APPLIED
CALCULUS

GENERAL INTEREST

APPLIED
CALCULUS

FOURTH EDITION

S. T. TAN

Stonehill College

BROOKS/COLE PUBLISHING COMPANY
An International Thomson Publishing Company I(T)P®

Pacific Grove • Albany • Belmont • Bonn • Boston • Cincinnati • Detroit • Johannesburg • London
Madrid • Melbourne • Mexico City • New York • Paris • Singapore • Tokyo • Toronto • Washington

To Pat, Bill, and Michael

Sponsoring Editor: *Margot Hanis*
Marketing Team: *Caroline Croley, Margaret Parks, Debra Johnston*
Editorial Assistant: *Kimberly Raburn*
Production Coordinator: *Marjorie Z. Sanders*
Production: *Cecile Joyner, The Cooper Company*
Manuscript Editor: *Carol Reitz*
Interior Design: *Julia Gecha*

Interior Illustration: *S T Associates*
Cover Design: *Lisa Henry*
Cover Illustration: *Judith L. Harkness*
Typesetting and Color Separation: *The PRD Group, Inc.*
Cover Printing: *Phoenix Color Corp.*
Printing and Binding: *R. R. Donnelley & Sons Company/ Crawfordsville*

Photo Credits: page xvi: (all) PhotoDisc, Inc. **xvii:** (top) © David Young-Wolff/PhotoEdit; (middle) PhotoDisc, Inc.; (bottom) © Elizabeth Zuckerman/PhotoEdit. **2:** © David Young-Wolff/PhotoEdit. **54:** © Tony Freeman/PhotoEdit. **172:** © 1996 R. J. Western. **256:** NASA/photo composite. **266:** PhotoDisc, Inc. **360:** © 1996 R. J. Western. **426:** PhotoDisc, Inc. **532:** © Juergen Gebhardt/The Image Bank. **580:** © Terje Rakke/The Image Bank. **672:** © Harald Sund/The Image Bank. **708:** PhotoDisc, Inc. **752:** © Kay Chernush/The Image Bank. **820:** © Barrie Rokeach/The Image Bank.

For more information, contact:

BROOKS/COLE PUBLISHING COMPANY
511 Forest Lodge Road
Pacific Grove, CA 93950
USA

International Thomson Publishing Europe
Berkshire House 168-173
High Holborn
London WC1V 7AA
England

Thomas Nelson Australia
102 Dodds Street
South Melbourne, 3205
Victoria, Australia

Nelson Canada
1120 Birchmount Road
Scarborough, Ontario
Canada M1K 5G4

International Thomson Editores
Seneca 53
Col. Polanco
11560 México, D. F., México

International Thomson Publishing GmbH
Königswinterer Strasse 418
53227 Bonn
Germany

International Thomson Publishing Japan
Hirakawacho Kyowa Building, 3F
2-2-1 Hirakawacho
Chiyoda-ku, Tokyo 102
Japan

International Thomson Publishing Asia
60 Albert Street
#15-01 Albert Complex
Singapore 189969

Printed in the United States of America

10 9 8 7 6 5 4

Library of Congress Cataloging-in-Publication Data

Tan, Soo Tang.
    Applied calculus / S. T. Tan.—4th ed.
        p. cm.
    Includes index.
    ISBN 0-534-95556-8
    1. Calculus. I. Title.
QA303.T14 1998
515—dc21   98-9936
                                  CIP

THIS BOOK IS PRINTED ON ACID-FREE RECYCLED PAPER

CONTENTS

* Sections marked with an asterisk are not prerequisites for later material.

CHAPTER 6   Integration   426

CHAPTER 7   Additional Topics in Integration   532

PREFACE

Applied Calculus, Fourth Edition, is suitable for use in a two-semester or three-quarter introductory calculus course for students in the managerial, life, and social sciences. As with the previous editions, our objective in *Applied Calculus* is twofold: (1) to write a textbook that is readable by students and (2) to make the book a useful teaching tool for instructors. We hope that with the present edition we have come one step closer to realizing our goal. The fourth edition of this text incorporates many suggestions by users of the earlier editions.

FEATURES

The following list includes some of the many important features of the book:

- **Coverage of Topics** The book contains more than enough material for the usual applied calculus course. Optional sections have been marked with an asterisk in the table of contents, thereby allowing the instructor to be flexible in choosing the topics most suitable for his or her course.

- **Approach** The problem-solving approach is stressed throughout the book. Numerous examples and solved problems are used to clarify each new concept or result in order to facilitate students' comprehension of new material. Figures are used extensively to help students visualize concepts and ideas.

- **Level of Presentation** Our approach is intuitive and we state the results informally. However, we have taken special care to ensure that this approach does not compromise the mathematical content and accuracy. Proofs of certain results are given, but they may be omitted if desired.

- **Applications** The text is application oriented. Many interesting, relevant, and up-to-date applications are drawn from the fields of business, economics, social and behavioral sciences, life sciences, physical sciences, and other fields of general interest. Some of these applications have their source in newspapers, weekly periodicals, and other magazines. Applications are found in the illustrative examples in the main body of the text as well as in the exercise sets. In fact, one goal of the text is to include at least one real-life application in each section (whenever feasible).

- **Exercises** Each section of the text is accompanied by an extensive set of exercises containing ample problems of a routine, computational nature that will help students master new techniques. The routine problems are followed by an extensive set of application-oriented problems that test students' mastery of the topics. Self-check exercises are also included at the end of each section. These exercises give the students a chance to test themselves on their understanding of the material.

- **Portfolios**   These interviews are designed to convey to the student the real-world experiences of professionals who have a background in mathematics and use it in their professions.

TECHNOLOGY

In this book our primary emphasis lies in the use of the graphing utility, which is employed in two ways:

1. It is used to explore mathematical concepts and also to shed further light on selected worked examples in the text. In this capacity, it serves to augment the analytic solution obtained in the text with a graphical and/or numerical solution, thus adding to a greater understanding of the problem. Exercises in this category appear under the heading:

 EXPLORING WITH TECHNOLOGY

2. It is used to solve problems whose solutions require a prodigious amount of calculations. Indeed, it is the availability of the graphing utility that makes it possible to include the many real-life problems in this book. Such exercises are found at the end of many subsections in the book under the heading:

/ / / USING TECHNOLOGY

NEW IN THE 4TH EDITION

Exploring with Technology Questions

These optional questions appear throughout the main body of the text and serve to enhance the student's understanding of the concepts and theory presented. Complete solutions to these exercises are given in the *Instructor's Complete Solutions Manual*.

/ / / Using Technology Subsections

These pages contain optional material and are placed at the end of the sections for which their use is appropriate. The subsections are written in the traditional example–exercise format with answers given at the back of the book. They may be used in the classroom if desired or as material for self-study by the student.

As many up-to-date and relevant applications have been introduced in these subsections, they provide students with an opportunity to interpret results in a real-life setting.

Group Discussion Questions

These are optional questions, appearing throughout the main body of the text, that can be discussed in class or assigned as homework. These questions generally require more thought and effort than the usual exercises. Complete solutions to these exercises are given in the *Instructor's Complete Solutions Manual*.

Critical Thinking Questions

These new exercises go beyond the usual exercises involving computation and provide the student with a little more challenge. Many of these questions call for the interpretation of results or the theory presented in the section.

New Internet-Linked Supplements

Students and instructors will now have access to these additional materials at the Brooks/Cole World Wide Web site: http://www.brookscole.com/math/authors/tans

- Study hints, review material, and practice chapter tests for students
- Group projects and extended problems for each chapter
- Instructions, including keystrokes, for the procedures referenced in the text for specific calculators (TI-82, TI-85, and other popular models)
- Modified "Using Technology" sections for CAS systems, including the command statements for Mathematica, Maple, and other popular systems

Sources

We have included sources for those applications that are based on real-life data.

New Exercises

A wealth of innovative and timely exercises have been added to pique student interest. More exercises (both rote and applied) calling for graphical interpretations have also been added.

Helpful Changes

A new real-life example—the motion of a magnetic levitation train—is used to motivate topics in limits, derivatives, and integration. The use of the motion of a familiar object in the different settings serves to enhance comprehension of the topics.

Chapter 2: A new illustrative example, "Spending by businesses on computer security equipment," has been added to the discussion of mathematical models in Section 2.3. The motion of a maglev is now used to motivate an intuitive discussion of the Intermediate Value Theorem.

Chapter 3: The use of the application "Estimating the Size of the Rings of Neptune" in Section 3.7, Differentials, is yet another example of the importance attached to our philosophy of motivating topics with real-life examples. A discussion of relative error has also been added to this section.

Chapter 4: The chapter has been reorganized and condensed. Sections 4.1 to 4.3 now cover the material previously covered in five sections. Topics are now organized into Applications of the First Derivative, Applications of the Second Derivative, and Curve Sketching.

Chapter 6: An early introduction to differential equations and a discussion of initial value problems have been added.

SUPPLEMENTS

- *Student's Solutions Manual,* available to both students and instructors, includes the solutions to odd-numbered exercises. ISBN 0-534-95558-4

- *Instructor's Complete Solutions Manual*, available only to instructors, includes solutions to all exercises. ISBN 0-534-95559-2

- *Test Bank with Chapter Tests,* free to adopters of the book, contains sample tests for each chapter. ISBN 0-534-35093-3

- *Thomson World Class Testing Tools* (Macintosh ISBN 0-534-35094-1; Windows ISBN 0-534-35096-8) This fully integrated suite of test creation, delivery, and classroom management tools includes World Class Test, Test Online, and World Class Management software. World Class Testing Tools allows professors to deliver tests via print, floppy, hard drive, LAN, or Internet. With these tools, professors can create cross-platform exam files from publisher files or existing WESTest 3.2 test banks, create and edit questions, and provide their own feedback to objective test questions—enabling the system to work as a tutorial or an examination. In addition, professors can generate questions algorithmically, creating tests that include multiple-choice, true/false, and matching questions. Professors can also track the progress of an entire class or an individual student. Testing and tutorial results can be integrated into the class management tool, which offers scoring, gradebook, and reporting capabilities.

ACKNOWLEDGMENTS

I wish to express my personal appreciation to each of the following reviewers, whose many suggestions have helped make a much improved book.

Reviewers of the Fourth Edition:

James V. Balch
Middle Tennessee State University

Janice Epstein
Texas A & M University

Gary J. Etgen
University of Houston

Matthew Gould
Vanderbilt University

Yvette Hester
Texas A & M University

Reviewers of the Previous Editions:

Michael Button
San Diego City College

Matthew P. Coleman
Fairfield University

William Coppage
Wright State University

Bruce Edwards
University of Florida at Gainesville

Charles S. Frady
Georgia State University

Howard Frisinger
Colorado State University

Larry Gerstein
University of California at Santa Barbara

John Haverhals
Bradley University

Frank Jenkins
John Carroll University

David E. Joyce
Clark University

Norman R. Martin
Northern Arizona University

Sandra Wray McAfee
University of Michigan

Maurice Monahan
South Dakota State University

Ralph J. Neuhaus
University of Idaho

Lloyd Olson
North Dakota State University

Wesley Orser
Clark College

Richard Porter
Northeastern University

Richard Quindley
Bridgewater State College

Thomas N. Roe
South Dakota State University

Donald R. Sherbert
University of Illinois

Lawrence V. Welch
Western Illinois University

I also wish to thank the editorial and production staff of Brooks/Cole: Margot Hanis, Jennifer Wilkinson, Caroline Croley, Margaret Parks, Debra Johnston, Jennifer Huber, Marjorie Sanders, and Vernon Boes for their thoughtful contributions and patient assistance and cooperation during the development and production of this book. Finally, I wish to thank Cecile Joyner of the Cooper Company and Carol Reitz, for doing an excellent job ensuring the accuracy and readability of this fourth edition, and Julie Gecha for her design of the interior of the book.

S. T. Tan
Stonehill College

APPLICATIONS

*In **Applied Calculus** we attempt to solve a wide variety of problems arising from many diverse fields of study. A small sample of the types of practical problems we will consider follows.*

POPULATION GROWTH   A study prepared for a Sunbelt town's Chamber of Commerce projected that the town's population in the next 3 years would grow according to the rule

$$P(x) = 50,000 + 30x^{3/2} + 20x$$

where $P(x)$ denotes the population x months from now. How fast will the population be increasing nine months from now? Sixteen months from now?

AIR POLLUTION   According to the South Coast Air Quality Management District, the level of nitrogen dioxide, a brown gas that impairs breathing, present in the atmosphere on a certain May day in downtown Los Angeles is approximated by

$$A(t) = 0.03t^3(t - 7)^4 + 60.2 \qquad (0 \le t \le 7)$$

where $A(t)$ is measured in pollutant standard index and t is measured in hours, with $t = 0$ corresponding to 7 A.M. How fast is the level of nitrogen dioxide increasing at 11 A.M.?

OPTIMAL DRIVING SPEED   A truck gets $400/x$ miles per gallon when driven at a constant speed of x miles per hour (between 50 and 70 miles per hour). If the price of fuel is $1 per gallon and the driver is paid $8 an hour, at what speed between 50 and 70 miles per hour is it most economical to drive?

LEARNING CURVES The Eastman Optical Company produces a 35-mm single-lens reflex camera. Eastman's training department determined that after completing the basic training program, a new, previously inexperienced employee would be able to assemble

$$Q(t) = 50 - 30e^{-0.5t}$$

model F cameras per day, t months after the employee began work on the assembly line. How many model F cameras can a new employee assemble per day after basic training? How many model F cameras can the average experienced employee assemble per day?

SUBWAY FARES A city's Metropolitan Transit Authority (MTA) operates a subway line for commuters from a certain suburb to the downtown metropolitan area. Currently, an average of 6000 passengers a day take the trains, paying a fare of $1.50 per ride. The board of the MTA, contemplating raising the fare to $1.75 per ride in order to generate a larger revenue, engaged the services of a consulting firm. The firm's study revealed that for each 25-cent increase in fare, the ridership would be reduced by an average of 1000 passengers a day. The consulting firm recommended that MTA stick to the current fare of $1.50 per ride, which already yields a maximum revenue. Show that the consultants' recommendations were correct.

ADVERTISING EXPENDITURES The Ross-Simons Company has a monthly advertising budget of $60,000. Their marketing department estimates that if they spend x dollars on newspaper advertising and y dollars on television advertising, then the monthly sales will be given by

$$z = f(x, y) = 90x^{1/4}y^{3/4}$$

dollars. Determine how much money Ross-Simons should spend on newspaper ads and on television ads per month to maximize its monthly sales.

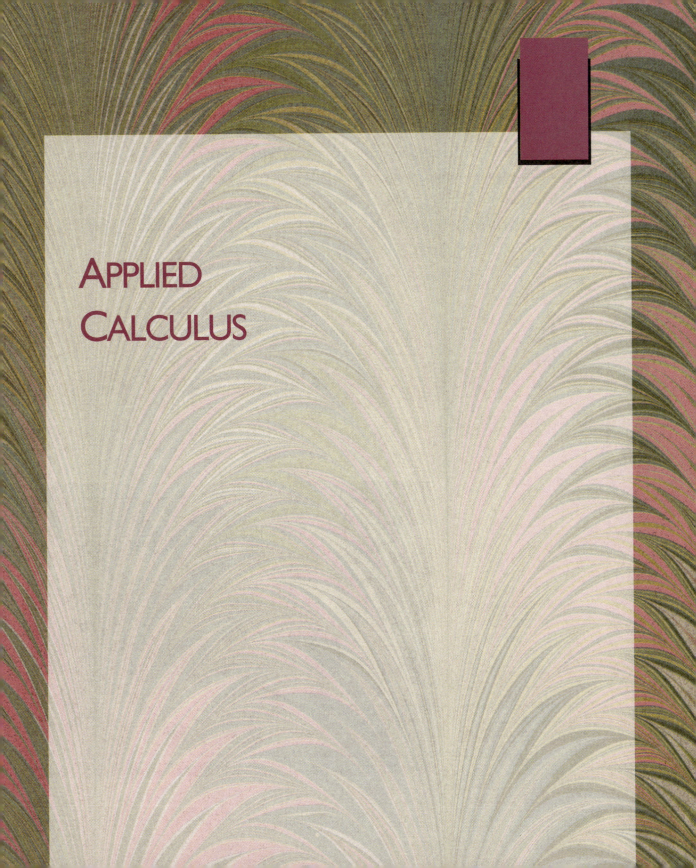

APPLIED
CALCULUS

The first two sections of this chapter contain a brief review of algebra. We then introduce the Cartesian coordinate system, which allows us to represent points in the plane in terms of ordered pairs of real numbers. This in turn enables us to compute the distance between two points algebraically. This chapter also covers straight lines. The slope of a straight line plays an important role in the study of calculus.

What sales figure can be predicted for next year? In Example 10, page 43, you will see how the manager of a local sporting goods store used sales figures from the previous years to predict the sales level for next year.

1

PRELIMINARIES

3

1.1 PRECALCULUS REVIEW I

Sections 1.1 and 1.2 review some of the basic concepts and techniques of algebra that are essential in the study of calculus. The material in this review will help you work through the examples and exercises in this book. You can read through this material now and do the exercises in areas where you feel a little "rusty," or you can review the material on an as-needed basis as you study the text. We begin our review with a discussion of real numbers.

The Real Number Line

The real number system is made up of the set of real numbers together with the usual operations of addition, subtraction, multiplication, and division.

Real numbers may be represented geometrically by points on a line. Such a line is called the **real number,** or **coordinate, line** and can be constructed as follows. Arbitrarily select a point on a straight line to represent the number zero. This point is called the **origin.** If the line is horizontal, then a point at a convenient distance to the right of the origin is chosen to represent the number 1. This determines the scale for the number line. Each positive real number lies at an appropriate distance to the right of the origin, and each negative real number lies at an appropriate distance to the left of the origin. See Figure 1.1.

Figure 1.1
The real number line.

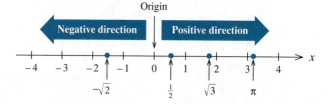

A *one-to-one correspondence* is set up between the set of all real numbers and the set of points on the number line. That is, exactly one point on the line is associated with each real number. Conversely, exactly one real number is associated with each point on the line. The real number that is associated with a point on the real number line is called the **coordinate** of that point.

Intervals

Throughout this book, we will often restrict our attention to certain subsets of the set of real numbers. For example, if x denotes the number of cars rolling off a plant assembly line each day, then x must be nonnegative—that is, $x \geq 0$. Further, suppose management decides that the daily production must not exceed 200 cars. Then x must satisfy the inequality $0 \leq x \leq 200$.

More generally, we will be interested in the following subsets of real numbers: open intervals, closed intervals, and half-open intervals. The set of all real numbers that lie *strictly* between two fixed numbers a and b is called an **open interval** (a, b). It consists of all real numbers x that satisfy the inequalities $a < x < b$, and it is called "open" because neither of its endpoints is included in the interval. A **closed interval** contains *both* of its endpoints. Thus, the set of all real numbers x that satisfy the inequalities $a \le x \le b$ is the closed interval $[a, b]$. Notice that square brackets are used to indicate that the endpoints are included in this interval. **Half-open intervals** contain only *one* of their endpoints. Thus, the interval $[a, b)$ is the set of all real numbers x that satisfy $a \le x < b$, whereas the interval $(a, b]$ is described by the inequalities $a < x \le b$. Examples of these **finite intervals** are illustrated in Table 1.1.

Table 1.1
Finite intervals.

Interval		Graph	Example	
Open	(a, b)		$(-2, 1)$	
Closed	$[a, b]$		$[-1, 2]$	
Half-open	$(a, b]$		$(\frac{1}{2}, 3]$	
Half-open	$[a, b)$		$[-\frac{1}{2}, 3)$	

In addition to finite intervals, we will encounter **infinite intervals.** Examples of infinite intervals are the half-lines (a, ∞), $[a, \infty)$, $(-\infty, a)$, and $(-\infty, a]$ defined by the set of all real numbers that satisfy $x > a$, $x \ge a$, $x < a$, and $x \le a$, respectively. The symbol ∞, called *infinity,* is not a real number. It is used here only for notational purposes in conjunction with the definition of infinite intervals. The notation $(-\infty, \infty)$ is used for the set of all real numbers x since, by definition, the inequalities $-\infty < x < \infty$ hold for any real number x. Infinite intervals are illustrated in Table 1.2.

Table 1.2
Infinite intervals.

Interval	Graph	Example	
(a, ∞)		$(2, \infty)$	
$[a, \infty)$		$[-1, \infty)$	
$(-\infty, a)$		$(-\infty, 1)$	
$(-\infty, a]$		$(-\infty, -\frac{1}{2}]$	

Properties of Inequalities

In practical applications, intervals are often found by solving one or more inequalities involving a variable. In such situations, the following properties may be used to advantage.

PROPERTIES OF INEQUALITIES

If a, b, and c are any real numbers, then

		Example
Property 1	If $a < b$ and $b < c$, then $a < c$.	$2 < 3$ and $3 < 8$, so $2 < 8$
Property 2	If $a < b$, then $a + c < b + c$.	$-5 < -3$, so $-5 + 2 < -3 + 2$; that is, $-3 < -1$
Property 3	If $a < b$ and $c > 0$, then $ac < bc$.	$-5 < -3$, and since $2 > 0$, we have $(-5)(2) < (-3)(2)$—that is, $-10 < -6$
Property 4	If $a < b$ and $c < 0$, then $ac > bc$.	$-2 < 4$, and since $-3 < 0$, we have $(-2)(-3) > (4)(-3)$—that is, $6 > -12$

Similar properties hold if each inequality sign, $<$, between a and b is replaced by $\geq$, $>$, or $\leq$.

A real number is a *solution of an inequality* involving a variable if a true statement is obtained when the variable is replaced by that number. The set of all real numbers satisfying the inequality is called the *solution set*.

EXAMPLE 1   Find the set of real numbers that satisfy $-1 \leq 2x - 5 < 7$.

Solution   Add 5 to each member of the given double inequality, obtaining

$$4 \leq 2x < 12$$

Next, multiply each member of the resulting double inequality by 1/2, yielding

$$2 \leq x < 6$$

Thus, the solution is the set of all values of x lying in the interval $[2, 6)$.

○ ○ ○

EXAMPLE 2   The management of Corbyco, a giant conglomerate, has esti-mated that x thousand dollars is needed to purchase

$$100{,}000(-1 + \sqrt{1 + 0.001x})$$

shares of common stock of the Starr Communications Company. Determine how much money Corbyco needs in order to purchase at least 100,000 shares of Starr's stock.

Solution   The amount of cash Corbyco needs to purchase at least 100,000 shares is found by solving the inequality

$$100{,}000(-1 + \sqrt{1 + 0.001x}) \geq 100{,}000$$

Proceeding, we find

$$-1 + \sqrt{1 + 0.001x} \geq 1$$
$$\sqrt{1 + 0.001x} \geq 2$$
$$1 + 0.001x \geq 4 \qquad \text{(Square both sides.)}$$
$$0.001x \geq 3$$
$$x \geq 3000$$

so Corbyco needs at least $3,000,000.   ○ ○ ○

Absolute Value

ABSOLUTE VALUE

The **absolute value** of a number a is denoted by $|a|$ and is defined by

$$|a| = \begin{cases} a & \text{if } a \geq 0 \\ -a & \text{if } a < 0 \end{cases}$$

Since $-a$ is a positive number when a is negative, it follows that the absolute value of a number is always nonnegative. For example, $|5| = 5$ and $|-5| = -(-5) = 5$. Geometrically, $|a|$ is the distance between the origin and the point on the number line that represents the number a (Figure 1.2).

Figure 1.2
The absolute value of a number.

(a)   (b)

ABSOLUTE VALUE PROPERTIES

If a and b are any real numbers, then

Example

Property 5 $|-a| = |a|$

$|-3| = -(-3) = 3 = |3|$

Property 6 $|ab| = |a|\,|b|$

$|(2)(-3)| = |-6| = 6$
$= |2|\,|-3|$

Property 7 $\left|\dfrac{a}{b}\right| = \dfrac{|a|}{|b|}, (b \neq 0)$

$\left|\dfrac{(-3)}{(-4)}\right| = \dfrac{|3|}{|4|} = \dfrac{3}{4} = \dfrac{|-3|}{|-4|}$

Property 8 $|a + b| \leq |a| + |b|$

$|8 + (-5)| = |3| = 3$
$\leq |8| + |-5|$
$= 13$

Property 8 is called the **triangle inequality.**

EXAMPLE 3   Evaluate each of the following expressions:

a. $|\pi - 5| + 3$ **b.** $|\sqrt{3} - 2| + |2 - \sqrt{3}|$

Solution

a. Since $\pi - 5 < 0$, we see that $|\pi - 5| = -(\pi - 5)$. Therefore,

$$|\pi - 5| + 3 = -(\pi - 5) + 3 = 8 - \pi$$

b. Since $\sqrt{3} - 2 < 0$, we see that $|\sqrt{3} - 2| = -(\sqrt{3} - 2)$. Next, observe that $2 - \sqrt{3} > 0$, so $|2 - \sqrt{3}| = 2 - \sqrt{3}$. Therefore,

$$|\sqrt{3} - 2| + |2 - \sqrt{3}| = -(\sqrt{3} - 2) + (2 - \sqrt{3})$$
$$= 4 - 2\sqrt{3} = 2(2 - \sqrt{3})$$

Exponents and Radicals

Recall that if b is any real number and n is a positive integer, then the expression b^n (read "b to the power n") is defined as the number

$$b^n = \underbrace{b \cdot b \cdot b \cdot \cdots \cdot b}_{n \text{ factors}}$$

The number b is called the **base,** and the superscript n is called the **power** of the exponential expression b^n. For example,

$$2^5 = 2 \cdot 2 \cdot 2 \cdot 2 \cdot 2 = 32 \quad \text{and} \quad \left(\frac{2}{3}\right)^3 = \left(\frac{2}{3}\right)\left(\frac{2}{3}\right)\left(\frac{2}{3}\right) = \frac{8}{27}$$

If $b \neq 0$, we define

$$b^0 = 1$$

For example, $2^0 = 1$ and $(-\pi)^0 = 1$, but the expression 0^0 is undefined.

Next, recall that if n is a positive integer, then the expression $b^{1/n}$ is defined to be the number that, when raised to the nth power, is equal to b. Thus,

$$(b^{1/n})^n = b$$

Such a number, if it exists, is called the **nth root of b,** also written $\sqrt[n]{b}$.

 Observe that the nth root of a negative number is not defined when n is even. For example, the square root of -2 is not defined because there is no real number b such that $b^2 = -2$. Also, given a number b, more than one number might satisfy our definition of the nth root. For example, both 3 and -3 squared equal 9, and each is a square root of 9. So, to avoid ambiguity, we define $b^{1/n}$ to be the positive nth root of b whenever it exists. Thus, $\sqrt{9} = 9^{1/2} = 3$.

Next, recall that if p/q (p, q, positive integers with $q \neq 0$) is a rational number in lowest terms, then the expression $b^{p/q}$ is defined as the number $(b^{1/q})^p$ or, equivalently, $\sqrt[q]{b^p}$, whenever it exists. Expressions involving negative rational exponents are taken care of by the definition

$$b^{-p/q} = \frac{1}{b^{p/q}}$$

Table 1.3

Definition of a^n ($a > 0$)	Example	Definition of a^n ($a > 0$)	Example
Integer exponent: If n is a positive integer, then $$a^n = a \cdot a \cdot a \cdots \cdot a$$ (n factors of a)	$2^5 = 2 \cdot 2 \cdot 2 \cdot 2 \cdot 2$ (5 factors) $= 32$	**Fractional exponent:** **a.** If n is a positive integer, then $$a^{1/n} \text{ or } \sqrt[n]{a}$$ denotes the **nth root** of a.	$16^{1/2} = \sqrt{16}$ $= 4$
Zero exponent: If n is equal to zero, then $$a^0 = 1$$ (0^0 is not defined.)	$7^0 = 1$	**b.** If m and n are positive integers, then $$a^{m/n} = \sqrt[n]{a^m} = (\sqrt[n]{a})^m$$	$8^{2/3} = (\sqrt[3]{8})^2$ $= 4$
Negative exponent: If n is a positive integer, then $$a^{-n} = \frac{1}{a^n} \quad (a \neq 0)$$	$6^{-2} = \frac{1}{6^2}$ $= \frac{1}{36}$	**c.** If m and n are positive integers, then $$a^{-m/n} = \frac{1}{a^{m/n}} \quad (a \neq 0)$$	$9^{-3/2} = \frac{1}{9^{3/2}}$ $= \frac{1}{27}$

Examples are

$$2^{3/2} = (2^{1/2})^3 \approx (1.4142)^3 \approx 2.8283$$

and

$$4^{-5/2} = \frac{1}{4^{5/2}} = \frac{1}{(4^{1/2})^5} = \frac{1}{2^5} = \frac{1}{32}$$

The rules defining the exponential expression a^n, where $a > 0$ for all rational values of n, are given in Table 1.3.

The first three definitions in Table 1.3 are also valid for negative values of a, whereas the fourth definition is valid only for negative values of a when n is odd. Thus,

$$(-8)^{1/3} = \sqrt[3]{-8} = -2 \qquad (n \text{ is odd.})$$
$$(-8)^{1/2} \text{ has no real value} \qquad (n \text{ is even.})$$

Finally, note that it can be shown that a^n has meaning for *all* real numbers n. For example, using a pocket calculator with a "y^x" key, we see that $2^{\sqrt{2}} \approx 2.665144$.

The five laws of exponents are listed in Table 1.4.

Table 1.4
The laws of exponents.

Law	Example
1. $a^m \cdot a^n = a^{m+n}$	$x^2 \cdot x^3 = x^{2+3} = x^5$
2. $\dfrac{a^m}{a^n} = a^{m-n} \qquad (a \neq 0)$	$\dfrac{x^7}{x^4} = x^{7-4} = x^3$
3. $(a^m)^n = a^{m \cdot n}$	$(x^4)^3 = x^{4 \cdot 3} = x^{12}$
4. $(ab)^n = a^n \cdot b^n$	$(2x)^4 = 2^4 \cdot x^4 = 16x^4$
5. $\left(\dfrac{a}{b}\right)^n = \dfrac{a^n}{b^n} \qquad (b \neq 0)$	$\left(\dfrac{x}{2}\right)^3 = \dfrac{x^3}{2^3} = \dfrac{x^3}{8}$

These laws are valid for any real numbers a, b, m, and n whenever the quantities are defined.

Remember, $(x^2)^3 \neq x^5$. The correct equation is $(x^2)^3 = x^{2 \cdot 3} = x^6$.

The next several examples illustrate the use of the laws of exponents.

EXAMPLE 4   Simplify the expression:

a. $(3x^2)(4x^3)$ **b.** $\dfrac{16^{5/4}}{16^{1/2}}$ **c.** $(6^{2/3})^3$ **d.** $(x^3y^{-2})^{-2}$ **e.** $\left(\dfrac{y^{3/2}}{x^{1/4}}\right)^{-2}$

Solution

a. $(3x^2)(4x^3) = 12x^{2+3} = 12x^5$ (Law 1)

b. $\dfrac{16^{5/4}}{16^{1/2}} = 16^{5/4 - 1/2} = 16^{3/4} = (\sqrt[4]{16})^3 = 2^3 = 8$ (Law 2)

c. $(6^{2/3})^3 = 6^{2/3 \cdot 3} = 6^{6/3} = 6^2 = 36$ (Law 3)

d. $(x^3 y^{-2})^{-2} = (x^3)^{-2}(y^{-2})^{-2} = x^{(3)(-2)}y^{(-2)(-2)} = x^{-6}y^4 = \dfrac{y^4}{x^6}$ (Law 4)

e. $\left(\dfrac{y^{3/2}}{x^{1/4}}\right)^{-2} = \dfrac{y^{(3/2)(-2)}}{x^{(1/4)(-2)}} = \dfrac{y^{-3}}{x^{-1/2}} = \dfrac{x^{1/2}}{y^3}$ (Law 5)

We can also use the laws of exponents to simplify expressions involving radicals, as illustrated in the next example.

EXAMPLE 5 Simplify the expression. (Assume x, y, and n are positive.)

a. $\sqrt[4]{16x^4 y^8}$ **b.** $\sqrt{12m^3 n} \cdot \sqrt{3m^5 n}$ **c.** $\dfrac{\sqrt[3]{-27x^6}}{\sqrt[3]{8y^3}}$

Solution

a. $\sqrt[4]{16x^4 y^8} = (16x^4 y^8)^{1/4} = 16^{1/4} \cdot x^{4/4} y^{8/4} = 2xy^2$

b. $\sqrt{12m^3 n} \cdot \sqrt{3m^5 n} = \sqrt{36m^8 n^2} = (36m^8 n^2)^{1/2} = 36^{1/2} \cdot m^{8/2} n^{2/2}$

$\qquad = 6m^4 n$

c. $\dfrac{\sqrt[3]{-27x^6}}{\sqrt[3]{8y^3}} = \dfrac{(-27x^6)^{1/3}}{(8y^3)^{1/3}} = \dfrac{-27^{1/3} x^{6/3}}{8^{1/3} y^{3/3}} = -\dfrac{3x^2}{2y}$

When a radical appears in the numerator or denominator of an algebraic expression, we often try to simplify the expression by eliminating the radical from the numerator or denominator. This process, called **rationalization,** is illustrated in the next two examples.

EXAMPLE 6 Rationalize the denominator of the expression $\dfrac{3x}{2\sqrt{x}}$.

Solution $\dfrac{3x}{2\sqrt{x}} = \dfrac{3x}{2\sqrt{x}} \cdot \dfrac{\sqrt{x}}{\sqrt{x}} = \dfrac{3x\sqrt{x}}{2\sqrt{x^2}} = \dfrac{3x\sqrt{x}}{2x} = \dfrac{3}{2}\sqrt{x}$

EXAMPLE 7 Rationalize the numerator of the expression $\dfrac{3\sqrt{x}}{2x}$.

Solution $\dfrac{3\sqrt{x}}{2x} = \dfrac{3\sqrt{x}}{2x} \cdot \dfrac{\sqrt{x}}{\sqrt{x}} = \dfrac{3\sqrt{x^2}}{2x\sqrt{x}} = \dfrac{3x}{2x\sqrt{x}} = \dfrac{3}{2\sqrt{x}}$

1.1 EXERCISES

In exercises 1–4, determine whether the statement is true or false.

1. $-3 < -20$ **2.** $-5 \le -5$ **3.** $\dfrac{2}{3} > \dfrac{5}{6}$

4. $-\dfrac{5}{6} < -\dfrac{11}{12}$

In exercises 5–10, show the given interval on a number line.

5. $(3, 6)$ **6.** $(-2, 5]$ **7.** $[-1, 4)$

8. $[-\frac{6}{5}, -\frac{1}{2}]$ **9.** $(0, \infty)$ **10.** $(-\infty, 5]$

In exercises 11–20, find the values of x that satisfy the inequality (inequalities).

11. $2x + 4 < 8$ **12.** $-6 > 4 + 5x$

13. $-4x \ge 20$ **14.** $-12 \le -3x$

15. $-6 < x - 2 < 4$ **16.** $0 \le x + 1 \le 4$

17. $x + 1 > 4$ or $x + 2 < -1$

18. $x + 1 > 2$ or $x - 1 < -2$

19. $x + 3 > 1$ and $x - 2 < 1$

20. $x - 4 \le 1$ and $x + 3 > 2$

In exercises 21–30, evaluate the expression.

21. $|-6 + 2|$ **22.** $4 + |-4|$

23. $\dfrac{|-12 + 4|}{|16 - 12|}$ **24.** $\left|\dfrac{0.2 - 1.4}{1.6 - 2.4}\right|$

25. $\sqrt{3}|-2| + 3|-\sqrt{3}|$ **26.** $|-1| + \sqrt{2}|-2|$

27. $|\pi - 1| + 2$ **28.** $|\pi - 6| - 3$

29. $|\sqrt{2} - 1| + |3 - \sqrt{2}|$

30. $|2\sqrt{3} - 3| - |\sqrt{3} - 4|$

In exercises 31–36, suppose that a and b are real numbers other than zero and that a > b. State whether the inequality is true or false.

31. $b - a > 0$ **32.** $\dfrac{a}{b} > 1$

33. $a^2 > b^2$ **34.** $\dfrac{1}{a} > \dfrac{1}{b}$

35. $a^3 > b^3$ **36.** $-a < -b$

In exercises 37–42, determine whether the statement is true for all real numbers a and b.

37. $|-a| = a$ **38.** $|b^2| = b^2$

39. $|a - 4| = |4 - a|$ **40.** $|a + 1| = |a| + 1$

41. $|a + b| = |a| + |b|$ **42.** $|a - b| = |a| - |b|$

In exercises 43–58, evaluate the expression.

43. $27^{2/3}$ **44.** $8^{-4/3}$

45. $\left(\dfrac{1}{\sqrt{3}}\right)^0$ **46.** $(7^{1/2})^4$

47. $\left[\left(\dfrac{1}{8}\right)^{1/3}\right]^{-2}$ **48.** $\left[\left(-\dfrac{1}{3}\right)^2\right]^{-3}$

49. $\left(\dfrac{7^{-5} \cdot 7^2}{7^{-2}}\right)^{-1}$ **50.** $\left(\dfrac{9}{16}\right)^{-1/2}$

51. $(125^{2/3})^{-1/2}$ **52.** $\sqrt[3]{2^6}$

53. $\dfrac{\sqrt{32}}{\sqrt{8}}$ **54.** $\sqrt[3]{\dfrac{-8}{27}}$

55. $\dfrac{16^{5/8}16^{1/2}}{16^{7/8}}$ **56.** $\left(\dfrac{9^{-3} \cdot 9^5}{9^{-2}}\right)^{-1/2}$

57. $16^{1/4} \cdot (8)^{-1/3}$ **58.** $\dfrac{6^{2.5} \cdot 6^{-1.9}}{6^{-1.4}}$

In exercises 59–68, determine whether the statement is true or false. Give a reason for your choice.

59. $x^4 + 2x^4 = 3x^4$ **60.** $3^2 \cdot 2^2 = 6^2$

61. $x^3 \cdot 2x^2 = 2x^6$ **62.** $3^3 + 3 = 3^4$

63. $\dfrac{2^{4x}}{1^{3x}} = 2^{4x-3x}$ **64.** $(2^2 \cdot 3^2)^2 = 6^4$

65. $\dfrac{1}{4^{-3}} = \dfrac{1}{64}$ **66.** $\dfrac{4^{3/2}}{2^4} = \dfrac{1}{2}$

67. $(1.2^{1/2})^{-1/2} = 1$ **68.** $5^{2/3} \cdot (25)^{2/3} = 25$

In exercises 69–74, rewrite the expression using positive exponents only.

69. $(xy)^{-2}$ **70.** $3s^{1/3} \cdot s^{-7/3}$

71. $\dfrac{x^{-1/3}}{x^{1/2}}$

72. $\sqrt{x^{-1}} \cdot \sqrt{9x^{-3}}$

73. $12^0(s + t)^{-3}$

74. $(x - y)(x^{-1} + y^{-1})$

In exercises 75–90, simplify the expression. (Assume x, y, r, s, and t are positive.)

75. $\dfrac{x^{7/3}}{x^{-2}}$

76. $(49x^{-2})^{-1/2}$

77. $(x^2y^{-3})(x^{-5}y^3)$

78. $\dfrac{5x^6y^3}{2x^2y^7}$

79. $\dfrac{x^{3/4}}{x^{-1/4}}$

80. $\left(\dfrac{x^3y^2}{z^2}\right)^2$

81. $\left(\dfrac{x^3}{-27y^{-6}}\right)^{-2/3}$

82. $\left(\dfrac{e^x}{e^{x-2}}\right)^{-1/2}$

83. $\left(\dfrac{x^{-3}}{y^{-2}}\right)^2\left(\dfrac{y}{x}\right)^4$

84. $\dfrac{(r^n)^4}{r^{5-2n}}$

85. $\sqrt[3]{x^{-2}} \cdot \sqrt{4x^5}$

86. $\sqrt{81x^6y^{-4}}$

87. $-\sqrt[4]{16x^4y^8}$

88. $\sqrt[3]{x^{3a+b}}$

89. $\sqrt[6]{64x^8y^3}$

90. $\sqrt[3]{27r^6} \cdot \sqrt{s^2t^4}$

In exercises 91–94, use the fact that $2^{1/2} \approx 1.414$ and $3^{1/2} \approx 1.732$ to evaluate the expression without using a calculator.

91. $2^{3/2}$ **92.** $8^{1/2}$ **93.** $9^{3/4}$ **94.** $6^{1/2}$

In exercises 95–98, use the fact that $10^{1/2} \approx 3.162$ and $10^{1/3} \approx 2.154$ to evaluate the expression without using a calculator.

95. $10^{3/2}$ **96.** $1000^{3/2}$ **97.** $10^{2.5}$

98. $(0.0001)^{-1/3}$

In exercises 99–104, rationalize the denominator of the expression.

99. $\dfrac{3}{2\sqrt{x}}$

100. $\dfrac{3}{\sqrt{xy}}$

101. $\dfrac{2y}{\sqrt{3y}}$

102. $\dfrac{5x^2}{\sqrt{3x}}$

103. $\dfrac{1}{\sqrt[3]{x}}$

104. $\sqrt{\dfrac{2x}{y}}$

In exercises 105–110, rationalize the numerator of the expression.

105. $\dfrac{2\sqrt{x}}{3}$

106. $\dfrac{\sqrt[3]{x}}{24}$

107. $\sqrt{\dfrac{2y}{x}}$

108. $\sqrt[3]{\dfrac{2x}{3y}}$

109. $\dfrac{\sqrt[3]{x^2z}}{y}$

110. $\dfrac{\sqrt[3]{x^2y}}{2x}$

111. Driving Range of a Car   An advertisement for a certain car states that the EPA fuel economy is 20 mpg city and 27 mpg highway and that the car's fuel-tank capacity is 18.1 gallons. Assuming ideal driving conditions, determine the driving range for the car from the foregoing data.

112. Find the minimum cost C (in dollars), given that

$$5(C - 25) \geq 1.75 + 2.5C$$

113. Find the maximum profit P (in dollars) given that

$$6(P - 2500) \leq 4(P + 2400)$$

114. Celsius and Fahrenheit Temperatures   The relationship between Celsius (°C) and Fahrenheit (°F) temperatures is given by the formula

$$C = \frac{5}{9}(F - 32)$$

a. If the temperature range for Montreal during the month of January is $-15° < °C < -5°$, find the range in degrees Fahrenheit in Montreal for the same period.
b. If the temperature range for New York City during the month of June is $63° < °F < 80°$, find the range in degrees Celsius in New York City for the same period.

115. Meeting Sales Targets   A salesman's monthly commission is 15% on all sales over $12,000. If his goal is to make a commission of at least $1000 per month, what minimum monthly sales figures must he attain?

116. Markup on a Car   The markup on a used car was at least 30% of its current wholesale price. If the car was sold for $2800, what was the maximum wholesale price?

117. Quality Control   The P.A.R. Manufacturing Company manufactures steel rods. Suppose the rods ordered by a customer are manufactured to a specification of 0.5 inch and are acceptable only if they are within the *tolerance limits* of 0.49 inch and 0.51 inch. Letting x denote the diameter of a rod, write an inequality using absolute value to express a criterion involving x that must be satisfied in order for a rod to be acceptable.

118. Quality Control   The diameter x (in inches) of a batch of ball bearings manufactured by the P.A.R. Manufacturing Company satisfies the inequality

$$|x - 0.1| \leq 0.01$$

What is the smallest diameter a ball bearing in the batch can have? The largest diameter?

119. Meeting Profit Goals   A manufacturer of a certain commodity has estimated that her profit in thousands of dollars is given by the expression

$$-6x^2 + 30x - 10$$

where x (in thousands) is the number of units produced. What production range will enable the manufacturer to realize a profit of at least $14,000 on the commodity?

120. Distribution of Incomes   The distribution of income in a certain city can be described by the exponential model $y = (2.8 \cdot 10^{11})(x)^{-1.5}$, where y is the number of families with an income of x or more dollars.

a. How many families in this city have an income of $20,000 or more?

b. How many families have an income of $40,000 or more?

c. How many families have an income of $100,000 or more?

1.2   PRECALCULUS REVIEW II

Operations with Algebraic Expressions

In calculus, we often work with algebraic expressions such as

$$2x^{4/3} - x^{1/3} + 1, \quad 2x^2 - x - \frac{2}{\sqrt{x}}, \quad \frac{3xy + 2}{x + 1}, \quad \text{and} \quad 2x^3 + 2x + 1$$

An algebraic expression of the form ax^n, where the coefficient a is a real number and n is a nonnegative integer, is called a **monomial,** meaning it consists of one term. For example, $7x^2$ is a monomial. A **polynomial** is a monomial or the sum of two or more monomials. For example,

$$x^2 + 4x + 4, \quad x^3 + 5, \quad x^4 + 3x^2 + 3, \quad \text{and} \quad x^2y + xy + y$$

are all polynomials.

Constant terms and terms containing the same variable factor are called **like,** or **similar, terms.** Like terms may be combined by adding or subtracting their numerical coefficients. For example,

$$3x + 7x = 10x \quad \text{and} \quad \frac{1}{2}xy + 3xy = \frac{7}{2}xy$$

The distributive property of the real number system,

$$ab + ac = a \cdot (b + c)$$

is used to justify this procedure.

To add or subtract two or more algebraic expressions, first remove the parentheses and then combine like terms. The resulting expression is written in order of decreasing degree from left to right.

$\circledR$* **EXAMPLE 1**

a. $(2x^4 + 3x^3 + 4x + 6) - (3x^4 + 9x^3 + 3x^2)$

$= 2x^4 + 3x^3 + 4x + 6 - 3x^4 - 9x^3 - 3x^2$ (Remove
parentheses.)

$= 2x^4 - 3x^4 + 3x^3 - 9x^3 - 3x^2 + 4x + 6$

$= -x^4 - 6x^3 - 3x^2 + 4x + 6$ (Combine like terms.)

b. $2t^3 - \{t^2 - [t - (2t - 1)] + 4\}$

$= 2t^3 - \{t^2 - [t - 2t + 1] + 4\}$

$= 2t^3 - \{t^2 - [-t + 1] + 4\}$ (Remove parentheses and combine
like terms within brackets.)

$= 2t^3 - \{t^2 + t - 1 + 4\}$ (Remove brackets.)

$= 2t^3 - \{t^2 + t + 3\}$ (Combine like terms within the braces.)

$= 2t^3 - t^2 - t - 3$ (Remove braces.)   ◐ ◐ ◐

An algebraic expression is said to be **simplified** if none of its terms are similar. Observe that when the algebraic expression in Example 1b was simplified, the innermost grouping symbols were removed first; that is, the parentheses ( ) were removed first, the brackets [ ] second, and the braces { } third.

When algebraic expressions are multiplied, each term of one algebraic expression is multiplied by each term of the other. The resulting algebraic expression is then simplified.

$\circledR$ **EXAMPLE 2**   Perform the indicated operations:

a. $(x^2 + 1)(3x^2 + 10x + 3)$ **b.** $(e^t + e^{-t})e^t - e^t(e^t - e^{-t})$

Solution

a. $(x^2 + 1)(3x^2 + 10x + 3)$

$= x^2(3x^2 + 10x + 3) + 1(3x^2 + 10x + 3)$

$= 3x^4 + 10x^3 + 3x^2 + 3x^2 + 10x + 3$

$= 3x^4 + 10x^3 + 6x^2 + 10x + 3$

b. $(e^t + e^{-t})e^t - e^t(e^t - e^{-t})$

$= e^{2t} + e^0 - e^{2t} + e^0$

$= e^{2t} - e^{2t} + e^0 + e^0$

$= 1 + 1$ (Recall that $e^0 = 1$.)

$= 2$ ◐ ◐ ◐

Certain product formulas that are frequently used in algebraic computations are given in Table 1.5.

◐ ◐ ◐

* The symbol $\circledR$ indicates that these examples were selected from the calculus portion of the text in order to help you review the algebraic computations you will *actually* be using in calculus.

Table 1.5

Formula	Example
$(a + b)^2 = a^2 + 2ab + b^2$	$(2x + 3y)^2 = (2x)^2 + 2(2x)(3y) + (3y)^2$
	$= 4x^2 + 12xy + 9y^2$
$(a - b)^2 = a^2 - 2ab + b^2$	$(4x - 2y)^2 = (4x)^2 - 2(4x)(2y) + (2y)^2$
	$= 16x^2 - 16xy + 4y^2$
$(a + b)(a - b) = a^2 - b^2$	$(2x + y)(2x - y) = (2x)^2 - (y)^2$
	$= 4x^2 - y^2$

Factoring

Factoring is the process of expressing an algebraic expression as a product of other algebraic expressions. For example, by applying the distributive property we may write

$$3x^2 - x = x(3x - 1)$$

The first step in factoring an algebraic expression is to check to see whether it contains any common terms. If it does, the greatest common term is then factored out. For example, the common factor of the algebraic expression $2a^2x + 4ax + 6a$ is $2a$, because

$$2a^2x + 4ax + 6a = \mathbf{2a} \cdot ax + \mathbf{2a} \cdot 2x + \mathbf{2a} \cdot 3 = \mathbf{2a}(ax + 2x + 3)$$

Ⓡ **EXAMPLE 3**   Factor out the greatest common factor in each of the following expressions:

a. $-0.3t^2 + 3t$ **b.** $2x^{3/2} - 3x^{1/2}$ **c.** $2ye^{xy^2} + 2xy^3e^{xy^2}$

d. $4x(x + 1)^{1/2} - 2x^2(\frac{1}{2})(x + 1)^{-1/2}$

Solution

a. $-0.3t^2 + 3t = -0.3t(t - 10)$

b. $2x^{3/2} - 3x^{1/2} = x^{1/2}(2x - 3)$

c. $2ye^{xy^2} + 2xy^3e^{xy^2} = 2ye^{xy^2}(1 + xy^2)$

d. $4x(x + 1)^{1/2} - 2x^2(\frac{1}{2})(x + 1)^{-1/2} = 4x(x + 1)^{1/2} - x^2(x + 1)^{-1/2}$
$$= x(x + 1)^{-1/2}[4(x + 1)^{1/2}(x + 1)^{1/2} - x]$$
$$= x(x + 1)^{-1/2}[4(x + 1) - x]$$
$$= x(x + 1)^{-1/2}(4x + 4 - x) = x(x + 1)^{-1/2}(3x + 4)$$

Here we select $(x + 1)^{-1/2}$ as the common factor because it is "contained" in each algebraic term. In particular, observe that

$$(x + 1)^{-1/2}(x + 1)^{1/2}(x + 1)^{1/2} = (x + 1)^{1/2}$$ ◦ ◦ ◦

Sometimes an algebraic expression may be factored by regrouping and rearranging its terms and then factoring out a common term. This technique is illustrated in Example 4.

EXAMPLE 4   Factor:

a. $2ax + 2ay + bx + by$ **b.** $3x\sqrt{y} - 4 - 2\sqrt{y} + 6x$

Solution

a. First, factor the common term $2a$ from the first two terms and the common term b from the last two terms. Thus,

$$2ax + 2ay + bx + by = 2a(x + y) + b(x + y)$$

Since $(x + y)$ is common to both terms of the polynomial, we may factor it out. Hence,

$$2a(x + y) + b(x + y) = (x + y)(2a + b)$$

b. $\begin{aligned}[t] 3x\sqrt{y} - 4 - 2\sqrt{y} + 6x &= 3x\sqrt{y} - 2\sqrt{y} + 6x - 4 \\ &= \sqrt{y}(3x - 2) + 2(3x - 2) \\ &= (3x - 2)(\sqrt{y} + 2) \end{aligned}$ ○ ○ ○

The first step in factoring a polynomial is to find the common factors. The next step is to express the polynomial as the product of a constant and/or one or more prime polynomials.

Certain product formulas that are useful in factoring binomials and trinomials are listed in Table 1.6.

Table 1.6

Formula	Example
Difference of two squares	
$x^2 - y^2 = (x + y)(x - y)$	$x^2 - 36 = (x + 6)(x - 6)$
	$\begin{aligned} 8x^2 - 2y^2 &= 2(4x^2 - y^2) \\ &= 2(2x + y)(2x - y) \end{aligned}$
	$9 - a^6 = (3 + a^3)(3 - a^3)$
Perfect-square trinomial	
$x^2 + 2xy + y^2 = (x + y)^2$	$x^2 + 8x + 16 = (x + 4)^2$
$x^2 - 2xy + y^2 = (x - y)^2$	$4x^2 - 4xy + y^2 = (2x - y)^2$
Sum of two cubes	
$x^3 + y^3 = (x + y)(x^2 - xy + y^2)$	$z^3 + 27 = z^3 + (3)^3$
	$\quad = (z + 3)(z^2 - 3z + 9)$
Difference of two cubes	
$x^3 - y^3 = (x - y)(x^2 + xy + y^2)$	$8x^3 - y^6 = (2x)^3 - (y^2)^3$
	$\quad = (2x - y^2)(4x^2 + 2xy^2 + y^4)$

The factors of the second-degree polynomial with integral coefficients

$$px^2 + qx + r$$

are $(ax + b)(cx + d)$, where $ac = p$, $ad + bc = q$, and $bd = r$. Since only a limited number of choices are possible, we use a trial-and-error method to factor polynomials having this form.

For example, to factor $x^2 - 2x - 3$, we first observe that the only possible first-degree terms are

$$(x \quad)(x \quad) \qquad \text{(Since the coefficient of } x^2 \text{ is 1)}$$

Next, we observe that the product of the constant terms is (-3). This gives us the following possible factors:

$$(x - 1)(x + 3)$$
$$(x + 1)(x - 3)$$

Looking once again at the polynomial $x^2 - 2x - 3$, we see that the coefficient of x is -2. Checking to see which set of factors yields -2 for the coefficient of x, we find that

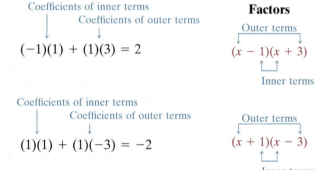

and we conclude that the correct factorization is

$$x^2 - 2x - 3 = (x + 1)(x - 3)$$

With practice, you will soon find that you can perform many of these steps mentally, and the need to write out each step will be eliminated.

Ⓡ **EXAMPLE 5**   Factor:

a. $3x^2 + 4x - 4$ **b.** $3x^2 - 6x - 24$

Solution

a. Using trial and error, we find that the correct factorization is

$$3x^2 + 4x - 4 = (3x - 2)(x + 2)$$

b. Since each term has the common factor 3, we have

$$3x^2 - 6x - 24 = 3(x^2 - 2x - 8)$$

Using the trial-and-error method of factorization, we find that

$$x^2 - 2x - 8 = (x - 4)(x + 2)$$

Thus, we have

$$3x^2 - 6x - 24 = 3(x - 4)(x + 2)$$

○ ○ ○

Roots of Polynomial Equations

A polynomial equation of degree n in the variable x is an equation of the form

$$a_n x^n + a_{n-1} x^{n-1} + \cdots + a_0 = 0$$

where n is a nonnegative integer and $a_0, a_1, \ldots, a_n$ are real numbers with $a_n \neq 0$. For example, the equation

$$-2x^5 + 8x^3 - 6x^2 + 3x + 1 = 0$$

is a polynomial equation of degree 5 in x.

The **roots** of a polynomial equation are precisely the values of x that satisfy the given equation.* One way of finding the roots of a polynomial equation is to first factor the polynomial and then solve the resulting equation. For example, the polynomial equation

$$x^3 - 3x^2 + 2x = 0$$

may be rewritten in the form

$$x(x^2 - 3x + 2) = 0$$

or
$$x(x - 1)(x - 2) = 0$$

Since the product of two real numbers can be equal to zero if and only if one (or both) of the factors is equal to zero, we have

$$x = 0, \quad x - 1 = 0, \quad \text{or} \quad x - 2 = 0$$

from which we see that the desired roots are $x = 0, 1,$ and 2.

The Quadratic Formula

In general, the problem of finding the roots of a polynomial equation is a difficult one. But the roots of a quadratic equation (a polynomial equation of degree 2) are easily found either by factoring or by using the following quadratic formula.

THE QUADRATIC FORMULA	The solutions of the equation $ax^2 + bx + c = 0$ $(a \neq 0)$ are given by $$x = \frac{-b \pm \sqrt{b^2 - 4ac}}{2a}$$

○ ○ ○

* In this book, we are interested only in the *real* roots of an equation.

EXAMPLE 6   Solve each of the following quadratic equations:

a. $2x^2 + 5x - 12 = 0$ **b.** $x^2 = -3x + 8$

Solution

a. The equation is in standard form, with $a = 2$, $b = 5$, and $c = -12$. Using the quadratic formula, we find

$$x = \frac{-b \pm \sqrt{b^2 - 4ac}}{2a} = \frac{-5 \pm \sqrt{5^2 - 4(2)(-12)}}{2(2)}$$

$$= \frac{-5 \pm \sqrt{121}}{4} = \frac{-5 \pm 11}{4}$$

$$= -4 \text{ or } \frac{3}{2}$$

This equation can also be solved by factoring. Thus

$$2x^2 + 5x - 12 = (2x - 3)(x + 4) = 0$$

from which we see that the desired roots are $x = 3/2$ or $x = -4$, as obtained earlier.

b. We first rewrite the given equation in the standard form $x^2 + 3x - 8 = 0$, from which we see that $a = 1$, $b = 3$, and $c = -8$. Using the quadratic formula, we find

$$x = \frac{-b \pm \sqrt{b^2 - 4ac}}{2a} = \frac{-3 \pm \sqrt{3^2 - 4(1)(-8)}}{2(1)}$$

$$= \frac{-3 \pm \sqrt{41}}{2}$$

That is, the solutions are $\dfrac{-3 + \sqrt{41}}{2} \approx 1.7$ and $\dfrac{-3 - \sqrt{41}}{2} \approx -4.7$. In this case, the quadratic formula proves quite handy!   ○ ○ ○

Rational Expressions

Quotients of polynomials are called **rational expressions.** Examples of rational expressions are

$$\frac{6x - 1}{2x + 3}, \quad \frac{3x^2y^3 - 2xy}{4x}, \quad \text{and} \quad \frac{2}{5ab}$$

Since rational expressions are quotients in which the variables represent real numbers, the properties of the real numbers apply to rational expressions as well, and operations with rational fractions are performed in the same manner as operations with arithmetic fractions. For example, using the properties of the real number system, we may write

$$\frac{ac}{bc} = \frac{a}{b} \cdot \frac{c}{c} = \frac{a}{b} \cdot 1 = \frac{a}{b}$$

where a, b, and c are any real numbers and b and c are not zero.

Similarly, using the same properties of real numbers, we may write

$$\frac{(x + 2)(x - 3)}{(x - 2)(x - 3)} = \frac{x + 2}{x - 2} \qquad (x \neq 2, 3)$$

after "canceling" the common factors.

 $\dfrac{\cancel{3} + 4x}{\cancel{3}} \neq 1 + 4x$ is an example of incorrect cancellation. Instead we need

to write $\dfrac{3 + 4x}{3} = \dfrac{3}{3} + \dfrac{4x}{3} = 1 + \dfrac{4x}{3}$.

A rational expression is simplified, or in lowest terms, when the numerator and denominator have no common factors other than 1 and −1 and the expression contains no negative exponents.

(R) **EXAMPLE 7**   Simplify the following expressions:

a. $\dfrac{x^2 + 2x - 3}{x^2 + 4x + 3}$ **b.** $\dfrac{[(t^2 + 4)(2t - 4) - (t^2 - 4t + 4)(2t)]}{(t^2 + 4)^2}$

Solution

a. $\dfrac{x^2 + 2x - 3}{x^2 + 4x + 3} = \dfrac{(x + 3)(x - 1)}{(x + 3)(x + 1)} = \dfrac{x - 1}{x + 1}$

b. $\dfrac{[(t^2 + 4)(2t - 4) - (t^2 - 4t + 4)(2t)]}{(t^2 + 4)^2}$

$$= \frac{2t^3 - 4t^2 + 8t - 16 - 2t^3 + 8t^2 - 8t}{(t^2 + 4)^2} \qquad \begin{array}{l}\text{(Carry out the}\\\text{indicated}\\\text{multiplication.)}\end{array}$$

$$= \frac{4t^2 - 16}{(t^2 + 4)^2} \qquad \text{(Combine like terms.)}$$

$$= \frac{4(t^2 - 4)}{(t^2 + 4)^2} \qquad \text{(Factor.)} \qquad \text{○ ○ ○}$$

The operations of multiplication and division are performed with algebraic fractions in the same manner as with arithmetic fractions (see Table 1.7).

Table 1.7

Operation	Example
If P, Q, R, and S are polynomials, then	
Multiplication	
$\dfrac{P}{Q} \cdot \dfrac{R}{S} = \dfrac{PR}{QS} \qquad (Q, S \neq 0)$	$\dfrac{2x}{y} \cdot \dfrac{(x + 1)}{(y - 1)} = \dfrac{2x(x + 1)}{y(y - 1)} = \dfrac{2x^2 + 2x}{y^2 - y}$
Division	
$\dfrac{P}{Q} \div \dfrac{R}{S} = \dfrac{P}{Q} \cdot \dfrac{S}{R} = \dfrac{PS}{QR} \qquad (Q, R, S \neq 0)$	$\dfrac{x^2 + 3}{y} \div \dfrac{y^2 + 1}{x} = \dfrac{x^2 + 3}{y} \cdot \dfrac{x}{y^2 + 1}$
	$= \dfrac{x^3 + 3x}{y^3 + y}$

When rational expressions are multiplied and divided, the resulting expressions should be simplified.

EXAMPLE 8   Perform the indicated operations and simplify:

$$\frac{2x-8}{x+2} \cdot \frac{x^2+4x+4}{x^2-16}$$

Solution $\dfrac{2x-8}{x+2} \cdot \dfrac{x^2+4x+4}{x^2-16}$

$$= \frac{2(x-4)}{x+2} \cdot \frac{(x+2)^2}{(x+4)(x-4)}$$

$$= \frac{2(x-4)(x+2)(x+2)}{(x+2)(x+4)(x-4)} \qquad \text{[Cancel the common factors } (x+2)(x-4).\text{]}$$

$$= \frac{2(x+2)}{x+4}$$

○ ○ ○

For rational expressions, the operations of addition and subtraction are performed by finding a common denominator of the fractions and then adding or subtracting the fractions. Table 1.8 shows the rules for fractions with equal denominators.

Table 1.8

Operation	Example
If P, Q, and R are polynomials, then	
Addition	
$\dfrac{P}{R} + \dfrac{Q}{R} = \dfrac{P+Q}{R} \qquad (R \neq 0)$	$\dfrac{2x}{x+2} + \dfrac{6x}{x+2} = \dfrac{2x+6x}{x+2} = \dfrac{8x}{x+2}$
Subtraction	
$\dfrac{P}{R} - \dfrac{Q}{R} = \dfrac{P-Q}{R} \qquad (R \neq 0)$	$\dfrac{3y}{y-x} - \dfrac{y}{y-x} = \dfrac{3y-y}{y-x} = \dfrac{2y}{y-x}$

To add or subtract fractions that have different denominators, first find a common denominator, preferably the least common denominator (LCD). Then carry out the indicated operations following the procedure described in Table 1.8.

To find the least common denominator (LCD) of two or more rational expressions:

I. *Find the prime factors of each denominator.*

2. *Form the product of the different prime factors that occur in the denominators.* Each prime factor in this product should be raised to the highest power of that factor appearing in the denominators.

$$\frac{x}{2+y} \neq \frac{x}{2} + \frac{x}{y}$$

 EXAMPLE 9   Simplify:

a. $\dfrac{2x}{x^2+1} + \dfrac{6(3x^2)}{x^3+2}$ **b.** $\dfrac{1}{x+h} - \dfrac{1}{x}$

Solution

a. $\dfrac{2x}{x^2+1} + \dfrac{6(3x^2)}{x^3+2}$

$= \dfrac{2x(x^3+2) + 6(3x^2)(x^2+1)}{(x^2+1)(x^3+2)}$ [LCD $= (x^2+1)(x^3+2)$.]

$= \dfrac{2x^4 + 4x + 18x^4 + 18x^2}{(x^2+1)(x^3+2)}$

$= \dfrac{20x^4 + 18x^2 + 4x}{(x^2+1)(x^3+2)}$

$= \dfrac{2x(10x^3 + 9x + 2)}{(x^2+1)(x^3+2)}$

b. $\dfrac{1}{x+h} - \dfrac{1}{x} = \dfrac{1}{x+h} \cdot \dfrac{x}{x} - \dfrac{1}{x} \cdot \dfrac{x+h}{x+h}$ [LCD $= (x)(x+h)$.]

$= \dfrac{x}{x(x+h)} - \dfrac{x+h}{x(x+h)}$

$= \dfrac{x - x - h}{x(x+h)}$

$= \dfrac{-h}{x(x+h)}$ ◦ ◦ ◦

Other Algebraic Fractions

The techniques used to simplify rational expressions may also be used to simplify algebraic fractions in which the numerator and denominator are not polynomials, as illustrated in Example 10.

EXAMPLE 10   Simplify:

a. $\dfrac{1 + \dfrac{1}{x+1}}{x - \dfrac{4}{x}}$ **b.** $\dfrac{x^{-1} + y^{-1}}{x^{-2} - y^{-2}}$

Solution

a.
$$\frac{1 + \dfrac{1}{x+1}}{x - \dfrac{4}{x}} = \frac{1 \cdot \dfrac{x+1}{x+1} + \dfrac{1}{x+1}}{x \cdot \dfrac{x}{x} - \dfrac{4}{x}} = \frac{\dfrac{x+1+1}{x+1}}{\dfrac{x^2-4}{x}}$$

$$= \frac{x+2}{x+1} \cdot \frac{x}{x^2-4} = \frac{x+2}{x+1} \cdot \frac{x}{(x+2)(x-2)}$$

$$= \frac{x}{(x+1)(x-2)}$$

b.
$$\frac{x^{-1} + y^{-1}}{x^{-2} - y^{-2}} = \frac{\dfrac{1}{x} + \dfrac{1}{y}}{\dfrac{1}{x^2} - \dfrac{1}{y^2}} = \frac{\dfrac{y+x}{xy}}{\dfrac{y^2-x^2}{x^2y^2}} \qquad \left(x^{-n} = \frac{1}{x^n} \right)$$

$$= \frac{y+x}{xy} \cdot \frac{x^2y^2}{y^2-x^2} = \frac{y+x}{xy} \cdot \frac{(xy)^2}{(y+x)(y-x)}$$

$$= \frac{xy}{y-x}$$

⊙ ⊙ ⊙

EXAMPLE 11   Perform the given operations and simplify:

a. $\dfrac{x^2(2x^2+1)^{1/2}}{x-1} \cdot \dfrac{4x^3 - 6x^2 + x - 2}{x(x-1)(2x^2+1)}$ **b.** $\dfrac{12x^2}{\sqrt{2x^2+3}} + 6\sqrt{2x^2+3}$

Solution

a. $\dfrac{x^2(2x^2+1)^{1/2}}{x-1} \cdot \dfrac{4x^3 - 6x^2 + x - 2}{x(x-1)(2x^2+1)}$

$$= \frac{x(4x^3 - 6x^2 + x - 2)}{(x-1)^2(2x^2+1)^{1-1/2}}$$

$$= \frac{x(4x^3 - 6x^2 + x - 2)}{(x-1)^2(2x^2+1)^{1/2}}$$

b. $\dfrac{12x^2}{\sqrt{2x^2+3}} + 6\sqrt{2x^2+3} = \dfrac{12x^2}{(2x^2+3)^{1/2}} + 6(2x^2+3)^{1/2}$

$$= \frac{12x^2 + 6(2x^2+3)^{1/2}(2x^2+3)^{1/2}}{(2x^2+3)^{1/2}}$$

$$= \frac{12x^2 + 6(2x^2+3)}{(2x^2+3)^{1/2}}$$

$$= \frac{24x^2 + 18}{(2x^2+3)^{1/2}} = \frac{6(4x^2+3)}{\sqrt{2x^2+3}}$$

⊙ ⊙ ⊙

Rationalizing Algebraic Fractions

When the denominator of an algebraic fraction contains sums or differences involving radicals, we may **rationalize the denominator**—that is, transform the fraction into an equivalent one with a denominator that does not contain radicals. In doing so, we make use of the fact that

$$(\sqrt{a} + \sqrt{b})(\sqrt{a} - \sqrt{b}) = (\sqrt{a})^2 - (\sqrt{b})^2$$
$$= a - b$$

This procedure is illustrated in Example 12.

EXAMPLE 12   Rationalize the denominator of $\dfrac{1}{1 + \sqrt{x}}$.

Solution   Upon multiplying the numerator and the denominator by $(1 - \sqrt{x})$, we obtain

$$\frac{1}{1 + \sqrt{x}} = \frac{1}{1 + \sqrt{x}} \cdot \frac{1 - \sqrt{x}}{1 - \sqrt{x}}$$
$$= \frac{1 - \sqrt{x}}{1 - (\sqrt{x})^2}$$
$$= \frac{1 - \sqrt{x}}{1 - x}$$ ◦ ◦ ◦

In other situations, it may be necessary to rationalize the numerator of an algebraic expression. In calculus, for example, one encounters the following problem.

EXAMPLE 13   Rationalize the numerator: $\dfrac{\sqrt{1 + h} - 1}{h}$

Solution

$$\frac{\sqrt{1 + h} - 1}{h} = \frac{\sqrt{1 + h} - 1}{h} \cdot \frac{\sqrt{1 + h} + 1}{\sqrt{1 + h} + 1}$$
$$= \frac{(\sqrt{1 + h})^2 - (1)^2}{h(\sqrt{1 + h} + 1)}$$
$$= \frac{1 + h - 1}{h(\sqrt{1 + h} + 1)} \qquad ((\sqrt{1 + h})^2 = \sqrt{1 + h} \cdot \sqrt{1 + h}$$
$$\qquad\qquad\qquad\qquad = 1 + h)$$
$$= \frac{h}{h(\sqrt{1 + h} + 1)}$$
$$= \frac{1}{\sqrt{1 + h} + 1}$$ ◦ ◦ ◦

1.2 EXERCISES

In exercises 1–22, perform the indicated operations and simplify each expression.

1. $(7x^2 - 2x + 5) + (2x^2 + 5x - 4)$

2. $(3x^2 + 5xy + 2y) + (4 - 3xy - 2x^2)$

3. $(5y^2 - 2y + 1) - (y^2 - 3y - 7)$

4. $3(2a - b) - 4(b - 2a)$

5. $x - \{2x - [-x - (1 - x)]\}$

6. $3x^2 - \{x^2 + 1 - x[x - (2x - 1)]\} + 2$

7. $\left(\dfrac{1}{3} - 1 + e\right) - \left(-\dfrac{1}{3} - 1 + e^{-1}\right)$

8. $-\dfrac{3}{4}y - \dfrac{1}{4}x + 100 + \dfrac{1}{2}x + \dfrac{1}{4}y - 120$

9. $3\sqrt{8} + 8 - 2\sqrt{y} + \dfrac{1}{2}\sqrt{x} - \dfrac{3}{4}\sqrt{y}$

10. $\dfrac{8}{9}x^2 + \dfrac{2}{3}x + \dfrac{16}{3}x^2 - \dfrac{16}{3}x - 2x + 2$

11. $(x + 8)(x - 2)$ **12.** $(5x + 2)(3x - 4)$

13. $(a + 5)^2$ **14.** $(3a - 4b)^2$

15. $(x + 2y)^2$ **16.** $(6 - 3x)^2$

17. $(2x + y)(2x - y)$ **18.** $(3x + 2)(2 - 3x)$

19. $(x^2 - 1)(2x) - x^2(2x)$

20. $(x^{1/2} + 1)\left(\dfrac{1}{2}x^{-1/2}\right) - (x^{1/2} - 1)\left(\dfrac{1}{2}x^{-1/2}\right)$

21. $2(t + \sqrt{t})^2 - 2t^2$ **22.** $2x^2 + (-x + 1)^2$

In exercises 23–30, factor out the greatest common factor from each expression.

23. $4x^5 - 12x^4 - 6x^3$

24. $4x^2y^2z - 2x^5y^2 + 6x^3y^2z^2$

25. $7a^4 - 42a^2b^2 + 49a^3b$

26. $3x^{2/3} - 2x^{1/3}$ **27.** $e^{-x} - xe^{-x}$

28. $2ye^{xy^2} + 2xy^3e^{xy^2}$ **29.** $2x^{-5/2} - \dfrac{3}{2}x^{-3/2}$

30. $\dfrac{1}{2}\left(\dfrac{2}{3}u^{3/2} - 2u^{1/2}\right)$

In exercises 31–44, factor each expression.

31. $6ac + 3bc - 4ad - 2bd$

32. $3x^3 - x^2 + 3x - 1$

33. $4a^2 - b^2$ **34.** $12x^2 - 3y^2$

35. $10 - 14x - 12x^2$ **36.** $x^2 - 2x - 15$

37. $3x^2 - 6x - 24$ **38.** $3x^2 - 4x - 4$

39. $12x^2 - 2x - 30$ **40.** $(x + y)^2 - 1$

41. $9x^2 - 16y^2$ **42.** $8a^2 - 2ab - 6b^2$

43. $x^6 + 125$ **44.** $x^3 - 27$

In exercises 45–52, perform the indicated operations and simplify each expression.

45. $(x^2 + y^2)x - xy(2y)$ **46.** $2kr(R - r) - kr^2$

47. $2(x - 1)(2x + 2)^3[4(x - 1) + (2x + 2)]$

48. $5x^2(3x^2 + 1)^4(6x) + (3x^2 + 1)^5(2x)$

49. $4(x - 1)^2(2x + 2)^3(2) + (2x + 2)^4(2)(x - 1)$

50. $(x^2 + 1)(4x^3 - 3x^2 + 2x) - (x^4 - x^3 + x^2)(2x)$

51. $(x^2 + 2)^2[5(x^2 + 2)^2 - 3](2x)$

52. $(x^2 - 4)(x^2 + 4)(2x + 8) - (x^2 + 8x - 4)(4x^3)$

In exercises 53–58, find the real roots of each equation by factoring.

53. $x^2 + x - 12 = 0$ **54.** $3x^2 - x - 4 = 0$

55. $4t^2 + 2t - 2 = 0$ **56.** $-6x^2 + x + 12 = 0$

57. $\frac{1}{4}x^2 - x + 1 = 0$ **58.** $\frac{1}{2}a^2 + a - 12 = 0$

In exercises 59–64, solve the equation by using the quadratic formula.

59. $4x^2 + 5x - 6 = 0$ **60.** $3x^2 - 4x + 1 = 0$

61. $8x^2 - 8x - 3 = 0$ **62.** $x^2 - 6x + 6 = 0$

63. $2x^2 + 4x - 3 = 0$ **64.** $2x^2 + 7x - 15 = 0$

In exercises 65–70, simplify the expression.

65. $\dfrac{x^2 + x - 2}{x^2 - 4}$ **66.** $\dfrac{2a^2 - 3ab - 9b^2}{2ab^2 + 3b^3}$

67. $\dfrac{12t^2 + 12t + 3}{4t^2 - 1}$ **68.** $\dfrac{x^3 + 2x^2 - 3x}{-2x^2 - x + 3}$

69. $\dfrac{(4x-1)(3)-(3x+1)(4)}{(4x-1)^2}$

70. $\dfrac{(1+x^2)^2(2)-2x(2)(1+x^2)(2x)}{(1+x^2)^4}$

In exercises 71–88, perform the indicated operations and simplify each expression.

71. $\dfrac{2a^2-2b^2}{b-a}\cdot\dfrac{4a+4b}{a^2+2ab+b^2}$

72. $\dfrac{x^2-6x+9}{x^2-x-6}\cdot\dfrac{3x+6}{2x^2-7x+3}$

73. $\dfrac{3x^2+2x-1}{2x+6}\div\dfrac{x^2-1}{x^2+2x-3}$

74. $\dfrac{3x^2-4xy-4y^2}{x^2y}\div\dfrac{(2y-x)^2}{x^3y}$

75. $\dfrac{58}{3(3t+2)}+\dfrac{1}{3}$

76. $\dfrac{a+1}{3a}+\dfrac{b-2}{5b}$

77. $\dfrac{2x}{2x-1}-\dfrac{3x}{2x+5}$

78. $\dfrac{-xe^x}{x+1}+e^x$

79. $\dfrac{4}{x^2-9}-\dfrac{5}{x^2-6x+9}$

80. $\dfrac{x}{1-x}+\dfrac{2x+3}{x^2-1}$

81. $\dfrac{1+\dfrac{1}{x}}{1-\dfrac{1}{x}}$

82. $\dfrac{\dfrac{1}{x}+\dfrac{1}{y}}{1-\dfrac{1}{xy}}$

83. $\dfrac{4x^2}{2\sqrt{2x^2+7}}+\sqrt{2x^2+7}$

84. $6(2x+1)^2\sqrt{x^2+x}+\dfrac{(2x+1)^4}{2\sqrt{x^2+x}}$

85. $\dfrac{2x(x+1)^{-1/2}-(x+1)^{1/2}}{x^2}$

86. $\dfrac{(x^2+1)^{1/2}-2x^2(x^2+1)^{-1/2}}{1-x^2}$

87. $\dfrac{(2x+1)^{1/2}-(x+2)(2x+1)^{-1/2}}{2x+1}$

88. $\dfrac{2(2x-3)^{1/3}-(x-1)(2x-3)^{-2/3}}{(2x-3)^{2/3}}$

In exercises 89–94, rationalize the denominator of each expression.

89. $\dfrac{1}{\sqrt{3}-1}$

90. $\dfrac{1}{\sqrt{x}+5}$

91. $\dfrac{1}{\sqrt{x}-\sqrt{y}}$

92. $\dfrac{a}{1-\sqrt{a}}$

93. $\dfrac{\sqrt{a}+\sqrt{b}}{\sqrt{a}-\sqrt{b}}$

94. $\dfrac{2\sqrt{a}+\sqrt{b}}{2\sqrt{a}-\sqrt{b}}$

In exercises 95–100, rationalize the numerator of each expression.

95. $\dfrac{\sqrt{x}}{3}$

96. $\dfrac{\sqrt[3]{y}}{x}$

97. $\dfrac{1-\sqrt{3}}{3}$

98. $\dfrac{\sqrt{x}-1}{x}$

99. $\dfrac{1+\sqrt{x+2}}{\sqrt{x+2}}$

100. $\dfrac{\sqrt{x+3}-\sqrt{x}}{3}$

1.3 THE CARTESIAN COORDINATE SYSTEM

The Cartesian Coordinate System

In Section 1.1 we saw how a one-to-one correspondence between the set of real numbers and the points on a straight line leads to a coordinate system on a line (a one-dimensional space).

A similar representation for points in a plane (a two-dimensional space) is realized through the **Cartesian coordinate system,** which is constructed as

Figure 1.3
The Cartesian coordinate system.

follows: Take two perpendicular lines, one of which is normally chosen to be horizontal. These lines intersect at a point O, called the **origin** (Figure 1.3). The horizontal line is called the **x-axis,** and the vertical line is called the **y-axis.** A number scale is set up along the x-axis, with the positive numbers lying to the right of the origin and the negative numbers lying to the left of it. Similarly, a number scale is set up along the y-axis, with the positive numbers lying above the origin and the negative numbers lying below it.

The number scales on the two axes need not be the same. Indeed, in many applications different quantities are represented by x and y. For example, x may represent the number of typewriters sold and y the total revenue resulting from the sales. In such cases it is often desirable to choose different number scales to represent the different quantities. Note, however, that the zeros of both number scales coincide at the origin of the two-dimensional coordinate system.

A point in the plane can now be represented uniquely in this coordinate system by an **ordered pair** of numbers—that is, a pair (x, y), where x is the first number and y the second. To see this, let P be any point in the plane (Figure 1.4). Draw perpendiculars from P to the x-axis and y-axis, respectively. Then the number x is precisely the number that corresponds to the point on the x-axis at which the perpendicular through P hits the x-axis. Similarly, y is the number that corresponds to the point on the y-axis at which the perpen- dicular through P crosses the y-axis.

Figure 1.4
An ordered pair (x, y).

Conversely, given an ordered pair (x, y) with x as the first number and y the second, a point P in the plane is uniquely determined as follows: Locate the point on the x-axis represented by the number x and draw a line through that point parallel to the y-axis. Next, locate the point on the y-axis represented by the number y and draw a line through that point parallel to the x-axis. The point of intersection of these two lines is the point P (see Figure 1.4).

In the ordered pair (x, y), x is called the **abscissa,** or **x-coordinate,** y is

Figure 1.5
Several points in the Cartesian plane.

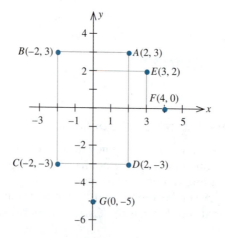

Figure 1.6
The four quadrants in the Cartesian plane.

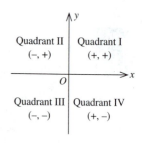

called the **ordinate,** or **y-coordinate,** and x and y together are referred to as the **coordinates** of the point P.

Letting (a, b) denote the point P with x-coordinate a and y-coordinate b, the points $A = (2, 3)$, $B = (-2, 3)$, $C = (-2, -3)$, $D = (2, -3)$, $E = (3, 2)$, $F = (4, 0)$, and $G = (0, -5)$ are plotted in Figure 1.5. The fact that, in general, $(x, y) \neq (y, x)$ is clearly illustrated by points A and E.

The axes divide the plane into four quadrants. Quadrant I consists of the points (x, y) that satisfy $x > 0$ and $y > 0$; Quadrant II, the points (x, y), where $x < 0$ and $y > 0$; Quadrant III, the points (x, y), where $x < 0$ and $y < 0$; and Quadrant IV, the points (x, y), where $x > 0$ and $y < 0$ (Figure 1.6).

The Distance Formula

One immediate benefit that arises from using the Cartesian coordinate system is that the distance between any two points in the plane may be expressed solely in terms of their coordinates. Suppose, for example, that (x_1, y_1) and (x_2, y_2) are any two points in the plane (Figure 1.7). Then the distance between these two points can be computed using the following formula.

Figure 1.7
The distance d between the points (x_1, y_1) and (x_2, y_2).

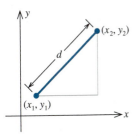

DISTANCE FORMULA

The distance d between two points $P_1(x_1, y_1)$ and $P_2(x_2, y_2)$ in the plane is given by

$$d = \sqrt{(x_2 - x_1)^2 + (y_2 - y_1)^2} \tag{1}$$

For a proof of this result, see exercise 35, page 34.

 EXAMPLE 1 Find the distance between the points $(-4, 3)$ and $(2, 6)$.

Solution Let $P_1(-4, 3)$ and $P_2(2, 6)$ be points in the plane. Then we have

$$x_1 = -4, \quad y_1 = 3, \quad x_2 = 2, \quad \text{and} \quad y_2 = 6$$

Using formula (1), we have

$$d = \sqrt{[2 - (-4)]^2 + (6 - 3)^2}$$
$$= \sqrt{6^2 + 3^2}$$
$$= \sqrt{45} = 3\sqrt{5}$$

○ ○ ○

Refer to Example 1.

Suppose we label the point $(2, 6)$ as P_1 and the point $(-4, 3)$ as P_2. (a) Show that the distance d between the two points is the same as that obtained earlier. (b) Prove that, in general, the distance d in formula (1) is independent of the way we label the two points.

Application

EXAMPLE 2   In the following diagram (Figure 1.8), S represents the position of a power relay station located on a straight coastal highway, and M shows the location of a marine biology experimental station on an island. A cable is to be laid connecting the relay station with the experimental station. If the cost of running the cable on land is $2 per running foot and the cost of running the cable underwater is $6 per running foot, find the total cost for laying the cable.

Figure 1.8
Cable connecting relay station S to experimental station M.

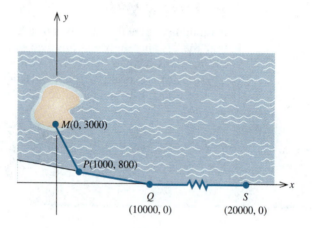

Solution   The length of cable required on land is given by the distance from P to Q plus the distance from Q to S. The distance is

$$\sqrt{(10{,}000 - 1000)^2 + (0 - 800)^2} + \sqrt{(20{,}000 - 10{,}000)^2 + (0 - 0)^2}$$
$$= \sqrt{9000^2 + 800^2} + 10{,}000$$
$$= \sqrt{81{,}640{,}000} + 10{,}000$$
$$\approx 19{,}035.49$$

or approximately 19,035.49 feet. Next, we see that the length of cable required underwater is given by the distance from M to P. This distance is

$$\sqrt{(0 - 1000)^2 + (3000 - 800)^2} = \sqrt{1000^2 + 2200^2}$$
$$= \sqrt{5{,}840{,}000}$$
$$\approx 2416.61$$

or approximately 2,416.61 feet. Therefore, the total cost for laying the cable is

$$2(19{,}035.49) + 6(2416.61) = 52{,}570.64$$

or approximately $52,571.   ⚬ ⚬ ⚬

In the Cartesian coordinate system, the two axes are perpendicular to each other. Consider a coordinate system in which the x- and y-axes are noncollinear (that is, the axes do not lie along a straight line) and are not perpendicular to each other (see the accompanying figure).

a. Describe how a point is represented in this coordinate system by an ordered pair (x, y) of real numbers. Conversely, show how an ordered pair (x, y) of real numbers uniquely determines a point in the plane.

b. Suppose you want to find a formula for the distance between two points $P_1(x_1, y_1)$ and $P_2(x_2, y_2)$ in the plane. What is the advantage that the Cartesian coordinate system has over the coordinate system under consideration? Comment on your answer.

SELF-CHECK EXERCISES 1.3

1. a. Plot the points $A(4, -2)$, $B(2, 3)$, and $C(-3, 1)$.

   b. Find the distance between the points A and B; between B and C; between A and C.

   c. Use the Pythagorean Theorem to show that the triangle with vertices A, B, and C is a right triangle.

2. The figure at the top of page 32 shows the location of cities A, B, and C. Suppose that a pilot wishes to fly from city A to city C but must make a mandatory stopover in city B. If the single-engine light plane has a range of 650 miles, can she make the trip without refueling in city B?

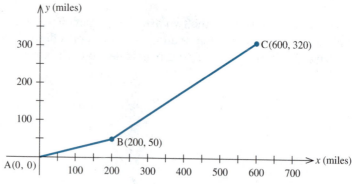

Solutions to Self-Check Exercises 1.3 can be found on page 34.

1.3 EXERCISES

In exercises 1–6, refer to the following figure and determine the coordinates of each point and the quadrant in which it is located.

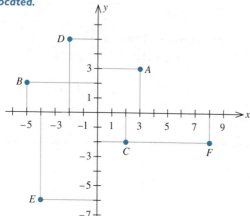

1. *A*   **2.** *B*   **3.** *C*

4. *D*   **5.** *E*   **6.** *F*

In exercises 7–12, refer to the following figure.

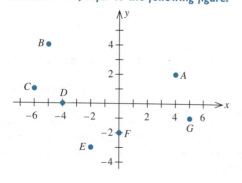

7. Which point has coordinates (4, 2)?

8. What are the coordinates of point *B*?

9. Which points have negative *y*-coordinates?

10. Which point has a negative *x*-coordinate and a negative *y*-coordinate?

11. Which point has an *x*-coordinate that is equal to zero?

12. Which point has a *y*-coordinate that is equal to zero?

In exercises 13–20, sketch a set of coordinate axes and plot each point.

13. $(-2, 5)$ **14.** $(1, 3)$

15. $(3, -1)$ **16.** $(3, -4)$

17. $(8, -7/2)$ **18.** $(-5/2, 3/2)$

19. $(4.5, -4.5)$ **20.** $(1.2, -3.4)$

In exercises 21–24, find the distance between the given points.

21. $(1, 3)$ and $(4, 7)$ **22.** $(1, 0)$ and $(4, 4)$

23. $(-1, 3)$ and $(4, 9)$ **24.** $(-2, 1)$ and $(10, 6)$

25. Find the coordinates of the points that are 10 units away from the origin and have a *y*-coordinate equal to -6.

26. Find the coordinates of the points that are 5 units away from the origin and have an *x*-coordinate equal to 3.

27. Show that the points $(3, 4), (-3, 7), (-6, 1),$ and $(0, -2)$ form the vertices of a square.

28. Show that the triangle with vertices $(-5, 2), (-2, 5),$ and $(5, -2)$ is a right triangle.

29. Distance Traveled A grand tour of four cities begins at city A and makes successive stops at cities B, C, and D before returning to city A. If the cities are located as shown in the following figure, find the total distance covered on the tour.

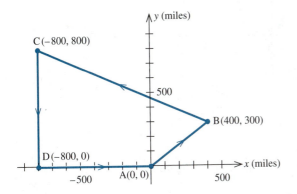

30. Delivery Charges A furniture store offers free setup and delivery services to all points within a 25-mile radius of its warehouse distribution center. If you live 20 miles east and 14 miles south of the warehouse, will you incur a delivery charge? Justify your answer.

31. Travel Time Towns A, B, C, and D are located as shown in the following figure. Two highways link town A to town D. Route 1 runs from town A to town D via town B, and route 2 runs from town A to town D via town C. If a salesman wishes to drive from town A to town D and traffic conditions are such that he could expect to average the same speed on either route, which highway should he take in order to arrive in the shortest time?

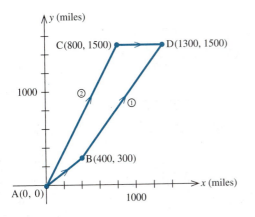

32. Minimizing Shipping Costs Refer to the figure for exercise 31. Suppose that a fleet of 100 automobiles are to be shipped from an assembly plant in town A to town D. They may be shipped either by freight train along route 1 at a cost of 11 cents per mile per automobile or by truck along route 2 at a cost of $10\frac{1}{2}$ cents per mile per automobile. Which means of transportation minimizes the shipping cost? What is the net savings?

33. Consumer Decisions Mr. Barclay wishes to determine which antenna he should purchase for his home. The TV store has supplied him with the following information:

Range in Miles			
VHF	UHF	Model	Price
30	20	A	$40.00
45	35	B	$50.00
60	40	C	$60.00
75	55	D	$70.00

Barclay wishes to receive channel 17 (VHF), which is located 25 miles east and 35 miles north of his home, and channel 38 (UHF), which is located 20 miles south and 32 miles west of his home. Which model will allow him to receive both channels at the least cost? (Assume that the terrain between Barclay's home and both broadcasting stations is flat.)

34. Calculating the Cost of Laying Cable In the following diagram, S represents the position of a power relay station located on a coastal highway, and M shows the location of a marine biology experimental station on an island. A cable is to be laid connecting the relay station with the experimental station. If the cost of running the cable on land is $2 per running foot and the cost of

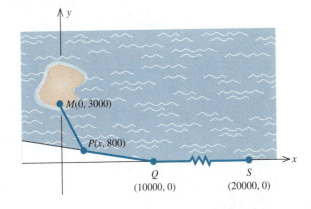

running cable under water is $6 per running foot, find an expression in terms of x that gives the total cost for laying the cable. Use this expression to find the total cost when $x = 900$. When $x = 1000$.

35. Let (x_1, y_1) and (x_2, y_2) be two points lying in the xy-plane. Show that the distance between the two points is given by

$$d = \sqrt{(x_2 - x_1)^2 + (y_2 - y_1)^2}$$

[*Hint:* Refer to the accompanying figure and use the Pythagorean Theorem.]

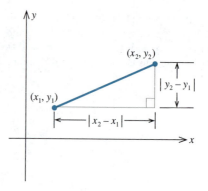

SOLUTIONS TO SELF–CHECK EXERCISES 1.3

1. a. The points are plotted in the following figure:

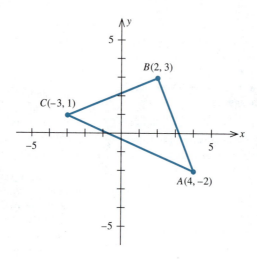

b. The distance between A and B is

$$d(A, B) = \sqrt{(2 - 4)^2 + [3 - (-2)]^2}$$
$$= \sqrt{(-2)^2 + 5^2} = \sqrt{4 + 25} = \sqrt{29}$$

The distance between B and C is

$$d(B, C) = \sqrt{(-3 - 2)^2 + (1 - 3)^2}$$
$$= \sqrt{(-5)^2 + (-2)^2} = \sqrt{25 + 4} = \sqrt{29}$$

The distance between A and C is

$$d(A, C) = \sqrt{(-3 - 4)^2 + [1 - (-2)]^2}$$
$$= \sqrt{(-7)^2 + 3^2} = \sqrt{49 + 9} = \sqrt{58}$$

c. We will show that

$$[d(A, C)]^2 = [d(A, B)]^2 + [d(B, C)]^2$$

From (b), we see that $[d(A, B)]^2 = 29$, $[d(B, C)]^2 = 29$, and $[d(A, C)]^2 = 58$, and the desired result follows.

2. The distance between city A and city B is

$$d(A, B) = \sqrt{200^2 + 50^2} \approx 206$$

or 206 miles. The distance between city B and city C is

$$d(B, C) = \sqrt{[600 - 200]^2 + [320 - 50]^2}$$
$$= \sqrt{400^2 + 270^2} \approx 483$$

or 483 miles. Therefore, the total distance the pilot would have to cover is 689 miles, so she must refuel in city B.

1.4 STRAIGHT LINES

In computing income tax, business firms are allowed by law to depreciate certain assets such as buildings, machines, furniture, automobiles, and so on, over a period of time. Linear depreciation, or the straight-line method, is often used for this purpose. The graph of the straight line shown in Figure 1.9 describes the book value V of a computer that has an initial value of $100,000 and that is being depreciated linearly over five years with a scrap value of $30,000. Note that only the solid portion of the straight line is of interest here.

The book value of the computer at the end of year t, where t lies between 0 and 5, can be read directly from the graph. But there is one shortcoming in this approach: The result depends on how accurately you draw and read the graph. A better and more accurate method is based on finding an *algebraic* representation of the depreciation line.

In order to see how a straight line in the xy-plane may be described algebraically, we need to first recall certain properties of straight lines.

Figure 1.9
Linear depreciation of an asset.

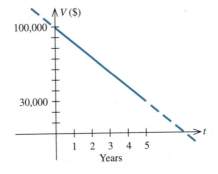

Figure 1.10
m is undefined.

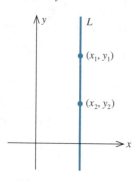

Slope of a Line

Let L denote the unique straight line that passes through the two distinct points (x_1, y_1) and (x_2, y_2). If $x_1 = x_2$, then L is a vertical line and the slope is undefined (Figure 1.10). If $x_1 \neq x_2$, we define the slope of L as follows:

SLOPE OF A NONVERTICAL LINE	If (x_1, y_1) and (x_2, y_2) are any two distinct points on a nonvertical line L, then the slope m of L is given by

$$m = \frac{\Delta y}{\Delta x} = \frac{y_2 - y_1}{x_2 - x_1} \tag{2}$$

Figure 1.11

Observe that the slope of a straight line is a constant whenever it is defined. The number $\Delta y = y_2 - y_1$ (Δy is read "delta y") is a measure of the vertical change in y, and $\Delta x = x_2 - x_1$ is a measure of the horizontal change in x, as shown in Figure 1.11. From this figure we can see that the slope m of a straight line L is a measure of the *rate of change of y with respect to x.*

Figure 1.12a shows a straight line L_1 with slope 2. Observe that L_1 has the property that a unit increase in x results in a 2-unit increase in y. To see this, let $\Delta x = 1$ in formula (2) so that $m = \Delta y$. Since $m = 2$, we conclude that $\Delta y = 2$. Similarly, Figure 1.12b shows a line L_2 with slope -1. Observe that a straight line with positive slope slants upward from left to right (y

Figure 1.12

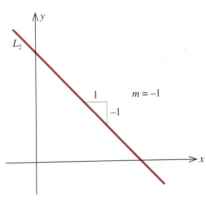

(a) The line rises ($m > 0$).

(b) The line falls ($m < 0$).

Figure 1.13
A family of straight lines.

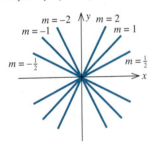

increases as x increases), whereas a line with negative slope slants downward from left to right (y decreases as x increases). Finally, Figure 1.13 shows a family of straight lines passing through the origin with indicated slopes.

EXAMPLE 1   Sketch the straight line that passes through the point $(-2, 5)$ and has slope $-4/3$.

Solution   First plot the point $(-2, 5)$ (Figure 1.14). Next, recall that a slope of $-4/3$ indicates that an increase of 1 unit in the x-direction produces a *decrease* of 4/3 units in the y-direction, or equivalently, a 3-unit increase in the x-direction produces a 3(4/3), or 4-unit, decrease in the y-direction. Using this information, we plot the point (1, 1) and draw the line through the two points.

Figure 1.14
L has slope $-4/3$ and passes through $(-2, 5)$.

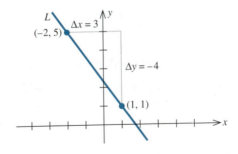

Show that the slope of a nonvertical line is independent of the two distinct points $P_1(x_1, y_1)$ and $P_2(x_2, y_2)$ used to compute it. *Hint:* Suppose we pick two other distinct points, $P_3(x_3, y_3)$ and $P_4(x_4, y_4)$ lying on L. Draw a picture and use similar triangles to demonstrate that using P_3 and P_4 gives the same value as that obtained using P_1 and P_2.

EXAMPLE 2 Find the slope m of the line that passes through the points $(-1, 1)$ and $(5, 3)$.

Solution Choose (x_1, y_1) to be the point $(-1, 1)$ and (x_2, y_2) to be the point $(5, 3)$. Then, with $x_1 = -1$, $y_1 = 1$, $x_2 = 5$, and $y_2 = 3$, we find

$$m = \frac{y_2 - y_1}{x_2 - x_1} = \frac{3 - 1}{5 - (-1)} = \frac{1}{3} \qquad \text{[Using (2)]}$$

Figure 1.15
L passes through (5, 3) and (−1, 1).

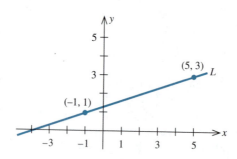

(Figure 1.15). Try to verify that the result obtained would have been the same had we chosen the point $(-1, 1)$ to be (x_2, y_2) and the point $(5, 3)$ to be (x_1, y_1).

Figure 1.16
The slope of the horizontal line L is 0.

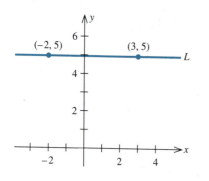

EXAMPLE 3 Find the slope of the line that passes through the points $(-2, 5)$ and $(3, 5)$.

Solution The slope of the required line is given by

$$m = \frac{5 - 5}{3 - (-2)} = \frac{0}{5} = 0$$

(Figure 1.16).

REMARK   In general, the slope of a horizontal line is zero.

We can use the slope of a straight line to determine whether a line is parallel to another line.

PARALLEL LINES

Two distinct lines are **parallel** if and only if their slopes are equal or their slopes are undefined.

EXAMPLE 4 Let L_1 be a line that passes through the points $(-2, 9)$ and $(1, 3)$, and let L_2 be the line that passes through the points $(-4, 10)$ and $(3, -4)$. Determine whether L_1 and L_2 are parallel.

Figure 1.17

L_1 and L_2 have the same slope and hence are parallel.

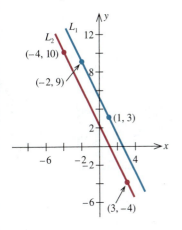

Solution   The slope m_1 of L_1 is given by

$$m_1 = \frac{3-9}{1-(-2)} = -2$$

The slope m_2 of L_2 is given by

$$m_2 = \frac{-4-10}{3-(-4)} = -2$$

Since $m_1 = m_2$, the lines L_1 and L_2 are in fact parallel (Figure 1.17).   ◦ ◦ ◦

Equations of Lines

We will now show that every straight line lying in the *xy*-plane may be represented by an equation involving the variables *x* and *y*. One immediate benefit of this is that problems involving straight lines may be solved algebraically.

Let *L* be a straight line parallel to the *y*-axis (perpendicular to the *x*-axis) (Figure 1.18). Then *L* crosses the *x*-axis at some point $(a, 0)$ with the *x*-coordinate given by $x = a$, where *a* is some real number. Any other point on *L* has the form $(a, \bar{y})$, where $\bar{y}$ is an appropriate number. Therefore, the vertical line *L* is described by the sole condition

$$x = a$$

and this is, accordingly, the equation of *L*. For example, the equation $x = -2$ represents a vertical line 2 units to the left of the *y*-axis, and the equation $x = 3$ represents a vertical line 3 units to the right of the *y*-axis (Figure 1.19).

Next, suppose that *L* is a nonvertical line so that it has a well-defined slope *m*. Suppose (x_1, y_1) is a fixed point lying on *L* and (x, y) is a variable point on *L* distinct from (x_1, y_1) (Figure 1.20).

Figure 1.18

The vertical line $x = a$.

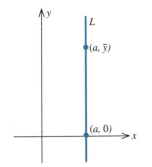

Figure 1.19

The vertical lines $x = -2$ and $x = 3$.

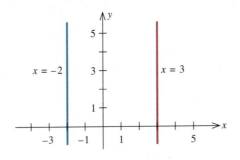

Figure 1.20

L passes through (x_1, y_1) and has slope m.

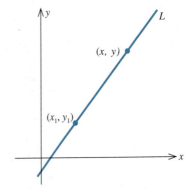

Using formula (2) with the point $(x_2, y_2) = (x, y)$, we find that the slope of L is given by

$$m = \frac{y - y_1}{x - x_1}$$

Upon multiplying both sides of the equation by $x - x_1$, we obtain formula (3).

POINT-SLOPE FORM

An equation of the line that has slope m and passes through the point (x_1, y_1) is given by

$$y - y_1 = m(x - x_1) \tag{3}$$

Equation (3) is called the **point-slope form of the equation of a line** since it utilizes a given point (x_1, y_1) on a line and the slope m of the line.

Figure 1.21

L passes through (1, 3) *and has slope 2.*

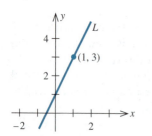

EXAMPLE 5 Find an equation of the line that passes through the point (1, 3) and has slope 2.

Solution Using the point-slope form of the equation of a line with the point (1, 3) and $m = 2$, we obtain

$$y - 3 = 2(x - 1) \qquad [(y - y_1) = m(x - x_1)]$$

which, when simplified, becomes

$$2x - y + 1 = 0$$

(Figure 1.21).   ◦ ◦ ◦

EXAMPLE 6 Find an equation of the line that passes through the points $(-3, 2)$ and $(4, -1)$.

Solution The slope of the line is given by

$$m = \frac{-1 - 2}{4 - (-3)} = -\frac{3}{7}$$

Using the point-slope form of the equation of a line with the point $(4, -1)$ and the slope $m = -3/7$, we have

$$y + 1 = -\frac{3}{7}(x - 4) \qquad [(y - y_1) = m(x - x_1)]$$

$$7y + 7 = -3x + 12$$

or $3x + 7y - 5 = 0$

(Figure 1.22).

Figure 1.22
L passes through $(-3, 2)$ *and*
$(4, -1)$.

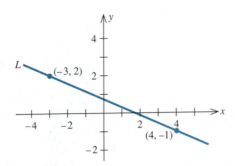

Consider the slope-intercept form of a straight line $y = mx + b$. Describe the family of straight lines obtained by keeping

a. the value of m fixed and allowing the value of b to vary
b. the value of b fixed and allowing the value of m to vary

We can use the slope of a straight line to determine whether a line is perpendicular to another line.

PERPENDICULAR LINES

If L_1 and L_2 are two distinct nonvertical lines that have slopes m_1 and m_2, respectively, then L_1 is **perpendicular** to L_2 (written $L_1 \perp L_2$) if and only if

$$m_1 = -\frac{1}{m_2}$$

If the line L_1 is vertical (so that its slope is undefined), then L_1 is perpendicular to another line, L_2, if and only if L_2 is horizontal (so that its slope is zero). For a proof of these results, see exercise 76, page 49.

Figure 1.23
L_2 *is perpendicular to* L_1 *and passes through* (3, 1).

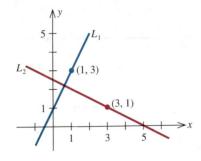

EXAMPLE 7 Find an equation of the line that passes through the point (3, 1) and is perpendicular to the line of Example 5.

Solution Since the slope of the line in Example 5 is 2, the slope of the required line is given by $m = -1/2$, the negative reciprocal of 2. Using the point-slope form of the equation of a line, we obtain

$$y - 1 = -\frac{1}{2}(x - 3) \qquad [(y - y_1) = m(x - x_1)]$$

$$2y - 2 = -x + 3$$

or $x + 2y - 5 = 0$

(Figure 1.23).

EXPLORING WITH TECHNOLOGY

1. Use a graphing utility to plot the straight lines L_1 and L_2 with equations $2x + y - 5 = 0$ and $41x + 20y - 11 = 0$ on the same set of axes using the standard viewing rectangle.
   a. Can you tell if the lines L_1 and L_2 are parallel to each other?
   b. Verify your observations by computing the slopes of L_1 and L_2 algebraically.
2. Use a graphing utility to plot the straight lines L_1 and L_2 with equations $x + 2y - 5 = 0$ and $5x - y + 5 = 0$ on the same set of axes using the standard viewing rectangle.
   a. Can you tell if the lines L_1 and L_2 are perpendicular to each other?
   b. Verify your observation by computing the slopes of L_1 and L_2 algebraically.

○   ○   ○

Figure 1.24

The line L has x-intercept a and y-intercept b.

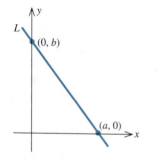

A straight line L that is neither horizontal nor vertical cuts the x-axis and the y-axis at, say, points $(a, 0)$ and $(0, b)$, respectively (Figure 1.24). The numbers a and b are called the **x-intercept** and **y-intercept,** respectively, of L.

Now, let L be a line with slope m and y-intercept b. Using formula (3), the point-slope form of the equation of a line, with the point $(0, b)$ and slope m, we have

$$y - b = m(x - 0)$$

or

$$y = mx + b$$

SLOPE-INTERCEPT FORM

The equation of the line that has slope m and intersects the y-axis at the point $(0, b)$ is given by

$$y = mx + b \tag{4}$$

EXAMPLE 8   Find an equation of the line that has slope 3 and y-intercept -4.

Solution   Using equation (4) with $m = 3$ and $b = -4$, we obtain the required equation

$$y = 3x - 4$$

○ ○ ○

EXAMPLE 9 Determine the slope and y-intercept of the line whose equation is $3x - 4y = 8$.

Solution Rewrite the given equation in the slope-intercept form and obtain

$$y = \frac{3}{4}x - 2$$

Comparing this result with equation (4), we find $m = 3/4$ and $b = -2$, and we conclude that the slope and y-intercept of the given line are 3/4 and -2, respectively.

○ ○ ○

EXPLORING WITH TECHNOLOGY

1. Use a graphing utility to plot the straight lines with equations $y = -2x + 3$, $y = -x + 3$, $y = x + 3$, and $y = 2.5x + 3$ on the same set of axes using the standard viewing rectangle. What effect does changing the coefficient m of x in the equation $y = mx + b$ have on its graph?

2. Use a graphing utility to plot the straight lines with equations $y = 2x - 2$, $y = 2x - 1$, $y = 2x$, $y = 2x + 1$, and $y = 2x + 4$ on the same set of axes using the standard viewing rectangle. What effect does changing the constant b in the equation $y = mx + b$ have on its graph?

3. Describe in words the effect of changing both m and b in the equation $y = mx + b$.

○ ○ ○

Applications

EXAMPLE 10 The sales manager of a local sporting goods store plotted sales versus time for the last five years and found the points to lie approximately along a straight line (Figure 1.25). By using the points corresponding to the first and fifth years, find an equation of the trend line. What sales figure can be predicted for the sixth year?

Figure 1.25
Sales of a sporting goods store.

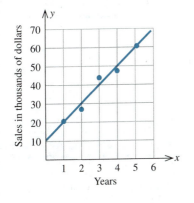

Solution Using formula (2) with the points (1, 20) and (5, 60), we find that the slope of the required line is given by

$$m = \frac{60 - 20}{5 - 1} = 10$$

Next, using the point-slope form of the equation of a line with the point (1, 20) and $m = 10$, we obtain

$$y - 20 = 10(x - 1) \qquad [(y - y_1) = m(x - x_1)]$$

or

$$y = 10x + 10$$

as the required equation.

The sales figure for the sixth year is obtained by letting $x = 6$ in the last equation, giving

$$y = 70$$

or $70,000.

o o o

EXAMPLE 11   Suppose that an art object purchased for $50,000 is expected to appreciate in value at a constant rate of $5000 per year for the next five years. Use formula (4) to write an equation predicting the value of the art object in the next several years. What will its value be three years from the date of purchase?

Solution   Let x denote the time (in years) that has elapsed since the date the object was purchased and let y denote the object's value (in dollars). Then $y = 50,000$ when $x = 0$. Furthermore, the slope of the required equation is given by $m = 5000$, since each unit increase in x (one year) implies an increase of 5000 units (dollars) in y. Using (4) with $m = 5000$ and $b = 50,000$, we obtain

$$y = 5000x + 50,000 \qquad (y = mx + b)$$

Three years from the date of purchase, the value of the object will be given by

$$y = 5000(3) + 50,000$$

or $65,000.

o o o

Refer to Example 11.

Can the equation predicting the value of the art object be used to predict long-term growth?

General Equation of a Line

We have considered several forms of the equation of a straight line in the plane. These different forms of the equation are equivalent to each other. In fact, each is a special case of the following equation.

GENERAL FORM OF A LINEAR EQUATION

The equation

$$Ax + By + C = 0 \qquad (5)$$

where A, B, and C are constants and A and B are not both zero, is called the general form of a linear equation in the variables x and y.

We will now state (without proof) an important result concerning the algebraic representation of straight lines in the plane.

THEOREM 1

An equation of a straight line is a linear equation; conversely, every linear equation represents a straight line.

This result justifies the use of the adjective *linear* describing Equation (5).

EXAMPLE 12 Sketch the straight line represented by the equation

$$3x - 4y - 12 = 0$$

Solution Since every straight line is uniquely determined by two distinct points, we need find only two such points through which the line passes in order to sketch it. For convenience let us compute the x- and y-intercepts. Setting $y = 0$, we find $x = 4$; thus the x-intercept is 4. Setting $x = 0$ gives $y = -3$, and the y-intercept is -3. A sketch of the line appears in Figure 1.26.

⚬ ⚬ ⚬

Following is a summary of the common forms of the equations of straight lines discussed in this section.

Figure 1.26
To sketch $3x - 4y - 12 = 0$, first find the x-intercept, 4, and the y-intercept, -3.

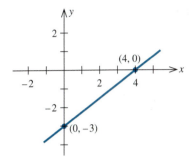

EQUATIONS OF STRAIGHT LINES		
Vertical line:	$x = a$	
Horizontal line:	$y = b$	
Point-slope form:	$y - y_1 = m(x - x_1)$	
Slope-intercept form:	$y = mx + b$	
General form:	$Ax + By + C = 0$	

SELF-CHECK EXERCISES 1.4

1. Determine the number a so that the line passing through the points $(a, 2)$ and $(3, 6)$ is parallel to a line with slope 4.

2. Find an equation of the line that passes through the point $(3, -1)$ and is perpendicular to a line with slope $-1/2$.

3. Does the point $(3, -3)$ lie on the line with equation $2x - 3y - 12 = 0$? Sketch the graph of the line.

4. The percentage of people over 65 who have high school diplomas is summarized in the following table:

Year (x)	1960	1965	1970	1975	1980	1985	1990
Percent with Diplomas (y)	20	25	30	36	42	47	52

Source: The World Almanac.

a. Plot the percentage of people over 65 who have high school diplomas (y) versus the year (x).

b. Draw the straight line L through the points (1960, 20) and (1990, 52).

c. Find an equation of the line L.

d. If the trend continues, estimate the percentage of people over 65 who will have high school diplomas by the year 2000.

Solutions to Self-Check Exercises 1.4 can be found on page 50.

 EXERCISES

In exercises 1–6, match the statement with one of the graphs (a)–(f).

1. The slope of the line is zero.

2. The slope of the line is undefined.

3. The slope of the line is positive and its y-intercept is positive.

4. The slope of the line is positive and its y-intercept is negative.

5. The slope of the line is negative and its x-intercept is negative.

6. The slope of the line is negative and its x-intercept is positive.

a.

b.

c.

d.

e.

f.

10.

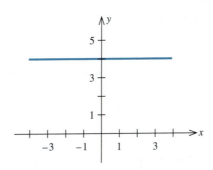

In exercises 7–10, find the slope of the line shown in each figure.

7.

8.

9.

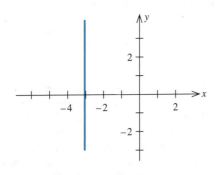

In exercises 11–16, find the slope of the line that passes through each pair of points.

11. (4, 3) and (5, 8)   **12.** (4, 5) and (3, 8)

13. (−2, 3) and (4, 8)   **14.** (−2, −2) and (4, −4)

15. (a, b) and (c, d)

16. (−a + 1, b − 1) and (a + 1, −b)

17. Given the equation $y = 4x - 3$, answer the following questions:
a. If x increases by 1 unit, what is the corresponding change in y?
b. If x decreases by 2 units, what is the corresponding change in y?

18. Given the equation $2x + 3y = 4$, answer the following questions:
a. Is the slope of the line described by this equation positive or negative?
b. As x increases in value, does y increase or decrease?
c. If x decreases by 2 units, what is the corresponding change in y?

In exercises 19 and 20, determine whether the line through each pair of points is parallel.

19. $A(1, -2)$, $B(-3, -10)$ and $C(1, 5)$, $D(-1, 1)$

20. $A(2, 3)$, $B(2, -2)$ and $C(-2, 4)$, $D(-2, 5)$

In exercises 21 and 22, determine whether the line through each pair of points is perpendicular.

21. $A(-2, 5)$, $B(4, 2)$ and $C(-1, -2)$, $D(3, 6)$

22. $A(2, 0)$, $B(1, -2)$ and $C(4, 2)$, $D(-8, 4)$

23. If the line passing through the points $(1, a)$ and $(4, -2)$ is parallel to the line passing through the points $(2, 8)$ and $(-7, a + 4)$, what is the value of a?

24. If the line passing through the points $(a, 1)$ and $(5, 8)$ is parallel to the line passing through the points $(4, 9)$ and $(a + 2, 1)$, what is the value of a?

25. Find an equation of the horizontal line that passes through $(-4, -3)$.

26. Find an equation of the vertical line that passes through $(0, 5)$.

In exercises 27–30, find an equation of the line that passes through the point and has the indicated slope m.

27. $(3, -4)$; $m = 2$ **28.** $(2, 4)$; $m = -1$

29. $(-3, 2)$; $m = 0$ **30.** $(1, 2)$; $m = -1/2$

In exercises 31–34, find an equation of the line that passes through the given points.

31. $(2, 4)$ and $(3, 7)$ **32.** $(2, 1)$ and $(2, 5)$

33. $(1, 2)$ and $(-3, -2)$ **34.** $(-1, -2)$ and $(3, -4)$

In exercises 35–38, find an equation of the line that has slope m and y-intercept b.

35. $m = 3$; $b = 4$ **36.** $m = -2$; $b = -1$

37. $m = 0$; $b = 5$ **38.** $m = -1/2$; $b = 3/4$

In exercises 39–44, write the equation in the slope-intercept form and then find the slope and y-intercept of the corresponding line.

39. $x - 2y = 0$ **40.** $y - 2 = 0$

41. $2x - 3y - 9 = 0$ **42.** $3x - 4y + 8 = 0$

43. $2x + 4y = 14$ **44.** $5x + 8y - 24 = 0$

45. Find an equation of the line that passes through the point $(-2, 2)$ and is parallel to the line $2x - 4y - 8 = 0$.

46. Find an equation of the line that passes through the point $(2, 4)$ and is perpendicular to the line $3x + 4y - 22 = 0$.

In exercises 47–52, find an equation of the line that satisfies the given condition.

47. The line parallel to the x-axis and 6 units below it

48. The line passing through the origin and parallel to the line joining the points $(2, 4)$ and $(4, 7)$

49. The line passing through the point (a, b) with slope equal to zero

50. The line passing through $(-3, 4)$ and parallel to the x-axis

51. The line passing through $(-5, -4)$ and parallel to the line joining $(-3, 2)$ and $(6, 8)$

52. The line passing through (a, b) with undefined slope

53. Given that the point $P(-3, 5)$ lies on the line $kx + 3y + 9 = 0$, find k.

54. Given that the point $P(2, -3)$ lies on the line $-2x + ky + 10 = 0$, find k.

In exercises 55–60, sketch the straight line defined by the given linear equation by finding the x- and y-intercepts. [Hint: See Example 12, page 45.]

55. $3x - 2y + 6 = 0$ **56.** $2x - 5y + 10 = 0$

57. $x + 2y - 4 = 0$ **58.** $2x + 3y - 15 = 0$

59. $y + 5 = 0$ **60.** $-2x - 8y + 24 = 0$

61. Show that an equation of a line through the points $(a, 0)$ and $(0, b)$ with $a \neq 0$ and $b \neq 0$ can be written in the form

$$\frac{x}{a} + \frac{y}{b} = 1$$

(Recall that the numbers a and b are the x- and y-intercepts, respectively, of the line. This form of an equation of a line is called the **intercept form.**)

In exercises 62–65, use the results of exercise 61 to find an equation of a line with the given x- and y-intercepts.

62. x-intercept 3; y-intercept 4

63. x-intercept -2; y-intercept -4

64. x-intercept $-\frac{1}{2}$; y-intercept $\frac{3}{4}$

65. x-intercept 4; y-intercept $-\frac{1}{2}$

In exercises 66 and 67, determine whether the given points lie on a straight line.

66. $A(-1, 7)$, $B(2, -2)$, and $C(5, -9)$

67. $A(-2, 1)$, $B(1, 7)$, and $C(4, 13)$

68. Social Security Contributions   For wages less than the maximum taxable wage base, social security contributions by employees are 7.65% of the employee's wages.

a. Find an equation that expresses the relationship between the wages earned (x) and the social security taxes paid (y) by an employee who earns less than the maximum taxable wage base.

b. For each additional dollar that an employee earns, by how much is his or her social security contribution increased? (Assume that the employee's wages are less than the maximum taxable wage base.)

c. What social security contributions will an employee who earns \$35,000 (which is less than the maximum taxable wage base) be required to make?

69. College Admissions Using data compiled by the Admissions Office at Faber University, college admissions officers estimate that 55% of the students who are offered admission to the freshman class at the university will actually enroll.

a. Find an equation that expresses the relationship between the number of students who actually enroll (y) and the number of students who are offered admission to the university (x).

b. If the desired freshman class size for the upcoming academic year is 1100 students, how many students should be admitted?

70. Weight of Whales The equation $W = 3.51L - 192$, expressing the relationship between the length L (in feet) and the expected weight W (in British tons) of adult blue whales, was adopted in the late 1960s by the International Whaling Commission.

a. What is the expected weight of an 80-foot blue whale?

b. Sketch the straight line that represents the equation.

71. Cost of a Commodity A manufacturer obtained the following data relating the cost y (in dollars) to the number of units (x) of a commodity produced:

Number of Units Produced (x)	0	20	40	60	80	100
Cost in Dollars (y)	200	208	222	230	242	250

a. Plot the cost (y) versus the quantity produced (x).

b. Draw the straight line through the points (0, 200) and (100, 250).

c. Derive an equation of the straight line of (b).

d. Taking this equation to be an approximation of the relationship between the cost and the level of production, estimate the cost of producing 54 units of the commodity.

72. Ideal Heights and Weights for Women The Venus Health Club for Women provides its members with the following table, which gives the average desirable weight for women of a certain height:

Height x (in inches)	60	63	66	69	72
Weight y (in pounds)	108	118	129	140	152

a. Plot the weight (y) versus the height (x).

b. Draw a straight line L through the points corresponding to heights of 5 feet and 6 feet.

c. Derive an equation of the line L.

d. Using the equation of (c), estimate the average desirable weight for a woman who is 5 feet 5 inches tall.

73. Sales Growth Metro Department Store's annual sales (in millions of dollars) during the past five years were:

Annual Sales (x)	5.8	6.2	7.2	8.4	9.0
Year (y)	1	2	3	4	5

a. Plot the annual sales (y) versus the year (x).

b. Draw a straight line L through the points corresponding to the first and fifth years.

c. Derive an equation of the line L.

d. Using the equation found in (c), estimate Metro's annual sales four years from now ($x = 9$).

74. Is there a difference between the statements "The slope of a straight line is zero" and "The slope of a straight line does not exist (is not defined)"? Explain your answer.

75. Show that two distinct lines with equations $a_1x + b_1y + c_1 = 0$ and $a_2x + b_2y + c_2 = 0$, respectively, are parallel if and only if $a_1b_2 - b_1a_2 = 0$.

[*Hint:* Write each equation in the slope-intercept form and compare.]

76. Prove that if a line L_1 with slope m_1 is perpendicular to a line L_2 with slope m_2, then $m_1m_2 = -1$.

[*Hint:* Refer to the following figure. Show that $m_1 = b$ and $m_2 = c$. Next, apply the Pythagorean Theorem to triangles OAC, OCB, and OBA to show that $1 = -bc$.]

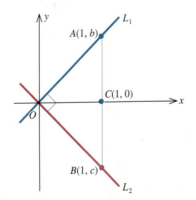

SOLUTIONS TO SELF-CHECK EXERCISES 1.4

I. The slope of the line that passes through the points $(a, 2)$ and $(3, 6)$ is

$$m = \frac{6 - 2}{3 - a}$$

$$= \frac{4}{3 - a}$$

Since this line is parallel to a line with slope 4, m must be equal to 4; that is,

$$\frac{4}{3 - a} = 4$$

or, upon multiplying both sides of the equation by $3 - a$,

$$4 = 4(3 - a)$$
$$4 = 12 - 4a$$
$$4a = 8$$

and $a = 2$

2. Since the required line L is perpendicular to a line with slope $-1/2$, the slope of L is

$$-\frac{1}{-1/2} = 2$$

Next, using the point-slope form of the equation of a line, we have

$$y - (-1) = 2(x - 3)$$
$$y + 1 = 2x - 6$$

or $y = 2x - 7$

3. Substituting $x = 3$ and $y = -3$ into the left-hand side of the given equation, we find

$$2(3) - 3(-3) - 12 = 3$$

which is not equal to zero (the right-hand side). Therefore, $(3, -3)$ does not lie on the line with equation $2x - 3y - 12 = 0$.

Setting $x = 0$, we find $y = -4$, the y-intercept. Next, setting $y = 0$ gives $x = 6$, the x-intercept. We now draw the line passing through the points $(0, -4)$ and $(6, 0)$ as shown.

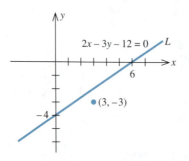

4. a. and **b.** See the accompanying figure.

c. The slope of L is

$$m = \frac{52 - 20}{1990 - 1960} = \frac{32}{30} = \frac{16}{15}$$

Using the point-slope form of the equation of a line with the point (1960, 20), we find

$$y - 20 = \frac{16}{15}(x - 1960) = \frac{16}{15}x - \frac{(16)(1960)}{15}$$

or

$$y = \frac{16}{15}x - \frac{6272}{3} + 20$$

$$= \frac{16}{15}x - \frac{6212}{3}$$

d. To estimate the percentage of people over 65 who will have high school diplomas by the year 2000, let $x = 2000$ in the equation obtained in (c). Thus, the required estimate is

$$y = \frac{16}{15}(2000) - \frac{6212}{3} \approx 62.67$$

or approximately 63%.

Group projects for this chapter can be found at the Brooks/Cole Web site: http://www.brookscole.com/math/authors/tans/

CHAPTER I SUMMARY OF PRINCIPAL FORMULAS AND TERMS

Formulas

I. Quadratic formula

$$x = \frac{-b \pm \sqrt{b^2 - 4ac}}{2a}$$

2. Distance between two points

$$d = \sqrt{(x_2 - x_1)^2 + (y_2 - y_1)^2}$$

3. Slope of a line $m = \dfrac{y_2 - y_1}{x_2 - x_1}$

4. Equation of a vertical line $x = a$

5. Equation of a horizontal line $y = b$

6. Point-slope form of the $y - y_1 = m(x - x_1)$
equation of a line

7. Slope-intercept form of the $y = mx + b$
equation of a line

8. General equation of a line $Ax + By + C = 0$

Terms

Real number (coordinate) line   Polynomial

Open interval   Roots of a polynomial equation

Closed interval   Cartesian coordinate system

Half-open interval   Ordered pair

Finite interval   Parallel lines

Infinite interval   Perpendicular lines

Absolute value

CHAPTER 1 REVIEW EXERCISES

In exercises 1–4, find the values of x that satisfy the inequality (inequalities).

1. $-x + 3 \le 2x + 9$ **2.** $-2 \le 3x + 1 \le 7$

3. $x - 3 > 2$ or $x + 3 < -1$

4. $2x^2 > 50$

In exercises 5–8, evaluate the expression.

5. $|-5 + 7| + |-2|$ **6.** $\left| \dfrac{5 - 12}{-4 - 3} \right|$

7. $|2\pi - 6| - \pi$

8. $|\sqrt{3} - 4| + |4 - 2\sqrt{3}|$

In exercises 9–14, evaluate the expression.

9. $\left(\dfrac{9}{4}\right)^{3/2}$ **10.** $\dfrac{5^6}{5^4}$

11. $(3 \cdot 4)^{-2}$ **12.** $(-8)^{5/3}$

13. $\dfrac{(3 \cdot 2^{-3})(4 \cdot 3^5)}{2 \cdot 9^3}$ **14.** $\dfrac{3\sqrt[3]{54}}{\sqrt[3]{18}}$

In exercises 15–19, simplify the expression.

15. $\dfrac{4(x^2 + y)^3}{x^2 + y}$ **16.** $\dfrac{a^6 b^{-5}}{(a^3 b^{-2})^{-3}}$

17. $\dfrac{\sqrt[4]{16x^5yz}}{\sqrt[4]{81xyz^5}}$, $(x, y,$ and z positive$)$

18. $(2x^3)(-3x^{-2})\left(\dfrac{1}{6}x^{-1/2}\right)$

19. $\left(\dfrac{3xy^2}{4x^3y}\right)^{-2}\left(\dfrac{3xy^3}{2x^2}\right)^3$

In exercises 20–23, factor the expression.

20. $-2\pi^2 r^3 + 100\pi r^2$ **21.** $2v^3w + 2vw^3 + 2u^2vw$

22. $16 - x^2$ **23.** $12t^3 - 6t^2 - 18t$

In exercises 24–27, solve the equation by factoring.

24. $8x^2 + 2x - 3 = 0$

25. $-6x^2 - 10x + 4 = 0$

26. $-x^3 - 2x^2 + 3x = 0$

27. $2x^4 + x^2 = 1$

In exercises 28 and 29, use the quadratic formula to solve the quadratic equation.

28. $x^2 - 2x - 5 = 0$

29. $2x^2 + 8x + 7 = 0$

In exercises 30–33, perform the indicated operations and simplify the expression.

30. $\dfrac{(t+6)(60) - (60t + 180)}{(t+6)^2}$

31. $\dfrac{6x}{2(3x^2 + 2)} + \dfrac{1}{4(x+2)}$

32. $\dfrac{2}{3}\left(\dfrac{4x}{2x^2 - 1}\right) + 3\left(\dfrac{3}{3x - 1}\right)$

33. $\dfrac{-2x}{\sqrt{x+1}} + 4\sqrt{x+1}$

34. Rationalize the numerator.

$$\frac{\sqrt{x} - 1}{x - 1}$$

35. Rationalize the denominator.

$$\frac{\sqrt{x} - 1}{2\sqrt{x}}$$

In exercises 36 and 37, find the distance between the two points.

36. $(-2, -3)$ and $(1, -7)$

37. $\left(\dfrac{1}{2}, \sqrt{3}\right)$ and $\left(-\dfrac{1}{2}, 2\sqrt{3}\right)$

In exercises 38–43, find an equation of the line L that passes through the point $(-2, 4)$ and satisfies the condition.

38. L is a vertical line.

39. L is a horizontal line.

40. L passes through the point $\left(3, \dfrac{7}{2}\right)$.

41. The x-intercept of L is 3.

42. L is parallel to the line $5x - 2y = 6$.

43. L is perpendicular to the line $4x + 3y = 6$.

44. Find an equation of the straight line that passes through the point $(2, 3)$ and is parallel to the line with equation $3x + 4y - 8 = 0$.

45. Find an equation of the straight line that passes through the point $(-1, 3)$ and is parallel to the line passing through the points $(-3, 4)$ and $(2, 1)$.

46. Find an equation of the line that passes through the point $(-2, -4)$ and is perpendicular to the line with equation $2x - 3y - 24 = 0$.

47. Sketch the graph of the equation $3x - 4y = 24$.

48. Sketch the graph of the line that passes through the point $(3, 2)$ and has slope $-2/3$.

49. Find the minimum cost C (in dollars) given that

$$2(1.5C + 80) \leq 2(2.5C - 20)$$

50. Find the maximum revenue R (in dollars) given that

$$12(2R - 320) \leq 4(3R + 240)$$

Additional study hints and sample chapter tests can be found at the Brooks/Cole Web site: http://www.brookscole.com/math/authors/tans/

In this chapter we define a *function*, a special relationship between two variables. The concept of a function enables us to describe many relationships that exist in applications. We also begin the study of differential calculus. Historically, differential calculus was developed in response to the problem of finding the tangent line to an arbitrary curve. But it quickly became apparent that solving this problem provided mathematicians with a method for solving many practical problems involving the rate of change of one quantity with respect to another. The basic tool used in differential calculus is the *derivative* of a function. The concept of the derivative is based, in turn, on a more fundamental notion— that of the *limit* of a function.

How does the change in the demand for a certain make of tires affect the unit price of the tires? The management of the Titan Tire Company has determined the demand function that relates the unit price of its Super Titan tires to the quantity demanded. In Example 7, page 155, you will see how this function can be used to compute the rate of change of the unit price of the Super Titan tires with respect to the quantity demanded.

2

FUNCTIONS, LIMITS, AND THE DERIVATIVE

2.1 FUNCTIONS AND THEIR GRAPHS

Functions

A manufacturer would like to know how his company's profit is related to its production level; a biologist would like to know how the size of the population of a certain culture of bacteria will change over time; a psychologist would like to know the relationship between the learning time of an individual and the length of a vocabulary list; and a chemist would like to know how the initial speed of a chemical reaction is related to the amount of substrate used. In each instance we are concerned with the same question: How does one quantity depend upon another? The relationship between two quantities is conveniently described in mathematics by using the concept of a function.

FUNCTION	A **function** is a rule that assigns to each element in a set A one and only one element in a set B.

The set A is called the **domain** of the function. It is customary to denote a function by a letter of the alphabet, such as the letter f. If x is an element in the domain of a function f, then the element in B that f associates with x is written $f(x)$ (read "f of x") and is called the value of f at x. The set comprising all the values assumed by $y = f(x)$ as x takes on all possible values in its domain is called the **range** of the function f.

We can think of a function f as a machine. The domain is the set of inputs (raw material) for the machine, the rule describes how the input is to be processed, and the value(s) of the function are the outputs of the machine (Figure 2.1).

We can also think of a function f as a mapping in which an element x in the domain of f is mapped onto a unique element $f(x)$ in B (Figure 2.2).

f(x) is unique

Figure 2.1
A function machine.

argument of function f

Value

Figure 2.2
The function f viewed as a mapping.

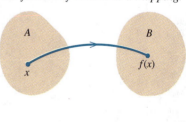

REMARKS

1. It is important to understand that the output $f(x)$ associated with an input x is unique. To appreciate the importance of this uniqueness property, consider a rule that associates with each item x in a department store its selling price y. Then, each x must correspond to *one and only one y*. Notice, however, that different x's may be associated with the same y. In the context of the present example, this says that different items may have the same price.

2. Although the sets A and B that appear in the definition of a function may be quite arbitrary, in this book they will denote sets of real numbers.

◦ ◦ ◦

An example of a function may be taken from the familiar relationship between the area of a circle and its radius. Letting x and y denote the radius and area of a circle, respectively, we have, from elementary geometry,

$$y = \pi x^2 \tag{1}$$

Equation (1) defines y as a function of x since for each admissible value of x (that is, for each nonnegative number representing the radius of a certain circle) there corresponds precisely one number $y = \pi x^2$ that gives the area of the circle. The rule defining this "area function" may be written as

$$f(x) = \pi x^2 \tag{2}$$

To compute the area of a circle of radius 5 inches, we simply replace x in equation (2) with the number 5. Thus, the area of the circle is

$$f(5) = \pi 5^2 = 25\pi$$

or 25π square inches.

In general, to evaluate a function at a specific value of x, we replace x with that value, as illustrated in Examples 1 and 2.

EXAMPLE 1 Let the function f be defined by the rule $f(x) = 2x^2 - x + 1$. Compute:

a. $f(1)$ **b.** $f(-2)$ **c.** $f(a)$ **d.** $f(a + h)$

Solution

a. $f(1) = 2(1)^2 - (1) + 1 = 2 - 1 + 1 = 2$
b. $f(-2) = 2(-2)^2 - (-2) + 1 = 8 + 2 + 1 = 11$
c. $f(a) = 2(a)^2 - (a) + 1 = 2a^2 - a + 1$
d. $f(a + h) = 2(a + h)^2 - (a + h) + 1 = 2a^2 + 4ah + 2h^2 - a - h + 1$

◦ ◦ ◦

EXAMPLE 2 The Thermo-Master Company manufactures an indoor-outdoor thermometer at its Mexican subsidiary. Management estimates that the profit (in dollars) realizable by Thermo-Master in the manufacture and sale of x thermometers per week is

$$P(x) = -0.001x^2 + 8x - 5000$$

Find Thermo-Master's weekly profit if its level of production is (a) 1000 thermometers per week and (b) 2000 thermometers per week.

Solution

a. The weekly profit realizable by Thermo-Master when the level of production is 1000 units per week is found by evaluating the profit function P at $x = 1000$. Thus,

$$P(1000) = -0.001(1000)^2 + 8(1000) - 5000 = 2000$$

or $2000.

b. When the level of production is 2000 units per week, the weekly profit is given by

$$P(2000) = -0.001(2000)^2 + 8(2000) - 5000 = 7000$$

or $7000.

Determining the Domain of a Function

Suppose we are given the function $y = f(x)$.* Then the variable x is called the **independent variable.** The variable y, whose value depends on x, is called the **dependent variable.**

In determining the domain of a function, we need to find what restrictions, if any, are to be placed on the independent variable x. In many practical applications, the domain of a function is dictated by the nature of the problem, as illustrated in Example 3.

EXAMPLE 3   An open box is to be made from a rectangular piece of cardboard 16 inches long and 10 inches wide by cutting away identical squares (x by x inches) from each corner and folding up the resulting flaps (Figure 2.3). Find an expression that gives the volume V of the box as a function of x. What is the domain of the function?

Figure 2.3

(a) The box is constructed by cutting x-by-x inch squares from each corner.

(b) The dimensions of the resulting box are $(10 - 2x)$ by $(16 - 2x)$ by x inches.

* It is customary to refer to a function f as $f(x)$ or by the equation $y = f(x)$ defining it.

Solution   The dimensions of the box are $(16 - 2x)$ inches long, $(10 - 2x)$ inches wide, and x inches high, so its volume (in cubic inches) is given by

$$V = f(x) = (16 - 2x)(10 - 2x)x \qquad \text{(Length · width · height)}$$
$$= (160 - 52x + 4x^2)x$$
$$= 4x^3 - 52x^2 + 160x$$

Since the length of each side of the box must be greater than or equal to zero, we see that

$$16 - 2x \geq 0, \quad 10 - 2x \geq 0, \quad \text{and} \quad x \geq 0$$

simultaneously; that is,

$$x \leq 8, \quad x \leq 5, \quad \text{and} \quad x \geq 0$$

All three inequalities are satisfied simultaneously provided that $0 \leq x \leq 5$. Thus, the domain of the function f is the interval $[0, 5]$.   ◦ ◦ ◦

In general, if a function is defined by a rule relating x to $f(x)$ without specific mention of its domain, it is understood that the domain will consist of all values of x for which $f(x)$ is a real number. In this connection, you should keep in mind that (1) division by zero is not permitted and (2) the square root of a negative number is not defined.

EXAMPLE 4   Find the domain of each of the functions defined by the following equations:

a. $f(x) = \sqrt{x - 1}$ **b.** $f(x) = \dfrac{1}{x^2 - 4}$ **c.** $f(x) = x^2 + 3$

Solution

a. Since the square root of a negative number is undefined, it is necessary that $x - 1 \geq 0$. The inequality is satisfied by the set of real numbers $x \geq 1$. Thus, the domain of f is the interval $[1, \infty)$.

b. The only restriction on x is that $x^2 - 4$ be different from zero since division by zero is not allowed. But, $(x^2 - 4) = (x + 2)(x - 2) = 0$ if $x = -2$ or $x = 2$. Thus, the domain of f in this case consists of the intervals $(-\infty, -2)$, $(-2, 2)$, and $(2, \infty)$.

c. Here, any real number satisfies the equation, so the domain of f is the set of all real numbers.   ◦ ◦ ◦

Graphs of Functions

If f is a function with domain A, then corresponding to each real number x in A there is precisely one real number $f(x)$. We can also express this fact by using **ordered pairs** of real numbers. Write each number x in A as the first member of an ordered pair and each number $f(x)$ corresponding to x as the second member of the ordered pair. This gives exactly one ordered pair $(x, f(x))$ for each x in A. This observation leads to an alternative definition of a function f:

A function f with domain A is the set of all ordered pairs $(x, f(x))$ where x belongs to A.

Observe that the condition that there be one and only one number $f(x)$ corresponding to each number x in A translates into the requirement that *no two ordered pairs have the same first number.*

Since ordered pairs of real numbers correspond to points in the plane, we have found a way to exhibit a function graphically.

GRAPH OF A FUNCTION OF ONE VARIABLE

The **graph of a function** f is the set of all points (x, y) in the xy-plane such that x is in the domain of f and $y = f(x)$.

Much information about the graph of a function can be gained by plotting a few points on its graph. Later on we will develop more systematic and sophisticated techniques for graphing functions.

Figure 2.4
The graph of $y = x^2 + 1$ is a parabola.

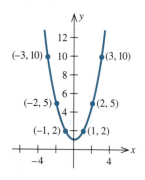

EXAMPLE 5   Sketch the graph of the function defined by the equation $y = x^2 + 1$.

Solution   The domain of the function is the set of all real numbers. By assigning several values to the variable x and computing the corresponding values for y, we obtain the following solutions to the equation $y = x^2 + 1$:

x	-3	-2	-1	0	1	2	3
y	10	5	2	1	2	5	10

By plotting these points and then connecting them with a smooth curve, we obtain the graph of $y = f(x)$, which is a parabola (Figure 2.4). ◦ ◦ ◦

Let $f(x) = x^2$.

a. Plot the graphs of $F(x) = x^2 + c$ on the same set of axes for $c = -2, -1, -\frac{1}{2}, 0, \frac{1}{2}, 1, 2$.
b. Plot the graphs of $G(x) = (x + c)^2$ on the same set of axes for $c = -2, -1, -\frac{1}{2}, 0, \frac{1}{2}, 1, 2$.
c. Plot the graphs of $H(x) = cx^2$ on the same set of axes for $c = -2, -1, -\frac{1}{2}, -\frac{1}{4}, 0, \frac{1}{4}, \frac{1}{2}, 1, 2$.
d. Study the family of graphs in parts (a), (b), and (c), and describe the relationship between the graph of a function f and the graphs of the functions defined by (i) $y = f(x) + c$, (ii) $y = f(x + c)$, and (iii) $y = cf(x)$, where c is a constant.

Some functions are defined in a piecewise fashion, as Examples 6 and 7 show.

EXAMPLE 6 The Madison Finance Company plans to open two branch offices two years from now in two separate locations: an industrial complex and a newly developed commercial center in the city. As a result of these expansion plans, Madison's total deposits during the next five years are expected to grow in accordance with the rule

$$f(x) = \begin{cases} \sqrt{2x} + 20 & 0 \le x \le 2 \\ \frac{1}{2}x^2 + 20 & 2 < x \le 5 \end{cases}$$

where $y = f(x)$ gives the total amount of money (in millions of dollars) on deposit with Madison in year x ($x = 0$ corresponds to the present). Sketch the graph of the function f.

Solution The function f is defined in a piecewise fashion on the interval $[0, 5]$. In the subdomain $[0, 2]$, the rule for f is given by $f(x) = \sqrt{2x} + 20$. The values of $f(x)$ corresponding to $x = 0, 1$, and 2 may be tabulated as follows:

x	0	1	2
$f(x)$	20	21.4	22

Next, in the subdomain $(2, 5]$, the rule for f is given by $f(x) = \frac{1}{2}x^2 + 20$. The values of $f(x)$ corresponding to $x = 3, 4$, and 5 are shown in the following table.

x	3	4	5
$f(x)$	24.5	28	32.5

Using the values of $f(x)$ in this table, we sketch the graph of the function f as shown in Figure 2.5.

Figure 2.5
We obtain the graph of the function $y = f(x)$ by graphing $y = \sqrt{2x} + 20$ over $[0, 2]$ and $y = \frac{1}{2}x^2 + 20$ over $(2, 5]$.

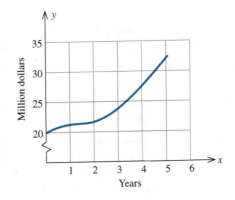

Figure 2.6
*The graph of y = f(x) is obtained
by graphing y = −x over (−∞, 0)
and y = √x over [0, ∞).*

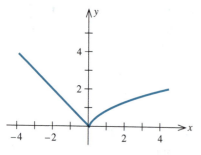

EXAMPLE 7   Sketch the graph of the function *f* defined by

$$f(x) = \begin{cases} -x & \text{if } x < 0 \\ \sqrt{x} & \text{if } x \geq 0 \end{cases}$$

Solution   The function *f* is defined in a piecewise fashion on the set of all real numbers. In the subdomain (−∞, 0), the rule for *f* is given by $f(x) = -x$. The equation $y = -x$ is a linear equation in the slope-intercept form (with slope −1 and intercept 0). Therefore, the graph of *f* corresponding to the subdomain (−∞, 0) is the half-line shown in Figure 2.6. Next, in the subdomain [0, ∞), the rule for *f* is given by $f(x) = \sqrt{x}$. The values of $f(x)$ corresponding to x = 0, 1, 2, 3, 4, 9, and 16 are shown in the following table.

x	0	1	2	3	4	9	16
f(x)	0	1	$\sqrt{2}$	$\sqrt{3}$	2	3	4

Using these values, we sketch the graph of the function *f* as shown in Figure 2.6.

● ● ●

Figure 2.7
*Since a vertical line passes through
the curve at more than one point,
we deduce that it is not the graph of
a function.*

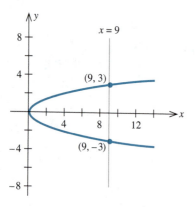

The Vertical-Line Test

Although it is true that every function *f* of a variable *x* has a graph in the *xy*-plane, it is important to realize that not every curve in the *xy*-plane is the graph of a function. For example, consider the curve depicted in Figure 2.7. This is the graph of the equation $x = y^2$. In general, the **graph of an equation** is the set of all ordered pairs (x, y) that satisfy the given equation. Observe that the points (9, −3) and (9, 3) both lie on the curve. This implies that the number x = 9 is associated with *two* numbers: y = −3 and y = 3. But this clearly violates the uniqueness property of a function. Thus, we conclude that the curve under consideration cannot be the graph of a function.

This example suggests the following test for determining when a curve is the graph of a function.

VERTICAL-LINE TEST   A curve in the *xy*-plane is the graph of a function $y = f(x)$ if and only if each vertical line intersects it in at most one point.

EXAMPLE 8   Determine which of the curves shown in Figure 2.8 are the graphs of functions of *x*.

Solution   The curves depicted in Figures 2.8a, c, and d are graphs of functions, because each curve satisfies the requirement that each vertical line intersects the curve in at most one point. Note that the vertical line shown in Figure 2.8c does not intersect the graph, because the point on the *x*-axis through

Figure 2.8
The vertical-line test can be used to determine which of these curves are graphs of functions.

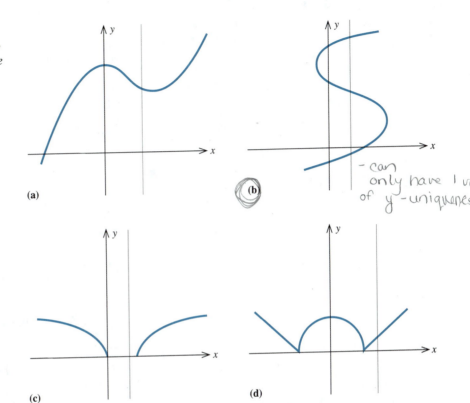

(a)

(b)

- can only have 1 value of y – uniqueness

(c)

(d)

which this line passes does not lie in the domain of the function. The curve depicted in Figure 2.8b is *not* the graph of a function because the vertical line shown there intersects the graph at three points.

SELF-CHECK EXERCISES 2.1

1. Let *f* be the function defined by

$$f(x) = \frac{\sqrt{x+1}}{x}$$

a. Find the domain of *f*.
b. Compute $f(3)$.
c. Compute $f(a + h)$.

2. Statistics obtained by Amoco Corporation show that more and more motorists are pumping their own gas. The following function gives self-serve sales as a percentage of all U.S. gas sales:

$$f(t) = \begin{cases} 6t + 17 & 0 \le t \le 6 \\ 15.98(t - 6)^{1/4} + 53 & 6 < t \le 20 \end{cases}$$

Here t is measured in years, with $t = 0$ corresponding to the beginning of 1974.

a. Sketch the graph of the function f.

b. What percentage of all gas sales at the beginning of 1978 were self-serve? At the beginning of 1994?

Source: Amoco Corporation

3. Let $f(x) = \sqrt{2x + 1} + 2$. Determine whether the point $(4, 6)$ lies on the graph of f.

Solutions to Self-Check Exercises 2.1 can be found on page 68.

2.1 EXERCISES

1. Let f be the function defined by $f(x) = 5x + 6$. Find $f(3), f(-3), f(a), f(-a),$ and $f(a + 3)$.

2. Let f be the function defined by $f(x) = 4x - 3$. Find $f(4), f(\frac{1}{4}), f(0), f(a),$ and $f(a + 1)$.

3. Let g be the function defined by $g(x) = 3x^2 - 6x - 3$. Find $g(0), g(-1), g(a), g(-a),$ and $g(x + 1)$.

4. Let h be the function defined by $h(x) = x^3 - x^2 + x + 1$. Find $h(-5), h(0), h(a),$ and $h(-a)$.

5. Let s be the function defined by $s(t) = 2t/(t^2 - 1)$. Find $s(4), s(0), s(a), s(2 + a),$ and $s(t + 1)$.

6. Let g be the function defined by $g(u) = (3u - 2)^{3/2}$. Find $g(1), g(6), g(\frac{11}{3}),$ and $g(u + 1)$.

7. Let f be the function defined by $f(t) = 2t^2/\sqrt{t - 1}$. Find $f(2), f(a), f(x + 1),$ and $f(x - 1)$.

8. Let f be the function defined by $f(x) = 2 + 2\sqrt{5 - x}$. Find $f(-4), f(1), f(\frac{11}{4}),$ and $f(x + 5)$.

9. Let f be the function defined by

$$f(x) = \begin{cases} x^2 + 1 & \text{if } x \leq 0 \\ \sqrt{x} & \text{if } x > 0 \end{cases}$$

Find $f(-2), f(0),$ and $f(1)$.

10. Let g be the function defined by

$$g(x) = \begin{cases} -\dfrac{1}{2}x + 1 & \text{if } x < 2 \\ \sqrt{x - 2} & \text{if } x \geq 2 \end{cases}$$

Find $g(-2), g(0), g(2),$ and $g(4)$.

11. Let f be the function defined by

$$f(x) = \begin{cases} -\dfrac{1}{2}x^2 + 3 & \text{if } x < 1 \\ 2x^2 + 1 & \text{if } x \geq 1 \end{cases}$$

Find $f(-1), f(0), f(1),$ and $f(2)$.

12. Let f be the function defined by

$$f(x) = \begin{cases} 2 + \sqrt{1 - x} & \text{if } x \leq 1 \\ \dfrac{1}{1 - x} & \text{if } x > 1 \end{cases}$$

Find $f(0), f(1),$ and $f(2)$.

In exercises 13–16, determine whether the point lies on the graph of the function.

13. $(2, \sqrt{3}); \quad g(x) = \sqrt{x^2 - 1}$

14. $(3, 3); \quad f(x) = \dfrac{x + 1}{\sqrt{x^2 + 7}} + 2$

15. $(-2, -3); \quad f(t) = \dfrac{|t - 1|}{t + 1}$

16. $\left(-3, -\dfrac{1}{13}\right); \quad h(t) = \dfrac{|t + 1|}{t^3 + 1}$

In exercises 17–30, find the domain of the function.

17. $f(x) = x^2 + 3$

18. $f(x) = 7 - x^2$

19. $f(x) = \dfrac{3x + 1}{x^2}$

20. $g(x) = \dfrac{2x + 1}{x - 1}$

21. $f(x) = \sqrt{x^2 + 1}$

22. $f(x) = \sqrt{x - 5}$

23. $f(x) = \sqrt{5 - x}$

24. $g(x) = \sqrt{2x^2 + 3}$

25. $f(x) = \dfrac{x}{x^2 - 1}$ **26.** $f(x) = \dfrac{1}{x^2 + x - 2}$

27. $f(x) = (x + 3)^{3/2}$ **28.** $g(x) = 2(x - 1)^{5/2}$

29. $f(x) = \dfrac{\sqrt{1 - x}}{x^2 - 4}$ **30.** $f(x) = \dfrac{\sqrt{x - 1}}{(x + 2)(x - 3)}$

31. Let f be a function defined by the rule $f(x) = x^2 - x - 6$.
  a. Find the domain of f.
  b. Compute $f(x)$ for $x = -3, -2, -1, 0, \frac{1}{2}, 1, 2, 3$.
  c. Use the results obtained in (a) and (b) to sketch the graph of f.

32. Let f be a function defined by the rule $f(x) = 2x^2 + x - 3$.
  a. Find the domain of f.
  b. Compute $f(x)$ for $x = -3, -2, -1, -\frac{1}{2}, 0, 1, 2, 3$.
  c. Use the results obtained in (a) and (b) to sketch the graph of f.

In exercises 33–44, sketch the graph of the function with the given rule.

33. $f(x) = 2x^2 + 1$ **34.** $f(x) = 9 - x^2$

35. $f(x) = 2 + \sqrt{x}$ **36.** $g(x) = 4 - \sqrt{x}$

37. $f(x) = \sqrt{1 - x}$ **38.** $f(x) = \sqrt{x - 1}$

39. $f(x) = |x| - 1$ **40.** $f(x) = |x| + 1$

41. $f(x) = \begin{cases} x & \text{if } x < 0 \\ 2x + 1 & \text{if } x \geq 0 \end{cases}$

42. $f(x) = \begin{cases} 4 - x & \text{if } x < 2 \\ 2x - 2 & \text{if } x \geq 2 \end{cases}$

43. $f(x) = \begin{cases} -x + 1 & \text{if } x \leq 1 \\ x^2 - 1 & \text{if } x > 1 \end{cases}$

44. $f(x) = \begin{cases} -x - 1 & \text{if } x < -1 \\ 0 & \text{if } -1 \leq x \leq 1 \\ x + 1 & \text{if } x > 1 \end{cases}$

In exercises 45–52, use the vertical-line test to determine whether the graph represents y as a function of x.

45.

46.

47.

48.

49.

50.

51.
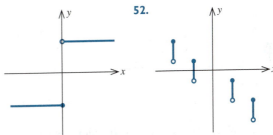

52.

53. The circumference of a circle is given by $C(r) = 2\pi r$, where r is the radius of the circle. What is the circumference of a circle with a 5-inch radius?

54. The volume of a sphere of radius r is given by $V(r) = \frac{4}{3}\pi r^3$. Compute $V(2.1)$ and $V(2)$. What does the quantity $V(2.1) - V(2)$ measure?

55. Sales of Prerecorded Music The following graphs show the sales y of prerecorded music (in billions of dollars) by format as a function of time t (in years) with $t = 0$ corresponding to 1985.

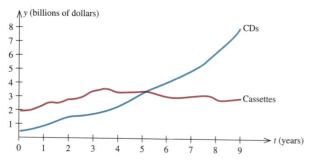

Source: Recording Industry Association of America

a. In what years were the sales of prerecorded cassettes greater than those of prerecorded CDs?
b. In what years were the sales of prerecorded CDs greater than those of prerecorded cassettes?
c. In what year were the sales of prerecorded cassettes the same as those of prerecorded CDs? Estimate the level of sales in each format at that time.

56. The Gender Gap The following graph shows the ratio of women's earnings to men's from 1960 through 1990.

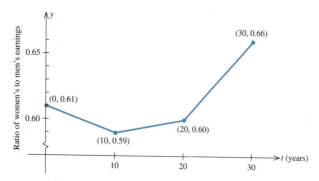

Source: U.S. Bureau of Labor Statistics

a. Write the rule for the function f giving the ratio of women's earnings to men's in year t, with $t = 0$ corresponding to 1960.

[*Hint:* The function f is defined piecewise and is linear over each of three subintervals.]

b. In what decade(s) was the gender gap expanding? Shrinking?
c. Refer to part (b). How fast was the gender gap (the ratio per year) expanding or shrinking in each of these decades?

57. Closing the Gender Gap in Education The following graph shows the ratio of bachelor's degrees earned by women to men from 1960 through 1990.

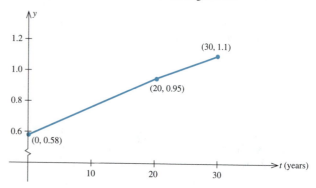

Source: Department of Education

a. Write the rule for the function f giving the ratio of bachelor's degrees earned by women to men in year t, with $t = 0$ corresponding to 1960.

[*Hint:* The function f is defined piecewise and is linear over each of two subintervals.]

b. How fast was the ratio changing in the period from 1960 through 1980? The decade from 1980 to 1990?
c. In what year (approximately) was the number of bachelor's degrees earned by women equal to that earned by men for the first time?

58. Consumption Function The consumption function in a certain economy is given by the equation

$$C(y) = 0.75y + 6$$

where $C(y)$ is the personal consumption expenditure, y is the disposable personal income, and both $C(y)$ and y are measured in billions of dollars. Find $C(0)$, $C(50)$, and $C(100)$.

59. Sales Taxes In a certain state, the sales tax T on the amount of taxable goods is 6% of the value of the goods purchased (x), where both T and x are measured in dollars.

a. Express T as a function of x.
b. Find $T(200)$ and $T(5.65)$.

60. Surface Area of a Single-Celled Organism The surface area S of a single-celled organism may be found by multiplying 4π times the square of the radius r of the cell. Express S as a function of r.

61. Friend's Rule   Friend's Rule, a method for calculating pediatric drug dosages, is based on a child's age. If a denotes the adult dosage (in mg) and if t is the age of the child (in years), then the child's dosage is given by

$$D(t) = \frac{2}{25} ta$$

If the adult dose of a substance is 500 mg, how much should a four-year-old child receive?

62. COLAs   Social security recipients receive an automatic cost-of-living adjustment (COLA) once each year. Their monthly benefit is increased by the amount that consumer prices increased during the preceding year. Suppose that consumer prices increased by 5.3% during the preceding year.
a. Express the adjusted monthly benefit of a social security recipient as a function of his or her current monthly benefit.
b. If Mr. Harrington's monthly social security benefit is now $620, what will his adjusted monthly benefit be?

63. Cost of Renting a Truck   The Ace Truck Leasing Company leases a certain size truck at $30 a day and 15 cents a mile, whereas the Acme Truck Leasing Company leases the same size truck at $25 a day and 20 cents a mile.
a. Find the daily cost of leasing from each company as a function of the number of miles driven.
b. Sketch the graphs of the two functions on the same set of axes.
c. Which company should a customer rent a truck from for one day if she plans to drive at most 70 miles and wishes to minimize her cost?

64. Linear Depreciation   A new machine was purchased by the National Textile company for $120,000. For income tax purposes, the machine is depreciated linearly over ten years; that is, the book value of the machine decreases at a constant rate, so that at the end of ten years the book value is zero.
a. Express the book value of the machine (V) as a function of the age, in years, of the machine (n).
b. Sketch the graph of the function in (a).
c. Find the book value of the machine at the end of the sixth year.
d. Find the rate at which the machine is being depreciated each year.

65. Linear Depreciation   Refer to exercise 64. An office building worth $1,000,000 when completed in 1980 was depreciated linearly over 50 years. What will the book value of the building be in 1999? In 2003? (Assume that the book value of the building will be zero at the end of the 50th year.)

66. Boyle's Law   As a consequence of Boyle's law, the pressure P of a fixed sample of gas held at a constant temperature is related to the volume V of the gas by the rule

$$P = f(V) = \frac{k}{V}$$

where k is a constant. What is the domain of the function f? Sketch the graph of the function f.

67. Poiseuille's Law   According to a law discovered by the nineteenth-century physician Poiseuille, the velocity (in centimeters per second) of blood r centimeters from the central axis of an artery is given by

$$v(r) = k(R^2 - r^2)$$

where k is a constant and R is the radius of the artery. Suppose that for a certain artery, $k = 1000$ and $R = 0.2$ so that $v(r) = 1000(0.04 - r^2)$.
a. What is the domain of the function $v(r)$?
b. Compute $v(0)$, $v(0.1)$, and $v(0.2)$ and interpret your results.

68. Population Growth   A study prepared for a certain Sunbelt town's Chamber of Commerce projected that the population of the town in the next three years will grow according to the rule

$$P(x) = 50{,}000 + 30x^{3/2} + 20x$$

where $P(x)$ denotes the population x months from now. By how much will the population increase during the next 9 months? During the next 16 months?

69. Worker Efficiency   An efficiency study conducted for the Elektra Electronics Company showed that the number of "Space Commander" walkie-talkies assembled by the average worker t hours after starting work at 8:00 A.M. is given by

$$N(t) = -t^3 + 6t^2 + 15t \qquad (0 \le t \le 4)$$

How many walkie-talkies can an average worker be expected to assemble between 8:00 and 9:00 A.M.? Between 9:00 and 10:00 A.M.?

70. Learning Curves   The Emory Secretarial School finds from experience that the average student taking advanced typing will progress according to the rule

$$N(t) = \frac{60t + 180}{t + 6}$$

where $N(t)$ measures the number of words per minute the student can type after t weeks in the course. How fast can the average student be expected to type after two weeks in the course? After four weeks in the course?

71. Politics   Political scientists have discovered the following empirical rule, known as the "cube rule," which gives the relationship between the proportion of seats in the House of Representatives won by Democratic candidates $s(x)$ and the proportion of popular votes x received by the Democratic presidential candidate:

$$s(x) = \frac{x^3}{x^3 + (1-x)^3} \qquad (0 \le x \le 1)$$

Compute $s(0.6)$ and interpret your result.

72. Home Shopping Industry   According to industry sources, revenue from the home shopping industry for the years since its inception may be approximated by the function

$$R(t) = \begin{cases} -0.03t^3 + 0.25t^2 - 0.12t & 0 \le t \le 3 \\ 0.57t - 0.63 & 3 < t \le 11 \end{cases}$$

where $R(t)$ measures the revenue in billions of dollars and t is measured in years, with $t = 0$ corresponding to the beginning of 1984. What was the revenue at the beginning of 1985? At the beginning of 1993?
Source: Paul Kagan Associates

73. Postal Regulations   The postage for first-class mail is 32 cents for the first ounce or fraction thereof, and 23 cents for each additional ounce or fraction thereof. Any parcel not exceeding 12 ounces may be sent by first-class mail. Letting x denote the weight of a parcel in ounces, and $f(x)$ the postage in cents, complete the following description of the "postage function" f:

$$f(x) = \begin{cases} 32 & \text{if } 0 < x \le 1 \\ 55 & \text{if } 1 < x \le 2 \\ \cdots & \\ ? & \text{if } 11 < x \le 12 \end{cases}$$

a. What is the domain of f?
b. Sketch the graph of f.

74. Harbor Cleanup   The amount of solids discharged from the MWRA (Massachusetts Water Resources Authority) sewage treatment plant on Deer Island (near Boston Harbor) is given by the function

$$f(t) = \begin{cases} 130 & 0 \le t \le 1 \\ -30t + 160 & 1 < t \le 2 \\ 100 & 2 < t \le 4 \\ -5t^2 + 25t + 80 & 4 < t \le 6 \\ 1.25t^2 - 26.25t + 162.5 & 6 < t \le 10 \end{cases}$$

where $f(t)$ is measured in tons per day and t is measured in years, with $t = 0$ corresponding to 1989.
Source: Metropolitan District Commission

a. What is the amount of solids discharged per day in 1989? In 1992? In 1996?
b. Sketch the graph of f.

SOLUTIONS TO SELF-CHECK EXERCISES 2.1

1. a. The expression under the radical sign must be nonnegative, so $x + 1 \ge 0$ or $x \ge -1$. Also, $x \ne 0$ because division by zero is not permitted. Therefore, the domain of f is $[-1, 0) \cup (0, \infty)$.

b. $f(3) = \dfrac{\sqrt{3+1}}{3} = \dfrac{\sqrt{4}}{3} = \dfrac{2}{3}$

c. $f(a + h) = \dfrac{\sqrt{(a+h)+1}}{a+h} = \dfrac{\sqrt{a+h+1}}{a+h}$

2. a. For t in the subdomain $[0, 6]$, the rule for f is given by $f(t) = 6t + 17$. The equation $y = 6t + 17$ is a linear equation, so that portion of the graph of f is the line segment joining the points $(0, 17)$ and $(6, 53)$. Next, in the subdomain $(6, 20]$, the rule for f is given by $f(t) = 15.98(t - 6)^{1/4} + 53$. Using a calculator, we construct the following table of values of $f(t)$ for selected values of t.

t	6	8	10	12	14	16	18	20
$f(t)$	53	72	75.6	78	79.9	81.4	82.7	83.9

We have included $t = 6$ in the table, although it does not lie in the subdomain of the function under consideration, in order to help us obtain a better sketch of that portion of the graph of f in the subdomain $(6, 20]$. The graph of f is as follows:

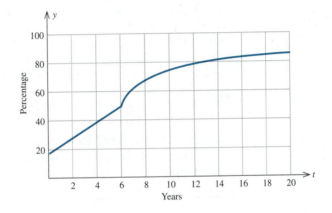

b. The percentage of all self-serve gas sales at the beginning of 1978 is found by evaluating f at $t = 4$. Since this point lies in the interval $[0, 6]$, we use the rule $f(t) = 6t + 17$ and find

$$f(4) = 6(4) + 17$$

giving 41% as the required figure. The percentage of all self-serve gas sales at the beginning of 1994 is given by

$$f(20) = 15.98(20 - 6)^{1/4} + 53$$

or approximately 83.9%.

3. A point (x, y) lies on the graph of the function f if and only if the coordinates satisfy the equation $y = f(x)$. Now

$$f(4) = \sqrt{2(4) + 1} + 2 = \sqrt{9} + 2 = 5 \neq 6$$

and we conclude that the given point does not lie on the graph of f.

Using Technology

Graphing a Function

Most of the graphs of functions in this book can be plotted with the help of a graphing utility. Furthermore, a graphing utility can be used to analyze the nature of a function. However, the amount and accuracy of the information obtained using a graphing utility depend on the experience and sophistication of the user. As you progress through this book, you will see that the more knowledge of calculus you gain, the more effective the graphing utility will prove as a tool in problem solving.

Finding a Suitable Viewing Rectangle

The first step in plotting the graph of a function with a graphing utility is to select a suitable viewing rectangle. We usually do this by experimenting. For example, you might first plot the graph using the *standard viewing rectangle* $[-10, 10]$ by $[-10, 10]$. If necessary, you then might adjust the viewing rectangle by enlarging it or reducing it to obtain a sufficiently complete view of the graph or at least the portion of the graph that is of interest.

EXAMPLE 1 Plot the graph of $f(x) = 2x^2 - 4x - 5$ in the standard viewing rectangle.

Solution The graph of f, shown in Figure T1, is a parabola. From our previous work (Example 5, Section 2.1), we know that the figure does give a good view of the graph.

Figure T1
The graph of $y = 2x^2 - 4x - 5$ in the viewing rectangle $[-10, 10] \times [-10, 10]$.

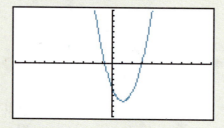

● ● ●

EXAMPLE 2 Let $f(x) = x^3(x - 3)^4$.

a. Plot the graph of f in the standard viewing rectangle.

b. Plot the graph of f in the rectangle $[-1, 5] \times [-40, 40]$.

Solution

a. The graph of f in the standard viewing rectangle is shown in Figure T2a. Since the graph does not appear to be complete, we need to adjust the viewing rectangle.

Figure T2
An incomplete sketch of $f(x) = x^3(x-3)^4$ is shown in (a) and a complete sketch is shown in (b).

(a)   **(b)**

b. The graph of f in the rectangle $[-1, 5] \times [-40, 40]$, shown in Figure T2b, is an improvement over the previous graph. (Later we will be able to show that the figure does, in fact, give a rather complete view of the graph of f.)

⊙ ⊙ ⊙

Evaluating a Function

A graphing utility can be used to find the value of a function with minimal effort, as the next example shows.

EXAMPLE 3   Let $f(x) = x^3 - 4x^2 + 4x + 2$.

a. Plot the graph of f in the standard viewing rectangle.

b. Find $f(3)$ and verify your result by direct computation.

c. Find $f(4.215)$.

Solution

a. The graph of f is shown in Figure T3.

Figure T3
The graph of $f(x) = x^3 - 4x^2 + 4x + 2$ in the standard viewing rectangle.

b. Using the evaluation function of the graphing utility and the value 3 for x, we find $y = 5$. This result is verified by computing

$$f(3) = 3^3 - 4(3^2) + 4(3) + 2 = 27 - 36 + 12 + 2 = 5$$

71

c. Using the evaluation function of the graphing utility and the value 4.215 for x, we find $y = 22.679738375$. Thus, $f(4.215) = 22.679738375$. The efficacy of the graphing utility is clearly demonstrated here! ◦ ◦ ◦

EXAMPLE 4 The anticipated rise in the number of Alzheimer's patients in the United States is given by

$$f(t) = -0.0277t^4 + 0.3346t^3 - 1.1261t^2 + 1.7575t + 3.7745 \qquad (0 \le t \le 6)$$

where $f(t)$ is measured in millions and t is measured in decades, with $t = 0$ corresponding to the beginning of 1990.

a. Use a graphing utility to plot the graph of f in the viewing rectangle $[0, 7] \times [0, 12]$.

b. What is the anticipated number of Alzheimer's patients in the United States at the beginning of the year 2000 ($t = 1$)? At the beginning of 2030 ($t = 4$)?
Source: Alzheimer's Association

Solution

a. The graph of f in the viewing rectangle $[0, 7] \times [0, 12]$ is shown in Figure T4.

Figure T4
The graph of f in the viewing rectangle $[0, 7] \times [0, 12]$.

b. Using the evaluation function of the graphing utility and the value 1 for x, we see that the anticipated number of Alzheimer's patients at the beginning of the year 2000 is given by

$$f(1) = 4.7128$$

or approximately 4.7 million. The anticipated number of Alzheimer's patients at the beginning of 2030 is given by

$$f(4) = 7.1101$$

or approximately 7.1 million. ◦ ◦ ◦

EXERCISES

In exercises 1–8, plot the graph of the function f in the standard viewing window.

1. $f(x) = 2x^2 - 16x + 29$ 2. $f(x) = -x^2 - 10x - 20$

3. $f(x) = x^3 - 2x^2 + x - 2$

4. $f(x) = -2.01x^3 + 1.21x^2 - 0.78x + 1$

5. $f(x) = 0.2x^4 - 2.1x^2 + 1$

6. $f(x) = -0.4x^4 + 1.2x - 1.2$

7. $f(x) = 2x\sqrt{x^2 + 1}$ 8. $f(x) = \dfrac{\sqrt{x} + 1}{\sqrt{x} - 1}$

In exercises 9–20, plot the graph of the function f in (a) the standard viewing window and (b) the indicated window.

9. $f(x) = 2x^2 - 32x + 125$; $[5, 15] \times [-5, 10]$

10. $f(x) = x^2 + 20x + 95$; $[-20, 10] \times [-15, -5]$

11. $f(x) = x^3 - 20x^2 + 8x - 10$; $[-20, 20] \times [-1200, 100]$

12. $f(x) = -2x^3 + 10x^2 - 15x - 5$; $[-10, 10] \times [-100, 100]$

13. $f(x) = x^4 - 2x^2 + 8$; $[-2, 2] \times [6, 10]$

14. $f(x) = x^4 - 2x^3$; $[-1, 3] \times [-2, 2]$

15. $f(x) = x + \dfrac{1}{x}$; $[-1, 3] \times [-5, 5]$

16. $f(x) = \dfrac{4}{x^2 - 8}$; $[-5, 5] \times [-5, 5]$

17. $f(x) = 2 - \dfrac{1}{x^2 + 1}$; $[-3, 3] \times [0, 3]$

18. $f(x) = x - 2\sqrt{x}$; $[0, 20] \times [-2, 10]$

19. $f(x) = x\sqrt{4 - x^2}$; $[-3, 3] \times [-2, 2]$

20. $f(x) = \dfrac{\sqrt{x} - 1}{x}$; $[0, 50] \times [-0.25, 0.25]$

In exercises 21–30, plot the graph of the function f in an appropriate viewing window. Note: The answer is not unique.

21. $f(x) = x^2 - 4x + 16$ 22. $f(x) = -x^2 + 2x - 11$

23. $f(x) = 2x^3 - 10x^2 + 5x - 10$

24. $f(x) = -x^3 + 5x^2 - 14x + 20$

25. $f(x) = 2x^4 - 3x^3 + 5x^2 - 20x + 40$

26. $f(x) = -2x^4 + 5x^2 - 4$

27. $f(x) = \dfrac{x^3}{x^3 + 1}$ 28. $f(x) = \dfrac{2x^4 - 3x}{x^2 - 1}$

29. $f(x) = 0.2\sqrt{x} - 0.3x^3$ 30. $f(x) = \sqrt{x}(2x - 1)^3$

In exercises 31–34, use the evaluation function of your graphing utility to find the value of f at the given value of x and verify your result by direct computation.

31. $f(x) = -3x^3 + 5x^2 - 2x + 8$; $x = -1$

32. $f(x) = 2x^4 - 3x^3 + 2x^2 + x - 5$; $x = 2$

33. $f(x) = \dfrac{x^4 - 3x^2}{x - 2}$; $x = 1$ 34. $f(x) = \dfrac{\sqrt{x^2 - 1}}{3x + 4}$; $x = 2$

In exercises 35–42, use the evaluation function of your graphing utility to find the value of f at the indicated value of x. Express your answer accurate to four decimal places.

35. $f(x) = 3x^3 - 2x^2 + x - 4$; $x = 2.145$

36. $f(x) = 2x^3 + 5x^2 + 3x + 1$; $x = -0.27$

37. $f(x) = 5x^4 - 2x^2 + 8x - 3$; $x = 1.28$

38. $f(x) = 4x^4 - 3x^3 + 1$; $x = -2.42$

39. $f(x) = \dfrac{2x^3 - 3x + 1}{3x - 2}$; $x = 2.41$

40. $f(x) = \dfrac{2x + 5}{3x^2 - 4x + 1}$; $x = -1.72$

41. $f(x) = \sqrt{2x^2 + 1} + \sqrt{3x^2 - 1}$; $x = 0.62$

42. $f(x) = 2x(3x^3 + 5)^{1/3}$; $x = -6.24$

43. **Manufacturing Capacity** Data obtained from the Federal Reserve show that the annual increase in manufacturing capacity between 1988 and 1994 is given by

$$f(t) = 0.0388889t^3 - 0.283333t^2 + 0.477778t$$
$$+ 2.04286 \qquad (0 \le t \le 6)$$

where $f(t)$ is a percentage and t is measured in years, with $t = 0$ corresponding to the beginning of 1988.

a. Use a graphing utility to plot the graph of f in the viewing rectangle $[0, 8] \times [0, 4]$.

b. What is the annual increase in manufacturing capacity at the beginning of 1990 ($t = 2$)? At the beginning of 1992 ($t = 4$)?

Source: Federal Reserve

44. Decline of Union Membership The total union membership as a percentage of the private work force is given by

$$f(t) = 0.00017t^4 - 0.00921t^3 + 0.15437t^2$$
$$- 1.360723t + 16.8028 \qquad (0 \le t \le 10)$$

where t is measured in years, with $t = 0$ corresponding to the beginning of 1983.

a. Plot the graph of f in the viewing rectangle $[0, 11] \times [8, 20]$.

b. What is the total union membership as a percentage of the private force at the beginning of 1986? At the beginning of 1993?

Source: American Federation of Labor and Congress of Industrial Organizations

45. Keeping with the Traffic Flow By driving at a speed to match the prevailing traffic speed, you decrease the chances of an accident. According to a University of Virginia School of Engineering and Applied Science study, the number of accidents per 100 million vehicle miles, y, is related to the deviation from the mean speed, x, in mph by the equation

$$y = 1.05x^3 - 21.95x^2 + 155.9x - 327.3$$
$$(6 \le x \le 11)$$

a. Plot the graph of y in the viewing rectangle $[6, 11] \times [20, 150]$.

b. What is the number of accidents per 100 million vehicle miles if the deviation from the mean speed is 6 mph? 8 mph? 11 mph?

Source: University of Virginia School of Engineering and Applied Science

2.2 THE ALGEBRA OF FUNCTIONS

The Sum, Difference, Product, and Quotient of Functions

Let $S(t)$ and $R(t)$ denote, respectively, the federal government's spending and revenue at any time t, measured in billions of dollars. The graphs of these functions for the period between 1981 and 1991 are shown in Figure 2.9.

Figure 2.9
$S(t) - R(t)$ gives the federal budget deficit at any time t.

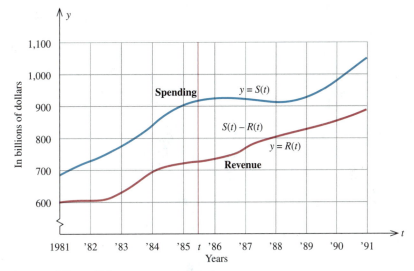

Source: Office of Management and Budget

The budget deficit at any time t is given by $S(t) - R(t)$ billion dollars. This observation suggests that we can define a function D whose value at any time t is given by $D(t) = S(t) - R(t)$. The function D, the *difference* of the two functions S and R, is written $D = S - R$, and may be called the "deficit function" since it gives the budget deficit at any time t. It has the same domain as the functions S and R.

Most functions are built up from other, generally simpler, functions. For example, we may view the function $f(x) = 2x + 4$ as the sum of the two functions $g(x) = 2x$ and $h(x) = 4$. The function $g(x) = 2x$ may, in turn, be viewed as the product of the functions $p(x) = 2$ and $q(x) = x$.

In general, given the functions f and g, we define the **sum, difference, product,** and **quotient** functions as $f + g$, $f - g$, fg, and f/g, respectively, with values defined by

$$(f + g)(x) = f(x) + g(x) \qquad \text{(Sum)}$$
$$(f - g)(x) = f(x) - g(x) \qquad \text{(Difference)}$$
$$(fg)(x) = f(x)g(x) \qquad \text{(Product)}$$
$$\left(\frac{f}{g}\right)(x) = \frac{f(x)}{g(x)} \qquad (g(x) \neq 0) \qquad \text{(Quotient)}$$

Thus, the rules for the sum, difference, product, and quotient of two functions f and g are obtained by simply adding, subtracting, multiplying, and dividing the respective expressions for $f(x)$ and $g(x)$.

EXAMPLE 1   Let f and g be functions defined by the rules $f(x) = \frac{1}{2}x^2$ and $g(x) = 2x + 1$, respectively. Find the rules for the sum s, the difference d, the product p, and the quotient q of the functions f and g.

Solution   The required rules are given by

$$s(x) = (f + g)(x) = f(x) + g(x) = \frac{1}{2}x^2 + 2x + 1$$

$$d(x) = (f - g)(x) = f(x) - g(x) = \frac{1}{2}x^2 - (2x + 1) = \frac{1}{2}x^2 - 2x - 1$$

$$p(x) = (fg)(x) = f(x)g(x) = \frac{1}{2}x^2 (2x + 1)$$

and $q(x) = \left(\frac{f}{g}\right)(x) = \frac{f(x)}{g(x)} = \frac{\frac{1}{2}x^2}{2x + 1} = \frac{x^2}{2(2x + 1)}$ $\left(x \neq -\frac{1}{2}\right)$ ○ ○ ○

Applications

The mathematical formulation of a problem arising from a practical situation often leads to an expression that involves the combination of functions. Consider, for example, the costs incurred in operating a business. Costs that remain more or less constant regardless of the firm's level of activity are called **fixed costs.** Examples of fixed costs are rental fees and executive salaries. On the other hand, costs that vary with production or sales are called **variable costs.** Examples of variable costs are wages and costs of raw materials. The **total cost** of operating a business is thus given by the *sum* of the variable costs and the fixed costs, as illustrated in the next example.

EXAMPLE 2   Suppose that Puritron, a manufacturer of water filters, has a monthly fixed cost of $10,000 and a variable cost of

$$-0.0001x^2 + 10x \qquad (0 \leq x \leq 40{,}000)$$

dollars, where x denotes the number of filters manufactured per month. Find a function C that gives the total cost incurred by Puritron in the manufacture of x filters.

Solution   Puritron's monthly fixed cost is always $10,000, regardless of the level of production, and it is described by the constant function $F(x) = 10{,}000$. Next, the variable cost is described by the function $V(x) = -0.0001x^2 + 10x$. Since the total cost incurred by Puritron at any level of production is the sum of the variable cost and the fixed cost, we see that the required total cost function is given by

$$C(x) = V(x) + F(x)$$
$$= -0.0001x^2 + 10x + 10{,}000 \qquad (0 \le x \le 40{,}000) \qquad \circ \; \circ \; \circ$$

Next, the **total profit** realized by a firm in operating a business is the *difference* between the total revenue realized and the total cost incurred; that is, $P(x) = R(x) - C(x)$.

EXAMPLE 3   Refer to Example 2. Suppose that the total revenue realized by Puritron from the sale of x water filters is given by the total revenue function

$$R(x) = -0.0005x^2 + 20x \qquad (0 \le x \le 40{,}000)$$

a. Find the total profit function—that is, the function that describes the total profit Puritron realizes in manufacturing and selling x water filters per month.

b. What is the profit when the level of production is 10,000 filters per month?

Solution

a. The total profit realized by Puritron in manufacturing and selling x water filters per month is the difference between the total revenue realized and the total cost incurred. Thus, the required total profit function is given by

$$P(x) = R(x) - C(x)$$
$$= (-0.0005x^2 + 20x) - (-0.0001x^2 + 10x + 10{,}000)$$
$$= -0.0004x^2 + 10x - 10{,}000$$

b. The profit realized by Puritron when the level of production is 10,000 filters per month is

$$P(10{,}000) = -0.0004(10{,}000)^2 + 10(10{,}000) - 10{,}000 = 50{,}000$$

or $50,000 per month. $\circ \; \circ \; \circ$

Composition of Functions

Another way to build up a function from other functions is through a process known as the *composition of functions.* Consider, for example, the function h, whose rule is given by $h(x) = \sqrt{x^2 - 1}$. Let f and g be functions defined by the rules $f(x) = x^2 - 1$ and $g(x) = \sqrt{x}$. Evaluating the function g at the point $f(x)$ [remember that for each real number x in the domain of f, $f(x)$ is simply a real number], we find that

$$g(f(x)) = \sqrt{f(x)} = \sqrt{x^2 - 1}$$

which is just the rule defining the function h!

In general, the **composition of a function g with a function f** is defined as the function h whose value is found by evaluating the function g at the point $f(x)$; that is,

$$h(x) = g(f(x))$$

Figure 2.10
The composite function $h = g \circ f$ is found by evaluating f at x and then g at $f(x)$, in that order.

Figure 2.11
The function $h = g \circ f$ viewed as a mapping.

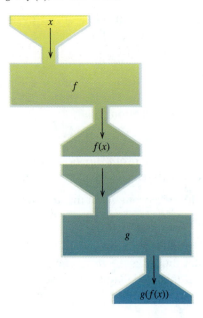

The function h is called a **composite function** and is commonly denoted by $h = g \circ f$ (read "g circle f" or "g composed with f"). The interpretation of the function h as a machine is illustrated in Figure 2.10, and its interpretation as a mapping is shown in Figure 2.11.

EXAMPLE 4   Let $f(x) = x^2 - 1$ and $g(x) = \sqrt{x} + 1$. Compute:

a. the rule for the composite function $g \circ f$

b. the rule for the composite function $f \circ g$

Solution

a. To find the rule for the composite function $g \circ f$, evaluate the function g at $f(x)$. Therefore,

$$(g \circ f)(x) = g(f(x)) = \sqrt{f(x)} + 1 = \sqrt{x^2 - 1} + 1$$

b. To find the rule for the composite function $f \circ g$, evaluate the function f at $g(x)$. Thus,

$$(f \circ g)(x) = f(g(x)) = (g(x))^2 - 1 = (\sqrt{x} + 1)^2 - 1$$
$$= x + 2\sqrt{x} + 1 - 1 = x + 2\sqrt{x}$$

⓿ ⓿ ⓿

Let $f(x) = \sqrt{x} + 1$ for $x \geq 0$ and let $g(x) = (x - 1)^2$ for $x \geq 1$.

a. Show that $(g \circ f)(x)$ and $(f \circ g)(x) = x$. [*Remark.* The function g is said to be the *inverse* of f and vice versa.]

b. Plot the graphs of f and g together with the straight line $y = x$. Describe the relationship between the graphs of f and g.

Example 4 reminds us that, in general, $g \circ f$ is different from $f \circ g$, and care must be taken regarding the order when computing a composite function.

EXAMPLE 5   An environmental impact study conducted for Oxnard's City Environmental Management Department indicates that, under existing environmental protection laws, the level of carbon monoxide present in the air due to pollution from automobile exhaust will be $0.01x^{2/3}$ parts per million when the number of motor vehicles is x thousand. A separate study conducted by a state government agency estimates that t years from now the number of motor vehicles in Oxnard will be $0.2t^2 + 4t + 64$ thousand.

a. Find an expression for the concentration of carbon monoxide in the air due to automobile exhaust t years from now.

b. What will the level of concentration be five years from now?

Solution

a. The level of carbon monoxide present in the air due to pollution from automobile exhaust is described by the function $g(x) = 0.01x^{2/3}$, where x is the number (in thousands) of motor vehicles. But the number of motor vehicles x (in thousands) t years from now may be estimated by the rule $f(t) = 0.2t^2 + 4t + 64$. Therefore, the concentration of carbon monoxide due to automobile exhaust t years from now is given by

$$C(t) = g(f(t)) = 0.01(0.2t^2 + 4t + 64)^{2/3}$$

parts per million.

b. The level of concentration five years from now will be

$$C(5) = 0.01[0.2(5)^2 + 4(5) + 64]^{2/3}$$
$$= (0.01)89^{2/3} \approx 0.20$$

or 0.20 parts per million.

○ ○ ○

Ebasco is a diversified engineering and construction company. In addition to designing and building electric generating facilities, the company provides clients with safety and risk assessment studies. Their studies include analyses of potential hazards, recommendations for safe work practices, and evaluations of responses to emergencies in chemical plants, fuel-storage depots, nuclear facilities, and hazardous waste sites. As a project manager, Marchlik deals directly with clients to help them understand and implement Ebasco's recommendations to ensure a safer environment.

Michael Marchlik

Title: Project Manager
Institution: Ebasco Services Incorporated

Calculus wasn't Mike Marchlik's favorite college subject. In fact, it was not until he started his first job that the "lights went on" and he realized how using calculus allowed him to solve real problems in his everyday work.

Marchlik emphasizes that he doesn't do "number crunching" himself. "I don't work out integrals, but in my work I use computer models that do that." The important issue is not computation but how the answers relate to client problems.

Marchlik's clients typically process highly toxic or explosive materials. Ebasco evaluates a client site, such as a chemical plant, to determine how safety systems might fail and what the probable consequences might be.

To avoid a major disaster like the one that occurred in Bhopal, India, in 1984, Marchlik and his team might be asked to determine how quickly a poisonous chemical would spread if a leak occurred. Or they might help avert a disaster like the one that rocked the Houston area in 1989. In that incident, hydrocarbon vapor exploded at a Phillips 66 chemical plant, shaking office buildings in Houston, 12 miles away. Several people were killed and hundreds were injured. Property damage totaled $1.39 billion.

In assessing risks for a fuel-storage depot, Ebasco considers variables such as weather conditions, including probable wind speed, the flow properties of a gas, and possible ignition sources. With today's more powerful computers, the models can involve a system of very complex equations to project likely scenarios.

Mathematical models vary, however. One model might forecast how much gas will flow out of a hole and how quickly it will disperse. Another model might project where the gas will go depending on local factors such as temperature and wind speed. Choosing the right model is essential. A model based on flat terrain when the client's storage depot is set among hills is going to produce the wrong answer.

Marchlik and his team run several models together to come up with their projections. Each model uses "equations that have to be integrated to come up with solutions." The bottom line? Marchlik stresses that "calculus is at the very heart" of Ebasco's risk-assessment work.

SELF-CHECK EXERCISES 2.2

1. Let f and g be functions defined by the rules $f(x) = \sqrt{x} + 1$ and $g(x) = \dfrac{x}{1+x}$, respectively. Find the rules for
  a. the sum s, the difference d, the product p, and the quotient q of f and g
  b. the composite functions $f \circ g$ and $g \circ f$

2. Health-care spending per person by the private sector comprises payments by individuals, corporations, and their insurance companies, and is approximated by the function

$$f(t) = 2.5t^2 + 31.3t + 406 \qquad (0 \le t \le 20)$$

where $f(t)$ is measured in dollars and t is measured in years, with $t = 0$ corresponding to the beginning of 1975. The corresponding government spending, comprising expenditures for medicaid, medicare, and other federal, state, and local government public health care, is

$$g(t) = 1.4t^2 + 29.6t + 251 \qquad (0 \le t \le 20)$$

where t has the same meaning as before.
  a. Find a function that gives the difference between private and government health-care spending per person at any time t.
  b. What was the difference between private and government expenditures per person at the beginning of 1985? At the beginning of 1993?
  Source: Health Care Financing Administration

Solutions to Self-Check Exercises 2.2 can be found on page 83.

2.2 EXERCISES

In exercises 1–8, let $f(x) = x^3 + 5$, $g(x) = x^2 - 2$, and $h(x) = 2x + 4$. Find the rule for each function.

1. $f + g$ **2.** $f - g$ **3.** fg **4.** gf

5. $\dfrac{f}{g}$ **6.** $\dfrac{f - g}{h}$ **7.** $\dfrac{fg}{h}$ **8.** fgh

In exercises 9–18, let $f(x) = x - 1$, $g(x) = \sqrt{x} + 1$, and $h(x) = 2x^3 - 1$. Find the rule for each function.

9. $f + g$ **10.** $g - f$ **11.** fg **12.** gf

13. $\dfrac{g}{h}$ **14.** $\dfrac{h}{g}$ **15.** $\dfrac{fg}{h}$ **16.** $\dfrac{fh}{g}$

17. $\dfrac{f - h}{g}$ **18.** $\dfrac{gh}{g - f}$

In exercises 19–24, find the rules for $f + g$, $f - g$, fg, and $\dfrac{f}{g}$.

19. $f(x) = x^2 + 5$; $g(x) = \sqrt{x} - 2$

20. $f(x) = \sqrt{x - 1}$; $g(x) = x^3 + 1$

21. $f(x) = \sqrt{x + 3}$; $g(x) = \dfrac{1}{x - 1}$

22. $f(x) = \dfrac{1}{x^2 + 1}$; $g(x) = \dfrac{1}{x^2 - 1}$

23. $f(x) = \dfrac{x + 1}{x - 1}$; $g(x) = \dfrac{x + 2}{x - 2}$

24. $f(x) = x^2 + 1$; $g(x) = \sqrt{x + 1}$

In exercises 25–30, find the rules for the composite functions
f ∘ g and g ∘ f.

25. $f(x) = x^2 + x + 1$; $g(x) = x^2$

26. $f(x) = 3x^2 + 2x + 1$; $g(x) = x + 3$

27. $f(x) = \sqrt{x} + 1$; $g(x) = x^2 - 1$

28. $f(x) = 2\sqrt{x} + 3$; $g(x) = x^2 + 1$

29. $f(x) = \dfrac{x}{x^2 + 1}$; $g(x) = \dfrac{1}{x}$

30. $f(x) = \sqrt{x + 1}$; $g(x) = \dfrac{1}{x - 1}$

In exercises 31–34, evaluate h(2), where h = g ∘ f.

31. $f(x) = x^2 + x + 1$; $g(x) = x^2$

32. $f(x) = \sqrt[3]{x^2 - 1}$; $g(x) = 3x^3 + 1$

33. $f(x) = \dfrac{1}{2x + 1}$; $g(x) = \sqrt{x}$

34. $f(x) = \dfrac{1}{x - 1}$; $g(x) = x^2 + 1$

In exercises 35–42, find functions f and g such that h =
g ∘ f. (Note: The answer is not unique.)

35. $h(x) = (2x^3 + x^2 + 1)^5$ **36.** $h(x) = (3x^2 - 4)^{-3}$

37. $h(x) = \sqrt{x^2 - 1}$ **38.** $h(x) = (2x - 3)^{3/2}$

39. $h(x) = \dfrac{1}{x^2 - 1}$ **40.** $h(x) = \dfrac{1}{\sqrt{x^2 - 4}}$

41. $h(x) = \dfrac{1}{(3x^2 + 2)^{3/2}}$

42. $h(x) = \dfrac{1}{\sqrt{2x + 1}} + \sqrt{2x + 1}$

In exercises 43–46, find f(a + h) − f(a) for each function.
Simplify your answer.

43. $f(x) = 3x + 4$ **44.** $f(x) = -\frac{1}{2}x + 3$

45. $f(x) = 4 - x^2$ **46.** $f(x) = x^2 - 2x + 1$

47. If $f(x) = x^2 + 1$, find and simplify

$$\frac{f(a + h) - f(a)}{h} \qquad (h \neq 0)$$

48. If $f(x) = 1/x$, find and simplify

$$\frac{f(a + h) - f(a)}{h} \qquad (h \neq 0)$$

49. Manufacturing Costs   TMI, Inc., a manufacturer of
blank audiocassette tapes, has a monthly fixed cost of

$12,100 and a variable cost of 60 cents per tape. Find a
function C that gives the total cost incurred by TMI in
the manufacture of x tapes per month.

50. Cost of Producing Word Processors   Apollo, Inc.,
manufactures its word processors at a variable cost of

$$V(x) = 0.000003x^3 - 0.03x^2 + 200x$$

dollars, where x denotes the number of units manufactured per month. The monthly fixed cost attributable
to the division that produces these word processors is
$100,000. Find a function C that gives the total cost
incurred by the manufacture of x word processors. What
is the total cost incurred in producing 2000 units per
month?

51. Cost of Producing Word Processors   Refer to exercise 50. Suppose the total revenue realized by Apollo,
Inc., from the sale of x word processors is given by the
total revenue function

$$R(x) = -0.1x^2 + 500x \qquad (0 \leq x \leq 5000)$$

where x is measured in dollars.
a. Find the total profit function.
b. What is the profit when 1500 units are produced and
sold each month?

52. Overcrowding of Prisons   The 1980s saw a trend toward old-fashioned punitive deterrence as opposed to
the more liberal penal policies and community-based
corrections popular in the 1960s and early 1970s. As a
result, prisons became more crowded and the gap between the number of people in prison and the prison
capacity widened. Based on figures from the U.S. Department of Justice, the number of prisoners (in thousands) in federal and state prisons is approximated by
the function

$$N(t) = 3.5t^2 + 26.7t + 436.2 \qquad (0 \leq t \leq 10)$$

where t is measured in years and $t = 0$ corresponds to
1983. The number of inmates for which prisons were
designed is given by

$$C(t) = 24.3t + 365 \qquad (0 \leq t \leq 10)$$

where $C(t)$ is measured in thousands and t has the same
meaning as before.
a. Find an expression that shows the gap between the
number of prisoners and the number of inmates for
which the prisons were designed at any time t.
b. Find the gap at the beginning of 1983. At the beginning of 1986.
Source: U.S. Dept. of Justice

53. Effect of Mortgage Rates on Housing Starts A study prepared for the National Association of Realtors estimated that the number of housing starts per year over the next five years will be

$$N(r) = \frac{7}{1 + 0.02r^2}$$

million units, where r (percent) is the mortgage rate. Suppose that the mortgage rate over the next t months is

$$r(t) = \frac{10t + 150}{t + 10} \qquad (0 \le t \le 24)$$

percent per year.
a. Find an expression for the number of housing starts per year as a function of t, t months from now.
b. Using the result from (a), determine the number of housing starts at present; 12 months from now; 18 months from now.

54. Hotel Occupancy Rate The occupancy rate of the all-suite Wonderland Hotel, located near an amusement park, is given by the function

$$r(t) = \frac{10}{81}t^3 - \frac{10}{3}t^2 + \frac{200}{9}t + 55 \qquad (0 \le t \le 11)$$

where t is measured in months and $t = 0$ corresponds to the beginning of January. Management has estimated

that the monthly revenue (in thousands of dollars) is approximated by the function

$$R(r) = -\frac{3}{5000}r^3 + \frac{9}{50}r^2 \qquad (0 \le r \le 100)$$

where r is the occupancy rate.
a. What is the hotel's occupancy rate at the beginning of January? At the beginning of June?
b. What is the hotel's monthly revenue at the beginning of January? At the beginning of June?
[*Hint:* Compute $R(r(0))$ and $R(r(5))$.]

55. Housing Starts and Construction Jobs The president of a major housing construction firm reports that the number of construction jobs (in millions) created is given by

$$N(x) = 1.42x$$

where x denotes the number of housing starts. Suppose that the number of housing starts in the next t months is expected to be

$$x(t) = \frac{7(t + 10)^2}{(t + 10)^2 + 2(t + 15)^2}$$

million units per year. Find an expression for the number of jobs created per month in the next t months. How many jobs will have been created 6 months from now? 12 months from now?

SOLUTIONS TO SELF-CHECK EXERCISES 2.2

1. a. $s(x) = f(x) + g(x) = \sqrt{x} + 1 + \dfrac{x}{1 + x}$

$d(x) = f(x) - g(x) = \sqrt{x} + 1 - \dfrac{x}{1 + x}$

$p(x) = f(x)g(x) = (\sqrt{x} + 1) \cdot \dfrac{x}{1 + x} = \dfrac{x(\sqrt{x} + 1)}{1 + x}$

$q(x) = \dfrac{f(x)}{g(x)} = \dfrac{\sqrt{x} + 1}{\dfrac{x}{1 + x}} = \dfrac{(\sqrt{x} + 1)(1 + x)}{x}$

b. $(f \circ g)(x) = f(g(x)) = \sqrt{\dfrac{x}{1 + x}} + 1$

$(g \circ f)(x) = g(f(x)) = \dfrac{\sqrt{x} + 1}{1 + (\sqrt{x} + 1)} = \dfrac{\sqrt{x} + 1}{\sqrt{x} + 2}$

2. a. The difference between private and government health-care spending per person at any time t is given by the function d with the rule

$$d(t) = f(t) - g(t) = (2.5t^2 + 31.3t + 406)$$
$$- (1.4t^2 + 29.6t + 251)$$
$$= 1.1t^2 + 1.7t + 155$$

b. The difference between private and government expenditures per person at the beginning of 1985 is given by

$$d(10) = 1.1(10)^2 + 1.7(10) + 155$$

or $282 per person.

The difference between private and government expenditures per person at the beginning of 1993 is given by

$$d(18) = 1.1(18)^2 + 1.7(18) + 155$$

or $542 per person.

2.3 FUNCTIONS AND MATHEMATICAL MODELS

Mathematical Models

Before we can apply mathematics to solving real-world problems, we must be able to formulate those problems in the language of mathematics. This process is referred to as **mathematical modeling.** A *mathematical model* may describe precisely the problem under consideration, or, more likely than not, it may provide only an acceptable approximation of the problem. For example, the accumulated amount A at the end of t years when a sum of P dollars is deposited in a fixed bank account, earning interest at the rate of r percent per year compounded m times a year, is given *exactly* by the formula (model)

$$A = P\left(1 + \frac{r}{m}\right)^{mt}$$

However, the size of a cancer tumor may be *approximated* by the volume of a sphere,

$$V = \frac{4}{3}\pi r^3$$

where r is the radius of the tumor in centimeters.

The many techniques used in constructing mathematical models of practical problems range from theoretical consideration of the problem on the one extreme to an interpretation of data associated with the problem on the other. The model for the accumulated amount of a fixed bank account mentioned earlier may be derived theoretically (see Chapter 5). Later we will see how linear equations (models) can be constructed from a given set of data points.

In calculus, we are concerned primarily with how one (dependent) variable depends on one or more (independent) variables. Consequently, most of our mathematical models will involve functions of one or more variables.* Once a function has been constructed to describe a specific real-world problem, a host of questions pertaining to the problem may be answered by analyzing the function (mathematical model). For example, if we have a function that gives the population of a certain culture of bacteria at any time t, then we can determine how fast the population is increasing or decreasing at any time t, and so on. Conversely, if we have a model that gives the rate of change of the cost of producing a certain item as a function of the level of production, and if we know the fixed cost incurred in producing this item, then we can find the total cost incurred in producing a certain number of those items.

Before going on, let us look at two mathematical models. The first one is used to estimate spending by business on computer security and the second is used to project the growth of the number of people enrolled in health maintenance organizations (HMOs). These models are derived from data using the least-squares technique. (We will not, however, pursue the actual construction of the models.)

EXAMPLE 1   The estimated spending by businesses on computer security equipment and services from 1987 to 1993 is given in the following table. The figures include spending for protection against computer criminals who steal, erase, or alter data, along with protection against fires, electrical failures, and natural disasters.

Year	1987	1988	1989	1990	1991	1992	1993
Spending (in billions of dollars)	0.49	0.59	0.66	0.73	0.81	0.93	1.02

A mathematical model approximating the amount of spending over the period in question is given by

$$S(t) = 0.0864t + 0.4879$$

where t is measured in years, with $t = 0$ corresponding to 1987.

a. Sketch the graph of the function S and the given data on the same set of axes.

b. Assuming that this trend continued, what was the spending by business on computer security equipment and services in 1995 ($t = 8$)?

c. What is the rate of increase of the annual expenditure over the period in question?

Source: Frost & Sullivan Inc.

○ ○ ○

* Functions of more than one variable will be studied later.

Figure 2.12
Estimated spending by businesses on computer security equipment and services.

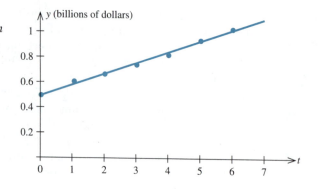

Solution

a. The graph of S is shown in Figure 2.12.

b. The estimated spending in 1995 is

$$S(8) = 0.0864(8) + 0.4879$$
$$\approx 1.1791$$

or approximately $1.18 billion.

c. The function S is linear, and so we conclude that the annual increase in the expenditure is given by the slope of the straight line represented by S, which is approximately $0.09 billion per year.   ◦ ◦ ◦

 EXAMPLE 2   The number of people enrolled in health maintenance organizations (HMOs) from 1984 to 1994 is given in the following table.

Year	1984	1986	1988	1990	1992	1994
Number of People (in millions)	15.8	25.8	33	37	41.6	50

A mathematical model approximating the number of people, $N(t)$, enrolled in HMOs during this period is

$$N(t) = 0.0514t^3 - 0.853t^2 + 6.8147t + 15.6524 \qquad (0 \le t \le 10)$$

where t is measured in years and $t = 0$ corresponds to 1984.

a. Sketch the graph of the function N to see how the model compares with the actual data.

b. Assuming that this trend continued, how many people were enrolled in HMOs at the beginning of 1998?

Source: Group Health Association of America

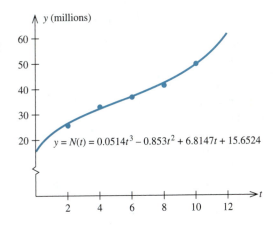

Figure 2.13
The graph of y = N(t) approximates the number of people enrolled in HMOs from 1984 to 1994.

Solution

a. The graph of the function N is shown in Figure 2.13.

b. The number of people enrolled in HMOs at the beginning of 1998 is given by

$$N(14) = 0.0514(14)^3 - 0.853(14)^2 + 6.8147(14) + 15.6524$$
$$= 84.9118$$

or approximately 84.91 million people.

⊙ ⊙ ⊙

We will discuss several mathematical models from the field of economics later in this section, but first we review some important functions that are the bases for many mathematical models.

Polynomial Functions

We begin by recalling a special class of functions, *polynomial functions*.

A POLYNOMIAL FUNCTION

A **polynomial function** of degree n is a function of the form

$$f(x) = a_0x^n + a_1x^{n-1} + \cdots + a_{n-1}x + a_n \qquad (a_0 \neq 0)$$

where $a_0, a_1, \ldots, a_n$ are constants and n is a nonnegative integer.

For example, the functions

$$f(x) = 4x^5 - 3x^4 + x^2 - x + 8$$

and

$$g(x) = 0.001x^3 - 2x^2 + 20x + 400$$

are polynomial functions of degrees 5 and 3, respectively. Observe that a polynomial function is defined everywhere so that it has domain $(-\infty, \infty)$.

A polynomial function of degree 1 ($n = 1$)

$$f(x) = a_0 x + a_1 \qquad (a_0 \neq 0)$$

is the equation of a straight line in the slope-intercept form with slope $m = a_0$ and y-intercept $b = a_1$ (see Section 1.4). For this reason, a polynomial function of degree 1 is called a **linear function.** For example, the linear function $f(x) = 2x + 3$ may be written as a linear equation in x and y—namely, $y = 2x + 3$ or $2x - y + 3 = 0$. Conversely, the linear equation $2x - 3y + 4 = 0$ can be solved for y in terms of x to yield the linear function $y = f(x) = \frac{2}{3}x + \frac{4}{3}$.

A polynomial function of degree 2 is referred to as a **quadratic function.** A polynomial function of degree 3 is called a **cubic function,** and so on. The mathematical model in Example 2 involves a cubic function.

Rational and Power Functions

Another important class of functions is rational functions. A **rational function** is simply the quotient of two polynomials. Examples of rational functions are

$$F(x) = \frac{3x^3 + x^2 - x + 1}{x - 2}$$

and

$$G(x) = \frac{x^2 + 1}{x^2 - 1}$$

In general, a rational function has the form

$$R(x) = \frac{f(x)}{g(x)}$$

where $f(x)$ and $g(x)$ are polynomial functions. Since division by zero is not allowed, we conclude that the domain of a rational function is the set of all real numbers except the zeros of g—that is, the roots of the equation $g(x) = 0$. Thus, the domain of the function F is the set of all numbers except $x = 2$, whereas the domain of the function G is the set of all numbers except those that satisfy $x^2 - 1 = 0$ or $x = \pm 1$.

Functions of the form

$$f(x) = x^r$$

where r is any real number, are called **power functions.** We encountered examples of power functions earlier in our work. For example, the functions

$$f(x) = \sqrt{x} = x^{1/2} \qquad \text{and} \qquad g(x) = \frac{1}{x^2} = x^{-2}$$

are power functions.

Many of the functions we will encounter later will involve combinations of the functions introduced here. For example, the following functions may be viewed as suitable combinations of such functions:

$$f(x) = \sqrt{\frac{1 - x^2}{1 + x^2}}$$

$$g(x) = \sqrt{x^2 - 3x + 4}$$

and
$$h(x) = (1 + 2x)^{1/2} + \frac{1}{(x^2 + 2)^{3/2}}$$

As with polynomials of degree 3 or greater, analyzing the properties of these functions is facilitated by using the tools of calculus, to be developed later.

 EXAMPLE 3   A study of driving costs based on 1992 model compact (six-cylinder) cars found that the average cost (car payments, gas, insurance, upkeep, and depreciation), measured in cents per mile, is approximated by the function

$$C(x) = \frac{2095}{x^{2.2}} + 20.08$$

where x (in thousands) denotes the number of miles the car is driven in one year. Using this model, estimate the average cost of driving a compact car 10,000 miles a year; 20,000 miles a year.
Source: Runzheimer International Study

Solution   The average cost of driving a compact car 10,000 miles a year is given by

$$C(10) = \frac{2095}{10^{2.2}} + 20.08$$
$$\approx 33.3$$

or approximately 33 cents per mile. The average cost of driving 20,000 miles a year is given by

$$C(20) = \frac{2095}{20^{2.2}} + 20.08$$
$$\approx 23.0$$

or approximately 23 cents per mile.   ⊙ ⊙ ⊙

Some Economic Models

In the remainder of this section we look at some economic models.

In a free market economy, consumer demand for a particular commodity depends on the commodity's unit price. A **demand equation** expresses the relationship between the unit price and the quantity demanded. The graph of the demand equation is called a **demand curve.** In general, the quantity demanded of a commodity decreases as the commodity's unit price increases, and vice versa. Accordingly, a **demand function** defined by $p = f(x)$, where p measures the unit price and x measures the number of units of the commodity in question, is generally characterized as a decreasing function of x; that is,

Figure 2.14
A demand curve.

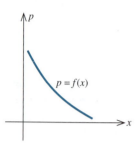

$p = f(x)$ decreases as x increases. Since both x and p assume only nonnegative values, the demand curve is that part of the graph of $f(x)$ that lies in the first quadrant (Figure 2.14).

In a competitive market, a relationship also exists between the unit price of a commodity and the commodity's availability in the market. In general, an increase in the commodity's unit price induces the producer to increase the supply of the commodity. Conversely, a decrease in the unit price generally leads to a drop in the supply. The equation that expresses the relation between the unit price and the quantity supplied is called a **supply equation,** and its graph is called a **supply curve.** A **supply function** defined by $p = f(x)$ is generally characterized as an increasing function of x; that is, $p = f(x)$ increases as x increases. Since both x and p assume only nonnegative values, the supply curve is that part of the graph of $f(x)$ that lies in the first quadrant (Figure 2.15).

Figure 2.15
A supply curve.

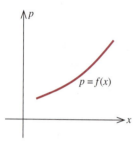

Under pure competition, the price of a commodity will eventually settle at a level dictated by the following condition: that the supply of the commodity be equal to the demand for it. If the price is too high, the consumer will not buy, and if the price is too low, the supplier will not produce. **Market equilibrium** prevails when the quantity produced is equal to the quantity demanded. The quantity produced at market equilibrium is called the **equilibrium quantity,** and the corresponding price is called the **equilibrium price.**

Market equilibrium corresponds to the point at which the demand curve and the supply curve intersect. In Figure 2.16, x_0 represents the equilibrium quantity and p_0 the equilibrium price. The point (x_0, p_0) lies on the supply curve and therefore satisfies the supply equation. At the same time, it also lies on the demand curve and therefore satisfies the demand equation. Thus, to find the point (x_0, p_0), and hence the equilibrium quantity and price, we solve the demand and supply equations simultaneously for x and p. For meaningful solutions, x and p must both be positive.

Figure 2.16
Market equilibrium corresponds to (x_0, p_0), the point at which the supply and demand curves intersect.

EXAMPLE 4   The demand function for a certain brand of videocassette is given by

$$p = d(x) = -0.01x^2 - 0.2x + 8$$

and the corresponding supply function is given by

$$p = s(x) = 0.01x^2 + 0.1x + 3$$

where p is expressed in dollars and x is measured in units of a thousand. Find the equilibrium quantity and price.

Solution   We solve the following system of equations:

$$p = -0.01x^2 - 0.2x + 8$$

and

$$p = 0.01x^2 + 0.1x + 3$$

Substituting the first equation into the second yields

$$-0.01x^2 - 0.2x + 8 = 0.01x^2 + 0.1x + 3$$

Figure 2.17
The supply curve and the demand
curve intersect at the point (10, 5).

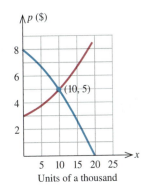

Units of a thousand

which is equivalent to

$$0.02x^2 + 0.3x - 5 = 0$$
$$2x^2 + 30x - 500 = 0$$
$$x^2 + 15x - 250 = 0$$
$$(x + 25)(x - 10) = 0$$

or

Thus, $x = -25$ or $x = 10$. Since x must be nonnegative, the root $x = -25$ is rejected. Therefore, the equilibrium quantity is 10,000 videocassettes. The equilibrium price is given by

$$p = 0.01(10)^2 + 0.1(10) + 3 = 5$$

or $5 per videocassette (Figure 2.17).

◦ ◦ ◦

EXPLORING WITH TECHNOLOGY

1. a. Use a graphing utility to plot the straight lines L_1 and L_2 with equations $y = 2x - 1$ and $y = 2.1x + 3$, respectively, on the same set of axes using the standard viewing rectangle. Do the lines appear to intersect?
   b. Plot the straight lines L_1 and L_2 using the viewing rectangle $[-100, 100] \times [-100, 100]$. Do the lines appear to intersect? Can you find the point of intersection using **TRACE** and **ZOOM**? Using the "intersection" function of your graphing utility?
   c. Find the point of intersection of L_1 and L_2 algebraically.
   d. Comment on the effectiveness of the methods of solutions in parts (b) and (c).

2. a. Use a graphing utility to plot the straight lines L_1 and L_2 with equations $y = 3x - 2$ and $y = -2x + 3$, respectively, on the same set of axes using the standard viewing rectangle. Then use **TRACE** and **ZOOM** to find the point of intersection of L_1 and L_2. Repeat using the "intersection" function of your graphing utility.
   b. Find the point of intersection of L_1 and L_2 algebraically.
   c. Comment on the effectiveness of the methods.

◦ ◦ ◦

SELF-CHECK EXERCISES 2.3

1. Thomas Young has suggested the following rule for calculating the dosage of medicine for children from ages 1 to 12 years. If a denotes the adult dosage (in mg), and t is the age of the child (in years), then the child's dosage is given by

$$D(t) = \frac{at}{t + 12}$$

If the adult dose of a substance is 500 mg, how much should a four-year-old child receive?

2. The demand function for Mrs. Baker's cookies is given by

$$d(x) = -\frac{2}{15}x + 4$$

where $d(x)$ is the wholesale price in dollars per pound and x is the quantity demanded each week, measured in thousands of pounds. The supply function for the cookies is given by

$$s(x) = \frac{1}{75}x^2 + \frac{1}{10}x + \frac{3}{2}$$

where $s(x)$ is the wholesale price in dollars per pound and x is the quantity, in thousands of pounds, that will be made available in the market per week by the supplier.

a. Sketch the graphs of the functions d and s.
b. Find the equilibrium quantity and price.

Solutions to Self-Check Exercises 2.3 can be found on page 98.

2.3 EXERCISES

In exercises 1–8, determine whether the equation defines y as a linear function of x. If so, write it in the form y = mx + b.

1. $2x + 3y = 6$

2. $-2x + 4y = 7$

3. $x = 2y - 4$

4. $2x = 3y + 8$

5. $2x - 4y + 9 = 0$

6. $3x - 6y + 7 = 0$

7. $2x^2 - 8y + 4 = 0$

8. $3\sqrt{x} + 4y = 0$

In exercises 9–14, determine whether the given function is a polynomial function, a rational function, or some other function. State the degree of each polynomial function.

9. $f(x) = 3x^6 - 2x^2 + 1$

10. $f(x) = \frac{x^2 - 9}{x - 3}$

11. $G(x) = 2(x^2 - 3)^3$

12. $H(x) = 2x^{-3} + 5x^{-2} + 6$

13. $f(t) = 2t^2 + 3\sqrt{t}$

14. $f(r) = \frac{6r}{(r^3 - 8)}$

15. Find the constants m and b in the linear function $f(x) = mx + b$ so that $f(0) = 2$ and $f(3) = -1$.

16. Find the constants m and b in the linear function $f(x) = mx + b$ so that $f(2) = 4$ and the straight line represented by f has slope -1.

17. A manufacturer has a monthly fixed cost of $40,000 and a production cost of $8 for each unit produced. The product sells for $12 per unit.
a. What is the cost function?
b. What is the revenue function?
c. What is the profit function?
d. Compute the profit (loss) corresponding to production levels of 8000 and 12,000 units.

18. A manufacturer has a monthly fixed cost of $100,000 and a production cost of $14 for each unit produced. The product sells for $20 per unit.
a. What is the cost function?
b. What is the revenue function?
c. What is the profit function?
d. Compute the profit (loss) corresponding to production levels of 12,000 and 20,000 units.

19. Disposable Income Economists define the *disposable annual income* for an individual by the equation $D = (1 - r)T$, where T is the individual's total income and r is the net rate at which he or she is taxed. What is the disposable income for an individual whose income is $40,000 and whose net tax rate is 28%?

20. Drug Dosages A method sometimes used by pediatricians to calculate the dosage of medicine for children is based on the child's surface area. If a denotes the adult dosage (in mg), and S is the surface area of the child (in square meters), then the child's dosage is given by

$$D(S) = \frac{Sa}{1.7}$$

If the adult dose of a substance is 500 mg, how much should a child whose surface area is 0.4 square meter receive?

21. Cowling's Rule Cowling's Rule is a method for calculating pediatric drug dosages. If a denotes the adult dosage (in mg) and t is the age of the child (in years), then the child's dosage is given by

$$D(t) = \left(\frac{t+1}{24}\right)a$$

If the adult dose of a substance is 500 mg, how much should a four-year-old child receive?

22. Worker Efficiency An efficiency study showed that the average worker at Delphi Electronics assembled cordless telephones at the rate of

$$f(t) = -\frac{3}{2}t^2 + 6t + 10 \qquad (0 \le t \le 4)$$

phones per hour, t hours after starting work during the morning shift. At what rate does the average worker assemble telephones 2 hours after starting work?

23. Revenue Functions The revenue (in dollars) realized by Apollo, Inc., from the sale of its electronic correcting typewriters is given by

$$R(x) = -0.1x^2 + 500x$$

where x denotes the number of units manufactured per month. What is Apollo's revenue when 1000 units are produced?

24. Effect of Advertising on Sales The quarterly profit of Cunningham Realty depends on the amount of money x spent on advertising per quarter according to the rule

$$P(x) = -\frac{1}{8}x^2 + 7x + 30 \qquad (0 \le x \le 50)$$

where $P(x)$ and x are measured in thousands of dollars. What is Cunningham's profit when its quarterly advertising budget is $28,000?

25. Reaction of a Frog to a Drug Experiments conducted by A. J. Clark suggest that the response $R(x)$ of a frog's heart muscle to the injection of x units of acetylcholine (as a percentage of the maximum possible effect of the drug) may be approximated by the rational function

$$R(x) = \frac{100x}{b+x} \qquad (x \ge 0)$$

where b is a positive constant that depends on the particular frog.
a. If a concentration of 40 units of acetylcholine produces a response of 50% for a certain frog, find the "response function" for this frog.
b. Using the model found in (a), find the response of the frog's heart muscle when 60 units of acetylcholine are administered.

26. Forecasting Sales The annual sales of the Crimson Drug Store are expected to be given by

$$S(t) = 2.3 + 0.4t$$

million dollars t years from now, whereas the annual sales of the Cambridge Drug Store are expected to be given by

$$S(t) = 1.2 + 0.6t$$

million dollars t years from now. When will the annual sales of the Cambridge Drug Store first surpass the annual sales of the Crimson Drug Store?

27. Cricket Chirping and Temperature Entomologists have discovered that there is a linear relationship between the number of chirps of crickets of a certain species and the air temperature. When the temperature is 70°F, the crickets chirp at the rate of 120 times per minute, and when the temperature is 80°F, they chirp at the rate of 160 times per minute.
a. Find an equation giving the relationship between the air temperature T and the number of chirps per minute, N, of the crickets.
b. Find N as a function of T and use this formula to determine the rate at which the crickets chirp when the temperature is 102°F.

28. Linear Depreciation In computing income tax, business firms are allowed by law to depreciate certain assets such as buildings, machines, furniture, automobiles, and so on, over a period of time. The linear depreciation, or straight-line method, is often used for this purpose. Suppose that an asset has an initial value of $$C$ and is to be depreciated linearly over n years with a scrap value of $$S$. Show that the book value of the asset at any time

t $(0 \leq t \leq n)$ is given by the linear function

$$V(t) = C - \frac{(C - S)}{n} t$$

[*Hint:* Find an equation of the straight line that passes through the points $(0, C)$ and (n, S). Then rewrite the equation in the slope-intercept form.]

29. Linear Depreciation   Using the linear depreciation model of exercise 28, find the book value of a printing machine at the end of the second year if its initial value is \$100,000 and it is depreciated linearly over five years with a scrap value of \$30,000.

30. Price of Ivory   According to the World Wildlife Fund, a group in the forefront of the fight against illegal ivory trade, the price of ivory (in dollars per kilo) compiled from a variety of legal and black market sources is approximated by the function

$$f(t) = \begin{cases} 8.37t + 7.44 & 0 \leq t \leq 8 \\ 2.84t + 51.68 & 8 < t \leq 30 \end{cases}$$

where t is measured in years and $t = 0$ corresponds to the beginning of 1970.
a. Sketch the graph of the function f.
b. What was the price of ivory at the beginning of 1970? At the beginning of 1990?

31. Price of Automobile Parts   For years, automobile manufacturers had a monopoly on the replacement-parts market, particularly for sheet metal parts such as fenders, doors, and hoods, the parts most often damaged in a crash. Beginning in the late 1970s, however, competition appeared on the scene. In a report conducted by an insurance company to study the effects of the competition, the price of an OEM (original equipment manufacturer) fender for a particular 1983 model car was found to be

$$f(t) = \frac{110}{\frac{1}{2}t + 1} \qquad (0 \leq t \leq 2)$$

where $f(t)$ is measured in dollars and t is in years. Over the same period of time, the price of a non-OEM fender for the car was found to be

$$g(t) = 26 \left(\frac{1}{4}t^2 - 1 \right)^2 + 52 \qquad (0 \leq t \leq 2)$$

where $g(t)$ is also measured in dollars. Find a function $h(t)$ that gives the difference in price between an OEM fender and a non-OEM fender. Compute $h(0)$, $h(1)$, and $h(2)$. What does the result of your computation

seem to say about the price gap between OEM and non-OEM fenders over the two years?

For the demand equations in exercises 32–35, where x represents the quantity demanded in units of a thousand and p is the unit price in dollars, (a) sketch the demand curve and (b) determine the quantity demanded when the unit price is set at \$p.

32. $p = -x^2 + 36$; $p = 11$ **33.** $p = -x^2 + 16$; $p = 7$

34. $p = \sqrt{9 - x^2}$; $p = 2$ **35.** $p = \sqrt{18 - x^2}$; $p = 3$

For the supply equations in exercises 36–39, where x is the quantity supplied in units of a thousand and p is the unit price in dollars, (a) sketch the supply curve and (b) determine the price at which the supplier will make 2000 units of the commodity available in the market.

36. $p = 2x^2 + 18$ **37.** $p = x^2 + 16x + 40$

38. $p = x^3 + x + 10$ **39.** $p = x^3 + 2x + 3$

40. Demand for Clock Radios   In the accompanying figure, L_1 is the demand curve for the model A clock radios manufactured by Ace Radio Inc., and L_2 is the demand curve for their model B clock radios.

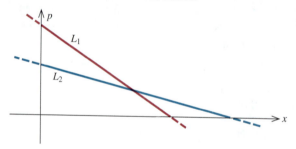

Which line has the greater slope? Interpret your results.

41. Supply of Clock Radios   In the accompanying figure, L_1 is the supply curve for the model A clock radios manufactured by Ace Radio Inc., and L_2 is the supply curve for their model B clock radios.

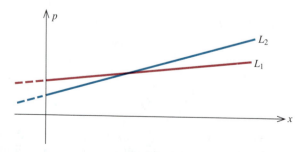

Which line has the greater slope? Interpret your results.

42. Demand for Smoke Alarms The demand function for the Sentinel smoke alarm is given by

$$p = \frac{30}{0.02x^2 + 1} \qquad (0 \le x \le 10)$$

where x (measured in units of a thousand) is the quantity demanded per week and p is the unit price in dollars. Sketch the graph of the demand function. What is the unit price that corresponds to a quantity demanded of 10,000 units?

43. Demand for Commodities Assume that the demand function for a certain commodity has the form

$$p = \sqrt{-ax^2 + b} \qquad (a \ge 0, b \ge 0)$$

where x is the quantity demanded, measured in units of a thousand, and p is the unit price in dollars. Suppose that the quantity demanded is 6000 ($x = 6$) when the unit price is $8 and 8000 ($x = 8$) when the unit price is $6. Determine the demand equation. What is the quantity demanded when the unit price is set at $7.50?

44. Supply Functions The supply function for the Luminar desk lamp is given by

$$p = 0.1x^2 + 0.5x + 15$$

where x is the quantity supplied (in thousands) and p is the unit price in dollars. Sketch the graph of the supply function. What unit price will induce the supplier to make 5000 lamps available in the marketplace?

45. Supply Functions Suppliers of transistor radios will market 10,000 units when the unit price is $20 and 62,500 units when the unit price is $35. Determine the supply function if it is known to have the form

$$p = a\sqrt{x} + b \qquad (a \ge 0, b \ge 0)$$

where x is the quantity supplied and p is the unit price in dollars. Sketch the graph of the supply function. What unit price will induce the supplier to make 40,000 transistor radios available in the marketplace?

46. Suppose the demand and supply equations for a certain commodity are given by $p = ax + b$ and $p = cx + d$, respectively, where $a < 0$, $c > 0$, and $b > d > 0$ (see the accompanying figure).
a. Find the equilibrium quantity and equilibrium price in terms of a, b, c, and d.

b. Use part (a) to determine what happens to the market equilibrium if c is increased while a, b, and d remain fixed. Interpret your answer in economic terms.
c. Use (a) to determine what happens to the market equilibrium if b is decreased while a, c, and d remain fixed. Interpret your answer in economic terms.

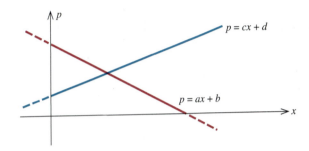

For each pair of supply and demand equations in exercises 47–50, where x represents the quantity demanded in units of a thousand and p the unit price in dollars, find the equilibrium quantity and the equilibrium price.

47. $p = -2x^2 + 80$ and $p = 15x + 30$

48. $p = -x^2 - 2x + 100$ and $p = 8x + 25$

49. $11p + 3x - 66 = 0$ and $2p^2 + p - x = 10$

50. $p = 60 - 2x^2$ and $p = x^2 + 9x + 30$

51. Market Equilibrium The weekly demand and supply functions for Sportsman 5×7 tents are given by

$$p = -0.1x^2 - x + 40$$

and

$$p = 0.1x^2 + 2x + 20$$

respectively, where p is measured in dollars and x is measured in units of a hundred. Find the equilibrium quantity and price.

52. Market Equilibrium The management of the Titan Tire Company has determined that the weekly demand and supply functions for their Super Titan tires are given by

$$p = 144 - x^2$$

and

$$p = 48 + \frac{1}{2}x^2$$

respectively, where p is measured in dollars and x is measured in units of a thousand. Find the equilibrium quantity and price.

USING TECHNOLOGY

Finding the Points of Intersection of Two Graphs

A graphing utility can be used to find the point(s) of intersection of the graphs of two functions.

EXAMPLE 1 Find the points of intersection of the graphs of

$$f(x) = 0.3x^2 - 1.4x - 3 \quad \text{and} \quad g(x) = -0.4x^2 + 0.8x + 6.4$$

Solution The graphs of both f and g in the standard viewing rectangle are shown in Figure T1. Using **TRACE** and **ZOOM** or the function for finding the points of intersection of two graphs on your graphing utility, we find the point(s) of intersection, accurate to four decimal places, to be $(-2.4158, 2.1329)$ and $(5.5587, -1.5125)$.

Figure T1
The graphs of f and g in the standard viewing window.

● ● ●

EXAMPLE 2 Consider the demand and supply functions

$$p = d(x) = -0.01x^2 - 0.2x + 8 \quad \text{and} \quad p = s(x) = 0.01x^2 + 0.1x + 3$$

of Example 4 in Section 2.3.

a. Plot the graphs of d and s in the viewing rectangle $[0, 15] \times [0, 10]$.

b. Verify that the equilibrium point is $(10, 5)$, as obtained in Example 4.

Solution

a. The graphs of d and s are shown in Figure T2.

Figure T2
The graphs of d and s in the window $[0, 15] \times [0, 10]$.

b. Using **TRACE** and **ZOOM** or the function for finding the point of intersection of two graphs, we see that $x = 10$ and $y = 5$, so the equilibrium point is $(10, 5)$, as obtained before.

○ ○ ○

EXERCISES

In exercises 1–6, find the points of intersection of the graphs of the functions. Express your answer accurate to four decimal places.

1. $f(x) = 1.2x + 3.8$; $g(x) = -0.4x^2 + 1.2x + 7.5$

2. $f(x) = 0.2x^2 - 1.3x - 3$; $g(x) = -1.3x + 2.8$

3. $f(x) = 0.3x^2 - 1.7x - 3.2$; $g(x) = -0.4x^2 + 0.9x + 6.7$

4. $f(x) = -0.3x^2 + 0.6x + 3.2$; $g(x) = 0.2x^2 - 1.2x - 4.8$

5. $f(x) = 0.3x^3 - 1.8x^2 + 2.1x - 2$; $g(x) = 2.1x - 4.2$

6. $f(x) = -0.2x^3 + 1.2x^2 - 1.2x + 2$; $g(x) = -0.2x^2 + 0.8x + 2.1$

7. The monthly demand and supply functions for a certain brand of wall clock are given by

$$p = -0.2x^2 - 1.2x + 50$$

and
$$p = 0.1x^2 + 3.2x + 25$$

respectively, where p is measured in dollars and x is measured in units of a hundred.
a. Plot the graphs of both functions in an appropriate viewing rectangle.
b. Find the equilibrium quantity and price.

8. The quantity demanded x (in units of a hundred) of Mikado miniature cameras per week is related to the unit price p (in dollars) by

$$p = -0.2x^2 + 80$$

The quantity x (in units of a hundred) that the supplier is willing to make available in the market is related to the unit price p (in dollars) by

$$p = 0.1x^2 + x + 40$$

a. Plot the graphs of both functions in an appropriate viewing rectangle.
b. Find the equilibrium quantity and price.

SOLUTIONS TO SELF-CHECK EXERCISES 2.3

1. Since the adult dose of the substance is 500 mg, $a = 500$; thus the rule in this case is

$$D(t) = \frac{500t}{t + 12}$$

A four-year-old should receive

$$D(4) = \frac{500(4)}{4 + 12}$$

or 125 mg of the substance.

2. a. The graphs of the functions d and s are shown in the following figure:

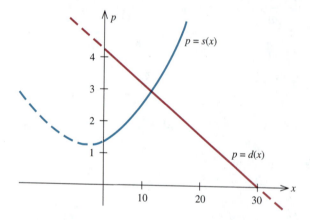

b. Solve the following system of equations:

$$p = -\frac{2}{15}x + 4$$

$$p = \frac{1}{75}x^2 + \frac{1}{10}x + \frac{3}{2}$$

Substituting the first equation into the second yields

$$\frac{1}{75}x^2 + \frac{1}{10}x + \frac{3}{2} = -\frac{2}{15}x + 4$$

$$\frac{1}{75}x^2 + \left(\frac{1}{10} + \frac{2}{15}\right)x - \frac{5}{2} = 0$$

or

$$\frac{1}{75}x^2 + \frac{7}{30}x - \frac{5}{2} = 0$$

Multiplying both sides of the last equation by 150, we have

$$2x^2 + 35x - 375 = 0$$

or

$$(2x - 15)(x + 25) = 0$$

Thus, $x = -25$ or $x = 15/2 = 7.5$. Since x must be nonnegative, we take $x = 7.5$, and the equilibrium quantity is 7500 pounds. The equilibrium price is given by

$$p = -\frac{2}{15}\left(\frac{15}{2}\right) + 4$$

or $3 per pound.

2.4   LIMITS

Introduction to Calculus

Historically, the development of calculus by Isaac Newton (1642–1727) and Gottfried Wilhelm Leibniz (1646–1716) resulted from the investigation of the following problems:

1. Finding the tangent line to a curve at a given point on the curve (Figure 2.18a)

2. Finding the area of a planar region bounded by an arbitrary curve (Figure 2.18b)

Figure 2.18

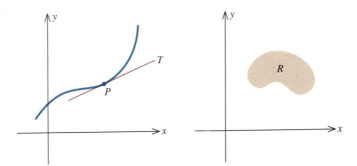

(a) What is the slope of the tangent line T at point P?

(b) What is the area of the region R?

    The tangent-line problem might appear to be unrelated to any practical applications of mathematics, but as you will see later, the problem of finding the *rate of change* of one quantity with respect to another is mathematically equivalent to the geometric problem of finding the slope of the *tangent line* to a curve at a given point on the curve. It is precisely the discovery of the relationship between these two problems that spurred the development of calculus in the seventeenth century and made it such an indispensable tool for solving practical problems. The following are a few examples of such problems.

- Finding the velocity of an object
- Finding the rate of change of a bacteria population with respect to time
- Finding the rate of change of a company's profit with respect to time
- Finding the rate of change of a travel agency's revenue with respect to the agency's expenditure for advertising

The study of the tangent-line problem led to the creation of *differential calculus,* which relies on the concept of the *derivative* of a function. The study of the area problem led to the creation of *integral calculus,* which relies on the concept of the *antiderivative,* or *integral,* of a function. (The derivative of a function and the integral of a function are intimately related, as you will see in Section 6.4.) Both the derivative of a function and the integral of a function are defined in terms of a more fundamental concept—the limit—our next topic.

A Real-Life Example

From data obtained in a test run conducted on a prototype of a maglev (magnetic levitation train), which moves along a straight monorail track, engineers have determined that the position of the maglev from the origin at time t is given by

$$\overset{\text{position}}{\underset{\text{of train}}{s}} = f(t) = 4t^2 \qquad (0 \le t \le 30) \qquad \textbf{(3)}$$

where f is called the **position function** of the maglev. The position of the maglev at time $t = 0, 1, 2, 3, \ldots, 10$, measured from its initial position, is

$$f(0) = 0, \quad f(1) = 4, \quad f(2) = 16, \quad f(3) = 36, \ldots, f(10) = 400$$

feet (see Figure 2.19).

Figure 2.19
A maglev moving along an elevated monorail track.

Suppose we want to find the velocity of the maglev at $t = 2$. This is just the velocity of the maglev as shown on its speedometer at that precise instant of time. Offhand, calculating this quantity using only equation (3) appears to be an impossible task; but consider what quantities we *can* compute using this relationship. Obviously, we can compute the position of the maglev at any time t as we did earlier for some selected values of t. Using these values, we can then compute the *average velocity* of the maglev over an interval of time. For example, the average velocity of the train over the time interval [2, 4] is given by

$$V = \frac{\text{Distance covered}}{\text{Time elapsed}} = \frac{f(4) - f(2)}{4 - 2}$$

$$= \frac{4(4^2) - 4(2^2)}{2}$$

$$= \frac{64 - 16}{2} = 24 \ \text{ft/sec}.$$

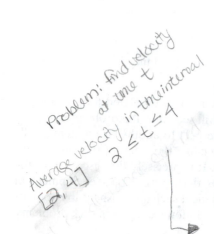

[handwritten margin note: Problem: find velocity at time t
Average velocity in the interval [2, 4]   2 ≤ t ≤ 4]

or 24 ft/sec. Although this is not quite the velocity of the maglev at $t = 2$, it does provide us with an approximation of its velocity at that time.

Can we do better? Intuitively, the smaller the time interval we pick (with $t = 2$ as the left endpoint), the better the average velocity over that time interval will approximate the actual velocity of the maglev at $t = 2$.*

Now, let us describe this process in general terms. Let $t > 2$. Then the average velocity of the maglev over the time interval $[2, t]$ is given by

[handwritten: average vel in interval from 2 to t = $\frac{f(t) - f(2)}{t - 2}$]

$$\frac{f(t) - f(2)}{t - 2} = \frac{4t^2 - 4(2^2)}{t - 2} = \frac{4(t^2 - 4)}{t - 2} \qquad \textbf{(4)}$$

By choosing the values of t closer and closer to 2, we obtain a sequence of numbers that gives the average velocities of the maglev over smaller and smaller time intervals. As we observed earlier, this sequence of numbers should approach the *instantaneous velocity* of the train at $t = 2$.

Let's try some sample calculations. Using equation (4) and taking the sequence $t = 2.5, 2.1, 2.01, 2.001,$ and 2.0001, which approaches 2, we find

The average velocity over $[2, 2.5]$ is $\dfrac{4(2.5^2 - 4)}{2.5 - 2} = 18$ ft/sec

The average velocity over $[2, 2.1]$ is $\dfrac{4(2.1^2 - 4)}{2.1 - 2} = 16.4$ ft/sec

and so forth. These results are summarized in Table 2.1.

Table 2.1

	t approaches 2 from the right.				
t	2.5	2.1	2.01	2.001	2.0001
Average Velocity over [2, t]	18	16.4	16.04	16.004	16.0004

Average velocity approaches 16 from the right.

From Table 2.1, we see that the average velocity of the maglev seems to approach the number 16 as it is computed over smaller and smaller time intervals. These computations suggest that the instantaneous velocity of the train at $t = 2$ is 16 ft/sec.

REMARK   Notice that we cannot obtain the instantaneous velocity for the maglev at $t = 2$ by substituting $t = 2$ into equation (2) because this value of t is not in the domain of the average velocity function.   ◦ ◦ ◦

◦ ◦ ◦

* Actually, any interval containing $t = 2$ will do.

Intuitive Definition of a Limit

Consider the function g defined by

$$g(t) = \frac{4(t^2 - 4)}{t - 2}$$

which gives the average velocity of the maglev [see equation (4)]. Suppose that we are required to determine the value that $g(t)$ approaches as t approaches the (fixed) number 2. If we take the sequence of values of t approaching 2 from the right-hand side, as we did earlier, we see that $g(t)$ approaches the number 16. Similarly, if we take a sequence of values of t approaching 2 from the left, such as $t = 1.5, 1.9, 1.999,$ and 1.9999, we obtain the results shown in Table 2.2.

Table 2.2

			t approaches 2 from the left.		
t	1.5	1.9	1.99	1.999	1.9999
g(t)	14	15.6	15.96	15.996	15.9996

$g(t)$ approaches 16 from the left.

Observe that $g(t)$ approaches the number 16 as t approaches 2—this time from the left-hand side. In other words, as t approaches 2 from *either* side of 2, $g(t)$ approaches 16. In this situation, we say that the limit of $g(t)$ as t approaches 2 is 16, written

$$\lim_{t \to 2} g(t) = \lim_{t \to 2} \frac{4(t^2 - 4)}{t - 2} = 16$$

The graph of the function g, shown in Figure 2.20, confirms this observation.

Figure 2.20
As t approaches t = 2 from either direction, g(t) approaches y = 16.

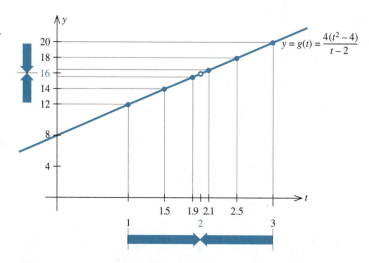

Observe that the point $t = 2$ is not in the domain of the function g [for this reason, the point $(2, 16)$ is missing from the graph of g]. This, however, is inconsequential because the value, if any, of $g(t)$ at $t = 2$ plays no role in computing the limit.

This example leads to the following informal definition.

LIMIT OF A FUNCTION

The function f has the limit L as x approaches a, written

$$\lim_{x \to a} f(x) = L$$

if the value $f(x)$ can be made as close to the number L as we please by taking x sufficiently close to (but not equal to) a.

EXPLORING WITH TECHNOLOGY

1. Use a graphing utility to plot the graph of $g(x) = \dfrac{4(x^2 - 4)}{x - 2}$ in the viewing rectangle $[0, 3] \times [0, 20]$.

2. Use **ZOOM** and **TRACE** to describe what happens to the values of $f(x)$ as x approaches 2, first from the right and then from the left.

3. What happens to the y-value when x takes on the value 2? Explain.

4. Reconcile your results with those of the preceding example.

Figure 2.21
$f(x)$ is close to 8 whenever x is close to 2.

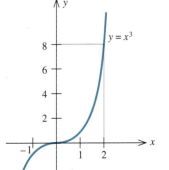

Evaluating the Limit of a Function

Let us now consider some examples involving the computation of limits.

EXAMPLE 1   Let $f(x) = x^3$ and evaluate $\lim_{x \to 2} f(x)$.

Solution   The graph of f is shown in Figure 2.21. You can see that $f(x)$ can be made as close to the number 8 as we please by taking x sufficiently close to 2. Therefore,

$$\lim_{x \to 2} x^3 = 8$$

Figure 2.22
$\lim\limits_{x \to 1} g(x) = 3$.

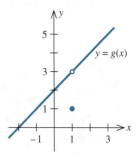

EXAMPLE 2 Let

$$g(x) = \begin{cases} x + 2 & \text{if } x \neq 1 \\ 1 & \text{if } x = 1 \end{cases}$$

Evaluate $\lim\limits_{x \to 1} g(x)$.

Solution The domain of g is the set of all real numbers. From the graph of g shown in Figure 2.22, we see that $g(x)$ can be made as close to 3 as we please by taking x sufficiently close to 1. Therefore

$$\lim_{x \to 1} g(x) = 3$$

Observe that $g(1) = 1$, which is not equal to the limit of the function g as x approaches 1. (Once again, the value of $g(x)$ at $x = 1$ has no bearing on the existence or value of the limit of g as x approaches 1.)   ○ ○ ○

EXAMPLE 3 Evaluate the limit of the following functions as x approaches the indicated point.

a. $f(x) = \begin{cases} -1 & \text{if } x < 0 \\ 1 & \text{if } x \geq 0 \end{cases}$; $x = 0$ **b.** $g(x) = \dfrac{1}{x^2}$; $x = 0$

Solution The graphs of the functions f and g are shown in Figure 2.23.

Figure 2.23

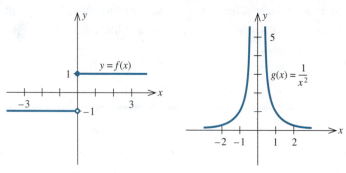

(a) $\lim\limits_{x \to 0} f(x)$ does not exist.   **(b)** $\lim\limits_{x \to 0} g(x)$ does not exist.

a. Referring to Figure 2.23a, we see that no matter how close x is to $x = 0$, $f(x)$ takes on the values 1 or -1 depending on whether x is positive or negative. Thus, there is no *single* real number L that $f(x)$ approaches as x approaches zero. We conclude that the limit of $f(x)$ does *not* exist as x approaches zero.

b. Referring to Figure 2.23b, we see that as x approaches $x = 0$ (from either side), $g(x)$ increases without bound and thus does not approach any specific real number. We conclude, accordingly, that the limit of $g(x)$ does not exist as x approaches zero.   ○ ○ ○

Consider the graph of the function h whose graph is depicted in the following figure:

It has the property that as x approaches 0 from either the right or the left, the curve oscillates more and more frequently between the lines $y = -1$ and $y = 1$.

a. Explain why $\lim\limits_{x \to 0} h(x)$ does not exist.

b. Compare this function with those in Example 3. More specifically, discuss the different ways the functions fail to have a limit at $x = 0$.

Until now, we have relied on knowing the actual values of a function or the graph of a function near $x = a$ to help us evaluate the limit of the function $f(x)$ as x approaches a. The following properties of limits, which we list without proof, enable us to evaluate limits of functions algebraically.

THEOREM I

Properties of Limits

Suppose that

$$\lim_{x \to a} f(x) = L \quad \text{and} \quad \lim_{x \to a} g(x) = M$$

Then

i. $\lim\limits_{x \to a}[f(x)]^r = [\lim\limits_{x \to a} f(x)]^r = L^r$ $\qquad$ (r, a real number)

ii. $\lim\limits_{x \to a} cf(x) = c \lim\limits_{x \to a} f(x) = cL$ $\qquad$ (c, a real number)

iii. $\lim\limits_{x \to a}[f(x) \pm g(x)] = \lim\limits_{x \to a} f(x) \pm \lim\limits_{x \to a} g(x) = L \pm M$

iv. $\lim\limits_{x \to a}[f(x)g(x)] = [\lim\limits_{x \to a} f(x)][\lim\limits_{x \to a} g(x)] = LM$

v. $\lim\limits_{x \to a} \dfrac{f(x)}{g(x)} = \dfrac{\lim\limits_{x \to a} f(x)}{\lim\limits_{x \to a} g(x)} = \dfrac{L}{M}$ $\qquad$ (Provided $M \neq 0$)

EXAMPLE 4   Use Theorem 1 to evaluate the following limits.

a. $\lim\limits_{x\to 2} x^3$ **b.** $\lim\limits_{x\to 4} 5x^{3/2}$ **c.** $\lim\limits_{x\to 1} (5x^4 - 2)$

d. $\lim\limits_{x\to 3} 2x^3 \sqrt{x^2 + 7}$ **e.** $\lim\limits_{x\to 2} \dfrac{2x^2 + 1}{x + 1}$

Solution

a. $\lim\limits_{x\to 2} x^3 = [\lim\limits_{x\to 2} x]^3$ [Property (i)]

$\qquad = 2^3 = 8 \qquad (\lim\limits_{x\to 2} x = 2)$

b. $\lim\limits_{x\to 4} 5x^{3/2} = 5[\lim\limits_{x\to 4} x^{3/2}]$ [Property (ii)]

$\qquad = 5(4)^{3/2} = 40$

c. $\lim\limits_{x\to 1}(5x^4 - 2) = \lim\limits_{x\to 1} 5x^4 - \lim\limits_{x\to 1} 2$ [Property (iii)]

To evaluate $\lim\limits_{x\to 1} 2$, observe that the constant function $g(x) = 2$ has value 2 for all values of x. Therefore, $g(x)$ must approach the limit 2 as x approaches $x = 1$ (or any other point for that matter!). Therefore,

$$\lim\limits_{x\to 1}(5x^4 - 2) = 5(1)^4 - 2$$

$$= 3$$

d. $\lim\limits_{x\to 3} 2x^3 \sqrt{x^2 + 7} = 2 \lim\limits_{x\to 3} x^3 \sqrt{x^2 + 7}$ [Property (ii)]

$\qquad = 2 \lim\limits_{x\to 3} x^3 \lim\limits_{x\to 3} \sqrt{x^2 + 7}$ [Property (iv)]

$\qquad = 2(3)^3 \sqrt{3^2 + 7}$ [Property (i)]

$\qquad = 2(27) \sqrt{16} = 216$

e. $\lim\limits_{x\to 2} \dfrac{2x^2 + 1}{x + 1} = \dfrac{\lim\limits_{x\to 2}(2x^2 + 1)}{\lim\limits_{x\to 2}(x + 1)}$ [Property (v)]

$= \dfrac{\lim\limits_{x\to 2} 2x^2 + \lim\limits_{x\to 2} 1}{\lim\limits_{x\to 2} x + \lim\limits_{x\to 2} 1} = \dfrac{2(2)^2 + 1}{2 + 1} = \dfrac{9}{3} = 3$

○ ○ ○

Indeterminate Forms

Let us emphasize once again that property (v) of limits is valid only when the limit of the function that appears in the denominator is not equal to zero at the point in question.

If the numerator has a limit different from zero and the denominator has a limit equal to zero, then the limit of the quotient does not exist at the point in question. This is the case with the function $g(x) = 1/x^2$ in Example 3b. Here, as x approaches zero, the numerator approaches 1 but the denominator approaches zero, so that the quotient becomes arbitrarily large. Thus, as observed earlier, the limit does not exist.

Next, consider

$$\lim_{x \to 2} \frac{4(x^2 - 4)}{x - 2}$$

which we evaluated earlier by looking at the values of the function for x near $x = 2$. If we attempt to evaluate this expression by applying property (v) of limits, we see that both the numerator and denominator of the function

$$\frac{4(x^2 - 4)}{x - 2}$$

approach zero as x approaches 2; that is, we obtain an expression of the form 0/0. In this event, we say that the limit of the quotient $f(x)/g(x)$ as x approaches 2 has the **indeterminate form 0/0.**

We will need to evaluate limits of this type when we discuss the derivative of a function, a fundamental concept in the study of calculus. As the name suggests, the meaningless expression 0/0 does not provide us with a solution to our problem. One strategy that can be used to solve this type of problem follows.

STRATEGY FOR EVALUATING INDETERMINATE FORMS

1. Replace the given function with an appropriate one that takes on the same values as the original function everywhere except at $x = a$.

2. Evaluate the limit of this function as x approaches a.

Examples 5 and 6 illustrate this strategy.

EXAMPLE 5 Evaluate

$$\lim_{x \to 2} \frac{4(x^2 - 4)}{x - 2}$$

Solution Since both the numerator and the denominator of this expression approach zero as x approaches 2, we have the indeterminate form 0/0. We rewrite

$$\frac{4(x^2 - 4)}{x - 2} = \frac{4(x - 2)(x + 2)}{(x - 2)}$$

which, upon canceling the common factors, is equivalent to $4(x + 2)$. Next, we replace $4(x^2 - 4)/(x - 2)$ with $4(x + 2)$ and take the limit as x approaches 2, obtaining

$$\lim_{x \to 2} \frac{4(x^2 - 4)}{x - 2} = \lim_{x \to 2} 4(x + 2) = 16$$

The graphs of the functions

$$f(x) = \frac{4(x^2 - 4)}{x - 2} \quad \text{and} \quad g(x) = 4(x + 2)$$

are shown in Figure 2.24a and b. Observe that the graphs are identical except when $x = 2$. The function g is defined for all values of x and, in particular, its value at $x = 2$ is $g(2) = 4(2 + 2) = 16$. Thus, the point $(2, 16)$ is on the graph of g. However, the function f is not defined at $x = 2$. Since $f(x) = g(x)$ for all values of x except $x = 2$, it follows that the graph of f must look exactly like the graph of g, with the exception that the point $(2, 16)$ is missing from the graph of f. This illustrates graphically why we can evaluate the limit of f by evaluating the limit of the "equivalent" function g.

Figure 2.24
The graphs of $f(x)$ and $g(x)$ are identical except at the point $(2, 16)$.

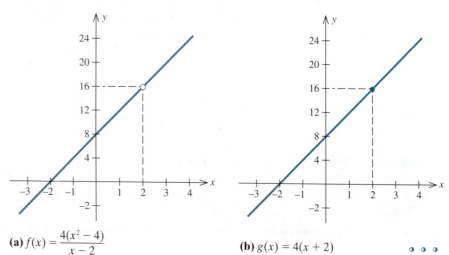

(a) $f(x) = \dfrac{4(x^2 - 4)}{x - 2}$ (b) $g(x) = 4(x + 2)$

EXPLORING WITH TECHNOLOGY

1. Use a graphing utility to plot the graph of $f(x) = \dfrac{4(x^2 - 4)}{x - 2}$ in the viewing rectangle $[0, 3] \times [0, 20]$.

Then use **ZOOM** and **TRACE** to find $\lim\limits_{x \to 2} \dfrac{4(x^2 - 4)}{x - 2}$.

2. Use a graphing utility to plot the graph of $g(x) = 4(x + 2)$ in the viewing rectangle $[0, 3] \times [0, 20]$. Then use **ZOOM** and **TRACE** to find $\lim\limits_{x \to 2} 4(x + 2)$. What happens to the y-value when x takes on the value 2? Explain.

3. Can you distinguish between the graphs of f and g?

4. Reconcile your results with those of Example 5.

EXPLORING WITH TECHNOLOGY

1. Use a graphing utility to plot the graph of $g(x) = \dfrac{\sqrt{1+x} - 1}{x}$ in the viewing rectangle $[-1, 2] \times [0, 1]$. Then use **ZOOM** and **TRACE** to find $\displaystyle\lim_{x \to 0} \dfrac{\sqrt{1+x} - 1}{x}$ by observing the values of $g(x)$ as x approaches 0 from the left and from the right.

2. Use a graphing utility to plot the graph of $f(x) = \dfrac{1}{\sqrt{1+x} + 1}$ in the viewing rectangle $[-1, 2] \times [0, 1]$. Then use **ZOOM** and **TRACE** to find $\displaystyle\lim_{x \to 0} \dfrac{1}{\sqrt{1+x} + 1}$. What happens to the y-value when x takes on the value 0? Explain.

3. Can you distinguish between the graphs of f and g?

4. Reconcile your results with those of Example 6.

◦ ◦ ◦

REMARK   Notice that the limit in Example 5 is the same limit that we evaluated earlier when we discussed the instantaneous velocity of a maglev at a specified time.

◦ ◦ ◦

EXAMPLE 6   Evaluate

$$\lim_{h \to 0} \frac{\sqrt{1+h} - 1}{h}$$

Solution   Letting h approach zero, we obtain the indeterminate form 0/0. Next, we rationalize the numerator of the quotient (see page 25) by multiplying both the numerator and the denominator by the expression $(\sqrt{1+h} + 1)$, obtaining

$$\frac{\sqrt{1+h} - 1}{h} = \frac{(\sqrt{1+h} - 1)(\sqrt{1+h} + 1)}{h(\sqrt{1+h} + 1)}$$

$$= \frac{1 + h - 1}{h(\sqrt{1+h} + 1)} \qquad [(\sqrt{a} - \sqrt{b})(\sqrt{a} + \sqrt{b}) = a - b]$$

$$= \frac{h}{h(\sqrt{1+h} + 1)}$$

$$= \frac{1}{\sqrt{1+h} + 1}$$

Therefore,

$$\lim_{h \to 0} \frac{\sqrt{1 + h} - 1}{h} = \lim_{h \to 0} \frac{1}{\sqrt{1 + h} + 1}$$

$$= \frac{1}{\sqrt{1} + 1} = \frac{1}{2}$$

o o o

Limits at Infinity

Up to now we have studied the limit of a function as x approaches a (finite) number a. There are occasions, however, when we want to know whether $f(x)$ approaches a unique number as x increases without bound. Consider, for example, the function P, giving the number of fruit flies (*Drosophila*) in a container under controlled laboratory conditions, as a function of a time t. The graph of P is shown in Figure 2.25. You can see from the graph of P that, as t increases without bound (gets larger and larger), $P(t)$ approaches the number 400. This number, called the *carrying capacity* of the environment, is determined by the amount of living space and food available, as well as other environmental factors.

Figure 2.25
The graph of P(t) gives the population of fruit flies in a laboratory experiment.

As another example, suppose that we are given the function

$$f(x) = \frac{2x^2}{1 + x^2}$$

and we want to determine what happens to $f(x)$ as x gets larger and larger. Picking the sequence of numbers 1, 2, 5, 10, 100, and 1000 and computing the corresponding values of $f(x)$, we obtain the following table of values:

x	1	2	5	10	100	1000
$f(x)$	1	1.6	1.92	1.98	1.9998	1.999998

Figure 2.26

The graph of $y = \dfrac{2x^2}{1+x^2}$ has a horizontal asymptote at $y = 2$.

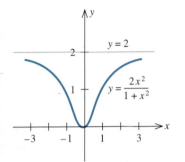

From the table, we see that as x gets larger and larger, $f(x)$ gets closer and closer to 2. The graph of the function f shown in Figure 2.26 confirms this observation. We call the line $y = 2$ a **horizontal asymptote.** *

In this situation we say that the limit of the function

$$f(x) = \frac{2x^2}{1+x^2}$$

as x increases without bound is 2, written

$$\lim_{x \to \infty} \frac{2x^2}{1+x^2} = 2$$

In the general case, the following definition is applicable:

LIMIT OF A FUNCTION AT INFINITY

The function f has the limit L as x increases without bound (or, as x approaches infinity), written

$$\lim_{x \to \infty} f(x) = L$$

if $f(x)$ can be made arbitrarily close to L by taking x large enough.

Similarly, the function f has the limit M as x decreases without bound (or as x approaches negative infinity), written

$$\lim_{x \to -\infty} f(x) = M$$

if $f(x)$ can be made arbitrarily close to M by taking x to be negative and sufficiently large in absolute value.

EXAMPLE 7   Let f and g be the functions

$$f(x) = \begin{cases} -1 & \text{if } x < 0 \\ 1 & \text{if } x \geq 0 \end{cases} \quad \text{and} \quad g(x) = \frac{1}{x^2}$$

Evaluate

a. $\lim\limits_{x \to \infty} f(x)$ and $\lim\limits_{x \to -\infty} f(x)$ **b.** $\lim\limits_{x \to \infty} g(x)$ and $\lim\limits_{x \to -\infty} g(x)$

Solution   The graphs of $f(x)$ and $g(x)$ are shown in Figure 2.27. Referring to the graphs of the respective functions, we see that

a. $\lim\limits_{x \to \infty} f(x) = 1$ and $\lim\limits_{x \to -\infty} f(x) = -1$

b. $\lim\limits_{x \to \infty} \dfrac{1}{x^2} = 0$ and $\lim\limits_{x \to -\infty} \dfrac{1}{x^2} = 0$.

○ ○ ○

* We will discuss asymptotes in greater detail in Section 4.3.

Figure 2.27

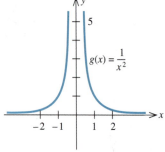

(a) $\lim\limits_{x \to \infty} f(x) = 1$ and $\lim\limits_{x \to -\infty} f(x) = -1$ **(b)** $\lim\limits_{x \to \infty} g(x) = 0$ and $\lim\limits_{x \to -\infty} g(x) = 0$ ○ ○ ○

All of the properties of limits listed in Theorem 1 are valid when a is replaced by ∞ or $-\infty$. In addition, we have the following property for the limit at infinity.

THEOREM 2

For all $n > 0$

$$\lim_{x \to \infty} \frac{1}{x^n} = 0 \quad \text{and} \quad \lim_{x \to -\infty} \frac{1}{x^n} = 0$$

provided that $\dfrac{1}{x^n}$ is defined.

EXPLORING WITH TECHNOLOGY

I. Use a graphing utility to plot the graphs of

$$y_1 = \frac{1}{x^{0.5}}, \quad y_2 = \frac{1}{x}, \quad \text{and} \quad y_3 = \frac{1}{x^{1.5}}$$

in the viewing rectangle $[0, 200] \times [0, 0.5]$. What can you say about $\lim\limits_{x \to \infty} \dfrac{1}{x^n}$ if $n = 0.5$, $n = 1$, and $n = 1.5$? Are these results predicted by Theorem 2?

2. Use a graphing utility to plot the graphs of

$$y_1 = \frac{1}{x} \quad \text{and} \quad y_2 = \frac{1}{x^{5/3}}$$

in the viewing rectangle $[-50, 0] \times [-0.5, 0]$. What can you say about $\lim\limits_{x \to -\infty} \dfrac{1}{x^n}$ if $n = 1$ and $n = \dfrac{5}{3}$? Are these results predicted by Theorem 2?
[*Hint:* To graph y_2, write it in the form $y2 = 1/(x^{(1/3)})^5$.]

○ ○ ○

In evaluating the limit at infinity of a rational function, the following technique is often used: *Divide the numerator and denominator of the expression by* x^n, *where n is the highest power present in the denominator of the expression.*

EXAMPLE 8   Evaluate

$$\lim_{x \to \infty} \frac{x^2 - x + 3}{2x^3 + 1}$$

Solution   Since the limits of both the numerator and the denominator do not exist as x approaches infinity, the property pertaining to the limit of a quotient [property (v)] is not applicable. Let us divide the numerator and denominator of the rational expression by x^3, obtaining

$$\lim_{x \to \infty} \frac{x^2 - x + 3}{2x^3 + 1} = \lim_{x \to \infty} \frac{\dfrac{1}{x} - \dfrac{1}{x^2} + \dfrac{3}{x^3}}{2 + \dfrac{1}{x^3}}$$

$$= \frac{0}{2} \qquad \text{(Using Theorem 2)}$$

$$= 0 \qquad \qquad \qquad \circ \ \circ \ \circ$$

EXAMPLE 9   Let

in class
possibly on test

$$f(x) = \frac{3x^2 + 8x - 4}{2x^2 + 4x - 5}$$

Compute $\lim_{x \to \infty} f(x)$ if it exists.

Solution   Again, we see that property (v) is not applicable. Dividing the numerator and the denominator by x^2, we obtain

$$\lim_{x \to \infty} \frac{3x^2 + 8x - 4}{2x^2 + 4x - 5} = \lim_{x \to \infty} \frac{3 + \dfrac{8}{x} - \dfrac{4}{x^2}}{2 + \dfrac{4}{x} - \dfrac{5}{x^2}}$$

$$= \frac{\lim\limits_{x \to \infty} 3 + 8 \lim\limits_{x \to \infty} \dfrac{1}{x} - 4 \lim\limits_{x \to \infty} \dfrac{1}{x^2}}{\lim\limits_{x \to \infty} 2 + 4 \lim\limits_{x \to \infty} \dfrac{1}{x} - 5 \lim\limits_{x \to \infty} \dfrac{1}{x^2}}$$

$$= \frac{3 + 0 - 0}{2 + 0 - 0}$$

$$= \frac{3}{2} \qquad \text{(Using Theorem 2)} \qquad \circ \ \circ \ \circ$$

EXAMPLE 10   Let $f(x) = \dfrac{2x^3 - 3x^2 + 1}{x^2 + 2x + 4}$ and evaluate

a. $\lim\limits_{x \to \infty} f(x)$ and   **b.** $\lim\limits_{x \to -\infty} f(x)$.

Solution

a. Dividing the numerator and the denominator of the rational expression by x^2, we obtain

$$\lim_{x \to \infty} \frac{2x^3 - 3x^2 + 1}{x^2 + 2x + 4} = \lim_{x \to \infty} \frac{2x - 3 + \dfrac{1}{x^2}}{1 + \dfrac{2}{x} + \dfrac{4}{x^2}}$$

Since the numerator becomes arbitrarily large whereas the denominator approaches 1 as x approaches infinity, we see that the quotient $f(x)$ gets larger and larger as x approaches infinity. In other words, the limit does not exist. We indicate this by writing

$$\lim_{x \to \infty} \frac{2x^3 - 3x^2 + 1}{x^2 + 2x + 4} = \infty$$

b. Once again, dividing both the numerator and the denominator by x^2, we obtain

$$\lim_{x \to -\infty} \frac{2x^3 - 3x^2 + 1}{x^2 + 2x + 4} = \lim_{x \to -\infty} \frac{2x - 3 + \dfrac{1}{x^2}}{1 + \dfrac{2}{x} + \dfrac{4}{x^2}}$$

In this case, the numerator becomes arbitrarily large in magnitude but negative in sign, whereas the denominator approaches 1 as x approaches negative infinity. Therefore, the quotient $f(x)$ decreases without bound and the limit does not exist. We indicate this by writing

$$\lim_{x \to -\infty} \frac{2x^3 - 3x^2 + 1}{x^2 + 2x + 4} = -\infty$$

●　●　●

Example 11 gives an application of the concept of the limit of a function at infinity.

EXAMPLE 11   The Custom Office Company makes a line of executive desks. It is estimated that the total cost of making x Senior Executive Model desks is $C(x) = 100x + 200{,}000$ dollars per year, so that the average cost of making x desks is given by

$$\overline{C}(x) = \frac{C(x)}{x} = \frac{100x + 200{,}000}{x} = 100 + \frac{200{,}000}{x}$$

dollars per desk. Evaluate $\lim\limits_{x \to \infty} \overline{C}(x)$ and interpret your results.

Figure 2.28

As the level of production increases, the average cost approaches $100 per desk.

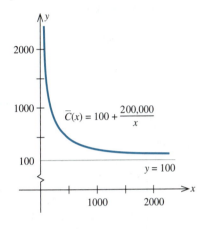

Solution

$$\lim_{x\to\infty} \overline{C}(x) = \lim_{x\to\infty}\left(100 + \frac{200{,}000}{x}\right)$$

$$= \lim_{x\to\infty} 100 + \lim_{x\to\infty}\frac{200{,}000}{x} = 100$$

A sketch of the graph of the function $\overline{C}(x)$ appears in Figure 2.28. The result we obtained is fully expected if we consider its economic implications. Note that as the level of production increases, the fixed cost per desk produced, represented by the term $(200{,}000/x)$, drops steadily. The average cost should approach a constant unit cost of production—$100 in this case.   ○ ○ ○

Consider the graph of the function f depicted in the following figure:

It has the property that the curve oscillates between $y = -1$ and $y = 1$ indefinitely in either direction.

a. Explain why $\lim_{x\to-\infty} f(x)$ and $\lim_{x\to\infty} f(x)$ do not exist.

b. Compare this function with those of Example 10. More specifically, discuss the different ways each function fails to have a limit at infinity or minus infinity.

SELF-CHECK EXERCISES 2.4

....

1. Find the indicated limit if it exists.

a. $\lim\limits_{x\to3} \dfrac{\sqrt{x^2 + 7} + \sqrt{3x - 5}}{x + 2}$

b. $\lim\limits_{x \to -1} \dfrac{x^2 - x - 2}{2x^2 - x - 3}$

2. The average cost per disc (in dollars) incurred by the Herald Record Company in pressing x compact audio discs is given by the average cost function

$$\overline{C}(x) = 1.8 + \dfrac{3000}{x}$$

Evaluate $\lim\limits_{x \to \infty} \overline{C}(x)$ and interpret your result.

Solutions to Self-Check Exercises 2.4 can be found on page 123.

2.4 EXERCISES

In exercises 1–8, use the graph of the given function f to determine $\lim\limits_{x \to a} f(x)$ at the indicated value of a, if it exists.

1.

2.

3.

4.

5.

6.

7.

8.
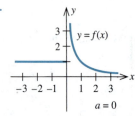

In exercises 9–16, complete the table by computing $f(x)$ at the given values of x. Use these results to estimate the indicated limit (if it exists).

9. $f(x) = x^2 + 1;\quad \lim\limits_{x \to 2} f(x)$

x	1.9	1.99	1.999	2.001	2.01	2.1
$f(x)$						

10. $f(x) = 2x^2 - 1;\quad \lim\limits_{x \to 1} f(x)$

x	0.9	0.99	0.999	1.001	1.01	1.1
$f(x)$						

11. $f(x) = \dfrac{|x|}{x};\quad \lim\limits_{x \to 0} f(x)$

x	−0.1	−0.01	−0.001	0.001	0.01	0.1
$f(x)$						

12. $f(x) = \dfrac{|x - 1|}{x - 1}$; $\displaystyle\lim_{x \to 1} f(x)$

x	0.9	0.99	0.999	1.001	1.01	1.1
$f(x)$						

13. $f(x) = \dfrac{1}{(x - 1)^2}$; $\displaystyle\lim_{x \to 1} f(x)$

x	0.9	0.99	0.999	1.001	1.01	1.1
$f(x)$						

14. $f(x) = \dfrac{1}{x - 2}$; $\displaystyle\lim_{x \to 2} f(x)$

x	1.9	1.99	1.999	2.001	2.01	2.1
$f(x)$						

15. $f(x) = \dfrac{x^2 + x - 2}{x - 1}$; $\displaystyle\lim_{x \to 1} f(x)$

x	0.9	0.99	0.999	1.001	1.01	1.1
$f(x)$						

16. $f(x) = \dfrac{x - 1}{x - 1}$; $\displaystyle\lim_{x \to 1} f(x)$

x	0.9	0.99	0.999	1.001	1.01	1.1
$f(x)$						

In exercises 17–22, sketch the graph of the function f and evaluate $\displaystyle\lim_{x \to a} f(x)$, if it exists, for the given values of a.

17. $f(x) = \begin{cases} x - 1 & \text{if } x \le 0 \\ -1 & \text{if } x > 0 \end{cases}$ $a = 0$

18. $f(x) = \begin{cases} x - 1 & \text{if } x \le 3 \\ -2x + 8 & \text{if } x > 3 \end{cases}$ $a = 3$

19. $f(x) = \begin{cases} x & \text{if } x < 1 \\ 0 & \text{if } x = 1 \\ -x + 2 & \text{if } x > 1 \end{cases}$ $a = 1$

20. $f(x) = \begin{cases} -2x + 4 & \text{if } x < 1 \\ 4 & \text{if } x = 1 \\ x^2 + 1 & \text{if } x > 1 \end{cases}$ $a = 1$

21. $f(x) = \begin{cases} |x| & \text{if } x \ne 0 \\ 1 & \text{if } x = 0 \end{cases}$ $a = 0$

22. $f(x) = \begin{cases} |x - 1| & \text{if } x \ne 1 \\ 0 & \text{if } x = 1 \end{cases}$ $a = 1$

In exercises 23–40, find the indicated limit.

23. $\displaystyle\lim_{x \to 2} 3$

24. $\displaystyle\lim_{x \to -2} -3$

25. $\displaystyle\lim_{x \to 3} x$

26. $\displaystyle\lim_{x \to -2} -3x$

27. $\displaystyle\lim_{x \to 1} (1 - 2x^2)$

28. $\displaystyle\lim_{t \to 3} (4t^2 - 2t + 1)$

29. $\displaystyle\lim_{x \to 1} (2x^3 - 3x^2 + x + 2)$

30. $\displaystyle\lim_{x \to 0} (4x^5 - 20x^2 + 2x + 1)$

31. $\displaystyle\lim_{s \to 0} (2s^2 - 1)(2s + 4)$

32. $\displaystyle\lim_{x \to 2} (x^2 + 1)(x^2 - 4)$

33. $\displaystyle\lim_{x \to 2} \dfrac{2x + 1}{x + 2}$

34. $\displaystyle\lim_{x \to 1} \dfrac{x^3 + 1}{2x^3 + 2}$

35. $\displaystyle\lim_{x \to 2} \sqrt{x + 2}$

36. $\displaystyle\lim_{x \to -2} \sqrt[3]{5x + 2}$

37. $\displaystyle\lim_{x \to -3} \sqrt{2x^4 + x^2}$

38. $\displaystyle\lim_{x \to 2} \sqrt{\dfrac{2x^3 + 4}{x^2 + 1}}$

39. $\displaystyle\lim_{x \to -1} \dfrac{\sqrt{x^2 + 8}}{2x + 4}$

40. $\displaystyle\lim_{x \to 3} \dfrac{x\sqrt{x^2 + 7}}{2x - \sqrt{2x + 3}}$

In exercises 41–48, find the indicated limit given that $\displaystyle\lim_{x \to a} f(x) = 3$ and $\displaystyle\lim_{x \to a} g(x) = 4$.

41. $\displaystyle\lim_{x \to a} [f(x) - g(x)]$

42. $\displaystyle\lim_{x \to a} 2f(x)$

43. $\displaystyle\lim_{x \to a} [2f(x) - 3g(x)]$

44. $\displaystyle\lim_{x \to a} [f(x)g(x)]$

45. $\displaystyle\lim_{x \to a} \sqrt{g(x)}$

46. $\displaystyle\lim_{x \to a} \sqrt[3]{5f(x) + 3g(x)}$

47. $\displaystyle\lim_{x \to a} \dfrac{2f(x) - g(x)}{f(x)g(x)}$

48. $\displaystyle\lim_{x \to a} \dfrac{g(x) - f(x)}{f(x) + \sqrt{g(x)}}$

In exercises 49–62, find the indicated limit, if it exists.

49. $\displaystyle\lim_{x \to 1} \dfrac{x^2 - 1}{x - 1}$

50. $\displaystyle\lim_{x \to -2} \dfrac{x^2 - 4}{x + 2}$

51. $\lim\limits_{x \to 0} \dfrac{x^2 - x}{x}$

52. $\lim\limits_{x \to 0} \dfrac{2x^2 - 3x}{x}$

53. $\lim\limits_{x \to -5} \dfrac{x^2 - 25}{x + 5}$

54. $\lim\limits_{b \to -3} \dfrac{b + 1}{b + 3}$

55. $\lim\limits_{x \to 1} \dfrac{x}{x - 1}$

56. $\lim\limits_{x \to 2} \dfrac{x + 2}{x - 2}$

57. $\lim\limits_{x \to -2} \dfrac{x^2 - x - 6}{x^2 + x - 2}$

58. $\lim\limits_{z \to 2} \dfrac{z^3 - 8}{z - 2}$

59. $\lim\limits_{x \to 1} \dfrac{\sqrt{x} - 1}{x - 1}$

$$\left[\text{Hint: Multiply by } \frac{\sqrt{x} + 1}{\sqrt{x} + 1}. \right]$$

60. $\lim\limits_{x \to 4} \dfrac{x - 4}{\sqrt{x} - 2}$

[*Hint:* See exercise 59.]

61. $\lim\limits_{x \to 1} \dfrac{x - 1}{x^3 + x^2 - 2x}$

62. $\lim\limits_{x \to -2} \dfrac{4 - x^2}{2x^2 + x^3}$

In exercises 63–68, use the graph of the function f to determine $\lim\limits_{x \to \infty} f(x)$ *and* $\lim\limits_{x \to -\infty} f(x)$, *if they exist.*

63.

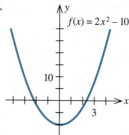

$f(x) = 2x^2 - 10$

64.

$f(x) = x^3 - x$

65.

$f(x) = \dfrac{1}{x^2}$

66.

$f(x) = \dfrac{x^2}{x^2 + 1}$

67.

$f(x) = 2 - |x|$

68.

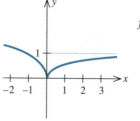

$$f(x) = \begin{cases} \sqrt{-x} & \text{if } x \le 0 \\ \dfrac{x}{x + 1} & \text{if } x > 0 \end{cases}$$

In exercises 69–72, complete the table by computing f(x) at the given values of x. Use the results to guess at the indicated limits, if they exist.

69. $f(x) = \dfrac{1}{x^2 + 1}$; $\lim\limits_{x \to \infty} f(x)$ and $\lim\limits_{x \to -\infty} f(x)$

x	1	10	100	1000
$f(x)$				

x	-1	-10	-100	-1000
$f(x)$				

70. $f(x) = \dfrac{2x}{x + 1}$; $\lim\limits_{x \to \infty} f(x)$ and $\lim\limits_{x \to -\infty} f(x)$

x	1	10	100	1000
$f(x)$				

x	-5	-10	-100	-1000
$f(x)$				

71. $f(x) = 3x^3 - x^2 + 10$; $\lim\limits_{x \to \infty} f(x)$ and $\lim\limits_{x \to -\infty} f(x)$

x	1	5	10	100	1000
$f(x)$					

x	-1	-5	-10	-100	-1000
$f(x)$					

72. $f(x) = \dfrac{|x|}{x}$; $\lim\limits_{x \to \infty} f(x)$ and $\lim\limits_{x \to -\infty} f(x)$

x	1	10	100	-1	-10	-100
$f(x)$						

In exercises 73–80, find the indicated limits, if they exist.

73. $\lim\limits_{x \to \infty} \dfrac{3x+2}{x-5}$

74. $\lim\limits_{x \to -\infty} \dfrac{4x^2-1}{x+2}$

75. $\lim\limits_{x \to -\infty} \dfrac{3x^3+x^2+1}{x^3+1}$

76. $\lim\limits_{x \to \infty} \dfrac{2x^2+3x+1}{x^4-x^2}$

77. $\lim\limits_{x \to -\infty} \dfrac{x^4+1}{x^3-1}$

78. $\lim\limits_{x \to \infty} \dfrac{4x^4-3x^2+1}{2x^4+x^3+x^2+x+1}$

79. $\lim\limits_{x \to \infty} \dfrac{x^5-x^3+x-1}{x^6+2x^2+1}$

80. $\lim\limits_{x \to \infty} \dfrac{2x^2-1}{x^3+x^2+1}$

81. Toxic Waste A city's main well was recently found to be contaminated with trichloroethylene, a cancer-causing chemical, as a result of an abandoned chemical dump leaching chemicals into the water. A proposal submitted to the selectmen of the city indicates that the cost, measured in millions of dollars, of removing x percent of the toxic pollutant is given by

$$C(x) = \frac{0.5x}{100-x} \qquad (0 < x < 100)$$

a. Find the cost of removing 50% of the pollutant; 60%; 70%; 80%; 90%; 95%.

b. Evaluate

$$\lim_{x \to 100} \frac{0.5x}{100-x}$$

and interpret your result.

82. A Doomsday Situation The population of a certain breed of rabbits introduced into an isolated island is given by

$$P(t) = \frac{72}{9-t} \qquad (0 < t < 9)$$

where t is measured in months.
a. Find the number of rabbits present in the island initially.
b. Show that the population of rabbits is increasing without bound.
c. Sketch the graph of the function P.

REMARK This phenomenon is referred to as a *doomsday* situation. ● ● ●

83. Average Cost The average cost per disc in dollars incurred by the Herald Record Company in pressing x video discs is given by the average cost function

$$\overline{C}(x) = 2.2 + \frac{2500}{x}$$

Evaluate $\lim\limits_{x \to \infty} \overline{C}(x)$ and interpret your result.

84. Concentration of a Drug in the Bloodstream The concentration of a certain drug in a patient's blood-stream t hours after injection is given by

$$C(t) = \frac{0.2t}{t^2+1}$$

milligrams per cubic centimeter. Evaluate $\lim\limits_{t \to \infty} C(t)$ and interpret your result.

85. Box Office Receipts The total worldwide box office receipts for a long-running blockbuster movie are approximated by the function

$$T(x) = \frac{120x^2}{x^2+4}$$

where $T(x)$ is measured in millions of dollars and x is the number of months since the movie's release.
a. What are the total box office receipts after the first month? The second month? The third month?
b. What will the movie gross in the long run?

86. Population Growth A major corporation is building a 4325-acre complex of homes, offices, stores, schools, and churches in the rural community of Glen Cove. As a result of this development, the planners have estimated that Glen Cove's population t years from now (in thousands) will be given by

$$P(t) = \frac{25t^2+125t+200}{t^2+5t+40}$$

a. What is the current population of Glen Cove?
b. What will the population be in the long run?

87. Driving Costs A study of driving costs of 1992 model subcompact (four-cylinder) cars found that the average cost (car payments, gas, insurance, upkeep, and depreciation), measured in cents per mile, is approximated by the function

$$C(x) = \frac{2010}{x^{2.2}} + 17.80$$

where x denotes the number of miles (in thousands) the car is driven in a year.

USING TECHNOLOGY

Finding the Limit of a Function

A graphing utility can be used to help us find the limit of a function, if it exists, as illustrated in the following examples.

EXAMPLE 1   Let $f(x) = \dfrac{x^3 - 1}{x - 1}$.

a. Plot the graph of f in the viewing rectangle $[-2, 2] \times [0, 4]$.

b. Use **ZOOM** to find $\lim\limits_{x \to 1} \dfrac{x^3 - 1}{x - 1}$.

c. Verify your result by evaluating the limit algebraically.

Solution

a. The graph of f in the viewing rectangle $[-2, 2] \times [0, 4]$ is shown in Figure T1.

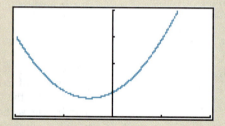

Figure T1

The graph of $f(x) = \dfrac{x^3 - 1}{x - 1}$ in the viewing rectangle $[-2, 2] \times [0, 4]$.

b. Using **ZOOM-IN** repeatedly, we see that the y-value approaches 3 as the x-value approaches 1. We conclude, accordingly, that

$$\lim_{x \to 1} \frac{x^3 - 1}{x - 1} = 3$$

c. We compute

$$\lim_{x \to 1} \frac{x^3 - 1}{x - 1} = \lim_{x \to 1} \frac{(x - 1)(x^2 + x + 1)}{x - 1}$$
$$= \lim_{x \to 1} (x^2 + x + 1) = 3$$

REMARK   If you attempt to find the limit in Example 1 by using the evaluation function of your graphing utility to find the value of $f(x)$ when $x = 1$, you will see that the graphing utility does not display the y-value. This happens because the point $x = 1$ is not in the domain of f.

EXAMPLE 2   Use **ZOOM** to find $\lim\limits_{x \to 0} (1 + x)^{1/x}$.

Solution   We first plot the graph of $f(x) = (1 + x)^{1/x}$ in a suitable viewing rectangle. Figure T2 shows a plot of f in the rectangle $[-1, 1] \times [0, 4]$. Using **ZOOM-IN** repeatedly, we see that $\lim\limits_{x \to 0} (1 + x)^{1/x} \approx 2.71828$.

Figure T2
The graph of $f(x) = (1 + x)^{1/x}$ in the viewing rectangle $[-1, 1] \times [0, 4]$.

The limit of $f(x) = (1 + x)^{1/x}$ as x approaches zero, denoted by the letter e, plays a very important role in the study of mathematics and its applications (see Section 5.6). Thus

$$\lim_{x \to 0} (1 + x)^{1/x} = e$$

where, as we have just seen, $e \approx 2.71828$.

EXAMPLE 3   When organic waste is dumped into a pond, the oxidation process that takes place reduces the pond's oxygen content. However, given time, nature will restore the oxygen content to its natural level. Suppose that the oxygen content t days after the organic waste has been dumped into the pond is given by

$$f(t) = 100 \left(\frac{t^2 + 10t + 100}{t^2 + 20t + 100} \right)$$

percent of its normal level.

a. Plot the graph of f in the viewing rectangle $[0, 200] \times [70, 100]$.

b. What can you say about $f(t)$ when t is very large?

c. Verify your observation in (b) by evaluating $\lim\limits_{t \to \infty} f(t)$.

Solution

a. The graph of f is shown in Figure T3.

b. From the graph of f it appears that $f(t)$ approaches 100 steadily as t gets

Figure T3
The graph of f in the viewing rectangle $[0, 200] \times [70, 100]$.

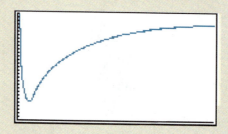

larger and larger. This observation tells us that eventually the oxygen content of the pond will be restored to its natural level.

c. To verify the observation made in part (b), we compute

$$\lim_{t \to \infty} f(t) = \lim_{t \to \infty} 100 \left(\frac{t^2 + 10t + 100}{t^2 + 20t + 100} \right)$$

$$= 100 \lim_{t \to \infty} \left(\frac{1 + \dfrac{10}{t} + \dfrac{100}{t^2}}{1 + \dfrac{20}{t} + \dfrac{100}{t^2}} \right) = 100$$

○ ○ ○

EXERCISES

*In exercises 1–10, use a graphing utility to find the indicated limit by first plotting the graph of the function in a suitable viewing rectangle and then using the **ZOOM-IN** feature of the calculator.*

1. $\lim_{x \to 1} \dfrac{2x^3 - 2x^2 + 3x - 3}{x - 1}$

2. $\lim_{x \to -2} \dfrac{2x^3 + 3x^2 - x + 2}{x + 2}$

3. $\lim_{x \to -1} \dfrac{x^3 + 1}{x + 1}$

4. $\lim_{x \to -1} \dfrac{x^4 - 1}{x - 1}$

5. $\lim_{x \to 1} \dfrac{x^3 - x^2 - x + 1}{x^3 - 3x + 2}$

6. $\lim_{x \to 2} \dfrac{x^3 + 2x^2 - 16}{2x^3 - x^2 + 2x - 16}$

7. $\lim_{x \to 0} \dfrac{\sqrt{x + 1} - 1}{x}$

8. $\lim_{x \to 0} \dfrac{(x + 4)^{3/2} - 8}{x}$

9. $\lim_{x \to 0} (1 + 2x)^{1/x}$

10. $\lim_{x \to 0} \dfrac{2^x - 1}{x}$

11. Use a graphing utility to show that $\lim_{x \to 3} \dfrac{2}{x - 3}$ does not exist.

12. Use a graphing utility to show that $\lim_{x \to 2} \dfrac{x^3 - 2x + 1}{x - 2}$ does not exist.

13. City Planning   A major developer is building a 5000-acre complex of homes, offices, stores, schools, and churches in the rural community of Marlboro. As a result of this development, the planners have estimated that Marlboro's population (in thousands) t years from now will be given by

$$P(t) = \frac{25t^2 + 125t + 200}{t^2 + 5t + 40}$$

a. Plot the graph of P in the viewing rectangle $[0, 50] \times [0, 30]$.

b. What will the population of Marlboro be in the long run?

[*Hint:* Find $\lim_{t \to \infty} P(t)$.]

a. What is the average cost of driving a subcompact car 5000 miles a year? 10,000 miles a year? 15,000 miles a year? 20,000 miles a year? 25,000 miles a year?

b. Use (a) to help sketch the graph of the function C.

c. What happens to the average cost as the number of miles driven increases without bound?

88. Speed of a Chemical Reaction Certain proteins, known as enzymes, serve as catalysts for chemical reactions in living things. In 1913, Leonor Michaelis and L. M. Menten discovered the following formula giving the initial speed V (in moles per liter per second) at which the reaction begins in terms of the amount of substrate x (the substance being acted upon, measured in moles per liter):

$$V = \frac{ax}{x + b}$$

where a and b are positive constants. Evaluate

$$\lim_{x \to \infty} \frac{ax}{x + b}$$

and interpret your result.

89. Show by means of an example that $\lim_{x \to a} [f(x) + g(x)]$ may exist even though neither $\lim_{x \to a} f(x)$ nor $\lim_{x \to a} g(x)$ exists. Does this example contradict Theorem 1?

90. Show by means of an example that $\lim_{x \to a} [f(x)g(x)]$ may exist even though neither $\lim_{x \to a} f(x)$ nor $\lim_{x \to a} g(x)$ exists. Does this example contradict Theorem 1?

SOLUTIONS TO SELF-CHECK EXERCISES 2.4

1. a. $\displaystyle \lim_{x \to 3} \frac{\sqrt{x^2 + 7} + \sqrt{3x - 5}}{x + 2} = \frac{\sqrt{9 + 7} + \sqrt{3(3) - 5}}{3 + 2}$

$$= \frac{\sqrt{16} + \sqrt{4}}{5}$$

$$= \frac{6}{5}$$

b. Letting x approach -1 leads to the indeterminate form $0/0$. Thus we proceed as follows:

$$\lim_{x \to -1} \frac{x^2 - x - 2}{2x^2 - x - 3} = \lim_{x \to -1} \frac{(x + 1)(x - 2)}{(x + 1)(2x - 3)}$$

$$= \lim_{x \to -1} \frac{x - 2}{2x - 3} \qquad \text{(Canceling the common factors)}$$

$$= \frac{-1 - 2}{2(-1) - 3}$$

$$= \frac{3}{5}$$

2. $\displaystyle \lim_{x \to \infty} \overline{C}(x) = \lim_{x \to \infty} \left(1.8 + \frac{3000}{x} \right)$

$$= \lim_{x \to \infty} 1.8 + \lim_{x \to \infty} \frac{3000}{x}$$

$$= 1.8$$

Our computation reveals that, as the production of audio discs increases "without bound," the average cost drops and approaches a unit cost of $1.80 per disc.

2.5 ONE-SIDED LIMITS AND CONTINUITY

One-Sided Limits

Consider the function f defined by

$$f(x) = \begin{cases} x - 1 & \text{if } x < 0 \\ x + 1 & \text{if } x \geq 0 \end{cases}$$

Figure 2.29

The function f does not have a limit as x approaches zero.

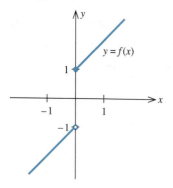

From the graph of f shown in Figure 2.29, we see that the function f does not have a limit as x approaches zero because no matter how close x is to zero, $f(x)$ takes on values that are close to 1 if x is positive and values that are close to -1 if x is negative. Therefore, $f(x)$ cannot be close to a single number L—no matter how close x is to zero. Now, if we restrict x to be greater than zero (to the right of zero), then we see that $f(x)$ can be made as close to the number 1 as we please by taking x sufficiently close to zero. In this situation, we say that the right-hand limit of f as x approaches zero (from the right) is 1, written

$$\lim_{x \to 0^+} f(x) = 1$$

Similarly, we see that $f(x)$ can be made as close to the number -1 as we please by taking x sufficiently close to, but to the left of, zero. In this situation, we say that the left-hand limit of f as x approaches zero (from the left) is -1, written

$$\lim_{x \to 0^-} f(x) = -1$$

These limits are called **one-sided limits.** More generally, we have the following definitions.

ONE-SIDED LIMITS

The function f has the **right-hand limit** L as x approaches a from the right, written

$$\lim_{x \to a^+} f(x) = L$$

if the values $f(x)$ can be made as close to L as we please by taking x sufficiently close to (but not equal to) a and to the right of a.

Similarly, the function f has the **left-hand limit** M as x approaches a from the left, written

$$\lim_{x \to a^-} f(x) = M$$

if the values $f(x)$ can be made as close to M as we please by taking x sufficiently close to (but not equal to) a and to the left of a.

The connection between one-sided limits and the two-sided limit defined earlier is given by the following theorem.

THEOREM 3

Let f be a function that is defined for all values of x close to $x = a$ with the possible exception of a itself. Then

$$\lim_{x \to a} f(x) = L \quad \text{if and only if} \quad \lim_{x \to a^+} f(x) = \lim_{x \to a^-} f(x) = L$$

Thus, the two-sided limit exists if and only if the one-sided limits exist and are equal.

EXAMPLE 1 Let

$$f(x) = \begin{cases} \sqrt{x} & \text{if } x > 0 \\ -x & \text{if } x \le 0 \end{cases} \quad \text{and} \quad g(x) = \begin{cases} -1 & \text{if } x < 0 \\ 1 & \text{if } x \ge 0 \end{cases}$$

a. Show that $\lim_{x \to 0} f(x)$ exists by studying the one-sided limits of f as x approaches $x = 0$.

b. Show that $\lim_{x \to 0} g(x)$ does not exist.

Solution

a. We find

$$\lim_{x \to 0^+} f(x) = \lim_{x \to 0^+} \sqrt{x} = 0$$

and

$$\lim_{x \to 0^-} f(x) = \lim_{x \to 0^-} (-x) = 0$$

so

$$\lim_{x \to 0} f(x) = 0 \text{ (Figure 2.30a)}$$

Figure 2.30

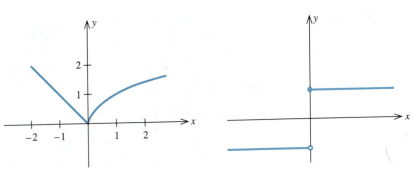

(a) $\lim_{x \to 0} f(x)$ exists.   **(b)** $\lim_{x \to 0} g(x)$ does not exist.

b. We have

$$\lim_{x \to 0^-} g(x) = -1 \qquad \text{and} \qquad \lim_{x \to 0^+} g(x) = 1$$

and since these one-sided limits are not equal, we conclude that $\lim_{x \to 0} g(x)$ does not exist (Figure 2.30b).

◗ ◗ ◗

Continuous Functions

Continuous functions will play an important role throughout most of our study of calculus. Loosely speaking, a function is continuous at a point if the graph of the function at that point is devoid of holes, gaps, jumps, or breaks. Consider, for example, the graph of the function f depicted in Figure 2.31.

Figure 2.31

The graph of this function is not continuous at $x = a$, $x = b$, $x = c$, and $x = d$.

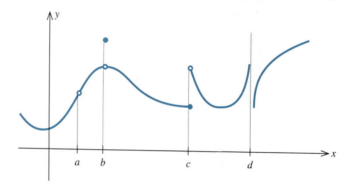

Let us take a closer look at the behavior of f at or near each of the points $x = a$, $x = b$, $x = c$, and $x = d$. First, note that f is not defined at $x = a$; that is, the point $x = a$ is not in the domain of f, thereby resulting in a "hole" in the graph of f. Next, observe that the value of f at b, $f(b)$, is not equal to the limit of $f(x)$ as x approaches b, resulting in a "jump" in the graph of f at that point. The function f does not have a limit at $x = c$ since the left-hand and right-hand limits of $f(x)$ are not equal, also resulting in a jump in the graph of f at that point. Finally, the limit of f does not exist at $x = d$, resulting in a break in the graph of f. The function f is *discontinuous* at each of these points. It is *continuous* at all other points.

CONTINUITY AT A POINT	A function f is **continuous at the point** $x = a$ if the following conditions are satisfied: **1.** $f(a)$ is defined. **2.** $\lim\limits_{x \to a} f(x)$ exists. **3.** $\lim\limits_{x \to a} f(x) = f(a)$.

Thus, a function f is continuous at the point $x = a$ if the limit of f at the point $x = a$ exists and has the value $f(a)$. Geometrically, f is continuous at the point $x = a$ if proximity of x to a implies the proximity of $f(x)$ to $f(a)$.

If f is not continuous at $x = a$, then f is said to be **discontinuous** at $x = a$. Also, f is **continuous on an interval** if f is continuous at every point in the interval.

Figure 2.32 depicts the graph of a continuous function on the interval (a, b). Notice that the graph of the function over the stated interval can be sketched without lifting one's pencil from the paper.

Figure 2.32
The graph of f is continuous on the interval (a, b).

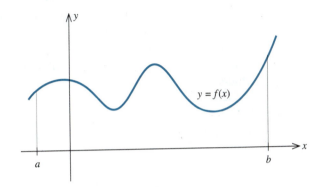

EXAMPLE 2   Find the values of x for which each of the following functions is continuous.

a. $f(x) = x + 2$ **b.** $g(x) = \dfrac{x^2 - 4}{x - 2}$ **c.** $h(x) = \begin{cases} x + 2 & \text{if } x \neq 2 \\ 1 & \text{if } x = 2 \end{cases}$

d. $F(x) = \begin{cases} -1 & \text{if } x < 0 \\ 1 & \text{if } x \geq 0 \end{cases}$ **e.** $G(x) = \begin{cases} \dfrac{1}{x} & \text{if } x > 0 \\ -1 & \text{if } x \leq 0 \end{cases}$

The graph of each function is shown in Figure 2.33.

Figure 2.33

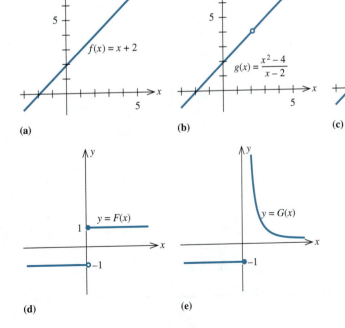

Solution

a. The function f is continuous everywhere, because the three conditions for continuity are satisfied for all values of x.

b. The function g is discontinuous at the point $x = 2$, because g is not defined at that point. It is continuous everywhere else.

c. The function h is discontinuous at $x = 2$ because the third condition for continuity is violated; the limit of $h(x)$ as x approaches 2 exists and has the value 4, but this limit is not equal to $h(2) = 1$. It is continuous for all other values of x.

d. The function F is continuous everywhere except at the point $x = 0$, where the limit of $F(x)$ fails to exist as x approaches zero (see Example 3a, Section 2.4).

e. Since the limit of $G(x)$ does not exist as x approaches zero, we conclude that G fails to be continuous at $x = 0$. The function G is continuous at all other points.

Properties of Continuous Functions

The following properties of continuous functions follow directly from the definition of continuity and the corresponding properties of limits. They are stated without proof.

PROPERTIES OF CONTINUOUS FUNCTIONS

1. The constant function $f(x) = c$ is continuous everywhere.
2. The identity function $f(x) = x$ is continuous everywhere.

If f and g are continuous at $x = a$, then

3. $[f(x)]^n$, where n is a real number, is continuous at $x = a$ whenever it is defined at that point.
4. $f \pm g$ is continuous at $x = a$.
5. fg is continuous at $x = a$.
6. $\dfrac{f}{g}$ is continuous at $x = a$ provided $g(a) \neq 0$.

Using these properties of continuous functions, we can prove the following results. (A proof is sketched in exercise 98, page 139.)

CONTINUITY OF POLYNOMIAL AND RATIONAL FUNCTIONS

1. A polynomial function $y = P(x)$ is continuous at every point x.
2. A rational function $R(x) = p(x)/q(x)$ is continuous at every point x where $q(x) \neq 0$.

EXAMPLE 3 Find the values of x for which each of the following functions is continuous.

a. $f(x) = 3x^3 + 2x^2 - x + 10$ **b.** $g(x) = \dfrac{8x^{10} - 4x + 1}{x^2 + 1}$

c. $h(x) = \dfrac{4x^3 - 3x^2 + 1}{x^2 - 3x + 2}$

Solution

a. The function f is a polynomial function of degree 3, so $f(x)$ is continuous for all values of x.

b. The function g is a rational function. Observe that the denominator of g—namely, $x^2 + 1$—is never equal to zero. Therefore, we conclude that g is continuous for all values of x.

c. The function h is a rational function. In this case, however, the denominator of h is equal to zero at $x = 1$ and $x = 2$, which can be seen by factoring it. Thus,

$$x^2 - 3x + 2 = (x - 2)(x - 1)$$

We conclude, therefore, that h is continuous everywhere except at $x = 1$ and $x = 2$, where it is discontinuous.   ○ ○ ○

Applications

Up to this point, most of the applications we have discussed involved functions that are continuous everywhere. In Example 4 we consider an application from the field of educational psychology that involves a discontinuous function.

EXAMPLE 4 Figure 2.34 depicts the learning curve associated with a certain individual. Beginning with no knowledge of the subject being taught, the individual makes steady progress toward understanding it

Figure 2.34
A learning curve that is discontinuous at $t = t_1$.

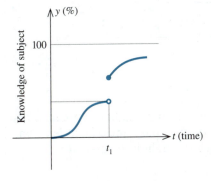

over the time interval $0 \leq t < t_1$. In this instance, the individual's progress slows down as we approach time t_1 because he fails to grasp a particularly difficult concept. All of a sudden, a breakthrough occurs at time t_1, propelling his knowledge of the subject to a higher level. The curve is discontinuous at t_1.

○ ○ ○

Intermediate Value Theorem

Let's look again at our model of the motion of the maglev on a straight stretch of track. We know that the train cannot vanish at any instant of time and it cannot skip portions of the track and reappear someplace else. To put it another way, the train cannot occupy the positions s_1 and s_2 without at least, at some time, occupying an intermediate position (Figure 2.35).

Figure 2.35
The position of the maglev.

In order to state this fact mathematically, recall that the position of the maglev as a function of time is described by

$$f(t) = 4t^2 \qquad (0 \leq t \leq 10)$$

Suppose the position of the maglev is s_1 at some time t_1 and its position is s_2 at some time t_2 (Figure 2.36).

Figure 2.36
If $s_1 \leq s_3 \leq s_2$, then there must be at least one t_3 ($t_1 \leq t_3 \leq t_2$) such that $f(t_3) = s_3$.

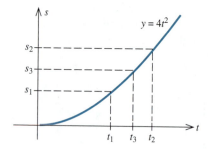

Then, if s_3 is any number between s_1 and s_2 giving an intermediate position of the maglev, there must be at least one t_3 between t_1 and t_2 giving the time at which the train is at s_3—that is, $f(t_3) = s_3$.

This discussion carries the gist of the Intermediate Value Theorem. The proof of this theorem can be found in most advanced calculus texts.

THEOREM 4

The Intermediate Value Theorem

If f is a continuous function on a closed interval $[a, b]$ and M is any number between $f(a)$ and $f(b)$, then there is at least one number c in $[a, b]$ such that $f(c) = M$ (see Figure 2.37).

(a) $f(c) = M$ **(b)** $f(c_1) = f(c_2) = f(c_3) = M$

Figure 2.37

To illustrate the Intermediate Value Theorem, let's look at the example involving the motion of the maglev again (see Figure 2.19, page 100). Notice that the initial position of the train is $f(0) = 0$, and the position at the end of its test run is $f(10) = 400$. Furthermore, the function f is continuous on $[0, 10]$. So, the Intermediate Value Theorem guarantees that if we arbitrarily pick a number between 0 and 400, say 100, giving the position of the maglev, there must be a $\bar{t}$ (read "t bar") between 0 and 10 at which time the train is at the position $s = 100$. To find the value of $\bar{t}$, we solve the equation $f(\bar{t}) = s$, or

$$4\bar{t}^2 = 100$$

giving $\bar{t} = 5$. (t must lie between 0 and 10.)

 It is important to remember when we use Theorem 4 that the function f must be continuous. The conclusion of the Intermediate Value Theorem may not hold if f is not continuous (see exercise 99).

The next theorem is an immediate consequence of the Intermediate Value Theorem. It not only tells us when a zero of a function f (root of the equation $f(x) = 0$) exists but also provides the basis for a method of approximating it.

THEOREM 5	**Existence of Zeros of a Continuous Function**

If f is a continuous function on a closed interval $[a, b]$, and if $f(a)$ and $f(b)$ have opposite signs, then there is at least one solution of the equation $f(x) = 0$ in the interval (a, b). (See Figure 2.38.)

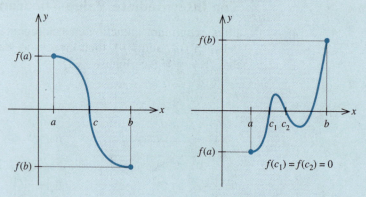

Figure 2.38
If $f(a)$ and $f(b)$ have opposite signs, there must be at least one number c $(a < c < b)$ such that $f(c) = 0$.

Figure 2.39
$f(a) < 0$ and $f(b) > 0$ but the graph of f does not cross the x-axis between a and b because f is discontinuous.

Geometrically, this property states that if the graph of a continuous function goes from above the x-axis to below the x-axis, or vice versa, it must *cross* the x-axis. This is not necessarily true if the function is discontinuous (Figure 2.39).

EXAMPLE 5   Let $f(x) = x^3 + x + 1$.

a. Show that f is continuous for all values of x.

b. Compute $f(-1)$ and $f(1)$ and use the results to deduce that there must be at least one point $x = c$, where c lies in the interval $(-1, 1)$ and $f(c) = 0$.

Solution

a. The function f is a polynomial function of degree 3 and is therefore continuous everywhere.

b. $f(-1) = (-1)^3 + (-1) + 1 = -1$
    $f(1) = 1^3 + 1 + 1 = 3$

Since $f(-1)$ and $f(1)$ have opposite signs, Theorem 5 tells us that there must be at least one point $x = c$ with $-1 < c < 1$ such that $f(c) = 0$.   ◦ ◦ ◦

The next example shows how the Intermediate Value Theorem can be used to help us find the zero of a function.

EXAMPLE 6   Let $f(x) = x^3 + x - 1$. Since f is a polynomial function it is continuous everywhere. Observe that $f(0) = -1$ and $f(1) = 1$ so that Theorem 5 guarantees the existence of at least one root of the equation $f(x) = 0$ in $(0, 1)$.*

We can locate the root more precisely by using Theorem 5 once again as follows: Evaluate $f(x)$ at the midpoint of $[0, 1]$, obtaining

$$f(0.5) = -0.375$$

Because $f(0.5) < 0$ and $f(1) > 0$, Theorem 5 now tells us that the root must lie in $(0.5, 1)$.

Table 2.3

Step	Root of $f(x) = 0$ Lies In
1	(0, 1)
2	(0.5, 1)
3	(0.5, 0.75)
4	(0.625, 0.75)
5	(0.625, 0.6875)
6	(0.65625, 0.6875)
7	(0.671875, 0.6875)
8	(0.6796875, 0.6875)
9	(0.6796875, 0.6835937)

Repeat the process: Evaluate $f(x)$ at the midpoint of $[0.5, 1]$, which is

$$\frac{0.5 + 1}{2} = 0.75$$

Thus,
$$f(0.75) = 0.1719$$

Because $f(0.5) < 0$ and $f(0.75) > 0$, Theorem 5 tells us that the root is in $(0.5, 0.75)$. This process can be continued. Table 2.3 summarizes the results of our computations through nine steps.

From Table 2.3 we see that the root is approximately 0.68, accurate to two decimal places. By continuing the process through a sufficient number of steps, we can obtain as accurate an approximation to the root as we please.

◑ ◑ ◑

REMARK   The process of finding the root of $f(x) = 0$ used in Example 6 is called the **method of bisection.** It is crude but effective.   ◑ ◑ ◑

SELF-CHECK EXERCISES 2.5

1. Evaluate $\lim\limits_{x \to -1^-} f(x)$ and $\lim\limits_{x \to -1^+} f(x)$, where

$$f(x) = \begin{cases} 1 & \text{if } x < -1 \\ 1 + \sqrt{x + 1} & \text{if } x \geq -1 \end{cases}$$

Does $\lim\limits_{x \to -1} f(x)$ exist?

◑ ◑ ◑

* It can be shown that f has precisely one zero in $(0, 1)$ (see exercise 90, Section 4.1.).

2. Determine the values of x for which the given function is discontinuous. At each point of discontinuity, indicate which condition(s) for continuity are violated. Sketch the graph of the function.

a. $f(x) = \begin{cases} -x^2 + 1 & \text{if } x \leq 1 \\ x - 1 & \text{if } x > 1 \end{cases}$

b. $g(x) = \begin{cases} -x + 1 & \text{if } x < -1 \\ 2 & \text{if } -1 < x \leq 1 \\ -x + 3 & \text{if } x > 1 \end{cases}$

Solutions to Self-Check Exercises 2.5 can be found on page 139.

2.5 EXERCISES

In exercises 1–8, use the graph of the function f to find $\lim\limits_{x \to a^-} f(x)$, $\lim\limits_{x \to a^+} f(x)$, *and* $\lim\limits_{x \to a} f(x)$ *at the indicated value of a, if the limit exists.*

1.

$a = 2$

2.

$a = 3$

7.

$a = 0$

8.

$a = 0$

3.

$a = -1$

4.

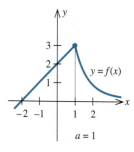

$a = 1$

In exercises 9–14, refer to the graph of the function f and determine whether each statement is true or false.

5.

$a = 1$

6.

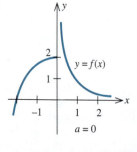

$a = 0$

9. $\lim\limits_{x \to -3^+} f(x) = 1$

10. $\lim\limits_{x \to 0} f(x) = f(0)$

11. $\lim\limits_{x \to 2^-} f(x) = 2$

12. $\lim\limits_{x \to 2^+} f(x) = 3$

13. $\lim\limits_{x \to 3} f(x)$ does not exist.

14. $\lim\limits_{x \to 5^-} f(x) = 3$

In exercises 15–20, refer to the graph of the function f and determine whether each statement is true or false.

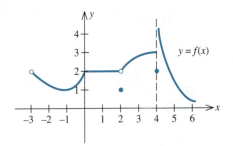

15. $\lim\limits_{x \to -3^+} f(x) = 2$

16. $\lim\limits_{x \to 0} f(x) = 2$

17. $\lim\limits_{x \to 2} f(x) = 1$

18. $\lim\limits_{x \to 4^-} f(x) = 3$

19. $\lim\limits_{x \to 4^+} f(x)$ does not exist.

20. $\lim\limits_{x \to 4} f(x) = 2$

In exercises 21–42 find the indicated one-sided limit, if it exists.

21. $\lim\limits_{x \to 1^+} (2x + 4)$

22. $\lim\limits_{x \to 1^-} (3x - 4)$

23. $\lim\limits_{x \to 2^-} \dfrac{x - 3}{x + 2}$

24. $\lim\limits_{x \to 1^+} \dfrac{x + 2}{x + 1}$

25. $\lim\limits_{x \to 0^+} \dfrac{1}{x}$

26. $\lim\limits_{x \to 0^-} \dfrac{1}{x}$

27. $\lim\limits_{x \to 0^+} \dfrac{x - 1}{x^2 + 1}$

28. $\lim\limits_{x \to 2^+} \dfrac{x + 1}{x^2 - 2x + 3}$

29. $\lim\limits_{x \to 0^+} \sqrt{x}$

30. $\lim\limits_{x \to 2^+} 2\sqrt{x - 2}$

31. $\lim\limits_{x \to -2^+} (2x + \sqrt{2 + x})$

32. $\lim\limits_{x \to -5^+} x(1 + \sqrt{5 + x})$

33. $\lim\limits_{x \to 1^-} \dfrac{1 + x}{1 - x}$

34. $\lim\limits_{x \to 1^+} \dfrac{1 + x}{1 - x}$

35. $\lim\limits_{x \to 2^-} \dfrac{x^2 - 4}{x - 2}$

36. $\lim\limits_{x \to -3^+} \dfrac{\sqrt{x + 3}}{x^2 + 1}$

37. $\lim\limits_{x \to 3^+} \dfrac{x^2 - 9}{x + 3}$

38. $\lim\limits_{x \to -2^-} \dfrac{\sqrt[3]{x + 10}}{2x^2 + 1}$

39. $\lim\limits_{x \to 0^+} f(x)$ and $\lim\limits_{x \to 0^-} f(x)$, where

$$f(x) = \begin{cases} 2x & \text{if } x < 0 \\ x^2 & \text{if } x \geq 0 \end{cases}$$

40. $\lim\limits_{x \to 0^+} f(x)$ and $\lim\limits_{x \to 0^-} f(x)$, where

$$f(x) = \begin{cases} -x + 1 & \text{if } x \leq 0 \\ 2x + 3 & \text{if } x > 0 \end{cases}$$

41. $\lim\limits_{x \to 1^+} f(x)$ and $\lim\limits_{x \to 1^-} f(x)$, where

$$f(x) = \begin{cases} \sqrt{x + 3} & \text{if } x \geq 1 \\ 2 + \sqrt{x} & \text{if } x < 1 \end{cases}$$

42. $\lim\limits_{x \to 1^+} f(x)$ and $\lim\limits_{x \to 1^-} f(x)$, where

$$f(x) = \begin{cases} x + 2\sqrt{x - 1} & \text{if } x \geq 1 \\ 1 - \sqrt{1 - x} & \text{if } x < 1 \end{cases}$$

In exercises 43–50, determine the values of x, if any, at which each function is discontinuous. At each point of discontinuity, state the condition(s) for continuity that are violated.

43.

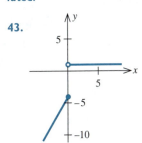

$$f(x) = \begin{cases} 2x - 4 & \text{if } x \leq 0 \\ 1 & \text{if } x > 0 \end{cases}$$

44.

$$f(x) = \begin{cases} x^2 + 1 & \text{if } x \neq 0 \\ 0 & \text{if } x = 0 \end{cases}$$

45.

$$f(x) = \begin{cases} x + 5 & \text{if } x \leq 0 \\ -x^2 + 5 & \text{if } x > 0 \end{cases}$$

46.

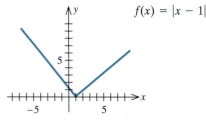

$f(x) = |x - 1|$

47.

$$f(x) = \begin{cases} x + 5 & \text{if } x < 0 \\ 2 & \text{if } x = 0 \\ -x^2 + 5 & \text{if } x > 0 \end{cases}$$

48.

$$f(x) = \begin{cases} \dfrac{x^2 - 1}{x + 1} & \text{if } x \neq -1 \\ 1 & \text{if } x = -1 \end{cases}$$

49.

$$f(x) = \begin{cases} -|x| + 1 & \text{if } x \neq 0 \\ 0 & \text{if } x = 0 \end{cases}$$

50.

$$f(x) = \begin{cases} \dfrac{1}{x^2} & \text{if } x \neq 0 \\ 1 & \text{if } x = 0 \end{cases}$$

In exercises 51–66, find the values of x for which each function is continuous.

51. $f(x) = 2x^2 + x - 1$

52. $f(x) = x^3 - 2x^2 + x - 1$

53. $f(x) = \dfrac{2}{x^2 + 1}$

54. $f(x) = \dfrac{x}{2x^2 + 1}$

55. $f(x) = \dfrac{2}{2x - 1}$

56. $f(x) = \dfrac{x + 1}{x - 1}$

57. $f(x) = \dfrac{2x + 1}{x^2 + x - 2}$

58. $f(x) = \dfrac{x - 1}{x^2 + 2x - 3}$

59. $f(x) = \begin{cases} x & \text{if } x \leq 1 \\ 2x - 1 & \text{if } x > 1 \end{cases}$

60. $f(x) = \begin{cases} -x + 1 & \text{if } x \leq -1 \\ x + 1 & \text{if } x > -1 \end{cases}$

61. $f(x) = \begin{cases} -2x + 1 & \text{if } x < 0 \\ x^2 + 1 & \text{if } x \geq 0 \end{cases}$

62. $f(x) = \begin{cases} x + 1 & \text{if } x \leq 1 \\ -x^2 + 1 & \text{if } x > 1 \end{cases}$

63. $f(x) = \begin{cases} \dfrac{x^2 - 1}{x - 1} & \text{if } x \neq 1 \\ 2 & \text{if } x = 1 \end{cases}$

64. $f(x) = \begin{cases} \dfrac{x^2 - 4}{x + 2} & \text{if } x \neq -2 \\ 1 & \text{if } x = -2 \end{cases}$

65. $f(x) = |x + 1|$ **66.** $f(x) = \dfrac{|x - 1|}{x - 1}$

In exercises 67–70 determine all values of x at which the function is discontinuous.

67. $f(x) = \dfrac{2x}{x^2 - 1}$ **68.** $f(x) = \dfrac{1}{(x - 1)(x - 2)}$

69. $f(x) = \dfrac{x^2 - 2x}{x^2 - 3x + 2}$ **70.** $f(x) = \dfrac{x^2 - 3x + 2}{x^2 - 2x}$

71. The Postage Function The graph of the "postage function"

$$f(x) = \begin{cases} 32 & \text{if } 0 < x \leq 1 \\ 55 & \text{if } 1 < x \leq 2 \\ \vdots & \\ 285 & \text{if } 11 < x \leq 12 \end{cases}$$

where x denotes the weight of a parcel in ounces and $f(x)$ the postage in cents, is shown in the figure on page 137. Determine the values of x for which f is discontinuous.

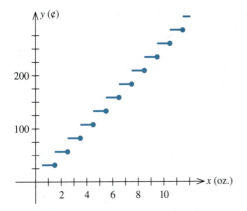

72. Inventory Control As part of an optimal inventory policy, the manager of an office supply company orders 500 reams of xerographic copy paper every 20 days. The accompanying graph shows the *actual* inventory level of copy paper in an office supply store during the first 60 business days of 1996. Determine the values of *t* for which the "inventory function" is discontinuous and give an interpretation of the graph.

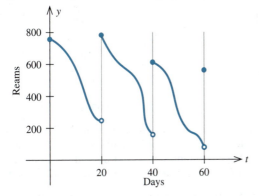

73. Learning Curves The following graph describes the progress Michael made in solving a problem correctly during a mathematics quiz. Here *y* denotes the percentage of work completed and *x* is measured in minutes. Give an interpretation of the graph.

74. Ailing Financial Institutions The Franklin Savings and Loan Company acquired two ailing financial institutions in 1992. One of them was acquired at time $t = T_1$, and the other was acquired at time $t = T_2$ ($t = 0$ corresponds to the beginning of 1992). The following graph shows the total amount of money on deposit with Franklin.

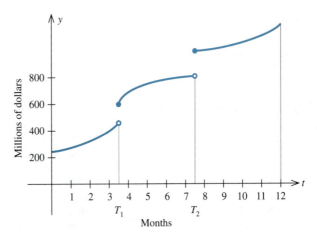

Explain the significance of the discontinuities of the function at T_1 and T_2.

75. Energy Consumption The following graph shows the amount of home heating oil remaining in a 200-gallon tank over a 120-day period ($t = 0$ corresponds to October 1). Explain why the function is discontinuous at $t = 40$, $t = 70$, $t = 95$, and $t = 110$.

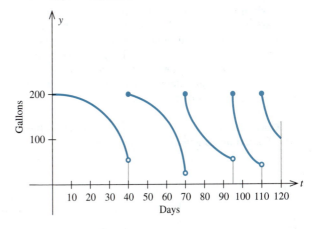

76. Prime Interest Rate The function *P*, whose graph follows, gives the prime rate (the interest rate banks charge their best corporate customers) as a function of time for the first 32 weeks in 1989. Determine the values of t for which P is discontinuous and interpret your results.

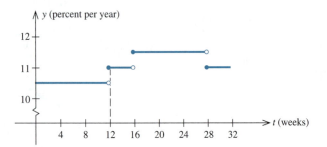

77. **Administration of an Intravenous Solution**   A dextrose solution is being administered to a patient intravenously. The 1-liter bottle holding the solution is removed and replaced by another as soon as the contents drop to approximately 5% of the initial (1-liter) amount. The rate of discharge is constant, and it takes 6 hours to discharge 95% of the contents of a full bottle. Draw a graph showing the amount of dextrose solution in a bottle in the IV system over a 24-hour period, assuming that we started with a full bottle.

78. **Commissions**   The base salary of a salesman working on commission is $12,000. For each $50,000 of sales beyond $100,000 he is paid a $1000 commission. Sketch a graph showing his earnings as a function of the level of his sales x. Determine the values of x for which the function f is discontinuous.

79. **Parking Fees**   The fee charged per car in a downtown parking lot is $1 for the first half hour and 50 cents for each additional half hour or part thereof, subject to a maximum of $5. Derive a function f relating the parking fee to the length of time a car is left in the lot. Sketch the graph of f and determine the values of x for which the function f is discontinuous.

80. **Commodity Prices**   The function that gives the cost of a certain commodity is defined by

$$C(x) = \begin{cases} 5x & \text{if } 0 < x < 10 \\ 4x & \text{if } 10 \le x < 30 \\ 3.5x & \text{if } 30 \le x < 60 \\ 3.25x & \text{if } x \ge 60 \end{cases}$$

where x is the number of pounds of a certain commodity sold and $C(x)$ is measured in dollars. Sketch the graph

of the function C and determine the values of x for which the function C is discontinuous.

81. **Energy Expended by a Fish**   Suppose that a fish swimming a distance of L ft at a speed of v ft/sec relative to the water and against a current flowing at the rate of u ft/sec $(u < v)$ expends a total energy given by

$$E(v) = \frac{aLv^3}{v - u}$$

where E is measured in foot-pounds (ft-lb), and a is a constant.
  a. Evaluate $\lim_{v \to u^+} E(v)$ and interpret your result.
  b. Evaluate $\lim_{v \to \infty} E(v)$ and interpret your result.

82. Let

$$f(x) = \begin{cases} x + 2 & \text{if } x \le 1 \\ kx^2 & \text{if } x > 1 \end{cases}$$

Find the value of k that will make f continuous on $(-\infty, \infty)$.

83. Let

$$f(x) = \begin{cases} \dfrac{x^2 - 4}{x + 2} & \text{if } x \ne -2 \\ k & \text{if } x = -2 \end{cases}$$

For what value of k will f be continuous on $(-\infty, \infty)$?

84. **a.** Suppose f is continuous at a and g is discontinuous at a. Is the sum $f + g$ discontinuous at a? Explain.
  b. Suppose f and g are both discontinuous at a. Is the sum $f + g$ necessarily discontinuous at a? Explain.

85. **a.** Suppose f is continuous at a and g is discontinuous at a. Is the product fg necessarily discontinuous at a? Explain.
  b. Suppose f and g are both discontinuous at a. Is the product fg necessarily discontinuous at a? Explain.

In exercises 86–89, (a) show that the function f is continuous for all values of x in the interval [a, b] and (b) prove that f must have at least one zero in the interval (a, b) by showing that f(a) and f(b) have opposite signs.

86. $f(x) = x^2 - 6x + 8$; $a = 1$, $b = 3$

87. $f(x) = x^3 - 2x^2 + 3x + 2$; $a = -1$, $b = 1$

88. $f(x) = 2x^3 - 3x^2 - 36x + 14$; $a = 0$, $b = 1$

89. $f(x) = 2x^{5/3} - 5x^{4/3}$; $a = 14$, $b = 16$

In exercises 90–91, use the Intermediate Value Theorem to find the value of c such that $f(c) = M$.

90. $f(x) = x^2 - x + 1$ on $[-1, 4]$; $M = 7$

91. $f(x) = x^2 - 4x + 6$ on $[0, 3]$; $M = 2$

92. Use the method of bisection (see Example 6) to find the root of the equation $x^3 - x + 1 = 0$ accurate to two decimal places.

93. Use the method of bisection to find the root of the equation $x^5 + 2x - 7 = 0$ accurate to two decimal places.

94. Joan is looking straight out a window of an apartment building at a height of 32 ft from the ground. A boy throws a tennis ball straight up by the side of the building where the window is located. Suppose the height of the ball (measured in feet) from the ground at time t is $h(t) = 4 + 64t - 16t^2$.
a. Show that $h(0) = 4$ and $h(2) = 68$.
b. Use the Intermediate Value Theorem to conclude that the ball must cross Joan's line of sight at least once.
c. At what time(s) does the ball cross Joan's line of sight? Interpret your results.

95. True or false. Suppose the function f is defined on the interval $[a, b]$. If $f(a)$ and $f(b)$ have the same sign, then f has no zero in $[a, b]$. Explain your answer.

96. Let $f(x) = x - \sqrt{1 - x^2}$.
a. Show that f is continuous for all values of x in the interval $[-1, 1]$.
b. Show that f has at least one zero in $[-1, 1]$.
c. Find the zeros of f in $[-1, 1]$ by solving the equation $f(x) = 0$.

97. Let $f(x) = \dfrac{x^2}{x^2 + 1}$.
a. Show that f is continuous for all values of x.
b. Show that $f(x)$ is nonnegative for all values of x.
c. Show that f has a zero at $x = 0$. Does this contradict Theorem 5?

98. a. Prove that a polynomial function $y = P(x)$ is continuous at every point x. Follow these steps:
1. Use properties 2 and 3 of continuous functions to establish that the function $g(x) = x^n$, where n is a positive integer, is continuous everywhere.
2. Use properties 1 and 5 to show that $f(x) = cx^n$, where c is a constant and n is a positive integer, is continuous everywhere.
3. Use property 4 to complete the proof of the result.
b. Prove that a rational function $R(x) = p(x)/q(x)$ is continuous at every point x, where $q(x) \neq 0$.
[*Hint:* Use the result of (a) and property 6.]

99. Show that the conclusion of the Intermediate Value Theorem does not hold if f is discontinuous on $[a, b]$.

SOLUTIONS TO SELF-CHECK EXERCISES 2.5

1. For $x < -1, f(x) = 1$, and so

$$\lim_{x \to -1^-} f(x) = \lim_{x \to -1^-} 1 = 1$$

For $x \geq -1, f(x) = 1 + \sqrt{x + 1}$, and so

$$\lim_{x \to -1^+} f(x) = \lim_{x \to -1^+} (1 + \sqrt{x + 1}) = 1$$

Since the left-hand and right-hand limits of f exist as x approaches $x = -1$ and both are equal to 1, we conclude that

$$\lim_{x \to -1} f(x) = 1$$

Finding the Points of Discontinuity of a Function

You can very often recognize the points of discontinuity of a function f by examining its graph. For example, Figure T1 shows the graph of $f(x) = \dfrac{x}{x^2 - 1}$ obtained using a graphing utility. It is evident that f is discontinuous at $x = -1$ and $x = 1$. This observation is also borne out by the fact that both these points are not in the domain of f.

Figure T1

The graph of $f(x) = \dfrac{x}{x^2 - 1}$ in the viewing rectangle $[-4, 4] \times [-10, 10]$.

Consider the function

$$g(x) = \frac{2x^3 + x^2 - 7x - 6}{x^2 - x - 2}$$

Using a graphing utility we obtain the graph of g shown in Figure T2. An examination of this graph does not reveal any points of discontinuity. However, if we factor both the numerator and the denominator of the rational expression, we see that

$$g(x) = \frac{(x + 1)(x - 2)(2x + 3)}{(x + 1)(x - 2)}$$
$$= 2x + 3$$

provided $x \neq -1$ and $x \neq 2$, so that its graph, in fact, looks like that shown in Figure T3.

Figure T2

The graph of $g(x) = \dfrac{2x^3 + x^2 - 7x - 6}{x^2 - x - 2}$ in the standard viewing rectangle.

Figure T3
The graph of g has holes at (−1, 1)
and (2, 7).

This example shows the limitation of the graphing utility and reminds us of the importance of studying functions analytically!

Graphing Functions Defined Piecewise

The following example illustrates how to plot the graphs of functions defined in a piecewise manner on a graphing utility.

EXAMPLE 1 Plot the graph of

$$f(x) = \begin{cases} x + 1 & \text{if } x \leq 1 \\ \dfrac{2}{x} & \text{if } x > 1 \end{cases}$$

Solution We enter the function

$$y1 = (x + 1)(x \leq 1) + (2/x)(x > 1)$$

The graph of the function in the viewing rectangle $[-5, 5] \times [-2, 4]$ is shown in Figure T4.

Figure T4
The graph of f in the viewing rectangle $[-5, 5] \times [-2, 4]$.

○ ○ ○

141

EXAMPLE 2 The percentage of U.S. households, $P(t)$, watching television during weekdays between the hours of 4 P.M. and 4 A.M. is given by

$$P(t) = \begin{cases} 0.01354t^4 - 0.49375t^3 + 2.58333t^2 + 3.8t + 31.60704 & 0 \le t \le 8 \\ 1.35t^2 - 33.05t + 208 & 8 < t \le 12 \end{cases}$$

where t is measured in hours, with $t = 0$ corresponding to 4 P.M. Plot the graph of P in the viewing rectangle $[0, 12] \times [0, 80]$.

Solution   We enter the function

$$y2 = (0.01354x^4 - 0.49375x^3 + 2.58333x^2 + 3.8x + 31.60704)(x \ge 0)(x \le 8)$$
$$+ (1.35x^2 - 33.05x + 208)(x > 8)(x \le 12)$$

The graph of P is shown in Figure T5.

Figure T5
The graph of P in the viewing rectangle $[0, 12] \times [0, 80]$.

Source: A. C. Nielsen Co.

EXERCISES

In exercises 1–10, use a graphing utility to plot the graph of f and to spot the points of discontinuity of f. Then use analytical means to verify your observation and find all points of discontinuity.

1. $f(x) = \dfrac{2}{x^2 - x}$

2. $f(x) = \dfrac{2x + 1}{x^2 + x - 2}$

3. $f(x) = \dfrac{\sqrt{x}}{x^2 - x - 2}$

4. $f(x) = \dfrac{3}{\sqrt{x}(x + 1)}$

5. $f(x) = \dfrac{6x^3 + x^2 - 2x}{2x^2 - x}$

6. $f(x) = \dfrac{2x^3 - x^2 - 13x - 6}{2x^2 - 5x - 3}$

7. $f(x) = \dfrac{2x^4 - 3x^3 - 2x^2}{2x^2 - 3x - 2}$

8. $f(x) = \dfrac{6x^4 - x^3 + 5x^2 - 1}{6x^2 - x - 1}$

9. $f(x) = \dfrac{x^3 + x^2 - 2x}{x^4 + 2x^3 - x - 2}$

[*Hint:* $x^4 + 2x^3 - x - 2 = (x^3 - 1)(x + 2)$]

10. $f(x) = \dfrac{x^3 - x}{x^{4/3} - x + x^{1/3} - 1}$

[*Hint:* $x^{4/3} - x + x^{1/3} - 1 = (x^{1/3} - 1)(x + 1)$]
Can you explain why part of the graph is missing?

In exercises 11–14, use a graphing utility to plot the graph of f in the indicated viewing rectangle.

11. $f(x) = \begin{cases} -1 & \text{if } x \leq 1 \\ x+1 & \text{if } x > 1; [-5,5] \times [-2,8] \end{cases}$

12. $f(x) = \begin{cases} \frac{1}{3}x^2 - 2x & \text{if } x \leq 3 \\ -x+6 & \text{if } x > 3; [0,7] \times [-5,5] \end{cases}$

13. $f(x) = \begin{cases} 2 & \text{if } x \leq 0 \\ \sqrt{4-x^2} & \text{if } x > 0; [-2,2] \times [-4,4] \end{cases}$

14. $f(x) = \begin{cases} -x^2 + x + 2 & \text{if } x \leq 1 \\ 2x^3 - x^2 - 4 & \text{if } x > 1; [-4,4] \times [-5,5] \end{cases}$

15. **Flight Path of a Plane** The function

$$f(x) = \begin{cases} 0 & 0 \leq x < 1 \\ -0.00411523x^3 + 0.0679012x^2 & \\ \quad - 0.123457x + 0.0596708 & 1 \leq x < 10 \\ 1.5 & 10 \leq x \leq 100 \end{cases}$$

where both x and $f(x)$ are measured in units of 1000 ft, describes the flight path of a plane taking off from the origin and climbing to an altitude of 15,000 ft. Plot the graph of f to visualize the trajectory of the plane.

16. **Home Shopping Industry** According to industry sources, revenue from the home shopping industry for the years since its inception may be approximated by the function

$$R(t) = \begin{cases} -0.03t^3 + 0.25t^2 - 0.12t & 0 \leq t \leq 3 \\ 0.57t - 0.63 & 3 < t \leq 11 \end{cases}$$

where $R(t)$ measures the revenue in billions of dollars and t is measured in years, with $t = 0$ corresponding to the beginning of 1984. Plot the graph of R.
Source: Paul Kagan Associates

143

2. a. The graph of *f* is as follows:

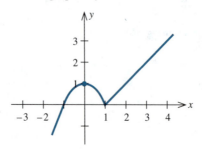

We see that *f* is continuous everywhere.
b. The graph of *g* is as follows:

Since *g* is not defined at $x = -1$, it is discontinuous there. It is continuous everywhere else.

2.6 THE DERIVATIVE

An Intuitive Example

We mentioned in Section 2.4 that the problem of finding the *rate of change* of one quantity with respect to another is mathematically equivalent to the problem of finding the *slope of the tangent line* to a curve at a given point on the curve. Before going on to establish this relationship, let us show its plausibility by looking at it from an intuitive point of view.

Consider the motion of the maglev discussed in Section 2.4. Recall that the position of the maglev at any time *t* is given by

$$s = f(t) = 4t^2 \qquad (0 \le t \le 30)$$

where *s* is measured in feet and *t* in seconds. The graph of the function *f* is sketched in Figure 2.40.

Observe that the graph of *f* rises slowly at first but more rapidly as *t* increases, reflecting the fact that the speed of the maglev is increasing with time. This observation suggests a relationship between the speed of the maglev at any time *t* and the *steepness* of the curve at the point corresponding to this value of *t*. Thus, it would appear that we can solve the problem of finding the speed of the maglev at any time if we can find a way to measure the steepness of the curve at any point on the curve.

Figure 2.40
Graph showing the position s of a maglev at time t.

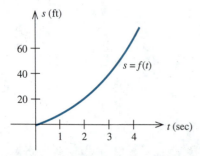

To discover a yardstick that will measure the steepness of a curve, consider the graph of a function f such as the one shown in Figure 2.41a. Think of the curve as representing a stretch of roller coaster track (Figure 2.41b). When the car is at the point P on the curve, a passenger sitting erect in the car and looking straight ahead will have a line of sight that is parallel to the line T, the tangent to the curve at P.

As Figure 2.41a suggests, the steepness of the curve—that is, the rate at which y is increasing or decreasing with respect to x—is given by the slope of the tangent line to the graph of f at the point $P(x, f(x))$. But for now we will show how this relationship can be used to estimate the rate of change of a function from its graph.

Figure 2.41

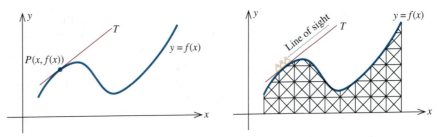

(a) T is the tangent line to the curve at P.

(b) T is parallel to the line of sight.

EXAMPLE I   The graph of the function $y = N(t)$, shown in Figure 2.42, gives the number of social security beneficiaries from the beginning of 1990 ($t = 0$) through the year 2045 ($t = 55$).

Figure 2.42
The number of social security bene-ficiaries from 1990 through 2045. We can use the slope of the tangent line at the indicated points to esti-mate the rate at which the number of social security beneficiaries will be changing.

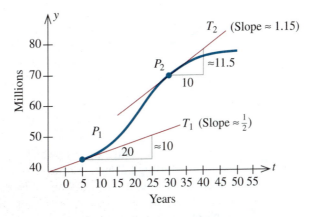

Use the graph of $y = N(t)$ to estimate the rate at which the number of social security beneficiaries will be growing at the beginning of the year 2000 ($t = 10$) and at the beginning of 2025 ($t = 35$). [Assume that the rate of change of the function N at any value of t is given by the slope of the tangent line at the point $P(t, N(t))$.]
Source: Social Security Administration

Solution   From the figure, we see that the slope of the tangent line T_1 to the graph of $y = N(t)$ at $P_1(10, 44.7)$ is approximately 0.5. This tells us that the quantity y is increasing at the rate of 1/2 unit per unit increase in t, when $t = 10$. In other words, at the beginning of the year 2000, the number of social security beneficiaries will be increasing at the rate of approximately 0.5 million, or 500,000, per year.

The slope of the tangent line T_2 at $P_2(35, 71.9)$ is approximately 1.15. This tells us that at the beginning of 2025, the number of social security beneficiaries will be growing at the rate of approximately 1.15 million, or 1,150,000, per year.

◑ ◑ ◑

Slope of a Tangent Line

In Example 1, we answered the questions raised by drawing the graph of the function N and estimating the position of the tangent lines. Ideally, however, we would like to solve a problem analytically whenever possible. In order to do this, we need a precise definition of the slope of a tangent line to a curve.

To define the tangent line to a curve C at a point P on the curve, fix P and let Q be any point on C distinct from P (Figure 2.43). The straight line passing through P and Q is called a **secant line.**

Figure 2.43

As Q approaches P along the curve C, the secant lines approach the tangent line T.

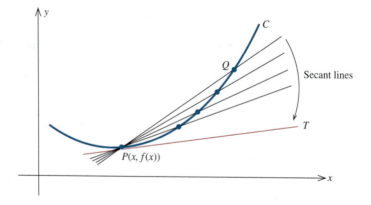

Now, as the point Q is allowed to move toward P along the curve, the secant line through P and Q rotates about the fixed point P and approaches a fixed line through P. This fixed line, which is the limiting position of the secant lines through P and Q as Q approaches P, is the **tangent line to the graph of f** at the point P.

We can describe the process more precisely as follows. Suppose the curve C is the graph of a function f defined by $y = f(x)$. Then the point P is described by $P(x, f(x))$ and the point Q by $Q(x + h, f(x + h))$, where h is some appropriate nonzero number (Figure 2.44a). Observe that we can make Q approach P along the curve C by letting h approach zero (Figure 2.44b).

Figure 2.44

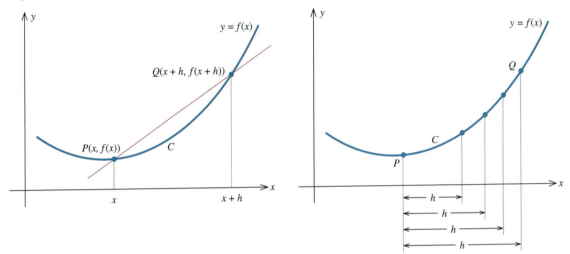

(a) The points $P(x, f(x))$ and $Q(x + h, f(x + h))$.   **(b)** As h approaches 0, Q approaches P.

Next, using the formula for the slope of a line, we can write the slope of the secant line passing through $P(x, f(x))$ and $Q(x + h, f(x + h))$ as

$$\frac{f(x + h) - f(x)}{(x + h) - x} = \frac{f(x + h) - f(x)}{h} \tag{5}$$

As observed earlier, Q approaches P and, therefore, the secant line through P and Q approaches the tangent line T as h approaches zero. Consequently, we might expect that the slope of the secant line would approach the slope of the tangent line T as h approaches zero. This leads to the following definition.

SLOPE OF A TANGENT LINE

The slope of the tangent line to the graph f at the point $P(x, f(x))$ is given by

$$\lim_{h \to 0} \frac{f(x + h) - f(x)}{h} \tag{6}$$

if it exists.

Rates of Change

We now show that the problem of finding the slope of the tangent line to the graph of a function f at the point $P(x, f(x))$ is mathematically equivalent to the problem of finding the rate of change of f at x. To see this, suppose that we are given a function f that describes the relationship between the two quantities x and y:

$$y = f(x)$$

Figure 2.45
*$f(x + h) - f(x)$ is the change in y
that corresponds to a change h in x.*

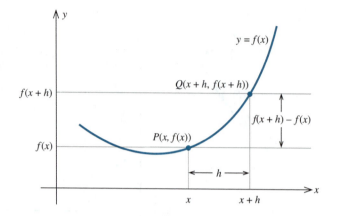

The number $f(x + h) - f(x)$ measures the change in y that corresponds to a change h in x (Figure 2.45).

Then, the **difference quotient**

$$\frac{f(x + h) - f(x)}{h} \tag{7}$$

measures the **average rate of change of y with respect to x** over the interval $[x, x + h]$. For example, if y measures the position of a car at time x, then quotient (7) gives the average velocity of the car over the time interval $[x, x + h]$.

Observe that the difference quotient (7) is the same as (5). We conclude that the difference quotient (7) also measures the slope of the secant line that passes through the two points $P(x, f(x))$ and $Q(x + h, f(x + h))$ lying on the graph of $y = f(x)$. Next, by taking the limit of the difference quotient (7) as h goes to zero—that is, by evaluating

$$\lim_{h \to 0} \frac{f(x + h) - f(x)}{h} \tag{8}$$

we obtain the **rate of change of f at x.** For example, if y measures the position of a car at time x, then the limit (8) gives the velocity of the car at time x. For emphasis, the rate of change of a function f at x is often called the **instantaneous rate of change of f at x.** This distinguishes it from the average rate of change of f, which is computed over an *interval* $[x, x + h]$ rather than at a *point x.*

Observe that the limit (8) is the same as (6). Therefore, the limit of the difference quotient also measures the slope of the tangent line to the graph of $y = f(x)$ at the point $(x, f(x))$.

The following summarizes this discussion.

AVERAGE AND INSTANTANEOUS RATES OF CHANGE	**Average rate of change** of f over the interval $[x, x + h]$ or **slope of the secant line** to the graph of f through the points $(x, f(x))$ and $(x + h, f(x + h))$ is

$$\frac{f(x + h) - f(x)}{h} \qquad (9)$$

Instantaneous rate of change of f at x or **slope of tangent line** to the graph of f at $(x, f(x))$ is

$$\lim_{h \to 0} \frac{f(x + h) - f(x)}{h} \qquad (10)$$

Explain the difference between the average rate of change of a function and the instantaneous rate of change of a function.

The Derivative

The limit (6), or (10), which measures both the slope of the tangent line to the graph of $y = f(x)$ at the point $P(x, f(x))$ and the (instantaneous) rate of change of f at x is given a special name: the **derivative of f at x.**

THE DERIVATIVE OF A FUNCTION	The derivative of a function f with respect to x is the function f' (read "f prime"), defined by

$$f'(x) = \lim_{h \to 0} \frac{f(x + h) - f(x)}{h} \qquad (11)$$

The domain of f' is the set of all x where the limit exists.

Thus, the derivative of a function f is a function f' that gives the slope of the tangent line to the graph of f at *any* point $(x, f(x))$, and also the rate of change of f at x (Figure 2.46).

Figure 2.46
The slope of the tangent line at
$P(x, f(x))$ *is* $f'(x)$; f *changes at the*
rate of $f'(x)$ *units per unit change*
in x at x.

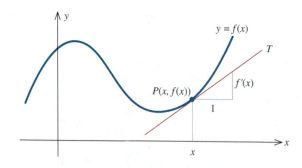

Other notations for the derivative of f include:

$$D_x f(x) \qquad \text{(Read "dee sub } x \text{ of } f \text{ of } x\text{")}$$

$$\frac{dy}{dx} \qquad \text{(Read "dee } y \text{ dee } x\text{")}$$

and $\qquad\qquad y' \qquad \text{(Read "} y \text{ prime")}$

The last two are used when the rule for f is written in the form $y = f(x)$.

The calculation of the derivative of f is facilitated using the following four-step process.

THE FOUR-STEP PROCESS FOR FINDING $f'(x)$

1. Compute $f(x + h)$.

2. Form the difference $f(x + h) - f(x)$.

3. Form the quotient $\dfrac{f(x + h) - f(x)}{h}$.

4. Compute $f'(x) = \lim\limits_{h \to 0} \dfrac{f(x + h) - f(x)}{h}$.

EXAMPLE 2   Find the slope of the tangent line to the graph of $f(x) = 3x + 5$ at any point $(x, f(x))$.

Solution   The slope of the tangent line at any point on the graph of f is given by the derivative of f at x. To find the derivative, we use the four-step process:

Step 1 $f(x + h) = 3(x + h) + 5 = 3x + 3h + 5$

Step 2 $f(x + h) - f(x) = (3x + 3h + 5) - (3x + 5) = 3h$

Step 3 $\dfrac{f(x + h) - f(x)}{h} = \dfrac{3h}{h} = 3$

Step 4 $f'(x) = \lim\limits_{h \to 0} \dfrac{f(x + h) - f(x)}{h} = \lim\limits_{h \to 0} 3 = 3$

We expect this result since the tangent line to any point on a straight line must coincide with the line itself and, therefore, must have the same slope as the line. In this case, the graph of f is a straight line with slope 3.   ◦ ◦ ◦

EXAMPLE 3   Let $f(x) = x^2$.

a. Compute $f'(x)$.

b. Compute $f'(2)$ and interpret your result.

Solution

Figure 2.47
The tangent line to the graph of $f(x) = x^2$ at (2, 4).

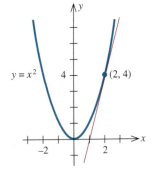

$y = x^2$

a. To find $f'(x)$, we use the four-step process:

Step 1 $f(x + h) = (x + h)^2 = x^2 + 2xh + h^2$

Step 2 $f(x + h) - f(x) = x^2 + 2xh + h^2 - x^2 = 2xh + h^2 = h(2x + h)$

Step 3 $\dfrac{f(x + h) - f(x)}{h} = \dfrac{h(2x + h)}{h} = 2x + h$

Step 4 $f'(x) = \lim\limits_{h \to 0} \dfrac{f(x + h) - f(x)}{h} = \lim\limits_{h \to 0} (2x + h) = 2x$

b. $f'(2) = 2(2) = 4$. This result tells us that the slope of the tangent line to the graph of f at the point (2, 4) is 4. It also tells us that the function f is changing at the rate of 4 units per unit change in x at $x = 2$. The graph of f and the tangent line at (2, 4) are shown in Figure 2.47.   ๐ ๐ ๐

EXPLORING WITH TECHNOLOGY

1. Consider the function $f(x) = x^2$ of Example 3. Suppose we want to compute $f'(2)$ using equation (11). Thus,

$$f'(2) = \lim_{h \to 0} \frac{f(2 + h) - f(2)}{h} = \lim_{h \to 0} \frac{(2 + h)^2 - 2^2}{h}$$

Use a graphing utility to plot the graph of

$$g(x) = \frac{(2 + x)^2 - 4}{x}$$

in the viewing rectangle $[-3, 3] \times [-2, 6]$.

2. Use **ZOOM** and **TRACE** to find $\lim\limits_{x \to 0} g(x)$.

3. Explain why the limit found in part (2) is $f'(2)$.

๐   ๐   ๐

EXAMPLE 4   Let $f(x) = x^2 - 4x$.

a. Compute $f'(x)$.

b. Find the point on the graph of f where the tangent line to the curve is horizontal.

c. Sketch the graph of f and the tangent line to the curve at the point found in (b).

d. What is the rate of change of f at this point?

Solution

a. To find $f'(x)$, we use the four-step process:

Step 1 $f(x + h) = (x + h)^2 - 4(x + h) = x^2 + 2xh + h^2 - 4x - 4h$

Step 2 $f(x + h) - f(x) = x^2 + 2xh + h^2 - 4x - 4h - (x^2 - 4x)$
$$= 2xh + h^2 - 4h = h(2x + h - 4)$$

Step 3 $\dfrac{f(x + h) - f(x)}{h} = \dfrac{h(2x + h - 4)}{h} = 2x + h - 4$

Step 4 $f'(x) = \lim\limits_{h \to 0} \dfrac{f(x + h) - f(x)}{h} = \lim\limits_{h \to 0} (2x + h - 4) = 2x - 4$

b. At a point on the graph of f where the tangent line to the curve is horizontal and hence has slope zero, the derivative f' of f is zero. Accordingly, to find such point(s) we set $f'(x) = 0$, which gives $2x - 4 = 0$, or $x = 2$. The corresponding value of y is given by $y = f(2) = -4$, and the required point is $(2, -4)$.

c. The graph of f and the tangent line are shown in Figure 2.48.

d. The rate of change of f at $x = 2$ is zero.

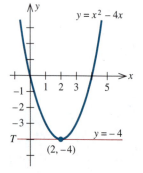

Figure 2.48
The tangent line to the graph of
$y = x^2 - 4x$ *at* $(2, -4)$ *is* $y = -4$.

● ● ●

EXAMPLE 5   Let $f(x) = 1/x$.

a. Compute $f'(x)$.

b. Find the slope of the tangent line T to the graph of f at the point where $x = 1$.

c. Find an equation of the tangent line T in (b).

Solution

a. To find $f'(x)$, we use the four-step process:

Step 1 $f(x + h) = \dfrac{1}{x + h}$

Step 2 $f(x + h) - f(x) = \dfrac{1}{x + h} - \dfrac{1}{x} = \dfrac{x - (x + h)}{x(x + h)} = -\dfrac{h}{x(x + h)}$

Can the tangent line to the graph of a function intersect the graph at more than one point? Explain your answer using illustrations.

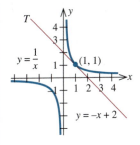

Figure 2.49
The tangent line to the graph of
$f(x) = 1/x$ *at* $(1, 1)$.

Step 3 $\dfrac{f(x + h) - f(x)}{h} = -\dfrac{h}{x(x + h)} \cdot \dfrac{1}{h} = -\dfrac{1}{x(x + h)}$

Step 4 $f'(x) = \lim\limits_{h \to 0} \dfrac{f(x + h) - f(x)}{h} = \lim\limits_{h \to 0} -\dfrac{1}{x(x + h)} = -\dfrac{1}{x^2}$

b. The slope of the tangent line T to the graph of f where $x = 1$ is given by $f'(1) = -1$.

c. When $x = 1$, $y = f(1) = 1$ and T is tangent to the graph of f at the point $(1, 1)$. From (b), we know that the slope of T is -1. Thus, an equation of T is

$$y - 1 = -1(x - 1)$$
$$y = -x + 2$$

or

See Figure 2.49.   ◦ ◦ ◦

EXPLORING WITH TECHNOLOGY

1. Use the results of Example 5 to draw the graph of $f(x) = 1/x$ and its tangent line at the point $(1, 1)$ by plotting the graphs of $y_1 = 1/x$ and $y_2 = -x + 2$ in the viewing rectangle $[-4, 4] \times [-4, 4]$.

2. Some graphing utilities draw the tangent line to the graph of a function at a given point automatically—you need only specify the function and give the x-coordinate of the point of tangency. If your graphing utility has this feature, verify the result of part (1) without finding an equation of the tangent line.

◦ ◦ ◦

Consider the following alternative approach to the definition of the derivative of a function: Let h be a positive number and suppose $P(x - h, f(x - h))$ and $Q(x + h, f(x + h))$ are any two points on the graph of f.

a. Give a geometric and a physical interpretation of the quotient

$$\frac{f(x + h) - f(x - h)}{2h}$$

Make a sketch to illustrate your answer.

(Continued on next page)

(*Continued from previous page*)

b. Give a geometric and a physical interpretation of the limit

$$\lim_{h \to 0} \frac{f(x + h) - f(x - h)}{2h}$$

Make a sketch to illustrate your answer.

c. Explain why it makes sense to define

$$f'(x) = \lim_{h \to 0} \frac{f(x + h) - f(x - h)}{2h}$$

d. Using the definition given in part (c), formulate a four-step process for finding $f'(x)$ similar to that given on page 150 and use it to find the derivative of $f(x) = x^2$. Compare your answer with that obtained in Example 3 on page 151.

Applications

EXAMPLE 6   Suppose that the distance (in feet) covered by a car moving along a straight road t seconds after starting from rest is given by the function $f(t) = 2t^2$ ($0 \le t \le 30$).

a. Calculate the average velocity of the car over the time intervals [22, 23], [22, 22.1], and [22, 22.01].

b. Calculate the (instantaneous) velocity of the car when $t = 22$.

c. Compare the results obtained in (a) with that obtained in (b).

Solution

a. We first compute the average velocity (average rate of change of f) over the interval $[t, t + h]$ using (9). We find

$$\frac{f(t + h) - f(t)}{h} = \frac{2(t + h)^2 - 2t^2}{h}$$

$$= \frac{2t^2 + 4th + 2h^2 - 2t^2}{h}$$

$$= 4t + 2h$$

Next, using $t = 22$ and $h = 1$, we find that the average velocity of the car over the time interval [22, 23] is

$$4(22) + 2(1) = 90$$

or 90 feet per second. Similarly, using $t = 22$, $h = 0.1$, and $h = 0.01$, we find that its average velocities over the time intervals [22, 22.1] and [22, 22.01] are 88.2 and 88.02 feet per second, respectively.

b. Using the limit (10), we see that the instantaneous velocity of the car at any time t is given by

$$\lim_{h \to 0} \frac{f(t + h) - f(t)}{h} = \lim_{h \to 0} (4t + 2h) \qquad \text{[Using the results from (a)]}$$

$$= 4t$$

In particular, the velocity of the car 22 seconds from rest ($t = 22$) is given by

$$v = 4(22)$$

Figure 2.50

The graph of the demand function
$p = 144 - x^2$.

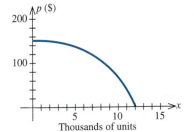

or 88 feet per second.

c. The computations in (a) show that as the time intervals over which the average velocity of the car are computed become smaller and smaller, the average velocities over these intervals do approach 88 ft/sec, the instantaneous velocity of the car at $t = 22$.   ○ ○ ○

EXAMPLE 7   The management of the Titan Tire Company has determined that the weekly demand function for their Super Titan tires is given by

$$p = f(x) = 144 - x^2$$

where p is measured in dollars and x is measured in units of a thousand (Figure 2.50).

a. Find the average rate of change in the unit price of a tire if the quantity demanded is between 5000 and 6000 tires. Between 5000 and 5100 tires. Betweeen 5000 and 5010 tires.

b. What is the instantaneous rate of change of the unit price when the quantity demanded is 5000 units?

Solution

a. The average rate of change of the unit price of a tire if the quantity demanded is between x and $x + h$ is

$$\frac{f(x + h) - f(x)}{h} = \frac{[144 - (x + h)^2] - (144 - x^2)}{h}$$

$$= \frac{144 - x^2 - 2x(h) - h^2 - 144 + x^2}{h}$$

$$= -2x - h$$

To find the average rate of change of the unit price of a tire when the quantity demanded is between 5000 and 6000 tires (that is, over the interval [5, 6]), we take $x = 5$ and $h = 1$, obtaining

$$-2(5) - 1 = -11$$

or $-\$11$ per 1000 tires. (Remember, x is measured in units of a thousand.) Similarly, taking $h = 0.1$ and $h = 0.01$ with $x = 5$, we find that the average rates of change of the unit price when the quantities demanded are between 5000 and 5100 and between 5000 and 5010 are $-\$10.10$ and $-\$10.01$ per 1000 tires, respectively.

b. The instantaneous rate of change of the unit price of a tire when the quantity demanded is x units is given by

$$\lim_{h \to 0} \frac{f(x + h) - f(x)}{h} = \lim_{h \to 0} (-2x - h) \quad \text{[Using the results from (a)]}$$

$$= -2x$$

In particular, the instantaneous rate of change of the unit price per tire when the quantity demanded is 5000 is given by $-2(5)$, or $-\$10$ per 1000 tires.

○ ○ ○

The derivative of a function provides us with a tool for measuring the rate of change of one quantity with respect to another. Table 2.4 lists several other applications involving this limit.

Table 2.4

x Stands for	*y* Stands for	$\dfrac{f(a + h) - f(a)}{h}$ Measures the	$\displaystyle\lim_{h \to 0} \dfrac{f(a + h) - f(a)}{h}$ Measures the
time	**concentration of a drug** in the bloodstream at time x	average rate of change in the concentration of the drug over the time interval $[a, a + h]$	instantaneous rate of change in the concentration of the drug in the bloodstream at time $x = a$
number of items sold	**revenue** at a sales level of x units	average rate of change in the revenue when the sales level is between $x = a$ and $x = a + h$	instantaneous rate of change in the revenue when the sales level is a units
time	**volume of sales** at time x	average rate of change in the volume of sales over the time interval $[a, a + h]$	instantaneous rate of change in the volume of sales at time $x = a$
time	**population** of *Drosophila* (fruit flies) at time x	average rate of growth of the fruit fly population over the time interval $[a, a + h]$	instantaneous rate of change of the fruit fly population at time $x = a$
temperature in a chemical reaction	**amount of product formed in the chemical reaction** when the temperature is x degrees	average rate of formation of chemical product over the temperature range $[a, a + h]$	instantaneous rate of formation of chemical product when the temperature is a degrees

Differentiability and Continuity

In practical applications, one encounters functions that fail to be **differentiable**—that is, do not have a derivative at certain values in the domain of the function f. It can be shown that a continuous function f fails to be differentiable at a point $x = a$ when the graph of f makes an abrupt change of direction at that point. We call such a point a "corner." A function also fails to be differentiable at a point where the tangent line is vertical since the slope of a vertical line is undefined. These cases are illustrated in Figure 2.51.

Figure 2.51

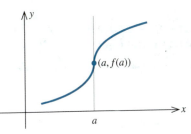

(a) The graph makes an abrupt change of direction at $x = a$.

(b) The slope at $x = a$ is undefined.

Figure 2.52
The function f is not differentiable at (8, 48).

The next example illustrates a function that is not differentiable at a point.

EXAMPLE 8   Mary works at the B&O department store, where, on a weekday, she is paid $6 per hour for the first 8 hours and $9 per hour for overtime. The function

$$f(x) = \begin{cases} 6x & 0 \le x \le 8 \\ 9x - 24 & 8 < x \end{cases}$$

gives Mary's earnings on a weekday in which she worked x hours. Sketch the graph of the function f and explain why it is not differentiable at $x = 8$.

Solution   The graph of f is shown in Figure 2.52. Observe that the graph of f has a corner at $x = 8$ and consequently is not differentiable at $x = 8$.

○ ○ ○

We close this section by mentioning the connection between the continuity and the differentiability of a function at a given value $x = a$ in the domain of f. By re-examining the function of Example 8, it becomes clear that f is continuous everywhere and, in particular, when $x = 8$. This shows that, in general, the continuity of a function at a point $x = a$ does not necessarily imply the differentiability of the function at that point. The converse, however, is true: If a function f is differentiable at a point $x = a$, then it is continuous there.

EXPLORING WITH TECHNOLOGY

1. Use a graphing utility to plot the graph of $f(x) = x^{1/3}$ in the viewing rectangle $[-2, 2] \times [-2, 2]$.

2. Use the graphing utility to draw the tangent line to the graph of f at the point $(0, 0)$. Can you explain why the process breaks down?

○ ○ ○

DIFFERENTIABILITY AND CONTINUITY	If a function is differentiable at $x = a$, then it is continuous at $x = a$.

For a proof of this result, see exercise 57, page 163.

EXAMPLE 9   Figure 2.53 depicts a portion of the graph of a function. Explain why the function fails to be differentiable at each of the points $x = a, b, c, d, e, f$, and g.

Figure 2.53
The graph of this function is not differentiable at the points a–g.

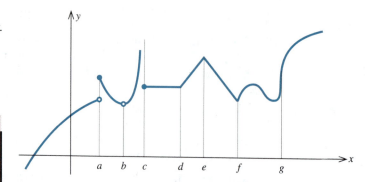

Suppose a function f is differentiable at $x = a$. Can there be two tangent lines to the graphs of f at the point $(a, f(a))$? Explain your answer.

Solution   The function fails to be differentiable at the points $x = a, b$, and c because it is discontinuous at each of these points. The derivative of the function does not exist at $x = d, e$, and f because it has a kink at each of these points. Finally, the function is not differentiable at $x = g$ because the tangent line is vertical at that point.

● ● ●

SELF–CHECK EXERCISES 2.6

1. Let $f(x) = -x^2 - 2x + 3$.
   a. Find the derivative f' of f using the definition of the derivative.
   b. Find the slope of the tangent line to the graph of f at the point $(0, 3)$.
   c. Find the rate of change of f when $x = 0$.
   d. Find an equation of the tangent line to the graph of f at the point $(0, 3)$.
   e. Sketch the graph of f and the tangent line to the curve at the point $(0, 3)$.

2. The losses (in millions of dollars) due to bad loans extended chiefly in agriculture, real estate, shipping, and energy by the Franklin Bank are estimated to be

$$A = f(t) = -t^2 + 10t + 30 \qquad (0 \le t \le 10)$$

where t is the time in years ($t = 0$ corresponds to the beginning of 1994). How fast were the losses mounting at the beginning of 1997? At the beginning of 1999? How fast will the losses be mounting at the beginning of 2001? Interpret your results.

Solutions to Self-Check Exercises 2.6 can be found on page 163.

2.6 EXERCISES

1. **Average Weight of an Infant** The following graph shows the weight measurements of the average infant from the time of birth ($t = 0$) through age 2 ($t = 24$). By computing the slopes of the respective tangent lines, estimate the rate of change of the average infant's weight when $t = 3$ and when $t = 18$. What is the average rate of change in the average infant's weight over its first year of life?

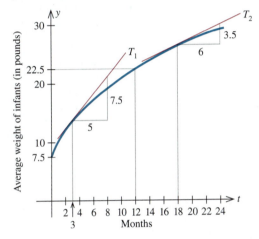

2. **Forestry** The following graph shows the volume of wood produced in a single-species forest. Here $f(t)$ is measured in cubic meters per hectare and t is measured in years. By computing the slopes of the respective tangent lines, estimate the rate at which the wood grown is changing at the beginning of the 10th year and at the beginning of the 30th year.

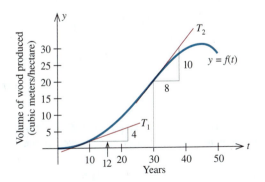

Source: The Random House Encyclopedia

3. **TV Viewing Patterns** The following graph, based on data supplied by the A. C. Nielsen Company, shows the percentage of U.S. households watching television during a 24-hr period on a weekday ($t = 0$ corresponds to 6 A.M.). By computing the slopes of the respective tangent lines, estimate the rate of change of the percentage of households watching television at 4 P.M. and 11 P.M.

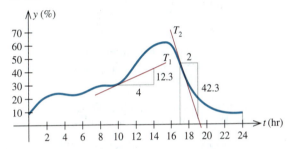

Source: A. C. Nielsen Company

4. **Crop Yield** Productivity and yield of cultivated crops are often reduced by insect pests. The following graph shows the relationship between the yield of a certain crop, $f(x)$, as a function of the density of aphids x. (Aphids are small insects that suck plant juices.) Here $f(x)$ is measured in kg/4000 sq meters, and x is measured in hundreds of aphids per bean stem. By computing the slopes of the respective tangent lines, estimate the rate of change of the crop yield with respect to the density of aphids when that density is 200 aphids per bean stem. When it is 800 aphids per bean stem.

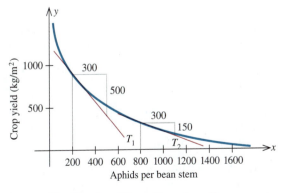

Source: The Random House Encyclopedia

5. The position of car A and car B, starting out side by side and traveling along a straight road, is given by $s = f(t)$ and $s = g(t)$, respectively, where s is measured in feet and t is measured in seconds (see accompanying figure).

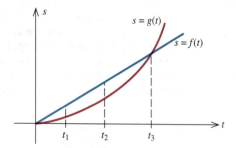

a. Which car is traveling faster at t_1?

b. What can you say about the speed of the cars at t_2?

[*Hint:* Compare tangent lines.]

c. Which car is traveling faster at t_3?

d. What can you say about the positions of the cars at t_3?

6. The velocity of car A and car B, starting out side by side and traveling along a straight road, is given by $v = f(t)$ and $v = g(t)$, respectively, where v is measured in ft/sec and t is measured in seconds (see accompanying figure).

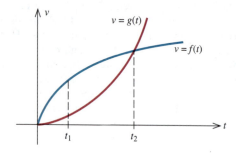

a. What can you say about the velocity and acceleration of the two cars at t_1? (Acceleration is the rate of change of velocity.)

b. What can you say about the velocity and acceleration of the two cars at t_2?

7. Effect of a Bactericide on Bacteria In the following figure, $f(t)$ gives the population P_1 of a certain bacteria culture at time t after a portion of bactericide A was introduced into the population at $t = 0$. The graph of g gives the population P_2 of a similar bacteria culture at time t after a portion of bactericide B was introduced into the population at $t = 0$.

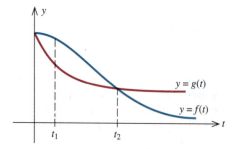

a. Which population is decreasing faster at t_1?

b. Which population is decreasing faster at t_2?

c. Which bactericide is more effective in reducing the population of bacteria in the short run? In the long run?

8. Market Share The following figure shows the devastating effect the opening of a new discount department store had on an established department store in a small town. The revenue of the discount store at time t (in months) is given by $f(t)$ million dollars, whereas the revenue of the established department store at time t is given by $g(t)$ million dollars. Answer the following questions by giving the value of t at which the specified event took place.

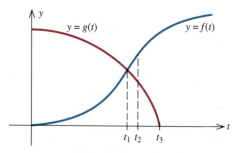

a. The revenue of the established department store is decreasing at the slowest rate.

b. The revenue of the established department store is decreasing at the fastest rate.

c. The revenue of the discount store first overtakes that of the established store.

d. The revenue of the discount store is increasing at the fastest rate.

In exercises 9–16, use the four-step process to find the slope of the tangent line to the graph of the given function at any point.

9. $f(x) = 13$ **10.** $f(x) = -6$

11. $f(x) = 2x + 7$ **12.** $f(x) = 8 - 4x$

13. $f(x) = 3x^2$

14. $f(x) = -\frac{1}{2}x^2$

15. $f(x) = -x^2 + 3x$

16. $f(x) = 2x^2 + 5x$

In exercises 17–22, find the slope of the tangent line to the graph of each function at the given point and determine an equation of the tangent line.

17. $f(x) = 2x + 7$ at $(2, 11)$

18. $f(x) = -3x + 4$ at $(-1, 7)$

19. $f(x) = 3x^2$ at $(1, 3)$

20. $f(x) = 3x - x^2$ at $(-2, -10)$

21. $f(x) = -\frac{1}{x}$ at $\left(3, -\frac{1}{3}\right)$

22. $f(x) = \frac{3}{2x}$ at $\left(1, \frac{3}{2}\right)$

23. Let $f(x) = 2x^2 + 1$.
   a. Find the derivative f' of f.
   b. Find an equation of the tangent line to the curve at the point $(1, 3)$.
   c. Sketch the graph of f.

24. Let $f(x) = x^2 + 6x$.
   a. Find the derivative f' of f.
   b. Find the point on the graph of f where the tangent line to the curve is horizontal.
   [*Hint:* Find the value of x for which $f'(x) = 0$.]

   c. Sketch the graph of f and the tangent line to the curve at the point found in (b).

25. Let $f(x) = x^2 - 2x + 1$.
   a. Find the derivative f' of f.
   b. Find the point on the graph of f where the tangent line to the curve is horizontal.
   c. Sketch the graph of f and the tangent line to the curve at the point found in (b).
   d. What is the rate of change of f at this point?

26. Let $f(x) = \dfrac{1}{x - 1}$.

   a. Find the derivative f' of f.
   b. Find an equation of the tangent line to the curve at the point $(-1, -\frac{1}{2})$.
   c. Sketch the graph of f.

27. Let $y = f(x) = x^2 + x$.
   a. Find the average rate of change of y with respect to x in the interval from $x = 2$ to $x = 3$. In the interval from $x = 2$ to $x = 2.5$. In the interval from $x = 2$ to $x = 2.1$.

   b. Find the (instantaneous) rate of change of y at $x = 2$.
   c. Compare the results obtained in (a) with that of (b).

28. Let $y = f(x) = x^2 - 4x$.
   a. Find the average rate of change of y with respect to x in the interval from $x = 3$ to $x = 4$. In the interval from $x = 3$ to $x = 3.5$. In the interval from $x = 3$ to $x = 3.1$.
   b. Find the (instantaneous) rate of change of y at $x = 3$.
   c. Compare the results obtained in (a) with that of (b).

29. **Velocity of a Car** Suppose that the distance s (in feet) covered by a car moving along a straight road t seconds after starting from rest is given by the function $f(t) = 2t^2 + 48t$.
   a. Calculate the average velocity of the car over the time intervals [20, 21], [20, 20.1], and [20, 20.01].
   b. Calculate the (instantaneous) velocity of the car when $t = 20$.
   c. Compare the results of (a) with that of (b).

30. **Velocity of a Ball Thrown into the Air** A ball is thrown straight up with an initial velocity of 128 ft/sec, so that its height (in feet) after t seconds is given by $s(t) = 128t - 16t^2$.
   a. What is the average velocity of the ball over the time intervals [2, 3], [2, 2.5], and [2, 2.1]?
   b. What is the instantaneous velocity at time $t = 2$?
   c. What is the instantaneous velocity at time $t = 5$? Is the ball rising or falling at this time?
   d. When will the ball hit the ground?

31. During the construction of a high-rise building, a worker accidentally dropped his portable electric screwdriver from a height of 400 ft. After t seconds, the screwdriver had fallen a distance of $s = 16t^2$ ft.
   a. How long did it take the screwdriver to reach the ground?
   b. What was the average velocity of the screwdriver between the time it was dropped and the time it hit the ground?
   c. What was the velocity of the screwdriver at the time it hit the ground?

32. A hot air balloon rises vertically from the ground so that its height after t seconds is $h = \frac{1}{2}t^2 + \frac{1}{2}t$ ft $(0 \le t \le 60)$.
   a. What is the height of the balloon at the end of 40 sec?
   b. What is the average velocity of the balloon between $t = 0$ and $t = 40$?
   c. What is the velocity of the balloon at the end of 40 sec?

33. At a temperature of 20°C, the volume V (in liters) of 1.33 g of O_2 is related to its pressure p (in atmospheres) by the formula $V = 1/p$.
a. What is the average rate of change of V with respect to p as p increases from $p = 2$ to $p = 3$?
b. What is the rate of change of V with respect to p when $p = 2$?

34. Cost of Producing Surfboards The total cost $C(x)$ (in dollars) incurred by the Aloha Company in manufacturing x surfboards a day is given by

$$C(x) = -10x^2 + 300x + 130 \qquad (0 \le x \le 15)$$

a. Find $C'(x)$.
b. What is the rate of change of the total cost when the level of production is ten surfboards a day?
c. What is the average cost Aloha incurs in manufacturing ten surfboards a day?

35. Effect of Advertising on Profit The quarterly profit of Cunningham Realty (in thousands of dollars) is given by

$$P(x) = -\frac{1}{3}x^2 + 7x + 30 \qquad (0 \le x \le 50)$$

where x (in thousands of dollars) is the amount of money Cunningham spends on advertising per quarter.
a. Find $P'(x)$.
b. What is the rate of change of Cunningham's quarterly profit if the amount it spends on advertising is $10,000 per quarter ($x = 10$)? $30,000 per quarter ($x = 30$)?

36. Demand for Tents The demand function for the Sportsman 5 × 7 tents is given by

$$p = f(x) = -0.1x^2 - x + 40$$

where p is measured in dollars and x is measured in units of a thousand.
a. Find the average rate of change in the unit price of a tent if the quantity demanded is between 5000 and 5050 tents. Between 5000 and 5010 tents.
b. What is the rate of change of the unit price if the quantity demanded is 5000?

37. A Country's GDP The gross domestic product (GDP) of a certain country is projected to be

$$N(t) = t^2 + 2t + 50 \qquad (0 \le t \le 5)$$

billion dollars t years from now. What will the rate of change of the country's GDP be two years from now? Four years from now?

38. Growth of Bacteria Under a set of controlled laboratory conditions, the size of the population of a certain bacteria culture at time t (in minutes) is described by the function

$$P = f(t) = 3t^2 + 2t + 1$$

Find the rate of population growth at $t = 10$ minutes.

In exercises 39–43, let x and f(x) represent the given quantities. Fix x = a and let h be a small positive number. Give an interpretation of the quantities

$$\frac{f(a+h) - f(a)}{h} \quad \text{and} \quad \lim_{h \to 0} \frac{f(a+h) - f(a)}{h}$$

39. x denotes time and $f(x)$ denotes the population of seals at time x

40. x denotes time and $f(x)$ denotes the prime interest rate at time x

41. x denotes time and $f(x)$ denotes a country's industrial production

42. x denotes the level of production of a certain commodity and $f(x)$ denotes the total cost incurred in producing x units of the commodity

43. x denotes altitude and $f(x)$ denotes atmospheric pressure

In each of exercises 44–49, the graph of a function is shown. For each function, state whether or not (a) f(x) has a limit at x = a, (b) f(x) is continuous at x = a and (c) f(x) is differentiable at x = a. Justify your answers.

44.

45.

46.

47.

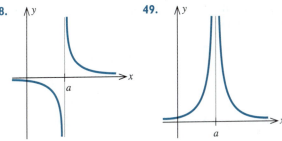

48.

49.

50. The distance s (in feet) covered by a motorcycle traveling in a straight line and starting from rest in t seconds is given by the function

$$s(t) = -0.1t^3 + 2t^2 + 24t$$

Calculate the motorcycle's average velocity over the time interval $[2, 2 + h]$ for $h = 1, 0.1, 0.01, 0.001, 0.0001$, and 0.00001, and use your results to guess at the motorcycle's instantaneous velocity at $t = 2$.

51. The daily total cost $C(x)$ incurred by Trappee and Sons, Inc., for producing x cases of Texa-Pep hot sauce is given by

$$C(x) = 0.000002x^3 + 5x + 400$$

Calculate

$$\frac{C(100 + h) - C(100)}{h}$$

for $h = 1, 0.1, 0.01, 0.001$, and 0.0001, and use your results to estimate the rate of change of the total cost function when the level of production is 100 cases a day.

52. Sketch the graph of the function $f(x) = |x + 1|$ and show that the function does not have a derivative at $x = -1$.

53. Sketch the graph of the function $f(x) = 1/(x - 1)$ and show that the function does not have a derivative at $x = 1$.

54. Let

$$f(x) = \begin{cases} x^2 & \text{if } x \le 1 \\ ax + b & \text{if } x > 1 \end{cases}$$

Find the values of a and b so that f is continuous and has a derivative at $x = 1$. Sketch the graph of f.

55. Sketch the graph of the function $f(x) = x^{2/3}$. Is the function continuous at $x = 0$? Does $f'(0)$ exist? Why or why not?

56. Prove that the derivative of the function $f(x) = |x|$ for $x \ne 0$ is given by

$$f'(x) = \begin{cases} 1 & \text{if } x > 0 \\ -1 & \text{if } x < 0 \end{cases}$$

[*Hint:* Recall the definition of the absolute value of a number.]

57. Show that if a function f is differentiable at a point $x = a$, then f must be continuous at that point.
[*Hint:* Write

$$f(x) - f(a) = \left[\frac{f(x) - f(a)}{x - a}\right](x - a)$$

Use the product rule for limits and the definition of the derivative to show that

$$\lim_{x \to a} [f(x) - f(a)] = 0]$$

SOLUTIONS TO SELF-CHECK EXERCISES 2.6

1. a. $f'(x) = \lim_{h \to 0} \dfrac{f(x + h) - f(x)}{h}$

$= \lim_{h \to 0} \dfrac{[-(x + h)^2 - 2(x + h) + 3] - (-x^2 - 2x + 3)}{h}$

$= \lim_{h \to 0} \dfrac{-x^2 - 2xh - h^2 - 2x - 2h + 3 + x^2 + 2x - 3}{h}$

$= \lim_{h \to 0} \dfrac{h(-2x - h - 2)}{h}$

$= \lim_{h \to 0} (-2x - h - 2) = -2x - 2$

b. From the result of (a), we see that the slope of the tangent line to the graph of f at any point $(x, f(x))$ is given by

$$f'(x) = -2x - 2$$

In particular, the slope of the tangent line to the graph of f at $(0, 3)$ is

$$f'(0) = -2$$

c. The rate of change of f when $x = 0$ is given by $f'(0) = -2$, or -2 units per unit change in x.

d. Using the result from (b), we see that an equation of the required tangent line is

$$y - 3 = -2(x - 0)$$

or

$$y = -2x + 3$$

e.

$y = -x^2 - 2x + 3$

$y = -2x + 3$

2. The rate of change of the losses at any time t is given by

$$
\begin{aligned}
f'(t) &= \lim_{h \to 0} \frac{f(t + h) - f(t)}{h} \\
&= \lim_{h \to 0} \frac{[-(t + h)^2 + 10(t + h) + 30] - (-t^2 + 10t + 30)}{h} \\
&= \lim_{h \to 0} \frac{-t^2 - 2th - h^2 + 10t + 10h + 30 + t^2 - 10t - 30}{h} \\
&= \lim_{h \to 0} \frac{h(-2t - h + 10)}{h} \\
&= \lim_{h \to 0} (-2t - h + 10) \\
&= -2t + 10
\end{aligned}
$$

Therefore, the rate of change of the losses suffered by the bank at the beginning of 1997 ($t = 3$) was

$$f'(3) = -2(3) + 10 = 4$$

That is, the losses were increasing at the rate of $4 million per year. At the beginning of 1999 ($t = 5$),

$$f'(5) = -2(5) + 10 = 0$$

USING TECHNOLOGY

Graphing a Function and Its Tangent Lines

We can use a graphing utility to plot the graph of a function f and the tangent line at any point on the graph.

EXAMPLE 1   Let $f(x) = x^2 - 4x$.

a. Find an equation of the tangent line to the graph of f at the point $(3, -3)$.

b. Plot both the graph of f and the tangent line found in (a) on the same set of axes.

Solution

a. The slope of the tangent line at any point on the graph of f is given by $f'(x)$. But from Example 4 (page 152) we find $f'(x) = 2x - 4$. Using this result, we see that the slope of the required tangent line is

$$f'(3) = 2(3) - 4 = 2$$

Finally, using the point-slope form of the equation of a line, we find that an equation of the tangent line is

$$y - (-3) = 2(x - 3)$$
$$y + 3 = 2x - 6$$

or $y = 2x - 9$.

b. The graph of f in the standard viewing rectangle and the tangent line of interest are shown in Figure T1.

Figure T1
The graph of $f(x) = x^2 - 4x$ and the tangent line $y = 2x - 9$ in the standard viewing rectangle.

○ ○ ○

REMARK   Some graphing utilities will draw both the graph of a function f and the tangent line to the graph of f at a specified point when the function and the specified value of x are entered.   ○ ○ ○

Finding the Derivative of a Function at a Given Point

The numerical derivative operation of a graphing utility can be used to give an approximate value of the derivative of a function for a given value of x.

EXAMPLE 2 Let $f(x) = \sqrt{x}$.

a. Use the numerical derivative operation of a graphing utility to find the derivative of f at $(4, 2)$.

b. Find an equation of the tangent line to the graph of f at $(4, 2)$.

c. Plot the graph of f and the tangent line on the same set of axes.

Solution

a. Using the numerical derivative operation of a graphing utility, we find that

$$f'(4) = \tfrac{1}{4}$$

b. An equation of the required tangent line is

$$y - 2 = \frac{1}{4}(x - 4)$$

or $y = \tfrac{1}{4}x + 1$.

c. The graph of f and the tangent line in the viewing rectangle $[0, 15] \times [0, 4]$ is shown in Figure T2.

Figure T2
The graph of $f(x) = \sqrt{x}$ and the tangent line $y = \dfrac{1}{4}x + 1$ in the viewing rectangle $[0, 15] \times [0, 4]$.

In exercises 1–10, (a) find an equation of the tangent line to the graph of f at the indicated point, and (b) use a graphing utility to plot the graph of f and the tangent line on the same set of axes. Use a suitable viewing rectangle.

1. $f(x) = 4x - 3$; $(2, 5)$

2. $f(x) = -2x + 5$; $(1, 3)$

3. $f(x) = 2x^2 + x$; $(-2, 6)$

4. $f(x) = -x^2 + 2x$; $(1, 1)$

5. $f(x) = 2x^2 + x - 3$; $(2, 7)$

6. $f(x) = -3x^2 + 2x - 1$; $(1, -2)$

7. $f(x) = x + \dfrac{1}{x}$; $(1, 2)$

8. $f(x) = x - \dfrac{1}{x}$; $(1, 0)$

9. $f(x) = \sqrt{x}$; $(4, 2)$

10. $f(x) = \dfrac{1}{\sqrt{x}}$; $\left(4, \dfrac{1}{2}\right)$

In exercises 11–20, (a) use the numerical derivative operation of a graphing utility to find the derivative of f for the given value of x (to two desired places of accuracy), (b) find an equation of the tangent line to the graph of f at the indicated point, and (c) plot the graph of f and the tangent line on the same set of axes. Use a suitable viewing rectangle.

11. $f(x) = x^3 + x + 1$; $x = 1$; $(1, 3)$

12. $f(x) = -2x^3 + 3x^2 + 2$; $x = -1$; $(-1, 7)$

13. $f(x) = x^4 - 3x^2 + 1$; $x = 2$; $(2, 5)$

14. $f(x) = -x^4 + 3x + 1$; $x = 1$; $(1, 3)$

15. $f(x) = x - \sqrt{x}$; $x = 4$; $(4, 2)$

16. $f(x) = x^{3/2} - x$; $x = 4$; $(4, 4)$

17. $f(x) = \dfrac{1}{x + 1}$; $x = 1$; $\left(1, \dfrac{1}{2}\right)$

18. $f(x) = \dfrac{x}{x + 1}$; $x = 3$; $\left(3, \dfrac{3}{4}\right)$

19. $f(x) = x\sqrt{x^2 + 1}$; $x = 2$; $(2, 2\sqrt{5})$

20. $f(x) = \dfrac{x}{\sqrt{x^2 + 1}}$; $x = 1$; $\left(1, \dfrac{\sqrt{2}}{2}\right)$

and we see that the growth in losses due to bad loans was zero at this point. At the beginning of 2001 ($t = 7$),

$$f'(7) = -2(7) + 10 = -4$$

and we conclude that the losses will be decreasing at the rate of $4 million per year.

Group projects for this chapter can be found at the Brooks/Cole Web site: http://www.brookscole.com/math/authors/tans/

CHAPTER 2 SUMMARY OF PRINCIPAL FORMULAS AND TERMS

Formulas

1. Average rate of change of f over $[x, x + h]$

or

Slope of the secant line to the graph of f through $(x, f(x))$ and $(x + h, f(x + h))$

or

Difference quotient

$$\frac{f(x + h) - f(x)}{h}$$

2. Instantaneous rate of change of f at $(x, f(x))$

or

Slope of tangent line to the graph of f at $(x, f(x))$ at x

or

Derivative of f

$$\lim_{h \to 0} \frac{f(x + h) - f(x)}{h}$$

Terms

Function	Demand function
Domain	Supply function
Independent variable	Market equilibrium
Dependent variable	Equilibrium quantity
Graph of a function	Equilibrium price
Graph of an equation	Limit of a function
Vertical-line test	Indeterminate form
Composite function	Limit of a function at infinity
Polynomial function	Right-hand limit of a function
Linear function	Left-hand limit of a function
Quadratic function	Continuity of a function at a point
Cubic function	Secant line
Rational function	Tangent line to the graph of f
Power function	Differentiable function

CHAPTER 2 REVIEW EXERCISES

1. Find the domain of each function:

a. $f(x) = \sqrt{9 - x}$ **b.** $f(x) = \dfrac{x + 3}{2x^2 - x - 3}$

2. Let $f(x) = 3x^2 + 5x - 2$. Find
a. $f(-2)$ **b.** $f(a + 2)$
c. $f(2a)$ **d.** $f(a + h)$

3. Let $y^2 = 2x + 1$.
a. Sketch the graph of this equation.
b. Is y a function of x? Why?
c. Is x a function of y? Why?

4. Sketch the graph of the function defined by

$$f(x) = \begin{cases} x + 1 & \text{if } x < 1 \\ -x^2 + 4x - 1 & \text{if } x \geq 1 \end{cases}$$

5. Let $f(x) = 1/x$ and $g(x) = 2x + 3$. Find
a. $f(x)g(x)$ **b.** $f(x)/g(x)$
c. $f(g(x))$ **d.** $g(f(x))$

In exercises 6–19, find the indicated limits, if they exist.

6. $\lim\limits_{x \to 0} (5x - 3)$ **7.** $\lim\limits_{x \to 1} (x^2 + 1)$

8. $\lim\limits_{x \to -1} (3x^2 + 4)(2x - 1)$

9. $\lim\limits_{x \to 3} \dfrac{x - 3}{x + 4}$ **10.** $\lim\limits_{x \to 2} \dfrac{x + 3}{x^2 - 9}$

11. $\lim\limits_{x \to -2} \dfrac{x^2 - 2x - 3}{x^2 + 5x + 6}$ **12.** $\lim\limits_{x \to 3} \sqrt{2x^3 - 5}$

13. $\lim\limits_{x \to 3} \dfrac{4x - 3}{\sqrt{x + 1}}$ **14.** $\lim\limits_{x \to 1^+} \dfrac{x - 1}{x(x - 1)}$

15. $\lim\limits_{x \to 1^-} \dfrac{\sqrt{x} - 1}{x - 1}$ **16.** $\lim\limits_{x \to \infty} \dfrac{x^2}{x^2 - 1}$

17. $\lim\limits_{x \to -\infty} \dfrac{x + 1}{x}$ **18.** $\lim\limits_{x \to \infty} \dfrac{3x^2 + 2x + 4}{2x^2 - 3x + 1}$

19. $\lim\limits_{x \to -\infty} \dfrac{x^2}{x + 1}$

20. Sketch the graph of the function

$$f(x) = \begin{cases} 2x - 3 & \text{if } x \leq 2 \\ -x + 3 & \text{if } x > 2 \end{cases}$$

and evaluate $\lim\limits_{x \to a^+} f(x)$, $\lim\limits_{x \to a^-} f(x)$, and $\lim\limits_{x \to a} f(x)$ at the point $a = 2$, if the limits exist.

21. Sketch the graph of the function

$$f(x) = \begin{cases} 4 - x & \text{if } x \leq 2 \\ x + 2 & \text{if } x > 2 \end{cases}$$

and evaluate $\lim\limits_{x \to a^+} f(x)$, $\lim\limits_{x \to a^-} f(x)$, and $\lim\limits_{x \to a} f(x)$ at the point $a = 2$, if the limits exist.

In exercises 22–25, determine all values of x for which each function is discontinuous.

22. $g(x) = \begin{cases} x + 3 & \text{if } x \neq 2 \\ 0 & \text{if } x = 2 \end{cases}$

23. $f(x) = \dfrac{3x + 4}{4x^2 - 2x - 2}$

24. $f(x) = \begin{cases} \dfrac{1}{(x + 1)^2} & \text{if } x \neq -1 \\ 2 & \text{if } x = -1 \end{cases}$

25. $f(x) = \dfrac{|2x|}{x}$

26. Let $y = x^2 + 2$.
a. Find the average rate of change of y with respect to x in the intervals $[1, 2]$, $[1, 1.5]$, and $[1, 1.1]$.
b. Find the (instantaneous) rate of change of y at $x = 1$.

27. Use the definition of the derivative to find the slope of the tangent line to the graph of the function $f(x) = 3x + 5$ at any point $P(x, f(x))$ on the graph.

28. Use the definition of the derivative to find the slope of the tangent line to the graph of the function $f(x) = -1/x$ at any point $P(x, f(x))$ on the graph.

29. Use the definition of the derivative to find the slope of the tangent line to the graph of the function $f(x) = \frac{3}{2}x + 5$ at the point $(-2, 2)$ and determine an equation of the tangent line.

30. Use the definition of the derivative to find the slope of the tangent line to the graph of the function $f(x) = -x^2$ at the point $(2, -4)$ and determine an equation of the tangent line.

31. The graph of the function f is shown in the accompanying figure.

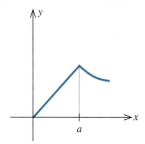

a. Is f continuous at $x = a$? Why?
b. Is f differentiable at $x = a$? Justify your answers.

32. Sales of a certain clock radio are approximated by the relationship $S(x) = 6000x + 30,000 \ (0 \le x \le 5)$, where $S(x)$ denotes the number of clock radios sold in year x ($x = 0$ corresponds to the year 1992). Find the number of clock radios expected to be sold in the year 1996.

33. A company's total sales (in millions of dollars) are approximately linear as a function of time (in years). Sales in 1991 were $2.4 million, whereas sales in 1996 amounted to $7.4 million.
a. Find an equation that gives the company's sales as a function of time.
b. What were the sales in 1994?

34. A company has a fixed cost of $30,000 and a production cost of $6 for each unit it manufactures. A unit sells for $10.
a. What is the cost function?
b. What is the revenue function?
c. What is the profit function?
d. Compute the profit (loss) corresponding to production levels of 6000, 8000, and 12,000 units, respectively.

35. Find the point of intersection of the two straight lines having the equations $y = \frac{3}{4}x + 6$ and $3x - 2y + 3 = 0$.

36. The cost and revenue functions for a certain firm are given by $C(x) = 12x + 20,000$ and $R(x) = 20x$, respectively. Find the company's break-even point.

37. Given the demand equation $3x + p - 40 = 0$ and the supply equation $2x - p + 10 = 0$, where p is the unit price in dollars and x represents the quantity in units of a thousand, determine the equilibrium quantity and the equilibrium price.

38. Clark's Rule is a method for calculating pediatric drug dosages based on a child's weight. If a denotes the adult dosage (in mg) and if w is the weight of the child (in pounds), then the child's dosage is given by

$$D(w) = \frac{aw}{150}$$

If the adult dose of a substance is 500 mg, how much should a child who weighs 35 pounds receive?

39. The monthly revenue R (in hundreds of dollars) realized in the sale of Royal electric shavers is related to the unit price p (in dollars) by the equation

$$R(p) = -\frac{1}{2}p^2 + 30p$$

Find the revenue when an electric shaver is priced at $30.

40. The membership of the newly opened Venus Health Club is approximated by the function

$$N(x) = 200(4 + x)^{1/2} \qquad (1 \le x \le 24)$$

where $N(x)$ denotes the number of members x months after the club's grand opening. Find $N(0)$ and $N(12)$ and interpret your results.

41. Psychologist L. L. Thurstone discovered the following model for the relationship between the learning time T and the length of a list n:

$$T = f(n) = An\sqrt{n - b}$$

where A and b are constants that depend on the person and the task. Suppose that for a certain person and a certain task, $A = 4$ and $b = 4$. Compute $f(4), f(5), \ldots, f(12)$ and use this information to sketch the graph of the function f. Interpret your results.

42. The monthly demand and supply functions for the Luminar desk lamp are given by

$$p = d(x) = -1.1x^2 + 1.5x + 40$$
and $$p = s(x) = 0.1x^2 + 0.5x + 15$$

respectively, where p is measured in dollars and x in units of a thousand. Find the equilibrium quantity and price.

43. The Photo-Mart transfers movie films to videocassettes. The fees charged for this service are shown in the following table. Find a function C relating the cost $C(x)$ to the number of feet x of film transferred. Sketch the graph of the function C and discuss its continuity.

Length of Film in Feet (x)	Price ($) for Conversion
$1 \le x \le 100$	5.00
$100 < x \le 200$	9.00
$200 < x \le 300$	12.50
$300 < x \le 400$	15.00
$x > 400$	$7 + 0.02x$

44. The average cost (in dollars) of producing x units of a certain commodity is given by

$$\overline{C}(x) = 20 + \frac{400}{x}$$

Evaluate $\lim_{x \to \infty} \overline{C}(x)$ and interpret your results.

 Additional study hints and sample chapter tests can be found at the Brooks/Cole Web site: http://www.brookscole.com/math/authors/tans/

This chapter gives several rules that will greatly simplify the task of finding the derivative of a function, thus enabling us to study how fast one quantity is changing with respect to another in many real-world situations. For example, we will be able to find how fast the population of an endangered species of whales grows after certain conservation measures have been implemented, how fast an economy's consumer price index (CPI) is changing at any time, and how fast the learning time changes with respect to the length of a list. We also see how these rules of differentiation facilitate the study of marginal analysis, the study of the rate of change of economic quantities. Finally, we introduce the notion of the differential of a function. Using differentials is a relatively easy way of approximating the change in one quantity due to a small change in a related quantity.

How is a pond's oxygen content affected by organic waste? In Example 7, page 193, you will see how to find the rate at which oxygen is being restored to the pond after organic waste has been dumped into it.

3

DIFFERENTIATION

3.1 BASIC RULES OF DIFFERENTIATION

Four Basic Rules

The method used in Chapter 2 for computing the derivative of a function is based on a faithful interpretation of the definition of the derivative as the limit of a quotient. Thus, to find the rule for the derivative f' of a function f, we first computed the difference quotient

$$\frac{f(x + h) - f(x)}{h}$$

and then evaluated its limit as h approached zero. As you have probably observed, this method is tedious even for relatively simple functions.

The main purpose of this chapter is to derive certain rules that will simplify the process of finding the derivative of a function. Throughout this book we will use the notation

$$\frac{d}{dx}[f(x)]$$

[read "dee, dee x of f of x"] to mean "the derivative of f with respect to x at x." In stating the rules of differentiation, we assume that the functions f and g are differentiable.

RULE 1: DERIVATIVE OF A CONSTANT $\dfrac{d}{dx}(c) = 0$ (c, a constant)

The derivative of a constant function is equal to zero.

We can see this from a geometric viewpoint by recalling that the graph of a constant function is a straight line parallel to the x-axis (Figure 3.1). Since the tangent line to a straight line at any point on the line coincides with the straight line itself, its slope (as given by the derivative of $f(x) = c$) must be zero. We can also use the definition of the derivative to prove this result by computing

$$f'(x) = \lim_{h \to 0} \frac{f(x + h) - f(x)}{h}$$

$$= \lim_{h \to 0} \frac{c - c}{h}$$

$$= \lim_{h \to 0} 0 = 0$$

Figure 3.1

The slope of the tangent line to the graph of $f(x) = c$, where c is a constant, is zero.

EXAMPLE 1

a. If $f(x) = 28$, then

$$f'(x) = \frac{d}{dx}(28) = 0$$

b. If $f(x) = -2$, then

$$f'(x) = \frac{d}{dx}(-2) = 0$$

ο ο ο

RULE 2: THE POWER RULE	If n is any real number, then $\dfrac{d}{dx}(x^n) = nx^{n-1}$.

Let us verify the Power Rule for the special case $n = 2$. If $f(x) = x^2$, then

$$f'(x) = \frac{d}{dx}(x^2) = \lim_{h \to 0} \frac{f(x+h) - f(x)}{h}$$

$$= \lim_{h \to 0} \frac{(x+h)^2 - x^2}{h}$$

$$= \lim_{h \to 0} \frac{x^2 + 2xh + h^2 - x^2}{h}$$

$$= \lim_{h \to 0} \frac{2xh + h^2}{h} = \lim_{h \to 0} (2x + h) = 2x$$

as we set out to show.

The proof of the Power Rule for the general case is not easy to prove and will be omitted. However, you will be asked to prove the rule for the special case $n = 3$ in exercise 67, page 186.

EXAMPLE 2

a. If $f(x) = x$, then

$$f'(x) = \frac{d}{dx}(x) = 1 \cdot x^{1-1} = x^0 = 1$$

b. If $f(x) = x^8$, then

$$f'(x) = \frac{d}{dx}(x^8) = 8x^7$$

c. If $f(x) = x^{5/2}$, then

$$f'(x) = \frac{d}{dx}(x^{5/2}) = \frac{5}{2}x^{3/2}$$

ο ο ο

In order to differentiate a function whose rule involves a radical, we first rewrite the rule using fractional powers. The resulting expression can then be differentiated using the Power Rule.

EXAMPLE 3   Find the derivative of the given function.

a. $f(x) = \sqrt{x}$ **b.** $g(x) = \dfrac{1}{\sqrt[3]{x}}$

Solution

a. Rewriting $\sqrt{x}$ in the form $x^{1/2}$, we obtain

$$f'(x) = \frac{d}{dx}(x^{1/2})$$

$$= \frac{1}{2}x^{-1/2} = \frac{1}{2x^{1/2}} = \frac{1}{2\sqrt{x}}$$

b. Rewriting $\dfrac{1}{\sqrt[3]{x}}$ in the form $x^{-1/3}$, we obtain

$$g'(x) = \frac{d}{dx}(x^{-1/3})$$

$$= -\frac{1}{3}x^{-4/3} = -\frac{1}{3x^{4/3}}$$ ◦ ◦ ◦

RULE 3: DERIVATIVE OF A CONSTANT MULTIPLE OF A FUNCTION	$\dfrac{d}{dx}[cf(x)] = c\dfrac{d}{dx}[f(x)]$ (*c*, a constant)

The derivative of a constant times a differentiable function is equal to the constant times the derivative of the function.

This result follows from the following computations.
    If $g(x) = cf(x)$, then

$$g'(x) = \lim_{h \to 0} \frac{g(x+h) - g(x)}{h} = \lim_{h \to 0} \frac{cf(x+h) - cf(x)}{h}$$

$$= c\lim_{h \to 0} \frac{f(x+h) - f(x)}{h}$$

$$= cf'(x)$$

EXAMPLE 4

a. If $f(x) = 5x^3$, then

$$f'(x) = \frac{d}{dx}(5x^3) = 5\frac{d}{dx}(x^3)$$
$$= 5(3x^2) = 15x^2$$

b. If $f(x) = \dfrac{3}{\sqrt{x}}$, then

$$f'(x) = \frac{d}{dx}(3x^{-1/2})$$

$$= 3\left(-\frac{1}{2}x^{-3/2}\right) = -\frac{3}{2x^{3/2}}$$

● ● ●

RULE 4: THE SUM RULE

$$\frac{d}{dx}[f(x) \pm g(x)] = \frac{d}{dx}[f(x)] \pm \frac{d}{dx}[g(x)]$$

The derivative of the sum or difference of two differentiable functions is equal to the sum or difference of their derivatives.

This result may be extended to the sum and difference of any finite number of differentiable functions. Let us verify the rule for a sum of two functions. If $s(x) = f(x) + g(x)$, then

$$s'(x) = \lim_{h \to 0} \frac{s(x + h) - s(x)}{h}$$

$$= \lim_{h \to 0} \frac{[f(x + h) + g(x + h)] - [f(x) + g(x)]}{h}$$

$$= \lim_{h \to 0} \frac{[f(x + h) - f(x)] + [g(x + h) - g(x)]}{h}$$

$$= \lim_{h \to 0} \frac{f(x + h) - f(x)}{h} + \lim_{h \to 0} \frac{g(x + h) - g(x)}{h}$$

$$= f'(x) + g'(x)$$

EXAMPLE 5   Find the derivatives of the following functions:

a. $f(x) = 4x^5 + 3x^4 - 8x^2 + x + 3$ **b.** $g(t) = \dfrac{t^2}{5} + \dfrac{5}{t^3}$

Solution

a. $f'(x) = \dfrac{d}{dx}(4x^5 + 3x^4 - 8x^2 + x + 3)$

$$= \dfrac{d}{dx}(4x^5) + \dfrac{d}{dx}(3x^4) - \dfrac{d}{dx}(8x^2) + \dfrac{d}{dx}(x) + \dfrac{d}{dx}(3)$$

$$= 20x^4 + 12x^3 - 16x + 1$$

b. Here the independent variable is t instead of x, so we differentiate with respect to t. Thus,

$$g'(t) = \dfrac{d}{dt}\left(\dfrac{1}{5}t^2 + 5t^{-3}\right) \qquad \left(\text{Rewriting } \dfrac{1}{t^3} \text{ as } t^{-3}\right)$$

$$= \dfrac{2}{5}t - 15t^{-4}$$

$$= \dfrac{2t^5 - 75}{5t^4} \qquad \left(\text{Rewriting } t^{-4} \text{ as } \dfrac{1}{t^4} \text{ and simplifying}\right) \qquad \text{◖ ◖ ◖}$$

EXAMPLE 6   Find the slope and an equation of the tangent line to the graph of $f(x) = 2x + \dfrac{1}{\sqrt{x}}$ at the point $(1, 3)$.

Solution   The slope of the tangent line at any point on the graph of f is given by

$$f'(x) = \dfrac{d}{dx}\left(2x + \dfrac{1}{\sqrt{x}}\right)$$

$$= \dfrac{d}{dx}(2x + x^{-1/2}) \qquad \left(\text{Rewriting } \dfrac{1}{\sqrt{x}} = \dfrac{1}{x^{1/2}} = x^{-1/2}\right)$$

$$= 2 - \dfrac{1}{2}x^{-3/2} \qquad \text{(Using the Sum Rule)}$$

$$= 2 - \dfrac{1}{2x^{3/2}}$$

In particular, the slope of the tangent line to the graph of f at $(1, 3)$ (where $x = 1$) is

$$f'(1) = 2 - \dfrac{1}{2(1^{3/2})} = 2 - \dfrac{1}{2} = \dfrac{3}{2}$$

Using the point-slope form of the equation of a line with slope 3/2 and the point $(1, 3)$, we see that an equation of the tangent line is

$$y - 3 = \dfrac{3}{2}(x - 1) \qquad [(y - y_1) = m(x - x_1)]$$

or, upon simplification,

$$y = \dfrac{3}{2}x + \dfrac{3}{2} \qquad \text{◖ ◖ ◖}$$

Applications

EXAMPLE 7 A group of marine biologists at the Neptune Institute of Oceanography recommended that a series of conservation measures be carried out over the next decade to save a certain species of whale from extinction. After implementing the conservation measures, the population of this species is expected to be

$$N(t) = 3t^3 + 2t^2 - 10t + 600 \qquad (0 \leq t \leq 10)$$

where $N(t)$ denotes the population at the end of year t. Find the rate of growth of the whale population when $t = 2$ and $t = 6$. How large will the whale population be eight years after implementing the conservation measures?

Solution The rate of growth of the whale population at any time t is given by

$$N'(t) = 9t^2 + 4t - 10$$

In particular, when $t = 2$ and $t = 6$, we have

$$N'(2) = 9(2)^2 + 4(2) - 10$$
$$= 34$$
$$N'(6) = 9(6)^2 + 4(6) - 10$$
$$= 338$$

so the whale population's rate of growth will be 34 whales per year after two years and 338 per year after six years.

The whale population at the end of the eighth year will be

$$N(8) = 3(8)^3 + 2(8)^2 - 10(8) + 600$$
$$= 2184$$

or 2184 whales. The graph of the function N appears in Figure 3.2. Note the rapid growth of the population in the later years, as the conservation measures begin to pay off, compared to the growth in the early years.   ○ ○ ○

Figure 3.2
The whale population after year t is given by N(t).

EXAMPLE 8 The altitude of a rocket (in feet) t seconds into flight is given by

$$s = f(t) = -t^3 + 96t^2 + 195t + 5 \qquad (t \geq 0)$$

a. Find an expression v for the rocket's velocity at any time t.

b. Compute the rocket's velocity when $t = 0, 30, 50, 65,$ and 70. Interpret your results.

c. Using the results from the solution to (b) and the observation that at the highest point in its trajectory the rocket's velocity is zero, find the maximum altitude attained by the rocket.

Solution

a. The rocket's velocity at any time t is given by

$$v = f'(t) = -3t^2 + 192t + 195$$

b. The rocket's velocity when $t = 0, 30, 50, 65,$ and 70 is given by

$$f'(0) = -3(0)^2 + 192(0) + 195 = 195$$
$$f'(30) = -3(30)^2 + 192(30) + 195 = 3255$$
$$f'(50) = -3(50)^2 + 192(50) + 195 = 2295$$
$$f'(65) = -3(65)^2 + 192(65) + 195 = 0$$
$$f'(70) = -3(70)^2 + 192(70) + 195 = -1065$$

Figure 3.3

The rocket's altitude t seconds into flight is given by f(t).

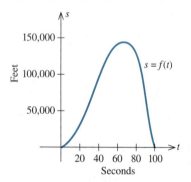

or 195, 3255, 2295, 0, and −1065 feet per second.

Thus, the rocket has an initial velocity of 195 ft/sec at $t = 0$ and accelerates to a velocity of 3255 ft/sec at $t = 30$. Fifty seconds into the flight, the rocket's velocity is 2295 ft/sec, which is less than the velocity at $t = 30$. This means that the rocket begins to decelerate after an initial period of acceleration. (Later on we will learn how to determine the rocket's maximum velocity.)

The deceleration continues: The velocity is 0 ft/sec at $t = 65$ and −1065 ft/sec when $t = 70$. This number tells us that 70 seconds into flight the rocket is heading back to Earth with a speed of 1065 ft/sec.

c. The results of (b) show that the rocket's velocity is zero when $t = 65$. At this instant, the rocket's maximum altitude is

$$s = f(65) = -(65)^3 + 96(65)^2 + 195(65) + 5$$
$$= 143,655$$

or 143,655 feet. A sketch of the graph of f appears in Figure 3.3.   ○ ○ ○

EXPLORING WITH TECHNOLOGY

Refer to Example 8.

a. Use a graphing utility to plot the graph of the velocity function

$$v = f'(t) = -3t^2 + 192t + 195$$

using the viewing rectangle $[0, 120] \times [-5000, 5000]$. Then using **ZOOM** and **TRACE** or the root-finding capability of your graphing utility, verify that $f'(65) = 0$.

b. Plot the graph of the position function of the rocket

$$s = f(t) = -t^3 + 96t^2 + 195t + 5$$

using the viewing rectangle $[0, 120] \times [0, 150,000]$. Then, using **ZOOM** and **TRACE** repeatedly, verify that the maximum altitude of the rocket is 143,655 ft.

(Continued on next page)

(Continued from previous page)
c. Use **ZOOM** and **TRACE** or the root-finding capability of your graphing utility to find when the rocket returns to Earth.

० ० ०

SELF-CHECK EXERCISES 3.1

1. Find the derivative of each of the following functions using the rules of differentiation.

a. $f(x) = 1.5x^2 + 2x^{1.5}$

b. $g(x) = 2\sqrt{x} + \dfrac{3}{\sqrt{x}}$

2. Let $f(x) = 2x^3 - 3x^2 + 2x - 1$.
a. Compute $f'(x)$.
b. What is the slope of the tangent line to the graph of f when $x = 2$?
c. What is the rate of change of the function f at $x = 2$?

3. A certain country's gross domestic product (GDP) (in millions of dollars) is described by the function

$$G(t) = -2t^3 + 45t^2 + 20t + 6000 \qquad (0 \le t \le 11)$$

where $t = 0$ corresponds to the beginning of 1986.
a. At what rate was the GDP changing at the beginning of 1991? At the beginning of 1993? At the beginning of 1996?
b. What was the average rate of growth of the GDP over the period 1991 to 1996?

Solutions to Self-Check Exercises 3.1 can be found on page 186.

3.1 EXERCISES

In exercises 1–34, find the derivative of the function f by using the rules of differentiation.

1. $f(x) = -3$

2. $f(x) = 365$

3. $f(x) = x^5$

4. $f(x) = x^7$

5. $f(x) = x^{2.1}$

6. $f(x) = x^{0.8}$

7. $f(x) = 3x^2$

8. $f(x) = -2x^3$

9. $f(r) = \pi r^2$

10. $f(r) = \frac{4}{3}\pi r^3$

11. $f(x) = 9x^{1/3}$

12. $f(x) = \frac{5}{4}x^{4/5}$

13. $f(x) = 3\sqrt{x}$

14. $f(u) = \dfrac{2}{\sqrt{u}}$

15. $f(x) = 7x^{-12}$

16. $f(x) = 0.3x^{-1.2}$

17. $f(x) = 5x^2 - 3x + 7$

18. $f(x) = x^3 - 3x^2 + 1$

19. $f(x) = -x^3 + 2x^2 - 6$

20. $f(x) = x^4 - 2x^2 + 5$

21. $f(x) = 0.03x^2 - 0.4x + 10$

22. $f(x) = 0.002x^3 - 0.05x^2 + 0.1x - 20$

23. $f(x) = \dfrac{x^3 - 4x^2 + 3}{x}$

USING TECHNOLOGY

FINDING THE RATE OF CHANGE OF A FUNCTION

We can use the numerical derivative operation of a graphing utility to obtain the value of the derivative at a given value of x. Since the derivative of a function $f(x)$ measures the rate of change of the function with respect to x, the numerical derivative operation can be used to answer questions pertaining to the rate of change of one quantity y with respect to another quantity x, where $y = f(x)$, for a specific value of x.

EXAMPLE 1 Let $y = 3t^3 + 2\sqrt{t}$.

a. Use the numerical derivative operation of a graphing utility to find how fast y is changing with respect to t when $t = 1$.

b. Verify the result of (a) using the rules of differentiation of this section.

Solution

a. Write $f(t) = 3t^3 + 2\sqrt{t}$. Using the numerical derivative operation of a graphing utility, we find that the rate of change of y with respect to t when $t = 1$ is given by $f'(1) = 10$.

b. Here $f(t) = 3t^3 + 2t^{1/2}$ and

$$f'(t) = 9t^2 + 2\left(\frac{1}{2}t^{-1/2}\right) = 9t^2 + \frac{1}{\sqrt{t}}$$

Using this result, we see that when $t = 1$, y is changing at the rate of

$$f'(1) = 9(1^2) + \frac{1}{\sqrt{1}} = 10$$

or 10 units per unit change in t, as obtained earlier.   ○ ○ ○

EXAMPLE 2 According to the U.S. Department of Energy and the Shell Development Company, a typical car's fuel economy depends on the speed it is driven and is approximated by the function

$$f(x) = 0.00000310315x^4 - 0.000455174x^3$$
$$+ 0.00287869x^2 + 1.25986x \qquad (0 \le x \le 75)$$

where x is measured in mph and $f(x)$ is measured in miles per gallon (mpg).

a. Use a graphing utility to graph the function f on the interval $[0, 75]$.

b. Find the rate of change of f when $x = 20$. When $x = 50$.

c. Interpret your results.

Source: U.S. Department of Energy and the Shell Development Company

Solution

a. The result is shown in Figure T1.

b. Using the numerical derivative operation of a graphing utility, we see that $f'(20) = 0.9280996$. The rate of change of f when $x = 50$ is given by $f'(50) = -0.314501$.

Figure T1

c. The results of part (b) tell us that when a typical car is being driven at 20 mph, its fuel economy increases at the rate of approximately 0.9 mpg per 1 mph increase in its speed. At a speed of 50 mph, its fuel economy decreases at the rate of approximately 0.3 mpg per 1 mph increase in its speed. ○ ○ ○

EXERCISES

In exercises 1–6, use the numerical derivative operation of a graphing utility to find the rate of change of $f(x)$ at the given value of x. Give your answer accurate to four decimal places.

1. $f(x) = 4x^5 - 3x^3 + 2x^2 + 1$; $x = 0.5$

2. $f(x) = -x^5 + 4x^2 + 3$; $x = 0.4$

3. $f(x) = x - 2\sqrt{x}$; $x = 3$

4. $f(x) = \dfrac{\sqrt{x} - 1}{x}$; $x = 2$

5. $f(x) = x^{1/2} - x^{1/3}$; $x = 1.2$

6. $f(x) = 2x^{5/4} + x$; $x = 2$

7. Carbon Monoxide in the Atmosphere The projected average global atmosphere concentration of carbon monoxide is approximated by the function

$$f(t) = 0.881443t^4 - 1.45533t^3 + 0.695876t^2$$
$$+ 2.87801t + 293 \qquad (0 \le t \le 4)$$

where t is measured in 40-year intervals, with $t = 0$ corresponding to the beginning of 1860 and $f(t)$ is measured in parts per million by volume.
a. Use a graphing utility to plot the graph of f in the viewing rectangle $[0, 4] \times [280, 400]$.
b. Use a graphing utility to estimate how fast the projected average global atmospheric concentration of carbon monoxide was changing at the beginning of the year 1900 ($t = 1$). At the beginning of 1990 ($t = 3.5$).
Source: "Beyond the Limits," Meadows et al.

8. Growth of HMOs Based on data compiled by the Group Health Association of America, the number of people receiving their care in an HMO (Health Maintenance Organization) from the beginning of 1984 through 1994 is approximated by the function

$$f(t) = 0.0514t^3 - 0.853t^2 + 6.8147t$$
$$+ 15.6524 \qquad (0 \le t \le 11)$$

where $f(t)$ gives the number of people in millions and t is measured in years, with $t = 0$ corresponding to the beginning of 1984.
a. Use a graphing utility to plot the graph of f in the viewing window $[0, 12] \times [0, 80]$.
b. How fast was the number of people receiving their care in an HMO growing at the beginning of 1992?
Source: Group Health Association of America

9. Home Sales According to the Greater Boston Real Estate Board—Multiple Listing Service, the average number of days a single-family home remains for sale from listing to accepted offer is approximated by the function

$$f(t) = 0.0171911t^4 - 0.662121t^3 + 6.18083t^2$$
$$- 8.97086t + 53.3357 \qquad (0 \le t \le 10)$$

where t is measured in years, with $t = 0$ corresponding to the beginning of 1984.
a. Use a graphing utility to plot the graph of f in the viewing rectangle $[0, 12] \times [0, 120]$.
b. How fast is the average number of days a single-family home remains for sale from listing to accepted offer changing at the beginning of 1984 ($t = 0$)? At the beginning of 1988 ($t = 4$)?
Source: Greater Boston Real Estate Board—Multiple Listing Service

10. Manufacturing Capacity Data obtained from the Federal Reserve shows that the annual change in manufacturing capacity between 1988 and 1994 is given by

$$f(t) = 0.0388889t^3 - 0.283333t^2$$
$$+ 0.477778t + 2.04286 \qquad (0 \le t \le 6)$$

where $f(t)$ is a percentage and t is measured in years, with $t = 0$ corresponding to the beginning of 1988.
a. Use a graphing utility to plot the graph of f in the viewing rectangle $[0, 8] \times [0, 4]$.
b. How fast was $f(t)$ changing at the beginning of 1990 ($t = 2$)? At the beginning of 1992 ($t = 4$)?
Source: Federal Reserve

24. $f(x) = \dfrac{x^3 + 2x^2 + x - 1}{x}$

25. $f(x) = 4x^4 - 3x^{5/2} + 2$

26. $f(x) = 5x^{4/3} - \frac{2}{3}x^{3/2} + x^2 - 3x + 1$

27. $f(x) = 3x^{-1} + 4x^{-2}$ **28.** $f(x) = -\frac{1}{3}(x^{-3} - x^6)$

29. $f(t) = \dfrac{4}{t^4} - \dfrac{3}{t^3} + \dfrac{2}{t}$

30. $f(x) = \dfrac{5}{x^3} - \dfrac{2}{x^2} - \dfrac{1}{x} + 200$

31. $f(x) = 2x - 5\sqrt{x}$ **32.** $f(t) = 2t^2 + \sqrt{t^3}$

33. $f(x) = \dfrac{2}{x^2} - \dfrac{3}{x^{1/3}}$ **34.** $f(x) = \dfrac{3}{x^3} + \dfrac{4}{\sqrt{x}} + 1$

35. Let $f(x) = 2x^3 - 4x$. Find:
   a. $f'(-2)$ **b.** $f'(0)$ **c.** $f'(2)$

36. Let $f(x) = 4x^{5/4} + 2x^{3/2} + x$. Find:
   a. $f'(0)$ **b.** $f'(16)$

In exercises 37–40, find the given limit by evaluating the derivative of a suitable function at an appropriate point. [Hint: Look at the definition of the derivative.]

37. $\displaystyle\lim_{h \to 0} \dfrac{(1 + h)^3 - 1}{h}$ **38.** $\displaystyle\lim_{x \to 1} \dfrac{x^5 - 1}{x - 1}$

[Hint: Let $h = x - 1$.]

39. $\displaystyle\lim_{h \to 0} \dfrac{3(2 + h)^2 - (2 + h) - 10}{h}$

40. $\displaystyle\lim_{t \to 0} \dfrac{1 - (1 + t)^2}{t(1 + t)^2}$

In exercises 41–44, find the slope and an equation of the tangent line to the graph of the function f at the specified point.

41. $f(x) = 2x^2 - 3x + 4$; $(2, 6)$

42. $f(x) = -\frac{5}{3}x^2 + 2x + 2$; $(-1, -\frac{5}{3})$

43. $f(x) = x^4 - 3x^3 + 2x^2 - x + 1$; $(1, 0)$

44. $f(x) = \sqrt{x} + \dfrac{1}{\sqrt{x}}$; $\left(4, \dfrac{5}{2}\right)$

45. Let $f(x) = x^3$.
   a. Find the point on the graph of f where the tangent line is horizontal.
   b. Sketch the graph of f and draw the horizontal tangent line.

46. Let $f(x) = x^3 - 4x^2$. Find the point(s) on the graph of f where the tangent line is horizontal.

47. Let $f(x) = x^3 + 1$.
   a. Find the point(s) on the graph of f where the slope of the tangent line is equal to 12.
   b. Find the equation(s) of the tangent line(s) of (a).
   c. Sketch the graph of f showing the tangent line(s).

48. Let $f(x) = \frac{2}{3}x^3 + x^2 - 12x + 6$. Find the values of x for which:
   a. $f'(x) = -12$ **b.** $f'(x) = 0$
   c. $f'(x) = 12$

49. Let $f(x) = \frac{1}{4}x^4 - \frac{1}{3}x^3 - x^2$. Find the point(s) on the graph of f where the slope of the tangent line is equal to:
   a. $-2x$ **b.** 0 **c.** $10x$

50. A straight line perpendicular to and passing through the point of tangency of the tangent line is called the *normal* to the curve. Find an equation of the tangent line and the normal to the curve $y = x^3 - 3x + 1$ at the point $(2, 3)$.

51. Growth of a Cancerous Tumor   The volume of a spherical cancer tumor is given by the function

$$V(r) = \frac{4}{3}\pi r^3$$

where r is the radius of the tumor in centimeters. Find the rate of change in the volume of the tumor when:
   a. $r = \frac{2}{3}$ cm   **b.** $r = \frac{5}{4}$ cm

52. Velocity of Blood in an Artery   The velocity (in centimeters per second) of blood r centimeters from the central axis of an artery is given by

$$v(r) = k(R^2 - r^2)$$

where k is a constant and R is the radius of the artery (see the accompanying figure). Suppose that $k = 1000$ and $R = 0.2$ cm. Find $v(0.1)$ and $v'(0.1)$ and interpret your results.

R

r

Blood vessel

53. Effect of Stopping on Average Speed   According to data from a study by General Motors, the average speed of your trip A (in mph) is related to the number of stops per mile you make on the trip x by the equation

$$A = \frac{26.5}{x^{0.45}}$$

Compute dA/dx for $x = 0.25$ and $x = 2$, and interpret your results.
Source: General Motors

54. Worker Efficiency   An efficiency study conducted for the Elektra Electronics Company showed that the number of "Space Commander" walkie-talkies assembled by the average worker t hours after starting work at 8 A.M. is given by

$$N(t) = -t^3 + 6t^2 + 15t$$

a. Find the rate at which the average worker will be assembling walkie-talkies t hours after starting work.
b. At what rate will the average worker be assembling walkie-talkies at 10 A.M.? At 11 A.M.?
c. How many walkie-talkies will the average worker assemble between 10 A.M. and 11 A.M.?

55. Consumer Price Index   An economy's consumer price index (CPI) is described by the function

$$I(t) = -0.2t^3 + 3t^2 + 100 \qquad (0 \le t \le 10)$$

where $t = 0$ corresponds to 1986.
a. At what rate was the CPI changing in 1991? In 1993? In 1996?
b. What was the average rate of increase in the CPI over the period from 1991 to 1996?

56. Effect of Advertising on Sales   The relationship between the amount of money x that the Cannon Precision Instruments Corporation spends on advertising and the company's total sales $S(x)$ is given by the function

$$S(x) = -0.002x^3 + 0.6x^2 + x + 500 \qquad (0 \le x \le 200)$$

where x is measured in thousands of dollars. Find the rate of change of the sales with respect to the amount of money spent on advertising. Are Cannon's total sales increasing at a faster rate when the amount of money spent on advertising is (a) $100,000 or (b) $150,000?

57. Population Growth   A study prepared for a Sunbelt town's chamber of commerce projected that the town's population in the next 3 years will grow according to the rule

$$P(t) = 50,000 + 30t^{3/2} + 20t$$

where $P(t)$ denotes the population t months from now. How fast will the population be increasing 9 months from now? Sixteen months from now?

58. Curbing Population Growth   Five years ago, the government of a Pacific island state launched an extensive propaganda campaign toward curbing the country's population growth. According to the Census Department, the population (measured in thousands of people) for the following 4 years was

$$P(t) = -\frac{1}{3}t^3 + 64t + 3000$$

where t is measured in years and $t = 0$ at the start of the campaign. Find the rate of change of the population at the end of years 1, 2, 3, and 4. Was the plan working?

59. Conservation of Species   A certain species of turtle faces extinction because dealers collect truckloads of turtle eggs to be sold as aphrodisiacs. After severe conservation measures are implemented, it is hoped that the turtle population will grow according to the rule

$$N(t) = 2t^3 + 3t^2 - 4t + 1000 \qquad (0 \le t \le 10)$$

where $N(t)$ denotes the population at the end of year t. Find the rate of growth of the turtle population when $t = 2$ and $t = 8$. What will the population be 10 years after the conservation measures are implemented?

60. Flight of a Rocket   The altitude (in feet) of a rocket t seconds into flight is given by

$$s = f(t) = -2t^3 + 114t^2 + 480t + 1 \qquad (t \ge 0)$$

a. Find an expression v for the rocket's velocity at any time t.
b. Compute the rocket's velocity when $t = 0, 20, 40,$ and 60. Interpret your results.
c. Using the results from the solution to (b), find the maximum altitude attained by the rocket.
[*Hint:* At its highest point, the velocity of the rocket is zero.]

61. Stopping Distance of a Racing Car   During a test by the editors of an auto magazine, the stopping distance s (in feet) of the MacPherson X-2 racing car conformed to the rule

$$s = f(t) = 120t - 15t^2 \qquad (t \ge 0)$$

where t was the time (in seconds) after the brakes were applied.
a. Find an expression for the car's velocity v at any time t.

b. What was the car's velocity when the brakes were first applied?

c. What was the car's stopping distance for that particular test?

[*Hint:* The stopping time is found by setting $v = 0$.]

62. Demand Functions The demand function for the Luminar desk lamp is given by

$$p = f(x) = -0.1x^2 - 0.4x + 35$$

where x is the quantity demanded (measured in thousands) and p is the unit price in dollars.

a. Find $f'(x)$.

b. What is the rate of change of the unit price when the quantity demanded is 10,000 units ($x = 10$)? What is the unit price at that level of demand?

63. Increase in Temporary Workers According to the Labor Department, the number of temporary workers (in millions) is estimated to be

$$N(t) = 0.025t^2 + 0.255t + 1.505 \qquad (0 \le t \le 5)$$

where t is measured in years, with $t = 0$ corresponding to 1991.

a. How many temporary workers were there at the beginning of 1994?

b. How fast was the number of temporary workers growing at the beginning of 1994?

Source: Labor Department

64. Supply Functions The supply function for a certain make of transistor radio is given by

$$p = f(x) = 0.1\sqrt{x} + 10$$

where x is the quantity supplied and p is the unit price in dollars.

a. Find $f'(x)$.

b. What is the rate of change of the unit price if the quantity supplied is 40,000 transistor radios?

65. Average Speed of a Vehicle on a Highway The average speed of a vehicle on a stretch of Route 134 between

6 A.M. and 10 A.M. on a typical weekday is approximated by the function

$$f(t) = 20t - 40\sqrt{t} + 50 \qquad (0 \le t \le 4)$$

where $f(t)$ is measured in miles per hour and t is measured in hours, $t = 0$ corresponding to 6 A.M.

a. Compute $f'(t)$.

b. Compute $f(0)$, $f(1)$, and $f(2)$, and interpret your results.

c. Compute $f'(\frac{1}{2})$, $f'(1)$, and $f'(2)$, and interpret your results.

66. Health-Care Spending Despite efforts at cost containment, the cost of the Medicare program is increasing at a high rate. Two major reasons for this increase are an aging population and the constant development and extensive use by physicians of new technologies. Based on data from the Health Care Financing Administration and the U.S. Census Bureau, health-care spending through the year 2000 may be approximated by the function

$$S(t) = 0.02836t^3 - 0.05167t^2 + 9.60881t$$
$$+ 41.9 \qquad (0 \le t \le 35)$$

where $S(t)$ is the spending in billions of dollars and t is measured in years, with $t = 0$ corresponding to the beginning of 1965.

a. Find an expression for the rate of change of health-care spending at any time t.

b. How fast was health-care spending changing at the beginning of 1980? How fast will health-care spending be changing at the beginning of 2000?

c. What was the amount of health-care spending at the beginning of 1980? What will the amount of health-care spending be at the beginning of 2000?

Source: Health Care Financing Administration

67. Prove the Power Rule (Rule 2) for the special case $n = 3$.

$$\left[\textit{Hint:} \text{ Compute } \lim_{h \to 0} \left[\frac{(x + h)^3 - x^3}{h} \right]. \right]$$

SOLUTIONS TO SELF-CHECK EXERCISES 3.1

I. a. $f'(x) = \dfrac{d}{dx}(1.5x^2) + \dfrac{d}{dx}(2x^{1.5})$

$$= (1.5)(2x) + (2)(1.5x^{0.5})$$

$$= 3x + 3\sqrt{x} = 3(x + \sqrt{x})$$

b. $g'(x) = \dfrac{d}{dx}(2x^{1/2}) + \dfrac{d}{dx}(3x^{-1/2})$

$\qquad = (2)\left(\dfrac{1}{2}x^{-1/2}\right) + (3)\left(-\dfrac{1}{2}x^{-3/2}\right)$

$\qquad = x^{-1/2} - \dfrac{3}{2}x^{-3/2}$

$\qquad = \dfrac{1}{2}x^{-3/2}(2x-3) = \dfrac{2x-3}{2x^{3/2}}$

2. a. $f'(x) = \dfrac{d}{dx}(2x^3) - \dfrac{d}{dx}(3x^2) + \dfrac{d}{dx}(2x) - \dfrac{d}{dx}(1)$

$\qquad = (2)(3x^2) - (3)(2x) + 2$

$\qquad = 6x^2 - 6x + 2$

b. The slope of the tangent line to the graph of f when $x = 2$ is given by

$$f'(2) = 6(2)^2 - 6(2) + 2 = 14$$

c. The rate of change of f at $x = 2$ is given by $f'(2)$. Using the results of (b), we see that $f'(2)$ is 14 units per unit change in x.

3. a. The rate at which the GDP was changing at any time t $(0 < t < 11)$ is given by

$$G'(t) = -6t^2 + 90t + 20$$

In particular, the rates of change of the GDP at the beginning of the years 1991 $(t = 5)$, 1993 $(t = 7)$, and 1996 $(t = 10)$ are given by

$$G'(5) = 320, \quad G'(7) = 356, \quad \text{and} \quad G'(10) = 320$$

respectively; that is, by $320 million per year, $356 million per year, and $320 million per year, respectively.

b. The average rate of growth of the GDP over the period from the beginning of 1991 $(t = 5)$ to the beginning of 1996 $(t = 10)$ is given by

$$\dfrac{G(10) - G(5)}{10 - 5} = \dfrac{[-2(10)^3 + 45(10)^2 + 20(10) + 6000]}{5}$$

$$- \dfrac{[-2(5)^3 + 45(5)^2 + 20(5) + 6000]}{5}$$

$$= \dfrac{8700 - 6975}{5}$$

or $345 million per year.

3.2 THE PRODUCT AND QUOTIENT RULES

In this section, we study two more rules of differentiation: the **Product Rule** and the **Quotient Rule.**

The Product Rule

The derivative of the product of two differentiable functions is given by the following rule:

RULE 5: THE PRODUCT RULE	$$\frac{d}{dx}[f(x)g(x)] = f(x)g'(x) + g(x)f'(x)$$

The derivative of the product of two functions is the first function times the derivative of the second plus the second function times the derivative of the first.

The Product Rule may be extended to the case involving the product of any finite number of functions (see exercise 59, page 197). We prove the Product Rule at the end of this section.

 The derivative of the product of two functions is not given by the product of the derivatives of the functions; that is, in general

$$\frac{d}{dx}[f(x)g(x)] \neq f'(x)g'(x)$$

EXAMPLE 1 Find the derivative of the function

$$f(x) = (2x^2 - 1)(x^3 + 3)$$

Solution By the Product Rule,

$$f'(x) = (2x^2 - 1)\frac{d}{dx}(x^3 + 3) + (x^3 + 3)\frac{d}{dx}(2x^2 - 1)$$
$$= (2x^2 - 1)(3x^2) + (x^3 + 3)(4x)$$
$$= 6x^4 - 3x^2 + 4x^4 + 12x$$
$$= 10x^4 - 3x^2 + 12x$$
$$= x(10x^3 - 3x + 12)$$

$\circ\ \circ\ \circ$

EXAMPLE 2 Differentiate (that is, find the derivative of) the function

$$f(x) = x^3(\sqrt{x} + 1)$$

Solution   First, we express the function in exponential form, obtaining

$$f(x) = x^3(x^{1/2} + 1)$$

By the Product Rule,

$$f'(x) = x^3 \frac{d}{dx}(x^{1/2} + 1) + (x^{1/2} + 1)\frac{d}{dx}x^3$$

$$= x^3\left(\frac{1}{2}x^{-1/2}\right) + (x^{1/2} + 1)(3x^2)$$

$$= \frac{1}{2}x^{5/2} + 3x^{5/2} + 3x^2$$

$$= \frac{7}{2}x^{5/2} + 3x^2$$

○ ○ ○

REMARK   We can also solve the problem by first expanding the product before differentiating f. Examples for which this is not possible will be considered in Section 3.3, where the true value of the Product Rule will be appreciated.

○ ○ ○

The Quotient Rule

The derivative of the quotient of two differentiable functions is given by the following rule:

RULE 6: THE QUOTIENT RULE

$$\frac{d}{dx}\left[\frac{f(x)}{g(x)}\right] = \frac{g(x)f'(x) - f(x)g'(x)}{[g(x)]^2} \qquad (g(x) \neq 0)$$

As an aid to remembering this expression, observe that it has the following form:

$$\frac{d}{dx}\left[\frac{f(x)}{g(x)}\right]$$

$$= \frac{(\text{Denominator})\left(\begin{array}{c}\text{Derivative of}\\ \text{numerator}\end{array}\right) - (\text{Numerator})\left(\begin{array}{c}\text{Derivative of}\\ \text{denominator}\end{array}\right)}{(\text{Square of denominator})}$$

For a proof of the Quotient Rule, see exercise 60, page 197.

 The derivative of a quotient is not equal to the quotient of the derivatives; that is,

$$\frac{d}{dx}\left[\frac{f(x)}{g(x)}\right] \neq \frac{f'(x)}{g'(x)}$$

For example, if $f(x) = x^3$ and $g(x) = x^2$, then

$$\frac{d}{dx}\left[\frac{f(x)}{g(x)}\right] = \frac{d}{dx}\left(\frac{x^3}{x^2}\right) = \frac{d}{dx}(x) = 1$$

which is not equal to

$$\frac{f'(x)}{g'(x)} = \frac{\dfrac{d}{dx}(x^3)}{\dfrac{d}{dx}(x^2)} = \frac{3x^2}{2x} = \frac{3}{2}x$$

EXAMPLE 3   Find $f'(x)$ if $f(x) = \dfrac{x}{2x - 4}$.

Solution   Using the Quotient Rule, we obtain

$$f'(x) = \frac{(2x - 4)\dfrac{d}{dx}(x) - x\dfrac{d}{dx}(2x - 4)}{(2x - 4)^2}$$

$$= \frac{(2x - 4)(1) - x(2)}{(2x - 4)^2}$$

$$= \frac{2x - 4 - 2x}{(2x - 4)^2} = -\frac{4}{(2x - 4)^2}$$ ○ ○ ○

EXAMPLE 4   Find $f'(x)$ if $f(x) = \dfrac{x^2 + 1}{x^2 - 1}$.

Solution   By the Quotient Rule,

$$f'(x) = \frac{(x^2 - 1)\dfrac{d}{dx}(x^2 + 1) - (x^2 + 1)\dfrac{d}{dx}(x^2 - 1)}{(x^2 - 1)^2}$$

$$= \frac{(x^2 - 1)(2x) - (x^2 + 1)(2x)}{(x^2 - 1)^2}$$

$$= \frac{2x^3 - 2x - 2x^3 - 2x}{(x^2 - 1)^2}$$

$$= -\frac{4x}{(x^2 - 1)^2}$$ ○ ○ ○

EXAMPLE 5   Find $h'(x)$ if

$$h(x) = \frac{\sqrt{x}}{x^2 + 1}$$

Solution   Rewrite $h(x)$ in the form $h(x) = \dfrac{x^{1/2}}{x^2 + 1}$. By the Quotient Rule, we find

$$h'(x) = \frac{(x^2 + 1)\dfrac{d}{dx}(x^{1/2}) - x^{1/2}\dfrac{d}{dx}(x^2 + 1)}{(x^2 + 1)^2}$$

$$= \frac{(x^2 + 1)\left(\dfrac{1}{2}x^{-1/2}\right) - x^{1/2}(2x)}{(x^2 + 1)^2}$$

$$= \frac{\dfrac{1}{2}x^{-1/2}(x^2 + 1 - 4x^2)}{(x^2 + 1)^2} \qquad \text{(Factoring out } \tfrac{1}{2}x^{-1/2} \text{ from the numerator)}$$

$$= \frac{1 - 3x^2}{2\sqrt{x}(x^2 + 1)^2} \qquad\qquad\qquad\qquad\qquad \bullet\ \bullet\ \bullet$$

Applications

EXAMPLE 6   The sales (in millions of dollars) of a laser disc recording of a hit movie t years from the date of release is given by

$$S(t) = \frac{5t}{t^2 + 1}$$

a. Find the rate at which the sales are changing at time t.

b. How fast are the sales changing at the time the laser discs are released $(t = 0)$? Two years from the date of release?

Solution

a. The rate at which the sales are changing at time t is given by $S'(t)$. Using the Quotient Rule, we obtain

$$S'(t) = \frac{d}{dt}\left[\frac{5t}{t^2 + 1}\right] = 5\frac{d}{dt}\left[\frac{t}{t^2 + 1}\right]$$

$$= 5\left[\frac{(t^2 + 1)(1) - t(2t)}{(t^2 + 1)^2}\right]$$

$$= 5\left[\frac{t^2 + 1 - 2t^2}{(t^2 + 1)^2}\right] = \frac{5(1 - t^2)}{(t^2 + 1)^2}$$

b. The rate at which the sales are changing at the time the laser discs are released is given by

$$S'(0) = \frac{5(1 - 0)}{(0 + 1)^2} = 5$$

That is, they are increasing at the rate of $5 million per year.

Two years from the date of release, the sales are changing at the rate of

$$S'(2) = \frac{5(1 - 4)}{(4 + 1)^2} = -\frac{3}{5} = -0.6$$

That is, they are decreasing at the rate of $600,000 per year.

The graph of the function S is shown in Figure 3.4.

Figure 3.4
After a spectacular rise, the sales begin to taper off.

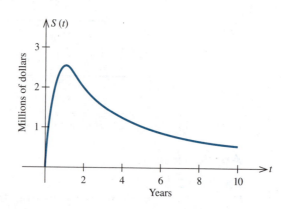

● ● ●

Exploring with Technology

Refer to Example 6.

a. Use a graphing utility to plot the graph of the function S using the viewing rectangle $[0, 10] \times [0, 3]$.

b. Use **TRACE** and **ZOOM** to determine the coordinates of the highest point on the graph of S in the interval $[0, 10]$. Interpret your results.

● ● ●

Suppose the revenue of a company is given by $R(x) = xp(x)$, where x is the number of units of the product sold at a unit price of $p(x)$ dollars.

a. Compute $R'(x)$ and explain, in words, the relationship between $R'(x)$ and $p(x)$ and/or its derivative.

b. What can you say about $R'(x)$ if $p(x)$ is constant? Is this expected?

EXAMPLE 7   When organic waste is dumped into a pond, the oxidation process that takes place reduces the pond's oxygen content. However, given time, nature will restore the oxygen content to its natural level. Suppose the oxygen content t days after organic waste has been dumped into the pond is given by

$$f(t) = 100 \left[\frac{t^2 - 4t + 4}{t^2 + 4} \right] \qquad (0 \leq t < \infty)$$

percent of its normal level.

a. Derive a general expression that gives the rate of change of the pond's oxygen level at any time t.

b. How fast is the pond's oxygen content changing one day after the organic waste has been dumped? Two days after? Three days after?

Solution

a. The rate of change of the pond's oxygen level at any time t is given by the derivative of the function f. Thus, the required expression is

$$f'(t) = 100 \frac{d}{dt} \left[\frac{t^2 - 4t + 4}{t^2 + 4} \right]$$

$$= 100 \left[\frac{(t^2 + 4)(2t - 4) - (t^2 - 4t + 4)(2t)}{(t^2 + 4)^2} \right]$$

$$= 100 \left[\frac{2t^3 - 4t^2 + 8t - 16 - 2t^3 + 8t^2 - 8t}{(t^2 + 4)^2} \right]$$

$$= \frac{100(4t^2 - 16)}{(t^2 + 4)^2} = \frac{400(t^2 - 4)}{(t^2 + 4)^2}$$

b. The rate at which the pond's oxygen content is changing one day after the organic waste has been dumped is given by

$$f'(1) = \frac{400(1 - 4)}{(1 + 4)^2} = -48$$

That is, it is dropping at the rate of 48% per day. Two days after, the rate is

$$f'(2) = \frac{400(4 - 4)}{(4 + 4)^2} = 0$$

That is, it is neither increasing nor decreasing. Three days after, the rate is

$$f'(3) = \frac{400(3^2 - 4)}{(3^2 + 4)^2} = 11.83$$

That is, the oxygen content is increasing at the rate of 11.83% per day, and the restoration process has indeed begun.   ⊙ ⊙ ⊙

Consider a particle moving along a straight line. Newton's second law of motion states that the external force F acting on the particle is equal to the rate of change of its momentum. Thus,

$$F = \frac{d}{dt}(mv)$$

where m, the mass of the particle, and v, its velocity, are both functions of time t.

a. Use the Product Rule to show that

$$F = m\frac{dv}{dt} + v\frac{dm}{dt}$$

and explain the expression on the right-hand side in words.

b. Use the results of (a) to show that if the mass of a particle is constant, then $F = ma$, where a is the acceleration of the particle.

Verification of the Product Rule

We will now verify the Product Rule. If $p(x) = f(x)g(x)$, then

$$p'(x) = \lim_{h \to 0} \frac{p(x+h) - p(x)}{h}$$

$$= \lim_{h \to 0} \frac{f(x+h)g(x+h) - f(x)g(x)}{h}$$

By adding $-f(x+h)g(x) + f(x+h)g(x)$ (which is zero!) to the numerator and factoring, we have

$$p'(x) = \lim_{h \to 0} \frac{f(x+h)[g(x+h) - g(x)] + g(x)[f(x+h) - f(x)]}{h}$$

$$= \lim_{h \to 0} \left\{ f(x+h)\left[\frac{g(x+h) - g(x)}{h}\right] + g(x)\left[\frac{f(x+h) - f(x)}{h}\right] \right\}$$

$$= \lim_{h \to 0} f(x+h)\left[\frac{g(x+h) - g(x)}{h}\right]$$

$$+ \lim_{h \to 0} g(x)\left[\frac{f(x+h) - f(x)}{h}\right] \qquad \text{(By property (iii) of limits)}$$

$$= \lim_{h \to 0} f(x+h) \cdot \lim_{h \to 0} \frac{g(x+h) - g(x)}{h}$$

$$+ \lim_{h \to 0} g(x) \cdot \lim_{h \to 0} \frac{f(x+h) - f(x)}{h} \qquad \text{(By property (iv) of limits)}$$

$$= f(x)g'(x) + g(x)f'(x)$$

Observe that in the second from the last link in the chain of equalities, we have used the fact that $\lim_{h \to 0} f(x + h) = f(x)$ because f is continuous at x.

SELF–CHECK EXERCISES 3.2

1. Find the derivative of $f(x) = \dfrac{2x + 1}{x^2 - 1}$.

2. What is the slope of the tangent line to the graph of

$$f(x) = (x^2 + 1)(2x^3 - 3x^2 + 1)$$

at the point $(2, 25)$? How fast is the function f changing when $x = 2$?

3. The total sales of the Security Products Corporation in its first 2 years of operation are given by

$$S = f(t) = \frac{0.3t^3}{1 + 0.4t^2} \qquad (0 \le t \le 2)$$

where S is measured in millions of dollars and $t = 0$ corresponds to the date Security Products began operations. How fast were the sales increasing at the beginning of the company's second year of operation?

Solutions to Self-Check Exercises 3.2 can be found on page 200.

3.2 EXERCISES

In exercises 1–30, find the derivative of the given function.

1. $f(x) = 2x(x^2 + 1)$

2. $f(x) = 3x^2(x - 1)$

3. $f(t) = (t - 1)(2t + 1)$

4. $f(x) = (2x + 3)(3x - 4)$

5. $f(x) = (3x + 1)(x^2 - 2)$

6. $f(x) = (x + 1)(2x^2 - 3x + 1)$

7. $f(x) = (x^3 - 1)(x + 1)$

8. $f(x) = (x^3 - 12x)(3x^2 + 2x)$

9. $f(w) = (w^3 - w^2 + w - 1)(w^2 + 2)$

10. $f(x) = \dfrac{1}{5}x^5 + (x^2 + 1)(x^2 - x - 1) + 28$

11. $f(x) = (5x^2 + 1)(2\sqrt{x} - 1)$

12. $f(t) = (1 + \sqrt{t})(2t^2 - 3)$

13. $f(x) = (x^2 - 5x + 2)\left(x - \dfrac{2}{x}\right)$

14. $f(x) = (x^3 + 2x + 1)\left(2 + \dfrac{1}{x^2}\right)$

15. $f(x) = \dfrac{1}{x - 2}$

16. $g(x) = \dfrac{3}{2x + 4}$

17. $f(x) = \dfrac{x - 1}{2x + 1}$

18. $f(t) = \dfrac{1 - 2t}{1 + 3t}$

19. $f(x) = \dfrac{1}{x^2 + 1}$

20. $f(u) = \dfrac{u}{u^2 + 1}$

21. $f(s) = \dfrac{s^2 - 4}{s + 1}$

22. $f(x) = \dfrac{x^3 - 2}{x^2 + 1}$

23. $f(x) = \dfrac{\sqrt{x}}{x^2 + 1}$

24. $f(x) = \dfrac{x^2 + 1}{\sqrt{x}}$

25. $f(x) = \dfrac{x^2 + 2}{x^2 + x + 1}$

26. $f(x) = \dfrac{x + 1}{2x^2 + 2x + 3}$

27. $f(x) = \dfrac{(x + 1)(x^2 + 1)}{x - 2}$

28. $f(x) = (3x^2 - 1)\left(x^2 - \dfrac{1}{x}\right)$

29. $f(x) = \dfrac{x}{x^2 - 4} - \dfrac{x - 1}{x^2 + 4}$

30. $f(x) = \dfrac{x + \sqrt{3x}}{3x - 1}$

In exercises 31–34, suppose f and g are functions that are differentiable at x = 1 and that f(1) = 2, f'(1) = −1, g(1) = −2, g'(1) = 3. Find the value of h'(1).

31. $h(x) = f(x)g(x)$ **32.** $h(x) = (x^2 + 1)g(x)$

33. $h(x) = \dfrac{xf(x)}{x + g(x)}$ **34.** $h(x) = \dfrac{f(x)g(x)}{f(x) - g(x)}$

In exercises 35–38, find the derivative of each of the given functions and evaluate f'(x) at the given value of x.

35. $f(x) = (2x - 1)(x^2 + 3); \; x = 1$

36. $f(x) = \dfrac{2x + 1}{2x - 1}; \; x = 2$

37. $f(x) = \dfrac{x}{x^4 - 2x^2 - 1}; \; x = -1$

38. $f(x) = (\sqrt{x} + 2x)(x^{3/2} - x); \; x = 4$

In exercises 39–42, find the slope and an equation of the tangent line to the graph of the function f at the specified point.

39. $f(x) = (x^3 + 1)(x^2 - 2); \; (2, 18)$

40. $f(x) = \dfrac{x^2}{x + 1}; \left(2, \dfrac{4}{3}\right)$

41. $f(x) = \dfrac{x + 1}{x^2 + 1}; \; (1, 1)$

42. $f(x) = \dfrac{1 + 2x^{1/2}}{1 + x^{3/2}}; \left(4, \dfrac{5}{9}\right)$

43. Find an equation of the tangent line to the graph of the function $f(x) = (x^3 + 1)(3x^2 - 4x + 2)$ at the point $(1, 2)$.

44. Find an equation of the tangent line to the graph of the function $f(x) = \dfrac{3x}{x^2 - 2}$ at the point $(2, 3)$.

45. Let $f(x) = (x^2 + 1)(2 - x)$. Fnd the point(s) on the graph of f where the tangent line is horizontal.

46. Let $f(x) = \dfrac{x}{x^2 + 1}$. Find the point(s) on the graph of f where the tangent line is horizontal.

47. Find the point(s) on the graph of the function $f(x) =$ $(x^2 + 6)(x - 5)$ where the slope of the tangent line is equal to -2.

48. Find the point(s) on the graph of the function $f(x) = \dfrac{x + 1}{x - 1}$ where the slope of the tangent line is equal to $-1/2$.

49. A straight line perpendicular to and passing through the point of tangency of the tangent line is called the *normal* to the curve. Find the equation of the tangent line and the normal to the curve $y = \dfrac{1}{1 + x^2}$ at the point $(1, \frac{1}{2})$.

50. Concentration of a Drug in the Bloodstream The concentration of a certain drug in a patient's bloodstream t hours after injection is given by

$$C(t) = \dfrac{0.2t}{t^2 + 1}$$

a. Find the rate at which the concentration of the drug is changing with respect to time.
b. How fast is the concentration changing $\frac{1}{2}$ hour after the injection? 1 hour after the injection? 2 hours after the injection?

51. Cost of Removing Toxic Waste A city's main well was recently found to be contaminated with trichloroethylene, a cancer-causing chemical, as a result of an abandoned chemical dump leaching chemicals into the water. A proposal submitted to the city's selectmen indicates that the cost, measured in millions of dollars, of removing x percent of the toxic pollutant is given by

$$C(x) = \dfrac{0.5x}{100 - x}$$

Find $C'(80)$, $C'(90)$, $C'(95)$, and $C'(99)$ and interpret your results.

52. Drug Dosages Thomas Young has suggested the following rule for calculating the dosage of medicine for children 1 to 12 years old. If a denotes the adult dosage (in mg), and if t is the child's age (in years), then the child's dosage is given by

$$D(t) = \dfrac{at}{t + 12}$$

Suppose that the adult dosage of a substance is 500 mg. Find an expression that gives the rate of change of a child's dosage with respect to the child's age. What is the rate of change of a child's dosage with respect to his or her age for a 6-year-old child? For a 10-year-old child?

53. Effect of Bactericide The number of bacteria $N(t)$ in a certain culture t minutes after an experimental bactericide is introduced obeys the rule

$$N(t) = \frac{10{,}000}{1 + t^2} + 2000$$

Find the rate of change of the number of bacteria in the culture 1 minute after and 2 minutes after the bactericide is introduced. What is the population of the bacteria in the culture 1 minute after the bactericide is introduced? 2 minutes after?

54. Demand Functions The demand function for the Sicard wristwatch is given by

$$d(x) = \frac{50}{0.01x^2 + 1} \qquad (0 \le x \le 20)$$

where x (measured in units of a thousand) is the quantity demanded per week and $d(x)$ is the unit price in dollars.
a. Find $d'(x)$.
b. Find $d'(5)$, $d'(10)$, and $d'(15)$ and interpret your results.

55. Learning Curves From experience, the Emory Secretarial School knows that the average student taking Advanced Typing will progress according to the rule

$$N(t) = \frac{60t + 180}{t + 6} \qquad (t \ge 0)$$

where $N(t)$ measures the number of words per minute the student can type after t weeks in the course.
a. Find an expression for $N'(t)$.
b. Compute $N'(t)$ for $t = 1, 3, 4,$ and 7 and interpret your results.
c. Sketch the graph of the function N. Does it confirm the results obtained in (b)?
d. What will the average student's typing speed be at the end of the 12-week course?

56. Box Office Receipts The total worldwide box office receipts for a long-running movie are approximated by the function

$$T(x) = \frac{120x^2}{x^2 + 4}$$

where $T(x)$ is measured in millions of dollars and x is the number of years since the movie's release. How fast are the total receipts changing one year after the movie's release? Three years after its release? Five years after its release?

57. Formaldehyde Levels A study on formaldehyde levels in 900 homes indicates that emissions of various chemicals can decrease over time. The formaldehyde level (parts per million) in an average home in the study is given by

$$f(t) = \frac{0.055t + 0.26}{t + 2} \qquad (0 \le t \le 12)$$

where t is the age of the house in years. How fast is the formaldehyde level of the average house dropping when it is new? At the beginning of its fourth year?
Source: Bonneville Power Administration

58. Population Growth A major corporation is building a 4325-acre complex of homes, offices, stores, schools, and churches in the rural community of Glen Cove. As a result of this development, the planners have estimated that Glen Cove's population (in thousands) t years from now will be given by

$$P(t) = \frac{25t^2 + 125t + 200}{t^2 + 5t + 40}$$

a. Find the rate at which Glen Cove's population is changing with respect to time.
b. What will the population be after 10 years? At what rate will the population be increasing when $t = 10$?

59. Extend the Product Rule for differentiation to the following case involving the product of three differentiable functions: Let $h(x) = u(x)v(x)w(x)$ and show that $h'(x) = u(x)v(x)w'(x) + u(x)v'(x)w(x) + u'(x)v(x)w(x)$.
[*Hint:* Let $f(x) = u(x)v(x)$, $g(x) = w(x)$, and $h(x) = f(x)g(x)$, and apply the Product Rule to the function h.]

60. Prove the Quotient Rule for differentiation (Rule 6).
Hint: Verify the following steps:
a. $\dfrac{k(x+h) - k(x)}{h} = \dfrac{f(x+h)g(x) - f(x)g(x+h)}{hg(x+h)g(x)}$
b. By adding $[-f(x)g(x) + f(x)g(x)]$ to the numerator and simplifying,

$$\frac{k(x+h) - k(x)}{h} = \frac{1}{g(x+h)g(x)}$$
$$\times \left\{ \left[\frac{f(x+h) - f(x)}{h} \right] \cdot g(x) \right.$$
$$\left. - \left[\frac{g(x+h) - g(x)}{h} \right] \cdot f(x) \right\}$$

c. $k'(x) = \lim\limits_{h \to 0} \dfrac{k(x+h) - k(x)}{h}$
$$= \frac{g(x)f'(x) - f(x)g'(x)}{[g(x)]^2}$$

USING TECHNOLOGY

THE PRODUCT AND QUOTIENT RULES

EXAMPLE 1 Let $f(x) = (2\sqrt{x} + 0.5x)(0.3x^3 + 2x - 0.3/x)$. Find $f'(0.2)$.

Solution Using the numerical derivative operation of a graphing utility, we find

$$f'(0.2) = 6.47974948127$$

EXAMPLE 2 **Importance of Time in Treating Heart Attacks**

According to the American Heart Association, the treatment benefit for heart attacks depends on the time to treatment and is described by the function

$$f(t) = \frac{-16.94t + 203.28}{t + 2.0328} \qquad (0 \le t \le 12)$$

where t is measured in hours and $f(t)$ is a percentage.

a. Use a graphing utility to graph the function f using the viewing rectangle $[0, 13] \times [0, 120]$.

b. Use a graphing utility to find the derivative of f when $t = 0$. When $t = 2$.

c. Interpret the results obtained in part (b).
Source: American Heart Association

Solution

a. The graph of f is shown in Figure T1.

Figure T1

b. Using the numerical derivative operation of a graphing utility, we find

$$f'(0) \approx -57.5266$$

and

$$f'(2) \approx -14.6165$$

c. The results of (b) show that the treatment benefit drops off at the rate of 58% per hour at the time when the heart attack first occurs and falls off at the rate of 15% per hour when the time to treatment is 2 hours. Thus, it is extremely urgent that a patient suffering a heart attack receive medical attention as soon as possible.   ○ ○ ○

EXERCISES

In exercises 1–6, use the numerical derivative operation of a graphing utility to find the rate of change of f(x) at the given value of x. Give your answer accurate to four decimal places.

1. $f(x) = (2x^2 + 1)(x^3 + 3x + 4); x = -0.5$

2. $f(x) = (\sqrt{x} + 1)(2x^2 + x - 3); x = 1.5$

3. $f(x) = \dfrac{\sqrt{x} - 1}{\sqrt{x} + 1}; x = 3$

4. $f(x) = \dfrac{\sqrt{x}(x^2 + 4)}{x^3 + 1}; x = 4$

5. $f(x) = \dfrac{\sqrt{x}(1 + x^{-1})}{x + 1}; x = 1$

6. $f(x) = \dfrac{x^2(2 + \sqrt{x})}{1 + \sqrt{x}}; x = 1$

7. New Construction Jobs The president of a major housing construction company claims that the number of construction jobs created in the next t months is given by

$$f(t) = 1.42 \left(\frac{7t^2 + 140t + 700}{3t^2 + 80t + 550} \right)$$

where $f(t)$ is measured in millions of jobs per year. At what rate will construction jobs be created 1 year from now, assuming her projection is correct?

8. Population Growth A major corporation is building a 4325-acre complex of homes, offices, stores, schools, and churches in the rural community of Glen Cove. As a result of this development, the planners have estimated that Glen Cove's population (in thousands) t years from now will be given by

$$P(t) = \frac{25t^2 + 125t + 200}{t^2 + 5t + 40}$$

a. What will the population be 10 years from now?
b. At what rate will the population be increasing 10 years from now?

SOLUTIONS TO SELF-CHECK EXERCISES 3.2

1. We use the Quotient Rule to obtain

$$f'(x) = \frac{(x^2 - 1)\dfrac{d}{dx}(2x + 1) - (2x + 1)\dfrac{d}{dx}(x^2 - 1)}{(x^2 - 1)^2}$$

$$= \frac{(x^2 - 1)(2) - (2x + 1)(2x)}{(x^2 - 1)^2}$$

$$= \frac{2x^2 - 2 - 4x^2 - 2x}{(x^2 - 1)^2}$$

$$= \frac{-2x^2 - 2x - 2}{(x^2 - 1)^2}$$

$$= -\frac{2(x^2 + x + 1)}{(x^2 - 1)^2}$$

2. The slope of the tangent line to the graph of f at any point is given by

$$f'(x) = (x^2 + 1)\frac{d}{dx}(2x^3 - 3x^2 + 1)$$

$$+ (2x^3 - 3x^2 + 1)\frac{d}{dx}(x^2 + 1)$$

$$= (x^2 + 1)(6x^2 - 6x) + (2x^3 - 3x^2 + 1)(2x)$$

In particular, the slope of the tangent line to the graph of f when $x = 2$ is

$$f'(2) = (2^2 + 1)[6(2^2) - 6(2)]$$
$$+ [2(2^3) - 3(2^2) + 1][2(2)]$$
$$= 60 + 20 = 80$$

Note that it is not necessary to simplify the expression for $f'(x)$ since we are required only to evaluate the expression at $x = 2$. We also conclude, from this result, that the function f is changing at the rate of 80 units per unit change in x when $x = 2$.

3. The rate at which the company's total sales are changing at any time t is given by

$$S'(t) = \frac{(1 + 0.4t^2)\dfrac{d}{dt}(0.3t^3) - (0.3t^3)\dfrac{d}{dt}(1 + 0.4t^2)}{(1 + 0.4t^2)^2}$$

$$= \frac{(1 + 0.4t^2)(0.9t^2) - (0.3t^3)(0.8t)}{(1 + 0.4t^2)^2}$$

Therefore, at the beginning of the second year of operation, Security Products' sales were increasing at the rate of

$$S'(1) = \frac{(1 + 0.4)(0.9) - (0.3)(0.8)}{(1 + 0.4)^2} = 0.520408$$

or $520,408 per year.

THE CHAIN RULE

This section introduces another rule of differentiation called the **Chain Rule.** When used in conjunction with the rules of differentiation developed in the last two sections, the Chain Rule enables us to greatly enlarge the class of functions we are able to differentiate.

The Chain Rule

Consider the function $h(x) = (x^2 + x + 1)^2$. If we were to compute $h'(x)$ using only the rules of differentiation from the previous sections, then our approach might be to expand $h(x)$. Thus,

$$h(x) = (x^2 + x + 1)^2 = (x^2 + x + 1)(x^2 + x + 1)$$
$$= x^4 + 2x^3 + 3x^2 + 2x + 1$$

from which we find

$$h'(x) = 4x^3 + 6x^2 + 6x + 2$$

But what about the function $H(x) = (x^2 + x + 1)^{100}$? The same technique may be used to find the derivative of the function H, but the amount of work involved in this case would be prodigious! Consider, also, the function $G(x) = \sqrt{x^2 + 1}$. For each of the two functions H and G, the rules of differentiation of the previous sections cannot be applied directly to compute the derivatives H' and G'.

Observe that both H and G are **composite** functions; that is, each is composed of, or built up from, simpler functions. For example, the function H is composed of the two simpler functions $f(x) = x^2 + x + 1$ and $g(x) = x^{100}$ as follows:

$$H(x) = g[f(x)] = [f(x)]^{100}$$
$$= (x^2 + x + 1)^{100}$$

In a similar manner, we see that the function G is composed of the two simpler functions $f(x) = x^2 + 1$ and $g(x) = \sqrt{x}$. Thus,

$$G(x) = g[f(x)] = \sqrt{f(x)}$$
$$= \sqrt{x^2 + 1}$$

As a first step toward finding the derivative h' of a composite function $h = g \circ f$ defined by $h(x) = g[f(x)]$, we write

$$u = f(x) \quad \text{and} \quad y = g[f(x)] = g(u)$$

The dependency of h on g and f is illustrated in Figure 3.5. Since u is a function of x, we may compute the derivative of u with respect to x, if f is a differentiable function, obtaining $du/dx = f'(x)$. Next, if g is a differentiable function of u, we may compute the derivative of g with respect to u, obtaining $dy/du = g'(u)$. Now, since the function h is composed of the function g and the function

Figure 3.5
The composite function h(x) =
g[f(x)].

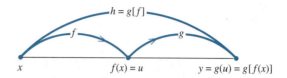

f, we might suspect that the rule $h'(x)$ for the derivative h' of h will be given by an expression that involves the rules for the derivatives of f and g. But how do we combine these derivatives to yield h'?

This question can be answered by interpreting the derivative of each function as giving the rate of change of that function. For example, suppose that $u = f(x)$ changes three times as fast as x—that is,

$$f'(x) = \frac{du}{dx} = 3$$

And suppose that $y = g(u)$ changes twice as fast as u—that is,

$$g'(u) = \frac{dy}{du} = 2$$

Then we would expect $y = h(x)$ to change six times as fast as x—that is,
$$h'(x) = g'(u)f'(x) = (2)(3) = 6$$
or equivalently,

$$\frac{dy}{dx} = \frac{dy}{du} \cdot \frac{du}{dx} = (2)(3) = 6$$

This observation suggests the following result, which we state without proof.

RULE 7: THE CHAIN RULE

If $h(x) = g[f(x)]$, then

$$h'(x) = \frac{d}{dx} g(f(x)) = g'(f(x))f'(x) \tag{1}$$

Equivalently, if we write $y = h(x) = g(u)$, where $u = f(x)$, then

$$\frac{dy}{dx} = \frac{dy}{du} \cdot \frac{du}{dx} \tag{2}$$

REMARKS

1. If we label the composite function h in the following manner

inside function
↓
$$h(x) = g[f(x)]$$
↑
outside function

then $h'(x)$ is just the *derivative* of the "outside function" *evaluated at the* "inside function" times the *derivative* of the "inside function."

2. Equation (2) can be remembered by observing that if we "cancel" the du's, then

$$\frac{dy}{dx} = \frac{dy}{du} \cdot \frac{du}{dx} = \frac{dy}{dx}$$

⊙ ⊙ ⊙

The Chain Rule for Powers of Functions

Many composite functions have the special form $h(x) = g(f(x))$, where g is defined by the rule $g(x) = x^n$ (n, a real number)—that is,

$$h(x) = [f(x)]^n$$

In other words, the function h is given by the power of a function f. The functions

$$h(x) = (x^2 + x + 1)^2, \quad H = (x^2 + x + 1)^{100}, \quad \text{and} \quad G = \sqrt{x^2 + 1}$$

discussed earlier are examples of this type of composite function. By using the following corollary of the Chain Rule, the General Power Rule, we are able to find the derivative of this type of function much more easily than by using the Chain Rule directly.

THE GENERAL POWER RULE

If the function f is differentiable and $h(x) = [f(x)]^n$ (n, a real number), then

$$h'(x) = \frac{d}{dx}[f(x)]^n = n[f(x)]^{n-1}f'(x) \tag{3}$$

To see this, we observe that $h(x) = g(f(x))$, where $g(x) = x^n$, so that, by virtue of the Chain Rule, we have

$$h'(x) = g'(f(x))f'(x)$$
$$= n[f(x)]^{n-1}f'(x)$$

since $g'(x) = nx^{n-1}$.

EXAMPLE 1   Find $H'(x)$, if

$$H(x) = (x^2 + x + 1)^{100}$$

Solution   Using the General Power Rule, we obtain

$$H'(x) = 100(x^2 + x + 1)^{99} \frac{d}{dx}(x^2 + x + 1)$$

$$= 100(x^2 + x + 1)^{99}(2x + 1)$$

Observe the bonus that comes from using the Chain Rule: The answer is completely factored.

⊙ ⊙ ⊙

EXAMPLE 2   Differentiate the function $G(x) = \sqrt{x^2 + 1}$.

Solution   We rewrite the function $G(x)$ as

$$G(x) = (x^2 + 1)^{1/2}$$

and apply the General Power Rule, obtaining

$$G'(x) = \frac{1}{2}(x^2 + 1)^{-1/2}\frac{d}{dx}(x^2 + 1)$$

$$= \frac{1}{2}(x^2 + 1)^{-1/2} \cdot 2x = \frac{x}{\sqrt{x^2 + 1}} \qquad \text{o o o}$$

EXAMPLE 3   Differentiate the function $f(x) = x^2(2x + 3)^5$.

Solution   Applying the Product Rule followed by the General Power Rule, we obtain

$$f'(x) = x^2\frac{d}{dx}(2x + 3)^5 + (2x + 3)^5\frac{d}{dx}(x^2)$$

$$= (x^2)5(2x + 3)^4 \cdot \frac{d}{dx}(2x + 3) + (2x + 3)^5(2x)$$

$$= 5x^2(2x + 3)^4(2) + 2x(2x + 3)^5$$

$$= 2x(2x + 3)^4(5x + 2x + 3) = 2x(7x + 3)(2x + 3)^4 \qquad \text{o o o}$$

EXAMPLE 4   Find $f'(x)$ if $f(x) = \dfrac{1}{(4x^2 - 7)^2}$.

Solution   Rewriting $f(x)$ and then applying the General Power Rule, we obtain

$$f'(x) = \frac{d}{dx}\left[\frac{1}{(4x^2 - 7)^2}\right] = \frac{d}{dx}(4x^2 - 7)^{-2}$$

$$= -2(4x^2 - 7)^{-3}\frac{d}{dx}(4x^2 - 7)$$

$$= -2(4x^2 - 7)^{-3}(8x) = -\frac{16x}{(4x^2 - 7)^3} \qquad \text{o o o}$$

EXAMPLE 5   Find the slope of the tangent line to the graph of the function

$$f(x) = \left(\frac{2x + 1}{3x + 2}\right)^3$$

at the point $(0, \frac{1}{8})$.

Solution   The slope of the tangent line to the graph of f at any point is given by $f'(x)$. To compute $f'(x)$, we use the General Power Rule followed by the Quotient Rule, obtaining

$$f'(x) = 3\left(\frac{2x+1}{3x+2}\right)^2 \frac{d}{dx}\left(\frac{2x+1}{3x+2}\right)$$

$$= 3\left(\frac{2x+1}{3x+2}\right)^2\left[\frac{(3x+2)(2) - (2x+1)(3)}{(3x+2)^2}\right]$$

$$= 3\left(\frac{2x+1}{3x+2}\right)^2\left[\frac{6x+4-6x-3}{(3x+2)^2}\right]$$

$$= \frac{3(2x+1)^2}{(3x+2)^4}$$

In particular, the slope of the tangent line to the graph of f at $(0, \frac{1}{8})$ is given by

$$f'(0) = \frac{3(0+1)^2}{(0+2)^4} = \frac{3}{16}$$

○ ○ ○

EXPLORING WITH TECHNOLOGY

Refer to Example 5.

a. Use a graphing utility to plot the graph of the function f using the viewing rectangle $[-2, 1] \times [-1, 2]$. Then draw the tangent line to the graph of f at the point $(0, \frac{1}{8})$.

b. For a better picture, repeat part (a) using the viewing rectangle $[-1, 1] \times [-0.1, 0.3]$.

c. Use the numerical differentiation capability of the graphing utility to verify that the slope of the tangent line at $(0, \frac{1}{8})$ is $\frac{3}{16}$.

○ ○ ○

Applications

EXAMPLE 6 The membership of The Fitness Center, which opened a few years ago, is approximated by the function

$$N(t) = 100(64 + 4t)^{2/3} \qquad (0 \le t \le 52)$$

where $N(t)$ gives the number of members at the beginning of week t.

a. Find $N'(t)$.

b. How fast was the center's membership increasing initially $(t = 0)$?

c. How fast was the membership increasing at the beginning of the 40th week?

d. What was the membership when the center first opened? At the beginning of the 40th week?

Solution

a. Using the General Power Rule, we obtain

$$N'(t) = \frac{d}{dt}[100(64 + 4t)^{2/3}]$$

$$= 100\frac{d}{dt}(64 + 4t)^{2/3}$$

$$= 100\left(\frac{2}{3}\right)(64 + 4t)^{-1/3}\frac{d}{dt}(64 + 4t)$$

$$= \frac{200}{3}(64 + 4t)^{-1/3}(4)$$

$$= \frac{800}{3(64 + 4t)^{1/3}}$$

b. The rate at which the membership was increasing when the center first opened is given by

$$N'(0) = \frac{800}{3(64)^{1/3}} \approx 66.7$$

or approximately 67 people per week.

c. The rate at which the membership was increasing at the beginning of the 40th week is given by

$$N'(40) = \frac{800}{3(64 + 160)^{1/3}} \approx 43.9$$

or approximately 44 people per week.

d. The membership when the center first opened is given by

$$N(0) = 100(64)^{2/3} = 100(16)$$

or approximately 1600 people. The membership at the beginning of the 40th week is given by

$$N(40) = 100(64 + 160)^{2/3} \approx 3688.3$$

or approximately 3688 people.   ○ ○ ○

The profit P of a one-product software manufacturer depends on the number of units of its products sold. The manufacturer estimates that it will sell x units of its product per week. Suppose $P = g(x)$ and $x = f(t)$, where g and f are differentiable functions.
a. Write an expression giving the rate of change of the profit with respect to the number of units sold.
b. Write an expression giving the rate of change of the number of units sold per week.
c. Write an expression giving the rate of change of the profit per week.

EXAMPLE 7 Arteriosclerosis begins during childhood when plaque (soft masses of fatty material) forms in the arterial walls, blocking the flow of blood through the arteries and leading to heart attacks, strokes, and gangrene. Suppose the idealized cross section of the aorta is circular with radius a cm, and that by year t the thickness of the plaque (assume that it is uniform) is $h = f(t)$ cm (Figure 3.6). Then the area of the opening is given by $A = \pi(a - h)^2$ square centimeters.

Figure 3.6
Cross section of the aorta.

Suppose that the radius of an individual's artery is 1 cm ($a = 1$) and the thickness of the plaque in year t is given by

$$h = g(t) = 1 - 0.01(10{,}000 - t^2)^{1/2} \text{ cm}$$

Since the area of the arterial opening is given by

$$A = f(h) = \pi(1 - h)^2$$

the rate at which A is changing with respect to time is given by

$$\frac{dA}{dt} = \frac{dA}{dh} \cdot \frac{dh}{dt} = f'(h) \cdot g'(t) \qquad \text{(By the Chain Rule)}$$

$$= 2\pi(1 - h)(-1)[-0.01(\tfrac{1}{2})(10{,}000 - t^2)^{-1/2}(-2t)] \qquad \text{(Using the Chain Rule twice)}$$

$$= -2\pi(1 - h)\left[\frac{0.01t}{(10{,}000 - t^2)^{1/2}}\right]$$

$$= -\frac{0.02\pi(1 - h)t}{\sqrt{10{,}000 - t^2}}$$

For example, when $t = 50$,

$$h = g(50) = 1 - 0.01(10{,}000 - 2{,}500)^{1/2} \approx 0.134$$

so that

$$\frac{dA}{dt} = -\frac{0.02\pi(1 - 0.134)50}{\sqrt{10{,}000 - 2{,}500}} \approx -0.03$$

that is, the area of the arterial opening is decreasing at the rate of 0.03 cm²/year.

○ ○ ○

Suppose the population P of a certain bacteria culture is given by $P = f(T)$, where T is the temperature of the medium. Further, suppose the temperature T is a function of time t in seconds—that is, $T = g(t)$. Give an interpretation of each of the following quantities:

a. $\dfrac{dP}{dT}$ **b.** $\dfrac{dT}{dt}$ **c.** $\dfrac{dP}{dt}$ **d.** $(f \circ g)(t)$ **e.** $f'(g(t))g'(t)$

Using Technology

Finding the Derivative
of a Composite Function

EXAMPLE 1   Find the rate of change of $f(x) = \sqrt{x}(1 + 0.02x^2)^{3/2}$ when $x = 2.1$.

Solution   Using the numerical derivative operation of a graphing utility, we find

$$f'(2.1) = 0.582146320119$$

or approximately 0.58 unit per unit change in x.   ○ ○ ○

EXAMPLE 2   The management of Astro World ("The Amusement Park of the Future") estimates that the total number of visitors (in thousands) to the amusement park t hours after opening time at 9 A.M. is given by

$$N(t) = \frac{30t}{\sqrt{2 + t^2}}$$

What is the rate at which visitors are admitted to the amusement park at 10:30 A.M.?

Solution   Using the numerical derivative operation of a graphing utility, we find

$$N'(1.5) \approx 6.8481$$

or approximately 6848 visitors per hour.   ○ ○ ○

EXERCISES

In exercises 1–6, use the numerical derivative operation of a graphing utility to find the rate of change of f(x) at the given value of x. Give your answer accurate to four decimal places.

1. $f(x) = \sqrt{x^2 - x^4}$; $x = 0.5$

2. $f(x) = x - \sqrt{1 - x^2}$; $x = 0.4$

3. $f(x) = x\sqrt{1 - x^2}$; $x = 0.2$

4. $f(x) = (x + \sqrt{x^2 + 4})^{3/2}$; $x = 1$

5. $f(x) = \dfrac{\sqrt{1 + x^2}}{x^3 + 2}$; $x = -1$

6. $f(x) = \dfrac{x^3}{1 + (1 + x^2)^{3/2}}$; $x = 3$

7. Worldwide Production of Vehicles According to *Automotive News*, the worldwide production of vehicles between 1960 and 1990 is given by the function

$$f(t) = 16.5\sqrt{1 + 2.2t} \qquad (0 \le t \le 3)$$

where $f(t)$ is measured in units of a million and t is measured in decades, with $t = 0$ corresponding to the beginning of 1960. What was the rate of change of the worldwide production of vehicles at the beginning of 1970? At the beginning of 1980?
Source: Automotive News

SELF-CHECK EXERCISES 3.3

1. Find the derivative of

$$f(x) = -\frac{1}{\sqrt{2x^2 - 1}}$$

2. Suppose that the life expectancy at birth (in years) of a female in a certain country is described by the function

$$g(t) = 50.02(1 + 1.09t)^{0.1} \qquad (0 \le t \le 150)$$

where t is measured in years and $t = 0$ corresponds to the beginning of 1900.
a. What is the life expectancy at birth of a female born at the beginning of 1980? At the beginning of the year 2000?
b. How fast is the life expectancy at birth of a female born at any time t changing?

Solutions to Self-Check Exercises 3.3 can be found on page 213.

3.3 EXERCISES

In exercises 1–46, find the derivative of the given function.

1. $f(x) = (2x - 1)^4$

2. $f(x) = (1 - x)^3$

3. $f(x) = (x^2 + 2)^5$

4. $f(t) = 2(t^3 - 1)^5$

5. $f(x) = (2x - x^2)^3$

6. $f(x) = 3(x^3 - x)^4$

7. $f(x) = (2x + 1)^{-2}$

8. $f(t) = \frac{1}{2}(2t^2 + t)^{-3}$

9. $f(x) = (x^2 - 4)^{3/2}$

10. $f(t) = (3t^2 - 2t + 1)^{3/2}$

11. $f(x) = \sqrt{3x - 2}$

12. $f(t) = \sqrt{3t^2 - t}$

13. $f(x) = \sqrt[3]{1 - x^2}$

14. $f(x) = \sqrt{2x^2 - 2x + 3}$

15. $f(x) = \dfrac{1}{(2x + 3)^3}$

16. $f(x) = \dfrac{2}{(x^2 - 1)^4}$

17. $f(t) = \dfrac{1}{\sqrt{2t - 3}}$

18. $f(x) = \dfrac{1}{\sqrt{2x^2 - 1}}$

19. $y = \dfrac{1}{(4x^4 + x)^{3/2}}$

20. $f(t) = \dfrac{4}{\sqrt[3]{2t^2 + t}}$

21. $f(x) = (3x^2 + 2x + 1)^{-2}$

22. $f(t) = (5t^3 + 2t^2 - t + 4)^{-3}$

23. $f(x) = (x^2 + 1)^3 - (x^3 + 1)^2$

24. $f(t) = (2t - 1)^4 + (2t + 1)^4$

25. $f(t) = (t^{-1} - t^{-2})^3$

26. $f(v) = (v^{-3} + 4v^{-2})^3$

27. $f(x) = \sqrt{x + 1} + \sqrt{x - 1}$

28. $f(u) = (2u + 1)^{3/2} + (u^2 - 1)^{-3/2}$

29. $f(x) = 2x^2(3 - 4x)^4$

30. $h(t) = t^2(3t + 4)^3$

31. $f(x) = (x - 1)^2(2x + 1)^4$

32. $g(u) = (1 + u^2)^5(1 - 2u^2)^8$

33. $f(x) = \left(\dfrac{x + 3}{x - 2}\right)^3$

34. $f(x) = \left(\dfrac{x + 1}{x - 1}\right)^5$

35. $s(t) = \left(\dfrac{t}{2t + 1}\right)^{3/2}$

36. $g(s) = \left(s^2 + \dfrac{1}{s}\right)^{3/2}$

37. $g(u) = \sqrt{\dfrac{u + 1}{3u + 2}}$

38. $g(x) = \sqrt{\dfrac{2x + 1}{2x - 1}}$

39. $f(x) = \dfrac{x^2}{(x^2 - 1)^4}$

40. $g(u) = \dfrac{2u^2}{(u^2 + u)^3}$

41. $h(x) = \dfrac{(3x^2 + 1)^3}{(x^2 - 1)^4}$

42. $g(t) = \dfrac{(2t - 1)^2}{(3t + 2)^4}$

43. $f(x) = \dfrac{\sqrt{2x + 1}}{x^2 - 1}$

44. $f(t) = \dfrac{4t^2}{\sqrt{2t^2 + 2t - 1}}$

45. $g(t) = \dfrac{\sqrt{t+1}}{\sqrt{t^2+1}}$ **46.** $f(x) = \dfrac{\sqrt{x^2+1}}{\sqrt{x^2-1}}$

In exercises 47–52, find dy/du, du/dx, and dy/dx.

47. $y = u^{4/3}$ and $u = 3x^2 - 1$

48. $y = \sqrt{u}$ and $u = 7x - 2x^2$

49. $y = u^{-2/3}$ and $u = 2x^3 - x + 1$

50. $y = 2u^2 + 1$ and $u = x^2 + 1$

51. $y = \sqrt{u} + \dfrac{1}{\sqrt{u}}$ and $u = x^3 - x$

52. $y = \dfrac{1}{u}$ and $u = \sqrt{x} + 1$

In exercises 53–56, find an equation of the tangent line to the graph of the function at the given point.

53. $f(x) = (1 - x)(x^2 - 1)^2$; $(2, -9)$

54. $f(x) = \left(\dfrac{x+1}{x-1}\right)^2$; $(3, 4)$

55. $f(x) = x\sqrt{2x^2 + 7}$; $(3, 15)$

56. $f(x) = \dfrac{8}{\sqrt{x^2 + 6x}}$; $(2, 2)$

57. Television Viewing The number of viewers of a television series introduced several years ago is approximated by the function

$$N(x) = (60 + 2x)^{2/3} \qquad (1 \le x \le 26)$$

where $N(x)$ (measured in millions) denotes the number of weekly viewers of the series in the xth week. Find the rate of increase of the weekly audience at the end of the 2nd week and at the end of the 12th week. How many viewers were there in the 2nd week? In the 24th week?

58. Male Life Expectancy Suppose that the life expectancy of a male at birth in a certain country is described by the function

$$f(t) = 46.9(1 + 1.09t)^{0.1} \qquad (0 \le t \le 150)$$

where t is measured in years and $t = 0$ corresponds to the beginning of 1900. How long can a male born at the beginning of the year 2000 in that country expect to live? What is the rate of change of the life expectancy of a male born in that country at the beginning of the year 2000?

59. Concentration of Carbon Monoxide in the Air According to a joint study conducted by Oxnard's Environmental Management Department and a state government agency, the concentration of carbon monoxide in the air due to automobile exhaust t years from now is given by

$$C(t) = 0.01(0.2t^2 + 4t + 64)^{2/3}$$

parts per million.
a. Find the rate at which the level of carbon monoxide is changing with respect to time.
b. Find the rate at which the level of carbon monoxide will be changing 5 years from now.

60. Continuing Education Enrollment The registrar of Kellogg University estimates that the total student enrollment in the Continuing Education division will be given by

$$N(t) = -\dfrac{20{,}000}{\sqrt{1 + 0.2t}} + 21{,}000$$

where $N(t)$ denotes the number of students enrolled in the division t years from now. Find an expression for $N'(t)$. How fast is the student enrollment increasing currently? How fast will it be increasing 5 years from now?

61. Air Pollution According to the South Coast Air Quality Management District, the level of nitrogen dioxide, a brown gas that impairs breathing, present in the atmosphere on a certain May day in downtown Los Angeles is approximated by

$$A(t) = 0.03t^3(t - 7)^4 + 60.2 \qquad (0 \le t \le 7)$$

where $A(t)$ is measured in pollutant standard index and t is measured in hours, with $t = 0$ corresponding to 7 A.M.
a. Find $A'(t)$.
b. Find $A'(1)$, $A'(3)$, and $A'(4)$ and interpret your results.

62. Effect of Luxury Tax on Consumption Government economists of a developing country determined that the purchase of imported perfume is related to a proposed "luxury tax" by the formula

$$N(x) = \sqrt{10{,}000 - 40x - 0.02x^2} \qquad (0 \le x \le 200)$$

where $N(x)$ measures the percentage of normal consumption of perfume when a "luxury tax" of x percent is imposed on it. Find the rate of change of $N(x)$ for taxes of 10%, 100%, and 150%.

63. Pulse Rate of an Athlete   The pulse rate (the number of heartbeats per minute) of a long-distance runner t seconds after leaving the starting line is given by

$$P(t) = \frac{300\sqrt{\frac{1}{2}t^2 + 2t + 25}}{t + 25} \qquad (t \geq 0)$$

Compute $P'(t)$. How fast is the athlete's pulse rate increasing 10 seconds into the run? 60 seconds into the run? 2 minutes into the run? What is her pulse rate 2 minutes into the run?

64. Thurstone Learning Model   Psychologist L. L. Thurstone suggested the following relationship between learning time T and the length of a list n:

$$T = f(n) = An\sqrt{n - b}$$

where A and b are constants that depend on the person and the task.
a. Compute dT/dn and interpret your result.
b. Suppose that for a certain person and a certain task, $A = 4$ and $b = 4$. Compute $f'(13)$ and $f'(29)$ and interpret your results.

65. Oil Spills   In calm waters, the oil spilling from the ruptured hull of a grounded tanker spreads in all directions. Assuming that the area polluted is a circle and that its radius is increasing at a rate of 2 ft/sec, determine how fast the area is increasing when the radius of the circle is 40 feet.

66. Arteriosclerosis   Refer to Example 7, page 207. Suppose that the radius of an individual's artery is 1 cm and the thickness of the plaque (in cm) t years from now is given by

$$h = g(t) = \frac{0.5t^2}{t^2 + 10} \qquad (0 \leq t \leq 10)$$

How fast will the arterial opening be decreasing 5 years from now?

67. Traffic Flow   Opened in the late 1950s, the Central Artery in downtown Boston was designed to move 75,000 vehicles a day. The number of vehicles moved per day is approximated by the function

$$x = f(t) = 6.25t^2 + 19.75t + 74.75 \qquad (0 \leq t \leq 5)$$

where x is measured in thousands and t in decades, with $t = 0$ corresponding to the beginning of 1959. Suppose the average speed of traffic flow in mph is given by

$$S = g(x) = -0.00075x^2 + 67.5 \qquad (75 \leq x \leq 350)$$

where x has the same meaning as before. What was the rate of change of the average speed of traffic flow at the beginning of 1999? What was the average speed of traffic flow at that time?

68. Hotel Occupancy Rates   The occupancy rate of the all-suite Wonderland Hotel, located near an amusement park, is given by the function

$$r(t) = \frac{10}{81}t^3 - \frac{10}{3}t^2 + \frac{200}{9}t + 60 \qquad (0 \leq t \leq 12)$$

where t is measured in months and $t = 0$ corresponds to the beginning of January. Management has estimated that the monthly revenue (in thousands of dollars per month) is approximated by the function

$$R(r) = -\frac{3}{5000}r^3 + \frac{9}{50}r^2 \qquad (0 \leq r \leq 100)$$

where r is the occupancy rate.
a. Find an expression that gives the rate of change of Wonderland's occupancy rate with respect to time.
b. Find an expression that gives the rate of change of Wonderland's monthly revenue with respect to the occupancy rate.
c. What is the rate of change of Wonderland's monthly revenue with respect to time at the beginning of January? At the beginning of June?

[*Hint:* Use the Chain Rule to find $R'(r(0))r'(0)$ and $R'(r(6))r'(6)$.]

69. Effect of Housing Starts on Jobs   The president of a major housing construction firm claims that the number of construction jobs created is given by

$$N(x) = 1.42x$$

where x denotes the number of housing starts. Suppose that the number of housing starts in the next t months is expected to be

$$x(t) = \frac{7t^2 + 140t + 700}{3t^2 + 80t + 550}$$

million units per year. Find an expression that gives the rate at which the number of construction jobs will be created t months from now. At what rate will construction jobs be created 1 year from now?

70. Demand for PCs   The quantity demanded per month, x, of a certain make of personal computer (PC) is related to the average unit price, p (in dollars), of personal computers by the equation

$$x = f(p) = \frac{100}{9}\sqrt{810{,}000 - p^2}$$

It is estimated that t months from now, the average price of a PC will be given by

$$p(t) = \frac{400}{1 + \frac{1}{8}\sqrt{t}} + 200 \qquad (0 \le t \le 60)$$

dollars. Find the rate at which the quantity demanded per month of the PCs will be changing 16 months from now.

71. **Cruise Ship Bookings** The management of Cruise World, operators of Caribbean luxury cruises, expects that the percentage of young adults booking passage on their cruises in the years ahead will rise dramatically. They have constructed the following model, which gives the percentage of young adult passengers in year t:

$$p = f(t) = 50\left(\frac{t^2 + 2t + 4}{t^2 + 4t + 8}\right) \qquad (0 \le t \le 5)$$

Young adults normally pick shorter cruises and generally spend less on their passage. The following model gives an approximation of the average amount of money R (in dollars) spent per passenger on a cruise when the

percentage of young adults is p:

$$R(p) = 1000\left(\frac{p + 4}{p + 2}\right)$$

Find the rate at which the price of the average passage will be changing 2 years from now.

72. In Section 3.1, we proved that

$$\frac{d}{dx}(x^n) = nx^{n-1}$$

for the special case when $n = 2$. Use the Chain Rule to show that

$$\frac{d}{dx}(x^{1/n}) = \frac{1}{n}x^{1/n - 1}$$

for any nonzero integer n, assuming that $f(x) = x^{1/n}$ is differentiable.

[*Hint:* Let $f(x) = x^{1/n}$ so that $[f(x)]^n = x$. Differentiate both sides with respect to x.]

73. With the aid of exercise 72, prove that

$$\frac{d}{dx}(x^r) = rx^{r-1}$$

for any rational number r.

[*Hint:* Let $r = m/n$, where m and n are integers with $n \neq 0$, and write $x^r = (x^m)^{1/n}$.]

SOLUTIONS TO SELF-CHECK EXERCISES 3.3

1. Rewriting, we have

$$f(x) = -(2x^2 - 1)^{-1/2}$$

Using the General Power Rule, we find

$$f'(x) = -\frac{d}{dx}(2x^2 - 1)^{-1/2}$$

$$= -\left(-\frac{1}{2}\right)(2x^2 - 1)^{-3/2}\frac{d}{dx}(2x^2 - 1)$$

$$= \frac{1}{2}(2x^2 - 1)^{-3/2}(4x)$$

$$= \frac{2x}{(2x^2 - 1)^{3/2}}$$

2. a. The life expectancy at birth of a female born at the beginning of 1980 is given by

$$g(80) = 50.02[1 + 1.09(80)]^{0.1} \approx 78.29$$

or approximately 78 years. Similarly, the life expectancy at birth of a female born at the beginning of the year 2000 is given by

$$g(100) = 50.02[1 + 1.09(100)]^{0.1} \approx 80.04$$

or approximately 80 years.

b. The rate of change of the life expectancy at birth of a female born at any time t is given by $g'(t)$. Using the General Power Rule, we have

$$g'(t) = 50.02 \frac{d}{dt}(1 + 1.09t)^{0.1}$$

$$= (50.02)(0.1)(1 + 1.09t)^{-0.9} \frac{d}{dt}(1 + 1.09t)$$

$$= (50.02)(0.1)(1.09)(1 + 1.09t)^{-0.9}$$

$$= 5.45218(1 + 1.09t)^{-0.9}$$

$$= \frac{5.45218}{(1 + 1.09t)^{0.9}}$$

3.4 MARGINAL FUNCTIONS IN ECONOMICS

Marginal analysis is the study of the rate of change of economic quantities. For example, an economist is not merely concerned with the value of an economy's gross domestic product (GDP) at a given time, but is equally concerned with the rate at which it is growing or declining. In the same vein, a manufacturer is not only interested in the total cost corresponding to a certain level of production of a commodity, but is also interested in the rate of change of the total cost with respect to the level of production, and so on. Let us begin with an example to explain the meaning of the adjective *marginal,* as used by economists.

Cost Functions

 EXAMPLE 1 Suppose that the total cost in dollars incurred per week by the Polaraire Company for manufacturing x refrigerators is given by the total cost function

$$C(x) = 8000 + 200x - 0.2x^2 \qquad (0 \le x \le 400)$$

a. What is the actual cost incurred for manufacturing the 251st refrigerator?

b. Find the rate of change of the total cost function with respect to x when $x = 250$.

c. Compare the results obtained in (a) and (b).

Solution

a. The actual cost incurred in producing the 251st refrigerator is the difference between the total cost incurred in producing the first 251 refrigerators and the total cost of producing the first 250 refrigerators:

$$
\begin{aligned}
C(251) - C(250) &= [8000 + 200(251) - 0.2(251)^2] \\
&\quad - [8000 + 200(250) - 0.2(250)^2] \\
&= 45{,}599.8 - 45{,}500 \\
&= 99.80
\end{aligned}
$$

or $99.80.

b. The rate of change of the total cost function C with respect to x is given by the derivative of C—that is, $C'(x) = 200 - 0.4x$. Thus, when the level of production is 250 refrigerators, the rate of change of the total cost with respect to x is given by

$$
\begin{aligned}
C'(250) &= 200 - 0.4(250) \\
&= 100
\end{aligned}
$$

or $100.

c. From the solution to (a), we know that the actual cost for producing the 251st refrigerator is $99.80. This answer is very closely approximated by the answer to (b), $100. To see why this is so, observe that the difference $C(251) - C(250)$ may be written in the form

$$
\frac{C(251) - C(250)}{1} = \frac{C(250 + 1) - C(250)}{1} = \frac{C(250 + h) - C(250)}{h}
$$

where $h = 1$. In other words, the difference $C(251) - C(250)$ is precisely the average rate of change of the total cost function C over the interval $[250, 251]$, or, equivalently, the slope of the secant line through the points $(250, 45{,}500)$ and $(251, 45{,}599.8)$. However, the number $C'(250) = 100$ is the instantaneous rate of change of the total cost function C at $x = 250$, or, equivalently, the slope of the tangent line to the graph of C at $x = 250$.

Now when h is small, the average rate of change of the function C is a good approximation to the instantaneous rate of change of the function C, or, equivalently, the slope of the secant line through the points in question is a good approximation to the slope of the tangent line through the point in question. Thus, we may expect

$$
\begin{aligned}
C(251) - C(250) &= \frac{C(251) - C(250)}{1} = \frac{C(250 + h) - C(250)}{h} \\
&\approx \lim_{h \to 0} \frac{C(250 + h) - C(250)}{h} = C'(250)
\end{aligned}
$$

which is precisely the case in this example.

The actual cost incurred in producing an additional unit of a certain commodity given that a plant is already at a certain level of operation is called the **marginal cost.** Knowing this cost is very important to management in their decision-making processes. As we saw in Example 1, the marginal cost is approximated by the rate of change of the total cost function evaluated at the appropriate point. For this reason, economists have defined the **marginal cost function** to be the derivative of the corresponding total cost function. In other words, if C is a total cost function, then the marginal cost function is defined to be its derivative C'. Thus, the adjective *marginal* is synonymous with *derivative of.*

 EXAMPLE 2   A subsidiary of the Elektra Electronics Company manufactures a programmable pocket calculator. Management determined that the daily total cost of producing these calculators (in dollars) is given by

$$C(x) = 0.0001x^3 - 0.08x^2 + 40x + 5000$$

where x stands for the number of calculators produced.

a. Find the marginal cost function.

b. What is the marginal cost when $x = 200, 300, 400$, and 600?

c. Interpret your results.

Solution

a. The marginal cost function C' is given by the derivative of the total cost function C. Thus,

$$C'(x) = 0.0003x^2 - 0.16x + 40$$

b. The marginal cost when $x = 200, 300, 400$, and 600 is given by

$$C'(200) = 0.0003(200)^2 - 0.16(200) + 40 = 20$$
$$C'(300) = 0.0003(300)^2 - 0.16(300) + 40 = 19$$
$$C'(400) = 0.0003(400)^2 - 0.16(400) + 40 = 24$$
$$C'(600) = 0.0003(600)^2 - 0.16(600) + 40 = 52$$

or $20, $19, 24, and 52, respectively.

c. From the results of (b), we see that Elektra's actual cost for producing the 201st calculator is approximately $20. The actual cost incurred for producing one additional calculator when the level of production is already 300 calculators is approximately $19, and so on. Observe that when the level of production is already 600 units, the actual cost of producing one additional unit is approximately $52. The higher cost for producing this additional unit when the level of production is 600 units may be the result of several factors, among them excessive costs incurred because of overtime or higher maintenance, production breakdown caused by greater stress and strain on the equipment, and so on. The graph of the total cost function appears in Figure 3.7.   ◐ ◐ ◐

Figure 3.7
The cost of producing x calculators is given by C(x).

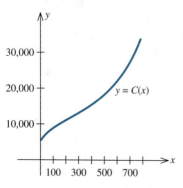

Average Cost Functions

Let us now introduce another marginal concept closely related to the marginal cost. Let $C(x)$ denote the total cost incurred in producing x units of a certain commodity. Then the **average cost** of producing x units of the commodity is obtained by dividing the total production cost by the number of units produced. This leads to the following definition.

AVERAGE COST FUNCTION	Suppose $C(x)$ is a total cost function. Then the average cost function, denoted by $\overline{C}(x)$ and read "C bar of x," is $$\frac{C(x)}{x} \tag{4}$$

The derivative $\overline{C}'(x)$ of the average cost function, called the **marginal average cost function,** measures the rate of change of the average cost function with respect to the number of units produced.

EXAMPLE 3 The total cost of producing x units of a certain commodity is given by

$$C(x) = 400 + 20x$$

dollars.

a. Find the average cost function $\overline{C}$.

b. Find the marginal average cost function $\overline{C}'$.

c. Interpret the results obtained in (a) and (b).

Solution

a. The average cost function is given by

$$\overline{C}(x) = \frac{C(x)}{x} = \frac{400 + 20x}{x}$$

$$= 20 + \frac{400}{x}$$

b. The marginal average cost function is

$$\overline{C}'(x) = -\frac{400}{x^2}$$

c. Since the marginal average cost function is negative for all admissible values of x, the rate of change of the average cost function is negative for all $x > 0$; that is, $\overline{C}(x)$ decreases as x increases. However, the graph of $\overline{C}$ always

Figure 3.8
*As the level of production increases,
the average cost approaches $20.*

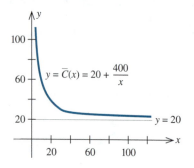

lies above the horizontal line $y = 20$, but it approaches the line since

$$\lim_{x \to \infty} \overline{C}(x) = \lim_{x \to \infty} \left(20 + \frac{400}{x} \right) = 20$$

A sketch of the graph of the function $\overline{C}(x)$ appears in Figure 3.8. This result is fully expected if we consider the economic implications. Note that as the level of production increases, the fixed cost per unit of production, represented by the term $(400/x)$, drops steadily. The average cost approaches the constant unit of production, which is $20 in this case.   ○ ○ ○

EXAMPLE 4   Once again consider the subsidiary of the Elektra Electronics Company. The daily total cost for producing its programmable calculators is given by

$$C(x) = 0.0001x^3 - 0.08x^2 + 40x + 5000$$

dollars, where x stands for the number of calculators produced (see Example 2).

a. Find the average cost function $\overline{C}$.

b. Find the marginal average cost function $\overline{C}'$. Compute $\overline{C}'(500)$.

c. Sketch the graph of the function $\overline{C}$ and interpret the results obtained in (a) and (b).

Solution

a. The average cost function is given by

$$\overline{C}(x) = \frac{C(x)}{x} = 0.0001x^2 - 0.08x + 40 + \frac{5000}{x}$$

b. The marginal average cost function is given by

$$\overline{C}'(x) = 0.0002x - 0.08 - \frac{5000}{x^2}$$

Also $$\overline{C}'(500) = 0.0002(500) - 0.08 - \frac{5000}{(500)^2} = 0$$

Figure 3.9
The average cost reaches a minimum of $35 when 500 calculators are produced.

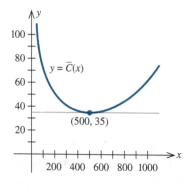

c. To sketch the graph of the function $\overline{C}$, observe that if x is a small positive number, then $\overline{C}(x) > 0$. Furthermore, $\overline{C}(x)$ becomes arbitrarily large as x approaches zero from the right, since the term $(5000/x)$ becomes arbitrarily large as x approaches zero. Next, the result $\overline{C}'(500) = 0$ obtained in (b) tells us that the tangent line to the graph of the function $\overline{C}$ is horizontal at the point $(500, 35)$ on the graph. Finally, plotting the points on the graph corresponding to, say, $x = 100, 200, 300, \ldots, 900$, we obtain the sketch in Figure 3.9. As expected, the average cost drops as the level of production increases. But in this case, as opposed to the case in Example 3, the average cost reaches a minimum value of $35, corresponding to a production level of 500, and *increases* thereafter.

This phenomenon is typical in situations where the marginal cost increases from some point on as production increases, as in Example 2. This situation is in contrast to that of Example 3, in which the marginal cost remains constant at any level of production.   ○ ○ ○

EXPLORING WITH TECHNOLOGY

Refer to Example 4.

a. Use a graphing utility to plot the graph of the average cost function

$$\overline{C}(x) = 0.0001x^2 - 0.08x + 40 + \frac{5000}{x}$$

using the viewing rectangle $[0, 1000] \times [0, 100]$. Then, using **ZOOM** and **TRACE**, show that the lowest point on the graph of $\overline{C}$ is (500, 35).

b. Draw the tangent line to the graph of $\overline{C}$ at (500, 35). What is its slope? Is this expected?

c. Plot the graph of the marginal average cost function

$$\overline{C}'(x) = 0.0002x - 0.08 - \frac{5000}{x^2}$$

using the viewing rectangle $[0, 2000] \times [-1, 1]$. Then use **ZOOM** and **TRACE** to show that the zero of the function $\overline{C}'$ occurs at $x = 500$. Verify this result using the root-finding capability of your graphing utility. Is this result compatible with that obtained in part (b)? Explain your answer.

○   ○   ○

Revenue Functions

Another marginal concept, the *marginal revenue function,* is associated with the revenue function R, given by

$$R(x) = px \tag{5}$$

where x is the number of units sold of a certain commodity and p is the unit selling price. In general, however, the unit selling price p of a commodity is related to the quantity x of the commodity demanded. This relationship, $p = f(x)$, is called a *demand equation* (see Section 2.3). Solving the demand equation for p in terms of x, we obtain the unit price function f, given by

$$p = f(x)$$

Thus, the revenue function R is given by

$$R(x) = px = xf(x)$$

where f is the unit price function. The derivative R' of the function R, called the **marginal revenue function,** measures the rate of change of the revenue function.

EXAMPLE 5   Suppose that the relationship between the unit price p in dollars and the quantity demanded x of the Acrosonic model F loudspeaker system is given by the equation

$$p = -0.02x + 400 \qquad (0 \leq x \leq 20{,}000)$$

a. Find the revenue function R.

b. Find the marginal revenue function R'.

c. Compute $R'(2000)$ and interpret your result.

Solution

a. The revenue function R is given by

$$\begin{aligned} R(x) &= px \\ &= x(-0.02x + 400) \\ &= -0.02x^2 + 400x \qquad (0 \leq x \leq 20{,}000) \end{aligned}$$

b. The marginal revenue function R' is given by

$$R'(x) = -0.04x + 400$$

c. $R'(2000) = -0.04(2000) + 400 = 320$

Thus, the actual revenue to be realized from the sale of the 2001st loudspeaker system is approximately \$320.   ◦ ◦ ◦

Profit Functions

Our final example of a marginal function involves the profit function. The profit function P is given by

$$P(x) = R(x) - C(x) \tag{6}$$

where R and C are the revenue and cost functions and x is the number of units of a commodity produced and sold. The **marginal profit function** $P'(x)$ measures the rate of change of the profit function P and provides us with a good approximation of the actual profit or loss realized from the sale of the $(x + 1)$st unit of the commodity (assuming the xth unit has been sold).

EXAMPLE 6   Refer to Example 5. Suppose that the cost of producing x units of the Acrosonic model F loudspeaker is

$$C(x) = 100x + 200{,}000$$

dollars.

a. Find the profit function P.

b. Find the marginal profit function P'.

c. Compute $P'(2000)$ and interpret your result.

d. Sketch the graph of the profit function P.

Solution

a. From the solution to Example 5a, we have

$$R(x) = -0.02x^2 + 400x$$

Thus, the required profit function P is given by

$$
\begin{aligned}
P(x) &= R(x) - C(x) \\
&= (-0.02x^2 + 400x) - (100x + 200{,}000) \\
&= -0.02x^2 + 300x - 200{,}000
\end{aligned}
$$

b. The marginal profit function P' is given by

$$P'(x) = -0.04x + 300$$

c.
$$P'(2000) = -0.04(2000) + 300 = 220$$

Thus, the actual profit realized from the sale of the 2001st loudspeaker system is approximately $220.

d. The graph of the profit function P appears in Figure 3.10.   ○ ○ ○

Figure 3.10

The total profit made when x loudspeakers are produced is given by P(x).

Elasticity of Demand

Finally, let us use the marginal concepts introduced in this section to derive an important criterion used by economists to analyze the demand function: *elasticity of demand.*

In what follows, it will be convenient to write the demand function f in the form $x = f(p)$; that is, we will think of the quantity demanded of a certain commodity as a function of its unit price. Since the quantity demanded of a commodity usually decreases as its unit price increases, the function f is typically a decreasing function of p (see Figure 3.11a).

Figure 3.11

(a) A demand function.

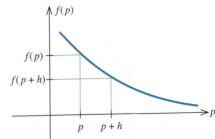

(b) $f(p + h)$ is the quantity demanded when the unit price increases from p to $p + h$ dollars.

Suppose that the unit price of a commodity is increased by h dollars from p dollars to $(p + h)$ dollars (Figure 3.11b). Then the quantity demanded drops from $f(p)$ units to $f(p + h)$ units, a change of $[f(p + h) - f(p)]$ units. The percentage change in the unit price is

$$\frac{h}{p}(100) \qquad \left(\frac{\text{Change in unit price}}{\text{Price } p}\right)(100)$$

and the corresponding percentage change in the quantity demanded is

$$100\left[\frac{f(p + h) - f(p)}{f(p)}\right] \qquad \left(\frac{\text{Change in quantity demanded}}{\text{Quantity demanded at price } p}\right)(100)$$

Now, one good way to measure the effect that a percentage change in price has on the percentage change in the quantity demanded is to look at the ratio of the latter to the former. We find

$$\frac{\text{Percentage change in the quantity demanded}}{\text{Percentage change in the unit price}} = \frac{100\left[\dfrac{f(p + h) - f(p)}{f(p)}\right]}{100\left(\dfrac{h}{p}\right)}$$

$$= \frac{\dfrac{f(p + h) - f(p)}{h}}{\dfrac{f(p)}{p}}$$

If f is differentiable at p, then

$$\frac{f(p + h) - f(p)}{h} \approx f'(p)$$

when h is small. Therefore, if h is small, then the ratio is approximately equal to

$$\frac{f'(p)}{\dfrac{f(p)}{p}} = \frac{pf'(p)}{f(p)}$$

Economists call the negative of this quantity the *elasticity of demand*.

ELASTICITY OF DEMAND

If f is a differentiable demand function defined by $x = f(p)$, then the **elasticity of demand** at price p is given by

$$E(p) = -\frac{pf'(p)}{f(p)} \qquad\qquad (7)$$

REMARK   It will be shown later (Section 4.1) that if f is decreasing on an interval, then $f'(p) < 0$ for p in that interval. In light of this, we see that since both p and $f(p)$ are positive, the quantity $\dfrac{pf'(p)}{f(p)}$ is negative. Because economists would rather work with a positive value, the elasticity of demand $E(p)$ is defined to be the negative of this quantity.

EXAMPLE 7   Consider the demand equation

$$p = -0.02x + 400 \qquad (0 \le x \le 20{,}000)$$

which describes the relationship between the unit price in dollars and the quantity demanded x of the Acrosonic model F loudspeaker systems.

a. Find the elasticity of demand $E(p)$.

b. Compute $E(100)$ and interpret your result.

c. Compute $E(300)$ and interpret your result.

Solution

a. Solving the given demand equation for x in terms of p, we find

$$x = f(p) = -50p + 20{,}000$$

from which we see that

$$f'(p) = -50$$

Therefore,

$$E(p) = -\frac{pf'(p)}{f(p)} = -\frac{p(-50)}{-50p + 20{,}000}$$

$$= \frac{p}{400 - p}$$

b. $E(100) = \dfrac{100}{400 - 100} = \dfrac{1}{3}$, which is the elasticity of demand when $p = 100$. To interpret this result, recall that $E(100)$ is the negative of the ratio of the percentage change in the quantity demanded to the percentage change in the unit price when $p = 100$. Therefore, our result tells us that when the unit price p is set at \$100 per speaker, an increase of 1% in the unit price will cause a decrease of approximately 0.33% in the quantity demanded.

c. $E(300) = \dfrac{300}{400 - 300} = 3$, which is the elasticity of demand when $p = 300$. It tells us that when the unit price is set at \$300 per speaker, an increase of 1% in the unit price will cause a decrease of approximately 3% in the quantity demanded.

Economists often use the following terminology to describe demand in terms of elasticity.

ELASTICITY OF DEMAND

The demand is said to be **elastic** if $E(p) > 1$.
The demand is said to be **unitary** if $E(p) = 1$.
The demand is said to be **inelastic** if $E(p) < 1$.

As an illustration, our computations in Example 7 revealed that demand for Acrosonic loudspeakers is elastic when $p = 300$ but inelastic when $p = 100$. These computations confirm that when demand is elastic, a small percentage change in the unit price will result in a greater percentage change in the quantity demanded; and when demand is inelastic, a small percentage change in the unit price will cause a smaller percentage change in the quantity demanded. Finally, when demand is unitary, a small percentage change in the unit price will result in the same percentage change in the quantity demanded.

We can describe the way revenue responds to changes in the unit price using the notion of elasticity. If the quantity demanded of a certain commodity is related to its unit price by the equation $x = f(p)$, then the revenue realized through the sale of x units of the commodity at a price of p dollars each is

$$R(p) = px = pf(p)$$

The rate of change of the revenue with respect to the unit price p is given by

$$R'(p) = f(p) + pf'(p)$$

$$= f(p)\left[1 + \frac{pf'(p)}{f(p)}\right]$$

$$= f(p)[1 - E(p)]$$

Now, suppose the demand is elastic when the unit price is set at a dollars. Then $E(a) > 1$, and so $1 - E(a) < 0$. Since $f(p)$ is positive for all values of p, we see that

$$R'(a) = f(a)[1 - E(a)] < 0$$

and so $R(p)$ is decreasing at $p = a$. This implies that a small increase in the unit price when $p = a$ results in a decrease in the revenue, whereas a small decrease in the unit price will result in an increase in the revenue. Similarly, you can show that if the demand is inelastic when the unit price is set at a dollars, then a small increase in the unit price will cause the revenue to increase, and a small decrease in the unit price will cause the revenue to decrease. Finally, if the demand is unitary when the unit price is set at a dollars, then $E(a) = 1$ and $R'(a) = 0$. This implies that a small increase or decrease in the unit price will not result in a change in the revenue. The following statements summarize this discussion.

1. If the demand is elastic at p $(E(p) > 1)$, then an increase in the unit price will cause the revenue to decrease, whereas a decrease in the unit price will cause the revenue to increase.

2. If the demand is inelastic at p $(E(p) < 1)$, then an increase in the unit price will cause the revenue to increase, and a decrease in the unit price will cause the revenue to decrease.

3. If the demand is unitary at p $(E(p) = 1)$, then an increase in the unit price will cause the revenue to stay about the same.

These results are illustrated in Figure 3.12.

Figure 3.12
The revenue is increasing on an interval where the demand is inelastic, decreasing on an interval where the demand is elastic, and stationary at the point where the demand is unitary.

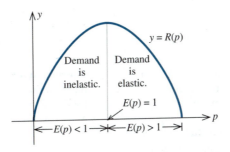

REMARK   As an aid to remembering this, note that

1. if demand is elastic, then the change in revenue and the change in the unit price move in opposite directions.

2. if demand is inelastic, then they move in the same direction.   ○ ○ ○

EXAMPLE 8   Refer to Example 7.

a. Is demand elastic, unitary, or inelastic when $p = 100$? When $p = 300$?

b. If the price is $100, will raising the unit price slightly cause the revenue to increase or decrease?

Solution

a. From the results of Example 7, we see that $E(100) = \frac{1}{3} < 1$ and $E(300) = 3 > 1$. We conclude accordingly that demand is inelastic when $p = 100$ and elastic when $p = 300$.

b. Since demand is inelastic when $p = 100$, raising the unit price slightly will cause the revenue to increase.   ○ ○ ○

SELF-CHECK EXERCISES 3.4

 1. The weekly demand for Pulsar VCRs (videocassette recorders) is given by the demand equation

$$p = -0.02x + 300 \qquad (0 \le x \le 15{,}000)$$

where p denotes the wholesale unit price in dollars and x denotes the quantity demanded. The weekly total cost function associated with manufacturing these VCRs is

$$C(x) = 0.000003x^3 - 0.04x^2 + 200x + 70{,}000$$

dollars.

a. Find the revenue function R and the profit function P.

b. Find the marginal cost function C', the marginal revenue function R', and the marginal profit function P'.

c. Find the marginal average cost function $\overline{C}'$.

d. Compute $C'(3000)$, $R'(3000)$, and $P'(3000)$ and interpret your results.

2. Refer to the preceding exercise. Determine whether the demand is elastic, unitary, or inelastic when $p = 100$ and when $p = 200$.

Solutions to Self-Check Exercises 3.4 can be found on page 229.

3.4 EXERCISES

1. Production Costs The graph of a typical total cost function $C(x)$ associated with the manufacture of x units of a certain commodity is shown in the following figure.
a. Explain why the function C is always increasing.
b. As the level of production x increases, the cost per unit drops, so that $C(x)$ increases but at a slower pace. However, a level of production is soon reached at which the cost per unit begins to increase dramatically (due to a shortage of raw material, overtime, breakdown of machinery due to excessive stress and strain) so that $C(x)$ continues to increase at a faster pace. Use the graph of C to find the approximate level of production x_0 where this occurs.

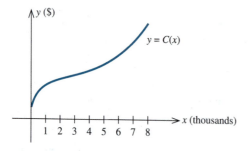

A calculator is recommended for exercises 2–33.

2. Marginal Cost The total weekly cost in dollars incurred by the Lincoln Record Company in pressing x long-playing records is

$$C(x) = 2000 + 2x - 0.0001x^2 \qquad (0 \le x \le 6000)$$

a. What is the actual cost incurred in producing the 1001st record? The 2001st record?
b. What is the marginal cost when $x = 1000$? When $x = 2000$?

3. Marginal Cost A division of Ditton Industries manufactures the Futura model microwave oven. The daily cost (in dollars) of producing these microwave ovens is

$$C(x) = 0.0002x^3 - 0.06x^2 + 120x + 5000$$

where x stands for the number of units produced.
a. What is the actual cost incurred in manufacturing the 101st oven? The 201st oven? The 301st oven?
b. What is the marginal cost when $x = 100$? When $x = 200$? When $x = 300$?

4. Marginal Average Cost The Custom Office Company makes a line of executive desks. It is estimated that the total cost for making x units of their Senior Executive Model is

$$C(x) = 100x + 200{,}000$$

dollars per year.
a. Find the average cost function $\overline{C}$.
b. Find the marginal average cost function $\overline{C}'$.
c. What happens to $\overline{C}(x)$ when x is very large? Interpret your results.

5. Marginal Average Cost The management of the Thermo-Master Company, whose Mexican subsidiary manufactures an indoor-outdoor thermometer, has estimated that the total weekly cost in dollars for producing x thermometers is

$$C(x) = 5000 + 2x$$

a. Find the average cost function $\overline{C}$.
b. Find the marginal average cost function $\overline{C}'$.
c. Interpret your results.

6. Find the average cost function $\overline{C}$ and the marginal average cost function $\overline{C}'$ associated with the total cost function C of exercise 2.

7. Find the average cost function $\overline{C}$ and the marginal average cost function $\overline{C}'$ associated with the total cost function C of exercise 3.

8. Marginal Revenue The Williams Commuter Air Service realizes a monthly revenue of

$$R(x) = 8000x - 100x^2$$

dollars when the price charged per passenger is x dollars.
a. Find the marginal revenue R'.
b. Compute $R'(39)$, $R'(40)$, and $R'(41)$. What do your results imply?

9. Marginal Revenue The management of the Acrosonic Company plans to market the Electro-Stat, an electrostatic speaker system. The marketing department has determined that the demand for these speakers is

$$p = -0.04x + 800 \qquad (0 \le x \le 20{,}000)$$

where p denotes the speaker's unit price (in dollars) and x denotes the quantity demanded.

a. Find the revenue function R.

b. Find the marginal revenue function R'.

c. Compute $R'(5000)$ and interpret your result.

10. Marginal Profit   Lynbrook West, an apartment complex, has 100 two-bedroom units. The monthly profit (in dollars) realized from renting x apartments is

$$P(x) = -10x^2 + 1760x - 50,000$$

a. What is the actual profit realized from renting the 51st unit, assuming that 50 units have already been rented?

b. Compute the marginal profit when $x = 50$ and compare your results with that obtained in (a).

11. Marginal Profit   Refer to exercise 9. Acrosonic's production department estimates that the total cost (in dollars) incurred in manufacturing x Electro-Stat speaker systems in the first year of production will be

$$C(x) = 200x + 300,000$$

a. Find the profit function P.

b. Find the marginal profit function P'.

c. Compute $P'(5000)$ and $P'(8000)$.

d. Sketch the graph of the profit function and interpret your results.

12. Marginal Cost, Revenue, and Profit   The weekly demand for the Pulsar 25 color console television is

$$p = 600 - 0.05x \quad (0 \le x \le 12,000)$$

where p denotes the wholesale unit price in dollars and x denotes the quantity demanded. The weekly total cost function associated with manufacturing the Pulsar 25 is given by

$$C(x) = 0.000002x^3 - 0.03x^2 + 400x + 80,000$$

where $C(x)$ denotes the total cost incurred in producing x sets.

a. Find the revenue function R and the profit function P.

b. Find the marginal cost function C', the marginal revenue function R', and the marginal profit function P'.

c. Compute $C'(2000)$, $R'(2000)$, and $P'(2000)$ and interpret your results.

d. Sketch the graphs of the functions C, R, and P and interpret (b) and (c) using the graphs obtained.

13. Marginal Cost, Revenue, and Profit   The Pulsar Corporation also manufactures a series of 19-inch color television sets. The quantity x of these sets demanded each week is related to the wholesale unit price p by the equation

$$p = -0.006x + 180$$

The weekly total cost incurred by Pulsar for producing x sets is

$$C(x) = 0.000002x^3 - 0.02x^2 + 120x + 60,000$$

dollars. Answer the questions in exercise 12 for these data.

14. Marginal Average Cost   Find the average cost function $\overline{C}$ associated with the total cost function C of exercise 12.

a. What is the marginal average cost function $\overline{C}'$?

b. Compute $\overline{C}'(5,000)$ and $\overline{C}'(10,000)$ and interpret your results.

c. Sketch the graph of $\overline{C}$.

15. Marginal Average Cost   Find the average cost function $\overline{C}$ associated with the total cost function C of exercise 13.

a. What is the marginal average cost function $\overline{C}'$?

b. Compute $\overline{C}'(5,000)$ and $\overline{C}'(10,000)$ and interpret your results.

16. Marginal Revenue   The quantity of Sicard wristwatches demanded per month is related to the unit price by the equation

$$p = \frac{50}{0.01x^2 + 1} \quad (0 \le x \le 20)$$

where p is measured in dollars and x in units of a thousand.

a. Find the revenue function R.

b. Find the marginal revenue function R'.

c. Compute $R'(2)$ and interpret your result.

17. Marginal Propensity to Consume   The consumption function of the U.S. economy for 1929 to 1941 is

$$C(x) = 0.712x + 95.05$$

where $C(x)$ is the personal consumption expenditure and x is the personal income, both measured in billions of dollars. Find the rate of change of consumption with respect to income, dC/dx. This quantity is called the *marginal propensity to consume*.

18. Marginal Propensity to Consume Refer to exercise 17. Suppose that a certain economy's consumption function is

$$C(x) = 0.873x^{1.1} + 20.34$$

where $C(x)$ and x are measured in billions of dollars. Find the marginal propensity to consume when $x = 10$.

19. Marginal Propensity to Save Suppose that $C(x)$ measures an economy's personal consumption expenditure and x the personal income, both in billions of dollars. Then

$$S(x) = x - C(x) \quad \text{(Income minus consumption)}$$

measures the economy's savings corresponding to an income of x billion dollars. Show that

$$\frac{dS}{dx} = 1 - \frac{dC}{dx}$$

The quantity dS/dx is called the *marginal propensity to save.*

20. Refer to exercise 19. For the consumption function of exercise 17, find the marginal propensity to save.

21. Refer to exercise 19. For the consumption function of exercise 18, find the marginal propensity to save when $x = 10$.

For each demand equation in exercises 22–27, compute the elasticity of demand and determine whether the demand is elastic, unitary, or inelastic at the indicated price.

22. $x = -\frac{3}{2}p + 9; \quad p = 2$

23. $x = -\frac{5}{4}p + 20; \quad p = 10$

24. $x + \frac{1}{3}p - 20 = 0; \quad p = 30$

25. $0.4x + p - 20 = 0; \quad p = 10$

26. $p = 144 - x^2; \quad p = 96$

27. $p = 169 - x^2; \quad p = 29$

28. Elasticity of Demand The management of the Titan Tire Company has determined that the quantity demanded x of their Super Titan tires per week is related to the unit price p by the equation

$$x = \sqrt{144 - p}$$

where p is measured in dollars and x in units of a thousand.

a. Compute the elasticity of demand when $p = 63$, when $p = 96$, and when $p = 108$.
b. Interpret the results obtained in (a).
c. Is the demand elastic, unitary, or inelastic when $p = 63$? When $p = 96$? When $p = 108$?

29. Elasticity of Demand The demand equation for the Roland portable hair dryer is given by

$$x = \frac{1}{5}(225 - p^2) \quad (0 \le p \le 15)$$

where x (measured in units of a hundred) is the quantity demanded per week and p is the unit price in dollars.
a. Is the demand elastic or inelastic when $p = 8$ and when $p = 10$?
b. When is the demand unitary?
[*Hint:* Solve $E(p) = 1$ for p.]

c. If the unit price is lowered slightly from $10, will the revenue increase or decrease?
d. If the unit price is increased slightly from $8, will the revenue increase or decrease?

30. Elasticity of Demand The quantity demanded per week x (in units of a hundred) of the Mikado miniature camera is related to the unit price p (in dollars) by the demand equation

$$x = \sqrt{400 - 5p} \quad (0 \le p \le 80)$$

a. Is the demand elastic or inelastic when $p = 40$? When $p = 60$?
b. When is the demand unitary?
c. If the unit price is lowered slightly from $60, will the revenue increase or decrease?
d. If the unit price is increased slightly from $40, will the revenue increase or decrease?

31. Elasticity of Demand The proprietor of the Showplace, a video club, has estimated that the rental price p (in dollars) of prerecorded videocassette tapes is related to the quantity x rented per week by the demand equation

$$x = \frac{2}{3}\sqrt{36 - p^2} \quad (0 \le p \le 6)$$

Currently, the rental price is $2 per tape.
a. Is the demand elastic or inelastic at this rental price?
b. If the rental price is increased, will the revenue increase or decrease?

32. **Elasticity of Demand**   The demand function for a certain make of exercise bicycle sold exclusively through cable television is

$$p = \sqrt{9 - 0.02x} \qquad (0 \le x \le 450)$$

where p is the unit price in hundreds of dollars and x is the quantity demanded per week. Compute the elasticity of demand and determine the range of prices corresponding to inelastic, unitary, and elastic demand.

[*Hint:* Solve the equation $E(p) = 1.$]

33. **Elasticity of Demand**   The demand equation for the Sicard wristwatch is given by

$$x = 10 \sqrt{\frac{50 - p}{p}} \qquad (0 < p \le 50)$$

where x (measured in units of a thousand) is the quantity demanded per week and p is the unit price in dollars. Compute the elasticity of demand and determine the range of prices corresponding to inelastic, unitary, and elastic demand.

SOLUTIONS TO SELF-CHECK EXERCISES 3.4

1. a.
$$\begin{aligned} R(x) &= px \\ &= x(-0.02x + 300) \\ &= -0.02x^2 + 300x \qquad (0 \le x \le 15{,}000) \end{aligned}$$

$$\begin{aligned} P(x) &= R(x) - C(x) \\ &= -0.02x^2 + 300x \\ &\quad - (0.000003x^3 - 0.04x^2 + 200x + 70{,}000) \\ &= -0.000003x^3 + 0.02x^2 + 100x - 70{,}000 \end{aligned}$$

b.
$$C'(x) = 0.000009x^2 - 0.08x + 200$$
$$R'(x) = -0.04x + 300$$

and

$$P'(x) = -0.000009x^2 + 0.04x + 100$$

c. The average cost function is

$$\begin{aligned} \overline{C}(x) &= \frac{C(x)}{x} \\ &= \frac{0.000003x^3 - 0.04x^2 + 200x + 70{,}000}{x} \\ &= 0.000003x^2 - 0.04x + 200 + \frac{70{,}000}{x} \end{aligned}$$

Therefore, the marginal average cost function is

$$\overline{C}'(x) = 0.000006x - 0.04 - \frac{70{,}000}{x^2}$$

d. Using the results from (b), we find

$$\begin{aligned} C'(3000) &= 0.000009(3000)^2 - 0.08(3000) + 200 \\ &= 41 \end{aligned}$$

That is, when the level of production is already 3000 VCRs, the actual cost of producing one additional VCR is approximately $41. Next,

$$R'(3000) = -0.04(3000) + 300 = 180$$

That is, the actual revenue to be realized from selling the 3001st VCR is approximately $180. Finally,

$$P'(3000) = -0.000009(3000)^2 + 0.04(3000) + 100$$
$$= 139$$

That is, the actual profit realized from selling the 3001st VCR is approximately $139.

2. We first solve the given demand equation for x in terms of p, obtaining

$$x = f(p) = -50p + 15{,}000$$

and

$$f'(p) = -50$$

Therefore,

$$E(p) = -\frac{pf'(p)}{f(p)} = -\frac{p}{-50p + 15{,}000}(-50)$$

$$= \frac{p}{300 - p} \qquad (0 \le p < 300)$$

Next, we compute

$$E(100) = \frac{100}{300 - 100} = \frac{1}{2} < 1$$

and we conclude that demand is inelastic when $p = 100$. Also,

$$E(200) = \frac{200}{300 - 200} = 2 > 1$$

and we see that demand is elastic when $p = 200$.

3.5 HIGHER-ORDER DERIVATIVES

Higher-Order Derivatives

The derivative f' of a function f is also a function. As such, the differentiability of f' may be considered. Thus, the function f' has a derivative f'' at a point x in the domain of f' if the limit of the quotient

$$\frac{f'(x + h) - f'(x)}{h}$$

exists as h approaches zero. In other words, it is the derivative of the first derivative.

The function f'' obtained in this manner is called the **second derivative** of the function f, just as the derivative f' of f is often called the first derivative of f. Continuing in this fashion, we are led to considering the third, fourth, and higher-order derivatives of f whenever they exist. Notations for the first, second, third, and, in general, nth derivatives of a function f at a point x are

$$f'(x), f''(x), f'''(x), \ldots, f^{(n)}(x)$$

or

$$D^1f(x), D^2f(x), D^3f(x), \ldots, D^nf(x)$$

If f is written in the form $y = f(x)$, then the notations for its derivatives are

$$y', y'', y''', \ldots, y^{(n)}$$

$$\frac{dy}{dx}, \frac{d^2y}{dx^2}, \frac{d^3y}{dx^3}, \ldots, \frac{d^ny}{dx^n}$$

or

$$D^1y, D^2y, D^3y, \ldots, D^ny$$

respectively.

EXAMPLE 1 Find the derivatives of all orders of the polynomial function $f(x) = x^5 - 3x^4 + 4x^3 - 2x^2 + x - 8$.

Solution We have

$$f'(x) = 5x^4 - 12x^3 + 12x^2 - 4x + 1$$

$$f''(x) = \frac{d}{dx} f'(x) = 20x^3 - 36x^2 + 24x - 4$$

$$f'''(x) = \frac{d}{dx} f''(x) = 60x^2 - 72x + 24$$

$$f^{(4)}(x) = \frac{d}{dx} f'''(x) = 120x - 72$$

$$f^{(5)}(x) = \frac{d}{dx} f^{(4)}(x) = 120$$

and, in general,

$$f^{(n)}(x) = 0 \quad \text{for } n > 5$$

EXAMPLE 2 Find the third derivative of the function f defined by $y = x^{2/3}$. What is its domain?

Solution We have

$$y' = \frac{2}{3} x^{-1/3}$$

$$y'' = \left(\frac{2}{3}\right)\left(-\frac{1}{3}\right) x^{-4/3} = -\frac{2}{9} x^{-4/3}$$

Figure 3.13
The graph of the function $y = x^{2/3}$.

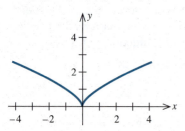

so the required derivative is

$$y''' = \left(-\frac{2}{9}\right)\left(-\frac{4}{3}\right)x^{-7/3} = \frac{8}{27}x^{-7/3} = \frac{8}{27x^{7/3}}$$

The common domain of the functions f', f'', and f''' is the set of all real numbers except $x = 0$. The domain of $y = x^{2/3}$ is the set of all real numbers. The graph of the function $y = x^{2/3}$ appears in Figure 3.13.   ◦ ◦ ◦

REMARK   Always simplify an expression before differentiating it to obtain the next order derivative.   ◦ ◦ ◦

EXAMPLE 3   Find the second derivative of the function $y = (2x^2 + 3)^{3/2}$.

Solution   We have, using the General Power Rule,

$$y' = \frac{3}{2}(2x^2 + 3)^{1/2}(4x) = 6x(2x^2 + 3)^{1/2}$$

Next, using the Product Rule and then the Chain Rule, we find

$$y'' = (6x) \cdot \frac{d}{dx}(2x^2 + 3)^{1/2} + \left[\frac{d}{dx}(6x)\right](2x^2 + 3)^{1/2}$$

$$= (6x)\left(\frac{1}{2}\right)(2x^2 + 3)^{-1/2}(4x) + 6(2x^2 + 3)^{1/2}$$

$$= 12x^2(2x^2 + 3)^{-1/2} + 6(2x^2 + 3)^{1/2}$$

$$= 6(2x^2 + 3)^{-1/2}[2x^2 + (2x^2 + 3)]$$

$$= \frac{6(4x^2 + 3)}{\sqrt{2x^2 + 3}}$$ ◦ ◦ ◦

Applications

Just as the derivative of a function f at a point x measures the rate of change of the function f at that point, the second derivative of f (the derivative of f') measures the rate of change of the derivative f' of the function f. The third derivative of the function f, f''', measures the rate of change of f'', and so on.

In Chapter 4 we will discuss applications involving the geometric interpretation of the second derivative of a function. The following example gives an interpretation of the second derivative in a familiar role.

EXAMPLE 4   Refer to the example on page 100. The distance s (in feet) covered by a maglev moving along a straight track t seconds after starting from rest is given by the function $s = 4t^2$ ($0 \le t \le 10$). What is the maglev's acceleration at the end of 30 seconds?

Solution   The velocity of the maglev t seconds from rest is given by

$$v = \frac{ds}{dt} = \frac{d}{dt}(4t^2) = 8t$$

The acceleration of the maglev t seconds from rest is given by the rate of change of the velocity of t—that is,

$$a = \frac{d}{dt}v = \frac{d}{dt}\left(\frac{ds}{dt}\right) = \frac{d^2s}{dt^2} = \frac{d}{dt}(8t) = 8$$

or 8 feet per second per second, normally abbreviated, 8 ft/sec².   **o o o**

EXAMPLE 5   A ball is thrown straight up into the air from the roof of a building. The height of the ball as measured from the ground is given by

$$s = -16t^2 + 24t + 120$$

where s is measured in feet and t in seconds. Find the velocity and acceleration of the ball 3 seconds after it is thrown into the air.

Solution   The velocity v and acceleration a of the ball at any time t are given by

$$v = \frac{ds}{dt} = \frac{d}{dt}(-16t^2 + 24t + 120) = -32t + 24$$

and

$$a = \frac{d^2t}{dt^2} = \frac{d}{dt}\left(\frac{ds}{dt}\right) = \frac{d}{dt}(-32t + 24) = -32$$

Therefore, the velocity of the ball 3 seconds after it is thrown into the air is

$$v = -32(3) + 24 = -72$$

That is, the ball is falling downward at a speed of 72 ft/sec. The acceleration of the ball is 32 ft/sec² downward at any time during the motion.   **o o o**

Figure 3.14
The CPI of a certain economy from year a to year b is given by I(t).

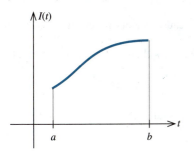

Another interpretation of the second derivative of a function—this time from the field of economics—follows. Suppose that the consumer price index (CPI) of an economy between the years a and b is described by the function $I(t)$ ($a \leq t \leq b$) (Figure 3.14). Then the first derivative of I, $I'(t)$, gives the rate of inflation of the economy at any time t. The second derivative of I, $I''(t)$, gives the *rate of change of the inflation rate* at any time t. Thus, when the economist or politician claims that "inflation is slowing," what he or she is saying is that the rate of inflation is decreasing. Mathematically, this is equivalent to noting that the second derivative $I''(t)$ is negative at the time t under consideration. Observe that $I'(t)$ could be positive at a time when $I''(t)$ is negative (see Example 6). Thus, one may not draw the conclusion from the aforementioned quote that prices of goods and services are about to drop!

EXAMPLE 6   An economy's consumer price index (CPI) is described by the function

$$I(t) = -0.2t^3 + 3t^2 + 100 \qquad (0 \le t \le 9)$$

where $t = 0$ corresponds to the year 1988. Compute $I'(6)$ and $I''(6)$ and use these results to show that even though the CPI was rising at the beginning of 1994, "inflation was moderating" at that time.

Figure 3.15
The CPI of an economy is given by $I(t)$.

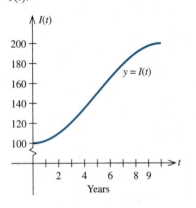

Solution   We find

$$I'(t) = -0.6t^2 + 6t$$

and

$$I''(t) = -1.2t + 6$$

so

$$I'(6) = -0.6(6)^2 + 6(6) = 14.4$$

and

$$I''(6) = -1.2(6) + 6 = -1.2$$

Our computations reveal that at the beginning of 1994 ($t = 6$), the CPI was increasing at the rate of 14.4 points per year, whereas the rate of the inflation rate was decreasing by 1.2 points per year. Thus, inflation was moderating at that time (Figure 3.15). In Section 4.2, we will see that relief actually began in early 1993.   ○ ○ ○

SELF-CHECK EXERCISES 3.5

1. Find the third derivative of

$$f(x) = 2x^5 - 3x^3 + x^2 - 6x + 10$$

2. Let

$$f(x) = \frac{1}{1+x}$$

Find $f'(x)$, $f''(x)$, and $f'''(x)$.

3. A certain species of turtle faces extinction because dealers collect truckloads of turtle eggs to be sold as aphrodisiacs. After severe conservation measures are implemented, it is hoped that the turtle population will grow according to the rule

$$N(t) = 2t^3 + 3t^2 - 4t + 1000 \qquad (0 \le t \le 10)$$

where $N(t)$ denotes the population at the end of year t. Compute $N''(2)$ and $N''(8)$ and interpret your results.

Solutions to Self-Check Exercises 3.5 can be found on page 238.

3.5 EXERCISES

In exercises 1–20, find the first and second derivatives of the given function.

1. $f(x) = 4x^2 - 2x + 1$

2. $f(x) = -0.2x^2 + 0.3x + 4$

3. $f(x) = 2x^3 - 3x^2 + 1$

4. $g(x) = -3x^3 + 24x^2 + 6x - 64$

5. $h(t) = t^4 - 2t^3 + 6t^2 - 3t + 10$

6. $f(x) = x^5 - x^4 + x^3 - x^2 + x - 1$

7. $f(x) = (x^2 + 2)^5$ **8.** $g(t) = t^2(3t + 1)^4$

9. $g(t) = (2t^2 - 1)^2(3t^2)$

10. $h(x) = (x^2 + 1)^2(x - 1)$

11. $f(x) = (2x^2 + 2)^{7/2}$

12. $h(w) = (w^2 + 2w + 4)^{5/2}$

13. $f(x) = x(x^2 + 1)^2$ **14.** $g(u) = u(2u - 1)^3$

15. $f(x) = \dfrac{x}{2x + 1}$ **16.** $g(t) = \dfrac{t^2}{t - 1}$

17. $f(s) = \dfrac{s - 1}{s + 1}$ **18.** $f(u) = \dfrac{u}{u^2 + 1}$

19. $f(u) = \sqrt{4 - 3u}$ **20.** $f(x) = \sqrt{2x - 1}$

In exercises 21–28, find the third derivative of the given function.

21. $f(x) = 3x^4 - 4x^3$

22. $f(x) = 3x^5 - 6x^4 + 2x^2 - 8x + 12$

23. $f(x) = \dfrac{1}{x}$ **24.** $f(x) = \dfrac{2}{x^2}$

25. $g(s) = \sqrt{3s - 2}$ **26.** $g(t) = \sqrt{2t + 3}$

27. $f(x) = (2x - 3)^4$ **28.** $g(t) = (\frac{1}{2}t^2 - 1)^5$

29. Acceleration of a Falling Object During the construction of an office building, a hammer is accidentally dropped from a height of 256 ft. The distance the hammer falls in t seconds is $s = 16t^2$. What is the hammer's velocity when it strikes the ground? What is its acceleration?

30. Acceleration of a Car The distance s (in feet) covered by a car t seconds after starting from rest is given by

$$s = 20t + 8t^2 - t^3 \qquad (0 \le t \le 6)$$

Find a general expression for the car's acceleration at any time t ($0 \le t \le 6$). Show that the car is decelerating $2\frac{2}{3}$ seconds after starting from rest.

31. Crime Rates The number of major crimes committed in Bronxville between 1988 and 1995 is approximated by the function

$$N(t) = -0.1t^3 + 1.5t^2 + 100 \qquad (0 \le t \le 7)$$

where $N(t)$ denotes the number of crimes committed in year t and $t = 0$ corresponds to the year 1988. Enraged by the dramatic increase in the crime rate, Bronxville's citizens, with the help of the local police, organized "Neighborhood Crime Watch" groups in early 1992 to combat this menace.

a. Verify that the crime rate was increasing from 1988 through 1995.

[*Hint:* Compute $N'(0), N'(1), \ldots, N'(7)$.]

b. Show that the Neighborhood Crime Watch program was working by computing $N''(4), N''(5), N''(6)$, and $N''(7)$.

32. GDP of a Developing Country A developing country's gross domestic product (GDP) from 1988 to 1996 is approximated by the function

$$G(t) = -0.2t^3 + 2.4t^2 + 60 \qquad (0 \le t \le 8)$$

where $G(t)$ is measured in billions of dollars and $t = 0$ corresponds to the year 1988.

a. Compute $G'(0), G'(1), \ldots, G'(8)$.

b. Compute $G''(0), G''(1), \ldots, G''(8)$.

c. Using the results obtained in (a) and (b), show that after a spectacular growth rate in the early years, the growth of the GDP cooled off.

33. Test Flight of a VTOL In a test flight of the McCord Terrier, McCord Aviation's experimental VTOL (vertical take-off and landing) aircraft, it was determined that t seconds after lift-off, when the craft was operated in

USING TECHNOLOGY

FINDING THE SECOND DERIVATIVE OF A FUNCTION AT A GIVEN POINT

Some graphing utilities have the capability of numerically computing the second derivative of a function at a point. If your graphing utility has this capability, use it to work through the examples and exercises of this section.

EXAMPLE 1 Use the (second) numerical derivative operation of a graphing utility to find the second derivative of $f(x) = \sqrt{x}$ when $x = 4$.

Solution Using the (second) numerical derivative operation of a graphing utility, we find

$$f''(4) = \text{der2 } (x^\wedge.5, x, 4) = -0.03125 \qquad \circ \; \circ \; \circ$$

EXAMPLE 2 The anticipated rise in Alzheimer's patients in the United States is given by

$$f(t) = -0.02765t^4 + 0.3346t^3 - 1.1261t^2$$
$$+ 1.7575t + 3.7745 \qquad (0 \le t \le 6)$$

where $f(t)$ is measured in millions and t is measured in decades, with $t = 0$ corresponding to the beginning of 1990.

a. How fast is the number of Alzheimer's patients in the United States anticipated to be changing at the beginning of 2030?

b. How fast is the rate of change of the number of Alzheimer's patients in the United States anticipated to be changing at the beginning of 2030?

c. Plot the graph of f in the viewing rectangle $[0, 7] \times [0, 12]$.
Source: Alzheimer's Association

Solution

a. Using the numerical derivative operation of a graphing utility, we find that the number of Alzheimer's patients at the beginning of 2030 can be anticipated to be changing at the rate of

$$f'(4) = 1.7311$$

That is, the number is increasing at the rate of approximately 1.7 million patients per decade.

b. Using the (second) numerical derivative operation of a graphing utility, we find that

$$f''(4) = 0.4694$$

That is, the rate of change of the number of Alzheimer's patients is increasing at the rate of approximately 0.5 million patients per decade per decade.

c. The graph is shown in Figure T1.

Figure T1
The graph of f in the viewing window $[0, 7] \times [0, 12]$.

EXERCISES

In exercises 1–8, find the value of the second derivative of f at the given value of x. Express your answer correct to four decimal places.

1. $f(x) = 2x^3 - 3x^2 + 1; x = -1$

2. $f(x) = 2.5x^5 - 3x^3 + 1.5x + 4; x = 2.1$

3. $f(x) = 2.1x^{3.1} - 4.2x^{1.7} + 4.2; x = 1.4$

4. $f(x) = 1.7x^{4.2} - 3.2x^{1.3} + 4.2x - 3.2; x = 2.2$

5. $f(x) = \dfrac{x^2 + 2x - 5}{x^3 + 1}; x = 2.1$

6. $f(x) = \dfrac{x^3 + x + 2}{2x^2 - 5x + 4}; x = 1.2$

7. $f(x) = \dfrac{x^{1/2} + 2x^{3/2} + 1}{2x^{1/2} + 3}; x = 0.5$

8. $f(x) = \dfrac{\sqrt{x} - 1}{2x + \sqrt{x} + 4}; x = 2.3$

9. **Rate of Bank Failures** The Federal Deposit Insurance Company (FDIC) estimates that the rate at which banks were failing between 1982 and 1994 is given by

$$f(t) = -0.063447t^4 - 1.953283t^3 + 14.632576t^2 - 6.684704t + 47.458874 \quad (0 \le t \le 12)$$

where $f(t)$ is measured in the number of banks per year and t is measured in years, with $t = 0$ corresponding to the beginning of 1982. Compute $f''(6)$ and interpret your results.
Source: Federal Deposit Insurance Corp.

10. **Multimedia Sales** According to the Electronic Industries Association, sales in the multimedia market (hardware and software) are expected to be

$$S(t) = -0.0094t^4 + 0.1204t^3 - 0.0868t^2 + 0.0195t + 3.3325 \quad (0 \le t \le 10)$$

where $S(t)$ is measured in billions of dollars and t is measured in years, with $t = 0$ corresponding to 1990. Compute $S''(7)$ and interpret your results.
Source: Electronics Industries Association

the vertical take-off mode, its altitude (in feet) was

$$h(t) = \frac{1}{16}t^4 - t^3 + 4t^2 \qquad (0 \le t \le 8)$$

a. Find an expression for the craft's velocity at time t.
b. Find the craft's velocity when $t = 0$ (the initial velocity), $t = 4$, and $t = 8$.
c. Find an expression for the craft's acceleration at time t.
d. Find the craft's acceleration when $t = 0$, $t = 4$, and $t = 8$.
e. Find the craft's height when $t = 0$, $t = 4$, and $t = 8$.

34. U.S. Census   According to the U.S. Census Bureau, the number of Americans aged 45 to 54 will be approximately

$$N(t) = -0.00233t^4 + 0.00633t^3 - 0.05417t^2$$
$$+ 1.3467t + 25$$

million people in year t, where $t = 0$ corresponds to the beginning of 1990. Compute $N'(10)$ and $N''(10)$ and interpret your results.
Source: U.S. Census Bureau

35. Air Purification   During testing of a certain brand of air purifier, it was determined that the amount of smoke remaining t minutes after the start of the test was

$$A(t) = 100 - 17.63t + 1.915t^2 - 0.1316t^3$$
$$+ 0.00468t^4 - 0.00006t^5$$

percent of the original amount. Compute $A'(10)$ and $A''(10)$ and interpret your results.
Source: Consumer Reports

36. Let f be the function defined by the rule $f(x) = x^{7/3}$. Show that f has first- and second-order derivatives at all points x, in particular at $x = 0$. Show also that the third derivative of f does not exist at $x = 0$.

37. Construct a function f that has derivatives of order up through and including n at a point a but fails to have the $(n + 1)$st derivative there.
[*Hint:* See exercise 36.]

38. Show that a polynomial function has derivatives of all orders.
[*Hint:* Let $P(x) = a_0 x^n + a_1 x^{n-1} + a_2 x^{n-2} + \cdots + a_n$ be a polynomial of degree n, where n is a positive integer and a_0, $a_1, \ldots, a_n$ are constants with $a_0 \ne 0$. Compute $P'(x), P''(x), \ldots$]

SOLUTIONS TO SELF-CHECK EXERCISES 3.5

1. $f'(x) = 10x^4 - 9x^2 + 2x - 6$
$f''(x) = 40x^3 - 18x + 2$
$f'''(x) = 120x^2 - 18$

2. We write $f(x) = (1 + x)^{-1}$ and use the General Power Rule, obtaining

$$f'(x) = (-1)(1 + x)^{-2}\frac{d}{dx}(1 + x) = -(1 + x)^{-2}(1)$$

$$= -(1 + x)^{-2} = -\frac{1}{(1 + x)^2}$$

Continuing, we find

$$f''(x) = -(-2)(1 + x)^{-3}$$
$$= 2(1 + x)^{-3} = \frac{2}{(1 + x)^3}$$

and

$$f'''(x) = 2(-3)(1 + x)^{-4}$$
$$= -6(1 + x)^{-4}$$
$$= -\frac{6}{(1 + x)^4}$$

3. $N'(t) = 6t^2 + 6t - 4$

$N''(t) = 12t + 6 = 6(2t + 1)$

Therefore, $N''(2) = 30$ and $N''(8) = 102$. The results of our computations reveal that at the end of year 2, the *rate* of growth of the turtle population is increasing at the rate of 30 turtles per year per year. At the end of year 8, the rate is increasing at the rate of 102 turtles per year per year. Clearly, the conservation measures are paying off handsomely.

3.6 IMPLICIT DIFFERENTIATION AND RELATED RATES

Differentiating Implicitly

Up to now we have dealt with functions expressed in the form $y = f(x)$; that is, the dependent variable y is expressed *explicitly* in terms of the independent variable x. However, not all functions are expressed in this form. Consider, for example, the equation

$$x^2 y + y - x^2 + 1 = 0 \qquad (8)$$

This equation does express y *implicitly* as a function of x. In fact, solving (8) for y in terms of x, we obtain

$$(x^2 + 1)y = x^2 - 1 \qquad \text{(Implicit equation)}$$

or $\qquad y = f(x) = \dfrac{x^2 - 1}{x^2 + 1} \qquad \text{(Explicit equation)}$

which gives an explicit representation of f.

Next, consider the equation

$$y^4 - y^3 - y + 2x^3 - x = 8$$

When certain restrictions are placed on x and y, this equation defines y as a function of x. But in this instance, we would be hard pressed to find y explicitly in terms of x. The following question arises naturally: How does one go about computing dy/dx in this case?

As it turns out, thanks to the Chain Rule, a method *does* exist for computing the derivative of a function directly from the implicit equation defining the function. This method is called **implicit differentiation** and is demonstrated in the next several examples.

EXAMPLE 1 Find dy/dx given the equation $y^2 = x$.

Solution Differentiating both sides of the equation with respect to x, we obtain

$$\frac{d}{dx}(y^2) = \frac{d}{dx}(x)$$

In order to carry out the differentiation of the term $\dfrac{d}{dx}(y^2)$, we note that y is a function of x. Writing $y = f(x)$ to remind us of this fact, we find that

$$\dfrac{d}{dx}(y^2) = \dfrac{d}{dx}[f(x)]^2 \qquad \text{[Writing } y = f(x)\text{]}$$

$$= 2f(x)f'(x) \qquad \text{(Using the Chain Rule)}$$

$$= 2y\dfrac{dy}{dx} \qquad \text{[Returning to using } y \text{ instead of } f(x)\text{]}$$

Therefore, the equation

$$\dfrac{d}{dx}(y^2) = \dfrac{d}{dx}(x)$$

is equivalent to

$$2y\dfrac{dy}{dx} = 1$$

Solving for $\dfrac{dy}{dx}$ yields

$$\dfrac{dy}{dx} = \dfrac{1}{2y}$$

Before considering other examples, let us summarize the important steps involved in implicit differentiation. (Here we assume that dy/dx exists.)

FINDING $\dfrac{dy}{dx}$ BY IMPLICIT DIFFERENTIATION

1. Differentiate both sides of the equation *with respect to x*. (Make sure that the derivative of any term involving y includes the factor dy/dx.)

2. Solve the resulting equation for $\dfrac{dy}{dx}$ in terms of x and y.

EXAMPLE 2   Find dy/dx given the equation

$$y^3 - y + 2x^3 - x = 8$$

Solution   Differentiating both sides of the given equation with respect to x, we obtain

$$\dfrac{d}{dx}(y^3 - y + 2x^3 - x) = \dfrac{d}{dx}(8)$$

or

$$\dfrac{d}{dx}(y^3) - \dfrac{d}{dx}(y) + \dfrac{d}{dx}(2x^3) - \dfrac{d}{dx}(x) = 0$$

Now, recalling that y is a function of x, we apply the Chain Rule to the first two terms on the left. Thus,

$$3y^2 \frac{dy}{dx} - \frac{dy}{dx} + 6x^2 - 1 = 0$$

$$(3y^2 - 1)\frac{dy}{dx} = 1 - 6x^2$$

$$\frac{dy}{dx} = \frac{1 - 6x^2}{3y^2 - 1}$$ ◦ ◦ ◦

Refer to Example 2.

Suppose we think of the equation $y^3 - y + 2x^3 - x = 8$ as defining x implicitly as a function of y. Find dx/dy and justify your method of solution.

EXAMPLE 3   Consider the equation $x^2 + y^2 = 4$.

a. Find dy/dx by implicit differentiation.

b. Find the slope of the tangent line to the graph of the function $y = f(x)$ at the point $(1, \sqrt{3})$.

c. Find an equation of the tangent line of (b).

Solution

a. Differentiating both sides of the equation with respect to x, we obtain

$$\frac{d}{dx}(x^2 + y^2) = \frac{d}{dx}(4)$$

$$\frac{d}{dx}(x^2) + \frac{d}{dx}(y^2) = 0$$

$$2x + 2y\frac{dy}{dx} = 0$$

and

$$\frac{dy}{dx} = -\frac{x}{y} \qquad (y \neq 0)$$

b. The slope of the tangent line to the graph of the function at the point $(1, \sqrt{3})$ is given by

$$\left.\frac{dy}{dx}\right|_{(1,\sqrt{3})} = -\frac{x}{y}\bigg|_{(1,\sqrt{3})} = -\frac{1}{\sqrt{3}}$$

(*Note:* This notation is read "dy/dx evaluated at the point $(1, \sqrt{3})$.")

c. An equation of the tangent line in question is found by using the point-slope form of the equation of a line with the slope $m = -1/\sqrt{3}$ and the point $(1, \sqrt{3})$. Thus,

$$y - \sqrt{3} = -\frac{1}{\sqrt{3}}(x - 1)$$

$$\sqrt{3}y - 3 = -x + 1$$

or $$x + \sqrt{3}y - 4 = 0$$

A sketch of this tangent line is shown in Figure 3.16.

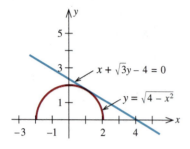

Figure 3.16
The line $x + \sqrt{3}y - 4 = 0$ is tangent to the graph of the function $y = f(x)$.

We can also solve the equation $x^2 + y^2 = 4$ explicitly for y in terms of x. If we do this, we obtain

$$y = \pm\sqrt{4 - x^2}$$

From this, we see that the equation $x^2 + y^2 = 4$ defines the two functions

$$y = f(x) = \sqrt{4 - x^2}$$

and

$$y = g(x) = -\sqrt{4 - x^2}$$

Since the point $(1, \sqrt{3})$ does not lie on the graph of $y = g(x)$, we conclude that

$$y = f(x) = \sqrt{4 - x^2}$$

is the required function. The graph of f is the upper semicircle shown in Figure 3.16.   ○ ○ ○

To find dy/dx at a *specific* point (a, b), differentiate the given equation implicitly with respect to x and then replace x and y by a and b, respectively, *before* solving the equation for dy/dx. This often simplifies the amount of algebra involved.

EXAMPLE 4   Find dy/dx given that x and y are related by the equation

$$x^2y^3 + 6x^2 = y + 12$$

and that $y = 2$ when $x = 1$.

Solution   Differentiating both sides of the given equation with respect to x, we obtain

$$\frac{d}{dx}(x^2y^3) + \frac{d}{dx}(6x^2) = \frac{d}{dx}(y) + \frac{d}{dx}(12)$$

$$x^2 \cdot \frac{d}{dx}(y^3) + y^3 \cdot \frac{d}{dx}(x^2) + 12x = \frac{dy}{dx} \qquad \text{[Using the Product Rule on } \frac{d}{dx}(x^2y^3)]$$

$$3x^2y^2\frac{dy}{dx} + 2xy^3 + 12x = \frac{dy}{dx}$$

Substituting $x = 1$ and $y = 2$ into this equation gives

$$3(1)^2(2)^2\frac{dy}{dx} + 2(1)(2)^3 + 12(1) = \frac{dy}{dx}$$

$$12\frac{dy}{dx} + 16 + 12 = \frac{dy}{dx}$$

and, solving for $\dfrac{dy}{dx}$,

$$\frac{dy}{dx} = -\frac{28}{11}$$

Note that it is not necessary to find an explicit expression for dy/dx.   ○ ○ ○

REMARK   In Examples 3 and 4, you can verify that the points at which we evaluated dy/dx actually lie on the curve in question by showing that the coordinates of the points satisfy the given equations.   ◦ ◦ ◦

EXAMPLE 5   Find dy/dx given that x and y are related by the equation

$$\sqrt{x^2 + y^2} - x^2 = 5$$

Solution   Differentiating both sides of the given equation with respect to x, we obtain

$$\frac{d}{dx}(x^2 + y^2)^{1/2} - \frac{d}{dx}(x^2) = \frac{d}{dx}(5)$$

[Writing $\sqrt{x^2 + y^2} = (x^2 + y^2)^{1/2}$]

$$\frac{1}{2}(x^2 + y^2)^{-1/2}\frac{d}{dx}(x^2 + y^2) - 2x = 0$$

(Using the General Power Rule on the first term)

$$\frac{1}{2}(x^2 + y^2)^{-1/2}\left(2x + 2y\frac{dy}{dx}\right) - 2x = 0$$

$$2x + 2y\frac{dy}{dx} = 4x(x^2 + y^2)^{1/2}$$

[Transposing $2x$ and multiplying both sides by $2(x^2 + y^2)^{1/2}$]

$$2y\frac{dy}{dx} = 4x(x^2 + y^2)^{1/2} - 2x$$

$$\frac{dy}{dx} = \frac{2x\sqrt{x^2 + y^2} - x}{y}$$

◦ ◦ ◦

Related Rates

Implicit differentiation is a useful technique for solving a class of problems known as **related rates** problems. For example, suppose that x and y are each functions of a third variable t. Here, x might denote the mortgage rate and y the number of single-family homes sold at any time t. Further, suppose that we have an equation that gives the relationship between x and y (the number of houses sold y is related to the mortgage rate x). Differentiating both sides of this equation implicitly with respect to t, we obtain an equation that gives a relationship between $\dfrac{dx}{dt}$ and $\dfrac{dy}{dt}$. In the context of our example, this equation gives us a relationship between the rate of change of the mortgage rate and the rate of change of the number of houses sold, as a function of time. Thus, knowing

$$\frac{dx}{dt} \qquad \text{(How fast the mortgage rate is changing at time } t\text{)}$$

we can determine

$$\frac{dy}{dt} \qquad \text{(How fast the sale of houses is changing at that instant of time)}$$

EXAMPLE 6   A study prepared for the National Association of Realtors estimates that the number of housing starts in the Southwest, $N(t)$ (in units of a million), over the next five years is related to the mortgage rate $r(t)$ (percent per year) by the equation

$$9N^2 + r = 36$$

What is the rate of change of the number of housing starts with respect to time when the mortgage rate is 11% per year and is increasing at the rate of 1.5% per year?

Solution   We are given that

$$r = 11 \qquad \text{and} \qquad \frac{dr}{dt} = 1.5$$

at a certain instant of time, and we are required to find $\dfrac{dN}{dt}$. First, by substituting $r = 11$ into the given equation, we find

$$9N^2 + 11 = 36$$

$$N^2 = \frac{25}{9}$$

or $N = 5/3$ (we reject the negative root). Next, differentiating the given equation implicitly on both sides with respect to t, we obtain

$$\frac{d}{dt}(9N^2) + \frac{d}{dt}(r) = \frac{d}{dt}(36)$$

$$18N\frac{dN}{dt} + \frac{dr}{dt} = 0 \qquad \color{red}{\text{(Use the Chain Rule on the first term.)}}$$

Then, substituting $N = \dfrac{5}{3}$ and $\dfrac{dr}{dt} = 1.5$ into this equation gives

$$18\left(\frac{5}{3}\right)\frac{dN}{dt} + 1.5 = 0$$

Solving this equation for dN/dt then gives

$$\frac{dN}{dt} = -\frac{1.5}{30} \approx -0.05$$

Thus, at the instant of time under consideration, the number of housing starts is decreasing at the rate of 50,000 units per year.   ◦ ◦ ◦

EXAMPLE 7   A major audio-tape manufacturer is willing to make x thousand 10-packs of metal alloy audiocassette tapes per week available in the marketplace when the wholesale price is $\$p$ per 10-pack. It is known that the relationship between x and p is governed by the supply equation

$$x^2 - 3xp + p^2 = 5$$

How fast is the supply of tapes changing when the price per 10-pack is $11, the quantity supplied is 4000 10-packs, and the wholesale price per 10-pack is increasing at the rate of 10 cents per 10-pack per week?

Solution   We are given that

$$p = 11, \quad x = 4, \quad \text{and} \quad \frac{dp}{dt} = 0.1$$

at a certain instant of time, and we are required to find $\dfrac{dx}{dt}$. Differentiating the given equation on both sides with respect to t, we obtain

$$\frac{d}{dt}(x^2) - \frac{d}{dt}(3xp) + \frac{d}{dt}(p^2) = \frac{d}{dt}(5)$$

$$2x\frac{dx}{dt} - 3\left(p\frac{dx}{dt} + x\frac{dp}{dt}\right) + 2p\frac{dp}{dt} = 0 \qquad \begin{array}{c}\text{(Use the Product Rule} \\ \text{on the second term.)}\end{array}$$

Substituting the given values of p, x, and $\dfrac{dp}{dt}$ into the last equation, we have

$$2(4)\frac{dx}{dt} - 3\left[(11)\frac{dx}{dt} + 4(0.1)\right] + 2(11)(0.1) = 0$$

$$8\frac{dx}{dt} - 33\frac{dx}{dt} - 1.2 + 2.2 = 0$$

$$25\frac{dx}{dt} = 1$$

or

$$\frac{dx}{dt} = 0.04$$

Thus, at the instant of time under consideration the supply of 10-pack audiocassettes is increasing at the rate of $(0.04)(1000)$, or 40, 10-packs per week.

● ● ●

In certain related problems, we need to formulate the problem mathematically before analyzing it. The following guidelines can be used to help solve problems of this type.

SOLVING RELATED RATES PROBLEMS

1. Assign a variable to each quantity. Draw a diagram if needed.
2. Write the *given* values of the variables and their rates of change with respect to t.
3. Find an equation giving the relationship between the variables.
4. Differentiate both sides of this equation implicitly with respect to t.
5. Replace the variables and their derivatives by the numerical data found in step 2, and solve the equation for the required rate of change.

EXAMPLE 8   At a distance of 4000 feet from the launch site, a spectator is observing a rocket being launched. If the rocket lifts off vertically and is rising at a speed of 600 ft/sec when it is at an altitude of 3000 ft, how fast is the distance between the rocket and the spectator changing at that instant?

Solution

Step 1   Let

$$y = \text{the altitude of the rocket}$$

and $x = \text{the distance between the rocket and the spectator}$

at any time t (see Figure 3.17).

Figure 3.17

The rate at which x is changing with respect to time is related to the rate of change of y with respect to time.

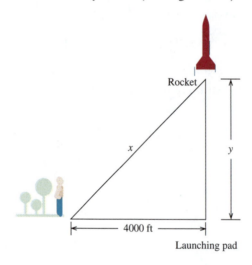

Step 2   We are given that at a certain instant of time

$$y = 3000 \quad \text{and} \quad \frac{dy}{dt} = 600$$

and are asked to find dx/dt at that instant.

Step 3   Applying the Pythagorean Theorem to the right triangle in Figure 3.17, we find that

$$x^2 = y^2 + 4000^2$$

Therefore, when $y = 3000$,

$$x = \sqrt{3000^2 + 4000^2} = 5000$$

Step 4   Next, we differentiate the equation $x^2 = y^2 + 4000^2$ with respect to t, obtaining

$$2x \frac{dx}{dt} = 2y \frac{dy}{dt}$$

(Remember that both x and y are functions of t.)

Step 5 Substituting $x = 5000$, $y = 3000$, and $dy/dt = 600$, we find

$$2(5000)\frac{dx}{dt} = 2(3000)(600)$$

or
$$\frac{dx}{dt} = 360$$

Therefore, the distance between the rocket and the spectator is changing at a rate of 360 ft/sec.   ○ ○ ○

 Be sure that you do *not* replace the variables in the equation found in step 3 by their numerical values before differentiating the equation.

SELF-CHECK EXERCISES 3.6

1. Given the equation $x^3 + 3xy + y^3 = 4$, find $\dfrac{dy}{dx}$ by implicit differentiation.

2. Find an equation of the tangent line to the graph of $16x^2 + 9y^2 = 144$ at the point $\left(2, -\dfrac{4\sqrt{5}}{3}\right)$.

Solutions to Self-Check Exercises 3.6 can be found on page 249.

3.6 EXERCISES

In exercises 1–8, find the derivative dy/dx (a) by solving each of the given implicit equations for y explicitly in terms of x and (b) by differentiating each of the given equations implicitly. Show that, in each case, the results are equivalent.

1. $x + 2y = 5$

2. $3x + 4y = 6$

3. $xy = 1$

4. $xy - y - 1 = 0$

5. $x^3 - x^2 - xy = 4$

6. $x^2y - x^2 + y - 1 = 0$

7. $\dfrac{x}{y} - x^2 = 1$

8. $\dfrac{y}{x} - 2x^3 = 4$

In exercises 9–30, find dy/dx by implicit differentiation.

9. $x^2 + y^2 = 16$

10. $2x^2 + y^2 = 16$

11. $x^2 - 2y^2 = 16$

12. $x^3 + y^3 + y - 4 = 0$

13. $x^2 - 2xy = 6$

14. $x^2 + 5xy + y^2 = 10$

15. $x^2y^2 - xy = 8$

16. $x^2y^3 - 2xy^2 = 5$

17. $x^{1/2} + y^{1/2} = 1$

18. $x^{1/3} + y^{1/3} = 1$

19. $\sqrt{x + y} = x$

20. $(2x + 3y)^{1/3} = x^2$

21. $\dfrac{1}{x^2} + \dfrac{1}{y^2} = 1$

22. $\dfrac{1}{x^3} + \dfrac{1}{y^3} = 5$

23. $\sqrt{xy} = x + y$

24. $\sqrt{xy} = 2x + y^2$

25. $\dfrac{x + y}{x - y} = 3x$

26. $\dfrac{x - y}{2x + 3y} = 2x$

27. $xy^{3/2} = x^2 + y^2$

28. $x^2y^{1/2} = x + 2y^3$

29. $(x + y)^3 + x^3 + y^3 = 0$

30. $(x + y^2)^{10} = x^2 + 25$

In exercises 31–34, find an equation of the tangent line to the graph of the function f defined by the given equation at the indicated point.

31. $4x^2 + 9y^2 = 36$; $(0, 2)$

32. $y^2 - x^2 = 16$; $(2, 2\sqrt{5})$

33. $x^2y^3 - y^2 + xy - 1 = 0$; $(1, 1)$

34. $(x - y - 1)^3 = x$; $(1, -1)$

In exercises 35–38, find the second derivative d^2y/dx^2 of each of the functions defined implicitly by the given equation.

35. $xy = 1$

36. $x^3 + y^3 = 28$

37. $y^2 - xy = 8$

38. $x^{1/3} + y^{1/3} = 1$

39. The volume of a right-circular cylinder of radius r and height h is $V = \pi r^2 h$. Suppose that the radius and height of the cylinder are changing with respect to time t.
   a. Find a relationship between dV/dt, dr/dt, and dh/dt.
   b. At a certain instant of time, the radius and height of the cylinder are 2 and 6 inches and are increasing at the rate of 0.1 and 0.3 in./sec, respectively. How fast is the volume of the cylinder increasing?

40. A car leaves an intersection traveling west. Its position 4 seconds later is 20 ft from the intersection. At the same time, another car leaves the same intersection heading north so that its position 4 seconds later is 28 ft from the intersection. If the speed of the cars at that instant of time is 9 ft/sec and 11 ft/sec, respectively, find the rate at which the distance between the two cars is changing.

41. Price-Demand   Suppose that the quantity demanded weekly of the Super Titan radial tires is related to its unit price by the equation

$$p + x^2 = 144$$

where p is measured in dollars and x is measured in units of a thousand. How fast is the quantity demanded changing when $x = 9$, $p = 63$, and the price per tire is increasing at the rate of $2 per week?

42. Price-Supply   Suppose the quantity x of Super Titan radial tires made available per week in the marketplace by the Titan Tire Company is related to the unit selling price by the equation

$$p - \frac{1}{2}x^2 = 48$$

where x is measured in units of a thousand and p is in dollars. How fast is the weekly supply of Super Titan radial tires being introduced into the marketplace when $x = 6$, $p = 66$, and the price per tire is decreasing at the rate of $3 per week?

43. Price-Demand   The demand equation for a certain brand of metal alloy audiocassette tape is

$$100x^2 + 9p^2 = 3600$$

where x represents the number (in thousands) of 10-packs demanded per week when the unit price is p. How fast is the quantity demanded increasing when the unit price per 10-pack is $14 and the selling price is dropping at the rate of 15 cents per 10-pack per week?
[*Hint:* To find the value of x when $p = 14$, solve the equation $100x^2 + 9p^2 = 3600$ for x when $p = 14$.]

44. Effect of Price on Supply   Suppose that the wholesale price of a certain brand of medium-size eggs p (in dollars per carton) is related to the weekly supply x (in thousands of cartons) by the equation

$$625p^2 - x^2 = 100$$

If 25,000 cartons of eggs are available at the beginning of a certain week and the price is falling at the rate of 2 cents per carton per week, at what rate is the supply falling?
[*Hint:* To find the value of p when $x = 25$, solve the supply equation for p when $x = 25$.]

45. Supply-Demand   Refer to exercise 44. If 25,000 cartons of eggs are available at the beginning of a certain week and the supply is falling at the rate of 1000 cartons per week, at what rate is the wholesale price changing?

46. Elasticity of Demand   The demand function for a certain make of cartridge typewriter ribbon is

$$p = -0.01x^2 - 0.1x + 6$$

where p is the unit price in dollars and x is the quantity demanded each week, measured in units of a thousand. Compute the elasticity of demand and determine whether the demand is inelastic, unitary, or elastic when $x = 10$.

47. Elasticity of Demand   The demand function for a certain brand of long-playing record is

$$p = -0.01x^2 - 0.2x + 8$$

where p is the wholesale unit price in dollars and x is the quantity demanded each week, measured in units of a thousand. Compute the elasticity of demand, and determine whether the demand is inelastic, unitary, or elastic when $x = 15$.

48. The volume V of a cube with sides of length x inches is changing with respect to time. At a certain instant of time, the sides of the cube are 5 inches long and increasing at the rate of 0.1 in./sec. How fast is the volume of the cube changing at that instant of time?

49. Oil Spills   In calm waters, oil spilling from the ruptured hull of a grounded tanker spreads in all directions. If the area polluted is a circle and its radius is increasing at a rate of 2 ft/sec, determine how fast the area is increasing when the radius of the circle is 40 feet.

50. Two ships leave the same port at noon. Ship A sails north at 15 mph and Ship B sails east at 12 mph. How fast is the distance between them changing at 1 P.M.?

51. A car leaves an intersection traveling east. Its position t seconds later is given by $x = t^2 + t$ feet. At the same time, another car leaves the same intersection heading north, traveling $y = t^2 + 3t$ feet in t seconds. Find the rate at which the distance between the two cars will be changing 5 seconds later.

52. At a distance of 50 feet from the pad, a man observes a helicopter taking off from a heliport. If the helicopter lifts off vertically and is rising at a speed of 44 ft/sec when it is at an altitude of 120 feet, how fast is the distance between the helicopter and the man changing at that instant?

53. A spectator watches a rowing race from the edge of a river bank. The lead boat is moving in a straight line that is 120 feet from the river bank. If the boat is moving at a constant speed of 20 ft/sec, how fast is the boat moving away from the spectator when it is 50 feet past her?

54. A man 6 feet tall is walking away from a street light 18 feet high at a speed of 6 ft/sec. How fast is the tip of his shadow moving along the ground?

55. A 20-foot ladder is leaning against a wall. If the bottom of the ladder is pulled away from the wall at a rate of 2 ft/sec, how fast is the top of the ladder sliding down the wall when the bottom of the ladder is 12 feet from the wall?

[*Hint:* Refer to the adjacent figure. By the Pythagorean Theorem, $x^2 + y^2 = 400$. Find dy/dt when $x = 12$ and $dx/dt = 2$.]

56. A 13-foot ladder is leaning against a wall. If the bottom of the ladder is pulled away from the wall at a rate of 2.5 ft/sec, how fast is the top of the ladder sliding down the wall when the bottom of the ladder is 12 feet from the wall?

SOLUTIONS TO SELF-CHECK EXERCISES 3.6

1. Differentiating both sides of the equation with respect to x, we have

$$3x^2 + 3y + 3xy' + 3y^2y' = 0$$
$$(x^2 + y) + (x + y^2)y' = 0$$

or

$$y' = -\frac{x^2 + y}{x + y^2}$$

2. To find the slope of the tangent line to the graph of the function at any point, we differentiate the equation implicitly with respect to x, obtaining

$$32x + 18yy' = 0$$

or

$$y' = -\frac{16x}{9y}$$

In particular, the slope of the tangent line at $\left(2, -\dfrac{4\sqrt{5}}{3}\right)$ is

$$m = -\frac{16(2)}{9\left(-\dfrac{4\sqrt{5}}{3}\right)} = \frac{8}{3\sqrt{5}}$$

Using the point-slope form of the equation of a line, we find

$$y - \left(-\frac{4\sqrt{5}}{3}\right) = \frac{8}{3\sqrt{5}}(x - 2)$$

or

$$y = \frac{8\sqrt{5}}{15}x - \frac{36\sqrt{5}}{15}$$

3.7 DIFFERENTIALS

The Millers are planning to buy a house in the near future and estimate that they will need a 30-year fixed-rate mortgage for $120,000. If the interest rate increases from the present rate of 9% per year to 9.4% per year between now and the time the Millers decide to secure the loan, approximately how much more per month will their mortgage be? (You will be asked to answer this question in exercise 43, page 262.)

Questions such as this, in which one wishes to *estimate* the change in the dependent variable (monthly mortgage payment) corresponding to a small change in the independent variable (interest rate per year), occur in many real-life applications. For example:

- An economist would like to know how a small increase in a country's capital expenditure will affect the country's gross domestic output.
- A sociologist would like to know how a small increase in the amount of capital investment in a housing project will affect the crime rate.
- A businesswoman would like to know how raising a product's unit price by a small amount will affect her profit.
- A bacteriologist would like to know how a small increase in the amount of a bactericide will affect a population of bacteria.

In order to calculate these changes and estimate their effects, we use the *differential* of a function, a concept that will be introduced shortly.

Increments

Let x denote a variable quantity and suppose that x changes from x_1 to x_2. This change in x is called the **increment in x** and is denoted by the symbol Δx (read "delta x"). Thus,

$$\Delta x = x_2 - x_1 \qquad \text{(Final value minus initial value)} \tag{9}$$

John Decker

Title: Mortgage Counselor
Institution: Shawmut Bank

Decker stresses that he and his colleagues "strive to grant loans. That's our job." But before allowing someone to file a formal application, Decker takes the person over several "prequalification hurdles" to gauge his or her ability to handle a mortgage.

To start, Decker relies on a two-tiered, debt-to-income ratio to see whether a person has sufficient gross monthly income to make payments. Under the first tier, the proposed monthly payment (principal and interest, property tax, homeowner's insurance, and, when applicable, a condo fee) cannot exceed 28% of an individual's gross monthly income. If Decker's initial calculations are positive, he then must determine the individual's ability to meet monthly mortgage payments while also repaying other debts, such as car and student loans, credit cards, alimony, and so on. These combined payments cannot exceed 36% percent of gross monthly income. A typical person with an $800 mortgage obligation and $550 in other payments would have to earn $3750 per month to clear these first two hurdles.

Contrary to popular belief, bankers *want* to lend money. "The idea is to grant mortgages," says Decker, which contribute substantially to a bank's profitability. But lending money means making sensible decisions about how much to lend as well as a person's ability to repay the loan.

Using a loan-to-value formula, banks might lend 80% of a property's value. In such cases, the applicant has to put 20% down to make the purchase. Or Shawmut may decide to lend up to 95% or as little as 75% of the appraised value.

Understandably, banks don't like to see bankruptcies, late payments, or liens on a personal credit history. Decker notes, however, that even this final hurdle doesn't necessarily mean failure in securing a mortgage. He works closely with each individual to overcome any stigma that might prompt a rejection.

Once Decker has put together a successful mortgage application, it is reviewed internally. Then, even though a mortgage is granted, it might come through at a slightly higher interest rate—10.4% instead of the expected 10% rate. Differentials would be used to compute the change in monthly payments, for although this might seem like a small change, on a large mortgage it can be enough of a variable to affect the new customer's ability to make monthly payments.

Using the debt-to-income ratio, Decker plugs the new variable into his formulas to determine whether a problem exists. These formulas used to compute the monthly payments involve the use of exponential functions. On a $100,000, 30-year, fixed-rate mortgage, payments will increase about $30 per month. Decker notes that such a small increase doesn't usually pose a significant problem.

If a customer can't make the new payment, however, Decker explores alternatives until he finds a solution. For Decker, it comes down to this: "If there is any possible way to give a loan, we're going to make it work."

Decker's job is simple: to lend money to people who want to buy a home. The hard part is deciding whether applicants qualify for one of Shawmut's 40 different mortgage plans. Sifting through income figures, current indebtedness, and credit history helps Decker determine which individuals make the best mortgage candidates.

EXAMPLE 1   Find the increment in x:

a. as x changes from 3 to 3.2   **b.** as x changes from 3 to 2.7

Solution

a. Here $x_1 = 3$ and $x_2 = 3.2$, so

$$\Delta x = x_2 - x_1 = 3.2 - 3 = 0.2$$

b. Here $x_1 = 3$ and $x_2 = 2.7$. Therefore,

$$\Delta x = x_2 - x_1 = 2.7 - 3 = -0.3$$ ◦ ◦ ◦

Observe that Δx plays the same role that h played in Section 2.4.

Now, suppose two quantities, x and y, are related by an equation $y = f(x)$, where f is a function. If x changes from x to $x + \Delta x$, then the corresponding change in y is called the *increment in y*. It is denoted by Δy and is defined in Figure 3.18 by

$$\Delta y = f(x + \Delta x) - f(x) \tag{10}$$

Figure 3.18
An increment of Δx in x induces an increment of $\Delta y = f(x + \Delta x) - f(x)$ in y.

EXAMPLE 2   Let $y = x^3$. Find Δx and Δy when

a. x changes from 2 to 2.01   **b.** x changes from 2 to 1.98

Solution   Let $f(x) = x^3$.

a. Here $\Delta x = 2.01 - 2 = 0.01$. Next,

$$\Delta y = f(x + \Delta x) - f(x) = f(2.01) - f(2)$$
$$= (2.01)^3 - 2^3 = 8.120601 - 8 = 0.120601$$

b. Here $\Delta x = 1.98 - 2 = -0.02$. Next,

$$\Delta y = f(x + \Delta x) - f(x) = f(1.98) - f(2)$$
$$= (1.98)^3 - 2^3 = 7.762392 - 8 = -0.237608$$ ◦ ◦ ◦

Differentials

We can obtain a relatively quick and simple way of approximating Δy, the change in y due to a small change Δx, by examining the graph of the function f shown in Figure 3.19.

Figure 3.19
If Δx is small, dy is a good approximation of Δy.

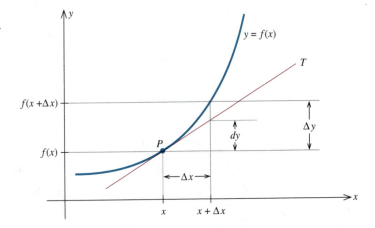

Observe that near the point of tangency P, the tangent line T is close to the graph of f. Therefore if Δx is small, then dy is a good approximation of Δy. We can find an expression for dy as follows: Notice that the slope of T is given by

$$\frac{dy}{\Delta x} \quad \text{(Rise divided by run)}$$

However, the slope of T is given by $f'(x)$. Therefore, we have

$$\frac{dy}{\Delta x} = f'(x)$$

or $dy = f'(x)\Delta x$. Thus, we have the approximation

$$\Delta y \approx dy = f'(x)\Delta x$$

in terms of the derivative of f at x. The quantity dy is called the *differential of y*.

THE DIFFERENTIAL

Let $y = f(x)$ define a differentiable function of x. Then:

1. The **differential dx** of the independent variable x is $dx = \Delta x$.
2. The **differential dy** of the dependent variable y is

$$dy = f'(x)\Delta x = f'(x)dx \tag{11}$$

REMARKS

1. For the independent variable x: There is no difference between Δx and dx—both measure the change in x from x to $x + \Delta x$.

2. For the dependent variable y: Δy measures the *actual* change in y as x changes from x to $x + \Delta x$, whereas dy measures the *approximate* change in y corresponding to the same change in x.

3. The differential dy depends on both x and dx, but for fixed x, dy is a linear function of dx.

◦ ◦ ◦

EXAMPLE 3   Let $y = x^3$.

a. Find the differential dy of y.

b. Use dy to approximate Δy when x changes from 2 to 2.01.

c. Use dy to approximate Δy when x changes from 2 to 1.98.

d. Compare the results of (b) with those of Example 2.

Solution

a. Let $f(x) = x^3$. Then

$$dy = f'(x)\, dx = 3x^2 dx$$

b. Here $x = 2$ and $dx = 2.01 - 2 = 0.01$. Therefore,

$$dy = 3x^2 dx = 3(2)^2(0.01) = 0.12$$

c. Here $x = 2$ and $dx = 1.98 - 2 = -0.02$. Therefore,

$$dy = 3x^2 dx = 3(2)^2(-0.02) = -0.24$$

d. As you can see, both approximations 0.12 and -0.24 are quite close to the actual changes of Δy obtained in Example 2: 0.120601 and -0.237608.

◦ ◦ ◦

Observe how much easier it is to find an approximation to the exact change in a function with the help of the differential, rather than calculating the exact change in the function itself. In the following examples, we take advantage of this fact.

EXAMPLE 4   Approximate the value of $\sqrt{26.5}$ using differentials. Verify your result using the $\boxed{\sqrt{}}$ key on your calculator.

Solution   Since we want to compute the square root of a number, let us consider the function $y = f(x) = \sqrt{x}$. Since 25 is the number nearest 26.5 whose square root is readily recognized, let us take $x = 25$. We want to know the change in y, Δy, as x changes from $x = 25$ to $x = 26.5$, an increase of

$\Delta x = 1.5$ units. Using equation (11), we find

$$\Delta y \approx dy = f'(x)\Delta x$$

$$= \left[\frac{1}{2\sqrt{x}}\Big|_{x=25}\right] \cdot (1.5) = \left(\frac{1}{10}\right)(1.5) = 0.15$$

Therefore,

$$\sqrt{26.5} - \sqrt{25} = \Delta y \approx 0.15$$

and

$$\sqrt{26.5} \approx \sqrt{25} + 0.15 = 5.15$$

The exact value of $\sqrt{26.5}$, rounded off to five decimal places, is 5.14782. Thus, the error incurred in the approximation is 0.00218.   ◦ ◦ ◦

Applications

EXAMPLE 5   The total cost incurred in operating a certain type of truck on a 500-mile trip, traveling at an average speed of v miles per hour, is estimated to be

$$C(v) = 125 + v + \frac{4500}{v}$$

dollars. Find the approximate change in the total operating cost when the average speed is increased from 55 mph to 58 mph.

Solution   With $v = 55$ and $\Delta v = dv = 3$, we find

$$\Delta C \approx dC = C'(v)dv = \left(1 - \frac{4500}{v^2}\right)\Bigg|_{v=55} \cdot 3$$

$$= \left(1 - \frac{4500}{3025}\right)(3) \approx -1.46$$

so the total operating cost is found to decrease by $1.46. This might explain why so many independent truckers often exceed the 55 mph speed limit.   ◦ ◦ ◦

EXAMPLE 6   The relationship between the amount of money x spent by Cannon Precision Instruments on advertising and Cannon's total sales $S(x)$ is given by the function

$$S(x) = -0.002x^3 + 0.6x^2 + x + 500 \qquad (0 \le x \le 200)$$

where x is measured in thousands of dollars. Use differentials to estimate the change in Cannon's total sales if advertising expenditures are increased from $100,000 ($x = 100$) to $105,000 ($x = 105$).

Solution   The required change in sales is given by

$$\Delta S \approx dS = S'(100)dx$$
$$= -0.006x^2 + 1.2x + 1\big|_{x=100} \cdot (5) \qquad (dx = 105 - 100 = 5)$$
$$= (-60 + 120 + 1)(5) = 305$$

that is, an increase of $305,000.

● ● ●

EXAMPLE 7   The Rings of Neptune

a. A ring has an inner radius of r units and an outer radius of R units, where $(R - r)$ is small in comparison to r (see Figure 3.20a). Use differentials to estimate the area of the ring.

Figure 3.20

$dr = R - r$

r

R

(a) The area of the ring is the circumference of the inner circle times the thickness.

(b) Neptune and its rings.

b. Recent observations, including those of Voyager I and II, showed that Neptune's ring system is considerably more complex than had been believed. For one thing, it is made up of a large number of distinguishable rings rather than one continuous great ring as previously thought (see Figure 3.20b). The outermost ring, 1989N1R, has an inner radius of approximately 62,900 km (measured from the center of the planet), and a radial width of approximately 50 km. Using these data, estimate the area of the ring.

Solution

a. Using the fact that the area of a circle of radius x is $A = f(x) = \pi x^2$, we find

$$\pi R^2 - \pi r^2 = f(R) - f(r)$$
$$= \Delta A \qquad \text{(Remember } \Delta A = \text{Change in } f \text{ when}$$
$$\approx dA \qquad \quad x \text{ changes from } x = r \text{ to } x = R.)$$
$$= f'(r)dr$$

where $dr = R - r$. So, we see that the area of the ring is approximately $2\pi r(R - r)$ square units. In words, the area of the ring is approximately equal to

Circumference of the inner circle × Thickness of the ring

b. Applying the results of (a) with $r = 62,900$ and $dr = 50$, we find that the area of the ring is approximately $2\pi(62,900)(50)$, or 19,760,618 sq. km, which is roughly 4% of Earth's surface.   ○ ○ ○

Before looking at the next example, we need to familiarize ourselves with some terminology. If a quantity with exact value q is measured or calculated with an error of Δq, then the quantity $\Delta q/q$ is called the relative error in the measurement or calculation of q. If the quantity $\Delta q/q$ is expressed as a percentage, it is then called the percentage error. Because Δq is approximated by dq, we normally approximate the relative error $\Delta q/q$ by dq/q.

EXAMPLE 8 Suppose the radius of a ball-bearing is measured to be 0.5 in., with a maximum error of ± 0.0002 in. Then the relative error in r is

$$\frac{dr}{r} = \frac{\pm 0.0002}{0.5} = \pm 0.0004$$

and the percentage error is $\pm 0.04\%$.   ○ ○ ○

EXAMPLE 9 Suppose the side of a cube is measured with a maximum percentage error of 2%. Use differentials to estimate the maximum percentage error in the calculated volume of the cube.

Solution Suppose the side of the cube is x, so that its volume is

$$V = x^3$$

We are given that $\left|\dfrac{dx}{x}\right| \le 0.02$. Now,

$$dV = 3x^2 dx$$

and so

$$\frac{dV}{V} = \frac{3x^2 dx}{x^3} = 3\frac{dx}{x}$$

Therefore,

$$\left|\frac{dV}{V}\right| = 3\left|\frac{dx}{x}\right| \le 3(0.02) = 0.06$$

and we see that the maximum percentage error in the measurement of the volume of the cube is 6%.   ○ ○ ○

Finally, we want to point out that if at some point in reading this section you have a sense of déjà vu, do not be surprised, because the notion of the differential was first used in Section 3.4 (see Example 1). There we took $\Delta x = 1$ since we were interested in finding the marginal cost when the level of production was increased from $x = 250$ to $x = 251$. If we had used differentials, we would have found

$$C(251) - C(250) \approx C'(250)dx$$

so that taking $dx = \Delta x = 1$, we have $C(251) - C(250) \approx C'(250)$, which agrees with the result obtained in Example 1. Thus, in Section 3.4, we touched upon the notion of the differential, albeit in the special case in which $dx = 1$.

SELF-CHECK EXERCISES 3.7

1. Find the differential of $f(x) = \sqrt{x} + 1$.

2. A certain country's government economists have determined that the demand equation for corn in that country is given by

$$p = f(x) = \frac{125}{x^2 + 1}$$

where p is expressed in dollars per bushel and x, the quantity demanded per year, is measured in billions of bushels. The economists are forecasting a harvest of 6 billion bushels for the year. If the actual production of corn were 6.2 billion bushels for the year instead, what would be the approximate drop in the predicted price of corn per bushel?

Solutions to Self-Check Exercises 3.7 can be found on page 263.

3.7 EXERCISES

In exercises 1–14, find the differential of the given function.

1. $f(x) = 2x^2$

2. $f(x) = 3x^2 + 1$

3. $f(x) = x^3 - x$

4. $f(x) = 2x^3 + x$

5. $f(x) = \sqrt{x + 1}$

6. $f(x) = \dfrac{3}{\sqrt{x}}$

7. $f(x) = 2x^{3/2} + x^{1/2}$

8. $f(x) = 3x^{5/6} + 7x^{2/3}$

9. $f(x) = x + \dfrac{2}{x}$

10. $f(x) = \dfrac{3}{x - 1}$

11. $f(x) = \dfrac{x - 1}{x^2 + 1}$

12. $f(x) = \dfrac{2x^2 + 1}{x + 1}$

13. $f(x) = \sqrt{3x^2 - x}$

14. $f(x) = (2x^2 + 3)^{1/3}$

15. Let f be a function defined by

$$y = f(x) = x^2 - 1$$

a. Find the differential of f.
b. Use your result from (a) to find the approximate change in y if x changes from 1 to 1.02.
c. Find the actual change in y if x changes from 1 to 1.02 and compare your result with that obtained in (b).

16. Let f be a function defined by

$$y = f(x) = 3x^2 - 2x + 6$$

a. Find the differential of f.

b. Use your result from (a) to find the approximate change in y if x changes from 2 to 1.97.

c. Find the actual change in y if x changes from 2 to 1.97 and compare your result with that obtained in (b).

17. Let f be a function defined by

$$y = f(x) = \frac{1}{x}$$

a. Find the differential of f.

b. Use your result from (a) to find the approximate change in y if x changes from -1 to -0.95.

c. Find the actual change in y if x changes from -1 to -0.95 and compare your result with that obtained in (b).

18. Let f be a function defined by

$$y = f(x) = \sqrt{2x + 1}$$

a. Find the differential of f.

b. Use your result from (a) to find the approximate change in y if x changes from 4 to 4.1.

c. Find the actual change in y if x changes from 4 to 4.1 and compare your result with that obtained in (b).

In exercises 19–26, use differentials to approximate the given quantity.

19. $\sqrt{10}$ **20.** $\sqrt{17}$ **21.** $\sqrt{49.5}$

22. $\sqrt{99.7}$ **23.** $\sqrt[3]{7.8}$ **24.** $\sqrt[4]{81.6}$

25. $\sqrt{0.089}$ **26.** $\sqrt[3]{0.00096}$

27. Use a differential to approximate $\sqrt{4.02} + \dfrac{1}{\sqrt{4.02}}$.

[*Hint:* Let $f(x) = \sqrt{x} + \dfrac{1}{\sqrt{x}}$ and compute dy with $x = 4$ and $dx = 0.02$.]

28. Use a differential to approximate $\dfrac{2(4.98)}{(4.98)^2 + 1}$.

[*Hint:* Study the hint for exercise 27.]

A calculator is recommended for the remainder of this exercise set.

29. Error Estimation The length of each edge of a cube is 12 cm, with a possible error in measurement of 0.02 cm. Use differentials to estimate the error that might occur when the volume of the cube is calculated.

30. Error Estimation A hemisphere-shaped dome of radius 60 ft is to be coated with a layer of rust-proofer before painting. Use differentials to estimate the amount of rust-proofer needed if the coat is to be 0.01 inch thick. [*Hint:* The volume of a hemisphere of radius r is $V = \frac{2}{3}\pi r^3$.]

31. Growth of a Cancerous Tumor The volume of a spherical cancer tumor is given by

$$V(r) = \frac{4}{3}\pi r^3$$

If the radius of a tumor is estimated at 1.1 cm, with a maximum error in measurement of 0.005 cm, determine the error that might occur when the volume of the tumor is calculated.

32. Unclogging Arteries Research done in the 1930s by the French physiologist Jean Poiseuille showed that the resistance R of a blood vessel of length l and radius r is $R = kl/r^4$, where k is a constant. Suppose a dose of the drug TPA increases r by 10%. How will this affect the resistance R? Assume that l is constant.

33. Gross Domestic Product An economist has determined that a certain country's gross domestic product (GDP) is approximated by the function $f(x) = 640x^{1/5}$, where $f(x)$ is measured in billions of dollars and x is the capital outlay in billions of dollars. Use differentials to estimate the change in the country's GDP if the country's capital expenditure changes from \$243 billion to \$248 billion.

34. Learning Curves The length of time (in seconds) a certain individual takes to learn a list of n items is approximated by

$$f(n) = 4n\sqrt{n - 4}$$

Use differentials to approximate the additional time it takes the individual to learn the items on a list when n is increased from 85 to 90 items.

35. Effect of Advertising on Profits The relationship between Cunningham Realty's quarterly profits, $P(x)$, and the amount of money x spent on advertising per quarter is described by the function

$$P(x) = -\frac{1}{8}x^2 + 7x + 30 \qquad (0 \le x \le 50)$$

where both $P(x)$ and x are measured in thousands of dollars. Use differentials to estimate the increase in profits when advertising expenditure each quarter is increased from \$24,000 to \$26,000.

Using Technology

Finding the Differential of a Function

The calculation of the differential of f at a given value of x involves the evaluation of the derivative of f at that point and can be facilitated through the use of the numerical derivative function.

EXAMPLE 1   Use dy to approximate Δy if $y = x^2(2x^2 + x + 1)^{2/3}$ and x changes from 2 to 1.98.

Solution   Let $f(x) = x^2(2x^2 + x + 1)^{2/3}$. Since $dx = 1.98 - 2 = -0.02$, we find the required approximation to be

$$dy = f'(2) \cdot (-0.02)$$

But using the numerical derivative operation, we find

$$f'(2) = 30.5758132855$$

and so $\qquad dy = (-0.02)(30.5758132855) = -0.611516266$ ○ ○ ○

EXAMPLE 2   The Meyers are considering the purchase of a house in the near future and estimate that they will need a loan of $120,000. Based on a 30-year conventional mortgage with an interest rate of r per year, their monthly repayment will be

$$P = \frac{10{,}000r}{1 - \left(1 + \dfrac{r}{12}\right)^{-360}}$$

dollars. If the interest rate increases from the present rate of 10% per year to 10.2% per year between now and the time the Meyers decide to secure the loan, approximately how much more per month will their mortgage payment be?

Solution   Let us write

$$P = f(r) = \frac{10{,}000r}{1 - \left(1 + \dfrac{r}{12}\right)^{-360}}$$

Then the increase in the mortgage payment will be approximately

$$dP = f'(0.1)dr = f'(0.1)(0.002) \qquad \text{(Since } dr = 0.102 - 0.1)$$
$$= (8867.59947979)(0.002) \approx 17.7352 \qquad \text{(Using the numerical derivative operation)}$$

or approximately $17.74 per month.   ○ ○ ○

EXERCISES

In exercises 1–6, use dy to approximate Δy for the function $y = f(x)$ when x changes from $x = a$ to $x = b$.

1. $f(x) = 0.21x^7 - 3.22x^4 + 5.43x^2 + 1.42x + 12.42$; $a = 3$, $b = 3.01$

2. $f(x) = \dfrac{0.2x^2 + 3.1}{1.2x + 1.3}$; $a = 2$, $b = 1.96$

3. $f(x) = \sqrt{2.2x^2 + 1.3x + 4}$; $a = 1$, $b = 1.03$

4. $f(x) = x\sqrt{2x^3 - x + 4}$; $a = 2$, $b = 1.98$

5. $f(x) = \dfrac{\sqrt{x^2 + 4}}{x - 1}$; $a = 4$, $b = 4.1$

6. $f(x) = 2.1x^2 + \dfrac{3}{\sqrt{x}} + 5$; $a = 3$, $b = 2.95$

7. **Calculating Mortgage Payments** Refer to Example 2. How much more per month will the Meyers' mortgage payment be if the interest rate increases from 10% to 10.3% per year? To 10.4% per year? To 10.5% per year?

8. **Estimating the Area of a Ring of Neptune** The ring 1989N2R of the planet Neptune has an inner radius of approximately 53,200 km (measured from the center of the planet) and a radial width of 15 km. Use differentials to estimate the area of the ring.

9. **Effect of Price Increase on Quantity Demanded** The quantity demanded per week of the Alpha Sports Watch, x (in thousands), is related to its unit price of p dollars by the equation

$$x = f(p) = 10\sqrt{\frac{50 - p}{p}} \qquad (0 \le p \le 50)$$

Use differentials to find the decrease in the quantity of the watches demanded per week if the unit price is increased from $40 to $42.

36. Effect of Mortgage Rates on Housing Starts A study prepared for the National Association of Realtors estimates that the number of housing starts per year over the next 5 years will be

$$N(r) = \frac{7}{1 + 0.02r^2}$$

million units, where r (percent) is the mortgage rate. Use differentials to estimate the decrease in the number of housing starts when the mortgage rate is increased from 12% to 12.5%.

37. Supply-Price The supply equation for a certain brand of transistor radio is given by

$$p = s(x) = 0.3 \sqrt{x} + 10$$

where x is the quantity supplied and p is the unit price in dollars. Use differentials to approximate the change in price when the quantity supplied is increased from 10,000 units to 10,500 units.

38. Demand-Price The demand function for the Sentinel smoke alarm is given by

$$p = d(x) = \frac{30}{0.02x^2 + 1}$$

where x is the quantity demanded (in units of a thousand) and p is the unit price in dollars. Use differentials to estimate the change in the price p when the quantity demanded changes from 5000 to 5500 units per week.

39. Surface Area of an Animal Animal physiologists use the formula

$$S = kW^{2/3}$$

to calculate an animal's surface area (in square meters) from its weight W (in kilograms), where k is a constant that depends on the animal under consideration. Suppose a physiologist calculates the surface area of a horse ($k = 0.1$). If the horse's weight is estimated at 300 kg, with a maximum error in measurement of 0.6 kg, determine the percentage error in the calculation of the horse's surface area.

40. Forecasting Profits The management of Trappee and Sons, Inc., forecast that they will sell 200,000 cases of their Texa-Pep hot sauce next year. Their annual profit is described by

$$P(x) = -0.000032x^3 + 6x - 100$$

thousand dollars, where x is measured in thousands of cases. If the maximum error in the forecast is 15%, determine the corresponding error in Trappee's profits.

41. Forecasting Commodity Prices A certain country's government economists have determined that the demand equation for soybeans in that country is given by

$$p = f(x) = \frac{55}{2x^2 + 1}$$

where p is expressed in dollars per bushel and x, the quantity demanded per year, is measured in billions of bushels. The economists are forecasting a harvest of 1.8 billion bushels for the year, with a maximum error of 15% in their forecast. Determine the corresponding maximum error in the predicted price per bushel of soybeans.

42. Crime Studies A sociologist has found that the number of serious crimes in a certain city per year is described by the function

$$N(x) = \frac{500(400 + 20x)^{1/2}}{(5 + 0.2x)^2}$$

where x (in cents per dollar deposited) is the level of reinvestment in the area in conventional mortgages by the city's ten largest banks. Use differentials to estimate the change in the number of crimes if the level of reinvestment changes from 20 cents per dollar deposited to 22 cents per dollar deposited.

43. Financing a Home The Millers are planning to buy a home in the near future and estimate that they will need a 30-year fixed-rate mortgage for $120,000. Their monthly payment P (in dollars) can be computed using the formula

$$P = \frac{10,000r}{1 - \left(1 + \dfrac{r}{12}\right)^{-360}}$$

where r is the interest rate per year.
a. Find the differential of P.
b. If the interest rate increases from the present rate of 9% per year to 9.2% per year between now and the time the Millers decide to secure the loan, approximately how much more will their monthly mortgage payment be? How much more will it be if the interest rate increases to 9.3% per year? To 9.4% per year? To 9.5% per year?

SOLUTIONS TO SELF-CHECK EXERCISES 3.7

1. We find

$$f'(x) = \frac{1}{2}x^{-1/2} = \frac{1}{2\sqrt{x}}$$

Therefore, the required differential of f is

$$dy = \frac{1}{2\sqrt{x}}\,dx$$

2. We first compute the differential

$$dp = -\frac{250x}{(x^2 + 1)^2}\,dx$$

Next, using equation (11) with $x = 6$ and $dx = 0.2$, we find

$$\Delta p \approx dp = -\frac{250(6)}{(36 + 1)^2}(0.2) = -0.22$$

or a drop in price of 22 cents per bushel.

CHAPTER 3 SUMMARY OF PRINCIPAL FORMULAS AND TERMS

Formulas

1. Derivative of a constant $\quad \dfrac{d}{dx}(c) = 0,\, c$ a constant

2. Power Rule $\quad \dfrac{d}{dx}(x^n) = nx^{n-1}$

3. Constant Multiple Rule $\quad \dfrac{d}{dx}(cu) = c\dfrac{du}{dx},\, c$ a constant

4. Sum Rule $\quad \dfrac{d}{dx}(u \pm v) = \dfrac{du}{dx} \pm \dfrac{dv}{dx}$

5. Product Rule $\quad \dfrac{d}{dx}(uv) = u\dfrac{dv}{dx} + v\dfrac{du}{dx}$

6. Quotient Rule $\quad \dfrac{d}{dx}\left(\dfrac{u}{v}\right) = \dfrac{v\dfrac{du}{dx} - u\dfrac{dv}{dx}}{v^2}$

7. Chain Rule $\quad \dfrac{dy}{dx} = \dfrac{dy}{du} \cdot \dfrac{du}{dx}$

8. General Power Rule $\quad \dfrac{d}{dx}(u^n) = nu^{n-1}\dfrac{du}{dx}$

9. Average cost function $\quad \overline{C}(x) = \dfrac{C(x)}{x}$

10. Revenue function $R(x) = px$

11. Profit function $P(x) = R(x) - C(x)$

12. Elasticity of demand $E(p) = -\dfrac{pf'(p)}{f(p)}$

13. Differential of y $dy = f'(x)dx$

Terms

Marginal cost function   Unitary demand

Marginal average cost function   Inelastic demand

Marginal revenue function   Second derivative of f

Marginal profit function   Implicit differentiation

Elastic demand   Related rates

CHAPTER 3 REVIEW EXERCISES

In exercises 1–30, find the derivative of the given function.

1. $f(x) = 3x^5 - 2x^4 + 3x^2 - 2x + 1$

2. $f(x) = 4x^6 + 2x^4 + 3x^2 - 2$

3. $g(x) = -2x^{-3} + 3x^{-1} + 2$

4. $f(t) = 2t^2 - 3t^3 - t^{-1/2}$

5. $g(t) = 2t^{-1/2} + 4t^{-3/2} + 2$

6. $h(x) = x^2 + \dfrac{2}{x}$

7. $f(t) = t + \dfrac{2}{t} + \dfrac{3}{t^2}$

8. $g(s) = 2s^2 - \dfrac{4}{s} + \dfrac{2}{\sqrt{s}}$

9. $h(x) = x^2 - \dfrac{2}{x^{3/2}}$

10. $f(x) = \dfrac{x+1}{2x-1}$

11. $g(t) = \dfrac{t^2}{2t^2 + 1}$

12. $h(t) = \dfrac{\sqrt{t}}{\sqrt{t} + 1}$

13. $f(x) = \dfrac{\sqrt{x} - 1}{\sqrt{x} + 1}$

14. $f(t) = \dfrac{t}{2t^2 + 1}$

15. $f(x) = \dfrac{x^2(x^2 + 1)}{x^2 - 1}$

16. $f(x) = (2x^2 + x)^3$

17. $f(x) = (3x^3 - 2)^8$

18. $h(x) = (\sqrt{x} + 2)^5$

19. $f(t) = \sqrt{2t^2 + 1}$

20. $g(t) = \sqrt[3]{1 - 2t^3}$

21. $s(t) = (3t^2 - 2t + 5)^{-2}$

22. $f(x) = (2x^3 - 3x^2 + 1)^{-3/2}$

23. $h(x) = \left(x + \dfrac{1}{x}\right)^2$

24. $h(x) = \dfrac{1 + x}{(2x^2 + 1)^2}$

25. $h(t) = (t^2 + t)^4(2t^2)$

26. $f(x) = (2x + 1)^3(x^2 + x)^2$

27. $g(x) = \sqrt{x}(x^2 - 1)^3$

28. $f(x) = \dfrac{x}{\sqrt{x^3 + 2}}$

29. $h(x) = \dfrac{\sqrt{3x + 2}}{4x - 3}$

30. $f(t) = \dfrac{\sqrt{2t + 1}}{(t + 1)^3}$

In exercises 31–36, find the second derivative of the given function.

31. $f(x) = 2x^4 - 3x^3 + 2x^2 + x + 4$

32. $g(x) = \sqrt{x} + \dfrac{1}{\sqrt{x}}$

33. $h(t) = \dfrac{t}{t^2 + 4}$

34. $f(x) = (x^3 + x + 1)^2$

35. $f(x) = \sqrt{2x^2 + 1}$

36. $f(t) = t(t^2 + 1)^3$

In exercises 37–42, find dy/dx by implicit differentiation.

37. $6x^2 - 3y^2 = 9$

38. $2x^3 - 3xy = 4$

39. $y^3 + 3x^2 = 3y$

40. $x^2 + 2x^2y^2 + y^2 = 10$

41. $x^2 - 4xy - y^2 = 12$

42. $3x^2y - 4xy + x - 2y = 6$

43. Let $f(x) = 2x^3 - 3x^2 - 16x + 3$.
  a. Find the points on the graph of f at which the slope of the tangent line is equal to -4.
  b. Find the equation(s) of the tangent line(s) of (a).

44. Let $f(x) = \frac{1}{3}x^3 + \frac{1}{2}x^2 - 4x + 1$.
  a. Find the points on the graph of f at which the slope of the tangent line is equal to -2.
  b. Find the equation(s) of the tangent line(s) of (a).

45. Find an equation of the tangent line to the graph of $y = \sqrt{4 - x^2}$ at the point $(1, \sqrt{3})$.

46. Find an equation of the tangent line to the graph of $y = x(x + 1)^5$ at the point $(1, 32)$.

47. Find the third derivative of the function

$$f(x) = \frac{1}{2x - 1}$$

What is its domain?

48. The number of subscribers to CNC Cable Television in the town of Randolph is approximated by the function

$$N(x) = 1000(1 + 2x)^{1/2} \qquad (1 \le x \le 30)$$

where $N(x)$ denotes the number of subscribers to the service in the xth week. Find the rate of increase in the number of subscribers at the end of the twelfth week.

49. The total weekly cost in dollars incurred by the Herald Record Company in pressing x video discs is given by the total cost function

$$C(x) = 2500 + 2.2x \qquad (0 \le x \le 8000)$$

a. What is the marginal cost when $x = 1000$? When $x = 2000$?
b. Find the average cost function $\overline{C}$ and the marginal average cost function $\overline{C}'$.
c. Using the results from (b), show that the average cost incurred by Herald in pressing a video disc approaches $2.20 per disc when the level of production is high enough.

50. The marketing department of Telecon Corporation has determined that the demand for their cordless phones obeys the relationship

$$p = -0.02x + 600 \qquad (0 \le x \le 30{,}000)$$

where p denotes the phone's unit price (in dollars) and x denotes the quantity demanded.
a. Find the revenue function R.
b. Find the marginal revenue function R'.
c. Compute $R'(10{,}000)$ and interpret your result.

51. The weekly demand for the Lectro-Copy photocopying machine is given by the demand equation

$$p = 2000 - 0.04x \qquad (0 \le x \le 50{,}000)$$

where p denotes the wholesale unit price in dollars and x denotes the quantity demanded. The weekly total cost function for manufacturing these copiers is given by

$$C(x) = 0.000002x^3 - 0.02x^2 + 1000x + 120{,}000$$

where $C(x)$ denotes the total cost incurred in producing x units.
a. Find the revenue function R, the profit function P, and the average cost function $\overline{C}$.
b. Find the marginal cost function C', the marginal revenue function R', the marginal profit function P', and the marginal average cost function $\overline{C}'$.
c. Compute $C'(3000)$, $R'(3000)$, and $P'(3000)$.
d. Compute $\overline{C}'(5000)$ and $\overline{C}'(8000)$ and interpret your results.

This chapter further explores the power of the derivative, which we use to help analyze the properties of functions. The information obtained can then be used to accurately sketch graphs of functions. We also see how the derivative is used in solving a large class of optimization problems, including finding what level of production will yield a maximum profit for a company; finding what level of production will result in minimal cost to a company; finding the maximum height attained by a rocket; finding the maximum velocity at which air is expelled when a person coughs; and a host of other problems.

What is the maximum altitude and what is the maximum velocity attained by the rocket? In Example 7, page 336, you will see how the techniques of calculus can be used to help answer these questions.

APPLICATIONS OF THE DERIVATIVE

4.1 APPLICATIONS OF THE FIRST DERIVATIVE

Determining the Intervals Where a Function Is Increasing or Decreasing

According to a study by the U.S. Department of Energy and the Shell Development Company, a typical car's fuel economy as a function of its speed is described by the graph shown in Figure 4.1. Observe that the fuel economy $f(x)$ in miles per gallon (mpg) improves as x, the vehicle's speed in miles per hour (mph), increases from 0 to 42, and then drops as the speed increases beyond 42 mph. We use the terms *increasing* and *decreasing* to describe the behavior of a function as we move from left to right along its graph.

Figure 4.1

A typical car's fuel economy improves as the speed at which it is driven increases from 0 mph to 42 mph and drops at speeds greater than 42 mph.

Source: U.S. Department of Energy and Shell Development Co.

More precisely, we have the following definitions.

INCREASING AND DECREASING FUNCTIONS

A function f is **increasing** on an interval (a, b) if for any two numbers x_1 and x_2 in (a, b), $f(x_1) < f(x_2)$ whenever $x_1 < x_2$ (Figure 4.2a).
A function f is **decreasing** on an interval (a, b) if for any two numbers x_1 and x_2 in (a, b), $f(x_1) > f(x_2)$ whenever $x_1 < x_2$ (Figure 4.2b).

Figure 4.2

(a) f is increasing on (a, b).   **(b)** f is decreasing on (a, b).

We say that f is *increasing at a point* c if there exists an interval (a, b) containing c such that f is increasing on (a, b). Similarly, we say that f is *decreasing at a point* c if there exists an interval (a, b) containing c such that f is decreasing on (a, b).

Since the rate of change of a function at a point $x = c$ is given by the derivative of the function at that point, the derivative lends itself naturally to being a tool for determining the intervals where a differentiable function is increasing or decreasing. Indeed, as we saw in Chapter 2, the derivative of a function at a point measures both the slope of the tangent line to the graph of the function at that point and the rate of change of the function at the same point. In fact, at a point where the derivative is positive, the slope of the tangent line to the graph is positive and the function is increasing. At a point where the derivative is negative, the slope of the tangent line to the graph is negative and the function is decreasing (Figure 4.3).

Figure 4.3

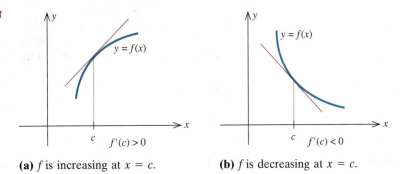

(a) f is increasing at $x = c$.   **(b)** f is decreasing at $x = c$.

These observations lead to the following important theorem, which we state without proof.

THEOREM 1

a. If $f'(x) > 0$ for each value of x in an interval (a, b), then f is increasing on (a, b).

b. If $f'(x) < 0$ for each value of x in an interval (a, b), then f is decreasing on (a, b).

c. If $f'(x) = 0$ for each value of x in an interval (a, b), then f is constant on (a, b).

EXAMPLE 1 Find the interval where the function $f(x) = x^2$ is increasing and the interval where it is decreasing.

Solution The derivative of $f(x) = x^2$ is $f'(x) = 2x$. Since

$$f'(x) = 2x > 0 \quad \text{if } x > 0$$

and

$$f'(x) = 2x < 0 \quad \text{if } x < 0$$

Figure 4.4
*The graph of f falls on $(-\infty, 0)$
where $f'(x) < 0$ and rises on $(0, \infty)$
where $f'(x) > 0$.*

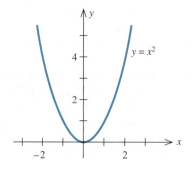

$y = x^2$

f is increasing on the interval $(0, \infty)$ and decreasing on the interval $(-\infty, 0)$ (Figure 4.4).

○ ○ ○

Recall that the graph of a continuous function cannot have any breaks. As a consequence, a continuous function cannot change sign unless it equals zero for some value of x. (See Theorem 5, page 132.) This observation suggests the following procedure for determining the sign of the derivative f' of a function f, and hence the intervals where the function f is increasing and where it is decreasing.

**DETERMINING THE
INTERVALS WHERE
A FUNCTION IS
INCREASING OR
DECREASING**

1. Find all values of x for which $f'(x) = 0$ or f' is discontinuous, and identify the open intervals determined by these points.
2. Select a test point c in each interval found in step 1 and determine the sign of $f'(c)$ in that interval.
 a. If $f'(c) > 0$, f is increasing on that interval.
 b. If $f'(c) < 0$, f is decreasing on that interval.

EXAMPLE 2   Determine the intervals where the function $f(x) = x^3 - 3x^2 - 24x + 32$ is increasing and where it is decreasing.

Solution

1. The derivative of f is

$$f'(x) = 3x^2 - 6x - 24 = 3(x + 2)(x - 4)$$

and it is continuous everywhere. The zeros of $f'(x)$ are $x = -2$ and $x = 4$, and these points divide the real line into the intervals $(-\infty, -2)$, $(-2, 4)$, and $(4, \infty)$.
2. To determine the sign of $f'(x)$ in the intervals $(-\infty, -2)$, $(-2, 4)$, and $(4, \infty)$, compute $f'(x)$ at a convenient test point in each interval. The results

Figure 4.5
Sign diagram for f'.

are shown in the following table:

Interval	Test Point c	$f'(c)$	Sign of $f'(x)$
$(-\infty, -2)$	-3	21	$+$
$(-2, 4)$	0	-24	$-$
$(4, \infty)$	5	21	$+$

Using these results, we obtain the sign diagram shown in Figure 4.5. We conclude that f is increasing on the intervals $(-\infty, -2)$ and $(4, \infty)$ and is decreasing on the interval $(-2, 4)$. Figure 4.6 shows the graph of f.

Figure 4.6
The graph of f rises on $(-\infty, -2)$, falls on $(-2, 4)$, and rises again on $(4, \infty)$.

EXPLORING WITH TECHNOLOGY

Refer to Example 2.
a. Use a graphing utility to plot the graphs of

$$f(x) = x^3 - 3x^2 - 24x + 32$$

and its derivative function

$$f'(x) = 3x^2 - 6x - 24$$

using the viewing rectangle $[-10, 10] \times [-50, 70]$.
b. By looking at the graph of f', determine the intervals where $f'(x) > 0$ and the intervals where $f'(x) < 0$. Next, look at the graph of f and determine the intervals where it is increasing and the intervals where it is decreasing. Describe the relationship. Is it what you expected?

REMARK   Do not be concerned with how the graphs in this section are obtained. We will learn how to sketch these graphs later. However, if you are familiar with the use of a graphing utility you may go ahead and verify each graph.   ○ ○ ○

EXAMPLE 3   Find the interval where the function $f(x) = x^{2/3}$ is increasing and the interval where it is decreasing.

Solution

1. The derivative of f is

$$f'(x) = \frac{2}{3}x^{-1/3} = \frac{2}{3x^{1/3}}$$

The function f' is not defined at $x = 0$, so f' is discontinuous there. It is continuous everywhere else. Furthermore, f' is not equal to zero anywhere. The point $x = 0$ divides the real line (the domain of f) into the intervals $(-\infty, 0)$ and $(0, \infty)$.

2. Pick a test point, say $x = -1$, in the interval $(-\infty, 0)$ and compute

$$f'(-1) = -\frac{2}{3}$$

Since $f'(-1) < 0$, we see that $f'(x) < 0$ on $(-\infty, 0)$. Next, we pick a test point, say $x = 1$, in the interval $(0, \infty)$ and compute

$$f'(1) = \frac{2}{3}$$

Since $f'(1) > 0$, we see that $f'(x) > 0$ on $(0, \infty)$. Figure 4.7 shows these results in the form of a sign diagram.

We conclude that f is decreasing on the interval $(-\infty, 0)$ and increasing on the interval $(0, \infty)$. The graph of f, shown in Figure 4.8, confirms these results.

True or false.
If f is continuous at c and f is increasing at c, then $f'(c) \neq 0$. Explain your answer. [*Hint:* Consider $f(x) = x^3$ and $c = 0$.]

Figure 4.7
Sign diagram for f'.

f' not defined at $x = 0$

Figure 4.8
f decreases on $(-\infty, 0)$ and increases on $(0, \infty)$.

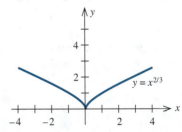

$y = x^{2/3}$

○ ○ ○

EXAMPLE 4 Find the intervals where the function $f(x) = x + 1/x$ is increasing and where it is decreasing.

Solution

1. The derivative of f is

$$f'(x) = 1 - \frac{1}{x^2} = \frac{x^2 - 1}{x^2}$$

Since f' is not defined at $x = 0$, it is discontinuous there. Furthermore, $f'(x)$ is equal to zero when $x^2 - 1 = 0$ or $x = \pm 1$. These values of x partition the domain of f' into the open intervals $(-\infty, -1)$, $(-1, 0)$, $(0, 1)$, and $(1, \infty)$, where the sign of f' is different from zero.

2. To determine the sign of f' in each of these intervals, we compute $f'(x)$ at the test points $x = -2, -\frac{1}{2}, \frac{1}{2}$, and 2, respectively, obtaining $f'(-2) = \frac{3}{4}$, $f'(-\frac{1}{2}) = -3$, $f'(\frac{1}{2}) = -3$, and $f'(2) = \frac{3}{4}$. From the sign diagram for f' (Figure 4.9), we conclude that f is increasing on $(-\infty, -1)$ and $(1, \infty)$ and decreasing on $(-1, 0)$ and $(0, 1)$.

The graph of f appears in Figure 4.10. Note that f' does not change sign as we move across the point of discontinuity, $x = 0$. (Compare this with Example 3.)

Figure 4.9
f' does not change sign as we move across $x = 0$.

f' is not defined at $x = 0$

Figure 4.10
The graph of f rises on $(-\infty, -1)$, falls on $(-1, 0)$ and $(0, 1)$, and rises again on $(1, \infty)$.

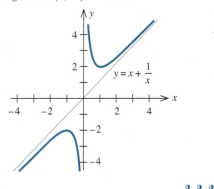

Example 4 reminds us that we must not automatically conclude that the derivative f' must change sign when we move across a point of discontinuity or a zero of f'.

Consider the profit function P associated with a certain commodity defined by

$$P(x) = R(x) - C(x) \qquad (x \geq 0)$$

where R is the revenue function, C is the total cost function, and x is the number of units of the product produced and sold.

a. Find an expression for $P'(x)$.
b. Find relationships in terms of the derivatives of R and C so that
   i. P is increasing at $x = a$
   ii. P is decreasing at $x = a$
   iii. P is neither increasing nor decreasing at $x = a$
   [*Hint:* Recall that the derivative of a function at $x = a$ measures the rate of change of the function at that point.]
c. Explain the results of part (b) in economic terms.

Relative Extrema

In addition to helping us determine where the graph of a function is increasing and decreasing, the first derivative may be used to help us locate certain "high points" and "low points" on the graph of f. Knowing these points is invaluable in sketching the graphs of functions and solving optimization problems. These "high points" and "low points" correspond to the *relative (local) maxima* and *relative minima* of a function. They are so called because they are the highest or the lowest points when compared to points nearby.

The graph shown in Figure 4.11 gives the U.S. budget deficit from 1980 ($t = 0$) through 1991. The relative maxima and the relative minima of the function f are indicated on the graph.

Figure 4.11
U.S. budget deficit from 1980 to 1991.

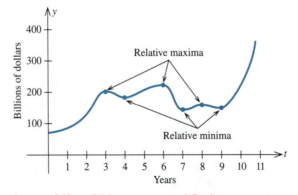

Source: Office of Management and Budget

More generally, we have the following definition:

| **RELATIVE MAXIMUM** | A function f has a **relative maximum** at $x = c$ if there exists an open interval (a, b) containing c such that $f(x) \leq f(c)$ for all x in (a, b). |

Geometrically, this means that there is *some* interval containing $x = c$ such that no point on the graph of f with its x-coordinate in that interval can lie above the point $(c, f(c))$; that is, $f(c)$ is the largest value of $f(x)$ in some interval around $x = c$. Figure 4.12 depicts the graph of a function f that has a relative maximum at $x = x_1$ and another at $x = x_3$.

Observe that all the points on the graph of f with x-coordinates in the interval I_1 containing x_1 (shown in blue) lie on or below the point $(x_1, f(x_1))$. This is also true for the point $(x_3, f(x_3))$ and the interval I_3. Thus, even though there are points on the graph of f that are "higher" than the points $(x_1, f(x_1))$ and $(x_3, f(x_3))$, the latter points are "highest" relative to points in their respective neighborhoods (intervals). Points on the graph of a function f that are "highest" and "lowest" with respect to *all* points in the domain of f will be studied in Section 4.4.

The definition of the relative minimum of a function parallels that of the relative maximum of a function.

| **RELATIVE MINIMUM** | A function f has a **relative minimum** at $x = c$ if there exists an open interval (a, b) containing c such that $f(x) \geq f(c)$ for all x in (a, b). |

The graph of the function f, depicted in Figure 4.12, has a relative minimum at $x = x_2$ and another at $x = x_4$.

Figure 4.12
f has a relative maximum at $x = x_1$ and at $x = x_3$.

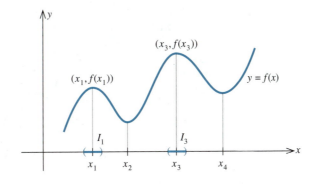

Finding the Relative Extrema

We refer to the relative maximum and relative minimum of a function as the **relative extrema** of that function. As a first step in our quest to find the relative extrema of a function, we consider functions that have derivatives at such points. Suppose that f is a function that is differentiable in some interval (a, b) that contains a point $x = c$, and that f has a relative maximum at $x = c$ (Figure 4.13a).

Figure 4.13

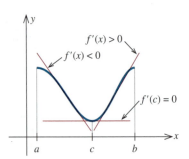

(a) f has a relative maximum at $x = c$.

(b) f has a relative minimum at $x = c$.

Observe that the slope of the tangent line to the graph of f must change from positive to negative as we move across the point $x = c$ from left to right. Therefore, the tangent line to the graph of f at the point $(c, f(c))$ must be horizontal; that is, $f'(c) = 0$ (Figure 4.13a).

Using a similar argument, it may be shown that the derivative f' of a differentiable function f must also be equal to zero at a point $x = c$, where f has a relative minimum (Figure 4.13b).

This analysis reveals an important characteristic of the relative extrema of a differentiable function f: *At any point c where f has a relative extremum, $f'(c) = 0$.*

Figure 4.14

$f'(0) = 0$, but f does not have a relative extremum at $(0, 0)$.

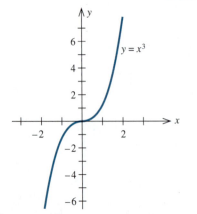

Before we develop a procedure for finding such points, a few words of caution are in order.

First, this result tells us that if a differentiable function f has a relative extremum at a point $x = c$, then $f'(c) = 0$. The converse of this statement—if $f'(c) = 0$ at some point $x = c$, then f must have a relative extremum at that point—is not true. Consider, for example, the function $f(x) = x^3$. Here $f'(x) = 3x^2$, so $f'(0) = 0$. Yet f has neither a relative maximum nor a relative minimum at $x = 0$ (Figure 4.14).

Second, our result assumes that the function is differentiable and thus has a derivative at a point that gives rise to a relative extremum. The functions $f(x) = |x|$ and $g(x) = x^{2/3}$ demonstrate that a relative extremum of a function may exist at a point at which the derivative does not exist. Both these functions fail to be differentiable at $x = 0$, but each has a relative minimum there. Figure 4.15 shows the graphs of these functions. Note that the slopes of the tangent lines change from negative to positive as we move across $x = 0$, just

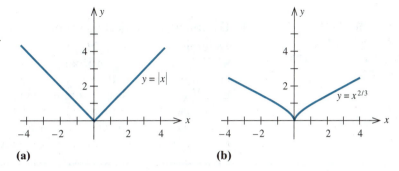

Figure 4.15
Each of these functions has a relative extremum at $(0, 0)$, but the derivative does not exist there.

as in the case of a function that is differentiable at a value of x that gives rise to a relative minimum.

We refer to a point in the domain of f that *may* give rise to a relative extremum as a *critical point.*

CRITICAL POINT OF f	A **critical point** of a function f is any point x in the domain of f such that $f'(x) = 0$ or $f'(x)$ does not exist.

Figure 4.16 depicts the graph of a function that has critical points at $x = a$, b, c, d, and e. Observe that $f'(x) = 0$ at $x = a$, b, and c. Next, since there is a corner at $x = d$, $f'(x)$ does not exist there. Finally, $f'(x)$ does not exist at $x = e$, because the tangent line there is vertical. Also, observe that the critical points $x = a$, b, and d give rise to relative extrema of f, whereas the critical points $x = c$ and $x = e$ do not.

Figure 4.16
Critical points of f.

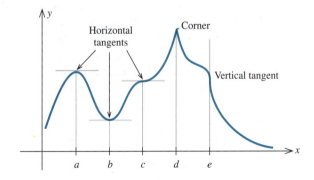

Having defined what a critical point is, we can now state a formal procedure for finding the relative extrema of a continuous function that is differentiable everywhere except at isolated values of x. Incorporated into the procedure is the so-called **First Derivative Test,** which enables us to determine whether or not a point gives rise to a relative maximum or a relative minimum of the function f.

PROCEDURE FOR FINDING RELATIVE EXTREMA (THE FIRST DERIVATIVE TEST)	**1.** Determine the critical points of f.
	 2. Determine the sign of $f'(x)$ to the left and right of each critical point.
	   a. If $f'(x)$ changes sign from *positive* to *negative* as we move across a critical point $x = c$, then $f(c)$ is a relative maximum.
	   b. If $f'(x)$ changes sign from *negative* to *positive* as we move across a critical point $x = c$, then $f(c)$ is a relative minimum.
	   c. If $f'(x)$ does not change sign as we move across a critical point $x = c$, then $f(c)$ is not a relative extremum.

EXAMPLE 5   Find the relative maxima and relative minima of the function $f(x) = x^2$.

Figure 4.17
f has a relative minimum at $x = 0$.

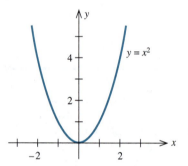

Solution   The derivative of $f(x) = x^2$ is given by $f'(x) = 2x$. Setting $f'(x) = 0$ yields $x = 0$ as the only critical point of f. Since

$$f'(x) < 0 \quad \text{if } x < 0$$

and
$$f'(x) > 0 \quad \text{if } x > 0$$

we see that $f'(x)$ changes sign from negative to positive as we move across the critical point $x = 0$. Thus we conclude that $f(0) = 0$ is a relative minimum of f (Figure 4.17).

EXAMPLE 6   Find the relative maxima and relative minima of the function $f(x) = x^{2/3}$ (see Example 3).

Solution   The derivative of f is $f'(x) = (2/3)x^{-1/3}$. As noted in Example 3, f' is not defined at $x = 0$, is continuous everywhere else, and is not equal to zero in its domain. Thus, $x = 0$ is the only critical point of the function f.

The sign diagram obtained in Example 3 is reproduced in Figure 4.18. We can see that the sign of $f'(x)$ changes from negative to positive as we move across $x = 0$ from left to right. Thus, an application of the First Derivative Test tells us that $f(0) = 0$ is a relative minimum of f (Figure 4.19).

Figure 4.18
Sign diagram for f'.

Figure 4.19
f has a relative minimum at $x = 0$.

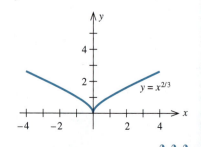

Recall that the average cost function $\overline{C}$ is defined by $\overline{C} = \dfrac{C(x)}{x}$, where $C(x)$ is the total

cost function and x is the number of units of a commodity manufactured (see Section 3.4).

a. Show that $\overline{C}'(x) = \dfrac{C'(x) - \overline{C}(x)}{x}$ $(x > 0)$.

b. Use the result of (a) to conclude that $\overline{C}$ is decreasing for values of x at which $C'(x) < \overline{C}(x)$. Find similar conditions for which $\overline{C}$ is increasing and for which $\overline{C}$ is constant.

c. Explain the results of (b) in economic terms.

EXAMPLE 7 Find the relative maxima and relative minima of the function

$$f(x) = x^3 - 3x^2 - 24x + 32$$

Solution The derivative of f is

$$f'(x) = 3x^2 - 6x - 24 = 3(x + 2)(x - 4)$$

and it is continuous everywhere. The zeros of $f'(x)$, $x = -2$ and $x = 4$, are the only critical points of the function f. The sign diagram for f' is shown in Figure 4.20. Examine the two critical points $x = -2$ and $x = 4$ for a relative extremum using the First Derivative Test and the sign diagram for f':

1. *The critical point* $x = -2$: Since the function $f'(x)$ changes sign from positive to negative as we move across $x = -2$ from left to right, we conclude that a relative maximum of f occurs at $x = -2$. The value of $f(x)$ when $x = -2$ is

$$f(-2) = (-2)^3 - 3(-2)^2 - 24(-2) + 32 = 60$$

2. *The critical point* $x = 4$: $f'(x)$ changes sign from negative to positive as we move across $x = 4$ from left to right, so $f(4) = -48$ is a relative minimum of f. The graph of f appears in Figure 4.21.

Figure 4.20
Sign diagram for f'.

Figure 4.21
f has a relative maximum at $x = -2$ and a relative minimum at $x = 4$.

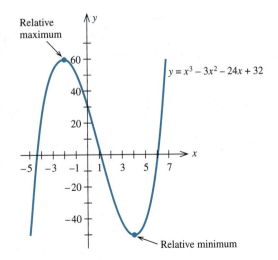

Figure 4.22
$x = 0$ *is not a critical point because f is not defined at* $x = 0$.

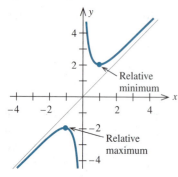

Figure 4.23
$f(x) = x + \dfrac{1}{x}$

EXAMPLE 8   Find the relative maxima and the relative minima of the function

$$f(x) = x + \frac{1}{x}$$

Solution   The derivative of f is

$$f'(x) = 1 - \frac{1}{x^2} = \frac{x^2 - 1}{x^2} = \frac{(x + 1)(x - 1)}{x^2}$$

Since f' is equal to zero at $x = -1$ and $x = 1$, these are critical points for the function f. Next, observe that f' is discontinuous at $x = 0$. However, because *f is not defined at that point,* the point $x = 0$ does not qualify as a critical point of f. Figure 4.22 shows the sign diagram for f'.

Since $f'(x)$ changes sign from positive to negative as we move across $x = -1$ from left to right, the First Derivative Test implies that $f(-1) = -2$ is a relative maximum of the function f. Next, $f'(x)$ changes sign from negative to positive as we move across $x = 1$ from left to right, so $f(1) = 2$ is a relative minimum of the function f. The graph of f appears in Figure 4.23. Note that this function has a relative maximum that lies below its relative minimum.

○ ○ ○

Exploring with Technology

Refer to Example 8.

a. Use a graphing utility to plot the graphs of $f(x) = x + \dfrac{1}{x}$ and its derivative function $f'(x) = 1 - \dfrac{1}{x^2}$ using the viewing rectangle $[-4, 4] \times [-8, 8]$.

b. By studying the graph of f', determine the critical points of f. Next, note the sign of $f'(x)$ immediately to the left and to the right of each critical point. What can you conclude about each critical point? Are your conclusions borne out by the graph of f?

○ ○ ○

Applications

EXAMPLE 9   The profit function of the Acrosonic Company is given by

$$P(x) = -0.02x^2 + 300x - 200,000$$

Figure 4.24
The profit function is increasing on (0, 7500) and decreasing on (7500, ∞).

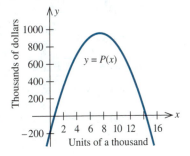

dollars, where x is the number of Acrosonic model F loudspeaker systems produced. Find where the function P is increasing and where it is decreasing.

Solution The derivative P' of the function P is

$$P'(x) = -0.04x + 300 = -0.04(x - 7500)$$

Thus, $P'(x) = 0$ when $x = 7500$. Furthermore, $P'(x) > 0$ for x in the interval $(0, 7500)$, and $P'(x) < 0$ for x in the interval $(7500, \infty)$. This means that the profit function P is increasing on $(0, 7500)$ and decreasing on $(7500, \infty)$ (Figure 4.24).   ◐ ◐ ◐

EXAMPLE 10 The number of major crimes committed in the city of Bronx-
.... ville from 1987 to 1994 is approximated by the function

$$N(t) = -0.1t^3 + 1.5t^2 + 100 \qquad (0 \le t \le 7)$$

where $N(t)$ denotes the number of crimes committed in year t and $t = 0$ corresponds to the beginning of 1987. Find where the function N is increasing and where it is decreasing.

Solution The derivative N' of the function N is

$$N'(t) = -0.3t^2 + 3t = -0.3t(t - 10)$$

Since $N'(t) > 0$ for t in the interval $(0, 7)$, the function N is increasing throughout that interval (Figure 4.25).   ◐ ◐ ◐

Figure 4.25
The number of crimes, $N(t)$, is increasing over the 7-year interval.

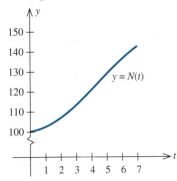

SELF-CHECK EXERCISES 4.1

1. Find the intervals where the function $f(x) = \frac{2}{3}x^3 - x^2 - 12x + 3$ is increasing and the intervals where it is decreasing.

2. Find the relative extrema of $f(x) = \dfrac{x^2}{1 - x^2}$.

Solutions to Self-Check Exercises 4.1 can be found on page 290.

4.1 EXERCISES

In exercises 1–8, you are given the graph of a function f. Determine the intervals where f is increasing, constant, or decreasing.

1.

2.

3.

4.

5.

6.

7.

8.

9. The Boston Marathon   The graph of the function f shown in the accompanying figure gives the elevation of that part of the Boston Marathon course that includes the notorious Heartbreak Hill. Determine the intervals (stretches of the course) where the function f is increasing (the runner is laboring), where it is constant (the runner is taking a breather), and where it is decreasing (the runner is coasting).

In exercises 10–35, find the interval(s) where each function is increasing and the interval(s) where it is decreasing.

10. $f(x) = 4 - 5x$

11. $f(x) = 3x + 5$

12. $f(x) = 2x^2 + x + 1$

13. $f(x) = x^2 - 3x$

14. $f(x) = x^3 - 3x^2$

15. $g(x) = x - x^3$

16. $f(x) = x^3 - 3x + 4$

17. $g(x) = x^3 + 3x^2 + 1$

18. $f(x) = \frac{2}{3}x^3 - 2x^2 - 6x - 2$

19. $f(x) = \frac{1}{3}x^3 - 3x^2 + 9x + 20$

20. $g(x) = x^4 - 2x^2 + 4$

21. $h(x) = x^4 - 4x^3 + 10$

22. $h(x) = \dfrac{1}{2x + 3}$

23. $f(x) = \dfrac{1}{x - 2}$

24. $g(t) = \dfrac{2t}{t^2 + 1}$

25. $h(t) = \dfrac{t}{t - 1}$

26. $f(x) = x^{2/3} + 5$

27. $f(x) = x^{3/5}$

28. $f(x) = (x - 5)^{2/3}$

29. $f(x) = \sqrt{x + 1}$

30. $g(x) = x\sqrt{x + 1}$

31. $f(x) = \sqrt{16 - x^2}$

32. $h(x) = \dfrac{x^2}{x - 1}$

33. $f(x) = \dfrac{x^2 - 1}{x}$

34. $g(x) = \dfrac{x}{(x + 1)^2}$

35. $f(x) = \dfrac{1}{(x - 1)^2}$

In exercises 36–43, you are given the graph of a function f. Determine the relative maxima and relative minima, if any.

36.

37.

38.

39.

40.

41.

42.

(a)

(b)

(c)

(d)

43.

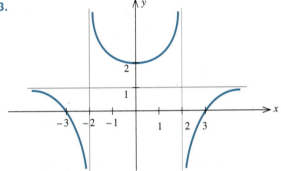

In exercises 48–71, find the relative maxima and relative minima, if any, of each function.

48. $g(x) = x^2 + 3x + 8$ **49.** $f(x) = x^2 - 4x$

50. $h(t) = -t^2 + 6t + 6$ **51.** $f(x) = \frac{1}{2}x^2 - 2x + 4$

52. $f(x) = x^{5/3}$ **53.** $f(x) = x^{2/3} + 2$

54. $f(x) = x^3 - 3x + 6$ **55.** $g(x) = x^3 - 3x^2 + 4$

56. $f(x) = \frac{1}{2}x^4 - x^2$

57. $F(x) = \frac{1}{3}x^3 - x^2 - 3x + 4$

58. $h(x) = \frac{1}{2}x^4 - 3x^2 + 4x - 8$

59. $g(x) = x^4 - 4x^3 + 8$

60. $F(t) = 3t^5 - 20t^3 + 20$

61. $f(x) = 3x^4 - 2x^3 + 4$

In exercises 44–47, match the graph of the function with the graph of its derivative in (a)–(d).

44.

45.

46.

47.

62. $h(x) = \dfrac{x}{x+1}$ **63.** $g(x) = \dfrac{x+1}{x}$

64. $g(x) = 2x^2 + \dfrac{4000}{x} + 10$

65. $f(x) = x + \dfrac{9}{x} + 2$

66. $g(x) = \dfrac{x}{x^2-1}$ **67.** $f(x) = \dfrac{x}{1+x^2}$

68. $g(t) = \dfrac{t^2}{1+t^2}$ **69.** $f(x) = \dfrac{x^2}{x^2-4}$

70. $g(x) = x\sqrt{x-4}$ **71.** $f(x) = (x-1)^{2/3}$

72. A stone is thrown straight up from the roof of an 80-foot building. The distance of the stone from the ground at any time t (in seconds) is given by

$$h(t) = -16t^2 + 64t + 80$$

feet. When is the stone rising and when is it falling? If the stone were to miss the building, when would it hit the ground? Sketch the graph of h.
[*Hint:* The stone is on the ground when $h(t) = 0$.]

73. Profit Functions The Mexican subsidiary of the Thermo-Master Company manufactures an indoor-outdoor thermometer. Management estimates that the profit realizable by the company for the manufacture and sale of x units of thermometers per week is

$$P(x) = -0.001x^2 + 8x - 5000$$

dollars. Find the intervals where the profit function P is increasing and the intervals where P is decreasing.

74. Flight of a Rocket The height (in feet) attained by a rocket t seconds into flight is given by the function

$$h(t) = -\frac{1}{3}t^3 + 16t^2 + 33t + 10$$

When is the rocket rising and when is it descending?

75. Environment of Forests Following the lead of the National Wildlife Federation, the Department of the Interior of a South American country began to record an index of environmental quality that measured progress and decline in the environmental quality of its forests. The index for the years 1984 through 1994 is approximated by the function

$$I(t) = \frac{1}{3}t^3 - \frac{5}{2}t^2 + 80 \qquad (0 \le t \le 10)$$

where $t = 0$ corresponds to the year 1984. Find the intervals where the function I is increasing and the intervals where it is decreasing. Interpret your results.
Source: World Almanac

76. Average Speed of a Highway Vehicle The average speed of a vehicle on a stretch of Route 134 between 6 A.M. and 10 A.M. on a typical weekday is approximated by the function

$$f(t) = 20t - 40\sqrt{t} + 50 \qquad (0 \le t \le 4)$$

where $f(t)$ is measured in miles per hour and t is measured in hours, with $t = 0$ corresponding to 6 A.M. Find the interval where f is increasing and the interval where f is decreasing and interpret your results.

77. Average Cost The average cost in dollars incurred by the Lincoln Record Company per week in pressing x long-playing records is given by

$$\overline{C}(x) = -0.0001x + 2 + \frac{2000}{x} \qquad (0 < x \le 6000)$$

Show that $\overline{C}(x)$ is always decreasing over the interval $(0, 6000)$.

78. Air Pollution According to the South Coast Air Quality Management District, the level of nitrogen dioxide, a brown gas that impairs breathing, present in the atmosphere on a certain May day in downtown Los Angeles is approximated by

$$A(t) = 0.03t^3(t-7)^4 + 60.2 \qquad (0 \le t \le 7)$$

where $A(t)$ is measured in pollutant standard index (PSI) and t is measured in hours with $t = 0$ corresponding to 7 A.M. At what time of day is the air pollution increasing and at what time is it decreasing?

79. Projected Retirement Funds Based on data from the Central Provident Fund of a certain country (a government agency similar to the Social Security Administration), the estimated cash in the fund in 1995 is given by

$$A(t) = -96.6t^4 + 403.6t^3 + 660.9t^2 + 250 \quad (0 \le t \le 5)$$

where $A(t)$ is measured in billions of dollars and t is measured in decades, with $t = 0$ corresponding to the year 1995. Find the interval where A is increasing and the interval where A is decreasing and interpret your results.
[*Hint:* Use the quadratic formula.]

80. Learning Curves The Emory Secretarial School finds from experience that the average student taking Advanced Typing will progress according to the rule

$$N(t) = \frac{60t + 180}{t + 6} \qquad (t \ge 0)$$

where $N(t)$ measures the number of words per minute the student can type after t weeks in the course. Compute $N'(t)$ and use this result to show that the function N is increasing on the interval $(0, \infty)$.

81. Drug Concentration in the Blood   The concentration of a certain drug in a patient's body t hours after injection is given by

$$C(t) = \frac{t^2}{2t^3 + 1} \qquad (0 \le t \le 4)$$

milligrams per cubic centimeter. When is the concentration of the drug increasing and when is it decreasing?

82. Air Pollution   The amount of nitrogen dioxide, a brown gas that impairs breathing, present in the atmosphere on a certain May day in the city of Long Beach is approximated by

$$A(t) = \frac{136}{1 + 0.25(t - 4.5)^2} + 28 \qquad (0 \le t \le 11)$$

where $A(t)$ is measured in pollutant standard index (PSI) and t is measured in hours, with $t = 0$ corresponding to 7 A.M. Find the intervals where A is increasing and where A is decreasing and interpret your results.

83. Prison Overcrowding   The 1980s saw a trend toward old-fashioned punitive deterrence as opposed to the more liberal penal policies and community-based corrections popular in the 1960s and early 1970s. As a result, prisons became more crowded and the gap between the number of people in prison and the prison capacity widened. Based on figures from the U.S. Department of Justice, the number of prisoners (in thousands) in federal and state prisons is approximated by the function

$$N(t) = 3.5t^2 + 26.7t + 436.2 \qquad (0 \le t \le 10)$$

where t is measured in years and $t = 0$ corresponds to 1984. The number of inmates for which prisons were designed is given by

$$C(t) = 24.3t + 365 \qquad (0 \le t \le 10)$$

where $C(t)$ is measured in thousands and t has the same meaning as before. Show that the gap between the number of prisoners and the number for which the prisons were designed has been widening at any time t.

[*Hint:* First write a function G that gives the gap between the number of prisoners and the number for which the prisons were designed at any time t. Then show that $G'(t) > 0$ for all values of t in the interval (0, 10).]

Source: U.S. Dept. of Justice

84. Using Theorem 1, verify that the linear function $f(x) = mx + b$ is (a) increasing everywhere if $m > 0$, (b) decreasing everywhere if $m < 0$, and (c) constant if $m = 0$.

85. In what interval is the quadratic function

$$f(x) = ax^2 + bx + c \qquad (a \ne 0)$$

increasing? In what interval is f decreasing?

86. Let

$$f(x) = \begin{cases} -3x & \text{if } x < 0 \\ 2x + 4 & \text{if } x \ge 0 \end{cases}$$

a. Compute $f'(x)$ and show that it changes sign from negative to positive as we move across $x = 0$.
b. Show that f does not have a relative minimum at $x = 0$. Does this contradict the First Derivative Test? Explain your answer.

87. Let

$$f(x) = \begin{cases} -x^2 + 3 & \text{if } x \ne 0 \\ 2 & \text{if } x = 0 \end{cases}$$

a. Compute $f'(x)$ and show that it changes sign from positive to negative as we move across $x = 0$.
b. Show that f does not have a relative maximum at $x = 0$. Does this contradict the First Derivative Test? Explain your answer.

88. Show that the quadratic function

$$f(x) = ax^2 + bx + c \qquad (a \ne 0)$$

has a relative extremum when $x = -b/2a$. Also, show that the relative extremum is a relative maximum if $a < 0$ and a relative minimum if $a > 0$.

89. Show that the cubic function

$$f(x) = ax^3 + bx^2 + cx + d \qquad (a \ne 0)$$

has no relative extremum if and only if $b^2 - 3ac \le 0$.

90. Refer to Example 6, page 133.
a. Show that f is increasing on the interval (0, 1).
b. Show that $f(0) = -1$ and $f(1) = 1$ and use the result of part (a) together with the Intermediate Value Theorem to conclude that there is exactly one root of $f(x) = 0$ in (0, 1).

91. Show that the function

$$f(x) = \frac{ax + b}{cx + d}$$

does not have a relative extremum if $ad - bc \ne 0$. What can you say about f if $ad - bc = 0$?

Using Technology

Using the First Derivative to Analyze a Function

A graphing utility is an effective tool for analyzing the properties of functions. This is especially true when we also bring into play the power of calculus, as the following examples show.

EXAMPLE 1   Let $f(x) = 2.4x^4 - 8.2x^3 + 2.7x^2 + 4x + 1$.

a. Use a graphing utility to plot the graph of f.

b. Find the intervals where f is increasing and the intervals where f is decreasing.

c. Find the relative extrema of f.

Solution

a. The graph of f in the viewing rectangle $[-2, 4] \times [-10, 10]$ is shown in Figure T1.

Figure T1
The graph of f in the viewing rectangle $[-2, 4] \times [-10, 10]$.

b. We compute

$$f'(x) = 9.6x^3 - 24.6x^2 + 5.4x + 4$$

and observe that f' is continuous everywhere. So the critical points of f occur at values of x where $f'(x) = 0$. In order to solve this last equation, observe that $f'(x)$ is a *polynomial function* of degree 3. The easiest way to solve the polynomial equation

$$9.6x^3 - 24.6x^2 + 5.4x + 4 = 0$$

is to use the function on a graphing utility for solving polynomial equations. (Not all graphing utilities have this function.) You can also use **TRACE** and **ZOOM**, but this will not give the same accuracy without a much greater effort. We find

$$x_1 \approx 2.22564943249, \quad x_2 \approx 0.63272944121, \quad x_3 \approx -0.295878873696$$

Referring to Figure T1, we conclude that f is decreasing on $(-\infty, -0.2959)$ and $(0.6327, 2.2256)$ (correct to four decimal places), and f is increasing on $(-0.2959, 0.6327)$ and $(2.2256, \infty)$.

c. Using the evaluation function of a graphing utility, we find the value of f at each of the critical points found in (b). Upon referring to Figure T1 once again, we see that $f(x_3) \approx 0.2836$ and $f(x_1) \approx -8.2366$ are relative minimum values of f, and $f(x_2) \approx 2.9194$ is a relative maximum value of f.   ○ ○ ○

REMARK   The equation $f'(x) = 0$ in Example 1 is a polynomial equation, and so it is easily solved using the function for solving polynomial equations. We could also solve the equation using the function for finding the roots of equations, but that would require much more work. For equations that are *not* polynomial equations, however, our only choice is to use the function for finding the roots of equations.   ○ ○ ○

If the derivative of a function is difficult to compute or simplify and we do not require great precision in the solution, we can find the relative extrema of the function using a combination of **ZOOM** and **TRACE**. This technique, which does not require the use of the derivative of f, is illustrated in the following example.

EXAMPLE 2   Let $f(x) = x^{1/3}(x^2 + 1)^{-3/2}3^{-x}$.

a. Use a graphing utility to plot the graph of f.

b. Find the relative extrema of f.

Solution

a. The graph of f in the viewing rectangle $[-4, 2] \times [-2, 1]$ is shown in Figure T2.

Figure T2
The graph of f in the viewing rectangle $[-4, 2] \times [-2, 1]$.

b. From the graph of f in Figure T2, we see that f has relative maxima when $x \approx -2$ and $x \approx 0.25$, and a relative minimum when $x \approx -0.75$. To obtain a better approximation of the first relative maximum, we zoom-in with the cursor at approximately the point on the graph corresponding to $x \approx -2$. Then, using **TRACE**, we see that a relative maximum occurs when $x \approx -1.76$ with value $y \approx -1.01$. Similarly, we find the other relative maximum where $x \approx 0.20$ with value $y \approx 0.44$. Repeating the procedure, we find the relative minimum at the point where $x \approx -0.86$ and $y \approx -1.07$.

Finally, we comment that if you have access to a computer and software such as Derive, Maple, or Mathematica, then symbolic differentiation will yield the derivative $f'(x)$ of any differentiable function. This software will also solve the equation $f'(x) = 0$ with ease. Thus, the use of a computer will simplify even more greatly the analysis of functions.

EXERCISES

In exercises 1–4, use a graphing utility to find (a) the intervals where f is increasing and the intervals where f is decreasing, and (b) the relative extrema of f. Express your answers accurate to four decimal places.

1. $f(x) = 3.4x^4 - 6.2x^3 + 1.8x^2 + 3x - 2$

2. $f(x) = 1.8x^4 - 9.1x^3 + 5x - 4$

3. $f(x) = 2x^5 - 5x^3 + 8x^2 - 3x + 2$

4. $f(x) = 3x^5 - 4x^2 + 3x - 1$

In exercises 5–8, use the ZOOM and TRACE features of a graphing utility to find (a) the intervals where f is increasing and the intervals where f is decreasing, and (b) the relative extrema of f. Express your answers accurate to two decimal places.

5. $f(x) = (2x + 1)^{1/3}(x^2 + 1)^{-2/3}$

6. $f(x) = [x^2(x^3 - 1)]^{1/3} + \dfrac{1}{x}$

7. $f(x) = x - \sqrt{1 - x^2}$

8. $f(x) = \dfrac{\sqrt{x}(x^2 - 1)^2}{x - 2}$

9. **Rate of Bank Failures**   The Federal Deposit Insurance Company (FDIC) estimates that the rate at which banks were failing between 1982 and 1994 is given by

$$f(t) = 0.063447t^4 - 1.953283t^3 + 14.632576t^2$$
$$- 6.684704t + 47.458874 \qquad (0 \leq t \leq 12)$$

where $f(t)$ is measured in the number of banks per year and t is measured in years, with $t = 0$ corresponding to the beginning of 1982.
a. Use a graphing utility to plot the graph of f in the viewing rectangle $[0, 13] \times [0, 220]$.
b. Determine the intervals where f is increasing and where f is decreasing and interpret your result.
c. Find the relative maximum of f and interpret your result.
Source: Federal Deposit Insurance Corporation

10. **Manufacturing Capacity**   Data obtained from the Federal Reserve show that the annual increase in manufacturing capacity between 1988 and 1994 is given by

$$f(t) = 0.0388889t^3 - 0.283333t^2 + 0.477778t$$
$$+ 2.04286 \qquad (0 \leq t \leq 6)$$

where $f(t)$ is a percentage and t is measured in years, with $t = 0$ corresponding to the beginning of 1988.

a. Use a graphing utility to plot the graph of f in the viewing rectangle $[0, 8] \times [0, 4]$.

b. Determine the intervals where f is increasing and where f is decreasing and interpret your result.

Source: Federal Reserve

11. **Home Sales** According to the Greater Boston Real Estate Board—Multiple Listing Service, the average number of days a single-family home remains for sale from listing to accepted offer is approximated by the function

$$f(t) = 0.0171911t^4 - 0.662121t^3 + 6.18083t^2$$
$$- 8.97086t + 53.3357 \qquad (0 \le t \le 10)$$

where t is measured in years, with $t = 0$ corresponding to the beginning of 1984.

a. Use a graphing utility to plot the graph of f in the viewing rectangle $[0, 12] \times [0, 120]$.

b. Use the graph of f to find, approximately, the intervals where f is increasing and the intervals where f is decreasing. What does this result tell us about the sales of single-family homes in the greater Boston area from 1984 through 1994?

Source: Greater Boston Real Estate Board—Multiple Listing Service

12. **Morning Traffic Rush** The speed of traffic flow on a certain stretch of Route 123 between 6 A.M. and 10 A.M. on a typical weekday is approximated by the function

$$f(t) = 20t - 40\sqrt{t} + 52 \qquad (0 \le t \le 4)$$

where $f(t)$ is measured in mph and t is measured in hours, with $t = 0$ corresponding to 6 A.M. Find the interval where f is increasing, the interval where f is decreasing, and the relative extrema of f. Interpret your results.

13. **Air Pollution** The amount of nitrogen dioxide, a brown gas that impairs breathing, present in the atmosphere on a certain May day in the city of Long Beach, is approximated by

$$A(t) = \frac{136}{1 + 0.25(t - 4.5)^2} + 28 \qquad (0 \le t \le 11)$$

where $A(t)$ is measured in pollutant standard index (PSI) and t is measured in hours, with $t = 0$ corresponding to 7 A.M. When is the PSI increasing and when is it decreasing? At what time is the PSI highest, and what is its value at that time?

SOLUTIONS TO SELF-CHECK EXERCISES 4.1

1. The derivative of f is

$$f'(x) = 2x^2 - 2x - 12 = 2(x + 2)(x - 3)$$

and it is continuous everywhere. The zeros of $f'(x)$ are $x = -2$ and $x = 3$. The sign diagram of f' is shown in the accompanying figure. We conclude that f is increasing on the intervals $(-\infty, -2)$ and $(3, \infty)$ and decreasing on the interval $(-2, 3)$.

2. The derivative of f is

$$f'(x) = \frac{(1 - x^2)\dfrac{d}{dx}(x^2) - x^2\dfrac{d}{dx}(1 - x^2)}{(1 - x^2)^2}$$

$$= \frac{(1 - x^2)(2x) - x^2(-2x)}{(1 - x^2)^2} = \frac{2x}{(1 - x^2)^2}$$

and it is continuous everywhere except at $x = \pm 1$. Since $f'(x)$ is equal to zero at $x = 0$, $x = 0$ is a critical point of f. Next, observe that $f'(x)$ is discontinuous at $x = \pm 1$, but since these points are not in the domain of f, they do not qualify as critical points of f. Finally, from the sign diagram of f' shown in the accompanying figure, we conclude that $f(0) = 0$ is a relative minimum of f.

4.2 APPLICATIONS OF THE SECOND DERIVATIVE

Determining the Intervals of Concavity

Consider the graphs shown in Figure 4.26, which give the estimated population of the world and of the United States through the year 2000. Both graphs are rising, indicating that both the U.S. population and the world population will continue to increase through the year 2000. But observe that the graph in Figure 4.26a opens upward, whereas the graph in Figure 4.26b opens down-

Figure 4.26

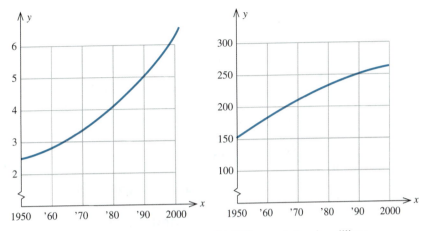

(a) World population in billions.   **(b)** U.S. population in millions.

Source: U.S. Dept. of Commerce and Worldwatch Institute

ward. What is the significance of this? To answer this question, let us look at the slopes of the tangent lines to various points on each graph (Figure 4.27).

In Figure 4.27a, we see that the slopes of the tangent lines to the graph are increasing as we move from left to right. Since the slope of the tangent line to the graph at a point on the graph measures the rate of change of the function at that point, we conclude that the world population is not only increasing through the year 2000 but is increasing at an *increasing* pace. A similar analysis of Figure 4.27b reveals that the U.S. population is increasing, but at a *decreasing* pace.

Figure 4.27

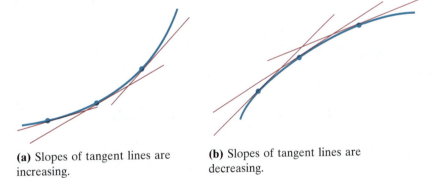

(a) Slopes of tangent lines are increasing.   **(b)** Slopes of tangent lines are decreasing.

The shape of a curve can be described using the notion of *concavity:*

CONCAVITY OF A FUNCTION f

Let the function f be differentiable on an interval (a, b). Then

1. f is **concave upward** on (a, b) if f' is increasing on (a, b).
2. f is **concave downward** on (a, b) if f' is decreasing on (a, b).

Figure 4.28

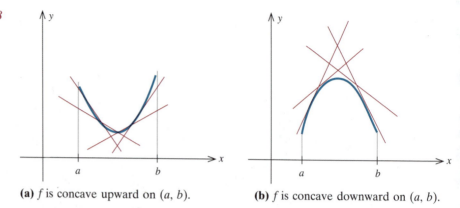

(a) f is concave upward on (a, b).   **(b)** f is concave downward on (a, b).

Geometrically, a curve is concave upward if it lies above its tangent lines (Figure 4.28a). Similarly, a curve is concave downward if it lies below its tangent lines (Figure 4.28b).

We also say that f is *concave upward at a point x = c* if there exists an interval (a, b) containing c in which f is concave upward. Similarly, we say that f is *concave downward at a point x = c* if there exists an interval (a, b) containing c in which f is concave downward.

If a function f has a second derivative f'', we can use f'' to determine the intervals of concavity of the function. Recall that $f''(x)$ measures the rate of change of the slope $f'(x)$ of the tangent line to the graph of f at the point $(x, f(x))$. Thus, if $f''(x) > 0$ on an interval (a, b), then the slopes of the tangent lines to the graph of f are increasing on (a, b) and so f is concave upward on (a, b). Similarly, if $f''(x) < 0$ on (a, b), then f is concave downward on (a, b). These observations suggest the following theorem.

THEOREM 2

a. If $f''(x) > 0$ for each value of x in (a, b), then f is concave upward on (a, b).

b. If $f''(x) < 0$ for each value of x in (a, b), then f is concave downward on (a, b).

The following procedure, based on the conclusions of Theorem 2, may be used to determine the intervals of concavity of a function.

DETERMINING THE INTERVALS OF CONCAVITY OF f

1. Determine the values of x for which f'' is zero or where f'' is not defined, and identify the open intervals determined by these points.

2. Determine the sign of f'' in each interval found in step 1. To do this, compute $f''(c)$, where c is any conveniently chosen test point in the interval.

a. If $f''(c) > 0$, f is concave upward on that interval.

b. If $f''(c) < 0$, f is concave downward on that interval.

Figure 4.29
Sign diagram for f".

Figure 4.30
f is concave downward on $(-\infty, 1)$ and concave upward on $(1, \infty)$.

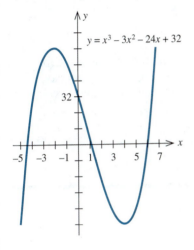

EXAMPLE 1 Determine where the function $f(x) = x^3 - 3x^2 - 24x + 32$ is concave upward and where it is concave downward.

Solution Here,

$$f'(x) = 3x^2 - 6x - 24$$

so

$$f''(x) = 6x - 6 = 6(x - 1)$$

and f'' is defined everywhere. Setting $f''(x) = 0$ gives $x = 1$. The sign diagram of f'' appears in Figure 4.29. We conclude that f is concave downward on the interval $(-\infty, 1)$ and is concave upward on the interval $(1, \infty)$. Figure 4.30 shows the graph of f. ○ ○ ○

EXAMPLE 2 Determine the intervals where the function $f(x) = x + 1/x$ is concave upward and where it is concave downward.

Solution We have

$$f'(x) = 1 - \frac{1}{x^2}$$

and

$$f''(x) = \frac{2}{x^3}$$

We deduce from the sign diagram for f'' (Figure 4.31) that the function f is concave downward on the interval $(-\infty, 0)$ and concave upward on the interval $(0, \infty)$. The graph of f is sketched in Figure 4.32.

Figure 4.31
The sign diagram for f".

Figure 4.32
f is concave downward on $(-\infty, 0)$ and concave upward on $(0, \infty)$.

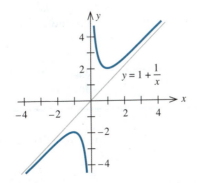

○ ○ ○

EXPLORING WITH TECHNOLOGY

Refer to Example 1.

a. Use a graphing utility to plot the graphs of

$$f(x) = x^3 - 3x^2 - 24x + 32$$

and its second derivative

$$f''(x) = 6x - 6$$

using the viewing rectangle $[-10, 10] \times [-80, 90]$.

b. By studying the graph of f'', determine the intervals where $f''(x) > 0$ and the intervals where $f''(x) < 0$. Next, look at the graph of f and determine the intervals where the graph of f is concave upward and the intervals where the graph of f is concave downward. Are these observations what you might have expected?

〇　〇　〇

Figure 4.33

The graph of S has a point of inflection at (50, 2700).

Inflection Points

Figure 4.33 shows the total sales S of a manufacturer of automobile air conditioners versus the amount of money x that the company spends on advertising its product. Notice that the graph of the continuous function $y = S(x)$ changes concavity—from upward to downward—at the point (50, 2700). This point is called an *inflection point* of S. To understand the significance of this inflection point, observe that the total sales increase rather slowly at first, but as more money is spent on advertising, the total sales increase rapidly. This rapid increase reflects the effectiveness of the company's ads. However, a point is soon reached after which any additional advertising expenditure results in increased sales but at a slower rate of increase. This point, commonly known as the *point of diminishing returns,* is the point of inflection of the function S. We will return to this example later.

Let us now state formally the definition of an inflection point.

INFLECTION POINT   A point on the graph of a differentiable function f at which the concavity changes is called an **inflection point.**

Observe that the graph of a function crosses its tangent line at a point of inflection (Figure 4.34).

Figure 4.34
At each point of inflection, the graph of a function crosses its tangent line.

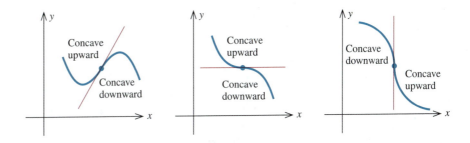

The following procedure may be used to find inflection points.

FINDING INFLECTION POINTS

1. Compute $f''(x)$.
2. Determine the points in the domain of f for which $f''(x) = 0$ or $f''(x)$ does not exist.
3. Determine the sign of $f''(x)$ to the left and right of each point $x = c$ found in step 2. If there is a change in the sign of $f''(x)$ as we move across the point $x = c$, then $(c, f(c))$ is an inflection point of f.

Figure 4.35
Sign diagram for f''.

The points determined in step 2 are only *candidates* for the inflection points of f. For example, you can easily verify that $f''(0) = 0$ if $f(x) = x^4$, but a sketch of the graph of f will show that $(0, 0)$ is not an inflection point of f.

EXAMPLE 3   Find the points of inflection of the function $f(x) = x^3$.

Figure 4.36
f has an inflection point at $(0, 0)$.

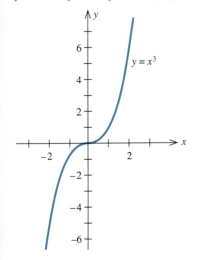

Solution

$$f'(x) = 3x^2$$

so

$$f''(x) = 6x$$

Observe that f'' is continuous everywhere and is zero if $x = 0$. The sign diagram of f'' is shown in Figure 4.35. From this diagram, we see that $f''(x)$ changes sign as we move across $x = 0$. Thus the point $(0, 0)$ is an inflection point of the function f (Figure 4.36).   ◦ ◦ ◦

EXAMPLE 4   Determine the intervals where the function $f(x) = (x - 1)^{5/3}$ is concave upward and where it is concave downward and find the inflection points of f.

Solution   The first derivative of f is

$$f'(x) = \frac{5}{3}(x - 1)^{2/3}$$

Figure 4.37
Sign diagram for f".

and the second derivative of f is

$$f''(x) = \frac{10}{9}(x-1)^{-1/3} = \frac{10}{9(x-1)^{1/3}}$$

We see that f'' is not defined at $x = 1$. Furthermore, $f''(x)$ is not equal to zero anywhere. The sign diagram of f'' is shown in Figure 4.37. From the sign diagram, we see that f is concave downward on $(-\infty, 1)$ and concave upward on $(1, \infty)$. Next, since $x = 1$ does lie in the domain of f, our computations also reveal that the point $(1, 0)$ is an inflection point of f (Figure 4.38).

● ● ●

Figure 4.38
f has an inflection point at (1, 0).

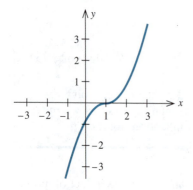

EXAMPLE 5 Determine the intervals where the function

$$f(x) = \frac{1}{x^2 + 1}$$

is concave upward and where it is concave downward and find the inflection points of f.

Solution The first derivative of f is

$$f'(x) = \frac{d}{dx}(x^2 + 1)^{-1} = -2x(x^2 + 1)^{-2} \qquad \text{(Using the General Power Rule)}$$

$$= -\frac{2x}{(x^2 + 1)^2}$$

Next, using the Quotient Rule, we find

$$f''(x) = \frac{(x^2 + 1)^2(-2) + (2x)2(x^2 + 1)(2x)}{(x^2 + 1)^4}$$

$$= \frac{(x^2 + 1)[-2(x^2 + 1) + 8x^2]}{(x^2 + 1)^4} = \frac{(x^2 + 1)(6x^2 - 2)}{(x^2 + 1)^4}$$

$$= \frac{2(3x^2 - 1)}{(x^2 + 1)^3} \qquad \text{(Canceling the common factors)}$$

Observe that f'' is continuous everywhere and is zero if

$$3x^2 - 1 = 0$$

$$x^2 = \frac{1}{3}$$

Figure 4.39
Sign diagram for f".

or $x = \pm\sqrt{3}/3$. The sign diagram for f'' is shown in Figure 4.39. From the sign diagram for f'', we see that f is concave upward on $(-\infty, -\sqrt{3}/3) \cup (\sqrt{3}/3, \infty)$ and concave downward on $(-\sqrt{3}/3, \sqrt{3}/3)$. Also, observe that $f''(x)$ changes sign as we move across the points $x = -\sqrt{3}/3$ and $x = \sqrt{3}/3$. Since

$$f\left(-\frac{\sqrt{3}}{3}\right) = \frac{1}{\frac{1}{3} + 1} = \frac{3}{4} \qquad \text{and} \qquad f\left(\frac{\sqrt{3}}{3}\right) = \frac{3}{4}$$

Figure 4.40

The graph of $f(x) = \dfrac{1}{x^2 + 1}$ is con-

cave upward on $\left(-\infty, -\dfrac{\sqrt{3}}{3}\right) \cup$

$\left(\dfrac{\sqrt{3}}{3}, \infty\right)$ and concave down-

ward on $\left(-\dfrac{\sqrt{3}}{3}, \dfrac{\sqrt{3}}{3}\right)$.

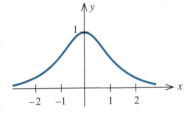

we see that the points $(-\sqrt{3}/3, 3/4)$ and $(\sqrt{3}/3, 3/4)$ are inflection points of f. The graph of f is shown in Figure 4.40.   ○ ○ ○

Applications

Examples 6 and 7 illustrate familiar interpretations of the significance of the inflection point of a function.

EXAMPLE 6   The total sales S (in thousands of dollars) of the Arctic Air Corporation, a manufacturer of automobile air conditioners, is related to the amount of money x the company spends on advertising its products by the formula

$$S = -0.01x^3 + 1.5x^2 + 200 \qquad (0 \le x \le 100)$$

where x is measured in thousands of dollars. Find the inflection point of the function S.

Solution   The first two derivatives of S are given by

$$S' = -0.03x^2 + 3x$$

and

$$S'' = -0.06x + 3$$

Setting $S'' = 0$ gives $x = 50$ as the only candidate for an inflection point of S. Moreover, since

$$S'' > 0 \quad \text{for} \quad x < 50$$

and

$$S'' < 0 \quad \text{for} \quad x > 50$$

the point $(50, 2700)$ is an inflection point of the function S. The graph of S appears in Figure 4.41. Notice that this is the graph of the function we discussed earlier.   ○ ○ ○

Figure 4.41
The graph of $S(x)$ has a point of inflection at $(50, 2700)$.

EXAMPLE 7   An economy's consumer price index (CPI) is described by the function

$$I(t) = -0.2t^3 + 3t^2 + 100 \qquad (0 \le t \le 9)$$

where $t = 0$ corresponds to the year 1988. Find the point of inflection of the function I and discuss its significance.

Figure 4.42
The graph of $I(t)$ has a point of inflection at $(5, 150)$.

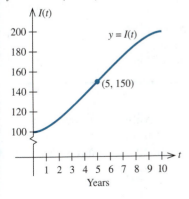

Solution   The first two derivatives of I are given by

$$I'(t) = -0.6t^2 + 6t$$

and

$$I''(t) = -1.2t + 6 = -1.2(t - 5)$$

Setting $I''(t) = 0$ gives $t = 5$ as the only candidate for an inflection point of I. Next, we observe that

$$I'' > 0 \quad \text{for} \quad t < 5$$

and

$$I'' < 0 \quad \text{for} \quad t > 5$$

so the point $(5, 150)$ is an inflection point of I. The graph of I is sketched in Figure 4.42.

Since the second derivative of I measures the rate of change of the inflation rate, our computations reveal that the rate of inflation had, in fact, peaked at $t = 5$. Thus, relief actually began at the beginning of 1993.   ○ ○ ○

The Second Derivative Test

We now show how the second derivative f'' of a function f can be used to help us determine whether a critical point of f is a relative extremum of f. Figure 4.43a shows the graph of a function that has a relative maximum at $x = c$. Observe that f is concave downward at that point. Similarly, Figure 4.43b shows that at a relative minimum of f the graph is concave upward. But from our previous work we know that f is concave downward at $x = c$ if $f''(c) < 0$ and f is concave upward at $x = c$ if $f''(c) > 0$. These observations suggest the following alternative procedure for determining whether a critical point of f gives rise to a relative extremum of f. This result is called the **Second Derivative Test** and is applicable when f'' exists.

Figure 4.43

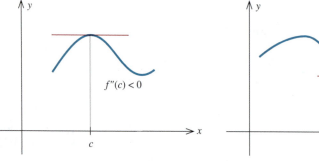

(a) f has a relative maximum at $x = c$.

(b) f has a relative minimum at $x = c$.

THE SECOND DERIVATIVE TEST

1. Compute $f'(x)$ and $f''(x)$.

2. Find all the critical points of f at which $f'(x) = 0$.

3. Compute $f''(c)$ for each such critical point c.
   a. If $f''(c) < 0$, then f has a relative maximum at c.
   b. If $f''(c) > 0$, then f has a relative minimum at c.
   c. If $f''(c) = 0$, the test fails; that is, it is inconclusive.

REMARK   As stated in step 3(c), the Second Derivative Test does not yield a conclusion if $f''(c) = 0$ or if $f''(c)$ does not exist. In other words, $x = c$ may give rise to a relative extremum or an inflection point (see exercise 86, page 308). In such cases, you should revert to the First Derivative Test.   ○ ○ ○

EXAMPLE 8   Determine the relative extrema of the function

$$f(x) = x^3 - 3x^2 - 24x + 32$$

using the Second Derivative Test. (See Example 7, Section 4.1.)

Solution   We have

$$f'(x) = 3x^2 - 6x - 24 = 3(x + 2)(x - 4)$$

so $f'(x) = 0$ gives $x = -2$ and $x = 4$, the critical points of f, as in Example 7. Next, we compute

$$f''(x) = 6x - 6 = 6(x - 1)$$

Since

$$f''(-2) = 6(-2 - 1) = -18 < 0$$

the Second Derivative Test implies that $f(-2) = 60$ is a relative maximum of f. Also,

$$f''(4) = 6(4 - 1) = 18 > 0$$

and the Second Derivative Test implies that $f(4) = -48$ is a relative minimum of f, which confirms the results obtained earlier.   ○ ○ ○

Suppose a function f has the following properties:

1. $f''(x) > 0$ for all x in an interval (a, b).

2. There is a point c between a and b such that $f'(c) = 0$.

What special property can you ascribe to the point $(c, f(c))$? Answer the question if property 1 is replaced by the property that $f''(x) < 0$ for all x in (a, b).

Comparing the First and Second Derivative Tests

Notice that both the First Derivative Test and the Second Derivative Test are used to classify the critical points of f. What are the pros and cons of the two tests? Since the Second Derivative Test is applicable only when f'' exists, it is less versatile than the First Derivative Test. For example, it cannot be used to locate the relative minimum $f(0) = 0$ of the function $f(x) = x^{2/3}$.

Furthermore, the Second Derivative Test is inconclusive when f'' is equal to zero at a critical point of f, whereas the First Derivative Test always yields positive conclusions. The Second Derivative Test is also inconvenient to use when f'' is difficult to compute. On the plus side, if f'' is computed easily, then we use the Second Derivative Test since it involves just the evaluation of f'' at the critical point(s) of f. Also, the conclusions of the Second Derivative Test are important in theoretical work.

We close this section by summarizing the different roles played by the first derivative f' and the second derivative f'' of a function f in determining the properties of the graph of f. The first derivative f' tells us where f is increasing and where f is decreasing, whereas the second derivative f'' tells us where f is concave upward and where f is concave downward. These different properties of f are reflected by the signs of f' and f'' in the interval of interest. The following table shows the general characteristics of the function f for various possible combinations of the signs of f' and f'' in the interval (a, b).

Signs of f' and f''	Properties of the Graph of f	General Shape of the Graph of f
$f'(x) > 0$ $f''(x) > 0$	f increasing f concave upward	⌣
$f'(x) > 0$ $f''(x) < 0$	f increasing f concave downward	⌒
$f'(x) < 0$ $f''(x) > 0$	f decreasing f concave upward	⌣
$f'(x) < 0$ $f''(x) < 0$	f decreasing f concave downward	⌒

SELF-CHECK EXERCISES 4.2

1. Determine where the function $f(x) = 4x^3 - 3x^2 + 6$ is concave upward and where it is concave downward.

2. Using the Second Derivative Test, if applicable, find the relative extrema of the function $f(x) = 2x^3 - \frac{1}{2}x^2 - 12x - 10$.

3. A certain country's gross domestic product (GDP) (in millions of dollars) in year t is described by the function

$$G(t) = -2t^3 + 45t^2 + 20t + 6000 \qquad (0 \le t \le 11)$$

where $t = 0$ corresponds to the beginning of the year 1985. Find the inflection point of the function G and discuss its significance.

Solutions to Self-Check Exercises 4.2 can be found on page 308.

4.2 EXERCISES

In exercises 1–8, you are given the graph of a function f. Determine the intervals where f is concave upward and where it is concave downward. Also, find all inflection points of f, if any.

1.

2.

3.

4.

5.

6.

7.

8.

In exercises 9–12, determine which graph—a, b, or c—is the graph of the function f with the specified properties.

9. $f(2) = 1$, $f'(2) > 0$, and $f''(2) < 0$

a.

b.

c.

10. $f(1) = 2$, $f'(x) > 0$ on $(-\infty, 1) \cup (1, \infty)$, and $f''(1) = 0$

a.

b.

c.

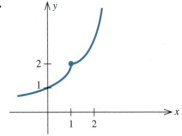

11. $f'(0)$ is undefined, f is decreasing on $(-\infty, 0)$, f is concave downward on $(0, 3)$, and f has an inflection point at $x = 3$.

a.

b.

c.

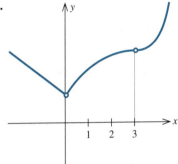

12. f is decreasing on $(-\infty, 2)$ and increasing on $(2, \infty)$, f is concave upward on $(1, \infty)$, and has inflection points at $x = 0$ and $x = 1$.

a.

b.

c.

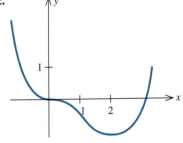

13. Effect of Advertising on Bank Deposits The following graphs were used by the CEO of the Madison Savings Bank to illustrate what effect a projected promotional campaign would have on its deposits over the next year. The functions D_1 and D_2 give the projected amount of money on deposit with the bank over the next 12 months with, and without, the proposed promotional campaign, respectively.
   a. Determine the signs of $D_1'(t)$, $D_2'(t)$, $D_1''(t)$, and $D_2''(t)$ on the interval $(0, 12)$.
   b. Explain the significance of your results in the context of the problem.

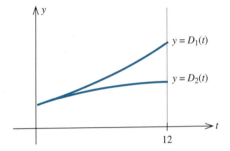

14. Assembly Time of a Worker In the following graph, $N(t)$ gives the number of transistor radios assembled by the average worker by the tth hour, where $t = 0$ corresponds to 8 A.M. and $0 \leq t \leq 4$. Explain the significance of the inflection point P shown on the graph.

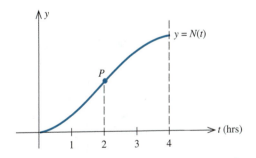

15. Water Pollution When organic waste is dumped into a pond, the oxidation process that takes place reduces the pond's oxygen content. However, given time, nature will restore the oxygen content to its natural level. In the graph on page 304, $P(t)$ gives the oxygen content (as a percentage of its normal level) t days after organic waste has been dumped into the pond. Explain the significance of the inflection point Q.

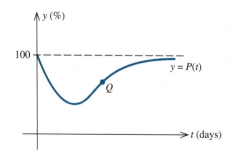

In exercises 16–21 show that the function is concave upward wherever it is defined.

16. $f(x) = 4x^2 - 12x + 7$

17. $g(x) = x^4 + \frac{1}{2}x^2 + 6x + 10$

18. $h(x) = \dfrac{1}{x^2}$

19. $f(x) = \dfrac{1}{x^4}$

20. $h(x) = \sqrt{x^2 + 4}$

21. $g(x) = -\sqrt{4 - x^2}$

In exercises 22–43, determine where the function is concave upward and where it is concave downward.

22. $g(x) = -x^2 + 3x + 4$

23. $f(x) = 2x^2 - 3x + 4$

24. $g(x) = x^3 - x$

25. $f(x) = x^3 - 1$

26. $f(x) = 3x^4 - 6x^3 + x - 8$

27. $f(x) = x^4 - 6x^3 + 2x + 8$

28. $f(x) = \sqrt[3]{x}$

29. $f(x) = x^{4/7}$

30. $g(x) = \sqrt{x - 2}$

31. $f(x) = \sqrt{4 - x}$

32. $g(x) = \dfrac{x}{x + 1}$

33. $f(x) = \dfrac{1}{x - 2}$

34. $g(x) = \dfrac{x}{1 + x^2}$

35. $f(x) = \dfrac{1}{2 + x^2}$

36. $f(x) = \dfrac{x + 1}{x - 1}$

37. $h(t) = \dfrac{t^2}{t - 1}$

38. $h(r) = -\dfrac{1}{(r - 2)^2}$

39. $g(x) = x + \dfrac{1}{x^2}$

40. $f(x) = (x - 2)^{2/3}$

41. $g(t) = (2t - 4)^{1/3}$

42. $f(x) = \dfrac{x^2 - 2}{x^3}$

43. $f(x) = \dfrac{x^2}{x^2 - 1}$

In exercises 44–55, find the inflection points, if any, of each function.

44. $g(x) = x^3 - 6x$

45. $f(x) = x^3 - 2$

46. $g(x) = 2x^3 - 3x^2 + 18x - 8$

47. $f(x) = 6x^3 - 18x^2 + 12x - 15$

48. $f(x) = x^4 - 2x^3 + 6$

49. $f(x) = 3x^4 - 4x^3 + 1$

50. $f(x) = \sqrt[5]{x}$

51. $g(t) = \sqrt[3]{t}$

52. $f(x) = (x - 2)^{4/3}$

53. $f(x) = (x - 1)^3 + 2$

54. $f(x) = 2 + \dfrac{3}{x}$

55. $f(x) = \dfrac{2}{1 + x^2}$

In exercises 56–73, find the relative extrema, if any, of each function. Use the Second Derivative Test, if applicable.

56. $g(x) = 2x^2 + 3x + 7$

57. $f(x) = -x^2 + 2x + 4$

58. $g(x) = x^3 - 6x$

59. $f(x) = 2x^3 + 1$

60. $f(x) = 2x^3 + 3x^2 - 12x - 4$

61. $f(x) = \frac{1}{3}x^3 - 2x^2 - 5x - 10$

62. $f(t) = 2t + \dfrac{3}{t}$

63. $g(t) = t + \dfrac{9}{t}$

64. $f(x) = \dfrac{2x}{x^2 + 1}$

65. $f(x) = \dfrac{x}{1 - x}$

66. $g(x) = x^2 + \dfrac{2}{x}$

67. $f(t) = t^2 - \dfrac{16}{t}$

68. $g(x) = \dfrac{1}{1 + x^2}$

69. $g(s) = \dfrac{s}{1 + s^2}$

70. $f(x) = \dfrac{x^2}{x^2 + 1}$

71. $f(x) = \dfrac{x^4}{x - 1}$

72. $f(x) = \dfrac{x^2 + 4}{x^2 - 1}$

73. $g(x) = \dfrac{2 - x}{(x + 2)^3}$

74. Effect of Budget Cuts on Drug-Related Crimes The following graphs were used by a police commissioner to illustrate what effect a budget cut would have on crime in the city. The number $N_1(t)$ gives the projected number of drug-related crimes in the next 12 months.

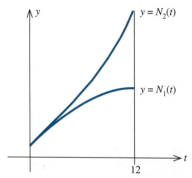

The number $N_2(t)$ gives the projected number of drug-related crimes in the same time frame if next year's budget is cut.

a. Explain why $N_1'(t)$ and $N_2'(t)$ are both positive on the interval $(0, 12)$.

b. What are the signs of $N_1''(t)$ and $N_2''(t)$ on the interval $(0, 12)$?

c. Interpret the results of (b).

75. Demand for RNs   The following graph gives the total number of "help wanted" ads for RNs (registered nurses) in 22 cities over the last 12 months as a function of time t (t measured in months).

a. Explain why $N'(t)$ is positive on the interval $(0, 12)$.

b. Determine the signs of $N''(t)$ on the interval $(0, 6)$ and the interval $(6, 12)$.

c. Interpret the results of (b).

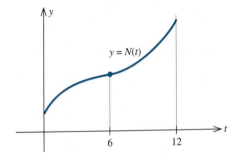

76. Effect of Advertising on Hotel Revenue   The total annual revenue R of the Miramar Resorts Hotel is related to the amount of money x the hotel spends on advertising its services by the function

$$R(x) = -0.003x^3 + 1.35x^2 + 2x + 8000$$

$$(0 \le x \le 400)$$

where both R and x are measured in thousands of dollars. Find the inflection point of R and discuss its significance.

77. Effect of Advertising on Sales   The total sales S (in thousands of dollars) of the Cannon Precision Instruments Corporation is related to the amount of money x that Cannon spends on advertising its products by the function

$$S(x) = -0.002x^3 + 0.6x^2 + x + 500$$

$$(0 \le x \le 200)$$

where x is measured in thousands of dollars. Find the inflection point of the function S and discuss its significance.

78. Forecasting Profits   As a result of increasing energy costs, the growth rate of the profit of the 4-year-old Venice Glassblowing Company has begun to decline. Venice's management, after consulting with energy experts, decides to implement certain energy-conservation measures aimed at cutting energy bills. The general manager reports that, according to his calculations, the growth rate of Venice's profit should be on the increase again within 4 years. If Venice's profit (in hundreds of dollars) x years from now is given by the function

$$P(x) = x^3 - 9x^2 + 40x + 50 \qquad (0 \le x \le 8)$$

determine whether the general manager's forecast will be accurate.

[*Hint:* Find the inflection point of the function P and study the concavity of P.]

79. Worker Efficiency   An efficiency study conducted for the Elektra Electronics Company showed that the number of Space Commander walkie-talkies assembled by the average worker t hours after starting work at 8 A.M. is given by

$$N(t) = -t^3 + 6t^2 + 15t \qquad (0 \le t \le 4)$$

At what time during the morning shift is the average worker performing at peak efficiency?

80. Cost of Producing Calculators   A subsidiary of Elektra Electronics manufactures programmable calculators. Management determines that the daily cost $C(x)$ of producing these calculators (in dollars) is

$$C(x) = 0.0001x^3 - 0.08x^2 + 40x + 5000$$

where x is the number of calculators produced. Find the inflection point of the function C and interpret your result.

81. Flight of a Rocket   The altitude (in feet) of a rocket t seconds into flight is given by

$$s = f(t) = -t^3 + 54t^2 + 480t + 6$$

Find the point of inflection of the function f and interpret your result. What is the maximum velocity attained by the rocket?

USING TECHNOLOGY

FINDING THE INFLECTION POINTS OF A FUNCTION

A graphing utility can be used to find the inflection points of a function and hence the intervals where the graph of the function is concave upward and the intervals where it is concave downward. Some graphing utilities have an operation for finding inflection points directly. If your graphing utility has this capability, use it to work through the example and exercises in this section.

EXAMPLE I Let $f(x) = 2.5x^5 - 12.4x^3 + 4.2x^2 - 5.2x + 4$.

a. Use a graphing utility to plot the graph of f.

b. Find the inflection points of f.

c. Find the intervals where f is concave upward and where it is concave downward.

Solution

a. The graph of f using the viewing rectangle $[-3, 3] \times [-25, 60]$ is shown in Figure T1.

Figure T1
The graph of f in the viewing rectangle $[-3, 3] \times [-25, 60]$.

b. From Figure T1, we see that f has three inflection points—one occurring at the point where the x-coordinate is approximately -1, another at the point where $x \approx 0$, and the third at the point where $x \approx 1$. To find the first inflection point, we use the inflection operation, moving the cursor to the point on the graph of f where $x \approx -1$. We obtain the point $(-1.2728, 34.6395)$ (accurate to four decimal places). Next, setting the cursor near the point $x = 0$ yields the inflection point $(0.1139, 3.4440)$. Finally, with the cursor set at $x = 1$, we obtain the third inflection point $(1.1589, -10.4594)$.

c. From the results of (b), we see that f is concave upward on the intervals $(-1.2728, 0.1139)$ and $(1.1589, \infty)$ and concave downward on $(-\infty, -1.2728)$ and $(0.1139, 1.1589)$.

EXERCISES

In exercises 1–8, use a graphing utility to find (a) the intervals where f is concave upward and the intervals where f is concave downward, and (b) the inflection points of f. Express your answers accurate to four decimal places.

1. $f(x) = 1.8x^4 - 4.2x^3 + 2.1x + 2$

2. $f(x) = -2.1x^4 + 3.1x^3 + 2x^2 - x + 1.2$

3. $f(x) = 1.2x^5 - 2x^4 + 3.2x^3 - 4x + 2$

4. $f(x) = -2.1x^5 + 3.2x^3 - 2.2x^2 + 4.2x - 4$

5. $f(x) = x^3(x^2 + 1)^{-1/3}$

6. $f(x) = x^2(x^3 - 1)^3$

7. $f(x) = \dfrac{x^2 - 1}{x^3}$

8. $f(x) = \dfrac{x + 1}{\sqrt{x}}$

9. **Growth of HMOs** Based on data compiled by the Group Health Association of America, the number of people receiving their care in an HMO (Health Maintenance Organization) from the beginning of 1984 through 1994 is approximated by the function

$$f(t) = 0.0514t^3 - 0.853t^2 + 6.8147t + 15.6524$$
$$(0 < t \le 11)$$

where $f(t)$ gives the number of people in millions and t is measured in years, with $t = 0$ corresponding to the beginning of 1984.
a. Use a graphing utility to plot the graph of f in the viewing rectangle $[0, 12] \times [0, 120]$.
b. Find the points of inflection of f.
c. At what time in the given time interval was the number of people receiving their care at an HMO increasing fastest?
Source: Group Health Association of America

10. **Manufacturing Capacity** Data obtained from the Federal Reserve show that the annual increase in manufacturing capacity between 1988 and 1994 is given by

$$f(t) = 0.0388889t^3 - 0.283333t^2 + 0.477778t$$
$$+ 2.04286 \qquad (0 \le t \le 6)$$

where $f(t)$ is a percentage and t is measured in years, with $t = 0$ corresponding to the beginning of 1988.
a. Use a graphing utility to plot the graph of f in the viewing rectangle $[0, 8] \times [0, 4]$.
b. Find the point of inflection and interpret your result.
Source: Federal Reserve

11. **Time on the Market** According to the Greater Boston Real Estate Board—Multiple Listing Service, the average number of days a single-family home remains for sale from listing to accepted offer is approximated by the function

$$f(t) = 0.0171911t^4 - 0.662121t^3 + 6.18083t^2$$
$$- 8.97086t + 53.3357 \qquad (0 \le t \le 10)$$

where t is measured in years, with $t = 0$ corresponding to the beginning of 1984.
a. Use a graphing utility to plot the graph of f in the viewing rectangle $[0, 12] \times [0, 120]$.
b. Find the points of inflection and interpret your result.
Source: Greater Boston Real Estate Board—Multiple Listing Service

12. **Multimedia Sales** According to the Electronic Industries Association, sales in the multimedia market (hardware and software) are expected to be

$$S(t) = -0.0094t^4 + 0.1204t^3 - 0.0868t^2$$
$$+ 0.0195t + 3.3325 \qquad (0 \le t \le 10)$$

where $S(t)$ is measured in billions of dollars and t is measured in years, with $t = 0$ corresponding to 1990.
a. Plot the graph of S in the viewing rectangle $[0, 12] \times [0, 25]$.
b. Find the inflection point of S and interpret your result.
Source: Electronic Industries Association

307

82. Air Pollution The level of ozone, an invisible gas that irritates and impairs breathing, present in the atmosphere on a certain May day in the city of Riverside was approximated by

$$A(t) = 1.0974t^3 - 0.0915t^4 \qquad (0 \le t \le 11)$$

where $A(t)$ is measured in pollutant standard index (PSI) and t is measured in hours, with $t = 0$ corresponding to 7 A.M. Use the Second Derivative Test to show that the function A has a relative maximum at approximately $t = 9$. Interpret your results.

83. Show that the quadratic function

$$f(x) = ax^2 + bx + c \qquad (a \ne 0)$$

is concave upward if $a > 0$ and concave downward if $a < 0$. Thus, by examining the sign of the coefficient of x^2, one can tell immediately whether the parabola opens upward or downward.

84. Suppose f has an inflection point at $(a, f(a))$. Must the function f' have a relative extremum at $x = a$? Explain your answer.

85. Show that the cubic function

$$f(x) = ax^3 + bx^2 + cx + d \qquad (a \ne 0)$$

has one and only one inflection point. Find the coordinates of this point.

86. Consider the functions $f(x) = x^3$, $g(x) = x^4$, and $h(x) = -x^4$.
a. Show that $x = 0$ is a critical point of each of the functions f, g, and h.
b. Show that the second derivative of each of the functions f, g, and h equals zero at $x = 0$.
c. Show that f has neither a relative maximum nor a relative minimum at $x = 0$, that g has a relative minimum at $x = 0$, and that h has a relative maximum at $x = 0$.

SOLUTIONS TO SELF-CHECK EXERCISES 4.2

1. We first compute

$$f'(x) = 12x^2 - 6x$$

and

$$f''(x) = 24x - 6 = 6(4x - 1)$$

Observe that f'' is continuous everywhere and has a zero at $x = \frac{1}{4}$. The sign diagram of f'' is shown in the accompanying figure.

From the sign diagram for f'' we see that f is concave upward on $(\frac{1}{4}, \infty)$ and concave downward on $(-\infty, \frac{1}{4})$.

2. First, we find the critical points of f by solving the equation

$$f'(x) = 6x^2 - x - 12 = 0$$

That is,

$$(3x + 4)(2x - 3) = 0$$

giving $x = -\frac{4}{3}$ and $x = \frac{3}{2}$. Next, we compute

$$f''(x) = 12x - 1$$

Since

$$f''\left(-\frac{4}{3}\right) = 12\left(-\frac{4}{3}\right) - 1 = -17 < 0$$

the Second Derivative Test implies that $f\left(-\frac{4}{3}\right) = \frac{10}{27}$ is a relative maximum of f. Also

$$f''\left(\frac{3}{2}\right) = 12\left(\frac{3}{2}\right) - 1 = 17 > 0$$

and we see that $f(\frac{3}{2}) = -\frac{179}{8}$ is a relative minimum.

3. We compute the second derivative of G. Thus,

$$G'(t) = -6t^2 + 90t + 20$$
$$G''(t) = -12t + 90$$

Now G'' is continuous everywhere and $G''(t) = 0$, where $t = \frac{15}{2}$, giving $t = \frac{15}{2}$ as the only candidate for an inflection point of G. Since $G''(t) > 0$ for $t < \frac{15}{2}$ and $G''(t) < 0$ for $t > \frac{15}{2}$, we see that the point $(\frac{15}{2}, \frac{15675}{2})$ is an inflection point of G. The results of our computations tell us that the country's GDP was increasing most rapidly at the beginning of July 1992.

4.3 CURVE SKETCHING

A Real-Life Example

As we have see on numerous occasions, the graph of a function is a useful aid for visualizing the function's properties. From a practical point of view, the graph of a function also gives, at one glance, a complete summary of all the information captured by the function.

Consider, for example, the graph of the function giving the Dow Jones Industrial Average (DJIA) on Black Monday, October 19, 1987 (Figure 4.44). Here $t = 0$ corresponds to 8:30 A.M., when the market was open for business, and $t = 7.5$ corresponds to 4 P.M., the closing time. The following information may be gleaned from studying the graph.

The graph is *decreasing* rapidly from $t = 0$ to $t = 1$, reflecting the sharp drop in the index in the first hour of trading. The point $(1, 2047)$ is a *relative minimum* point of the function, and this turning point coincides with the start of an aborted recovery. The short-lived rally, represented by the portion of the graph that is *increasing* on the interval $(1, 2)$, quickly fizzled out at $t = 2$ (10:30 A.M.). The *relative maximum* point $(2, 2150)$ marks the highest point of the recovery. The function is decreasing in the rest of the interval. The point $(4, 2006)$ is an *inflection point* of the function; it shows that there was a temporary respite at $t = 4$ (12:30 P.M.). However, selling pressure continued unabated, and the Dow Jones Industrial Index continued to fall until the closing bell. Finally, the graph also shows that the index opened at the high

Figure 4.44
The Dow Jones Industrial Average on Black Monday.

Source: *Wall Street Journal*

of the day ($f(0) = 2247$ is the *absolute maximum* of the function) and closed at the low of the day ($f(15/2) = 1739$ is the *absolute minimum* of the function), a drop of 508 points!*

Before we turn our attention to the actual task of sketching the graph of a function, let us look at some properties of graphs that will be helpful in this connection.

Vertical Asymptotes

Before going on, you might want to review the material on one-sided limits and the limit at infinity of a function (Sections 2.4 and 2.5).

Consider the graph of the function

$$f(x) = \frac{x + 1}{x - 1}$$

Figure 4.45
The graph of f has a vertical asymptote at x = 1.

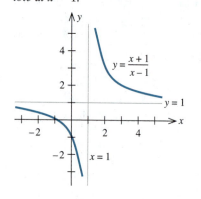

shown in Figure 4.45. Observe that $f(x)$ increases without bound (tends to infinity) as x approaches $x = 1$ from the right; that is,

$$\lim_{x \to 1^+} \frac{x + 1}{x - 1} = \infty$$

You can verify this by taking a sequence of values of x approaching $x = 1$ from the right and looking at the corresponding values of $f(x)$.

Here is another way of looking at the situation: Observe that if x is a number that is a little larger than 1, then both $(x + 1)$ and $(x - 1)$ are positive, so that $(x + 1)/(x - 1)$ is also positive. As x approaches $x = 1$, the numerator

---

* Absolute maxima and absolute minima of functions are covered in Section 4.4.

$(x + 1)$ approaches the number 2, but the denominator $(x - 1)$ approaches zero, so the quotient $(x + 1)/(x - 1)$ approaches infinity, as observed earlier. The line $x = 1$ is called a *vertical asymptote* of the graph of f.

For the function $f(x) = (x + 1)/(x - 1)$, you can show that

$$\lim_{x \to 1^-} \frac{x + 1}{x - 1} = -\infty$$

and this tells us how $f(x)$ approaches the asymptote $x = 1$ from the left. More generally, we have the following definition:

VERTICAL ASYMPTOTES

The line $x = a$ is a **vertical asymptote** of the graph of a function f if either

$$\lim_{x \to a^+} f(x) = \infty \quad \text{or} \quad -\infty$$

or

$$\lim_{x \to a^-} f(x) = \infty \quad \text{or} \quad -\infty$$

REMARK   Although a vertical asymptote of a graph is not part of the graph, it serves as a useful aid for sketching the graph.   ◦ ◦ ◦

For rational functions

$$f(x) = \frac{P(x)}{Q(x)}$$

there is a simple criterion for determining whether the graph of f has any vertical asymptotes.

FINDING VERTICAL ASYMPTOTES OF RATIONAL FUNCTIONS

Suppose that f is a rational function

$$f(x) = \frac{P(x)}{Q(x)}$$

where P and Q are polynomial functions. Then the line $x = a$ is a vertical asymptote of the graph of f if $Q(a) = 0$ but $P(a) \neq 0$.

For the function

$$f(x) = \frac{x + 1}{x - 1}$$

considered earlier, $P(x) = x + 1$ and $Q(x) = x - 1$. Observe that $Q(1) = 0$ but $P(1) = 2 \neq 0$, so $x = 1$ is a vertical asymptote of the graph of f.

EXAMPLE 1   Find the vertical asymptotes of the graph of the function

$$f(x) = \frac{x^2}{4 - x^2}$$

Solution   The function f is a rational function with $P(x) = x^2$ and $Q(x) = 4 - x^2$. The zeros of Q are found by solving

$$4 - x^2 = 0$$

that is,

$$(2 - x)(2 + x) = 0$$

giving $x = -2$ and $x = 2$. These are candidates for the vertical asymptotes of the graph of f. Examining $x = -2$, we compute $P(-2) = (-2)^2 = 4 \neq 0$, and we see that $x = -2$ is indeed a vertical asymptote of the graph of f. Similarly, we find $P(2) = 2^2 = 4 \neq 0$, and so $x = 2$ is also a vertical asymptote of the graph of f. The graph of f sketched in Figure 4.46 confirms these results.

Figure 4.46
$x = -2$ and $x = 2$ are vertical
asymptotes of the graph of f.

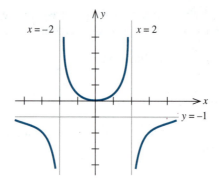

Recall that in order for the line $x = a$ to be a vertical asymptote of the graph of a rational function f, *only* the denominator of $f(x)$ must be equal to zero at $x = 0$. If *both* $P(a)$ and $Q(a)$ are equal to zero, then $x = a$ need *not* be a vertical asymptote. For example, look at the function $f(x) = \dfrac{4(x^2 - 4)}{x - 2}$, whose graph appears in Figure 2.24a, page 108.

Horizontal Asymptotes

Let us return to the function f defined by

$$f(x) = \frac{x + 1}{x - 1}$$

(Figure 4.47).

Figure 4.47
The graph of f has a horizontal asymptote at y = 1.

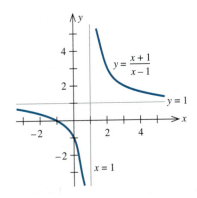

Observe that $f(x)$ approaches the horizontal line $y = 1$ as x approaches infinity, and, in this case, $f(x)$ approaches $y = 1$ as x approaches minus infinity as well. The line $y = 1$ is called a *horizontal asymptote* of the graph of f. More generally, we have the following definition.

HORIZONTAL ASYMPTOTES	The line $y = b$ is a **horizontal asymptote** of the graph of a function f if either $$\lim_{x \to \infty} f(x) = b \quad \text{or} \quad \lim_{x \to -\infty} f(x) = b$$

For the function

$$f(x) = \frac{x + 1}{x - 1}$$

we see that

$$\lim_{x \to \infty} \frac{x + 1}{x - 1} = \lim_{x \to \infty} \frac{1 + \dfrac{1}{x}}{1 - \dfrac{1}{x}} \qquad \text{(Divide numerator and denominator by } x.)$$

$$= 1$$

Also,

$$\lim_{x \to -\infty} \frac{x + 1}{x - 1} = \lim_{x \to -\infty} \frac{1 + \dfrac{1}{x}}{1 - \dfrac{1}{x}}$$

$$= 1$$

In either case, we conclude that $y = 1$ is a horizontal asymptote of the graph of f, as observed earlier.

EXAMPLE 2   Find the horizontal asymptotes of the graph of the function

$$f(x) = \frac{x^2}{4 - x^2}$$

Solution   We compute

$$\lim_{x \to \infty} \frac{x^2}{4 - x^2} = \lim_{x \to \infty} \frac{1}{\dfrac{4}{x^2} - 1} \qquad \text{(Dividing numerator and denominator by } x^2\text{)}$$

$$= -1$$

and so $y = -1$ is a horizontal asymptote, as before. The graph of f sketched in Figure 4.48 confirms this result.

Figure 4.48
The graph of f has a horizontal asymptote at $y = -1$.

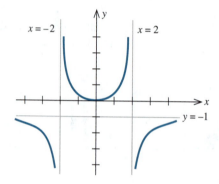

We next state an important property of polynomial functions.

> A polynomial function has no vertical or horizontal asymptotes.

To see this, note that a polynomial function $P(x)$ can be written as a rational function with denominator equal to 1. Thus,

$$P(x) = \frac{P(x)}{1}$$

Since the denominator is never equal to zero, P has no vertical asymptotes. Next, if P is a polynomial of degree greater than or equal to 1, then

$$\lim_{x \to \infty} P(x) \qquad \text{and} \qquad \lim_{x \to -\infty} P(x)$$

are either infinity or minus infinity; that is, they do not exist. Therefore, P has no horizontal asymptotes.

In the last two sections we saw how the first and second derivatives of a function are used to reveal various properties of the graph of a function f. We now show how this information can be used to help us sketch the graph of f. We begin by giving a general procedure for curve sketching.

A GUIDE TO CURVE SKETCHING

1. Determine the domain of f.
2. Find the x- and y-intercepts of f.*
3. Determine the behavior of f for large absolute values of x.
4. Find all horizontal and vertical asymptotes of f.
5. Determine the intervals where f is increasing and where f is decreasing.
6. Find the relative extrema of f.
7. Determine the concavity of f.
8. Find the inflection points of f.
9. Plot a few additional points to help further identify the shape of the graph of f, and sketch the graph.

We now illustrate the techniques of curve sketching with several examples.

Two Step-by-Step Examples

EXAMPLE 3   Sketch the graph of the function

$$y = f(x) = x^3 - 6x^2 + 9x + 2$$

Solution   Obtain the following information on the graph of f.

1. The domain of f is the interval $(-\infty, \infty)$.
2. By setting $x = 0$, we find that the y-intercept is 2. The x-intercept is found by setting $y = 0$, which, in this case, leads to a cubic equation. Since the solution is not readily found, we will not use this information.
3. Since

$$\lim_{x \to -\infty} f(x) = \lim_{x \to -\infty} (x^3 - 6x^2 + 9x + 2) = -\infty$$

and $$\lim_{x \to \infty} f(x) = \lim_{x \to \infty} (x^3 - 6x^2 + 9x + 2) = \infty$$

we see that f decreases without bound as x decreases without bound, and f increases without bound as x increases without bound.

◦ ◦ ◦

* The equation $f(x) = 0$ may be difficult to solve, in which case one may decide against finding the x-intercepts or use technology, if available, for assistance.

Figure 4.49
Sign diagram for f'.

Figure 4.50
Sign diagram for f''.

Figure 4.51
We first plot the intercept, the relative extrema, and the inflection point.

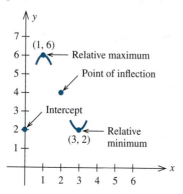

Figure 4.52
The graph of $y = x^3 - 6x^2 + 9x + 2$.

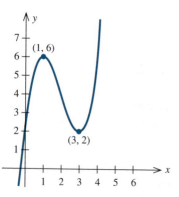

4. Since f is a polynomial function, there are no asymptotes.
$$f'(x) = 3x^2 - 12x + 9 = 3(x^2 - 4x + 3)$$
$$= 3(x - 3)(x - 1)$$

5. Setting $f'(x) = 0$ gives $x = 1$ or $x = 3$. The sign diagram for f' shows that f is increasing on the intervals $(-\infty, 1)$ and $(3, \infty)$ and decreasing on the interval $(1, 3)$ (Figure 4.49).

6. From the results of step 5, we see that $x = 1$ and $x = 3$ are critical points of f. Furthermore, f' changes sign from positive to negative as we move across $x = 1$, so a relative maximum of f occurs at $x = 1$. Similarly, we see that a relative minimum of f occurs at $x = 3$. Now,
$$f(1) = 1 - 6 + 9 + 2 = 6$$
and
$$f(3) = 3^3 - 6(3)^2 + 9(3) + 2 = 2$$
so $f(1) = 6$ is a relative maximum of f and $f(3) = 2$ is a relative minimum of f.

7.
$$f''(x) = 6x - 12 = 6(x - 2)$$
which is equal to zero when $x = 2$. The sign diagram of f'' shows that f is concave downward on the interval $(-\infty, 2)$ and concave upward on the interval $(2, \infty)$ (Figure 4.50).

8. From the results of step 7, we see that f'' changes sign as we move across the point $x = 2$. Next,
$$f(2) = 2^3 - 6(2)^2 + 9(2) + 2 = 4$$
and so the required inflection point of f is $(2, 4)$.

9. Summarizing, we have

Domain	$(-\infty, \infty)$
Intercept	$(0, 2)$
$\lim\limits_{x \to -\infty} f(x)$; $\lim\limits_{x \to \infty} f(x)$	$-\infty$; ∞
Asymptotes	None
Intervals where f is ↗ or ↘	↗ on $(-\infty, 1) \cup (3, \infty)$; ↘ on $(1, 3)$
Relative extrema	Rel. max. at $(1, 6)$; rel. min. at $(3, 2)$
Concavity	Downward on $(-\infty, 2)$; upward on $(2, \infty)$
Point of inflection	$(2, 4)$

In general, it is a good idea to start graphing by plotting the intercept, relative extrema, and inflection point (Figure 4.51). Then, using the rest of the information, we complete the graph of f, as sketched in Figure 4.52.

○ ○ ○

The average price of gasoline at the pump over a period of 3 months during which there was a temporary shortage of oil is described by the function f defined on the interval $[0, 3]$. During the first month, the price was increasing at an increasing rate. Starting with the second month, the good news was that the rate of increase was slowing down, although the price of gas was still increasing. This pattern continued until the end of the second month. The price of gas peaked at the end of $t = 2$ and began to fall at an increasing rate until $t = 3$.

a. Describe the signs of $f'(t)$ and $f''(t)$ over each of the intervals $(0, 1)$, $(1, 2)$, and $(2, 3)$.

b. Make a sketch showing a plausible graph of f over $[0, 3]$.

EXAMPLE 4   Sketch the graph of the function

$$y = f(x) = \frac{x + 1}{x - 1}$$

Solution   Obtain the following information:

1. f is undefined when $x = 1$, so the domain of f is the set of all real numbers other than $x = 1$.

2. Setting $y = 0$ gives -1, the x-intercept of f. Next, setting $x = 0$ gives -1 as the y-intercept of f.

3. Earlier we found that

$$\lim_{x \to \infty} \frac{x + 1}{x - 1} = 1 \quad \text{and} \quad \lim_{x \to -\infty} \frac{x + 1}{x - 1} = 1$$

(see page 313). Consequently, we see that $f(x)$ approaches the line $y = 1$ as $|x|$ becomes arbitrarily large. For $x > 1$, $f(x) > 1$ and $f(x)$ approaches the line $y = 1$ from above. For $x < 1$, $f(x) < 1$, so $f(x)$ approaches the line $y = 1$ from below.

4. The straight line $x = 1$ is a vertical asymptote of the graph of f. Also, from the results of step 3, we conclude that $y = 1$ is a horizontal asymptote of the graph of f.

Figure 4.53
The sign diagram for f'.

f' is not defined here

5.
$$f'(x) = \frac{(x - 1)(1) - (x + 1)(1)}{(x - 1)^2} = -\frac{2}{(x - 1)^2}$$

and is discontinuous at $x = 1$. The sign diagram of f' shows that $f'(x) < 0$ whenever it is defined. Thus, f is decreasing on the intervals $(-\infty, 1)$ and $(1, \infty)$ (Figure 4.53).

6. From the results of step 5, we see that there are no critical points of f since $f'(x)$ is never equal to zero for any value of x in the domain of f.

Figure 4.54
The sign diagram for f''.

f'' is not defined here

7.
$$f''(x) = \frac{d}{dx}[-2(x - 1)^{-2}] = 4(x - 1)^{-3} = \frac{4}{(x - 1)^3}$$

The sign diagram of f'' shows immediately that f is concave downward on the interval $(-\infty, 1)$ and concave upward on the interval $(1, \infty)$ (Figure 4.54).

8. From the results of step 7, we see that there are no candidates for inflection points of f since $f''(x)$ is never equal to zero for any value of x in the domain of f. Hence, f has no inflection points.

9. Summarizing, we have

Domain	$(-\infty, 1) \cup (1, \infty)$
Intercepts	$(-1, 0); (0, -1)$
$\lim_{x \to -\infty} f(x); \lim_{x \to \infty} f(x)$	$1; 1$
Asymptotes	$x = 1$ is a vertical asymptote; $y = 1$ is a horizontal asymptote
Intervals where f is ↗ or ↘	↘ on $(-\infty, 1) \cup (1, \infty)$
Relative extrema	None
Concavity	Downward on $(-\infty, 1)$; upward on $(1, \infty)$
Points of inflection	None

The graph of f is sketched in Figure 4.55.

Figure 4.55
The graph of f has a horizontal asymptote at y = 1 and a vertical asymptote at x = 1.

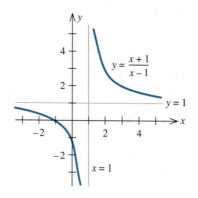

SELF-CHECK EXERCISES 4.3

1. Find the horizontal and vertical asymptotes of the graph of the function

$$f(x) = \frac{2x^2}{x^2 - 1}$$

2. Sketch the graph of the function

$$f(x) = \tfrac{2}{3}x^3 - 2x^2 - 6x + 4$$

Solutions to Self-Check Exercises 4.3 can be found on page 327.

In exercises 1–10, find the horizontal and vertical asymptotes of the graph.

1.

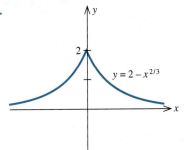

$y = 2 - x^{2/3}$

2.

$y = \dfrac{1}{(x+1)^2}$

3.

$y = \dfrac{1}{x^3}$

4.

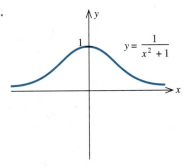

$y = \dfrac{1}{x^2 + 1}$

5.

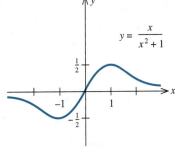

$y = \dfrac{x}{x^2 - 1}$

6.

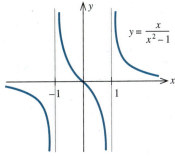

$y = \dfrac{x}{x^2 + 1}$

7.

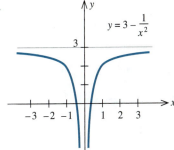

$y = 3 - \dfrac{1}{x^2}$

8.

$y = \dfrac{1}{(x+2)^2}$

9.

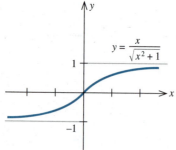

$$y = \frac{x}{\sqrt{x^2 + 1}}$$

10.

$$y = \frac{x^2 - 4}{x^2 - 1}$$

In exercises 11–28, find the horizontal and vertical asymptotes of the graph of the function. (You need not sketch the graph.)

11. $f(x) = \dfrac{1}{x}$

12. $f(x) = \dfrac{1}{x + 2}$

13. $f(x) = -\dfrac{2}{x^2}$

14. $g(x) = \dfrac{1}{1 + 2x^2}$

15. $f(x) = \dfrac{x - 1}{x + 1}$

16. $g(t) = \dfrac{t + 1}{2t - 1}$

17. $h(x) = x^3 - 3x^2 + x + 1$

18. $g(x) = 2x^3 + x^2 + 1$

19. $f(t) = \dfrac{t^2}{t^2 - 9}$

20. $g(x) = \dfrac{x^3}{x^2 - 4}$

21. $f(x) = \dfrac{3x}{x^2 - x - 6}$

22. $g(x) = \dfrac{2x}{x^2 + x - 2}$

23. $g(t) = 2 + \dfrac{5}{(t - 2)^2}$

24. $f(x) = 1 + \dfrac{2}{x - 3}$

25. $f(x) = \dfrac{x^2 - 2}{x^2 - 4}$

26. $h(x) = \dfrac{2 - x^2}{x^2 + x}$

27. $g(x) = \dfrac{x^3 - x}{x(x + 1)}$

28. $f(x) = \dfrac{x^4 - x^2}{x(x - 1)(x + 2)}$

In exercises 29 and 30, you are given the graphs of two functions f and g. One function is the derivative function of the other. Identify each of them.

29.

30.

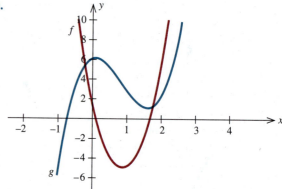

31. Terminal Velocity   A skydiver leaps from the gondola of a hot-air balloon. As she free-falls, air resistance, which is proportional to her velocity, builds up to a point where it balances the force due to gravity. The resulting motion may be described in terms of her velocity as follows: Starting at rest (zero velocity), her velocity increases and approaches a constant velocity, called the *terminal velocity*. Sketch a graph of her velocity v versus time t.

In exercises 32–35, use the information summarized in the table to sketch the graph of f.

32. $f(x) = x^3 - 3x^2 + 1$

Domain	$(-\infty, \infty)$
Intercept	y-intercept: 1
Asymptotes	None
Intervals where f is ↗ and ↘	↗ on $(-\infty, 0) \cup (2, \infty)$; ↘ on $(0, 2)$
Relative extrema	Rel. max. at $(0, 1)$; rel. min. at $(2, -3)$
Concavity	Downward on $(-\infty, 1)$; upward on $(1, \infty)$
Point of inflection	$(1, -1)$

33. $f(x) = \frac{1}{9}(x^4 - 4x^3)$

Domain	$(-\infty, \infty)$
Intercepts	x-intercepts: 0, 4; y-intercept: 0
Asymptotes	None
Intervals where f is ↗ and ↘	↗ on $(3, \infty)$; ↘ on $(-\infty, 3)$
Relative extrema	Rel. min. at $(3, -3)$
Concavity	Downward on $(0, 2)$; upward on $(-\infty, 0) \cup (2, \infty)$
Points of inflection	$(0, 0)$ and $(2, -16/9)$

34. $f(x) = \frac{4x - 4}{x^2}$

Domain	$(-\infty, 0) \cup (0, \infty)$
Intercept	x-intercept: 1
Asymptotes	x-axis and y-axis
Intervals where f is ↗ and ↘	↗ on $(0, 2)$; ↘ on $(-\infty, 0) \cup (2, \infty)$
Relative extrema	Rel. max. at $(2, 1)$
Concavity	Downward on $(-\infty, 0) \cup (0, 3)$; upward on $(3, \infty)$
Points of inflection	$(3, 8/9)$

35. $f(x) = x - 3x^{1/3}$

Domain	$(-\infty, \infty)$
Intercepts	x-intercepts: $\pm 3\sqrt{3}$
Asymptotes	None
Intervals where f is ↗ and ↘	↗ on $(-\infty, -1) \cup (1, \infty)$; ↘ on $(-1, 1)$
Relative extrema	Rel. max. at $(-1, 2)$; rel. min. at $(1, -2)$
Concavity	Downward on $(-\infty, 0)$; upward on $(0, \infty)$
Points of inflection	$(0, 0)$

In exercises 36–59, sketch the graph of the function using the curve-sketching guide of this section.

36. $f(x) = x^2 - 2x + 3$

37. $g(x) = 4 - 3x - 2x^3$

38. $f(x) = 2x^3 + 1$

39. $h(x) = x^3 - 3x + 1$

40. $f(t) = 2t^3 - 15t^2 + 36t - 20$

41. $f(x) = -2x^3 + 3x^2 + 12x + 2$

42. $f(t) = 3t^4 + 4t^3$

43. $h(x) = \frac{3}{2}x^4 - 2x^3 - 6x^2 + 8$

44. $f(x) = \sqrt{x^2 + 5}$

45. $f(t) = \sqrt{t^2 - 4}$

46. $f(x) = \sqrt[3]{x^2}$

47. $g(x) = \frac{1}{2}x - \sqrt{x}$

48. $f(x) = \frac{1}{x + 1}$

49. $g(x) = \frac{2}{x - 1}$

50. $g(x) = \frac{x}{x - 1}$

51. $h(x) = \frac{x + 2}{x - 2}$

52. $g(x) = \frac{x}{x^2 - 4}$

53. $f(t) = \frac{t^2}{1 + t^2}$

54. $f(x) = \frac{x^2 - 9}{x^2 - 4}$

55. $g(t) = -\frac{t^2 - 2}{t - 1}$

56. $h(x) = \frac{1}{x^2 - x - 2}$

57. $g(t) = \frac{t + 1}{t^2 - 2t - 1}$

58. $g(x) = (x + 2)^{3/2} + 1$

59. $h(x) = (x - 1)^{2/3} + 1$

60. Cost of Removing Toxic Pollutants A city's main well was recently found to be contaminated with trichloroethylene (a cancer-causing chemical) as a result of an abandoned chemical dump leaching chemicals into the water. A proposal submitted to the city's selectmen indicated that the cost, measured in millions of dollars, of removing $x\%$ of the toxic pollutants is given by

$$C(x) = \frac{0.5x}{100 - x}$$

a. Find the vertical asymptote of $C(x)$.
b. Is it possible to remove 100% of the toxic pollutant from the water?

61. Average Cost of Producing Video Discs The average cost per disc (in dollars) incurred by the Herald Record Company in pressing x video discs is given by the average cost function

$$\overline{C}(x) = 2.2 + \frac{2500}{x}$$

a. Find the horizontal asymptote of $\overline{C}(x)$.
b. What is the limiting value of the average cost?

USING TECHNOLOGY

ANALYZING THE PROPERTIES OF A FUNCTION

One of the main purposes of studying Section 4.3 is to see how the many concepts of calculus come together to paint a picture of a function. The techniques of graphing also play a very practical role. For example, using the techniques of graphing developed in Section 4.3, you can tell if the graph of a function generated by a graphing utility is reasonably complete. Furthermore, these techniques can often reveal details that are missing from a graph.

EXAMPLE 1 Consider the function $f(x) = 2x^3 - 3.5x^2 + x - 10$. A plot of the graph of f in the standard viewing rectangle is shown in Figure T1. Since the domain of f is the interval $(-\infty, \infty)$, we see that Figure T1 does not reveal the part of the graph to the left of the y-axis. This suggests that we enlarge the viewing rectangle accordingly. Figure T2 shows the graph of f in the viewing rectangle $[-10, 10] \times [-20, 10]$.

Figure T1
The graph of f in the standard viewing rectangle.

Figure T2
The graph of f in the viewing rectangle $[-10, 10] \times [-20, 10]$.

The behavior of f for large values of f ($\lim\limits_{x \to -\infty} f(x) = -\infty$ and $\lim\limits_{x \to \infty} f(x) = \infty$) suggests that this viewing rectangle has captured a sufficiently complete picture of f. Next, an analysis of the first derivative of f,

$$f'(x) = 6x^2 - 7x + 1 = (6x - 1)(x - 1)$$

reveals that f has critical values at $x = 1/6$ and $x = 1$. In fact, a sign diagram of f' shows that f has a relative maximum at $x = 1/6$ and a relative minimum at $x = 1$, details that are not revealed in the graph of f shown in Figure T2. In order to examine this portion of the graph of f, we use, say, the viewing rectangle $[-1, 2] \times [-11, -9]$. The resulting graph of f is shown in Figure T3, which certainly reveals the hitherto missing details! Thus, through an interaction of calculus and a graphing utility, we are able to obtain a good picture of the properties of f.

Figure T3
The graph of f in the viewing rectangle $[-1, 2] \times [-11, -9]$.

o o o

Finding x-Intercepts

As noted in Section 4.3, it is not always easy to find the x-intercepts of the graph of a function. But this information is very important in applications. By using the function for solving polynomial equations or the function for finding the roots of an equation, we can solve the equation $f(x) = 0$ quite easily and hence yield the x-intercepts of the graph of a function.

EXAMPLE 2   Let $f(x) = x^3 - 3x^2 + x + 1.5$.

a. Use the function for solving polynomial equations on a graphing utility to find the x-intercepts of the graph of f.

b. Use the function for finding the roots of an equation on a graphing utility to find the x-intercepts of the graph of f.

Solution

a. Observe that f is a polynomial function of degree 3, and so we may use the function for solving polynomial equations to solve the equation $x^3 - 3x^2 + x + 1.5 = 0$ $(f(x) = 0)$. We find that the solutions (x-intercepts) are

$$x_1 \approx -0.525687120865, \quad x_2 \approx 1.2586520225, \quad x_3 \approx 2.26703509836$$

b. Using the graph of f (Figure T4), we see that $x_1 \approx -0.5$, $x_2 \approx 1$, and $x_3 \approx 2$. Using the function for finding the roots of an equation on a graphing utility, and these values of x as initial guesses, we find

$$x_1 \approx -0.5256871209, \quad x_2 \approx 1.2586520225, \quad x_3 \approx 2.2670350984 \quad \bullet \bullet \bullet$$

REMARK   The function for solving polynomial equations on a graphing utility will solve a polynomial equation $f(x) = 0$, where f is a polynomial function. The function for finding the roots of a polynomial, however, will solve equations $f(x) = 0$ even if f is not a polynomial. $\bullet \bullet \bullet$

EXAMPLE 3   Unless payroll taxes are increased significantly and/or benefits are scaled back drastically, it is a matter of time before the current Social Security system goes broke. Based on data from the Board of Trustees of the Social Security Administration, the assets of the system—the Social Security "trust fund"—may be approximated by

$$f(t) = -0.0129t^4 + 0.3087t^3 + 2.1760t^2 + 62.8466t + 506.2955 \qquad (0 \le t \le 35)$$

where $f(t)$ is measured in millions of dollars and t is measured in years, with $t = 0$ corresponding to 1995.

a. Use a graphing calculator to sketch the graph of f.

b. Based on this model, when can the Social Security system be expected to go broke?
Source: Social Security Administration

Solution

a. The graph of f in the window $[0, 35] \times [-1000, 3500]$ is shown in Figure T5.

b. Using the function for finding the roots on a graphing utility, we find that $y = 0$ when $t \approx 34.1$, and this tells us that the system is expected to go broke around 2029.

Figure T5

○ ○ ○

EXERCISES

In exercises 1–4, use the method of Example 1 to analyze the function. (Note: Your answers will not be unique.)

1. $f(x) = 4x^3 - 4x^2 + x + 10$

2. $f(x) = x^3 + 2x^2 + x - 12$

3. $f(x) = \frac{1}{2}x^4 + x^3 + \frac{1}{2}x^2 - 10$

4. $f(x) = 2.25x^4 - 4x^3 + 2x^2 + 2$

In exercises 5–10, find the x-intercepts of the graph of f. Give your answer accurate to four decimal places.

5. $f(x) = 0.2x^3 - 1.2x^2 + 0.8x + 2.1$

6. $f(x) = -0.5x^3 + 1.7x^2 - 1.2$

7. $f(x) = 0.3x^4 - 1.2x^3 + 0.8x^2 + 1.1x - 2$

8. $f(x) = -0.2x^4 + 0.8x^3 - 2.1x + 1.2$

9. $f(x) = 2x^2 - \sqrt{x+1} - 3$

10. $f(x) = x - \sqrt{1 - x^2}$

62. Concentration of a Drug in the Bloodstream   The concentration of a certain drug in a patient's bloodstream t hours after injection is given by

$$C(t) = \frac{0.2t}{t^2 + 1}$$

milligrams per cubic centimeter.
a. Find the horizontal asymptote of $C(t)$.
b. Interpret your result.

63. Effect of Enzymes on Chemical Reactions   Certain proteins, known as enzymes, serve as catalysts for chemical reactions in living things. In 1913, Leonor Michaelis and L. M. Menten discovered the following formula giving the initial speed V (in moles per liter per second) at which the reaction begins in terms of the amount of substrate x (the substance that is being acted upon, measured in moles per liter):

$$V = \frac{ax}{x + b}$$

where a and b are positive constants.
a. Find the horizontal asymptote of V.
b. Interpret your result.

64. Worker Efficiency   An efficiency study showed that the total number of cordless telephones assembled by an average worker at Delphi Electronics t hours after starting work at 8 A.M. is given by

$$N(t) = -\frac{1}{2}t^3 + 3t^2 + 10t \qquad (0 \le t \le 4)$$

Sketch the graph of the function N and interpret your results.

65. GDP of a Developing Country   A developing country's gross domestic product (GDP) from 1988 to 1996 is approximated by the function

$$G(t) = -0.2t^3 + 2.4t^2 + 60 \qquad (0 \le t \le 8)$$

where $G(t)$ is measured in billions of dollars and $t = 0$ corresponds to the year 1988. Sketch the graph of the function G and interpret your results.

66. Concentration of a Drug in the Bloodstream   The concentration of a certain drug in a patient's bloodstream t hours after injection is given by

$$C(t) = \frac{0.2t}{t^2 + 1}$$

milligrams per cubic centimeter. Sketch the graph of the function C and interpret your results.

67. Oxygen Content of a Pond   When organic waste is dumped into a pond, the oxidation process that takes place reduces the pond's oxygen content. However, given time, nature will restore the oxygen content to its natural level. Suppose that the oxygen content t days after organic waste has been dumped into the pond is given by

$$f(t) = 100 \left(\frac{t^2 - 4t + 4}{t^2 + 4} \right) \qquad (0 \le t < \infty)$$

percent of its normal level. Sketch the graph of the function f and interpret your results.

68. Box Office Receipts   The total worldwide box office receipts for a long-running movie are approximated by the function

$$T(x) = \frac{120x^2}{x^2 + 4}$$

where $T(x)$ is measured in millions of dollars and x is the number of years since the movie's release. Sketch the graph of the function T and interpret your results.

69. Cost of Removing Toxic Pollutants   A city's main well was recently found to be contaminated with trichloroethylene, a cancer-causing chemical, as a result of an abandoned chemical dump leaching chemicals into the water. A proposal submitted to the city's selectmen indicates that the cost, measured in millions of dollars, of removing $x\%$ of the toxic pollutant is given by

$$C(x) = \frac{0.5x}{100 - x}$$

Sketch the graph of the function C and interpret your results.

SOLUTIONS TO SELF-CHECK EXERCISES 4.3

1. Since

$$\lim_{x \to \infty} \frac{2x^2}{x^2 - 1} = \lim_{x \to \infty} \frac{2}{1 - \dfrac{1}{x^2}} \qquad \text{(Divide the numerator and denominator by } x^2.)$$

$$= 2$$

we see that $y = 2$ is a horizontal asymptote. Next, since

$$x^2 - 1 = (x + 1)(x - 1) = 0$$

implies $x = -1$ or $x = 1$, these are candidates for the vertical asymptotes of f. Since the numerator of f is not equal to zero for $x = -1$ or $x = 1$, we conclude that $x = -1$ and $x = 1$ are vertical asymptotes of the graph of f.

2. We obtain the following information on the graph of f.

(1) The domain of f is the interval $(-\infty, \infty)$.

(2) By setting $x = 0$, we find the y-intercept is 4.

(3) Since

$$\lim_{x \to -\infty} f(x) = \lim_{x \to -\infty} \left(\frac{2}{3} x^3 - 2x^2 - 6x + 4 \right) = -\infty$$

and

$$\lim_{x \to \infty} f(x) = \lim_{x \to \infty} \left(\frac{2}{3} x^3 - 2x^2 - 6x + 4 \right) = \infty$$

we see that $f(x)$ decreases without bound as x decreases without bound, and $f(x)$ increases without bound as x increases without bound.

(4) Since f is a polynomial function, there are no asymptotes.

$$f'(x) = 2x^2 - 4x - 6 = 2(x^2 - 2x - 3)$$
$$= 2(x + 1)(x - 3)$$

(5) Setting $f'(x) = 0$ gives $x = -1$ or $x = 3$. The accompanying sign diagram for f' shows that f is increasing on the intervals $(-\infty, -1)$ and $(3, \infty)$ and decreasing on $(-1, 3)$.

Sign diagram for f'

(6) From the results of step 5, we see that $x = -1$ and $x = 3$ are critical points of f. Furthermore, the sign diagram of f' tells us that $x = -1$ gives rise to a relative maximum of f and $x = 3$ gives rise to a relative minimum of f. Now,

$$f(-1) = \frac{2}{3}(-1)^3 - 2(-1)^2 - 6(-1) + 4 = \frac{22}{3}$$

and

$$f(3) = \frac{2}{3}(3)^3 - 2(3)^2 - 6(3) + 4 = -14$$

so $f(-1) = 22/3$ is a relative maximum of f and $f(3) = -14$ is a relative minimum of f.

(7) $$f''(x) = 4x - 4 = 4(x - 1)$$

which is equal to zero when $x = 1$. The accompanying sign diagram of f'' shows that f is concave downward on the interval $(-\infty, 1)$ and concave upward on the interval $(1, \infty)$.

Sign diagram for f''

(8) From the results of step 7, we see that $x = 1$ is the only candidate for an inflection point of f. Since $f''(x)$ changes sign as we move across the point $x = 1$ and

$$f(1) = \frac{2}{3}(1)^3 - 2(1)^2 - 6(1) + 4 = -\frac{10}{3}$$

we see that the required inflection point is $(1, -10/3)$.
(9) Summarizing this information, we have

Domain	$(-\infty, \infty)$
Intercept	$(0, 4)$
Intervals where f is ↗ or ↘	↗ on $(-\infty, -1) \cup (3, \infty)$; ↘ on $(-1, 3)$
Relative extrema	Rel. max. at $(-1, 22/3)$; rel. min. at $(3, -14)$
Concavity	Downward on $(-\infty, 1)$; upward on $(1, \infty)$
Point of inflection	$(1, -10/3)$
$\lim\limits_{x \to -\infty} f(x); \lim\limits_{x \to \infty} f(x)$	$-\infty$; ∞
Asymptotes	None

The graph of f is sketched in the accompanying figure.

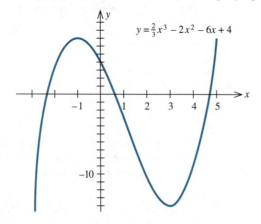

$$y = \frac{2}{3}x^3 - 2x^2 - 6x + 4$$

4.4 OPTIMIZATION I

Absolute Extrema

The graph of the function f in Figure 4.56 shows the average age of cars in use in the United States from the beginning of 1946 ($t = 0$) to the beginning of 1990 ($t = 44$). Observe that the highest average age of cars in use during this period is 9 years, whereas the lowest average age of cars in use during the same period is $5\frac{1}{2}$ years. The number 9, the largest value of $f(t)$ for all values of t in the interval [0, 44] (the domain of f) is called the *absolute maximum value of f* on that interval. The number $5\frac{1}{2}$, the smallest value of $f(t)$ for all values of t in [0, 44], is called the *absolute minimum value of f* on that interval. Notice, too, that the absolute maximum value of f is attained at the endpoint $t = 0$ of the interval, whereas the absolute minimum value of f is attained at the two interior points $t = 12$ (corresponding to 1958) and $t = 23$ (corresponding to 1969).

Figure 4.56
$f(t)$ gives the average age of cars in use in year t, t in [0, 44].

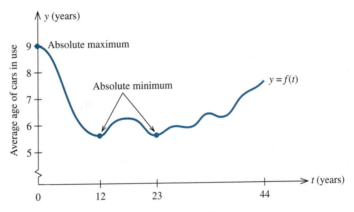

Source: American Automobile Association

A precise definition of the **absolute extrema** (absolute maximum or absolute minimum) of a function follows.

| **THE ABSOLUTE EXTREMA OF A FUNCTION f** | If $f(x) \leq f(c)$ for all x in the domain of f, then $f(c)$ is called the **absolute maximum value** of f. |
| | If $f(x) \geq f(c)$ for all x in the domain of f, then $f(c)$ is called the **absolute minimum value** of f. |

Figure 4.57 shows the graphs of several functions and gives the absolute maximum and absolute minimum of each function, if they exist.

Figure 4.57

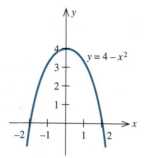

(a) $f(0) = 0$ is the absolute minimum of f; f has no absolute maximum.

(b) $f(0) = 4$ is the absolute maximum of f; f has no absolute minimum.

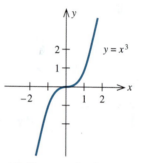

(c) $f\left(\dfrac{\sqrt{2}}{2}\right) = \dfrac{1}{2}$ is the absolute maximum of f. $f\left(-\dfrac{\sqrt{2}}{2}\right) = -\dfrac{1}{2}$ is the absolute minimum of f.

(d) f has no absolute extrema.

Absolute Extrema on a Closed Interval

As the preceding examples show, a continuous function defined on an arbitrary interval does not always have an absolute maximum or an absolute minimum. But an important case arises often in practical applications in which both the absolute maximum and the absolute minimum of a function are guaranteed to exist. This occurs when a continuous function is defined on a *closed* interval. Let us state this important result in the form of a theorem, whose proof we will omit.

THEOREM 3

If a function f is continuous on a closed interval $[a, b]$, then f has both an absolute maximum value and an absolute minimum value on $[a, b]$.

Observe that if an absolute extremum of a continuous function f occurs at a point in an open interval (a, b), then it must be a relative extremum of f and hence its x-coordinate must be a critical point of f. Otherwise, the

Figure 4.58
The relative minimum of f at x_3 is the absolute minimum of f. The right endpoint b gives rise to the absolute maximum value f(b) of f.

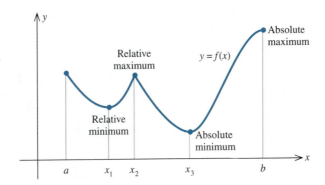

absolute extremum of f must occur at one or both of the endpoints of the interval $[a, b]$. A typical situation is illustated in Figure 4.58.

Here x_1, x_2, and x_3 are critical points of f. The absolute minimum of f occurs at x_3, which lies in the open interval (a, b) and is a critical point of f. The absolute maximum of f occurs at b, an endpoint. This observation suggests the following procedure for finding the absolute extrema of a continuous function on a closed interval.

FINDING THE ABSOLUTE EXTREMA OF f ON A CLOSED INTERVAL	1. Find the critical points of f that lie in (a, b). 2. Compute the value of f at each of these critical points of f and compute $f(a)$ and $f(b)$. 3. The absolute maximum value and absolute minimum value of f will correspond to the largest and smallest numbers, respectively, found in step 2.

Figure 4.59
F has an absolute minimum value of 0 and an absolute maximum value of 4.

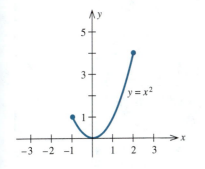

EXAMPLE 1   Find the absolute extrema of the function $F(x) = x^2$ defined on the interval $[-1, 2]$.

Solution   The function F is continuous on the closed interval $[-1, 2]$ and differentiable on the open interval $(-1, 2)$. The derivative of F is

$$F'(x) = 2x$$

so $x = 0$ is the only critical point of F. Next, evaluate $F(x)$ at $x = -1$, $x = 0$, and $x = 2$. Thus,

$$F(-1) = 1, \quad F(0) = 0, \quad F(2) = 4$$

It follows that 0 is the absolute minimum value of F and 4 is the absolute maximum value of F. The graph of F, which appears in Figure 4.59, confirms our results.   ο ο ο

EXAMPLE 2   Find the absolute extrema of the function

$$f(x) = x^3 - 2x^2 - 4x + 4$$

defined on the interval $[0, 3]$.

Solution   The function f is continuous on the closed interval $[0, 3]$ and differentiable on the open interval $(0, 3)$. The derivative of f is

$$f'(x) = 3x^2 - 4x - 4 = (3x + 2)(x - 2)$$

and it is equal to zero when $x = -\frac{2}{3}$ and $x = 2$. Since the point $x = -\frac{2}{3}$ lies outside the interval $[0, 3]$, it is dropped from further consideration and $x = 2$ is seen to be the sole critical point of f. Next, we evaluate $f(x)$ at the critical point of f as well as the endpoints of f, obtaining

$$f(0) = 4, \quad f(2) = -4, \quad f(3) = 1$$

From these results, we conclude that -4 is the absolute minimum value of f and 4 is the absolute maximum value of f. The graph of f, which appears in Figure 4.60, confirms our results. Observe that the absolute maximum of f occurs at the endpoint $x = 0$ of the interval $[0, 3]$, while the absolute minimum of f occurs at $x = 2$, which is a point in the interval $(0, 3)$.

Figure 4.60
f has an absolute maximum value of 4 and an absolute minimum value of −4.

EXPLORING WITH TECHNOLOGY

Let $f(x) = x^3 - 2x^2 - 4x + 4$. (This is the function of Example 2.)
a. Use a graphing utility to plot the graph of f using the viewing rectangle $[0, 3] \times [-5, 5]$. Use **TRACE** to find the absolute extrema of f on the interval $[0, 3]$ and thus verify the results obtained analytically in Example 2.
b. Plot the graph of f using the viewing rectangle $[-2, 1] \times [-5, 6]$. Use **ZOOM** and **TRACE** to find the absolute extrema of f on the interval $[-2, 1]$. Verify your results analytically.

EXAMPLE 3 Find the absolute maximum and absolute minimum values of the function $f(x) = x^{2/3}$ on the interval $[-1, 8]$.

Solution The derivative of f is

$$f'(x) = \frac{2}{3}x^{-1/3} = \frac{2}{3x^{1/3}}$$

Note that f' is not defined at $x = 0$, is continuous everywhere else, and does not equal zero for all x. Therefore, $x = 0$ is the only critical point of f. Evaluating $f(x)$ at $x = -1$, 0, and 8, we obtain

$$f(-1) = 1, \quad f(0) = 0, \quad f(8) = 4$$

We conclude that the absolute minimum value of f is 0, attained at $x = 0$, and the absolute maximum value of f is 4, attained at $x = 8$ (Figure 4.61).

● ● ●

Figure 4.61
f has an absolute minimum value of
$f(0) = 0$ and an absolute maximum
value of $f(8) = 4$.

Applications

Many real-world applications call for finding the absolute maximum value or the absolute minimum value of a given function. For example, management is interested in finding what level of production will yield the maximum profit for a company; a farmer is interested in finding the right amount of fertilizer to maximize crop yield; a doctor is interested in finding the maximum concentration of a drug in a patient's body and the time at which it occurs; and an engineer is interested in finding the dimension of a container with a specified shape and volume that can be constructed at a minimum cost.

EXAMPLE 4 The Acrosonic Company's total profit from manufacturing and selling x units of their model F loudspeaker systems is given by

$$P(x) = -0.02x^2 + 300x - 200,000 \qquad (0 \le x \le 20,000)$$

dollars. How many units of the loudspeaker system must Acrosonic produce to maximize its profits?

Solution To find the absolute maximum of P on $[0, 20,000]$, first find the critical points of P on the interval $(0, 20,000)$. To do this, compute

$$P'(x) = -0.04x + 300$$

Solving the equation $P'(x) = 0$ gives $x = 7500$. Next, evaluate $P(x)$ at $x = 7500$ as well as the endpoints $x = 0$ and $x = 20,000$ of the interval $[0, 20,000]$, obtaining

$$P(0) = -200,000$$
$$P(7500) = 925,000$$
and
$$P(20,000) = -2,200,000$$

Figure 4.62
P has an absolute maximum at
(7500, 925,000).

From these computations we see that the absolute maximum value of the function P is 925,000. Thus, by producing 7500 units, Acrosonic will realize a maximum profit of $925,000. The graph of P is sketched in Figure 4.62.

EXAMPLE 5 When a person coughs, the trachea, or windpipe, contracts, allowing air to be expelled at a maximum velocity. It can be shown that, during a cough, the velocity v of airflow is given by the function

$$v = f(r) = kr^2(R - r)$$

where r is the trachea's radius (in centimeters) during a cough, R is the trachea's normal radius (in centimeters), and k is a positive constant that depends on the length of the trachea. Find the radius r for which the velocity of airflow is greatest.

Figure 4.63
The velocity of airflow is greatest
when the radius of the contracted
trachea is $\frac{2}{3}R$.

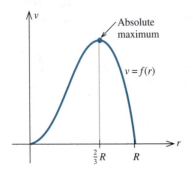

Solution To find the absolute maximum of f on $[0, R]$, first find the critical points of f on the interval $(0, R)$. We compute

$$f'(r) = 2kr(R - r) - kr^2 \quad \text{(Using the Product Rule)}$$
$$= -3kr^2 + 2kRr = kr(-3r + 2R)$$

Setting $f'(r) = 0$ gives $r = 0$ or $r = \frac{2}{3}R$, and so $r = \frac{2}{3}R$ is the sole critical point of f ($r = 0$ is an endpoint). Evaluating $f(r)$ at $r = \frac{2}{3}R$, as well as at the endpoints $r = 0$ and $r = R$, we obtain

$$f(0) = 0$$
$$f\left(\frac{2}{3}R\right) = \frac{4k}{27}R^3$$

and
$$f(R) = 0$$

from which we deduce that the velocity of airflow is greatest when the radius of the contracted trachea is $\frac{2}{3}R$—that is, when the radius is contracted by approximately 33%. The graph of the function f is shown in Figure 4.63.

Recall that the total profit function P is defined as $P(x) = R(x) - C(x)$, where R is the total revenue function, C is the total cost function, and x is the number of units of a product produced and sold. (Assume all derivatives exist.)

a. Show that at the level of production x_0 that yields the maximum profit for the company, the following two conditions are satisfied:

$$R'(x_0) = C'(x_0) \quad \text{and} \quad R''(x_0) < C''(x_0)$$

b. Interpret the two conditions in part (a) in economic terms and explain why they make sense.

Prove that if a cost function $C(x)$ is concave upward ($C''(x) > 0$), then the level of production that will result in the smallest average production cost occurs when

$$\overline{C}(x) = C'(x)$$

that is, when the average cost $\overline{C}(x)$ is equal to the marginal cost $C'(x)$.

[*Hints:*
1. Show that

$$\overline{C}'(x) = \frac{xC'(x) - C(x)}{x^2}$$

so that the critical point of the function $\overline{C}$ occurs when

$$xC'(x) - C(x) = 0$$

2. Show that at a critical point of $\overline{C}$

$$\overline{C}''(x) = \frac{C''(x)}{x}$$

Use the Second Derivative Test to reach the desired conclusion.]

EXAMPLE 6   The daily average cost function (in dollars per unit) of the Elektra Electronics Company is given by

$$\overline{C}(x) = 0.0001x^2 - 0.08x + 40 + \frac{5000}{x} \qquad (x > 0)$$

where x stands for the number of programmable calculators Elektra produces. Show that a production level of 500 units per day results in a minimum average cost for the company.

Solution   The domain of the function $\overline{C}$ is the interval $(0, \infty)$, which is not closed. In order to solve the problem, we resort to the graphic method. Using the techniques of graphing from the last section, we sketch the graph of $\overline{C}$ (Figure 4.64).

Now,

$$\overline{C}'(x) = 0.0002x - 0.08 - \frac{5000}{x^2}$$

Substituting the given value of x, 500, into $\overline{C}'(x)$ gives $\overline{C}'(500) = 0$, so $x = 500$ is a critical point of $\overline{C}$. Next,

$$\overline{C}''(x) = 0.0002 + \frac{10,000}{x^3}$$

Thus,

$$\overline{C}''(500) = 0.0002 + \frac{10,000}{(500)^3} > 0$$

and by the Second Derivative Test, a relative minimum of the function $\overline{C}$ occurs at the point $x = 500$. Furthermore, $\overline{C}''(x) > 0$ for $x > 0$, which implies

Figure 4.64
The minimum average cost is $35/unit.

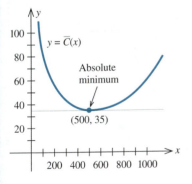

that the graph of $\overline{C}$ is concave upward everywhere, so the relative minimum of $\overline{C}$ must be the absolute minimum of $\overline{C}$. The minimum average cost is given by

$$\overline{C}(500) = 0.0001(500)^2 - 0.08(500) + 40 + \frac{5000}{500}$$

$$= 35$$

or \$35 per unit.

○ ○ ○

EXPLORING WITH TECHNOLOGY

Refer to the preceding Group Discussion question and Example 6.
a. Using a graphing utility, plot the graphs of

$$\overline{C}(x) = 0.0001x^2 - 0.08x + 40 + \frac{5000}{x}$$

and
$$C'(x) = 0.0003x^2 - 0.16x + 40$$

using the viewing rectangle $[0, 1000] \times [0, 150]$.
[*Note:* $C(x) = 0.0001x^3 - 0.08x^2 + 40x + 5000$ (Why?)]
b. Find the point of intersection of the graphs of $\overline{C}$ and C' and thus verify the assertion in the Group Discussion question for the special case studied in Example 6.

○ ○ ○

EXAMPLE 7   The altitude (in feet) of a rocket t seconds into flight is given by

$$s = f(t) = -t^3 + 96t^2 + 195t + 5 \qquad (t \geq 0)$$

a. Find the maximum altitude attained by the rocket.

b. Find the maximum velocity attained by the rocket.

Solution

a. The maximum altitude attained by the rocket is given by the largest value of the function f in the closed interval $[0, T]$, where T denotes the time the rocket impacts Earth. We know that such a number exists because the dominant term in the expression for the continuous function f is $-t^3$. So for t large enough, the value of $f(t)$ must change from positive to negative and, in particular, it must attain the value 0 for some T.

Figure 4.65
The maximum altitude of the rocket is 143,655 feet.

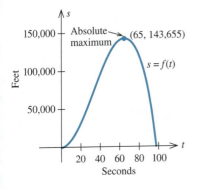

To find the absolute maximum of f, compute
$$f'(t) = -3t^2 + 192t + 195$$
$$= -3(t - 65)(t + 1)$$
and solve the equation $f'(t) = 0$, obtaining $t = -1$ and $t = 65$. Ignore $t = -1$ since it lies outside the interval $[0, T]$. This leaves the critical point $t = 65$ of f. Continuing, we compute
$$f(0) = 5, \quad f(65) = 143{,}655, \quad \text{and} \quad f(T) = 0$$
and conclude, accordingly, that the absolute maximum value of f is 143,655. Thus, the maximum altitude of the rocket is 143,655 feet, attained 65 seconds into flight. The graph of f is sketched in Figure 4.65.

b. To find the maximum velocity attained by the rocket, find the largest value of the function that describes the rocket's velocity at any time t—namely,
$$v = f'(t) = -3t^2 + 192t + 195 \qquad (t \geq 0)$$
We find the critical point of v by setting $v' = 0$. But
$$v' = -6t + 192$$
and the critical point of v is $t = 32$. Since
$$v'' = -6 < 0$$

Figure 4.66
The maximum velocity of the rocket is 3267 ft/sec.

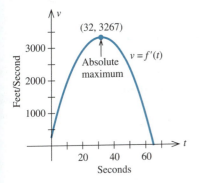

the Second Derivative Test implies that a relative maximum of v occurs at $t = 32$. Our computation has, in fact, clarified the property of the "velocity curve." Since $v'' < 0$ everywhere, the velocity curve is concave downward everywhere. With this observation, we assert that the relative maximum must, in fact, be the absolute maximum of v. The maximum velocity of the rocket is given by evaluating v at $t = 32$,
$$f'(32) = -3(32)^2 + 192(32) + 195$$
or 3267 ft/sec. The graph of the velocity function v is sketched in Figure 4.66.

○ ○ ○

SELF-CHECK EXERCISES 4.4

1. Let $f(x) = x - 2\sqrt{x}$.
   a. Find the absolute extrema of f on the interval $[0, 9]$.
   b. Find the absolute extrema of f.

2. Find the absolute extrema of $f(x) = 3x^4 + 4x^3 + 1$ on $[-2, 1]$.

3. The operating rate (expressed as a percentage) of factories, mines, and utilities in a certain region of the country on the tth day of the year 1996 is given by the function
$$f(t) = 80 + \frac{1{,}200t}{t^2 + 40{,}000} \qquad (0 \leq t \leq 250)$$
On which day of the first 250 days of 1996 was the manufacturing capacity operating rate highest?

Solutions to Self-Check Exercises 4.4 can be found on page 344.

4.4 EXERCISES

In exercises 1–8, you are given the graph of some function f defined on the indicated interval. Find the absolute maximum and the absolute minimum of f, if they exist.

1.

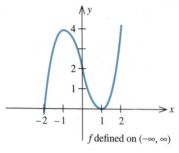

f defined on (−∞, ∞)

2.

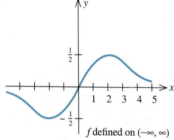

f defined on (−∞, ∞)

3.

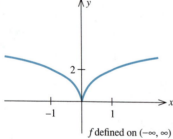

f defined on (−∞, ∞)

4.

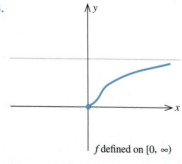

f defined on [0, ∞)

5.

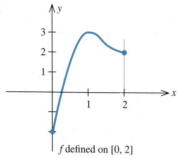

f defined on [0, 2]

6.

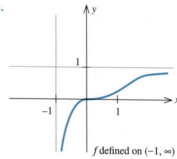

f defined on (−1, ∞)

7.

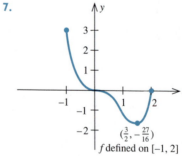

$(\frac{3}{2}, -\frac{27}{16})$

f defined on [−1, 2]

8.

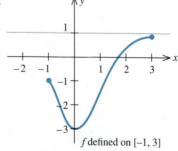

f defined on [−1, 3]

In exercises 9–38, find the absolute maximum value and the absolute minimum value, if any, of the given function.

9. $f(x) = 2x^2 + 3x - 4$

10. $g(x) = -x^2 + 4x + 3$

11. $h(x) = x^{1/3}$

12. $f(x) = x^{2/3}$

13. $f(x) = \dfrac{1}{1 + x^2}$

14. $f(x) = \dfrac{x}{1 + x^2}$

15. $f(x) = x^2 - 2x - 3$ on $[-2, 3]$

16. $g(x) = x^2 - 2x - 3$ on $[0, 4]$

17. $f(x) = -x^2 + 4x + 6$ on $[0, 5]$

18. $f(x) = -x^2 + 4x + 6$ on $[3, 6]$

19. $f(x) = x^3 + 3x^2 - 1$ on $[-3, 2]$

20. $g(x) = x^3 + 3x^2 - 1$ on $[-3, 1]$

21. $g(x) = 3x^4 + 4x^3$ on $[-2, 1]$

22. $f(x) = \frac{1}{2}x^4 - \frac{2}{3}x^3 - 2x^2 + 3$ on $[-2, 3]$

23. $f(x) = \dfrac{x + 1}{x - 1}$ on $[2, 4]$

24. $g(t) = \dfrac{t}{t - 1}$ on $[2, 4]$

25. $f(x) = 4x + \dfrac{1}{x}$ on $[1, 3]$

26. $f(x) = 9x - \dfrac{1}{x}$ on $[1, 3]$

27. $f(x) = \frac{1}{2}x^2 - 2\sqrt{x}$ on $[0, 3]$

28. $g(x) = \frac{1}{8}x^2 - 4\sqrt{x}$ on $[0, 9]$

29. $f(x) = \dfrac{1}{x}$ on $(0, \infty)$

30. $g(x) = \dfrac{1}{x + 1}$ on $(0, \infty)$

31. $f(x) = 3x^{2/3} - 2x$ on $[0, 3]$

32. $g(x) = x^2 + 2x^{2/3}$ on $[-2, 2]$

33. $f(x) = x^{2/3}(x^2 - 4)$ on $[-1, 2]$

34. $f(x) = x^{2/3}(x^2 - 4)$ on $[-1, 3]$

35. $f(x) = \dfrac{x}{x^2 + 2}$ on $[-1, 2]$

36. $f(x) = \dfrac{1}{x^2 + 2x + 5}$ on $[-2, 1]$

37. $f(x) = \dfrac{x}{\sqrt{x^2 + 1}}$ on $[-1, 1]$

38. $g(x) = x\sqrt{4 - x^2}$ on $[0, 2]$

39. A stone is thrown straight up from the roof of an 80-foot building. The height of the stone at any time t (in seconds), measured from the ground, is given by

$$h(t) = -16t^2 + 64t + 80$$

feet. What is the maximum height the stone reaches?

40. Maximizing Profits Lynbrook West, an apartment complex, has 100 two-bedroom units. The monthly profit realized from renting out x apartments is given by

$$P(x) = -10x^2 + 1{,}760x - 50{,}000$$

dollars. Find how many units should be rented out in order to maximize the monthly rental profit. What is the maximum monthly profit realizable?

41. Maximizing Profits The estimated monthly profit realizable by the Cannon Precision Instruments Corporation for manufacturing and selling x units of its model M1 camera is

$$P(x) = -0.04x^2 + 240x - 10{,}000$$

dollars. Determine how many cameras Cannon should produce per month in order to maximize its profits.

42. Flight of a Rocket The altitude (in feet) attained by a model rocket t seconds into flight is given by the function

$$h(t) = -\frac{1}{3}t^3 + 4t^2 + 20t + 2$$

Find the maximum altitude attained by the rocket.

43. Female Self-Employed Work Force Based on data obtained from the U.S. Department of Labor, the number of nonfarm, full-time, self-employed women can be approximated by

$$N(t) = 0.81t - 1.14\sqrt{t} + 1.53 \qquad (0 \le t \le 6)$$

where $N(t)$ is measured in millions and t is measured in 5-year intervals, with $t = 0$ corresponding to the beginning of 1963. Determine the absolute extrema of the function N on the interval $[0, 6]$. Interpret your results.

44. Maximizing Profits The management of Trappee and Sons, Inc., producers of the famous Texa-Pep hot sauce, estimate that their profit from the daily production and sale of x cases (each case consisting of 24 bottles) of the hot sauce is given by

$$P(x) = -0.000002x^3 + 6x - 400$$

dollars. What is the largest possible profit Trappee can make in one day?

45. Maximizing Profits The quantity demanded per month of the Walter Serkin recording of Beethoven's *Moonlight Sonata,* manufactured by Phonola Record Industries, is related to the price per record. The equation

$$p = -0.00042x + 6 \qquad (0 \le x \le 12,000)$$

where p denotes the unit price in dollars and x is the number of records demanded, relates the demand to the price. The total monthly cost for pressing and packaging x copies of this classical recording is given by

$$C(x) = 600 + 2x - 0.00002x^2 \qquad (0 \le x \le 20,000)$$

dollars. Determine how many copies Phonola should produce per month in order to maximize its profits. [*Hint:* The revenue is $R(x) = px$ and the profit is $P(x) = R(x) - C(x)$.]

46. Maximizing Profit A manufacturer of tennis rackets finds that the total cost $C(x)$ (in dollars) of manufacturing x rackets per day is given by $C(x) = 400 + 4x + 0.0001x^2$. Each racket can be sold at a price of p dollars, where p is related to x by the demand equation $p = 10 - 0.0004x$. If all the rackets that are manufactured can be sold, find the daily level of production that will yield a maximum profit for the manufacturer.

 47. Maximizing Profit The weekly demand for the Pulsar 25-inch color console television is given by the demand equation

$$p = -0.05x + 600 \qquad (0 \le x \le 12,000)$$

where p denotes the wholesale unit price in dollars and x denotes the quantity demanded. The weekly total cost function associated with manufacturing these sets is given by

$$C(x) = 0.000002x^3 - 0.03x^2 + 400x + 80,000$$

where $C(x)$ denotes the total cost incurred in producing x sets. Find the level of production that will yield a maximum profit for the manufacturer. [*Hint:* Use the quadratic formula.]

 48. Minimizing Average Costs Suppose that the total cost function for manufacturing a certain product is $C(x) = 0.2(0.01x^2 + 120)$ dollars, where x represents the number of units produced. Find the level of production that will minimize the average cost.

49. Minimizing Production Costs The total monthly cost in dollars incurred by Cannon Precision Instruments Corporation for manufacturing x units of the model M1

camera is given by the function

$$C(x) = 0.0025x^2 + 80x + 10,000$$

a. Find the average cost function $\overline{C}$.
b. Find the level of production that results in the smallest average production cost.
c. Find the level of production for which the average cost is equal to the marginal cost.
d. Compare the result of (c) with that of (b).

50. Minimizing Production Costs The daily total cost in dollars incurred by Trappee and Sons, Inc., for producing x cases of Texa-Pep hot sauce is given by the function

$$C(x) = 0.000002x^3 + 5x + 400$$

Using this function, answer the questions posed in exercise 49.

51. Maximizing Revenue Suppose that the quantity demanded per week of a certain dress is related to the unit price p by the demand equation $p = \sqrt{800 - x}$, where p is in dollars and x is the number of dresses made. How many dresses should be made and sold per week in order to maximize the revenue? [*Hint:* $R(x) = px$.]

52. Maximizing Revenue The quantity demanded per month of the Sicard wristwatch is related to the unit price by the equation

$$p = \frac{50}{0.01x^2 + 1} \qquad (0 \le x \le 20)$$

where p is measured in dollars and x is measured in units of a thousand. How many watches must be sold to yield a maximum revenue?

53. Oxygen Content of a Pond When organic waste is dumped into a pond, the oxidation process that takes place reduces the pond's oxygen content. However, given time, nature will restore the oxygen content to its natural level. Suppose that the oxygen content t days after organic waste has been dumped into the pond is given by

$$f(t) = 100 \left[\frac{t^2 - 4t + 4}{t^2 + 4} \right] \qquad (0 \le t < \infty)$$

percent of its normal level.
a. When is the level of oxygen content lowest?
b. When is the rate of oxygen regeneration greatest?

 54. Air Pollution The amount of nitrogen dioxide, a brown gas that impairs breathing, present in the atmosphere on a certain May day in the city of Long Beach is approximated by

$$A(t) = \frac{136}{1 + 0.25(t - 4.5)^2} + 28 \qquad (0 \le t \le 11)$$

where $A(t)$ is measured in pollutant standard index (PSI) and t is measured in hours, with $t = 0$ corresponding to 7 A.M. Determine the time of day when the pollution is at its highest level.

55. Maximizing Revenue The average revenue is defined as the function

$$\overline{R}(x) = \frac{R(x)}{x} \qquad (x > 0)$$

Prove that if a revenue function $R(x)$ is concave downward ($R''(x) < 0$), then the level of sales that will result in the largest average revenue occurs when $\overline{R}(x) = R'(x)$.

56. Velocity of Blood According to a law discovered by the nineteenth-century physician Jean Louis Marie Poiseuille, the velocity (in centimeters per second) of blood r centimeters from the central axis of an artery is given by

$$v(r) = k(R^2 - r^2)$$

where k is a constant and R is the radius of the artery. Show that the velocity of blood is greatest along the central axis.

57. GDP of a Developing Country A developing country's gross domestic product (GDP) from 1988 to 1996 is approximated by the function

$$G(t) = -0.2t^3 + 2.4t^2 + 60 \qquad (0 \le t \le 8)$$

where $G(t)$ is measured in billions of dollars and $t = 0$ corresponds to the year 1988. Show that the growth rate of the country's GDP was maximal in 1992.

58. Crime Rates The number of major crimes committed in the city of Bronxville between 1987 and 1994 is approximated by the function

$$N(t) = -0.1t^3 + 1.5t^2 + 100 \qquad (0 \le t \le 7)$$

where $N(t)$ denotes the number of crimes committed in year t ($t = 0$ corresponds to the year 1987). Enraged by the dramatic increase in the crime rate, the citizens of Bronxville, with the help of the local police, organized "Neighborhood Crime Watch" groups in early 1991 to combat this menace. Show that the growth in the crime rate was maximal in 1992, giving credence to the claim

that the Neighborhood Crime Watch program was working.

59. Social Security Surplus Based on data from the Social Security Administration, the estimated cash in the Social Security retirement and disability trust funds may be approximated by

$$f(t) = -0.0129t^4 + 0.3087t^3 + 2.1760t^2 + 62.8466t$$
$$+ 506.2955 \qquad (0 \le t \le 35)$$

where $f(t)$ is measured in billions of dollars and t is measured in years, with $t = 0$ corresponding to the year 1995. Show that the Social Security surplus will be at its highest level at approximately the middle of the year 2018.

[*Hint:* Show that $t = 23.6811$ is an approximate critical point of $f'(t)$.]

Source: Social Security Administration

60. Energy Expended by a Fish It has been conjectured that a fish swimming a distance of L ft at a speed of v ft/sec relative to the water and against a current flowing at the rate of u ft/sec ($u < v$) expends a total energy given by

$$E(v) = \frac{aLv^3}{v - u}$$

where E is measured in foot-pounds (ft-lbs) and a is a constant. Find the speed v at which the fish must swim in order to minimize the total energy expended. (*Note:* This result has been verified by biologists.)

61. Let f be a constant function—that is, let $f(x) = c$, where c is some real number. Show that every point $x = a$ is an absolute maximum and, at the same time, an absolute minimum of f.

62. Show that a polynomial function defined on the interval $(-\infty, \infty)$ cannot have both an absolute maximum and an absolute minimum unless it is a constant function.

63. One condition that must be satisfied before Theorem 3 (page 330) is applicable is that the function f must be continuous on the closed interval $[a, b]$. Define a function f on the closed interval $[-1, 1]$ by

$$f(x) = \begin{cases} \dfrac{1}{x} & \text{if } x \in [-1, 1] \qquad (x \ne 0) \\ 0 & \text{if } x = 0 \end{cases}$$

a. Show that f is not continuous at $x = 0$.
b. Show that $f(x)$ does not attain an absolute maximum or an absolute minimum on the interval $[-1, 1]$.
c. Confirm your results by sketching the function f.

USING TECHNOLOGY

FINDING THE ABSOLUTE EXTREMA OF A FUNCTION

Some graphing utilities have a function for finding the absolute maximum and the absolute minimum values of a continuous function on a closed interval. If your graphing utility has this capability, use it to work through the example and exercises of this section.

EXAMPLE 1 Let $f(x) = \dfrac{2x + 4}{(x^2 + 1)^{3/2}}$.

a. Use a graphing utility to plot the graph of f in the viewing rectangle $[-3, 3] \times [-1, 5]$.

b. Find the absolute maximum and absolute minimum values of f on the interval $[-3, 3]$. Express your answers accurate to four decimal places.

Solution

a. The graph of f is shown in Figure T1.

Figure T1

The graph of f in the viewing rectangle $[-3, 3] \times [-1, 5]$.

b. Using the function on a graphing utility for finding the absolute minimum value of a continuous function on a closed interval, we find the absolute minimum value of f to be -0.0632. Similarly, using the function for finding the absolute maximum value, we find the absolute maximum value to be 4.1593.

REMARK   Some graphing utilities will enable you to find the absolute minimum and absolute maximum values of a continuous function on a closed interval without having to graph the function.

EXERCISES

In exercises 1–6, use a graphing utility to find the absolute maximum and the absolute minimum values of f in the given interval using the method of Example 1. Express your answers accurate to four decimal places.

1. $f(x) = 3x^4 - 4.2x^3 + 6.1x - 2; [-2, 3]$

2. $f(x) = 2.1x^4 - 3.2x^3 + 4.1x^2 + 3x - 4; [-1, 2]$

3. $f(x) = \dfrac{2x^3 - 3x^2 + 1}{x^2 + 2x - 8}; [-3, 1]$

4. $f(x) = \sqrt{x}(x^3 - 4)^2; [0.5, 1]$

5. $f(x) = \dfrac{x^3 - 1}{x^2}$ on $[1, 3]$

6. $f(x) = \dfrac{x^3 - x^2 + 1}{x - 2}$ on $[1, 3]$

7. **Rate of Bank Failures**   The Federal Deposit Insurance Company (FDIC) estimates that the rate at which banks were failing between 1982 and 1994 is given by

$$f(t) = 0.063447t^4 - 1.953283t^3 + 14.632576t^2$$
$$- 6.684704t + 47.458874 \qquad (0 \le t \le 12)$$

where $f(t)$ is the number of banks per year and t is measured in years, with $t = 0$ corresponding to the beginning of 1982.
a. Use a graphing utility to plot the graph of f in the viewing rectangle $[0, 12] \times [0, 220]$.
b. What is the highest rate of bank failures during the period in question?
Source: Federal Deposit Insurance Corporation

8. **Boston's Daily Temperature for 1993**   The average daily temperature in Boston for 1993 is given by

$$f(t) = 0.0434841t^4 - 1.18523t^3 + 9.38548t^2$$
$$- 17.7553t + 38.9272 \qquad (0 \le t \le 12)$$

where $f(t)$ is measured in degrees Fahrenheit and t is in months, with $t = 0$ corresponding to the beginning

of 1993. Find the absolute extrema for f and interpret your results.
Source: Robert Lautzenheiser, Climatologist

9. **Time on the Market**   According to the Greater Boston Real Estate Board-Multiple Listing Service, the average number of days a single-family home remains for sale from listing to accepted offer is approximated by the function

$$f(t) = 0.0171911t^4 - 0.662121t^3 + 6.18083t^2$$
$$- 8.97086t + 53.3357 \qquad (0 \le t \le 10)$$

where t is measured in years, with $t = 0$ corresponding to the beginning of 1984.
a. Use a graphing utility to plot the graph of f in the viewing rectangle $[0, 12] \times [0, 120]$.
b. Find the absolute maximum value and the absolute minimum value of f in the interval $[0, 12]$. Interpret your results.
Source: Greater Boston Real Estate Board-Multiple Listing Service

10. **Why SSS Benefits May Exceed Payroll Taxes**   Unless payroll taxes are increased significantly and/or benefits are scaled back drastically, it is a matter of time before the current Social Security system goes broke. Based on data from the Board of Trustees of the Social Security Administration, the assets of the system—the Social Security "trust fund"—may be approximated by

$$f(t) = -0.0129t^4 + 0.3087t^3 + 2.1760t^2$$
$$+ 62.8466t + 506.2955 \qquad (0 \le t \le 35)$$

where $f(t)$ is measured in millions of dollars and t is measured in years, with $t = 0$ corresponding to 1995.
Source: Social Security Administration
a. Use a graphing calculator to sketch the graph of f.
b. Based on this model, when will the Social Security system start to pay out more benefits than it gets in payroll taxes?

SOLUTIONS TO SELF-CHECK EXERCISES 4.4

1. a. The function f is continuous in its domain and differentiable in the interval $(0, 9)$. The derivative of f is

$$f'(x) = 1 - x^{-1/2} = \frac{x^{1/2} - 1}{x^{1/2}}$$

and it is equal to zero when $x = 1$. Evaluating $f(x)$ at the endpoints $x = 0$ and $x = 9$ and at the critical point $x = 1$ of f, we have

$$f(0) = 0, \quad f(1) = -1, \quad \text{and} \quad f(9) = 3$$

From these results, we see that -1 is the absolute minimum value of f and 3 is the absolute maximum value of f.

b. In this case, the domain of f is the interval $[0, \infty)$, which is not closed. Therefore, we resort to the graphic method. Using the techniques of graphing, we sketch in the accompanying figure the graph of f.

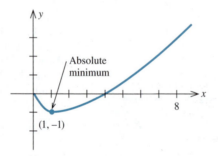

The graph of f shows that -1 is the absolute minimum value of f, but f has no absolute maximum since $f(x)$ increases without bound as x increases without bound.

2. The function f is continuous on the interval $[-2, 1]$. It is also differentiable on the open interval $(-2, 1)$. The derivative of f is

$$f'(x) = 12x^3 + 12x^2 = 12x^2(x + 1)$$

and it is continuous on $(-2, 1)$. Setting $f'(x) = 0$ gives $x = -1$ and $x = 0$ as critical points of f. Evaluating $f(x)$ at these critical points of f as well as at the endpoints of the interval $[-2, 1]$, we obtain

$$f(-2) = 17, \quad f(-1) = 0, \quad f(0) = 1, \quad \text{and} \quad f(1) = 8$$

From these results, we see that 0 is the absolute minimum value of f and 17 is the absolute maximum value of f.

3. The problem is solved by finding the absolute maximum of the function f on $[0, 250]$. Differentiating $f(t)$, we obtain

$$f'(t) = \frac{(t^2 + 40{,}000)(1{,}200) - 1{,}200t(2t)}{(t^2 + 40{,}000)^2}$$

$$= \frac{-1{,}200\,(t^2 - 40{,}000)}{(t^2 + 40{,}000)^2}$$

Upon setting $f'(t) = 0$ and solving the resulting equation, we obtain $t = -200$ or 200. Since -200 lies outside the interval $[0, 250]$, we are interested only in the critical point $t = 200$ of f. Evaluating $f(t)$ at $t = 0$, $t = 200$, and $t = 250$, we find

$$f(0) = 80, \quad f(200) = 83, \quad \text{and} \quad f(250) = 82.93$$

We conclude that the manufacturing capacity operating rate was the highest on the 200th day of 1996—that is, a little past the middle of July of 1996.

4.5 OPTIMIZATION II

Section 4.4 outlined how to find the solution to certain optimization problems in which the objective function is given. In this section, we consider problems in which we are required to first find the appropriate function to be optimized. The following guidelines will be useful for solving these problems.

GUIDELINES FOR SOLVING OPTIMIZATION PROBLEMS

1. Assign a letter to each variable mentioned in the problem. If appropriate, draw and label a figure.

2. Find an expression for the quantity to be optimized.

3. Use the conditions given in the problem to write the quantity to be optimized as a function f of *one* variable. Note any restrictions to be placed on the domain of f from physical considerations of the problem.

4. Optimize the function f over its domain using the methods of Section 4.4.

REMARK   In carrying out step 4, remember that if the function f to be optimized is continuous on a closed interval, then the absolute maximum and absolute minimum of f are, respectively, the largest and smallest values of $f(x)$ on the set comprising the critical points of f and the endpoints of the interval. If the domain of f is not a closed interval, then we resort to the graphical method.   ○ ○ ○

Maximization Problems

EXAMPLE I   A man wishes to have a rectangular-shaped garden in his backyard. He has 50 feet of fencing material with which to enclose his garden. Find the dimensions for the largest garden he can have if he uses all of the fencing material.

Solution

Step I   Let x and y denote the dimensions (in feet) of two adjacent sides of the garden (Figure 4.67) and let A denote its area.

Figure 4.67
What is the maximum rectangular area that can be enclosed with 50 feet of fencing?

Step 2   The area of the garden,

$$A = xy \qquad (1)$$

is the quantity to be maximized.

Step 3   The perimeter of the rectangle, $(2x + 2y)$ feet, must equal 50 feet. Therefore, we have the equation

$$2x + 2y = 50$$

Next, solving this equation for y in terms of x yields

$$y = 25 - x \qquad (2)$$

which, when substituted into equation (1), gives

$$A = x(25 - x)$$
$$= -x^2 + 25x$$

(Remember, the function to be optimized must involve just one variable.) Since the sides of the rectangle must be nonnegative, we must have $x \geq 0$ and $y = 25 - x \geq 0$; that is, we must have $0 \leq x \leq 25$. Thus, the problem is reduced to that of finding the absolute maximum of $A = f(x) = -x^2 + 25x$ on the closed interval $[0, 25]$.

Step 4   Observe that f is continuous on $[0, 25]$. So the absolute maximum value of f must occur at the endpoint(s) or at the critical point(s) of f. The derivative of the function A is given by

$$A' = f'(x) = -2x + 25$$

Setting $A' = 0$ gives

$$-2x + 25 = 0$$

or $x = 12.5$, as the critical point of A. Next, we evaluate the function $A = f(x)$ at $x = 12.5$ and at the endpoints $x = 0$ and $x = 25$ of the interval $[0, 25]$, obtaining

$$f(0) = 0, \quad f(12.5) = 156.25, \quad f(25) = 0$$

We see that the absolute maximum value of the function f is 156.25. From equation (2) we see that when $x = 12.5$, the value of y is given by $y = 12.5$. Thus, the garden would be of maximum area (156.25 square feet) if it were in the form of a square with sides 12.5 feet long.

⚬ ⚬ ⚬

EXAMPLE 2   By cutting away identical squares from each corner of a rectangular piece of cardboard and folding up the resulting flaps, the cardboard may be turned into an open box. If the cardboard is 16 inches long and 10 inches wide, find the dimensions of the box that will yield the maximum volume.

Solution

Step 1 Let x denote the length (in inches) of one side of each of the identical squares to be cut out of the cardboard (Figure 4.68), and let V denote the volume of the resulting box.

Figure 4.68
The dimensions of the open box are $(16 - 2x)$ by $(10 - 2x)$ by x inches.

Step 2 The dimensions of the box are $(16 - 2x)$ inches long, $(10 - 2x)$ inches wide, and x inches high. Therefore, its volume,

$$V = (16 - 2x)(10 - 2x)x$$
$$= 4(x^3 - 13x^2 + 40x) \qquad \text{(Expanding the expression)}$$

cubic inches, is the quantity to be maximized.

Step 3 Since each side of the box must be nonnegative, x must satisfy the inequalities $x \geq 0$, $16 - 2x \geq 0$, and $10 - 2x \geq 0$. This set of inequalities is satisfied if $0 \leq x \leq 5$. Thus, the problem at hand is equivalent to that of finding the absolute maximum of

$$V = f(x) = 4(x^3 - 13x^2 + 40x)$$

on the closed interval $[0, 5]$.

Step 4 Observe that f is continuous on $[0, 5]$, so the absolute maximum value of f must be attained at the endpoint(s) or at the critical point(s) of f.

Differentiating $f(x)$, we obtain

$$f'(x) = 4(3x^2 - 26x + 40)$$
$$= 4(3x - 20)(x - 2)$$

Upon setting $f'(x) = 0$ and solving the resulting equation for x, we obtain $x = 20/3$ or $x = 2$. Since 20/3 lies outside the interval $[0, 5]$, it is no longer considered and we are interested only in the critical point $x = 2$ of f. Next, evaluating $f(x)$ at $x = 0$, $x = 5$ (the endpoints of the interval $[0, 5]$), and $x = 2$, we obtain

$$f(0) = 0, \quad f(2) = 144, \quad f(5) = 0$$

Thus, the volume of the box is maximized by taking $x = 2$. The dimensions of the box are $12'' \times 6'' \times 2''$, and the volume is 144 cubic inches.

○ ○ ○

EXPLORING WITH TECHNOLOGY

Refer to Example 2.
a. Use a graphing utility to plot the graph of

$$f(x) = 4(x^3 - 13x^2 + 40x)$$

using the viewing rectangle $[0, 5] \times [0, 150]$. Explain what happens to $f(x)$ as x increases from $x = 0$ to $x = 5$ and give a physical interpretation.
b. Using **ZOOM** and **TRACE,** find the absolute maximum of f on the interval $[0, 5]$, and thus verify the solution for Example 2 obtained analytically.

○ ○ ○

EXAMPLE 3 A city's Metropolitan Transit Authority (MTA) operates a subway line for commuters from a certain suburb to the downtown metropolitan area. Currently, an average of 6000 passengers a day take the trains, paying a fare of $1.50 per ride. The Board of the MTA, contemplating raising the fare to $1.75 per ride in order to generate a larger revenue, engages the services of a consulting firm. The firm's study reveals that for each 25-cent increase in fare, the ridership will be reduced by an average of 1000 passengers a day. Thus the consulting firm recommends that MTA stick to the current fare of $1.50 per ride, which already yields a maximum revenue. Show that the consultants are correct.

Solution

Step 1 Let x denote the number of passengers per day, p denote the fare per ride, and R be MTA's revenue.

Step 2 In order to find a relationship between x and p, observe that the given data imply that when $x = 6000$, $p = 1.5$, and when $x = 5000$, $p = 1.75$. Therefore, the points $(6000, 1.5)$ and $(5000, 1.75)$ lie on a straight line. (Why?) To find the linear relationship between p and x, use the point-slope form of the equation of a straight line. Now, the slope of the line is

$$m = \frac{1.75 - 1.50}{5000 - 6000} = -0.00025$$

Therefore, the required equation is

$$p - 1.5 = -0.00025(x - 6000)$$
$$= -0.00025x + 1.5$$
$$p = -0.00025x + 3$$

Therefore, the revenue,

$$R = f(x) = xp = -0.00025x^2 + 3x \qquad \text{(Number of riders times unit fare)}$$

is the quantity to be maximized.

Step 3 Since both p and x must be nonnegative, we see that $0 \le x \le 12{,}000$, and the problem is that of finding the absolute maximum of the function f on the closed interval $[0, 12{,}000]$.

Step 4 Observe that f is continuous on $[0, 12{,}000]$. To find the critical point of R, we compute

$$f'(x) = -0.0005x + 3$$

and set it equal to zero, giving $x = 6000$. Evaluating the function f at $x = 6000$, as well as at the endpoints $x = 0$ and $x = 12{,}000$, yields

$$f(0) = 0$$
$$f(6000) = 9000$$
$$f(12{,}000) = 0$$

We conclude that a maximum revenue of $9000 per day is realized when the ridership is 6000 per day. The optimum price of the fare per ride is therefore $1.50, as recommended by the consultants. The graph of the revenue function R is shown in Figure 4.69.   ◦ ◦ ◦

Figure 4.69
f has an absolute maximum of 9000 when x = 6000.

Minimization Problems

EXAMPLE 4 The Betty Moore Company requires that its corned beef hash containers have a capacity of 54 cubic inches, have the shape of right-circular cylinders, and be made of tin. Determine the radius and height of the container that requires the least amount of metal.

Figure 4.70
We want to minimize the amount of material used to construct the container.

Solution

Step 1 Let the radius and height of the container be r and h inches, respectively, and let S denote the surface area of the container (Figure 4.70).

Step 2 The amount of tin used to construct the container is given by the total surface area of the cylinder. Now, the area of the base and the top of the cylinder are each πr^2 square inches and the area of the side is $2\pi rh$ square inches. Therefore,

$$S = 2\pi r^2 + 2\pi rh \qquad \qquad \textbf{(3)}$$

the quantity to be minimized.

Step 3 The requirement that the volume of a container be 54 cubic inches implies that

$$\pi r^2 h = 54 \tag{4}$$

Solving equation (4) for h, we obtain

$$h = \frac{54}{\pi r^2} \tag{5}$$

which, when substituted into (3), yields

$$S = 2\pi r^2 + 2\pi r \left(\frac{54}{\pi r^2}\right)$$

$$= 2\pi r^2 + \frac{108}{r}$$

Clearly, the radius r of the container must satisfy the inequality $r > 0$. The problem now is reduced to finding the absolute minimum of the function $S = f(r)$ on the interval $(0, \infty)$.

Step 4 Using the curve-sketching techniques of Section 4.3, we obtain the graph of f in Figure 4.71.

To find the critical point of f, we compute

$$S' = 4\pi r - \frac{108}{r^2}$$

and solve the equation $S' = 0$ for r:

$$4\pi r - \frac{108}{r^2} = 0$$

$$4\pi r^3 - 108 = 0$$

$$r^3 = \frac{27}{\pi}$$

or

$$r = \frac{3}{\sqrt[3]{\pi}} \approx 2 \tag{6}$$

Next, let us show that this value of r gives rise to the absolute minimum of f. To show this, we first compute

$$S'' = 4\pi + \frac{216}{r^3}$$

Since $S'' > 0$ if $r = 3/\sqrt[3]{\pi}$, the Second Derivative Test implies that the value of r in equation (6) gives rise to a relative minimum of f. Finally, this relative minimum of f is also the absolute minimum of f since f is always concave upward ($S'' > 0$ for all $r > 0$). To find the height of the given container, we substitute the value of r given in

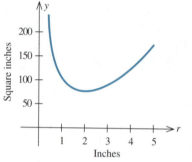

Figure 4.71
The total surface area of the right cylindrical container is graphed as a function of r.

(6) into (5). Thus

$$h = \frac{54}{\pi r^2} = \frac{54}{\pi \left(\dfrac{3}{\pi^{1/3}}\right)^2}$$

$$= \frac{54\pi^{2/3}}{(\pi)9}$$

$$= \frac{6}{\pi^{1/3}} = \frac{6}{\sqrt[3]{\pi}}$$

$$= 2r$$

We conclude that the required container has a radius of approximately 2 inches and a height of approximately 4 inches, or twice the size of the radius.   ๐ ๐ ๐

An Inventory Problem

One problem faced by many companies is that of controlling the inventory of goods carried. Ideally, the manager must ensure that the company has sufficient stock to meet customer demand at all times. At the same time, she must make sure that this is accomplished without overstocking (incurring unnecessary storage costs) and also without having to place orders too frequently (incurring reordering costs).

EXAMPLE 5   The Dixie Import-Export Company is the sole agent for the Excalibur 250 cc motorcycle. Management estimates that the demand for these motorcycles is 10,000 per year and that they will sell at a uniform rate throughout the year. The cost incurred in ordering each shipment of motorcycles is $10,000 and the cost per year of storing each motorcycle is $200.

Dixie's management faces the following problem: Ordering too many motorcycles at one time ties up valuable storage space and increases the storage cost. On the other hand, placing orders too frequently increases the ordering costs. How large should each order be, and how often should orders be placed, to minimize ordering and storage costs?

Solution   Let x denote the number of motorcycles in each order (the lot size). Then, assuming that each shipment arrives just as the previous shipment has been sold, the average number of motorcycles in storage during the year is $x/2$. You can see that this is the case by examining Figure 4.72. Thus, Dixie's storage cost for the year is given by $200(x/2)$, or $100x$ dollars.

Next, since the company requires 10,000 motorcycles for the year and since each order is for x motorcycles, the number of orders required is

$$\frac{10,000}{x}$$

Figure 4.72

As each lot is depleted, the new lot arrives. The average inventory level is x/2 if x is the lot size.

This gives an ordering cost of

$$10,000 \left(\frac{10,000}{x} \right) = \frac{100,000,000}{x}$$

dollars for the year. Thus, the total yearly cost incurred by Dixie, which includes the ordering and storage costs attributed to the sale of these motorcycles, is given by

$$C(x) = 100x + \frac{100,000,000}{x}$$

The problem is reduced to finding the absolute minimum of the function C in the interval $(0, 10,000]$. To accomplish this, we compute

$$C'(x) = 100 - \frac{100,000,000}{x^2}$$

Setting $C'(x) = 0$ and solving the resulting equation, we obtain $x = \pm 1000$. Since the number -1000 is outside the domain of the function C, it is rejected, leaving $x = 1000$ as the only critical point of C. Next, we find

$$C''(x) = \frac{200,000,000}{x^3}$$

Since $C''(1000) > 0$, the Second Derivative Test implies that the critical point $x = 1000$ is a relative minimum of the function C (Figure 4.73). Also, since $C''(x) > 0$ for all x in $(0, 10,000]$, the function C is concave upward everywhere so that the point $x = 1000$ also gives the absolute minimum of C. Thus, in order to minimize the ordering and storage costs, Dixie should place $10,000/1000$, or 10, orders a year, each for a shipment of 1000 motorcycles.

Figure 4.73

C has an absolute minimum at (1000, 200,000).

SELF-CHECK EXERCISES 4.5

 1. A man wishes to have an enclosed vegetable garden in his backyard. If the garden is to be a rectangular area of 300 square feet, find the dimensions of the garden that will minimize the amount of fencing material needed.

 2. The demand for the Super Titan tires is 1,000,000 per year. The set-up cost for each production run is $4000, and the manufacturing cost is $20 per tire. The cost of storing each tire over the year is $2. Assuming uniformity of demand throughout the year and instantaneous production, determine how many tires should be manufactured per production run in order to keep the production cost to a minimum.

Solutions to Self-Check Exercises 4.5 can be found on page 356.

4.5 EXERCISES

1. Enclosing the Largest Area   The owner of the Rancho Los Feliz has 3000 yards of fencing material with which to enclose a rectangular piece of grazing land along the straight portion of a river. If fencing is not required along the river, what are the dimensions of the largest area he can enclose? What is this area?

2. Enclosing the Largest Area   Refer to exercise 1. As an alternative plan, the owner of the Rancho Los Feliz might use the 3000 yards of fencing material to enclose the rectangular piece of grazing land along the straight portion of the river and then subdivide it by means of a fence running parallel to the sides. Again, no fencing is required along the river. What are the dimensions of the largest area that can be enclosed? What is this area? (See the accompanying figure.)

River

3. Minimizing Construction Costs   The management of the UNICO department store has decided to enclose an 800-square-foot area outside the building for displaying potted plants and flowers. One side will be formed by the external wall of the store, two sides will be constructed of pine boards, and the fourth side will be made of galvanized steel fencing material. If the pine board fencing costs $6 per running foot and the steel fencing costs $3 per running foot, determine the dimensions of the enclosure that can be erected at minimum cost.

4. Packaging   By cutting away identical squares from each corner of a rectangular piece of cardboard and folding up the resulting flaps, an open box may be made. If the cardboard is 15 inches long and 8 inches wide, find the dimensions of the box that will yield the maximum volume.

5. Metal Fabrication   If an open box is made from a tin sheet 8 inches square by cutting out identical squares from each corner and bending up the resulting flaps, determine the dimensions of the largest box that can be made.

6. **Minimizing Construction Costs** If an open box has a square base and a volume of 108 cubic inches, and is constructed from a tin sheet, find the dimensions of the box, assuming a minimum amount of material is used in its construction.

7. **Minimizing Construction Costs** What are the dimensions of a closed rectangular box that has a square cross section, a capacity of 128 cubic inches, and is constructed using the least amount of material?

8. **Minimizing Construction Costs** A rectangular box is to have a square base and a volume of 20 cubic feet. If the material for the base costs 30 cents per square foot, the material for the sides costs 10 cents per square foot, and the material for the top costs 20 cents per square foot, determine the dimensions of the box that can be constructed at minimum cost.

9. **Parcel Post Regulations** Postal regulations specify that a parcel sent by parcel post may have a combined length and girth of no more than 108 inches. Find the dimensions of a rectangular package that has a square cross section and the largest volume that may be sent through the mail. What is the volume of such a package?
[*Hint:* The length plus the girth is $4x + h$ (see the accompanying figure).]

10. **Book Design** A production editor at Saunders-Roe Publishing decided that the pages of a book should have

one-inch margins at the top and bottom and half-inch margins on the sides. She further stipulated that each page should have an area of 50 square inches (see the accompanying figure). Determine the page dimensions that will result in the maximum printed area on the page.

11. **Parcel Post Regulations** Postal regulations specify that a parcel sent by parcel post may have a combined length and girth of no more than 108 inches. Find the dimensions of the cylindrical package of greatest volume that may be sent through the mail. What is the volume of such a package? Compare with exercise 9.
[*Hint:* The length plus the girth is $2\pi r + l$.]

12. **Minimizing Costs** For its beef stew, the Betty Moore Company uses tin containers that have the form of right circular cylinders. Find the radius and height of a container if it has a capacity of 36 cubic inches and is constructed using the least amount of metal.

13. **Product Design** The cabinet that will enclose the Acrosonic model D loudspeaker system will be rectangular and will have an internal volume of 2.4 cubic feet. For aesthetic reasons, it has been decided that the height of the cabinet is to be 1.5 times its width. If the top, bottom, and sides of the cabinet are constructed of ve-

neer costing 40 cents per square foot, and the front (ignore the cutouts in the baffle) and rear are constructed of particle board costing 20 cents per square foot, what are the dimensions of the enclosure that can be constructed at a minimum cost?

14. **Designing a Norman Window** A Norman window has the shape of a rectangle surmounted by a semicircle (see the accompanying figure). If a Norman window is to have a perimeter of 28 ft, what should its dimensions be in order to allow the maximum amount of light through the window?

15. **Optimal Charter Flight Fare** If exactly 200 people sign up for a charter flight, the Leisure World Travel Agency charges $300 per person. However, if more than 200 people sign up for the flight (assume this is the case), then each fare is reduced by $1 for each additional person. Determine how many passengers will result in a maximum revenue for the travel agency. What is the maximum revenue? What would the fare per passenger be in this case?

[*Hint:* Let x denote the number of passengers beyond 200. Show that the revenue function R is given by $R(x) = (200 + x) \cdot (300 - x)$.]

16. **Maximizing Yield** An apple orchard has an average yield of 36 bushels of apples per tree if tree density is 22 trees per acre. For each unit increase in tree density, the yield decreases by 2 bushels. Find how many trees should be planted in order to maximize the yield.

17. **Strength of a Beam** A wooden beam has a rectangular cross section of height h inches and width w inches (see the accompanying figure). The strength S of the beam is directly proportional to its width and the square of its height. What are the dimensions of the cross section of the strongest beam that can be cut from a round log of diameter 24 inches?

[*Hint:* $S = kh^2w$, where k is a constant of proportionality.]

18. **Designing a Grain Silo** A grain silo has the shape of a right circular cylinder surmounted by a hemisphere (see the accompanying figure). If the silo is to have a capacity of 504π cubic feet, find the radius and height of the silo that requires the least amount of material to construct.

[*Hint:* The volume of the silo is $\pi r^2h + \frac{2}{3}\pi r^3$, and the surface area (including the floor) is $\pi(3r^2 + 2rh)$.]

19. **Minimizing Construction Costs** In the following diagram, S represents the position of a power relay station located on a straight coast and E shows the location of a marine biology experimental station on an island. A cable is to be laid connecting the relay station with the

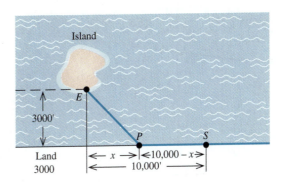

experimental station. If the cost of running the cable on land is $1 per running foot, and the cost of running the cable under water is $3 per running foot, locate the point P that will result in a minimum cost (solve for x).

20. **Flights of Birds**   During daylight hours, some birds fly more slowly over water than over land because some of their energy is expended in overcoming the down-drafts of air over open bodies of water. Suppose a bird that flies at a constant speed of 4 mph over water and 6 mph over land starts its journey at the point E on an island and ends at its nest N on the shore of the mainland, as shown in the accompanying figure. Find the location of the point P that allows the bird to complete its journey in the minimum time (solve for x).

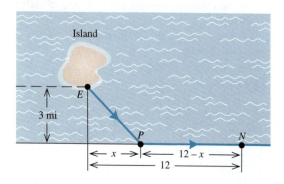

21. **Optimal Speed of a Truck**   A truck gets $400/x$ miles per gallon when driven at a constant speed of x miles per hour (between 50 and 70 miles per hour). If the price of fuel is $1 per gallon and the driver is paid $8 an hour, at what speed between 50 and 70 miles per hour is it most economical to drive?

22. **Inventory Control and Planning**   The demand for motorcycle tires imported by the Dixie Import-Export

Company is 40,000 per year and may be assumed to be uniform throughout the year. The cost of ordering a shipment of tires is $400, and the cost of storing each tire for a year is $2. Determine how many tires should be in each shipment if the ordering and storage costs are to be minimized. (Assume that each shipment arrives just as the previous one has been sold.)

23. **Inventory Control and Planning**   The McDuff Preserves Company expects to bottle and sell 2,000,000 32-ounce jars of jam. The company orders its containers from the Consolidated Bottle Company. The cost of ordering a shipment of bottles is $200, and the cost of storing each empty bottle for a year is 40 cents. Determine how many orders McDuff should place per year and how many bottles should be in each shipment if the ordering and storage costs are to be minimized. (Assume that each shipment of bottles is used up before the next shipment arrives.)

24. **Inventory Control and Planning**   The Neilsen Cookie Company sells its assorted butter cookies in containers that have a net content of 1 pound. The estimated demand for the cookies is 1,000,000 units. The set-up cost for each production run is $500, and the manufacturing cost is 50 cents for each container of cookies. The cost of storing each container of cookies over the year is 40 cents. Assuming uniformity of demand throughout the year and instantaneous production, determine how many containers of cookies Neilsen should produce per production run in order to minimize the production cost.

[*Hint:* Following the method of Example 5, show that the total production cost is given by the function

$$C(x) = \frac{500,000,000}{x} + 0.2x + 500,000$$

Then minimize the function C on the interval $(0, 1,000,000)$.]

SOLUTIONS TO SELF-CHECK EXERCISES 4.5

1. Let x and y (measured in feet) denote the length and width of the rectangular garden. Since the area is to be 300 square feet, we have

$$xy = 300$$

Next, the amount of fencing to be used is given by the perimeter, and this quantity is to be minimized. Thus, we want to minimize

$$2x + 2y$$

or, since $y = \dfrac{300}{x}$ (obtained by solving for y in the first equation), we see that the expression to be minimized is

$$f(x) = 2x + 2\left(\frac{300}{x}\right)$$

$$= 2x + \frac{600}{x}$$

for positive values of x. Now

$$f'(x) = 2 - \frac{600}{x^2}$$

Setting $f'(x) = 0$ yields $x = -\sqrt{300}$ or $x = \sqrt{300}$. We consider only the critical point $x = \sqrt{300}$ since $-\sqrt{300}$ lies outside the interval $(0, \infty)$. We then compute

$$f''(x) = \frac{1200}{x^3}$$

Since

$$f''(\sqrt{300}) > 0$$

the Second Derivative Test implies that a relative minimum of f occurs at $x = \sqrt{300}$. In fact, since $f''(x) > 0$ for all x in $(0, \infty)$, we conclude that $x = \sqrt{300}$ gives rise to the absolute minimum of f. The corresponding value of y, obtained by substituting this value of x into the equation $xy = 300$, is $y = \sqrt{300}$. Therefore, the required dimensions of the vegetable garden are approximately 17.3 ft $\times$ 17.3 ft.

2. Let x denote the number of tires in each production run. Then the average number of tires in storage is $x/2$, so the storage cost incurred by the company is $2(x/2)$, or x dollars. Next, since the company needs to manufacture 1,000,000 tires for the year in order to meet the demand, the number of production runs is $1,000,000/x$. This gives set-up costs amounting to

$$4000\left(\frac{1,000,000}{x}\right) = \frac{4,000,000,000}{x}$$

dollars for the year. The total manufacturing cost is $20,000,000. Thus, the total yearly cost incurred by the company is given by

$$C(x) = x + \frac{4,000,000,000}{x} + 20,000,000$$

Differentiating $C(x)$, we find

$$C'(x) = 1 - \frac{4,000,000,000}{x^2}$$

Setting $C'(x) = 0$ gives $x = 63{,}246$ as the critical point in the interval $(0, 1{,}000{,}000)$. Next, we find

$$C''(x) = \frac{8,000,000,000}{x^3}$$

Since $C''(x) > 0$ for all $x > 0$, we see that C is concave upward for all $x > 0$. Furthermore, $C''(63{,}246) > 0$ implies that $x = 63{,}246$ gives rise to a relative minimum of C (by the Second Derivative Test). Since C is always concave upward for $x > 0$, $x = 63{,}246$ gives the absolute minimum of C. Therefore, the company should manufacture 63,246 tires in each production run.

CHAPTER 4 SUMMARY OF PRINCIPAL TERMS

Terms

Increasing function   Concave downward

Decreasing function   Inflection point

Relative maximum   Second Derivative Test

Relative minimum   Vertical asymptote

Relative extrema   Horizontal asymptote

Critical point   Absolute extrema

First Derivative Test   Absolute maximum value

Concave upward   Absolute minimum value

CHAPTER 4 REVIEW EXERCISES

In exercises 1–10, (a) find the intervals where the given function f is increasing and where it is decreasing, (b) find the relative extrema of f, (c) find the intervals where f is concave upward and where it is concave downward, and (d) find the inflection points, if any, of f.

1. $f(x) = \frac{1}{3}x^3 - x^2 + x - 6$

2. $f(x) = (x - 2)^3$

3. $f(x) = x^4 - 2x^2$

4. $f(x) = x + \frac{4}{x}$

5. $f(x) = \frac{x^2}{x - 1}$

6. $f(x) = \sqrt{x - 1}$

7. $f(x) = (1 - x)^{1/3}$

8. $f(x) = x\sqrt{x - 1}$

9. $f(x) = \frac{2x}{x + 1}$

10. $f(x) = \frac{-1}{1 + x^2}$

In exercises 11–18, obtain as much information as possible on each of the given functions. Then use this information to sketch the graph of the function.

11. $f(x) = x^2 - 5x + 5$

12. $f(x) = -2x^2 - x + 1$

13. $g(x) = 2x^3 - 6x^2 + 6x + 1$

14. $g(x) = \frac{1}{3}x^3 - x^2 + x - 3$

15. $h(x) = x\sqrt{x - 2}$

16. $h(x) = \frac{2x}{1 + x^2}$

17. $f(x) = \frac{x - 2}{x + 2}$

18. $f(x) = x - \frac{1}{x}$

In exercises 19–22, find the horizontal and vertical asymptotes of the graphs of the given functions. Do not sketch the graphs.

19. $f(x) = \frac{1}{2x + 3}$

20. $f(x) = \frac{2x}{x + 1}$

21. $f(x) = \frac{5x}{x^2 - 2x - 8}$

22. $f(x) = \frac{x^2 + x}{x(x - 1)}$

In exercises 23–32, find the absolute maximum value and the absolute minimum value, if any, of the given function.

23. $f(x) = 2x^2 + 3x - 2$

24. $g(x) = x^{2/3}$

25. $g(t) = \sqrt{25 - t^2}$

26. $f(x) = \frac{1}{3}x^3 - x^2 + x + 1$ on $[0, 2]$

27. $h(t) = t^3 - 6t^2$ on $[2, 5]$

28. $g(x) = \frac{x}{x^2 + 1}$ on $[0, 5]$

29. $f(x) = x - \dfrac{1}{x}$ on $[1, 3]$

30. $h(t) = 8t - \dfrac{1}{t^2}$ on $[1, 3]$

31. $f(s) = s\sqrt{1 - s^2}$ on $[-1, 1]$

32. $f(x) = \dfrac{x^2}{x - 1}$ on $[-1, 3]$

33. Odyssey Travel Agency's monthly profit (in thousands of dollars) depends on the amount of money x spent on advertising per month according to the rule

$$P(x) = -x^2 + 8x + 20$$

where x is also measured in thousands of dollars. What should Odyssey's monthly advertising budget be in order to maximize its monthly profits?

34. The Department of the Interior of an African country began to record an index of environmental quality to measure progress or decline in the environmental quality of its wildlife. The index for the years 1984 through 1994 is approximated by the function

$$I(t) = \dfrac{50t^2 + 600}{t^2 + 10} \qquad (0 \le t \le 10)$$

a. Compute $I'(t)$ and show that $I(t)$ is decreasing on the interval $(0, 10)$.
b. Compute $I''(t)$. Study the concavity of the graph of I.
c. Sketch the graph of I.
d. Interpret your results.

35. The weekly demand for video discs manufactured by the Herald Record Company is given by

$$p = -0.0005x^2 + 60$$

where p denotes the unit price in dollars and x denotes the quantity demanded. The weekly total cost function associated with producing these discs is given by

$$C(x) = -0.001x^2 + 18x + 4000$$

where $C(x)$ denotes the total cost incurred in pressing x discs. Find the production level that will yield a maximum profit for the manufacturer.
[*Hint:* Use the quadratic formula.]

36. The total monthly cost (in dollars) incurred by the Carlota Music Company in manufacturing x units of its Professional Series guitars is given by the function

$$C(x) = 0.001x^2 + 100x + 4000$$

a. Find the average cost function $\overline{C}$.
b. Determine the production level that will result in the smallest average production cost.

37. The average worker at Wakefield Avionics, Inc., can assemble

$$N(t) = -2t^3 + 12t^2 + 2t \qquad (0 \le t \le 4)$$

ready-to-fly radio-controlled model airplanes t hours into the 8 A.M. to 12 noon morning shift. At what time during this shift is the average worker performing at peak efficiency?

38. You wish to construct a closed rectangular box that has a volume of 4 cubic feet. The length of the base of the box will be twice as long as its width. The material for the top and bottom of the box costs 30 cents per square foot. The material for the sides of the box costs 20 cents per square foot. Find the dimensions of the least expensive box that can be constructed.

39. The Lehen Vinters Company imports a certain brand of beer. The demand, which may be assumed to be uniform, is 800,000 cases per year. The cost of ordering a shipment of beer is $500, and the cost of storing each case of beer for a year is $2. Determine how many cases of beer should be in each shipment if the ordering and storage costs are to be kept to a minimum. (Assume that each shipment of beer arrives just as the previous one has been sold.)

40. Let

$$f(x) = \begin{cases} x^3 + 1 & \text{if } x \ne 0 \\ 2 & \text{if } x = 0 \end{cases}$$

a. Compute $f'(x)$ and show that it does not change sign as we move across $x = 0$.
b. Show that f has a relative maximum at $x = 0$. Does this contradict the First Derivative Test? Explain your answer.

The exponential function is, without doubt, the most important function in mathematics and its applications. After a brief introduction to the exponential function and its *inverse*, the logarithmic function, we learn how to differentiate such functions. This lays the foundation for exploring the many applications involving exponential functions. For example, we look at the role played by exponential functions in computing earned interest on a bank account, in studying the growth of a bacteria population in the laboratory, in studying the way radioactive matter decays, in studying the rate at which a factory worker learns a certain process, and in studying the rate at which a communicable disease is spread over time.

How many bacteria will there be in a culture at the end of a certain period of time? How fast will the bacteria population be growing at the end of that time? Example 1, page 414, answers these questions.

EXPONENTIAL AND LOGARITHMIC FUNCTIONS

Exponential Functions and Their Graphs

Suppose that you deposit a sum of $1000 in an account earning interest at the rate of 10% per year *compounded continuously* (the way most financial institutions compute interest). Then the accumulated amount at the end of t years ($0 \leq t \leq 20$) is described by the function f, whose graph appears in Figure 5.1.* Such a function is called an *exponential function*. Observe that the graph of f rises rather slowly at first but very rapidly as time goes by. For purposes of comparison, we have also shown the graph of the function $y = g(t) = 1000(1 + 0.10t)$, giving the accumulated amount for the same principal ($1000) but earning *simple* interest at the rate of 10% per year. The moral of the story: It is never too early to save.

Exponential functions play an important role in many real-world applications, as you will see throughout this chapter.

Figure 5.1
Under continuous compounding, a sum of money grows exponentially.

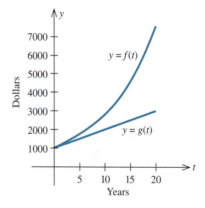

Observe that whenever b is a positive number and n is any real number, the expression b^n is a real number. This enables us to define an *exponential function* as follows:

EXPONENTIAL FUNCTION

The function defined by

$$f(x) = b^x \qquad (b > 0, b \neq 1)$$

is called an **exponential function with base b and exponent x.** The domain of f is the set of all real numbers.

○ ○ ○

* We will derive the rule for f in Section 5.3.

For example, the exponential function with base 2 is the function

$$f(x) = 2^x$$

with domain $(-\infty, \infty)$. The values of $f(x)$ for selected values of x follow:

$$f(3) = 2^3 = 8, \quad f\left(\frac{3}{2}\right) = 2^{3/2} = 2 \cdot 2^{1/2} = 2\sqrt{2}, \quad f(0) = 2^0 = 1,$$

$$f(-1) = 2^{-1} = \frac{1}{2}, \quad \text{and} \quad f\left(-\frac{2}{3}\right) = 2^{-2/3} = \frac{1}{2^{2/3}} = \frac{1}{\sqrt[3]{4}}$$

Computations involving exponentials are facilitated by the laws of exponents. These laws were stated in Section 1.1, and you might want to review the material there. For convenience, however, we will restate these laws.

THE LAWS OF EXPONENTS

Let a and b be positive numbers and let x and y be real numbers. Then,

1. $b^x \cdot b^y = b^{x+y}$

2. $\dfrac{b^x}{b^y} = b^{x-y}$

3. $(b^x)^y = b^{xy}$

4. $(ab)^x = a^x b^x$

5. $\left(\dfrac{a}{b}\right)^x = \dfrac{a^x}{b^x}$

The use of the laws of exponents is illustrated in the next example.

EXAMPLE 1

a. $16^{7/4} \cdot 16^{-1/2} = 16^{7/4 - 1/2} = 16^{5/4} = 2^5 = 32$ (Law 1)

b. $\dfrac{8^{5/3}}{8^{-1/3}} = 8^{5/3 - (-1/3)} = 8^2 = 64$ (Law 2)

c. $(64^{4/3})^{-1/2} = 64^{(4/3)(-1/2)} = 64^{-2/3}$

$$= \frac{1}{64^{2/3}} = \frac{1}{(64^{1/3})^2} = \frac{1}{4^2} = \frac{1}{16}$$ (Law 3)

d. $(16 \cdot 81)^{-1/4} = 16^{-1/4} \cdot 81^{-1/4} = \dfrac{1}{16^{1/4}} \cdot \dfrac{1}{81^{1/4}} = \dfrac{1}{2} \cdot \dfrac{1}{3} = \dfrac{1}{6}$ (Law 4)

e. $\left(\dfrac{3^{1/2}}{2^{1/3}}\right)^4 = \dfrac{3^{4/2}}{2^{4/3}} = \dfrac{9}{2^{4/3}}$ (Law 5)   ○ ○ ○

EXAMPLE 2   Let $f(x) = 2^{2x-1}$. Find the value of x for which $f(x) = 16$.

Solution   We want to solve the equation

$$2^{2x-1} = 16 = 2^4$$

But this equation holds if and only if

$$2x - 1 = 4 \qquad (b^m = b^n \Rightarrow m = n)$$

giving $x = \frac{5}{2}$.   ○ ○ ○

Exponential functions play an important role in mathematical analysis. Because of their special characteristics, they are some of the most useful functions and are found in virtually every field where mathematics is applied. To mention a few examples: Under ideal conditions the number of bacteria present at any time t in a culture may be described by an exponential function of t; radioactive substances decay over time in accordance with an "exponential" law of decay; money left on fixed deposit and earning compound interest grows exponentially; and some of the most important distribution functions encountered in statistics are exponential.

Let us begin our investigation into the properties of exponential functions by studying their graphs.

EXAMPLE 3   Sketch the graph of the exponential function $y = 2^x$.

Solution   First, as discussed earlier, the domain of the exponential function $y = f(x) = 2^x$ is the set of real numbers. Next, putting $x = 0$ gives $y = 2^0 = 1$, the y-intercept of f. There is no x-intercept since there is no value of x for which $y = 0$. To find the range of f, consider the following table of values:

x	-5	-4	-3	-2	-1	0	1	2	3	4	5
y	1/32	1/16	1/8	1/4	1/2	1	2	4	8	16	32

Figure 5.2
The graph of $y = 2^x$.

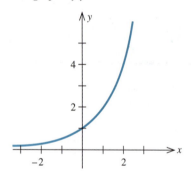

We see from these computations that 2^x decreases and approaches zero as x decreases without bound and that 2^x increases without bound as x increases without bound. Thus, the range of f is the interval $(0, \infty)$—that is, the set of positive real numbers. Finally, we sketch the graph of $y = f(x) = 2^x$ in Figure 5.2.

EXAMPLE 4   Sketch the graph of the exponential function $y = (1/2)^x$.

Figure 5.3
The graph of $y = (1/2)^x$.

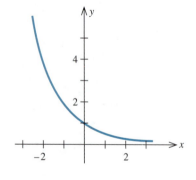

Solution   The domain of the exponential function $y = (1/2)^x$ is the set of all real numbers. The y-intercept is $(1/2)^0 = 1$; there is no x-intercept since there is no value of x for which $y = 0$. From the following table of values

x	-5	-4	-3	-2	-1	0	1	2	3	4	5
y	32	16	8	4	2	1	1/2	1/4	1/8	1/16	1/32

we deduce that $(1/2)^x = 1/2^x$ increases without bound as x decreases without bound and that $(1/2)^x$ decreases and approaches zero as x increases without bound. Thus, the range of f is the interval $(0, \infty)$. The graph of $y = f(x) = (1/2)^x$ is sketched in Figure 5.3.

The functions $y = 2^x$ and $y = (1/2)^x$, whose graphs you studied in Examples 3 and 4, are special cases of the exponential function $y = f(x) = b^x$, obtained by setting $b = 2$ and $b = 1/2$, respectively. In general, the exponential function $y = b^x$ with $b > 1$ has a graph similar to $y = 2^x$, whereas the graph of $y = b^x$ for $0 < b < 1$ is similar to that of $y = (1/2)^x$ (exercises 27 and 28). When $b = 1$, the function $y = b^x$ reduces to the constant function $y = 1$. For comparison, the graphs of all three functions are sketched in Figure 5.4.

Figure 5.4
$y = b^x$ is an increasing function of x if $b > 1$, a constant function if $b = 1$, and a decreasing function if $0 < b < 1$.

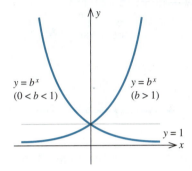

PROPERTIES OF THE EXPONENTIAL FUNCTION	The exponential function $y = b^x$ $(b > 0, b \neq 1)$ has the following properties:

1. Its domain is $(-\infty, \infty)$.

2. Its range is $(0, \infty)$.

3. Its graph passes through the point $(0, 1)$.

4. It is continuous on $(-\infty, \infty)$.

5. It is increasing on $(-\infty, \infty)$ if $b > 1$ and decreasing on $(-\infty, \infty)$ if $b < 1$.

The Base e

Exponential functions to the base e, where e is an irrational number whose value is 2.7182818 . . . , play an important role in both theoretical and applied problems. It can be shown, although we will not do so here, that

$$e = \lim_{m \to \infty} \left(1 + \frac{1}{m}\right)^m \tag{1}$$

However, you may convince yourself of the plausibility of this definition of the number e by examining Table 5.1, which may be constructed with the help of a calculator.

Table 5.1

m	10	100	1000	10,000	100,000	1,000,000
$\left(1 + \dfrac{1}{m}\right)^m$	2.59374	2.70481	2.71692	2.71815	2.71827	2.71828

EXPLORING WITH TECHNOLOGY

In order to obtain a visual confirmation of the fact that the expression $(1 + 1/m)^m$ approaches the number $e = 2.71828\ldots$ as m increases without bound, plot the graph of $f(x) = (1 + 1/x)^x$ in a suitable viewing rectangle and observe that $f(x)$ approaches $2.71828\ldots$ as x increases without bound. Use **ZOOM** and **TRACE** to find the value of $f(x)$ for large values of x.

○ ○ ○

Figure 5.5
The graph of $y = e^x$.

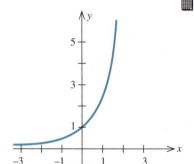

EXAMPLE 5   Sketch the graph of the function $y = e^x$.

Solution   Since $e > 1$, it follows from our previous discussion that the graph of $y = e^x$ is similar to the graph of $y = 2^x$ (see Figure 5.2). With the aid of a calculator we obtain the following table.

x	-3	-2	-1	0	1	2	3
y	0.05	0.14	0.37	1	2.72	7.39	20.09

The graph of $y = e^x$ is sketched in Figure 5.5.   ◦ ◦ ◦

Next, we consider another exponential function to the base e that is closely related to the previous function and is particularly useful in constructing models that describe "exponential decay."

Figure 5.6
The graph of $y = e^{-x}$.

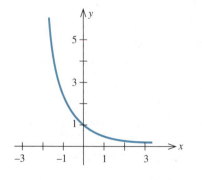

EXAMPLE 6   Using a calculator, sketch the graph of the function $y = e^{-x}$.

Solution   Since $e > 1$, it follows that $0 < 1/e < 1$, so $f(x) = e^{-x} = 1/e^x = (1/e)^x$ is an exponential function with base less than 1. Therefore it has a graph similar to that of the exponential function $y = (1/2)^x$. As before, we construct the following table of values of $y = e^{-x}$ for selected values of x.

x	-3	-2	-1	0	1	2	3
y	20.09	7.39	2.72	1	0.37	0.14	0.05

Using this table, we sketch the graph of $y = e^{-x}$ in Figure 5.6.   ◦ ◦ ◦

SELF-CHECK EXERCISES 5.1

1. Solve the equation $2^{2x+1} \cdot 2^{-3} = 2^{x-1}$.

2. Sketch the graph of $y = e^{0.4x}$.

Solutions to Self-Check Exercises 5.1 can be found on page 370.

5.1 EXERCISES

In exercises 1–8, evaluate the expression.

1. a. $4^{-3} \cdot 4^5$ **b.** $3^{-3} \cdot 3^6$

2. a. $(2^{-1})^3$ **b.** $(3^{-2})^3$

3. a. $9(9)^{-1/2}$ **b.** $5(5)^{-1/2}$

4. a. $[(-\frac{1}{2})^3]^{-2}$ **b.** $[(-\frac{1}{3})^2]^{-3}$

5. a. $\dfrac{(-3)^4(-3)^5}{(-3)^8}$ **b.** $\dfrac{(2^{-4})(2^6)}{2^{-1}}$

6. a. $3^{1/4} \cdot (9)^{-5/8}$ **b.** $2^{3/4} \cdot (4)^{-3/2}$

7. a. $\dfrac{5^{3.3} \cdot 5^{-1.6}}{5^{-0.3}}$ **b.** $\dfrac{4^{2.7} \cdot 4^{-1.3}}{4^{-0.4}}$

8. a. $(\frac{1}{16})^{-1/4}(\frac{27}{64})^{-1/3}$ **b.** $(\frac{8}{27})^{-1/3}(\frac{81}{256})^{-1/4}$

In exercises 9–16, simplify the expression.

9. a. $(64x^9)^{1/3}$ **b.** $(25x^3y^4)^{1/2}$

10. a. $(2x^3)(-4x^{-2})$ **b.** $(4x^{-2})(-3x^5)$

11. a. $\dfrac{6a^{-5}}{3a^{-3}}$ **b.** $\dfrac{4b^{-4}}{12b^{-6}}$

12. a. $y^{-3/2}y^{5/3}$ **b.** $x^{-3/5}x^{8/3}$

13. a. $(2x^3y^2)^3$ **b.** $(4x^2y^2z^3)^2$

14. a. $(x^{r/s})^{s/r}$ **b.** $(x^{-b/a})^{-a/b}$

15. a. $\dfrac{5^0}{(2^{-3}x^{-3}y^2)^2}$ **b.** $\dfrac{(x+y)(x-y)}{(x-y)^0}$

16. a. $\dfrac{(a^m \cdot a^{-n})^{-2}}{(a^{m+n})^2}$ **b.** $\left(\dfrac{x^{2n-2}y^{2n}}{x^{5n+1}y^{-n}}\right)^{1/3}$

In exercises 17–26, solve the equation for x.

17. $6^{2x} = 6^4$ **18.** $5^{-x} = 5^3$

19. $3^{3x-4} = 3^5$ **20.** $10^{2x-1} = 10^{x+3}$

21. $(2.1)^{x+2} = (2.1)^5$ **22.** $(-1.3)^{x-2} = (-1.3)^{2x+1}$

23. $8^x = \left(\dfrac{1}{32}\right)^{x-2}$ **24.** $3^{x-x^2} = \dfrac{1}{9^x}$

25. $3^{2x} - 12 \cdot 3^x + 27 = 0$

26. $2^{2x} - 4 \cdot 2^x + 4 = 0$

In exercises 27–35, sketch the graphs of the given functions on the same axes. A calculator is recommended for these exercises.

27. $y = 2^x$, $y = 3^x$, and $y = 4^x$

28. $y = (\frac{1}{2})^x$, $y = (\frac{1}{3})^x$, and $y = (\frac{1}{4})^x$

29. $y = 2^{-x}$, $y = 3^{-x}$, and $y = 4^{-x}$

30. $y = 4^{0.5x}$ and $y = 4^{-0.5x}$

31. $y = 4^{0.5x}$, $y = 4^x$, and $y = 4^{2x}$

32. $y = e^x$, $y = 2e^x$, and $y = 3e^x$

33. $y = e^{0.5x}$, $y = e^x$, and $y = e^{1.5x}$

34. $y = e^{-0.5x}$, $y = e^{-x}$, and $y = e^{-1.5x}$

35. $y = 0.5e^{-x}$, $y = e^{-x}$, and $y = 2e^{-x}$

USING TECHNOLOGY

Although the proof is outside the scope of this book, it can be proved that an exponential function of the form $f(x) = b^x$, where $b > 1$, will ultimately grow faster than the power function $g(x) = x^n$ for *any* positive real number n. To give a visual demonstration of this result for the special case of the exponential function $f(x) = e^x$, we can use a graphing calculator to plot the graphs of both f and g (for selected values of n) on the same set of axes in an appropriate viewing rectangle and observe that the graph of f ultimately lies above that of g.

EXAMPLE I Use a graphing utility to plot the graphs of (a) $f(x) = e^x$ and $g(x) = x^3$ on the same set of axes in the viewing rectangle $[0, 6] \times [0, 250]$ and (b) $f(x) = e^x$ and $g(x) = x^5$ in the viewing rectangle $[0, 20] \times [0, 1,000,000]$.

Solution

a. The graphs of $f(x) = e^x$ and $g(x) = x^3$ in the viewing rectangle $[0, 6] \times [0, 250]$ are shown in Figure T1a.

b. The graphs of $f(x) = e^x$ and $g(x) = x^5$ in the viewing rectangle $[0, 20] \times [0, 1,000,000]$ are shown in Figure T1b.

Figure T1

(a) The graphs of $f(x) = e^x$ and $g(x) = x^3$ in the viewing rectangle $[0, 6] \times [0, 250]$.

(b) The graphs of $f(x) = e^x$ and $g(x) = x^5$ in the viewing rectangle $[0, 20] \times [0, 1,000,000]$.

In the exercises that follow, you are asked to use a graphing utility to reveal the properties of exponential functions.

EXERCISES

In exercises 1 and 2, use a graphing utility to plot the graphs of the functions f and g on the same set of axes in the specified viewing rectangle.

1. $f(x) = e^x$ and $g(x) = x^2$; $[0, 4] \times [0, 30]$

2. $f(x) = e^x$ and $g(x) = x^4$; $[0, 15] \times [0, 20{,}000]$

In exercises 3 and 4, use a graphing utility to plot the graphs of the functions f and g on the same set of axes in an appropriate viewing rectangle to demonstrate that f ultimately grows faster than g. [Note: Your answer will not be unique.]

3. $f(x) = 2^x$ and $g(x) = x^{2.5}$

4. $f(x) = 3^x$ and $g(x) = x^3$

5. Use a graphing utility to plot the graphs of $f(x) = 2^x$, $g(x) = 3^x$, and $h(x) = 4^x$ on the same set of axes in the viewing rectangle $[0, 5] \times [0, 100]$. Comment on the relationship between the base b and the growth of the function $f(x) = b^x$.

6. Use a graphing utility to plot the graphs of $f(x) = (1/2)^x$, $g(x) = (1/3)^x$, and $h(x) = (1/4)^x$ on the same set of axes in the viewing rectangle $[0, 4] \times [0, 1]$. Comment

on the relationship between the base b and the growth of the function $f(x) = b^x$.

7. Use a graphing utility to plot the graphs of $f(x) = e^x$, $g(x) = 2e^x$, and $h(x) = 3e^x$ on the same set of axes in the viewing rectangle $[-3, 3] \times [0, 10]$. Comment on the role played by the constant k in the graph of $f(x) = ke^x$.

8. Use a graphing utility to plot the graphs of $f(x) = -e^x$, $g(x) = -2e^x$, and $h(x) = -3e^x$ on the same set of axes in the viewing rectangle $[-3, 3] \times [-10, 0]$. Comment on the role played by the constant k in the graph of $f(x) = ke^x$.

9. Use a graphing utility to plot the graphs of $f(x) = e^{0.5x}$, $g(x) = e^x$, and $h(x) = e^{1.5x}$ on the same set of axes in the viewing rectangle $[-2, 2] \times [0, 4]$. Comment on the role played by the constant k in the graph of $f(x) = e^{kx}$.

10. Use a graphing utility to plot the graphs of $f(x) = e^{-0.5x}$, $g(x) = e^{-x}$, and $h(x) = e^{-1.5x}$ on the same set of axes in the viewing rectangle $[-2, 2] \times [0, 4]$. Comment on the role played by the constant k in the graph of $f(x) = e^{kx}$.

SOLUTIONS TO SELF-CHECK EXERCISES 5.1

1. $2^{2x+1} \cdot 2^{-3} = 2^{x-1}$

$\dfrac{2^{2x+1}}{2^{x-1}} \cdot 2^{-3} = 1$ (Dividing both sides by 2^{x-1})

$2^{(2x+1)-(x-1)-3} = 1$

$2^{x-1} = 1$

This is true if and only if $x - 1 = 0$ or $x = 1$.

2. We first construct the following table of values.

x	-3	-2	-1	0	1	2	3	4
$y = e^{0.4x}$	0.3	0.5	0.7	1	1.5	2.2	3.3	5

Next, we plot these points and join them by a smooth curve to obtain the graph of f shown in the accompanying figure.

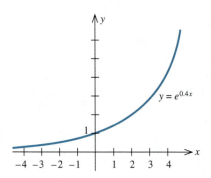

5.2 **LOGARITHMIC FUNCTIONS**

Logarithms

You are already familiar with exponential equations of the form

$$b^y = x \qquad (b > 0, b \neq 1)$$

where the variable x is expressed in terms of a real number b and a variable y. But what about solving this same equation for y? You may recall from your study of algebra that the number y is called the **logarithm of x to the base b** and is denoted by **$\log_b x$.** It is the exponent to which the base b must be raised in order to obtain the number x.

THE LOGARITHM OF x TO THE BASE b	$y = \log_b x$ if and only if $x = b^y$ $(x > 0)$

 Observe that the logarithm $\log_b x$ is defined only for positive values of x.

EXAMPLE 1

a. $\log_{10} 100 = 2$ since $100 = 10^2$

b. $\log_5 125 = 3$ since $125 = 5^3$

c. $\log_3 \dfrac{1}{27} = -3$ since $\dfrac{1}{27} = \dfrac{1}{3^3} = 3^{-3}$

d. $\log_{20} 20 = 1$ since $20 = 20^1$ ◦ ◦ ◦

EXAMPLE 2   Solve each of the following equations for x.

a. $\log_3 x = 4$ **b.** $\log_{16} 4 = x$ **c.** $\log_x 8 = 3$

Solution

a. By definition, $\log_3 x = 4$ implies $x = 3^4 = 81$.

b. $\log_{16} 4 = x$ is equivalent to $4 = 16^x = (4^2)^x = 4^{2x}$, from which we deduce that $x = \frac{1}{2}$.

c. Referring once again to the definition, we see that the equation $\log_x 8 = 3$ is equivalent to the equation $8 = x^3$, so $x = 2$.   ◦ ◦ ◦

The two widely used systems of logarithms are the system of **common logarithms,** which uses the number 10 as the base, and the system of **natural logarithms,** which uses the irrational number $e = 2.71828\ldots$ as the base. Also, it is standard practice to write **log** for $\log_{10}$ and **ln** for $\log_e$.

LOGARITHMIC NOTATION	$\log x = \log_{10} x$ (Common logarithm) $\ln x = \log_e x$ (Natural logarithm)

The system of natural logarithms is widely used in theoretical work. Using natural logarithms rather than logarithms to other bases often leads to simpler expressions.

Laws of Logarithms

Computations involving logarithms are facilitated by the following **laws of logarithms.**

LAWS OF LOGARITHMS

If m and n are positive numbers, then

1. $\log_b mn = \log_b m + \log_b n$

2. $\log_b \dfrac{m}{n} = \log_b m - \log_b n$

3. $\log_b m^n = n \log_b m$

4. $\log_b 1 = 0$

5. $\log_b b = 1$

You will be asked to prove these laws in exercises 49–51. Their derivations are based on the definition of a logarithm and the corresponding laws of exponents. The following examples illustrate the properties of logarithms.

EXAMPLE 3

a. $\log(2 \cdot 3) = \log 2 + \log 3$

b. $\ln \dfrac{5}{3} = \ln 5 - \ln 3$ **c.** $\log \sqrt{7} = \log 7^{1/2} = \dfrac{1}{2} \log 7$

d. $\log_5 1 = 0$ **e.** $\log_{45} 45 = 1$ ○ ○ ○

EXAMPLE 4
Given that $\log 2 \approx 0.3010$, $\log 3 \approx 0.4771$, and $\log 5 \approx 0.6990$, use the laws of logarithms to find

a. $\log 15$ **b.** $\log 7.5$ **c.** $\log 81$ **d.** $\log 50$

Solution

a. Note that $15 = 3 \cdot 5$, so by Law 1 for logarithms,

$$\log 15 = \log 3 \cdot 5$$
$$= \log 3 + \log 5$$
$$\approx 0.4771 + 0.6990$$
$$= 1.1761$$

b. Observing that $7.5 = 15/2 = (3 \cdot 5)/2$, we apply Laws 1 and 2, obtaining

$$\log 7.5 = \log \frac{(3)(5)}{2}$$
$$= \log 3 + \log 5 - \log 2$$
$$\approx 0.4771 + 0.6990 - 0.3010$$
$$= 0.8751$$

c. Since $81 = 3^4$, we apply Law 3 to obtain

$$\log 81 = \log 3^4$$
$$= 4 \log 3$$
$$\approx 4(0.4771)$$
$$= 1.9084$$

d. We write $50 = 5 \cdot 10$ and find

$$\log 50 = \log(5)(10)$$
$$= \log 5 + \log 10$$
$$\approx 0.6990 + 1 \qquad \text{(Using Law 5)}$$
$$= 1.6990$$

EXAMPLE 5   Expand and simplify the following expressions:

a. $\log_3 x^2 y^3$ **b.** $\log_2 \dfrac{x^2 + 1}{2^x}$ **c.** $\ln \dfrac{x^2 \sqrt{x^2 - 1}}{e^x}$

Solution

a. $\log_3 x^2 y^3 = \log_3 x^2 + \log_3 y^3$ (Law 1)
$\qquad\qquad\quad = 2 \log_3 x + 3 \log_3 y$ (Law 3)

b. $\log_2 \dfrac{x^2 + 1}{2^x} = \log_2(x^2 + 1) - \log_2 2^x$ (Law 2)

$\qquad\qquad\quad = \log_2(x^2 + 1) - x \log_2 2$ (Law 3)
$\qquad\qquad\quad = \log_2(x^2 + 1) - x$ (Law 5)

c. $\ln \dfrac{x^2 \sqrt{x^2 - 1}}{e^x} = \ln \dfrac{x^2(x^2 - 1)^{1/2}}{e^x}$ (Rewriting)

$\qquad\qquad = \ln x^2 + \ln(x^2 - 1)^{1/2} - \ln e^x$ (Laws 1 and 2)
$\qquad\qquad = 2 \ln x + \tfrac{1}{2} \ln(x^2 - 1) - x \ln e$ (Law 3)
$\qquad\qquad = 2 \ln x + \tfrac{1}{2} \ln(x^2 - 1) - x$ (Law 5)

Logarithmic Functions and Their Graphs

The definition of the logarithm implies that if b and n are positive numbers and b is different from 1, then the expression $\log_b n$ is a real number. This enables us to define a *logarithmic function* as follows:

LOGARITHMIC FUNCTION

The function defined by

$$f(x) = \log_b x \qquad (b > 0, \, b \neq 1)$$

is called the **logarithmic function with base** b. The domain of f is the set of all positive numbers.

Figure 5.7
The points (u, v) and (v, u) are mirror reflections of each other.

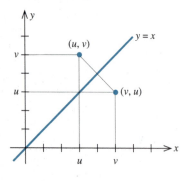

One easy way to obtain the graph of the logarithmic function $y = \log_b x$ is to construct a table of values of the logarithm (base b). However, another method—and a more instructive one—is based on exploiting the intimate relationship between logarithmic and exponential functions.

If a point (u, v) lies on the graph of $y = \log_b x$, then

$$v = \log_b u$$

But we can also write this equation in exponential form as

$$u = b^v$$

So the point (v, u) also lies on the graph of the function $y = b^x$. Let us look at the relationship between the points (u, v) and (v, u) and the line $y = x$ (Figure 5.7). If we think of the line $y = x$ as a mirror, then the point (v, u) is the mirror reflection of the point (u, v). Similarly, the point (u, v) is a mirror reflection of the point (v, u). We can take advantage of this relationship to help us draw the graph of logarithmic functions. For example, if we wish to draw the graph of $y = \log_b x$, where $b > 1$, then we need only draw the mirror reflection of the graph of $y = b^x$ with respect to the line $y = x$ (Figure 5.8).

Figure 5.8
The graphs of $y = b^x$ and $y = \log_b x$ are mirror reflections of each other.

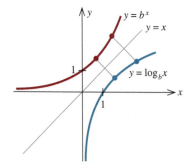

You may discover the following properties of the logarithmic function by taking the reflection of the graph of an appropriate exponential function (exercises 31 and 32).

PROPERTIES OF THE LOGARITHMIC FUNCTION

The logarithmic function $y = \log_b x$ ($b > 0$, $b \neq 1$) has the following properties:

1. Its domain is $(0, \infty)$.

2. Its range is $(-\infty, \infty)$.

3. Its graph passes through the point $(1, 0)$.

4. It is continuous on $(0, \infty)$.

5. It is increasing on $(0, \infty)$ if $b > 1$ and decreasing on $(0, \infty)$ if $b < 1$.

EXAMPLE 6 Sketch the graph of the function $y = \ln x$.

Solution We first sketch the graph of $y = e^x$. Then the required graph is obtained by tracing the mirror reflection of the graph of $y = e^x$ with respect to the line $y = x$ (Figure 5.9).

Figure 5.9
The graph of $y = \ln x$ is the mirror reflection of the graph of $y = e^x$.

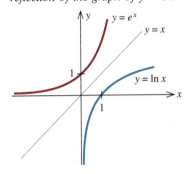

Properties Relating the Exponential and Logarithmic Functions

We made use of the relationship that exists between the exponential function $f(x) = e^x$ and the logarithmic function $g(x) = \ln x$ when we sketched the graph of g in Example 6. This relationship is further described by the following properties, which are an immediate consequence of the definition of the logarithm of a number.

| PROPERTIES RELATING e^x AND $\ln x$ | $e^{\ln x} = x \quad x > 0$ | (2) |
| | $\ln e^x = x \quad$ for any real number x | (3) |

(Try to verify these properties.)
From properties 2 and 3, we conclude that the composite function

$$(f \circ g)(x) = f[g(x)]$$
$$= e^{\ln x} = x$$

and
$$(g \circ f)(x) = g[f(x)]$$
$$= \ln e^x = x$$

Thus,

$$f[g(x)] = g[f(x)]$$
$$= x$$

Any two functions f and g that satisfy this relationship are said to be **inverses** of each other. Note that the function f undoes what the function g does, and vice versa, so that the composition of the two functions in any order results in the identity function $F(x) = x$.

The relationships expressed in equations (2) and (3) are useful in solving equations that involve exponentials and logarithms.

 EXAMPLE 7   Solve the equation $2e^{x+2} = 5$.

Solution   We first divide both sides of the equation by 2 to obtain

$$e^{x+2} = \frac{5}{2} = 2.5$$

Next, taking the natural logarithm of each side of the equation and using equation (3), we have

$$\ln e^{x+2} = \ln 2.5$$
$$x + 2 = \ln 2.5$$
$$x = -2 + \ln 2.5$$
$$\approx -1.08$$

○ ○ ○

 EXAMPLE 8   Solve the equation $5 \ln x + 3 = 0$.

Solution   Adding -3 to both sides of the equation leads to

$$5 \ln x = -3$$
$$\ln x = -\frac{3}{5} = -0.6$$

and so

$$e^{\ln x} = e^{-0.6}$$

Using equation (2), we conclude that

$$x = e^{-0.6}$$
$$\approx 0.55$$

○ ○ ○

Consider the equation $y = y_0 b^{kx}$, where y_0 and k are positive constants and $b > 0$, $b \neq 1$. Suppose we want to express y in the form $y = y_0 e^{px}$. Use the laws of logarithms to show that $p = k \ln b$ and hence that $y = y_0 e^{(k \ln b)x}$ is an alternative form of $y = y_0 b^{kx}$ using the base e.

SELF-CHECK EXERCISES 5.2

1. Sketch the graph of $y = 3^x$ and $y = \log_3 x$ on the same set of axes.

2. Solve the equation $3e^{x+1} - 2 = 4$.

Solutions to Self-Check Exercises 5.2 can be found on page 378.

5.2 EXERCISES

In exercises 1–10, express the given equation in logarithmic form.

1. $2^6 = 64$

2. $3^5 = 243$

3. $3^{-2} = \frac{1}{9}$

4. $5^{-3} = \frac{1}{125}$

5. $\left(\frac{1}{3}\right)^1 = \frac{1}{3}$

6. $\left(\frac{1}{2}\right)^{-4} = 16$

7. $32^{3/5} = 8$

8. $81^{3/4} = 27$

9. $10^{-3} = 0.001$

10. $16^{-1/4} = 0.5$

In exercises 11–16, use the facts that log 3 = 0.4771 and log 4 = 0.6021 to find the value of the given logarithm.

11. $\log 12$

12. $\log \frac{3}{4}$

13. $\log 16$

14. $\log \sqrt{3}$

15. $\log 48$

16. $\log \frac{1}{300}$

In exercises 17–26, use the laws of logarithms to simplify the given expression.

17. $\log x(x + 1)^4$

18. $\log x(x^2 + 1)^{-1/2}$

19. $\log \dfrac{\sqrt{x + 1}}{x^2 + 1}$

20. $\ln \dfrac{e^x}{1 + e^x}$

21. $\ln xe^{-x^2}$

22. $\ln x(x + 1)(x + 2)$

23. $\ln \dfrac{x^{1/2}}{x^2\sqrt{1 + x^2}}$

24. $\ln \dfrac{x^2}{\sqrt{x}(1 + x)^2}$

25. $\ln x^x$

26. $\ln x^{x^2+1}$

In exercises 27–30, sketch the graph of the given equation.

27. $y = \log_3 x$

28. $y = \log_{1/3} x$

29. $y = \ln 2x$

30. $y = \ln \frac{1}{2}x$

In exercises 31 and 32, sketch the graphs of the given equations on the same coordinate axes.

31. $y = 2^x$ and $y = \log_2 x$

32. $y = e^{3x}$ and $y = \ln 3x$

In exercises 33–42, use logarithms to solve the given equation for t.

33. $e^{0.4t} = 8$

34. $\frac{1}{3}e^{-3t} = 0.9$

35. $5e^{-2t} = 6$

36. $4e^{t-1} = 4$

37. $2e^{-0.2t} - 4 = 6$

38. $12 - e^{0.4t} = 3$

39. $\dfrac{50}{1 + 4e^{0.2t}} = 20$

40. $\dfrac{200}{1 + 3e^{-0.3t}} = 100$

41. $A = Be^{-t/2}$

42. $\dfrac{A}{1 + Be^{t/2}} = C$

43. Blood Pressure   A normal child's systolic blood pressure may be approximated by the function

$$p(x) = m(\ln x) + b$$

where $p(x)$ is measured in millimeters of mercury, x is measured in pounds, and m and b are constants. Given that $m = 19.4$ and $b = 18$, determine the systolic blood pressure of a child who weighs 92 pounds.

44. Magnitude of Earthquakes   On the Richter scale, the magnitude R of an earthquake is given by the formula

$$R = \log \frac{I}{I_0}$$

where I is the intensity of the earthquake being measured and I_0 is the standard reference intensity.
a. Express the intensity I of an earthquake of magnitude $R = 5$ in terms of the standard intensity I_0.
b. Express the intensity I of an earthquake of magnitude $R = 8$ in terms of the standard intensity I_0. How many times greater is the intensity of an earthquake of magnitude 8 than one of magnitude 5?
c. In modern times, the greatest loss of life attributable to an earthquake occurred in eastern China in 1976. Known as the Tangshan earthquake, it registered 8.2 on the Richter scale. How does the intensity of this

earthquake compare with the intensity of an earthquake of magnitude $R = 5$?

45. Sound Intensity   The relative loudness of a sound D of intensity I is measured in decibels, where

$$D = 10 \log \frac{I}{I_0}$$

and I_0 is the standard threshold of audibility.
a. Express the intensity I of a 30-decibel sound (the sound level of normal conversation) in terms of I_0.
b. Determine how many times greater the intensity of an 80-decibel sound (rock music) is than that of a 30-decibel sound.
c. Prolonged noise above 150 decibels causes immediate and permanent deafness. How does the intensity of a 150-decibel sound compare with the intensity of an 80-decibel sound?

46. Barometric Pressure   Halley's Law states that the barometric pressure (in inches of mercury) at an altitude of x miles above sea level is approximately given by the equation

$$p(x) = 29.92e^{-0.2x} \qquad (x \geq 0)$$

If the barometric pressure as measured by a hot-air balloonist is 20 inches of mercury, what is the balloonist's altitude?

47. Forensic Science   Forensic scientists use the following law to determine the time of death of accident or murder victims. If T denotes the temperature of a body t hours after death, then

$$T = T_0 + (T_1 - T_0)(0.97)^t$$

where T_0 is the air temperature and T_1 is the body temperature at the time of death. John Doe was found murdered at midnight in his house, when the room temperature was 70°F and his body temperature was 80°F. When was he killed? Assume that the normal body temperature is 98.6°F.

48. a. Given that $2^x = e^{kx}$, find k.
b. Show that, in general, if b is a nonnegative real number, then any equation of the form $y = b^x$ may be written in the form $y = e^{kx}$, for some real number k.

49. Use the definition of a logarithm to prove
a. $\log_b mn = \log_b m + \log_b n$
b. $\log_b \dfrac{m}{n} = \log_b m - \log_b n$

[*Hint:* Let $\log_b m = p$ and $\log_b n = q$. Then $b^p = m$ and $b^q = n$.]

50. Use the definition of a logarithm to prove

$$\log_b m^n = n \log_b m$$

51. Use the definition of a logarithm to prove
a. $\log_b 1 = 0$ **b.** $\log_b b = 1$

SOLUTIONS TO SELF-CHECK EXERCISES 5.2

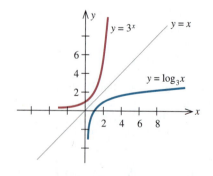

1. First, sketch the graph of $y = 3^x$ with the help of the following table of values.

x	−3	−2	−1	0	1	2	3
$y = 3^x$	1/27	1/9	1/3	0	3	9	27

Next, take the mirror reflection of this graph with respect to the line $y = x$ to obtain the graph of $y = \log_3 x$.

2.
$$3e^{x+1} - 2 = 4$$
$$3e^{x+1} = 6$$
$$e^{x+1} = 2$$
$$\ln e^{x+1} = \ln 2$$
$$(x + 1)\ln e = \ln 2 \qquad \text{(Law 3)}$$
$$x + 1 = \ln 2 \qquad \text{(Law 5)}$$
$$x = \ln 2 - 1$$
$$\approx -0.3069$$

5.3 COMPOUND INTEREST

Compound Interest

Compound interest is a natural application of the exponential function to the business world. We begin by recalling that simple interest is interest that is computed only on the original principal. Thus, if I denotes the interest on a principal P (in dollars) at an interest rate of r per year for t years, then we have

$$I = Prt$$

The **accumulated amount** A, the sum of the principal and interest after t years, is given by

$$A = P + I = P + Prt$$
$$= P(1 + rt) \quad \text{(Simple interest formula)} \tag{4}$$

Frequently, interest earned is periodically added to the principal and thereafter earns interest itself at the same rate. This is called **compound interest.** In order to find a formula for the accumulated amount, let us consider a numerical example. Suppose that $1000 (the principal) is deposited in a bank for a **term** of three years earning interest at the rate of 8% per year (called the **nominal,** or **stated, rate**) compounded annually. Then, using formula (4) with $P = 1000$, $r = 0.08$, and $t = 1$, we see that the accumulated amount at the end of the first year is

$$A_1 = P(1 + rt)$$
$$= 1000[1 + 0.08(1)] = 1000(1.08) = 1080$$

or $1080.

To find the accumulated amount A_2 at the end of the second year, we use (4) once again, this time with $P = A_1$. (Remember, the principal *and* interest now earn interest over the second year.) We obtain

$$A_2 = P(1 + rt) = A_1(1 + rt)$$
$$= 1000[1 + 0.08(1)][1 + 0.08(1)]$$
$$= 1000[1 + 0.08]^2 = 1000(1.08)^2 \approx 1166.40$$

or approximately $1166.40.

Finally, the accumulated amount A_3 at the end of the third year is found using (4) with $P = A_2$, giving

$$A_3 = P(1 + rt) = A_2(1 + rt)$$
$$= 1000[1 + 0.08(1)]^2[1 + 0.08(1)]$$
$$= 1000[1 + 0.08]^3 = 1000(1.08)^3 \approx 1259.71$$

or approximately $1259.71.

If you reexamine our calculations in this example, you will see that the accumulated amounts at the end of each year have the following form:

First year:	$A_1 = 1000(1 + 0.08)$	or	$A_1 = P(1 + r)$
Second year:	$A_2 = 1000(1 + 0.08)^2$	or	$A_2 = P(1 + r)^2$
Third year:	$A_3 = 1000(1 + 0.08)^3$	or	$A_3 = P(1 + r)^3$

These observations suggest the following general result: If P dollars are invested over a term of t years earning interest at the rate of r per year compounded annually, then the accumulated amount is

$$A = P(1 + r)^t \tag{5}$$

Formula (5) was derived under the assumption that interest was compounded *annually*. In practice, however, interest is usually compounded more than once a year. The interval of time between successive interest calculations is called the **conversion period.**

If interest at a nominal rate of r per year is compounded m times a year on a principal of P dollars, then the simple interest rate per conversion period is

$$i = \frac{r}{m} \qquad \frac{\text{(Annual interest rate)}}{\text{(Periods per year)}}$$

For example, if the nominal interest rate is 8% per year ($r = 0.08$) and interest is compounded quarterly ($m = 4$), then

$$i = \frac{r}{m} = \frac{0.08}{4} = 0.02$$

or 2% per period.

In order to find a general formula for the accumulated amount when a principal of P dollars is deposited in a bank for a term of t years and earns interest at the (nominal) rate of r per year compounded m times per year, we proceed as before using formula (5) repeatedly with the interest rate $i = r/m$. We see that the accumulated amount at the end of each period is as follows:

First period:	$A_1 = P(1 + i)$
Second period:	$A_2 = A_1(1 + i) = [P(1 + i)](1 + i) = P(1 + i)^2$
Third period:	$A_3 = A_2(1 + i) = [P(1 + i)^2](1 + i) = P(1 + i)^3$
$\vdots$	$\vdots$
nth period:	$A_n = A_{n-1}(1 + i) = [P(1 + i)^{n-1}](1 + i) = P(1 + i)^n$

But there are $n = mt$ periods in t years (number of conversion periods times the term). Therefore, the accumulated amount at the end of t years is given by

$$A = P(1 + i)^n$$

COMPOUND INTEREST FORMULA

$$A = P(1 + i)^n \qquad \text{(6)}$$

where $i = r/m$, $n = mt$, and

A = Accumulated amount at the end of n conversion periods
P = Principal
r = Nominal interest rate per year
m = Number of conversion periods per year
t = Term (number of years)

 EXAMPLE I Find the accumulated amount after three years if $1000 is invested at 8% per year compounded (a) annually, (b) semi-annually, (c) quarterly, and (d) monthly.

Solution

a. Here $P = 1000$, $r = 0.08$, and $m = 1$. Thus, $i = r = 0.08$ and $n = 3$, so formula (6) gives

$$A = 1000(1.08)^3$$
$$= 1259.71$$

or $1259.71.

b. Here $P = 1000$, $r = 0.08$, and $m = 2$. Thus, $i = 0.08/2 = 0.04$ and $n = (3)(2) = 6$, so that (6) gives

$$A = 1000(1.04)^6$$
$$= 1265.32$$

or $1265.32.

c. In this case, $P = 1000$, $r = 0.08$, and $m = 4$. Thus, $i = 0.08/4 = 0.02$ and $n = (3)(4) = 12$, so (6) gives

$$A = 1000(1.02)^{12}$$
$$= 1268.24$$

or $1268.24.

d. Here $P = 1000$, $r = 0.08$, and $m = 12$. Thus, $i = 0.08/12 = 0.0067$ and $n = (3)(12) = 36$, so (6) gives

$$A = 1000(1.0067)^{36}$$
$$= 1271.75$$

or $1271.75. These results are summarized in Table 5.2.

Table 5.2

Nominal Rate (r)	Conversion Period	Interest Rate/ Conversion Period	Initial Investment	Accumulated Amount
8%	Annual ($m = 1$)	8%	$1000	$1259.71
8%	Semiannual ($m = 2$)	4%	$1000	$1265.32
8%	Quarterly ($m = 4$)	2%	$1000	$1268.24
8%	Monthly ($m = 12$)	2/3%	$1000	$1271.75

Effective Rate of Interest

In the last example we saw that the interest actually earned on an investment depends on the frequency with which the interest is compounded. Thus, the stated, or nominal, rate of 8% per year does not reflect the actual rate at which interest is earned. This suggests that we need to find a common basis for comparing interest rates. One such way of comparing interest rates is provided by using the *effective rate*. The **effective rate** is the *simple* interest rate that would produce the same accumulated amount in one year as the nominal rate compounded m times a year. The effective rate is also called the **true rate.**

To derive a relation between the nominal interest rate, r per year compounded m times, and its corresponding effective rate, r_{eff} per year, let us assume an initial investment of P dollars. Then the accumulated amount after 1 year at a simple interest rate of r_{eff} per year is

$$A = P(1 + r_{eff})$$

Also, the accumulated amount after 1 year at an interest rate of r per year compounded m times a year is

$$A = P(1 + i)^n = P\left(1 + \frac{r}{m}\right)^m \qquad \text{(Since } i = r/m\text{)}$$

Equating the two expressions gives

$$P(1 + r_{eff}) = P\left(1 + \frac{r}{m}\right)^m$$

$$1 + r_{eff} = \left(1 + \frac{r}{m}\right)^m \qquad \text{(Dividing both sides by } P\text{)}$$

or, upon solving for r_{eff}, we obtain the formula for computing the effective rate of interest:

EFFECTIVE RATE OF INTEREST FORMULA

$$r_{\text{eff}} = \left(1 + \frac{r}{m}\right)^m - 1 \qquad (7)$$

where
$r_{\text{eff}} = $ Effective rate of interest
$r = $ Nominal interest rate per year
$m = $ Number of conversion periods per year

 EXAMPLE 2 Find the effective rate of interest corresponding to a nominal rate of 8% per year compounded (a) annually, (b) semiannually, (c) quarterly, and (d) monthly.

Solution

a. The effective rate of interest corresponding to a nominal rate of 8% per year compounded annually is, of course, given by 8% per year. This result is also confirmed by using formula (7) with $r = 0.08$ and $m = 1$.

Thus, $\qquad r_{\text{eff}} = (1 + 0.08) - 1 = 0.08$

b. Let $r = 0.08$ and $m = 2$. Then (7) yields

$$r_{\text{eff}} = \left(1 + \frac{0.08}{2}\right)^2 - 1$$
$$= (1.04)^2 - 1$$
$$= 0.0816$$

so the required effective rate is 8.16% per year.

c. Let $r = 0.08$ and $m = 4$. Then (7) yields

$$r_{\text{eff}} = \left(1 + \frac{0.08}{4}\right)^4 - 1$$
$$= (1.02)^4 - 1$$
$$= 0.08243$$

so the corresponding effective rate in this case is 8.243% per year.

d. Let $r = 0.08$ and $m = 12$. Then (7) yields

$$r_{\text{eff}} = \left(1 + \frac{0.08}{12}\right)^{12} - 1$$
$$= (1.0067)^{12} - 1$$
$$= 0.08343$$

so the corresponding effective rate in this case is 8.343% per year. ○ ○ ○

Now, if the effective rate of interest r_{eff} is known, then the accumulated amount after t years on an investment of P dollars may be more readily computed by using the formula

$$A = P(1 + r_{\text{eff}})^t$$

Table 5.3

Nominal Rate	Frequency of Interest Payment	Effective Rate	Initial Investment	Accumulated Amount After 3 Years
8%	Annually	8%	$1000	$1000(1 + 0.08)^3$ = \$1259.71
8%	Semiannually	8.16%	$1000	$1000(1 + 0.0816)^3$ = \$1265.32
8%	Quarterly	8.243%	$1000	$1000(1 + 0.08243)^3$ = \$1268.23
8%	Monthly	8.343%	$1000	$1000(1 + 0.08343)^3$ = \$1271.75

The Truth in Lending Act passed by Congress in 1968 requires that the effective rate of interest be disclosed in all contracts involving interest charges. The passage of this act has benefited consumers because they now have a common basis for comparing the various nominal rates quoted by different financial institutions. Furthermore, knowing the effective rate enables consumers to compute the actual charges involved in a transaction. Thus, if the effective rates of interest found in Example 2 were known, the accumulated values of Example 1, shown in Table 5.3, could have been readily found.

Present Value

Let us return to the compound interest formula (6), which expresses the accumulated amount at the end of n periods when interest at the rate of r is compounded m times a year. The principal P in (6) is often referred to as the **present value,** and the accumulated value A is called the **future value** since it is realized at a future date. In certain instances, an investor may wish to determine how much money he should invest now, at a fixed rate of interest, so that he will realize a certain sum at some future date. This problem may be solved by expressing P in terms of A. Thus, from (6) we find

$$P = A(1 + i)^{-n}$$

Here, as before, $i = r/m$, where m is the number of conversion periods per year.

PRESENT VALUE FORMULA FOR COMPOUND INTEREST

$$P = A(1 + i)^{-n} \tag{8}$$

 EXAMPLE 3   Find how much money should be deposited in a bank paying interest at the rate of 6% per year compounded monthly so that at the end of 3 years the accumulated amount will be $20,000.

Solution   Here, $r = 0.06$ and $m = 12$, so $i = 0.06/12 = 0.005$ and $n = (3)(12) = 36$. Thus, the problem is to determine P given that $A = 20,000$.

Using formula (8), we obtain

$$P = 20{,}000(1.005)^{-36}$$
$$= 16{,}713$$

or $16,713.

EXAMPLE 4   Find the present value of $49,158.60 due in 5 years at an interest rate of 10% per year compounded quarterly.

Solution   Using formula (8) with $r = 0.1$ and $m = 4$, so that $i = 0.1/4 = 0.025$, $n = (4)(5) = 20$, and $A = 49{,}158.6$, we obtain

$$P = (49{,}158.6)(1.025)^{-20} = 30{,}000$$

or $30,000.

Continuous Compounding of Interest

One question that arises naturally in the study of compound interest is: What happens to the accumulated amount over a fixed period of time if the interest is computed more and more frequently?

Intuition suggests that the more often interest is compounded, the larger the accumulated amount will be. This is confirmed by the results of Example 1, where we found that the accumulated amounts did, in fact, increase when we increased the number of conversion periods per year.

This leads us to another question: Does the accumulated amount approach a limit when the interest is computed more and more frequently over a fixed period of time?

To answer this question, let us look again at the compound interest formula:

$$A = P\left(1 + \frac{r}{m}\right)^{mt} \tag{9}$$

Recall that m is the number of conversion periods per year. So to find an answer to our problem, we should let m approach infinity (get larger and larger) in (9). But first we will rewrite this equation in the form

$$A = P\left[\left(1 + \frac{r}{m}\right)^{m}\right]^{t} \qquad \text{[Since } b^{xy} = (b^x)^y\text{]}$$

Now, letting $m \to \infty$, we find that

$$\lim_{m \to \infty}\left[P\left(1 + \frac{r}{m}\right)^{m}\right]^{t} = P\left[\lim_{m \to \infty}\left(1 + \frac{r}{m}\right)^{m}\right]^{t} \qquad \text{(Why?)}$$

Next, upon making the substitution $u = m/r$ and observing that $u \to \infty$ as $m \to \infty$, the foregoing expression reduces to

$$P\left[\lim_{u \to \infty}\left(1 + \frac{1}{u}\right)^{ur}\right]^{t} = P\left[\lim_{u \to \infty}\left(1 + \frac{1}{u}\right)^{u}\right]^{rt}$$

But
$$\lim_{u \to \infty}\left(1 + \frac{1}{u}\right)^{u} = e \qquad \text{[Using (1)]}$$

so
$$\lim_{m \to \infty} P\left[\left(1 + \frac{r}{m}\right)^{m}\right]^{t} = Pe^{rt}$$

Our computations tell us that as the frequency with which interest is compounded increases without bound, the accumulated amount approaches Pe^{rt}. In this situation, we say that interest is *compounded continuously*. Let us summarize this important result.

CONTINUOUS COMPOUND INTEREST FORMULA

$$A = Pe^{rt} \qquad\qquad (10)$$

where

$P = $ Principal
$r = $ Annual interest rate compounded continuously
$t = $ Time in years
$A = $ Accumulated amount at the end of t years

EXPLORING WITH TECHNOLOGY

In the opening paragraph of Section 5.1, we pointed out that the accumulated amount of an account earning interest *compounded continuously* will eventually outgrow by far the accumulated amount of an account earning interest at the same nominal rate but earning simple interest. Illustrate this fact using the following example.

Suppose you deposit $1000 in account I, earning interest at the rate of 10% per year compounded continuously so that the accumulated amount at the end of t years is $A_1(t) = 1000e^{0.1t}$. Suppose you also deposit $1000 in account II, earning simple interest at the rate of 10% per year so that the accumulated amount at the end of t years is $A_2(t) = 1000(1 + 0.1t)$. Use a graphing utility to sketch the graphs of the functions A_1 and A_2 in the viewing rectangle $[0, 20] \times [0, 10{,}000]$ to see the accumulated amounts $A_1(t)$ and $A_2(t)$ over a 20-year period.

○ ○ ○

EXAMPLE 5   Find the accumulated amount after three years if $1000 is invested at 8% per year compounded (a) daily (take the number of days in a year to be 365) and (b) continuously.

Solution

a. Using formula (6) with $P = 1000$, $r = 0.08$, $m = 365$, and $n = (365)(3) = 1095$, we find

$$A = 1000 \left(1 + \frac{0.08}{365}\right)^{1095} \approx 1271.22$$

or $1271.22.

b. Here we use formula (10) with $P = 1000$, $r = 0.08$, and $t = 3$, obtaining

$$A = 1000e^{(0.08)(3)}$$
$$\approx 1271.25 \qquad \text{(Using the ``e^x'' key)}$$

or $1271.25.

Observe that the accumulated amounts corresponding to interest compounded daily and interest compounded continuously differ by very little. The continuous compound interest formula is a very important tool in theoretical work in financial analysis.

If we solve formula (10) for P, we obtain

$$P = Ae^{-rt} \tag{11}$$

which gives the present value in terms of the future (accumulated) value for the case of continuous compounding.

EXAMPLE 6 The Blakely Investment Company owns an office building located in the commercial district of a city. As a result of the continued success of an urban renewal program, local business is enjoying a miniboom. The market value of Blakely's property is

$$V(t) = 300,000e^{\sqrt{t}/2}$$

where $V(t)$ is measured in dollars and t is the time in years from the present. If the expected rate of inflation is 9% compounded continuously for the next 10 years, find an expression for the present value $P(t)$ of the market price of the property valid for the next 10 years. Compute $P(7)$, $P(8)$, and $P(9)$, and interpret your results.

Solution Using formula (11) with $A = V(t)$ and $r = 0.09$, we find that the present value of the market price of the property t years from now is

$$P(t) = V(t)e^{-0.09t}$$
$$= 300,000e^{-0.09t + \sqrt{t}/2} \qquad (0 \le t \le 10)$$

Letting $t = 7$, 8, and 9, respectively, we find that

$$P(7) = 300,000e^{-0.09(7) + \sqrt{7}/2} = 599,837 \quad \text{or} \quad \$599,837$$
$$P(8) = 300,000e^{-0.09(8) + \sqrt{8}/2} = 600,640 \quad \text{or} \quad \$600,640$$
$$P(9) = 300,000e^{-0.09(9) + \sqrt{9}/2} = 598,115 \quad \text{or} \quad \$598,115$$

From the results of these computations, we see that the present value of the property's market price seems to decrease after a certain period of growth. This suggests that there is an optimal time for the owners to sell. Later we will show that the highest present value of the property's market price is $600,779, which occurs at time $t = 7.72$ years.   ○ ○ ○

EXPLORING WITH TECHNOLOGY

The effective rate of interest is given by

$$r_{eff} = \left(1 + \frac{r}{m}\right)^m - 1$$

where the number of conversion periods per year is m. In exercise 27 you will be asked to show that the effective rate of interest r_{eff} corresponding to a nominal interest rate r per year compounded continuously is given by

$$\hat{r}_{eff} = e^r - 1$$

In order to obtain a visual confirmation of this result, consider the special case where $r = 0.1$ (10% per year).

a. Use a graphing utility to plot the graph of both

$$y_1 = \left(1 + \frac{0.1}{x}\right)^x - 1 \qquad \text{and} \qquad y_2 = e^{0.1} - 1$$

in the viewing rectangle $[0, 3] \times [0, 0.12]$.

b. Does your result seem to imply that

$$\left(1 + \frac{r}{m}\right)^m - 1$$

approaches

$$\hat{r}_{eff} = e^r - 1$$

as m increases without bound for the special case $r = 0.1$?

○   ○   ○

SELF-CHECK EXERCISES 5.3

 1. Find the present value of $20,000 due in 3 years at an interest rate of 12% per year compounded monthly.

2. Mr. Baker is a retiree living on social security and the income from his investment. Currently, his $100,000 investment in a 1-year CD is yielding 11.6% interest compounded daily. If he reinvests the principal ($100,000) on the due date of the CD in another 1-year CD paying 9.2% interest compounded daily, find the net decrease in his yearly income from his investment.

3. a. What is the accumulated amount after 5 years if $10,000 is invested at 10% per year compounded continuously?
b. Find the present value of $10,000 due in 5 years at an interest rate of 10% per year compounded continuously.

Solutions to Self-Check Exercises 5.3 can be found on page 392.

5.3 EXERCISES

A calculator is recommended for these exercises.

In exercises 1–4, find the accumulated amount A if the principal P is invested at an interest rate of r per year for t years.

1. $P = \$2500$, $r = 7\%$, $t = 10$, compounded semiannually

2. $P = \$12,000$, $r = 8\%$, $t = 10$, compounded quarterly

3. $P = \$150,000$, $r = 10\%$, $t = 4$, compounded monthly

4. $P = \$150,000$, $r = 9\%$, $t = 3$, compounded daily

In exercises 5 and 6, find the effective rate corresponding to the given nominal rate.

5. a. 10% per year compounded semiannually
b. 9% per year compounded quarterly

6. a. 8% per year compounded monthly
b. 8% per year compounded daily

In exercises 7 and 8, find the present value of $40,000 due in four years at the given rate of interest.

7. a. 8% per year compounded semiannually
b. 8% per year compounded quarterly

8. a. 7% per year compounded monthly
b. 9% per year compounded daily

9. Find the accumulated amount after 4 years if $5000 is invested at 8% per year compounded continuously.

10. An amount of $25,000 is deposited in a bank that pays interest at the rate of 7% per year, compounded annually. What is the total amount on deposit at the end of 6 years, assuming there are no deposits or withdrawals during those 6 years? What is the interest earned in that period of time?

11. Housing Prices The Brennans are planning to buy a house 4 years from now. Housing experts in their area have estimated that the cost of a home will increase at a rate of 9% per year during that 4-year period. If this economic prediction holds true, how much can they expect to pay for a house that currently costs $80,000?

12. Energy Consumption A metropolitan utility company in a western city of the United States expects the consumption of electricity to increase by 8% per year during the next decade, due mainly to the expected population increase. If consumption does increase at this rate, find the amount by which the utility company will have to increase its generating capacity in order to meet the area's needs at the end of the decade.

13. Pension Funds The managers of a pension fund have invested $1.5 million in U.S. government certificates of deposit (CDs) that pay interest at the rate of 9.5% per year compounded semiannually over a period of 10 years. At the end of this period, how much will the investment be worth?

14. Savings Accounts Mr. Kaplan invested a sum of money 5 years ago in a savings account, which has since paid interest at the rate of 8% compounded quarterly. His investment is now worth $22,289.22. How much did he originally invest?

15. Loan Consolidation The proprietors of the Coachmen Inn secured two loans from the Union Bank: one for $8000 due in 3 years and one for $15,000 due in 6 years, both at an interest rate of 10% compounded semiannually. The bank agreed to allow the two loans to be consolidated into one loan payable in 5 years at the same interest rate. How much will the proprietors have to pay the bank at the end of 5 years?

16. Tax-Deferred Annuities   Mrs. Bennett is in the 28% tax bracket and has $25,000 available for investment during her current tax year. Assume that she remains in the same tax bracket over the next 10 years and determine the accumulated amount of her investment if

a. she puts the $25,000 into a tax-deferred annuity that pays 12% per year, tax deferred for 10 years.
b. she puts the $25,000 into a taxable instrument that pays 12% per year for 10 years.

[*Hint:* In this case the yield after taxes is 8.64% per year.]

17. Consumer Price Index   At an annual inflation rate of 7.5%, how long will it take the Consumer Price Index (CPI) to double?

18. Investment Returns   Ms. Collins purchased a house in 1987 for $80,000. In 1993 she sold the house and made a net profit of $28,000. Find the effective annual rate of return on her investment over the 6-year period.

19. Investment Returns   Mr. Stevens purchased 1000 shares of a certain stock for $25,250 (including commissions). He sold the shares 2 years later and received $32,100 after deducting commissions. Find the effective annual rate of return on his investment over the 2-year period.

20. Investment Options   Investment A offers a 10% return compounded semiannually, and investment B offers a 9.75% return compounded continuously. Which investment has a higher rate of return over a 4-year period?

21. Present Value   Find the present value of $59,673 due in 5 years at an interest rate of 8% per year compounded continuously.

22. Real Estate Investments   A condominium complex was purchased by a group of private investors for $1.4 million and sold 6 years later for $3.6 million. Find the annual rate of return (compounded continuously) on their investment.

23. Saving for College   Having received a large inheritance, a child's parents wish to establish a trust for the child's college education. If they need an estimated $70,000 seven years from now, how much should they set aside in trust now, if they invest the money at 10.5% compounded (a) quarterly? (b) continuously?

24. Effect of Inflation on Salaries   Mr. Lyons's current annual salary is $25,000. Ten years from now, how much will he need to earn in order to retain his present purchasing power if the rate of inflation over that period

is 6% per year? Assume that inflation is continuously compounded.

25. Pensions   Ms. Lindstrom, who is now 50 years old, is employed by a firm that guarantees her a pension of $40,000 per year at age 65. What is the present value of her first year's pension if inflation over the next 15 years is (a) 6%? (b) 8%? (c) 12%? Assume that inflation is continuously compounded.

26. Real Estate Investments   An investor purchased a piece of waterfront property. Because of the development of a marina in the vicinity, the market value of the property is expected to increase according to the rule

$$V(t) = 80,000e^{\sqrt{t}/2}$$

where $V(t)$ is measured in dollars and t is the time in years from the present. If the rate of inflation is expected to be 9% compounded continuously for the next 8 years, find an expression for the present value $P(t)$ of the property's market price valid for the next 8 years. What is $P(t)$ expected to be in 4 years?

27. Show that the effective rate of interest $\hat{r}_{\text{eff}}$ that corresponds to a nominal interest rate r per year compounded continuously is given by

$$\hat{r}_{\text{eff}} = e^r - 1$$

[*Hint:* From formula (7) we see that the effective rate $\hat{r}_{\text{eff}}$ corresponding to a nominal interest rate r per year compounded m times a year is given by

$$\hat{r}_{\text{eff}} = \left(1 + \frac{r}{m}\right)^m - 1$$

Let m tend to infinity in this expression.]

28. Refer to exercise 27. Find the effective rate of interest that corresponds to a nominal rate of 10% per year compounded (a) quarterly, (b) monthly, and (c) continuously.

29. Investment Analysis   Refer to exercise 27. Bank A pays interest on deposits at a 7% annual rate compounded quarterly, and Bank B pays interest on deposits at a $7\frac{1}{8}$% annual rate compounded continuously. Which bank has the higher effective rate of interest?

30. Investment Analysis   Find the nominal rate of interest that, when compounded monthly, yields an effective rate of interest of 10% per year.
[*Hint:* Use equation (7).]

31. Investment Analysis   Find the nominal rate of interest that, when compounded continuously, yields an effective rate of interest of 10% per year.
[*Hint:* See exercise 27.]

Misato Nakazaki

Title: Assistant Vice President
Institution: A large investment corporation

In the securities industry, buying and selling stocks and bonds has always required a mastery of concepts and formulas that outsiders find confusing. As a bond seller, Nakazaki routinely uses terms such as *issue, maturity, current yield, callable* and *convertible bonds,* and so on.

These terms, however, are easily defined. When corporations issue bonds, they are borrowing money at a fixed rate of interest. The bonds are scheduled to mature—to be paid back—on a specific date as much as 30 years into the future. Callable bonds allow the issuer to pay off the loans prior to their expected maturity, reducing overall interest payments. In its simplest terms, current yield is the price of a bond multiplied by the interest rate at which the bond is issued. For example, a bond with a face value of $1000 and an interest rate of 10% yields $100 per year in interest payments. When that same bond is resold at a premium on the secondary market for $1200, its current yield nets only an 8.3% rate of return based on the higher purchase price.

Bonds attract investors for many reasons. A key variable is the sensitivity of the bond's price to future changes in interest rates. If investors get locked into a low-paying bond when future bonds pay higher yields, they lose money. Nakazaki stresses that "no one knows for sure what rates will be over time." Employing differentials allows her to calculate interest-rate sensitivity for clients as they ponder purchase decisions.

Computerized formulas, "whose basis is calculus," says Nakazaki, help her factor the endless stream of numbers flowing across her desk.

On a typical day, Nakazaki might be given a bid on "10 million, GMAC, 8.5%, January 2003." Translation: Her customer wants her to buy General Motors Acceptance Corporation bonds with a face value of $10 million and an interest rate of 8.5%, maturing in January 2003.

After she calls her firm's trader to find out the yield on the bond in question, Nakazaki enters the price and other variables, such as the interest rate and date of maturity, and the computer prints out the answers. Nakazaki can then relay to her client the bond's current yield, accrued interest, and so on. In Nakazaki's rapid-fire work environment, such speed is essential. Nakazaki cautions that "computer users have to understand what's behind the formulas." The software "relies on the basics of calculus. If people don't understand the formula, it's useless for them to use the calculations."

With an MBA from New York University, Nakazaki typifies the younger generation of Japanese women who have chosen to succeed in the business world. Since earning her degree, she has sold bonds for a global securities firm in New York City.

Nakazaki's client list reads like a who's who of the leading Japanese banks, insurance companies, mutual funds, and corporations. As institutional buyers, her clients purchase large blocks of American corporate bonds and mortgage-backed securities such as Ginnie Maes.

SOLUTIONS TO SELF-CHECK EXERCISES 5.3

1. Using formula (8) with $r = 0.12$ and $m = 12$, so that

$$i = \frac{0.12}{12} = 0.01, \quad n = (12)(3) = 36, \quad \text{and} \quad A = 20{,}000$$

we find the required present value to be

$$P = 20{,}000(1.01)^{-36}$$
$$= 13{,}978.50$$

or $13,978.50.

2. The accumulated amount of Mr. Baker's current investment is found by using formula (6) with $P = 100{,}000$, $r = 0.116$, and $m = 360$. Thus,

$$i = \frac{0.116}{360} = 0.0003222 \quad \text{and} \quad n = 360$$

so the required accumulated amount is

$$A = 100{,}000(1.0003222)^{360}$$
$$= 112{,}296.59$$

or $112,296.59. Next, we compute the accumulated amount of Mr. Baker's reinvestment. Once again, using (6) with $P = 100{,}000$, $r = 0.092$, and $m = 360$ so that

$$i = \frac{0.092}{360} = 0.0002556 \quad \text{and} \quad n = 360,$$

we find the required accumulated amount in this case to be

$$\overline{A} = 100{,}000(1.0002556)^{360}$$

or $109,636.95. Therefore, Mr. Baker can expect to experience a net decrease in yearly income of

$$112{,}296.59 - 109{,}636.95$$

or $2,659.64.

3. a. Using formula (10) with $P = 10{,}000$, $r = 0.1$, and $t = 5$, we find that the required accumulated amount is given by

$$A = 10{,}000e^{(0.1)(5)}$$
$$= 16{,}487.21$$

or $16,487.21.

b. Using formula (11) with $A = 10{,}000$, $r = 0.1$, and $t = 5$, we see that the required present value is given by

$$P = 10{,}000e^{-(0.1)(5)}$$
$$= 6065.31$$

or $6065.31.

5.4 DIFFERENTIATION OF EXPONENTIAL FUNCTIONS

The Derivative of the Exponential Function

In order to study the effects of budget deficit-reduction plans at different income levels, it is important to know the income distribution of American families. Based on data from the House Budget Committee, the House Ways and Means Committee, and the U.S. Census Bureau, the graph of f shown in Figure 5.10 gives the number of American families y (in millions) as a function of their annual income x (in thousands of dollars) in 1990.

Figure 5.10
The graph of f shows the number of families versus their annual income.

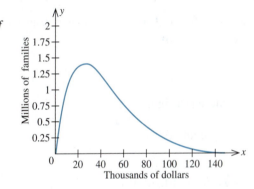

Source: House Budget Committee, House Ways and Means Committee, and U.S. Census Bureau

Observe that the graph of f rises very quickly and then tapers off. From the graph of f, you can see that the bulk of American families earned less than \$100,000 per year. In fact, 95% of U.S. families earned less than \$102,358 per year in 1990. (We will refer to this model again in Using Technology at the end of this section.)

To analyze mathematical models involving exponential and logarithmic functions in greater detail, we need to develop rules for computing the derivative of these functions. We begin by looking at the rule for computing the derivative of the exponential function.

RULE I   DERIVATIVE OF THE EXPONENTIAL FUNCTION

$$\frac{d}{dx}e^x = e^x$$

Thus, the derivative of the exponential function with base e is equal to the function itself. To demonstrate the validity of this rule, we compute

$$f'(x) = \lim_{h \to 0} \frac{f(x + h) - f(x)}{h}$$

$$= \lim_{h \to 0} \frac{e^{x+h} - e^x}{h}$$

$$= \lim_{h \to 0} \frac{e^x(e^h - 1)}{h} \qquad \text{(Writing } e^{x+h} = e^x e^h \text{ and factoring)}$$

$$= e^x \lim_{h \to 0} \frac{e^h - 1}{h} \qquad \text{(Why?)}$$

To evaluate

$$\lim_{h \to 0} \frac{e^h - 1}{h}$$

let us refer to Table 5.4, which is constructed with the aid of a calculator. From the table, we see that

$$\lim_{h \to 0} \frac{e^h - 1}{h} = 1$$

(Although a rigorous proof of this fact is possible, it is beyond the scope of this book. Also see Example 1, Using Technology, page 402.) Using this result, we conclude that

$$f'(x) = e^x \cdot 1 = e^x$$

as we set out to show.

Table 5.4

h	0.1	0.01	0.001	-0.1	-0.01	-0.001
$\dfrac{e^h - 1}{h}$	1.0517	1.0050	1.0005	0.9516	0.9950	0.9995

EXAMPLE 1   Compute the derivative of each of the following functions:

a. $f(x) = x^2 e^x$ **b.** $g(t) = (e^t + 2)^{3/2}$

Solution

a. The Product Rule gives

$$f'(x) = \frac{d}{dx}(x^2 e^x)$$

$$= x^2 \frac{d}{dx}(e^x) + e^x \frac{d}{dx}(x^2)$$

$$= x^2 e^x + e^x(2x)$$

$$= xe^x(x + 2)$$

b. Using the General Power Rule, we find

$$g'(t) = \frac{3}{2}(e^t + 2)^{1/2}\frac{d}{dt}(e^t + 2)$$

$$= \frac{3}{2}(e^t + 2)^{1/2}e^t = \frac{3}{2}e^t(e^t + 2)^{1/2}$$

● ● ●

EXPLORING WITH TECHNOLOGY

Consider the exponential function $f(x) = b^x$ ($b > 0$, $b \neq 1$).

a. Use the definition of the derivative of a function to show that

$$f'(x) = b^x \cdot \lim_{h \to 0}\frac{b^h - 1}{h}$$

b. Use the result of part (a) to show that

$$\frac{d}{dx}(2^x) = 2^x \cdot \lim_{h \to 0}\frac{2^h - 1}{h}$$

and

$$\frac{d}{dx}(3^x) = 3^x \cdot \lim_{h \to 0}\frac{3^h - 1}{h}$$

c. Use the technique in Using Technology, page 402, to show that (to two decimal places)

$$\lim_{h \to 0}\frac{2^h - 1}{h} = 0.69 \qquad \text{and} \qquad \lim_{h \to 0}\frac{3^h - 1}{h} = 1.10$$

d. Conclude from the results of parts (b) and (c) that

$$\frac{d}{dx}(2^x) \approx (0.69)2^x \qquad \text{and} \qquad \frac{d}{dx}(3^x) \approx (1.10)3^x$$

Thus,

$$\frac{d}{dx}(b^x) = k \cdot b^x$$

where k is an appropriate constant.

e. The results of part (d) suggest that, for convenience, we pick the base b, where $2 < b < 3$, so that $k = 1$. This value of b is $e \approx 2.718281828\ldots$. Thus

$$\frac{d}{dx}(e^x) = e^x$$

This is why we prefer to work with the exponential function $f(x) = e^x$.

● ● ●

Applying the Chain Rule to Exponential Functions

To enlarge the class of exponential functions to be differentiated, we appeal to the Chain Rule to obtain the following rule for differentiating composite functions of the form $h(x) = e^{f(x)}$. An example of such a function is $h(x) = e^{x^2 - 2x}$. Here $f(x) = x^2 - 2x$.

RULE 2 THE CHAIN RULE FOR EXPONENTIAL FUNCTIONS	If $f(x)$ is a differentiable function, then $$\frac{d}{dx}(e^{f(x)}) = e^{f(x)}f'(x)$$

To see this, observe that if $h(x) = g[f(x)]$, where $g(x) = e^x$, then by virtue of the Chain Rule,

$$h'(x) = g'(f(x))f'(x) = e^{f(x)}f'(x)$$

since $g'(x) = e^x$.

As an aid to remembering the Chain Rule for exponential functions, observe that it has the following form:

$$\frac{d}{dx}(e^{f(x)}) = e^{f(x)} \cdot \text{derivative of exponent}$$
$$\underset{\text{same}}{\underline{\quad\quad\quad}}$$

EXAMPLE 2   Find the derivative of each of the following functions.

a. $f(x) = e^{2x}$

b. $y = e^{-3x}$

c. $g(t) = e^{2t^2 + t}$

Solution

a. $f'(x) = e^{2x}\dfrac{d}{dx}(2x) = e^{2x} \cdot 2 = 2e^{2x}$

b. $\dfrac{dy}{dx} = e^{-3x}\dfrac{d}{dx}(-3x) = -3e^{-3x}$

c. $g'(t) = e^{2t^2 + t} \cdot \dfrac{d}{dt}(2t^2 + t) = (4t + 1)e^{2t^2 + t}$

EXAMPLE 3 Differentiate the function $y = xe^{-2x}$.

Solution Using the Product Rule, followed by the Chain Rule, we find

$$\frac{dy}{dx} = x \frac{d}{dx} e^{-2x} + e^{-2x} \frac{d}{dx}(x)$$

$$= xe^{-2x} \frac{d}{dx}(-2x) + e^{-2x} \qquad \text{(Using the Chain Rule on the first term)}$$

$$= -2xe^{-2x} + e^{-2x}$$

$$= e^{-2x}(1 - 2x) \qquad\qquad\qquad\qquad\qquad \text{◖ ◖ ◖}$$

EXAMPLE 4 Differentiate the function $g(t) = \dfrac{e^t}{e^t + e^{-t}}$.

Solution Using the Quotient Rule, followed by the Chain Rule, we find

$$g'(t) = \frac{(e^t + e^{-t}) \dfrac{d}{dt}(e^t) - e^t \dfrac{d}{dt}(e^t + e^{-t})}{(e^t + e^{-t})^2}$$

$$= \frac{(e^t + e^{-t})e^t - e^t(e^t - e^{-t})}{(e^t + e^{-t})^2}$$

$$= \frac{e^{2t} + 1 - e^{2t} + 1}{(e^t + e^{-t})^2} \qquad (e^\circ = 1)$$

$$= \frac{2}{(e^t + e^{-t})^2} \qquad\qquad\qquad\qquad \text{◖ ◖ ◖}$$

EXAMPLE 5 In Section 5.6 we will discuss some practical applications of the exponential function

$$Q(t) = Q_0 e^{kt}$$

where Q_0 and k are positive constants and $t \in [0, \infty)$. A quantity $Q(t)$ growing according to this law experiences *exponential growth*. Show that for a quantity $Q(t)$ experiencing exponential growth, the rate of growth of the quantity $Q'(t)$ at any time t is directly proportional to the amount of the quantity present.

Solution Using the Chain Rule for exponential functions, we compute the derivative Q' of the function Q. Thus,

$$Q'(t) = Q_0 e^{kt} \frac{d}{dt}(kt)$$

$$= Q_0 e^{kt}(k)$$

$$= kQ_0 e^{kt}$$

$$= kQ(t) \qquad (Q(t) = Q_0 e^{kt})$$

which is the desired conclusion. ◖ ◖ ◖

EXAMPLE 6   Find the points of inflection of the function $f(x) = e^{-x^2}$.

Solution   The first derivative of f is

$$f'(x) = -2xe^{-x^2}$$

Differentiating $f'(x)$ with respect to x yields

$$f''(x) = (-2x)(-2xe^{-x^2}) - 2e^{-x^2}$$
$$= 2e^{-x^2}(2x^2 - 1)$$

Setting $f''(x) = 0$ gives

$$2e^{-x^2}(2x^2 - 1) = 0$$

Figure 5.11
Sign diagram for f".

Since e^{-x^2} never equals zero for any real value of x, we see that $x = \pm 1/\sqrt{2}$ are the only candidates for inflection points of f. The sign diagram of f'', shown in Figure 5.11, tells us that both $x = -1/\sqrt{2}$ and $x = 1/\sqrt{2}$ give rise to inflection points of f.
Next,

$$f\left(-\frac{1}{\sqrt{2}}\right) = f\left(\frac{1}{\sqrt{2}}\right) = e^{-1/2}$$

and the inflection points of f are $(-1/\sqrt{2}, e^{-1/2})$ and $(1/\sqrt{2}, e^{-1/2})$. The graph of f appears in Figure 5.12.

○ ○ ○

Figure 5.12
The graph of $y = e^{-x^2}$ has two in-flection points.

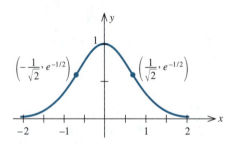

Application

Our final example involves finding the absolute maximum of an exponential function.

EXAMPLE 7   Refer to Example 6, Section 5.3. The present value of the market price of the Blakely Office Building is given by

$$P(t) = 300,000e^{-0.09t + \sqrt{t}/2} \qquad (0 \le t \le 10)$$

Find the optimal present value of the building's market price.

Solution To find the maximum value of P over $[0, 10]$, we compute

$$P'(t) = 300{,}000e^{-0.09t+\sqrt{t}/2}\frac{d}{dt}\left(-0.09t + \frac{1}{2}t^{1/2}\right)$$

$$= 300{,}000e^{-0.09t+\sqrt{t}/2}\left(-0.09 + \frac{1}{4}t^{-1/2}\right)$$

Setting $P'(t) = 0$ gives

$$-0.09 + \frac{1}{4t^{1/2}} = 0$$

since $e^{-0.09t+\sqrt{t}/2}$ is never zero for any value of t. Solving this equation, we find

$$\frac{1}{4t^{1/2}} = 0.09$$

$$t^{1/2} = \frac{1}{4(0.09)}$$

$$= \frac{1}{0.36}$$

or

$$t \approx 7.72$$

the sole critical point of the function P. Finally, evaluating $P(t)$ at the critical point as well as at the endpoints of $[0, 10]$, we have

t	$P(t)$
0	300,000
7.72	600,779
10	592,838

We conclude, accordingly, that the optimal present value of the property's market price is \$600,779 and that this will occur 7.72 years from now.

○ ○ ○

SELF-CHECK EXERCISES 5.4

1. Let $f(x) = xe^{-x}$.
   a. Find the first and second derivatives of f.
   b. Find the relative extrema of f.
   c. Find the inflection points of f.

2. An industrial asset is being depreciated at a rate so that its book value t years from now will be

$$V(t) = 50{,}000e^{-0.4t}$$

dollars. How fast will the book value of the asset be changing 3 years from now?

Solutions to Self-Check Exercises 5.4 can be found on page 404.

5.4 EXERCISES

In exercises 1–28, find the derivative of the function.

1. $f(x) = e^{3x}$

2. $f(x) = 3e^x$

3. $g(t) = e^{-t}$

4. $f(x) = e^{-2x}$

5. $f(x) = e^x + x$

6. $f(x) = 2e^x - x^2$

7. $f(x) = x^3 e^x$

8. $f(u) = u^2 e^{-u}$

9. $f(x) = \dfrac{2e^x}{x}$

10. $f(x) = \dfrac{x}{e^x}$

11. $f(x) = 3(e^x + e^{-x})$

12. $f(x) = \dfrac{e^x + e^{-x}}{2}$

13. $f(w) = \dfrac{e^w + 1}{e^w}$

14. $f(x) = \dfrac{e^x}{e^x + 1}$

15. $f(x) = 2e^{3x-1}$

16. $f(t) = 4e^{3t+2}$

17. $h(x) = e^{-x^2}$

18. $f(x) = e^{x^2-1}$

19. $f(x) = 3e^{-1/x}$

20. $f(x) = e^{1/(2x)}$

21. $f(x) = (e^x + 1)^{25}$

22. $f(x) = (4 - e^{-3x})^3$

23. $f(x) = e^{\sqrt{x}}$

24. $f(t) = -e^{-\sqrt{2t}}$

25. $f(x) = (x - 1)e^{3x+2}$

26. $f(s) = (s^2 + 1)e^{-s^2}$

27. $f(x) = \dfrac{e^x - 1}{e^x + 1}$

28. $g(t) = \dfrac{e^{-t}}{1 + t^2}$

In exercises 29–32, find the second derivative of the function.

29. $f(x) = e^{-4x} + 2e^{3x}$

30. $f(t) = 3e^{-2t} - 5e^{-t}$

31. $f(x) = 2xe^{3x}$

32. $f(t) = t^2 e^{-2t}$

33. Find an equation of the tangent line to the graph of $y = e^{2x-3}$ at the point $(\frac{3}{2}, 1)$.

34. Find an equation of the tangent line to the graph of $y = e^{-x^2}$ at the point $(1, 1/e)$.

35. Determine the intervals where the function $f(x) = e^{-x^2/2}$ is increasing and where it is decreasing.

36. Determine the intervals where the function $f(x) = x^2 e^{-x}$ is increasing and where it is decreasing.

37. Determine the intervals of concavity for the function $f(x) = \dfrac{e^x - e^{-x}}{2}$.

38. Determine the intervals of concavity for the function $f(x) = xe^x$.

39. Find the inflection point of the function $f(x) = xe^{-2x}$.

40. Find the inflection point(s) of the function $f(x) = 2e^{-x^2}$.

In exercises 41–44, find the absolute extrema of the function.

41. $f(x) = e^{-x^2}$ on $[-1, 1]$

42. $h(x) = e^{x^2-4}$ on $[-2, 2]$

43. $g(x) = (2x - 1)e^{-x}$ on $[0, \infty)$

44. $f(x) = xe^{-x^2}$ on $[0, 2]$

In exercises 45–48, use the curve-sketching guidelines of Chapter 4, page 315, to sketch the graph of the function.

45. $f(t) = e^t - t$

46. $h(x) = \dfrac{e^x + e^{-x}}{2}$

47. $f(x) = 2 - e^{-x}$

48. $f(x) = \dfrac{3}{1 + e^{-x}}$

A calculator is recommended for the remainder of these exercises.

49. Sales Promotion   The Lady Bug, a women's clothing chain store, found that t days after the end of a sales promotion the volume of sales was given by

$$S(t) = 20{,}000(1 + e^{-0.5t}) \qquad (0 \le t \le 5)$$

dollars. Find the rate of change of The Lady Bug's sales volume when $t = 1$, $t = 2$, $t = 3$, and $t = 4$.

50. Energy Consumption of Appliances   The average energy consumption of the typical refrigerator/freezer manufactured by York Industries is approximately

$$C(t) = 1486e^{-0.073t} + 500 \qquad (0 \le t \le 20)$$

kilowatt hours per year, where t is measured in years, with $t = 0$ corresponding to 1972.
a. What was the average energy consumption of the York refrigerator/freezer at the beginning of 1972?
b. Prove that the average energy consumption of the York refrigerator/freezer is decreasing over the years in question.
c. All refrigerator/freezers manufactured as of January 1, 1990, must meet the 950-kilowatt-hours-per-year

maximum energy-consumption standard set by the National Appliance Conservation Act. Show that the York refrigerator/freezer satisfies this requirement.

51. Polio Immunization   Polio, a once-feared killer, declined markedly in the United States in the 1950s after Jonas Salk developed the inactivated polio vaccine and mass immunization of children took place. The number of polio cases in the United States from the beginning of 1959 to the beginning of 1963 is approximated by the function

$$N(t) = 5.3e^{0.095t^2 - 0.85t} \qquad (0 \le t \le 4)$$

where $N(t)$ gives the number of polio cases (in thousands) and t is measured in years, with $t = 0$ corresponding to the beginning of 1959.
a. Show that the function N is decreasing over the time interval under consideration.
b. How fast was the number of polio cases decreasing at the beginning of 1959? At the beginning of 1962? [*Comment:* Following the introduction of the oral vaccine developed by Dr. Albert B. Sabin in 1963, polio in the United States has, for all practical purposes, been eliminated.]

52. Price of Perfume   The monthly demand for a certain brand of perfume is given by the demand equation

$$p = 100e^{-0.0002x} + 150$$

where p denotes the retail unit price in dollars and x denotes the quantity (in 1-ounce bottles) demanded.
a. Find the rate of change of the price per bottle when $x = 1000$; when $x = 2000$.
b. What is the price per bottle when $x = 1000$? When $x = 2000$?

53. Price of Wine   The monthly demand for a certain brand of table wine is given by the demand equation

$$p = 240\left(1 - \frac{3}{3 + e^{-0.0005x}}\right)$$

where p denotes the wholesale price per case (in dollars) and x denotes the number of cases demanded.
a. Find the rate of change of the price per case when $x = 1000$.
b. What is the price per case when $x = 1000$?

54. Spread of an Epidemic   During a flu epidemic, the total number of students on a state university campus who had contracted influenza by the xth day was given

by

$$N(x) = \frac{3000}{1 + 99e^{-x}} \qquad (x \ge 0)$$

a. How many students had influenza initially?
b. Derive an expression for the rate at which the disease was being spread and prove that the function N is increasing on the interval $(0, \infty)$.
c. Sketch the graph of N. What was the total number of students who contracted influenza during that particular epidemic?

55. Maximum Oil Production   It has been estimated that the total production of oil from a certain oil well is given by

$$T(t) = -1{,}000(t + 10)e^{-0.1t} + 10{,}000$$

thousand barrels t years after production has begun. Determine the year when the oil well will be producing at maximum capacity.

56. Optimal Selling Time   Refer to exercise 26, page 390. The present value of a piece of waterfront property purchased by an investor is given by the function

$$P(t) = 80{,}000e^{\sqrt{t}/2 - 0.09t} \qquad (0 \le t \le 8)$$

Determine the optimal time (based on present value) for the investor to sell the property. What is the property's optimal present value?

57. Oil Used to Fuel Productivity   A study on worldwide oil use was prepared for a major oil company. The study predicted that the amount of oil used to fuel productivity in a certain country is given by

$$f(t) = 1.5 + 1.8te^{-1.2t} \qquad (0 \le t \le 4)$$

where $f(t)$ denotes the number of barrels per $1000 of economic output and t is measured in decades ($t = 0$ corresponds to 1965). Compute $f'(0)$, $f'(1)$, $f'(2)$, and $f'(3)$, and interpret your results.

58. Percentage of Population Relocating   Based on data obtained from the Census Bureau, the manager of Plymouth Van Lines estimates that the percentage of the total population relocating in year t ($t = 0$ corresponds to the year 1960) may be approximated by the formula

$$P(t) = 20.6e^{-0.009t} \qquad (0 \le t \le 35)$$

Compute $P'(10)$, $P'(20)$, and $P'(30)$ and interpret your results.

USING TECHNOLOGY

EXAMPLE 1 At the beginning of Section 5.4, we demonstrated via a table of values of $(e^h - 1)/h$ for selected values of h the plausibility of the result

$$\lim_{h \to 0} \frac{e^h - 1}{h} = 1$$

In order to obtain a visual confirmation of this result, we plot the graph of

$$f(x) = \frac{e^x - 1}{x}$$

in the viewing rectangle $[-1, 1] \times [0, 2]$ (see Figure T1). From the graph of f, we see that $f(x)$ appears to approach 1 as x approaches 0.

Figure T1
The graph of f in the viewing rectangle $[-1, 1] \times [0, 2]$.

The numerical derivative function of a graphing utility will yield the derivative of an exponential or logarithmic function for any value of x, just as it did for algebraic functions.*

EXERCISES

In exercises 1–6, use the numerical derivative operation of a graphing utility to find the rate of change of f(x) at the given value of x. Give your answer accurate to four decimal places.

1. $f(x) = x^3 e^{-1/x}$; $x = -1$

2. $f(x) = (\sqrt{x} + 1)^{3/2} e^{-x}$; $x = 0.5$

3. $f(x) = x^3 \sqrt{\ln x}$; $x = 2$

4. $f(x) = \dfrac{\sqrt{x} \ln x}{x + 1}$; $x = 3.2$

5. $f(x) = e^{-x} \ln(2x + 1)$; $x = 0.5$

6. $f(x) = \dfrac{e^{-\sqrt{x}}}{\ln(x^2 + 1)}$; $x = 1$

7. **An Extinction Situation** The number of saltwater crocodiles in a certain area of northern Australia is given by

$$P(t) = \frac{300e^{-0.024t}}{5e^{-0.024t} + 1}$$

a. How many crocodiles were in the population initially?

b. Show that $\lim\limits_{t \to \infty} P(t) = 0$.

* The rules for differentiating logarithmic functions will be covered in Section 5.5. However, the exercises given here can be done without using these rules.

c. Use a graphing calculator to plot the graph of P in the viewing rectangle $[0, 200] \times [0, 70]$.

REMARK   This phenomenon is referred to as an *extinction* situation. ● ● ●

8. **Income of American Families**   Based on data compiled by the House Budget Committee, the House Ways and Means Committee, and the U.S. Census Bureau, it is estimated that the number of American families y (in millions) who earned x thousand dollars in 1990 is related by the equation

$$y = 0.1584xe^{-0.0000016x^3 + 0.00011x^2 - 0.04491x} \qquad (x > 0)$$

a. Use a graphing utility to plot the graph of the equation in the viewing rectangle $[0, 150] \times [0, 2]$.
b. How fast is y changing with respect to x when $x = 10$? When $x = 50$? Interpret your results.
Source: House Budget Committee, House Ways and Means Committee, and U.S. Census Bureau

9. **World Population Growth**   According to a study conducted by the United Nations Population Division, the world population in billions is approximated by the function

$$f(t) = \frac{12}{1 + 3.74914e^{-1.42804t}} \qquad (0 \le t \le 4)$$

where t is measured in half-centuries, with $t = 0$ corresponding to the beginning of 1950.

a. Use a graphing utility to plot the graph of f in the viewing rectangle $[0, 5] \times [0, 14]$.
b. How fast is the world population expected to increase at the beginning of the year 2000?
Source: United Nations Population Division

10. **Increase in Juvenile Offenders**   The number of youths aged 15 to 19 will increase by 21% between 1994 and 2005, pushing the crime rate up. According to the National Council on Crime and Delinquency, the number of violent crime arrests of juveniles under age 18 in year t is given by

$$f(t) = -0.438t^2 + 9.002t + 107 \qquad (0 \le t \le 13)$$

where $f(t)$ is measured in thousands and t in years, with $t = 0$ corresponding to 1989. According to the same source, if trends like inner-city drug use and wider availability of guns continues, then the number of violent crime arrests of juveniles under age 18 in year t will be

given by

$$g(t) = \begin{cases} -0.438t^2 + 9.002t + 107 & \text{if } 0 \le t < 4 \\ 99.456e^{0.07824t} & \text{if } 4 \le t \le 13 \end{cases}$$

where $g(t)$ is measured in thousands and $t = 0$ corresponds to 1989.
a. Compute $f(11)$ and $g(11)$ and interpret your results.
b. Compute $f'(11)$ and $g'(11)$ and interpret your results.
Source: National Council on Crime and Delinquency

11. **Increasing Crop Yields**   If left untreated on bean stems, aphids (small insects that suck plant juices) will multiply at an increasing rate during the summer months and reduce productivity and crop yield of cultivated crops. But if the aphids are treated in mid-June, the numbers decrease sharply to less than 100 per bean stem, allowing for steep rises in crop yield. The function

$$F(t) = \begin{cases} 62e^{1.152t} & \text{if } 0 \le t < 1.5 \\ 349e^{-1.324(t-1.5)} & \text{if } 1.5 \le t \le 3 \end{cases}$$

gives the number of aphids in a typical bean stem at time t, where t is measured in months, with $t = 0$ corresponding to the beginning of May.
Source: The Random House Encyclopedia
a. How many aphids are there on a typical bean stem at the beginning of June $(t = 1)$? At the beginning of July $(t = 2)$?
b. How fast is the population of aphids changing at the beginning of June? At the beginning of July?

12. **Percentage of Females in the Labor Force**   Based on data from the United States Census Bureau, the chief economist of Manpower, Inc., constructed the following formula, giving the percentage of the total female population in the civilian labor force, $P(t)$, at the beginning of the tth decade ($t = 0$ corresponds to the year 1900):

$$P(t) = \frac{74}{1 + 2.6e^{-0.166t + 0.04536t^2 - 0.0066t^3}} \qquad (0 \le t \le 11)$$

Assume this trend continues for the rest of the 20th century.
a. What will the percentage of the total female population in the civilian labor force be at the beginning of the year 2000?
b. What will the growth rate of the percentage of the total female population in the civilian labor force be at the beginning of the year 2000?
Source: U.S. Census Bureau

SOLUTIONS TO SELF-CHECK EXERCISES 5.4

1. a. Using the Product Rule, we obtain

$$f'(x) = x\frac{d}{dx}e^{-x} + e^{-x}\frac{d}{dx}x$$

$$= -xe^{-x} + e^{-x} = (1-x)e^{-x}$$

Using the Product Rule once again, we obtain

$$f''(x) = (1-x)\frac{d}{dx}e^{-x} + e^{-x}\frac{d}{dx}(1-x)$$

$$= (1-x)(-e^{-x}) + e^{-x}(-1)$$

$$= -e^{-x} + xe^{-x} - e^{-x} = (x-2)e^{-x}$$

b. Setting $f'(x) = 0$ gives

$$(1-x)e^{-x} = 0$$

Since $e^{-x} \neq 0$, we see that $1 - x = 0$, and this gives $x = 1$ as the only critical point of f. The sign diagram of f' shown in the accompanying figure tells us that the point $(1, e^{-1})$ is a relative maximum of f.

c. Setting $f''(x) = 0$ gives $x - 2 = 0$, so $x = 2$ is a candidate for an inflection point of f. The sign diagram of f'' (accompanying figure) shows that $(2, 2e^{-2})$ is an inflection point of f.

2. The rate of change of the book value of the asset t years from now is

$$V'(t) = 50,000\frac{d}{dt}e^{-0.4t}$$

$$= 50,000(-0.4)e^{-0.4t} = -20,000e^{-0.4t}$$

Therefore, 3 years from now the book value of the asset will be changing at the rate of

$$V'(3) = -20,000e^{-0.4(3)} = -20,000e^{-1.2} \approx -6023.88$$

that is, decreasing at the rate of approximately $6024 per year.

5.5 DIFFERENTIATION OF LOGARITHMIC FUNCTIONS

The Derivative of ln x

Let us now turn our attention to the differentiation of logarithmic functions.

RULE 3 DERIVATIVE OF ln x	$\dfrac{d}{dx} \ln |x| = \dfrac{1}{x}$ $(x \neq 0)$

To derive Rule 3, suppose that $x > 0$ and write $f(x) = \ln x$ in the equivalent form

$$x = e^{f(x)}$$

Differentiating both sides of the equation with respect to x, we find, using the Chain Rule,

$$1 = e^{f(x)} \cdot f'(x)$$

from which we see that

$$f'(x) = \frac{1}{e^{f(x)}}$$

or, since $e^{f(x)} = x$,

$$f'(x) = \frac{1}{x}$$

as we set out to show. You are asked to prove the rule for the case $x < 0$ in exercise 59.

EXAMPLE 1   Compute the derivative of each of the following functions:

a. $f(x) = x \ln x$ **b.** $g(x) = \dfrac{\ln x}{x}$

Solution

a. Using the Product Rule, we obtain

$$f'(x) = \frac{d}{dx}(x \ln x) = x \frac{d}{dx}(\ln x) + (\ln x)\frac{d}{dx}(x)$$

$$= x\left(\frac{1}{x}\right) + \ln x = 1 + \ln x$$

b. Using the Quotient Rule, we obtain

$$g'(x) = \frac{x\dfrac{d}{dx}(\ln x) - (\ln x)\dfrac{d}{dx}(x)}{x^2} = \frac{x\left(\dfrac{1}{x}\right) - \ln x}{x^2} = \frac{1 - \ln x}{x^2}$$

◦ ◦ ◦

You can derive the formula for the derivative of $f(x) = \ln x$ directly from the definition of the derivative as follows.

a. Show that

$$f'(x) = \lim_{h \to 0} \frac{f(x + h) - f(x)}{h} = \lim_{h \to 0} \ln \left(1 + \frac{h}{x}\right)^{1/h}$$

b. Put $m = x/h$ and note that $m \to \infty$ as $h \to 0$. Furthermore, $f'(x)$ can be written in the form

$$f'(x) = \lim_{m \to \infty} \ln \left(1 + \frac{1}{m}\right)^{m/x}$$

c. Finally, use both the fact that the natural logarithmic function is continuous and the definition of the number e to show that

$$f'(x) = \frac{1}{x} \ln \left[\lim_{m \to \infty} \left(1 + \frac{1}{m}\right)^{m}\right] = \frac{1}{x}$$

The Chain Rule for Logarithmic Functions

To enlarge the class of logarithmic functions to be differentiated, we appeal once more to the Chain Rule to obtain the following rule for differentiating composite functions of the form $h(x) = \ln f(x)$, where $f(x)$ is assumed to be a positive differentiable function.

RULE 4   THE CHAIN RULE FOR LOGARITHMIC FUNCTIONS

If $f(x)$ is a differentiable function, then

$$\frac{d}{dx}[\ln f(x)] = \frac{f'(x)}{f(x)} \qquad (f(x) > 0)$$

To see this, observe that $h(x) = g[f(x)]$, where $g(x) = \ln x \ (x > 0)$. Since $g'(x) = 1/x$, we have, using the Chain Rule,

$$h'(x) = g'(f(x))f'(x)$$
$$= \frac{1}{f(x)} f'(x) = \frac{f'(x)}{f(x)}$$

Observe that in the special case $f(x) = x$, $h(x) = \ln x$, so the derivative of h is, by Rule 3, given by $h'(x) = 1/x$.

EXAMPLE 2 Find the derivative of the function $f(x) = \ln(x^2 + 1)$.

Solution Using Rule 4, we see immediately that

$$f'(x) = \frac{\frac{d}{dx}(x^2 + 1)}{x^2 + 1} = \frac{2x}{x^2 + 1}$$

When differentiating functions involving logarithms, the rules of logarithms may be used to advantage, as shown in Examples 3 and 4.

EXAMPLE 3 Differentiate the function $y = \ln[(x^2 + 1)(x^3 + 2)^6]$.

Solution We first rewrite the given function using the properties of logarithms:

$$\begin{aligned}
y &= \ln[(x^2 + 1)(x^3 + 2)^6] \\
&= \ln(x^2 + 1) + \ln(x^3 + 2)^6 \qquad (\ln mn = \ln m + \ln n) \\
&= \ln(x^2 + 1) + 6\ln(x^3 + 2) \qquad (\ln m^n = n\ln m)
\end{aligned}$$

Differentiating and using Rule 4, we obtain

$$\begin{aligned}
y' &= \frac{\frac{d}{dx}(x^2 + 1)}{x^2 + 1} + \frac{6\frac{d}{dx}(x^3 + 2)}{x^3 + 2} \\
&= \frac{2x}{x^2 + 1} + \frac{6(3x^2)}{x^3 + 2} = \frac{2x}{x^2 + 1} + \frac{18x^2}{x^3 + 2}
\end{aligned}$$

EXPLORING WITH TECHNOLOGY

Use a graphing utility to plot the graphs of $f(x) = \ln x$; its first derivative function, $f'(x) = 1/x$; and its second derivative function, $f''(x) = -1/x^2$, using the same viewing rectangle $[0, 4] \times [-3, 3]$.

a. Describe the properties of the graph of f revealed by studying the graph of $f'(x)$. What can you say about the rate of increase of f for large values of x?

b. Describe the properties of the graph of f revealed by studying the graph of $f''(x)$. What can you say about the concavity of f for large values of x?

EXAMPLE 4 Find the derivative of the function $g(t) = \ln(t^2 e^{-t^2})$.

Solution Here again, to save a lot of work, we first simplify the given expression using the properties of logarithms. We have

$$g(t) = \ln(t^2 e^{-t^2})$$

$$= \ln t^2 + \ln e^{-t^2} \qquad (\ln mn = \ln m + \ln n)$$

$$= 2 \ln t - t^2 \qquad (\ln m^n = n \ln m \text{ and } \ln e = 1)$$

Therefore,
$$g'(t) = \frac{2}{t} - 2t = \frac{2(1 - t^2)}{t}$$

Logarithmic Differentiation

As we saw in the last two examples, the task of finding the derivative of a given function can be made easier by first applying the laws of logarithms to simplify the function. We now illustrate a process called **logarithmic differentiation,** which not only simplifies the calculation of the derivatives of certain functions but also enables us to compute the derivatives of functions we could not otherwise differentiate using the techniques developed thus far.

EXAMPLE 5 Differentiate $y = x(x + 1)(x^2 + 1)$ using logarithmic differentiation.

Solution First we take the natural logarithm on both sides of the given equation, obtaining

$$\ln y = \ln x(x + 1)(x^2 + 1)$$

Next, we use the properties of logarithms to rewrite the right-hand side of this equation, obtaining

$$\ln y = \ln x + \ln(x + 1) + \ln(x^2 + 1)$$

If we differentiate both sides of this equation, we have

$$\frac{d}{dx} \ln y = \frac{d}{dx}[\ln x + \ln(x + 1) + \ln(x^2 + 1)]$$

$$= \frac{1}{x} + \frac{1}{x + 1} + \frac{2x}{x^2 + 1} \qquad \text{(Using Rule 4)}$$

In order to evaluate the expression on the left-hand side, note that y is a function of x. Therefore, writing $y = f(x)$ to remind us of this fact, we have

$$\frac{d}{dx} \ln y = \frac{d}{dx} \ln[f(x)] \qquad [\text{Writing } y = f(x)]$$

$$= \frac{f'(x)}{f(x)} \qquad \text{(Using Rule 4)}$$

$$= \frac{y'}{y} \qquad [\text{Returning to using } y \text{ instead of } f(x)]$$

Therefore, we have

$$\frac{y'}{y} = \frac{1}{x} + \frac{1}{x+1} + \frac{2x}{x^2+1}$$

Finally, solving for y', we have

$$y' = y\left(\frac{1}{x} + \frac{1}{x+1} + \frac{2x}{x^2+1}\right)$$

$$= x(x+1)(x^2+1)\left(\frac{1}{x} + \frac{1}{x+1} + \frac{2x}{x^2+1}\right) \qquad \circ \ \circ \ \circ$$

Before considering other examples, let us summarize the important steps involved in logarithmic differentiation.

FINDING $\frac{dy}{dx}$ BY LOGARITHMIC DIFFERENTIATION

1. Take the natural logarithm on both sides of the equation and use the properties of logarithms to write any "complicated expression" as a sum of simpler terms.

2. Differentiate both sides of the equation with respect to x.

3. Solve the resulting equation for $\frac{dy}{dx}$.

EXAMPLE 6   Differentiate $y = x^2(x-1)(x^2+4)^3$.

Solution   Taking the natural logarithm on both sides of the given equation and using the laws of logarithms, we obtain

$$\ln y = \ln x^2(x-1)(x^2+4)^3$$

$$= \ln x^2 + \ln(x-1) + \ln(x^2+4)^3$$

$$= 2\ln x + \ln(x-1) + 3\ln(x^2+4)$$

Differentiating both sides of the equation with respect to x, we have

$$\frac{d}{dx}\ln y = \frac{y'}{y} = \frac{2}{x} + \frac{1}{x-1} + 3\cdot\frac{2x}{x^2+4}$$

Finally, solving for y', we have

$$y' = y\left(\frac{2}{x} + \frac{1}{x-1} + \frac{6x}{x^2+4}\right)$$

$$= x^2(x-1)(x^2+4)^3\left(\frac{2}{x} + \frac{1}{x-1} + \frac{6x}{x^2+4}\right) \qquad \circ \ \circ \ \circ$$

EXAMPLE 7   Find the derivative of $f(x) = x^x$ $(x > 0)$.

Solution   A word of caution! This function is neither a power function nor an exponential function. Taking the natural logarithm on both sides of the equation gives

$$\ln f(x) = \ln x^x = x \ln x$$

Differentiating both sides of the equation with respect to x, we obtain

$$\frac{f'(x)}{f(x)} = x \frac{d}{dx} \ln x + (\ln x) \frac{d}{dx} x$$

$$= x \left(\frac{1}{x} \right) + \ln x$$

$$= 1 + \ln x$$

Therefore,

$$f'(x) = f(x)(1 + \ln x) = x^x(1 + \ln x) \qquad \circ \ \circ \ \circ$$

EXPLORING WITH TECHNOLOGY

Refer to Example 7.

a. Use a graphing utility to plot the graph of $f(x) = x^x$ using the viewing rectangle $[0, 2] \times [0, 2]$. Then use **ZOOM** and **TRACE** to show that

$$\lim_{x \to 0^+} f(x) = 1$$

b. Use the results of part (a) and Example 7 to show that $\lim\limits_{x \to 0^+} f'(x) = -\infty$. Justify your answer.

$$\circ \quad \circ \quad \circ$$

SELF-CHECK EXERCISES 5.5

1. Find an equation of the tangent line to the graph of $f(x) = x \ln(2x + 3)$ at the point $(-1, 0)$.

2. Use logarithmic differentiation to compute y' given $y = (2x + 1)^3(3x + 4)^5$.

Solutions to Self-Check Exercises 5.5 can be found on page 412.

5.5 EXERCISES

In exercises 1–32, find the derivative of the function.

1. $f(x) = 5 \ln x$

2. $f(x) = \ln 5x$

3. $f(x) = \ln(x + 1)$

4. $g(x) = \ln(2x + 1)$

5. $f(x) = \ln x^8$

6. $h(t) = 2 \ln t^5$

7. $f(x) = \ln \sqrt{x}$

8. $f(x) = \ln(\sqrt{x} + 1)$

9. $f(x) = \ln \dfrac{1}{x^2}$

10. $f(x) = \ln \dfrac{1}{2x^3}$

11. $f(x) = \ln(4x^2 - 6x + 3)$

12. $f(x) = \ln(3x^2 - 2x + 1)$

13. $f(x) = \ln \dfrac{2x}{x + 1}$

14. $f(x) = \ln \dfrac{x + 1}{x - 1}$

15. $f(x) = x^2 \ln x$

16. $f(x) = 3x^2 \ln 2x$

17. $f(x) = \dfrac{2 \ln x}{x}$

18. $f(x) = \dfrac{3 \ln x}{x^2}$

19. $f(u) = \ln(u - 2)^3$

20. $f(x) = \ln(x^3 - 3)^4$

21. $f(x) = \sqrt{\ln x}$

22. $f(x) = \sqrt{\ln x + x}$

23. $f(x) = (\ln x)^3$

24. $f(x) = 2(\ln x)^{3/2}$

25. $f(x) = \ln(x^3 + 1)$

26. $f(x) = \ln \sqrt{x^2 - 4}$

27. $f(x) = e^x \ln x$

28. $f(x) = e^x \ln \sqrt{x + 3}$

29. $f(t) = e^{2t} \ln(t + 1)$

30. $g(t) = t^2 \ln(e^{2t} + 1)$

31. $f(x) = \dfrac{\ln x}{x}$

32. $g(t) = \dfrac{t}{\ln t}$

In exercises 33–36, find the second derivative of the function.

33. $f(x) = \ln 2x$

34. $f(x) = \ln(x + 5)$

35. $f(x) = \ln(x^2 + 2)$

36. $f(x) = (\ln x)^2$

In exercises 37–46, use logarithmic differentiation to find the derivative of the function.

37. $y = (x + 1)^2(x + 2)^3$

38. $y = (3x + 2)^4(5x - 1)^2$

39. $y = (x - 1)^2(x + 1)^3(x + 3)^4$

40. $y = \sqrt{3x + 5}(2x - 3)^4$

41. $y = \dfrac{(2x^2 - 1)^5}{\sqrt{x + 1}}$

42. $y = \dfrac{\sqrt{4 + 3x^2}}{\sqrt[3]{x^2 + 1}}$

43. $y = 3^x$

44. $y = x^{x+2}$

45. $y = (x^2 + 1)^x$

46. $y = x^{\ln x}$

47. Find an equation of the tangent line to the graph of $y = x \ln x$ at the point $(1, 0)$.

48. Find an equation of the tangent line to the graph of $y = \ln x^2$ at the point $(2, \ln 4)$.

49. Determine the intervals where the function $f(x) = \ln x^2$ is increasing and where it is decreasing.

50. Determine the intervals where the function $f(x) = \dfrac{\ln x}{x}$ is increasing and where it is decreasing.

51. Determine the intervals of concavity for the function $f(x) = x^2 + \ln x^2$.

52. Determine the intervals of concavity for the function $f(x) = \dfrac{\ln x}{x}$.

53. Find the inflection points of the function $f(x) = \ln(x^2 + 1)$.

54. Find the inflection points of the function $f(x) = x^2 \ln x$.

55. Find the absolute extrema of the function $f(x) = x - \ln x$ on $[\frac{1}{2}, 3]$.

56. Find the absolute extrema of the function $g(x) = x/\ln x$ on $[2, \infty)$.

In exercises 57 and 58, use the guidelines on page 315 to sketch the graph of the given function.

57. $f(x) = \ln(x - 1)$

58. $f(x) = 2x - \ln x$

59. Prove that $\dfrac{d}{dx} \ln |x| = \dfrac{1}{x}$ $(x \neq 0)$ for the case $x < 0$.

SOLUTIONS TO SELF-CHECK EXERCISES 5.5

1. The slope of the tangent line to the graph of f at any point $(x, f(x))$ lying on the graph of f is given by $f'(x)$. Using the Product Rule, we find

$$f'(x) = \frac{d}{dx}[x \ln(2x + 3)]$$

$$= x \frac{d}{dx} \ln(2x + 3) + \ln(2x + 3) \cdot \frac{d}{dx}(x)$$

$$= x \left(\frac{2}{2x + 3} \right) + \ln(2x + 3) \cdot 1$$

$$= \frac{2x}{2x + 3} + \ln(2x + 3)$$

In particular, the slope of the tangent line to the graph of f at the point $(-1, 0)$ is

$$f'(-1) = \frac{-2}{-2 + 3} + \ln 1 = -2$$

Therefore, using the point-slope form of the equation of a line, we see that a required equation is

$$y - 0 = -2(x + 1)$$

or

$$y = -2x - 2$$

2. Taking the logarithm on both sides of the equation gives

$$\ln y = \ln(2x + 1)^3(3x + 4)^5$$

$$= \ln(2x + 1)^3 + \ln(3x + 4)^5$$

$$= 3 \ln(2x + 1) + 5 \ln(3x + 4)$$

Differentiating both sides of the equation with respect to x, keeping in mind that y is a function of x, we obtain

$$\frac{d}{dx}(\ln y) = \frac{y'}{y} = 3 \cdot \frac{2}{2x + 1} + 5 \cdot \frac{3}{3x + 4}$$

$$= 3 \left[\frac{2}{2x + 1} + \frac{5}{3x + 4} \right]$$

$$= \left(\frac{6}{2x + 1} + \frac{15}{3x + 4} \right)$$

and

$$y' = (2x + 1)^3(3x + 4)^5 \cdot \left(\frac{6}{2x + 1} + \frac{15}{3x + 4} \right)$$

5.6 EXPONENTIAL FUNCTIONS AS MATHEMATICAL MODELS

Exponential Growth

Figure 5.13
Exponential growth.

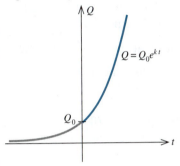

Many problems arising from practical situations can be described mathematically in terms of exponential functions or functions closely related to the exponential function. In this section, we look at some applications involving exponential functions from the fields of the life and social sciences.

In Section 5.1, we saw that the exponential function $f(x) = b^x$ is an increasing function when $b > 1$. In particular, the function $f(x) = e^x$ shares this property. From this result one may deduce that the function $Q(t) = Q_0 e^{kt}$, where Q_0 and k are positive constants, has the following properties:

1. $Q(0) = Q_0$
2. $Q(t)$ increases "rapidly" without bound as t increases without bound (Figure 5.13).

Property 1 follows from the computation

$$Q(0) = Q_0 e^0 = Q_0$$

Next, to study the rate of change of the function $Q(t)$, we differentiate it with respect to t, obtaining

$$Q'(t) = \frac{d}{dt}(Q_0 e^{kt})$$

$$= Q_0 \frac{d}{dt}(e^{kt})$$

$$= kQ_0 e^{kt}$$

$$= kQ(t) \tag{12}$$

Since $Q(t) > 0$ (because Q_0 is assumed to be positive) and $k > 0$, we see that $Q'(t) > 0$ and so $Q(t)$ is an increasing function of t. Our computation has, in fact, shed more light on an important property of the function $Q(t)$. Equation (12) says that the rate of increase of the function $Q(t)$ is proportional to the amount $Q(t)$ of the quantity present at time t. The implication is that as $Q(t)$ increases, so does the *rate of increase* of $Q(t)$, resulting in a very rapid increase in $Q(t)$ as t increases without bound.

Thus, the exponential function

$$Q(t) = Q_0 e^{kt} \qquad (0 \le t < \infty) \tag{13}$$

provides us with a mathematical model of a quantity $Q(t)$ that is initially present in the amount of $Q(0) = Q_0$ and whose rate of growth at any time t is directly proportional to the amount of the quantity present at time t. Such a quantity is said to exhibit **exponential growth,** and the constant k is called the **growth constant.** Interest earned on a fixed deposit when compounded continuously exhibits exponential growth. Other examples of exponential growth follow.

 EXAMPLE 1 Under ideal laboratory conditions, the number of bacteria in a culture grows in accordance with the law $Q(t) = Q_0 e^{kt}$, where Q_0 denotes the number of bacteria initially present in the culture, k is some constant determined by the strain of bacteria under consideration, and t is the elapsed time measured in hours. Suppose that there are 10,000 bacteria present initially in the culture and 60,000 present 2 hours later.

a. How many bacteria will there be in the culture at the end of 4 hours?

b. What is the rate of growth of the population after 4 hours?

Solution

a. We are given that $Q(0) = Q_0 = 10,000$, so $Q(t) = 10,000 e^{kt}$. Next, the fact that 60,000 bacteria are present 2 hours later translates into $Q(2) = 60,000$. Thus

$$60,000 = 10,000 e^{2k}$$

$$e^{2k} = 6$$

Taking the natural logarithm on both sides of the equation, we obtain

$$\ln e^{2k} = \ln 6$$

$$2k = \ln 6 \qquad \text{(Since } \ln e = 1\text{)}$$

or $\qquad\qquad k \approx 0.8959$

Thus, the number of bacteria present at any time t is given by

$$Q(t) = 10,000 e^{0.8959t}$$

In particular, the number of bacteria present in the culture at the end of 4 hours is given by

$$Q(4) = 10,000 e^{0.8959(4)}$$

$$= 360,029$$

b. The rate of growth of the bacteria population at any time t is given by

$$Q'(t) = kQ(t)$$

Thus, using the result from (a), we find that the rate at which the population is growing at the end of 4 hours is

$$Q'(4) = kQ(4)$$

$$\approx (0.8959)(360,029)$$

$$\approx 322,550$$

or approximately 322,550 bacteria per hour.   ○ ○ ○

Figure 5.14
Exponential decay.

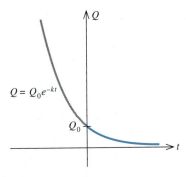

Exponential Decay

In contrast to exponential growth, a quantity exhibits **exponential decay** if it decreases at a rate that is directly proportional to its size. Such a quantity may be described by the exponential function

$$Q(t) = Q_0 e^{-kt} \qquad (t \in [0, \infty)) \tag{14}$$

where the positive constant Q_0 measures the amount present initially ($t = 0$) and k is some suitable positive number, called the **decay constant.** The choice of this number is determined by the nature of the substance under consideration. The graph of this function is sketched in Figure 5.14.

To verify the properties ascribed to the function $Q(t)$, we simply compute

$$Q(0) = Q_0 e^0 = Q_0$$

$$Q'(t) = \frac{d}{dt}(Q_0 e^{-kt})$$

and

$$= Q_0 \frac{d}{dt}(e^{-kt})$$

$$= -kQ_0 e^{-kt} = -kQ(t)$$

EXAMPLE 2   Radioactive substances decay exponentially. For example, the amount of radium present at any time t obeys the law $Q(t) = Q_0 e^{-kt}$, where Q_0 is the initial amount present and k is a suitable positive constant. The **half-life of a radioactive substance** is the time required for a given amount to be reduced by one-half. Now, it is known that the half-life of radium is approximately 1600 years. Suppose that initially there are 200 milligrams of pure radium. Find the amount left after t years. What is the amount left after 800 years?

Solution   The initial amount of radium present is 200 milligrams, so $Q(0) = Q_0 = 200$. Thus $Q(t) = 200e^{-kt}$. Next, the datum concerning the half-life of radium implies that $Q(1600) = 100$, and this gives

$$100 = 200e^{-1600k}$$

$$e^{-1600k} = \frac{1}{2}$$

Taking the natural logarithm on both sides of this equation yields

$$-1600k \ln e = \ln \frac{1}{2}$$

$$-1600k = \ln \frac{1}{2} \qquad (\ln e = 1)$$

$$k = -\frac{1}{1600} \ln \left(\frac{1}{2}\right) = 0.0004332$$

Therefore, the amount of radium left after t years is

$$Q(t) = 200e^{-0.0004332t}$$

In particular, the amount of radium left after 800 years is

$$Q(800) = 200e^{-0.0004332(800)} \approx 141.42$$

or approximately 141 milligrams.   ○ ○ ○

EXAMPLE 3   Carbon 14, a radioactive isotope of carbon, has a half-life of 5770 years. What is its decay constant?

Solution   We have $Q(t) = Q_0 e^{-kt}$. Since the half-life of the element is 5770 years, half of the substance is left at the end of that period. That is,

$$Q(5770) = Q_0 e^{-5770k} = \frac{1}{2} Q_0$$

or

$$e^{-5770k} = \frac{1}{2}$$

Taking the natural logarithm on both sides of this equation, we have

$$\ln e^{-5770k} = \ln \frac{1}{2}$$

$$-5770k = -0.693147$$

$$\approx 0.00012$$ ○ ○ ○

Carbon-14 dating is a well-known method used by anthropologists to establish the age of animal and plant fossils. This method assumes that the proportion of carbon 14 (C-14) present in the atmosphere has remained constant over the past 50,000 years. Professor Willard Libby, recipient of the Nobel Prize in chemistry in 1960, proposed this theory.

The amount of C-14 in the tissues of a living plant or animal is constant. However, when an organism dies, it stops absorbing new quantities of C-14 and the amount of C-14 in the remains diminishes because of the natural decay of the radioactive substance. Thus, the approximate age of a plant or animal fossil can be determined by measuring the amount of C-14 present in the remains.

EXAMPLE 4   A skull from an archeological site has one-tenth the amount of C-14 that it originally contained. Determine the approximate age of the skull.

Solution   Here

$$Q(t) = Q_0 e^{-kt}$$

$$= Q_0 e^{-0.00012t}$$

where Q_0 is the amount of C-14 present originally and k, the decay constant, is equal to 0.00012 (see Example 3). Since $Q(t) = (1/10)Q_0$, we have

$$\frac{1}{10}Q_0 = Q_0 e^{-0.00012t}$$

$$\ln\frac{1}{10} = -0.00012t \qquad \text{(Taking the natural logarithm on both sides)}$$

$$t = \frac{\ln\dfrac{1}{10}}{-0.00012}$$

$$\approx 19{,}200$$

or approximately 19,200 years.   ○ ○ ○

Learning Curves

The next example shows how the exponential function may be applied to describe certain types of learning processes. Consider the function

$$Q(t) = C - Ae^{-kt}$$

where C, A, and k are positive constants. To sketch the graph of the function Q, observe that its y-intercept is given by $Q(0) = C - A$. Next, we compute

$$Q'(t) = kAe^{-kt}$$

Since both k and A are positive, we see that $Q'(t) > 0$ for all values of t. Thus $Q(t)$ is an increasing function of t. Also,

$$\lim_{t\to\infty} Q(t) = \lim_{t\to\infty}(C - Ae^{-kt})$$

$$= \lim_{t\to\infty} C - \lim_{t\to\infty} Ae^{-kt}$$

$$= C$$

Figure 5.15
A learning curve.

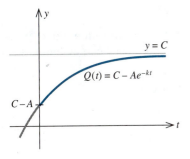

so $y = C$ is a horizontal asymptote of Q. Thus, $Q(t)$ increases and approaches the number C as t increases without bound. The graph of the function Q is shown in Figure 5.15, where that part of the graph corresponding to the negative values of t is drawn with a gray line since, in practice, one normally restricts the domain of the function to the interval $[0, \infty)$.

Observe that $Q(t)$ $(t > 0)$ increases rather rapidly initially but that the rate of increase slows down considerably after a while. To see this, we compute

$$\lim_{t\to\infty} Q'(t) = \lim_{t\to\infty} kAe^{-kt} = 0$$

This behavior of the graph of the function Q closely resembles the learning pattern experienced by workers engaged in highly repetitive work. For example, the productivity of an assembly-line worker increases very rapidly in the early stages of the training period. This productivity increase is a direct result of the worker's training and accumulated experience. But the rate of increase of productivity slows down as time goes by and the worker's productivity level approaches some fixed level due to the limitations of the worker and the machine. Because of this characteristic, the graph of the function $Q(t) = C - Ae^{-kt}$ is often called a **learning curve.**

 EXAMPLE 5   The Camera Division of the Eastman Optical Company produces a 35-mm single-lens reflex camera. Eastman's training department determines that after completing the basic training program, a new, previously inexperienced employee will be able to assemble

$$Q(t) = 50 - 30e^{-0.5t}$$

model F cameras per day, t months after the employee starts work on the assembly line.

a. How many model F cameras can a new employee assemble per day after basic training?

b. How many model F cameras can an employee with 1 month of experience assemble per day? An employee with 2 months of experience? An employee with 6 months of experience?

c. How many model F cameras can the average experienced employee assemble per day?

Solution

a. The number of model F cameras a new employee can assemble is given by

$$Q(0) = 50 - 30 = 20$$

b. The number of model F cameras that an employee with 1 month of experience, 2 months of experience, and 6 months of experience can assemble per day is given by

$$Q(1) = 50 - 30e^{-0.5} \approx 31.80$$
$$Q(2) = 50 - 30e^{-1} \approx 38.96$$
$$Q(6) = 50 - 30e^{-3} \approx 48.51$$

or approximately 32, 39, and 49, respectively.

c. As t increases without bound, $Q(t)$ approaches 50. Hence, the average experienced employee can ultimately be expected to assemble 50 model F cameras per day.   ◖ ◖ ◖

Other applications of the learning curve are found in models that describe the dissemination of information about a product or the velocity of an object dropped into a viscous medium.

Logistic Growth Functions

Our last example of an application of exponential functions to the description of natural phenomena involves the **logistic** (also called the **S-shaped,** or **sigmoi-**

Figure 5.16
A logistic curve.

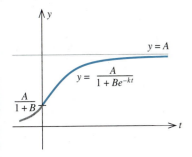

dal) curve, which is the graph of the function

$$Q(t) = \frac{A}{1 + Be^{-kt}}$$

where A, B, and k are positive constants. The function Q is called a **logistic growth function,** and the graph of the function Q is sketched in Figure 5.16.

Observe that $Q(t)$ increases rather rapidly for small values of t. In fact, for small values of t, the logistic curve resembles an exponential growth curve. However, the *rate of growth* of $Q(t)$ decreases quite rapidly as t increases and $Q(t)$ approaches the number A as t increases without bound.

Thus, the logistic curve exhibits both the property of rapid growth of the exponential growth curve as well as the "saturation" property of the learning curve. Because of these characteristics, the logistic curve serves as a suitable mathematical model for describing many natural phenomena. For example, if a small number of rabbits were introduced to a tiny island in the South Pacific, the rabbit population might be expected to grow very rapidly at first, but the growth rate would decrease quickly as overcrowding, scarcity of food, and other environmental factors affected it. The population would eventually stabilize at a level compatible with the life-support capacity of the environment. Models describing the spread of rumors and epidemics are other examples of the application of the logistic curve.

EXAMPLE 6   The number of soldiers at Fort MacArthur who contracted influenza after t days during a flu epidemic is approximated by the exponential model

$$Q(t) = \frac{5000}{1 + 1249e^{-kt}}$$

If 40 soldiers contracted the flu by the seventh day, find how many soldiers contracted the flu by the fifteenth day.

Solution   The given information implies that

$$Q(7) = 40$$

and

$$Q(7) = \frac{5000}{1 + 1249e^{-7k}}$$

$$= 40$$

Thus,

$$40(1 + 1249e^{-7k}) = 5000$$

$$1 + 1249e^{-7k} = \frac{5000}{40} = 125$$

or

$$e^{-7k} = \frac{124}{1249}$$

$$-7k = \ln\frac{124}{1249}$$

$$k = -\frac{\ln\dfrac{124}{1249}}{7} \approx 0.33$$

Therefore, the number of soldiers who contracted the flu after t days is given by

$$Q(t) = \frac{5000}{1 + 1249e^{-0.33t}}$$

In particular, the number of soldiers who contracted the flu by the fifteenth day is given by

$$Q(15) = \frac{5000}{1 + 1249e^{-15(0.33)}}$$

$$\approx 508$$

or approximately 508 soldiers.

○ ○ ○

EXPLORING WITH TECHNOLOGY

Refer to Example 6.

a. Use a graphing utility to plot the graph of the function Q using the viewing rectangle $[0, 40] \times [0, 5000]$.
b. Find how long it takes for the first 1000 soldiers to contract the flu. [*Suggestion:* Plot the graphs of $y_1 = Q(t)$ and $y_2 = 1000$ and find the point of intersection of the two graphs.]

○   ○   ○

SELF-CHECK EXERCISE 5.6

Suppose that the population of a country (in millions) at any time t grows in accordance with the rule

$$P = \left(P_0 + \frac{I}{k}\right)e^{kt} - \frac{I}{k}$$

where P denotes the population at any time t, k is a constant reflecting the natural growth rate of the population, I is a constant giving the (constant) rate of immigration into the country, and P_0 is the total population of the country at time $t = 0$. The population of the United States in the year 1980 ($t = 0$) was 226.5 million. If the natural growth rate is 0.8% annually ($k = 0.008$) and net immigration is allowed at the rate of half a million people per year ($I = 0.5$) until the end of the century, what will the population of the United States be in the year 2000?

Solutions to Self-Check Exercise 5.6 can be found on page 423.

5.6 EXERCISES

A calculator is recommended for this exercise set.

1. Exponential Growth   Given that a quantity $Q(t)$ is described by the exponential growth function

$$Q(t) = 400e^{0.05t}$$

where t is measured in minutes, answer the following questions.
a. What is the growth constant?
b. What quantity is present initially?
c. Using Table I in the appendix or a calculator, complete the following table of values:

t	0	10	20	100	1000
Q					

2. Exponential Decay   Given that a quantity $Q(t)$ exhibiting exponential decay is described by the function

$$Q(t) = 2000e^{-0.06t}$$

where t is measured in years, answer the following questions.
a. What is the decay constant?
b. What quantity is present initially?
c. Using Table I in the appendix or a calculator, complete the following table of values:

t	0	5	10	20	100
Q					

3. Growth of Bacteria   The growth rate of the bacterium *Escherichia coli*, a common bacterium found in the human intestine, is proportional to its size. Under ideal laboratory conditions, when this bacterium is grown in a nutrient broth medium, the number of cells in a culture doubles approximately every 20 minutes.
a. If the initial cell population is 100, determine the function $Q(t)$ that expresses the exponential growth of the number of cells of this bacterium as a function of time t (in minutes).
b. How long will it take for a colony of 100 cells to increase to a population of one million?
c. If the initial cell population were 1000, how would this alter our model?

4. World Population   The world population at the beginning of 1990 was 5.3 billion. Assume that the population continues to grow at its present rate of approximately 2% per year, and find the function $Q(t)$ that expresses the world population (in billions) as a function of time t (in years) where $t = 0$ corresponds to the beginning of 1990.
a. Using this function, complete the following table of values and sketch the graph of the function Q.

Year	1990	1995	2000	2005
World Population				

Year	2010	2015	2020	2025
World Population				

b. Find the estimated rate of growth in the year 2000.

5. World Population   Refer to exercise 4.
a. If the world population continues to grow at its present rate of approximately 2% per year, find the length of time t_0 required for the world population to triple in size.
b. Using the time t_0 found in (a), what would the world population be if the growth rate were reduced to 1.8%?

6. Resale Value   A certain piece of machinery was purchased 3 years ago by the Garland Mills Company for $500,000. Its present resale value is $320,000. Assuming that the machine's resale value decreases exponentially, what will it be 4 years from now?

7. Atmospheric Pressure   If the temperature is constant, then the atmospheric pressure P (in pounds per square inch) varies with the altitude above sea level h in accordance with the law

$$P = p_0 e^{-kh}$$

where p_0 is the atmospheric pressure at sea level and k is a constant. If the atmospheric pressure is 15 pounds per square inch at sea level and 12.5 pounds per square inch at 4000 feet, find the atmospheric pressure at an altitude of 12,000 feet.

8. **Radioactive Decay** The radioactive element polonium decays according to the law

$$Q(t) = Q_0 \cdot 2^{-(t/140)}$$

where Q_0 is the initial amount and the time t is measured in days. If the amount of polonium left after 280 days is 20 milligrams, what was the initial amount present?

9. **Radioactive Decay** Phosphorus 32 has a half-life of 14.2 days. If 100 grams of this substance are present initially, find the amount present after t days. What amount will be left after 7.1 days?

10. **Nuclear Fallout** Strontium 90, a radioactive isotope of strontium, is present in the fallout resulting from nuclear explosions. It is especially hazardous to animal life, including humans, because, upon ingestion of contaminated food, it is absorbed into the bone structure. Its half-life is 27 years. If the amount of strontium 90 in a certain area is found to be four times the "safe" level, find how much time must elapse before an "acceptable level" is reached.

11. **Carbon-14 Dating** Wood deposits recovered from an archeological site contain 20% of the carbon 14 they originally contained. How long ago did the tree from which the wood was obtained die?

12. **Carbon-14 Dating** Skeletal remains of the so-called "Pittsburgh Man," unearthed in Pennsylvania, had lost 82% of the carbon 14 they originally contained. Determine the approximate age of the bones.

13. **Learning Curves** The American Stenographic Institute finds that the average student taking Advanced Shorthand, an intensive 20-week course, progresses according to the function

$$Q(t) = 120(1 - e^{-0.05t}) + 60 \qquad (0 \leq t \leq 20)$$

where $Q(t)$ measures the number of words (per minute) of dictation that the student can take in shorthand after t weeks in the course. Sketch the graph of the function Q and answer the following questions.
a. What is the beginning shorthand speed for the average student in this course?
b. What shorthand speed does the average student attain halfway through the course?
c. How many words per minute can the average student take after completing this course?

14. **Effect of Advertising on Sales** The Metro Department Store found that t weeks after the end of a sales promotion the volume of sales was given by a function of the form

$$S(t) = B + Ae^{-kt} \qquad (0 \leq t \leq 4)$$

where $B = 50{,}000$ and is equal to the average weekly volume of sales before the promotion. The sales volumes at the end of the first and third weeks were $83,515 and $65,055, respectively. Assume that the sales volume is decreasing exponentially. Find:
a. the decay constant k
b. the sales volume at the end of the fourth week

15. **Demand for Computers** The Universal Instruments Company found that the monthly demand for its new line of Galaxy Home Computers t months after placing the line on the market was given by

$$D(t) = 2000 - 1500e^{-0.05t} \qquad (t > 0)$$

Graph this function and answer the following questions.
a. What is the demand after 1 month? One year? Two years? Five years?
b. At what level is the demand expected to stabilize?
c. Find the rate of growth of the demand after the tenth month.

16. **Newton's Law of Cooling** Newton's law of cooling states that the rate at which the temperature of an object changes is proportional to the difference in temperature between the object and that of the surrounding medium. Thus, the temperature $F(t)$ of an object that is greater than the temperature of its surrounding medium is given by

$$F(t) = T + Ae^{-kt}$$

where t is the time expressed in minutes, T is the temperature of the surrounding medium, and A and k are constants. Suppose that a cup of instant coffee is prepared with boiling water (212°F) and left to cool on the counter in a room where the temperature is 72°F. If $k = 0.1865$, determine when the coffee will be cool enough to drink (say, 110°F).

17. **Spread of an Epidemic** During a flu epidemic, the number of children in the Woodbridge Community School System who contracted influenza after t days was given by

$$Q(t) = \frac{1000}{1 + 199e^{-0.8t}}$$

a. How many children were stricken by the flu after the first day?
b. How many children had the flu after 10 days?
c. How many children eventually contracted the disease?

18. Growth of a Fruit-Fly Population On the basis of data collected during an experiment, a biologist found that the growth of the fruit fly (*Drosophila*) with a limited food supply could be approximated by the exponential model

$$N(t) = \frac{400}{1 + 39e^{-0.16t}}$$

where t denotes the number of days since the beginning of the experiment.

a. What was the initial fruit-fly population in the experiment?

b. What was the maximum fruit-fly population that could be expected under this laboratory condition?

c. What was the population of the fruit-fly colony on the twentieth day?

19. Percentage of Households with VCRs According to estimates by Paul Kroger Associates, the percentage of households that own videocassette recorders (VCRs) is given by

$$P(t) = \frac{68}{1 + 21.67e^{-0.62t}} \qquad (0 \le t \le 12)$$

where t is measured in years, with $t = 0$ corresponding to the beginning of 1985. What percentage of households owned VCRs at the beginning of 1985? At the beginning of 1995?

20. Population Growth in the 21st Century The population of the United States is approximated by the function

$$P(t) = \frac{616.5}{1 + 4.02e^{-0.5t}}$$

where $P(t)$ is measured in millions of people and t is measured in 30-year intervals, with $t = 0$ corresponding to 1930. What is the expected population of the United States in 2020 ($t = 3$)?

21. Spread of a Rumor Three hundred students attended the dedication ceremony of a new building on a college campus. The president of the traditionally female col-

lege announced a new expansion program, which included plans to make the college coeducational. The number of students who learned of the new program t hours later is given by the function

$$f(t) = \frac{3000}{1 + Be^{-kt}}$$

If 600 students on campus had heard about the new program 2 hours after the ceremony, how many students had heard about the policy after 4 hours?

22. Concentration of Glucose in the Bloodstream A glucose solution is administered intravenously into the bloodstream at a constant rate of r mg/hr. As the glucose is being administered, it is converted into other substances and removed from the bloodstream. Suppose the concentration of the glucose solution at time t is given by

$$C(t) = \frac{r}{k} - \left[\left(\frac{r}{k} \right) - C_0 \right] e^{-kt}$$

where C_0 is the concentration at time $t = 0$ and k is a constant.

a. Assuming that $C_0 < r/k$, evaluate

$$\lim_{t \to \infty} C(t)$$

and interpret your result.

b. Sketch the graph of the function C.

23. Gompertz Growth Curve Consider the function

$$Q(t) = Ce^{-Ae^{-kt}}$$

where $Q(t)$ is the size of a quantity at time t and A, C, and k are positive constants. The graph of this function, called the **Gompertz growth curve,** is used by biologists to describe restricted population growth.

a. Show that the function Q is always increasing.

b. Find the time t at which the growth rate $Q'(t)$ is increasing most rapidly.

[*Hint:* Find the inflection point of Q.]

c. Show that $\lim_{t \to \infty} Q(t) = C$ and interpret your result.

SOLUTION TO SELF-CHECK EXERCISE 5.6

We are given that $P_0 = 226.5$, $k = 0.008$, and $I = 0.5$. So

$$P = \left(226.5 + \frac{0.5}{0.008} \right) e^{0.008t} - \frac{0.5}{0.008}$$

$$= 289e^{0.008t} - 62.5$$

Therefore, the population in the year 2000 will be given by

$$P(20) = 289e^{0.16} - 62.5$$
$$\approx 276.6$$

or approximately 276.6 million.

CHAPTER 5 SUMMARY OF PRINCIPAL FORMULAS AND TERMS

Formulas

1. Exponential function with base b

$$y = b^x$$

2. The number e

$$e = \lim_{m \to \infty} \left(1 + \frac{1}{m}\right)^m = 2.71828$$

3. Exponential function with base e

$$y = e^x$$

4. Logarithmic function with base b

$$y = \log_b x$$

5. Logarithmic function with base e

$$y = \ln x$$

6. Inverse properties of $\ln x$ and e

$$\ln e^x = x \quad \text{and} \quad e^{\ln x} = x$$

7. Compound interest (accumulated amount)

$$A = P(1 + i)^n, \text{ where } i = r/m, \text{ and } n = mt$$

8. Effective rate of interest

$$r_{\text{eff}} = \left(1 + \frac{r}{m}\right)^m - 1$$

9. Compound interest (present value)

$$P = A(1 + i)^{-n}, \text{ where } i = r/m, \text{ and } n = mt$$

10. Continuous compound interest

$$A = Pe^{rt}$$

11. Derivative of the exponential function

$$\frac{d}{dx}(e^x) = e^x$$

12. Chain rule for exponential functions

$$\frac{d}{dx}(e^u) = e^u \frac{du}{dx}$$

13. Derivative of the logarithmic function

$$\frac{d}{dx} \ln |x| = \frac{1}{x}$$

14. Chain rule for logarithmic functions

$$\frac{d}{dx}(\ln u) = \frac{1}{u}\frac{du}{dx}$$

Terms

Common logarithm
Natural logarithm
Logarithmic differentiation
Exponential growth
Growth constant

Exponential decay
Decay constant
Half-life of a radioactive element
Logistic growth function

CHAPTER 5 REVIEW EXERCISES

1. Sketch on the same set of coordinate axes the graphs of the exponential functions defined by the equations.
   a. $y = 2^{-x}$ **b.** $y = (\frac{1}{2})^x$

In exercises 2 and 3, express each in logarithmic form.

2. $(\frac{2}{3})^{-3} = \frac{27}{8}$ 3. $16^{-3/4} = 0.125$

In exercises 4 and 5, solve each equation for x.

4. $\log_4(2x + 1) = 2$

5. $\ln(x - 1) + \ln 4 = \ln(2x + 4) - \ln 2$

In exercises 6–8, given that $\ln 2 = x$, $\ln 3 = y$, and $\ln 5 = z$, express each of the given logarithmic values in terms of x, y, and z.

6. $\ln 30$ 7. $\ln 3.6$ 8. $\ln 75$

9. Sketch the graph of the function $y = \log_2(x + 3)$.

10. Sketch the graph of the function $y = \log_3(x + 1)$.

In exercises 11–28, find the derivative of the function.

11. $f(x) = xe^{2x}$

12. $f(t) = \sqrt{t}e^t + t$

13. $g(t) = \sqrt{t}e^{-2t}$

14. $g(x) = e^x\sqrt{1 + x^2}$

15. $y = \dfrac{e^{2x}}{1 + e^{-2x}}$

16. $f(x) = e^{2x^2 - 1}$

17. $f(x) = xe^{-x^2}$

18. $g(x) = (1 + e^{2x})^{3/2}$

19. $f(x) = x^2e^x + e^x$

20. $g(t) = t \ln t$

21. $f(x) = \ln(e^{x^2} + 1)$

22. $f(x) = \dfrac{x}{\ln x}$

23. $f(x) = \dfrac{\ln x}{x + 1}$

24. $y = (x + 1)e^x$

25. $y = \ln(e^{4x} + 3)$

26. $f(r) = \dfrac{re^r}{1 + r^2}$

27. $f(x) = \dfrac{\ln x}{1 + e^x}$

28. $g(x) = \dfrac{e^{x^2}}{1 + \ln x}$

29. Find the second derivative of the function $y = \ln(3x + 1)$.

30. Find the second derivative of the function $y = x \ln x$.

31. Find $h'(0)$ if $h(x) = g(f(x))$, $g(x) = x + \dfrac{1}{x}$, and $f(x) = e^x$.

32. Find $h'(1)$ if $h(x) = g(f(x))$, $g(x) = \dfrac{x + 1}{x - 1}$, and $f(x) = \ln x$.

33. Use logarithmic differentiation to find the derivative of $f(x) = (2x^3 + 1)(x^2 + 2)^3$.

34. Use logarithmic differentiation to find the derivative of $f(x) = \dfrac{x(x^2 - 2)^2}{(x - 1)}$.

35. Find an equation of the tangent line to the graph of $y = e^{-2x}$ at the point $(1, e^{-2})$.

36. Find an equation of the tangent line to the graph of $y = xe^{-x}$ at the point $(1, e^{-1})$.

37. Sketch the graph of the function $f(x) = xe^{-2x}$.

38. Sketch the graph of the function $f(x) = x^2 - \ln x$.

39. Find the absolute extrema of the function $f(t) = te^{-t}$.

40. Find the absolute extrema of the function $g(t) = \dfrac{\ln t}{t}$ on $[1, 2]$.

41. A hotel was purchased by a conglomerate for $4.5 million and sold 5 years later for $8.2 million. Find the annual rate of return (compounded continuously).

42. Find the present value of $119,346 due in 4 years at an interest rate of 10% per year compounded continuously.

43. A culture of bacteria that initially contained 2000 bacteria has a count of 18,000 bacteria after 2 hours.
   a. Determine the function $Q(t)$ that expresses the exponential growth of the number of cells of this bacterium as a function of time t (in minutes).
   b. Find the number of bacteria present after 4 hours.

44. The radioactive element radium has a half-life of 1600 years. What is its decay constant?

45. The V.C.A. Television Company found that the monthly demand for its new line of video disc players t months after placing the players on the market is given by

$$D(t) = 4000 - 3000e^{-0.06t} \qquad (t \geq 0)$$

Graph this function and answer the following questions.
   a. What was the demand after 1 month? After 1 year? After 2 years?
   b. At what level is the demand expected to stabilize?

46. During a flu epidemic, the number of students at a certain university who contracted influenza after t days could be approximated by the exponential model

$$Q(t) = \dfrac{3000}{1 + 499e^{-kt}}$$

If 90 students contracted the flu by the tenth day, how many students contracted the flu by the twentieth day?

Differential calculus is concerned with the problem of finding the rate of change of one quantity with respect to another. In this chapter we begin the study of the other branch of calculus, known as integral calculus. Here we are interested in precisely the opposite problem: If we know the rate of change of one quantity with respect to another, can we find the relationship between the two quantities? The principal tool used in the study of integral calculus is the *antiderivative* of a function, and we develop rules for antidifferentiation, or *integration*, as the process of finding the antiderivative is called. We also show that a link is established between differential and integral calculus—via the Fundamental Theorem of Calculus.

How much will the solar cell panels cost? The head of Soloron Corporation's research and development department has projected that the cost of producing solar cell panels will drop at a certain rate in the next several years. In Example 7, page 448, you will see how this information can be used to predict the cost of solar cell panels in the coming years.

6

INTEGRATION

6.1 ANTIDERIVATIVES AND THE RULES OF INTEGRATION

Antiderivatives

Let us return, once again, to the example involving the motion of the maglev (Figure 6.1).

Figure 6.1
A maglev moving along an elevated monorail track.

In Chapter 2, we discussed the following problem:

> *If we know the position of the maglev at any time t, can we find its velocity at time t?*

As it turns out, if the position of the maglev is described by the position function f, then its velocity at any time t is given by $f'(t)$. Here f', the velocity function of the maglev, is just the derivative of f.

Now, in Chapters 6 and 7, we will consider precisely the opposite problem:

> *If we know the velocity of the maglev at any time t, can we find its position at time t?*

Stated another way, if we know the velocity function f' of the maglev, can we find its position function f?

In order to solve this problem, we need the concept of an *antiderivative* of a function.

ANTIDERIVATIVE   A function F is an **antiderivative** of f on an interval I if $F'(x) = f(x)$ for all x in I.

Thus, an antiderivative of a function f is a function F whose derivative is f. For example, $F(x) = x^2$ is an antiderivative of $f(x) = 2x$ because

$$F'(x) = \frac{d}{dx}(x^2) = 2x = f(x)$$

and $F(x) = x^3 + 2x + 1$ is an antiderivative of $f(x) = 3x^2 + 2$ because

$$F'(x) = \frac{d}{dx}(x^3 + 2x + 1) = 3x^2 + 2 = f(x)$$

EXAMPLE 1 Let $F(x) = \frac{1}{3}x^3 - 2x^2 + x - 1$. Show that F is an antiderivative of $f(x) = x^2 - 4x + 1$.

Solution Differentiating the function F, we obtain

$$F'(x) = x^2 - 4x + 1 = f(x)$$

and the desired result follows.   ❍ ❍ ❍

EXAMPLE 2 Let $F(x) = x$, $G(x) = x + 2$, and $H(x) = x + C$, where C is a constant. Show that F, G, and H are all antiderivatives of the function f defined by $f(x) = 1$.

Solution Since

$$F'(x) = \frac{d}{dx}(x) = 1 = f(x)$$

$$G'(x) = \frac{d}{dx}(x + 2) = 1 = f(x)$$

and $$H'(x) = \frac{d}{dx}(x + C) = 1 = f(x)$$

we see that F, G, and H are indeed antiderivatives of f.   ❍ ❍ ❍

Example 2 shows that once an antiderivative G of a function f is known, then another antiderivative of f may be found by adding an arbitrary constant to the function G. The following theorem states that no function other than one obtained in this manner can be an antiderivative of f. (We omit the proof.)

THEOREM 1

Let G be an antiderivative of a function f. Then every antiderivative F of f must be of the form $F(x) = G(x) + C$, where C is a constant.

Returning to Example 2, we see that there are infinitely many antiderivatives of the function $f(x) = 1$. We obtain each one by specifying the constant C in the function $F(x) = x + C$. Figure 6.2 shows the graphs of some of these

Figure 6.2
The graphs of some antiderivatives
of $f(x) = 1$.

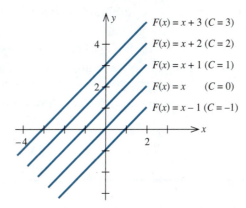

$F(x) = x + 3 \ (C = 3)$

$F(x) = x + 2 \ (C = 2)$

$F(x) = x + 1 \ (C = 1)$

$F(x) = x \qquad (C = 0)$

$F(x) = x - 1 \ (C = -1)$

antiderivatives for selected values of C. These graphs constitute part of a family of infinitely many parallel straight lines, each having a slope equal to 1. This result is expected since there are infinitely many curves (straight lines) with a given slope equal to 1. The antiderivatives $F(x) = x + C$ (C, a constant) are precisely the functions representing this family of straight lines.

EXAMPLE 3   Prove that the function $G(x) = x^2$ is an antiderivative of the function $f(x) = 2x$. Write a general expression for the antiderivatives of f.

Solution   Since $G'(x) = 2x = f(x)$, we have shown that $G(x) = x^2$ is an antiderivative of $f(x) = 2x$. By Theorem 1, every antiderivative of the function $f(x) = 2x$ has the form $F(x) = x^2 + C$, where C is some constant. The graphs of a few of the antiderivatives of f are shown in Figure 6.3.

Figure 6.3
The graphs of some antiderivatives of $f(x) = 2x$.

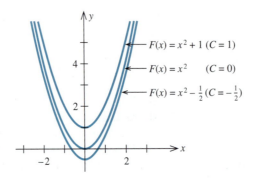

$F(x) = x^2 + 1$ ($C = 1$)

$F(x) = x^2$ ($C = 0$)

$F(x) = x^2 - \frac{1}{2}$ ($C = -\frac{1}{2}$)

EXPLORING WITH TECHNOLOGY

Let $f(x) = x^2 - 1$.

1. Show that $F(x) = \frac{1}{3}x^3 - x + C$, where C is an arbitrary constant, is an antiderivative of f.

2. Use a graphing utility to plot the graphs of the antiderivatives of f corresponding to $C = -2$, $C = -1$, $C = 0$, $C = 1$, and $C = 2$ on the same set of axes using the viewing rectangle $[-4, 4] \times [-4, 4]$.

3. If your graphing utility has the capability, draw the tangent line to each of the graphs in part (2) at the point whose x-coordinate is 2. What can you say about this family of tangent lines?

4. What is the slope of a tangent line in this family? Explain how you obtained your answer.

The Indefinite Integral

The process of finding all antiderivatives of a function is called **antidifferentiation, or integration.** We use the symbol $\int$, called an **integral sign,** to indicate that the operation of integration is to be performed on some function f. Thus,

$$\int f(x)\,dx = F(x) + C$$

(read "the indefinite integral of $f(x)$ with respect to x equals $F(x)$ plus C") tells us that the **indefinite integral** of f is the family of functions given by $F(x) + C$, where $F'(x) = f(x)$. The function f to be integrated is called the **integrand,** and the constant C is called a **constant of integration.** The expression dx following the integrand $f(x)$ reminds us that the operation is performed with respect to x. If the independent variable is t, we write $\int f(t)\,dt$ instead. In this sense both t and x are "dummy variables."

Using this notation, we can write the results of Examples 2 and 3 as

$$\int 1\,dx = x + C \qquad \text{and} \qquad \int 2x\,dx = x^2 + K$$

where C and K are arbitrary constants.

Basic Integration Rules

Our next task is to develop some rules for finding the indefinite integral of a given function f. Because integration and differentiation are reverse operations, we discover many of the rules of integration by first making an "educated guess" at the antiderivative F of the function f to be integrated. Then this result is verified by demonstrating that $F' = f$.

RULE I INDEFINITE INTEGRAL OF A CONSTANT

$$\int k\,dx = kx + C \qquad (k,\text{ a constant})$$

To prove this result, observe that

$$F'(x) = \frac{d}{dx}(kx + C) = k$$

EXAMPLE 4 Evaluate each of the following indefinite integrals.

a. $\int 2\,dx$ **b.** $\int \pi^2\,dx$

Solution Each of the integrands has the form $f(x) = k$, where k is a constant. Applying Rule 1 in each case yields

a. $\int 2\,dx = 2x + C$ **b.** $\int \pi^2\,dx = \pi^2 x + C$

○ ○ ○

Next, from the rule of differentiation,

$$\frac{d}{dx}x^n = nx^{n-1}$$

we obtain the following rule of integration.

RULE 2 THE POWER RULE	$\displaystyle\int x^n \, dx = \frac{1}{n+1}x^{n+1} + C \qquad (n \neq -1)$

An antiderivative of a power function is another power function obtained from the integrand by increasing its power by 1 and dividing the resulting expression by the new power.

To prove this result, observe that

$$\begin{aligned}
F'(x) &= \frac{d}{dx}\left[\frac{1}{n+1}x^{n+1} + C\right] \\
&= \frac{n+1}{n+1}x^n \\
&= x^n \\
&= f(x)
\end{aligned}$$

EXAMPLE 5   Evaluate each of the following indefinite integrals:

a. $\displaystyle\int x^3 \, dx$ **b.** $\displaystyle\int x^{3/2} \, dx$ **c.** $\displaystyle\int \frac{1}{x^{3/2}} \, dx$

Solution   Each integrand is a power function with exponent $n \neq -1$. Applying Rule 2 in each case yields the following results:

a. $\displaystyle\int x^3 \, dx = \frac{1}{4}x^4 + C$

b. $\displaystyle\int x^{3/2} \, dx = \frac{1}{5/2}x^{5/2} + C = \frac{2}{5}x^{5/2} + C$

c. $\displaystyle\int \frac{1}{x^{3/2}} \, dx = \int x^{-3/2} \, dx = \frac{1}{-1/2}x^{-1/2} + C = -2x^{-1/2} + C$

These results may be verified by differentiating each of the antiderivatives and showing that the result is equal to the corresponding integrand.

○ ○ ○

The next rule tells us that a constant factor may be moved through an integral sign.

RULE 3 INDEFINITE INTEGRAL OF A CONSTANT MULTIPLE OF A FUNCTION

$$\int cf(x)\,dx = c\int f(x)\,dx \qquad (c,\text{ a constant})$$

The indefinite integral of a constant multiple of a function is equal to the constant multiple of the indefinite integral of the function.

This result follows from the corresponding rule of differentiation (see Rule 3, Section 3.1).

Only a constant can be "moved out" of an integral sign. For example, it would be incorrect to write

$$\int x^2\,dx = x^2\int 1\,dx$$

In fact, $\int x^2\,dx = \frac{1}{3}x^3 + C$, whereas $x^2\int 1\,dx = x^2(x + C) = x^3 + Cx^2$.

EXAMPLE 6   Evaluate each of the following indefinite integrals.

a. $\displaystyle\int 2t^3\,dt$ **b.** $\displaystyle\int -3x^{-2}\,dx$

Solution   Each integrand has the form $cf(x)$, where c is a constant. Applying Rule 3, we obtain

a. $\displaystyle\int 2t^3\,dt = 2\int t^3\,dt = 2\left[\frac{1}{4}t^4 + K\right] = \frac{1}{2}t^4 + 2K = \frac{1}{2}t^4 + C$

where $C = 2K$. From now on, we will write the constant of integration as C, since any nonzero multiple of an arbitrary constant is an arbitrary constant.

b. $\displaystyle\int -3x^{-2}\,dx = -3\int x^{-2}\,dx = (-3)(-1)x^{-1} + C = \frac{3}{x} + C$

RULE 4 THE SUM RULE

$$\int [f(x) + g(x)]\,dx = \int f(x)\,dx + \int g(x)\,dx$$

and

$$\int [f(x) - g(x)]\,dx = \int f(x)\,dx - \int g(x)\,dx$$

The indefinite integral of a sum (difference) of two integrable functions is equal to the sum (difference) of their indefinite integrals.

This result is easily extended to the case involving the sum and difference of any finite number of functions. As in Rule 3, the proof of Rule 4 follows from the corresponding rule of differentiation (see Rule 4, Section 3.1).

EXAMPLE 7   Evaluate the indefinite integral

$$\int (3x^5 + 4x^{3/2} - 2x^{-1/2})\, dx$$

Solution   Applying the extended version of Rule 4, we find that

$$\int (3x^5 + 4x^{3/2} - 2x^{-1/2})\, dx$$

$$= \int 3x^5\, dx + \int 4x^{3/2}\, dx - \int 2x^{-1/2}\, dx$$

$$= 3 \int x^5\, dx + 4 \int x^{3/2}\, dx - 2 \int x^{-1/2}\, dx \qquad \text{(Rule 3)}$$

$$= (3) \left(\frac{1}{6} \right) x^6 + (4) \left(\frac{2}{5} \right) x^{5/2} - (2)(2)x^{1/2} + C \qquad \text{(Rule 2)}$$

$$= \frac{1}{2} x^6 + \frac{8}{5} x^{5/2} - 4x^{1/2} + C$$

Observe that we have combined the three constants of integration, which arise from evaluating the three indefinite integrals, to obtain one constant C. After all, the sum of three arbitrary constants is also an arbitrary constant.

RULE 5 THE INDEFINITE INTEGRAL OF THE EXPONENTIAL FUNCTION

$$\int e^x\, dx = e^x + C$$

The indefinite integral of the exponential function with base e is equal to the function itself (except, of course, for the constant of integration).

EXAMPLE 8   Evaluate the indefinite integral

$$\int (2e^x - x^3)\, dx$$

Solution   We have

$$\int (2e^x - x^3)\, dx = \int 2e^x\, dx - \int x^3\, dx$$

$$= 2 \int e^x\, dx - \int x^3\, dx$$

$$= 2e^x - \frac{1}{4} x^4 + C$$

The last rule of integration in this section covers the integration of the function $f(x) = x^{-1}$. Remember that this function constituted the only exceptional case in the integration of the power function $f(x) = x^n$ (see Rule 2).

RULE 6 THE INDEFINITE INTEGRAL OF THE FUNCTION $f(x) = x^{-1}$

$$\int x^{-1}\,dx = \int \frac{1}{x}\,dx = \ln|x| + C \qquad (x \neq 0)$$

To prove Rule 6, observe that

$$\frac{d}{dx}\ln|x| = \frac{1}{x} \qquad \text{(See Rule 3, Section 5.5.)}$$

EXAMPLE 9   Evaluate the indefinite integral

$$\int \left(2x + \frac{3}{x} + \frac{4}{x^2}\right) dx$$

Solution

$$\int \left(2x + \frac{3}{x} + \frac{4}{x^2}\right) dx = \int 2x\,dx + \int \frac{3}{x}\,dx + \int \frac{4}{x^2}\,dx$$

$$= 2\int x\,dx + 3\int \frac{1}{x}\,dx + 4\int x^{-2}\,dx$$

$$= 2\left(\frac{1}{2}\right)x^2 + 3\ln|x| + 4(-1)x^{-1} + C$$

$$= x^2 + 3\ln|x| - \frac{4}{x} + C$$

Differential Equations

Let us return to the problem posed at the beginning of the section: *Given the derivative of a function f', can we find the function f?* As an example, suppose we are given the function

$$f'(x) = 2x - 1 \tag{1}$$

and we wish to find $f(x)$. From what we now know, we can find f by integrating equation (1). Thus,

$$f(x) = \int f'(x)\,dx = \int (2x - 1)\,dx = x^2 - x + C \tag{2}$$

where C is an arbitrary constant. So there are infinitely many functions that have the derivative f', each differing from the other by a constant.

Equation (1) is called a *differential equation*. In general, a **differential equation** is an equation that involves the derivative or differential of an unknown function. (In the case of equation (1), the unknown function is f.) A **solution** of a differential equation is any function that satisfies the differential equation. Thus, equation (2) gives *all* the solutions of the differential equation (1) and, accordingly, it is called the **general solution** of the differential equation $f'(x) = 2x - 1$.

Figure 6.4
The graphs of some of the functions having the derivative $f'(x) = 2x - 1$. Observe that the slopes of the tangent lines to the graphs are the same for a fixed value of x.

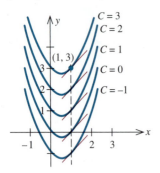

The graphs of $f(x) = x^2 - x + C$ for selected values of C are shown in Figure 6.4. These graphs have one property in common: For any fixed value of x, the tangent lines to these graphs have the same slope. This follows because any member of the family $f(x) = x^2 - x + C$ must have the same slope at x—namely, $2x - 1$!

Although there are infinitely many solutions to the differential equation $f'(x) = 2x - 1$, we can obtain a **particular solution** by specifying the value the function must assume at a certain value of x. For example, suppose we stipulate that the function f under consideration must satisfy the condition $f(1) = 3$ or, equivalently, the graph of f must pass through the point $(1, 3)$. Then, using the condition on the general solution $f(x) = x^2 - x + C$, we find that

$$f(1) = 1 - 1 + C = 3$$

and $C = 3$. Thus, the particular solution is $f(x) = x^2 - x + 3$ (see Figure 6.4).

The condition $f(1) = 3$ is an example of an *initial condition*. More generally, an **initial condition** is a condition imposed on the value of f at a point $x = a$.

Initial Value Problems

An **initial value problem** is one in which we are required to find a function satisfying (1) a differential equation and (2) one or more initial conditions. The following are examples of initial value problems.

EXAMPLE 10   Find the function f if it is known that

$$f'(x) = 3x^2 - 4x + 8 \quad \text{and} \quad f(1) = 9$$

Solution   We are required to solve the initial value problem

$$\left. \begin{array}{l} f'(x) = 3x^2 - 4x + 8 \\ f(1) = 9 \end{array} \right\}$$

Integrating the function f', we find

$$f(x) = \int f'(x)\, dx$$
$$= \int (3x^2 - 4x + 8)\, dx$$
$$= x^3 - 2x^2 + 8x + C$$

Using the condition $f(1) = 9$, we have

$$9 = f(1) = 1^3 - 2(1)^2 + 8(1) + C = 7 + C$$

or

$$C = 2$$

Therefore, the required function f is given by $f(x) = x^3 - 2x^2 + 8x + 2$.

Applications

EXAMPLE 11   In a test run of a maglev along a straight elevated monorail track, data obtained from reading its speedometer indicate that the velocity of the maglev at time t can be described by the velocity function

$$v(t) = 8t \qquad (0 \le t \le 30)$$

Find the position function of the maglev. Assume that initially the maglev is located at the origin of a coordinate line.

Solution   Let $s(t)$ denote the position of the maglev at any time t $(0 \le t \le 30)$. Then $s'(t) = v(t)$. So, we have the initial value problem

$$\left.\begin{array}{l} s'(t) = 8t \\ s(0) = 0 \end{array}\right\}$$

Integrating both sides of the differential equation $s'(t) = 8t$, we obtain

$$s(t) = \int s'(t)\, dt = \int 8t\, dt = 4t^2 + C$$

where C is an arbitrary constant. To evaluate C, we use the initial condition $s(0) = 0$ to write

$$s(0) = 4(0) + C = 0 \qquad \text{or} \qquad C = 0$$

Therefore, the required position function is $s(t) = 4t^2$ $(0 \le t \le 30)$.

○ ○ ○

EXAMPLE 12   The current circulation of the *Investor's Digest* is 3000 copies per week. The managing editor of the weekly projects a growth rate of

$$4 + 5t^{2/3}$$

copies per week, t weeks from now, for the next 3 years. Based on her projection, what will the circulation of the digest be 125 weeks from now?

Solution   Let $S(t)$ denote the circulation of the digest t weeks from now. Then $S'(t)$ is the rate of change in the circulation in the tth week and is given by

$$S'(t) = 4 + 5t^{2/3}$$

Furthermore, the current circulation of 3000 copies per week translates into the initial condition $S(0) = 3000$. Integrating the differential equation with respect to t gives

$$S(t) = \int S'(t)\, dt = \int (4 + 5t^{2/3})\, dt$$

$$= 4t + 5\left(\frac{t^{5/3}}{\frac{5}{3}}\right) + C = 4t + 3t^{5/3} + C$$

To determine the value of C, we use the condition $S(0) = 3000$ to write

$$S(0) = 4(0) + 3(0) + C = 3000$$

which gives $C = 3000$. Therefore, the circulation of the digest t weeks from now will be

$$S(t) = 4t + 3t^{5/3} + 3000$$

In particular, the circulation 125 weeks from now will be

$$S(125) = 4(125) + 3(125)^{5/3} + 3000 = 12,875$$

or 12,875 copies per week.   ○ ○ ○

SELF-CHECK EXERCISES 6.1

1. Evaluate $\int \left(\dfrac{1}{\sqrt{x}} - \dfrac{2}{x} + 3e^x \right) dx.$

2. Find the rule for the function f given that (1) the slope of the tangent line to the graph of f at any point $P(x, f(x))$ is given by the expression $3x^2 - 6x + 3$ and (2) the graph of f passes through the point $(2, 9)$.

3. Suppose that United Motors' share of the new cars sold in a certain country is changing at the rate of

$$f(t) = -0.01875t^2 + 0.15t - 1.2 \qquad (0 \le t \le 12)$$

percent at year t ($t = 0$ corresponds to the beginning of 1987). The company's market share at the beginning of 1987 was 48.4%. What was United Motors' market share at the beginning of 1999?

Solutions to Self-Check Exercises 6.1 can be found on page 442.

6.1 EXERCISES

In exercises 1–4, verify directly that F is an antiderivative of f.

1. $F(x) = \frac{1}{3}x^3 + 2x^2 - x + 2; f(x) = x^2 + 4x - 1$

2. $F(x) = xe^x + \pi; f(x) = e^x(1 + x)$

3. $F(x) = \sqrt{2x^2 - 1}; f(x) = \dfrac{2x}{\sqrt{2x^2 - 1}}$

4. $F(x) = x \ln x - x; f(x) = \ln x$

In exercises 5–8, (a) verify that G is an antiderivative of f, (b) find all antiderivatives of f, and (c) sketch the graphs of a few of the family of antiderivatives found in (b).

5. $G(x) = 2x; f(x) = 2$

6. $G(x) = 2x^2; f(x) = 4x$

7. $G(x) = \frac{1}{3}x^3; f(x) = x^2$

8. $G(x) = e^x; f(x) = e^x$

In exercises 9–50, find the indefinite integral.

9. $\int 6 \, dx$

10. $\int \sqrt{2} \, dx$

11. $\int x^3 \, dx$

12. $\int 2x^5 \, dx$

13. $\int x^{-4} \, dx$

14. $\int 3t^{-7} \, dt$

15. $\int x^{2/3} \, dx$

16. $\int 2u^{3/4} \, du$

17. $\int x^{-5/4} \, dx$

18. $\int 3x^{-2/3} \, dx$

19. $\int \dfrac{2}{x^2} \, dx$

20. $\int \dfrac{1}{3x^5} \, dx$

21. $\int \pi \sqrt{t}\, dt$

22. $\int \dfrac{3}{\sqrt{t}}\, dt$

23. $\int (3 - 2x)\, dx$

24. $\int (1 + u + u^2)\, du$

25. $\int (x^2 + x + x^{-3})\, dx$

26. $\int (0.3t^2 + 0.02t + 2)\, dt$

27. $\int 4e^x\, dx$

28. $\int (1 + e^x)\, dx$

29. $\int (1 + x + e^x)\, dx$

30. $\int (2 + x + 2x^2 + e^x)\, dx$

31. $\int \left(4x^3 - \dfrac{2}{x^2} - 1\right) dx$

32. $\int \left(6x^3 + \dfrac{3}{x^2} - x\right) dx$

33. $\int (x^{5/2} + 2x^{3/2} - x)\, dx$

34. $\int (t^{3/2} + 2t^{1/2} - 4t^{-1/2})\, dt$

35. $\int \left(\sqrt{x} + \dfrac{3}{\sqrt{x}}\right) dx$

36. $\int \left(\sqrt[3]{x^2} - \dfrac{1}{x^2}\right) dx$

37. $\int \left(\dfrac{u^3 + 2u^2 - u}{3u}\right) du$

$\left[\text{Hint: } \dfrac{u^3 + 2u^2 - u}{3u} = \dfrac{1}{3}u^2 + \dfrac{2}{3}u - \dfrac{1}{3}\right]$

38. $\int \dfrac{x^4 - 1}{x^2}\, dx$

$\left[\text{Hint: } \dfrac{x^4 - 1}{x^2} = x^2 - x^{-2}\right]$

39. $\int (2t + 1)(t - 2)\, dt$

40. $\int u^{-2}(1 - u^2 + u^4)\, du$

41. $\int \dfrac{1}{x^2}(x^4 - 2x^2 + 1)\, dx$

42. $\int \sqrt{t}(t^2 + t - 1)\, dt$

43. $\int \dfrac{ds}{(s + 1)^{-2}}$

44. $\int \left(\sqrt{x} + \dfrac{3}{x} - 2e^x\right) dx$

45. $\int (e^t + t^e)\, dt$

46. $\int \left(\dfrac{1}{x^2} - \dfrac{1}{\sqrt[3]{x^2}} + \dfrac{1}{\sqrt{x}}\right) dx$

47. $\int \left(\dfrac{x^3 + x^2 - x + 1}{x^2}\right) dx$

[Hint: Simplify the integrand first.]

48. $\int \dfrac{t^3 + \sqrt[3]{t}}{t^2}\, dt$

[Hint: Simplify the integrand first.]

49. $\int \dfrac{(\sqrt{x} - 1)^2}{x^2}\, dx$

[Hint: Simplify the integrand first.]

50. $\int (x + 1)^2 \left(1 - \dfrac{1}{x}\right) dx$

[Hint: Simplify the integrand first.]

In exercises 51–58, find f(x) by solving the initial value problem.

51. $f'(x) = 2x + 1; f(1) = 3$

52. $f'(x) = 3x^2 - 6x; f(2) = 4$

53. $f'(x) = 3x^2 + 4x - 1; f(2) = 9$

54. $f'(x) = \dfrac{1}{\sqrt{x}}; f(4) = 2$

55. $f'(x) = 1 + \dfrac{1}{x^2}; f(1) = 2$

56. $f'(x) = e^x - 2x; f(0) = 2$

57. $f'(x) = \dfrac{x + 1}{x}; f(1) = 1$

58. $f'(x) = 1 + e^x + \dfrac{1}{x}; f(1) = 3 + e$

In exercises 59–62, find the function f given that the slope of the tangent line to the graph of f at any point (x, f(x)) is f'(x) and that the graph of f passes through the given point.

59. $f'(x) = \frac{1}{2}x^{-1/2}; (2, \sqrt{2})$

60. $f'(t) = t^2 - 2t + 3; (1, 2)$

61. $f'(x) = e^x + x; (0, 3)$

62. $f'(x) = \dfrac{2}{x} + 1; (1, 2)$

63. Velocity of a Car   The velocity of a car (in ft/sec) t seconds after starting from rest is given by the function

$$f(t) = 2\sqrt{t} \qquad (0 \le t \le 30)$$

Find the car's position at any time t.

64. Velocity of a Maglev   The velocity in ft/sec of a maglev is

$$v(t) = 0.2t + 3 \qquad (0 \le t \le 120)$$

At $t = 0$, it is at the station. Find the function giving the position of the maglev at time t, assuming that the motion takes place along a straight stretch of track.

65. Cost of Producing Clocks   The Lorimar Watch Company manufactures travel clocks. The daily marginal cost function associated with producing these clocks is

$$C'(x) = 0.000009x^2 - 0.009x + 8$$

where $C'(x)$ is measured in dollars per unit and x denotes the number of units produced. Management has determined that the daily fixed cost incurred in producing these clocks is $120. Find the total cost incurred by Lorimar in producing the first 500 travel clocks per day.

66. Revenue Functions   The management of the Lorimar Watch Company has determined that the daily marginal revenue function associated with producing and selling their travel clocks is given by

$$R'(x) = -0.009x + 12$$

where x denotes the number of units produced and sold and $R'(x)$ is measured in dollars per unit.
a. Determine the revenue function $R(x)$ associated with producing and selling these clocks.
b. What is the demand equation that relates the wholesale unit price with the quantity of travel clocks demanded?

67. Profit Functions   The Cannon Precision Instruments Corporation makes an automatic electronic flash with Thyrister circuitry. The estimated marginal profit associated with producing and selling these electronic flashes is

$$(-0.004x + 20)$$

dollars per unit per month when the production level is x units per month. Cannon's fixed cost for producing and selling these electronic flashes is $16,000 per month. At what level of production does Cannon realize a maximum profit? What is the maximum monthly profit?

68. Cost of Producing Guitars   The Carlota Music Company estimates that the marginal cost of manufacturing its Professional Series guitars is

$$C'(x) = 0.002x + 100$$

dollars per month when the level of production is x guitars per month. The fixed costs incurred by Carlota are $4000 per month. Find the total monthly cost incurred by Carlota in manufacturing x guitars per month.

69. Quality Control   As part of a quality-control program, the chess sets manufactured by the Jones Brothers Company are subjected to a final inspection before packing. The rate of increase in the number of sets checked per hour by an inspector t hours into the 8 A.M. to 12 noon morning shift is approximately

$$N'(t) = -3t^2 + 12t + 45 \qquad (0 \le t \le 4)$$

Find an expression $N(t)$ that approximates the number of sets inspected at the end of t hours.
[*Hint: N(0) = 0.*]

How many sets does the average inspector check during a morning shift?

70. Ballast Dropped from a Balloon   A ballast is dropped from a stationary hot-air balloon that is hovering at an altitude of 400 ft. Its velocity after t seconds is $-32t$ ft/sec.
a. Find the height $h(t)$ of the ballast from the ground at time t.
[*Hint: h'(t) = -32t and h(0) = 400.*]

b. When will the ballast strike the ground?
c. Find the velocity of the ballast when it hits the ground.

Ballast

71. Cable TV Subscribers   A study conducted by Tele-Cable, Inc., estimates that the number of cable television subscribers will grow at the rate of

$$100 + 210t^{3/4}$$

new subscribers per month t months from the start date of the service. If 5000 subscribers signed up for the service before the starting date, how many subscribers will there be 16 months from that date?

72. Air Pollution   On an average summer day, the level of carbon monoxide in a city's air is 2 parts per million. An environmental protection agency's study predicts that, unless more stringent measures are taken to protect the city's atmosphere, the concentration of carbon monoxide present in the air will increase at the rate of

$$0.003t^2 + 0.06t + 0.1$$

parts per million per year t years from now. If no further pollution-control efforts are made, what will the concentration of carbon monoxide be on an average summer day five years from now?

73. **Population Growth** The development of Astro World ("The Amusement Park of the Future") on the outskirts of a city will increase the city's population at the rate of

$$4500\sqrt{t} + 1000$$

people per year t years from the start of construction. The population before construction is 30,000. Determine the projected population 9 years after construction of the park has begun.

74. **Ozone Pollution** The rate of change of the level of ozone, an invisible gas that is an irritant and impairs breathing, present in the atmosphere on a certain May day in the city of Riverside is given by

$$R(t) = 3.2922t^2 - 0.366t^3 \qquad (0 < t < 11)$$

(measured in pollutant standard index per hour). Here t is measured in hours, with $t = 0$ corresponding to 7 A.M. Find the ozone level $A(t)$ at any time t assuming that at 7 A.M. it is zero.

[*Hint:* $A'(t) = R(t)$ and $A(0) = 0$.]

75. **Surface Area of a Human** Empirical data suggest that the surface area of a 180-cm-tall human body changes at the rate of

$$S'(W) = 0.131773W^{-0.575}$$

square meters per kilogram, where W is the weight of the body in kg. If the surface area of a 180-cm-tall human body weighing 70 kg is 1.886277 square meters, what is the surface area of a human body of the same height weighing 75 kg?

76. **Flight of a Rocket** The velocity, in feet per second, of a rocket t seconds into vertical flight is given by

$$v(t) = -3t^2 + 192t + 120$$

Find an expression $h(t)$ that gives the rocket's altitude, in feet, t seconds after lift-off. What is the altitude of the rocket 30 seconds after lift-off?

[*Hint:* $h'(t) = v(t)$; $h(0) = 0$.]

77. **Flow of Blood in an Artery** Nineteenth-century physician Jean Louis Marie Poiseuille discovered that the rate of change of the velocity of blood r cm from the central axis of an artery (in cm/sec/cm) is given by

$$a(r) = -kr$$

where k is a constant. If the radius of an artery is R cm, find an expression for the velocity of blood as a function of r (see the accompanying figure).

[*Hint:* $v'(r) = a(r)$ and $v(R) = 0$. (Why?)]

Blood vessel

78. **Acceleration of a Car** A car traveling along a straight road at 66 ft/sec accelerated to a speed of 88 ft/sec over a distance of 440 ft. What was the acceleration of the car, assuming it was constant?

79. **Deceleration of a Car** What constant deceleration would a car moving along a straight road have to be subjected to if it were brought to rest from a speed of 88 ft/sec in 9 seconds? What would the stopping distance be?

80. A tank has a constant cross-sectional area of 50 sq. ft and an orifice of constant cross-sectional area of 1/2 sq. ft located at the bottom of the tank (see the accompanying figure).

If the tank is filled with water to a height of h ft and allowed to drain, then the height of the water decreases at a rate that is described by the equation

$$\frac{dh}{dt} = -\frac{1}{25}\left(\sqrt{20} - \frac{t}{50}\right) \qquad (0 \le t \le 50\sqrt{20})$$

Find an expression for the height of the water at any time t if its height initially is 20 ft.

81. **Launching a Fighter Aircraft** A fighter aircraft is launched from the deck of a Nimitz class aircraft carrier with the help of a steam catapult. If the aircraft is to attain a take-off speed of at least 240 ft/sec after traveling 800 ft along the flight deck, find the minimum acceleration it must be subjected to, assuming it is constant.

SOLUTIONS TO SELF-CHECK EXERCISES 6.1

1. $\int \left(\dfrac{1}{\sqrt{x}} - \dfrac{2}{x} + 3e^x \right) dx = \int \left(x^{-1/2} - \dfrac{2}{x} + 3e^x \right) dx$

$$= \int x^{-1/2}\, dx - 2\int \dfrac{1}{x}\, dx + 3\int e^x\, dx$$

$$= 2x^{1/2} - 2\ln|x| + 3e^x + C$$

$$= 2\sqrt{x} - 2\ln|x| + 3e^x + C$$

2. The slope of the tangent line to the graph of the function f at any point $P(x, f(x))$ is given by the derivative f' of f. Thus, the first condition implies that

$$f'(x) = 3x^2 - 6x + 3$$

which, upon integration, yields

$$f(x) = \int (3x^2 - 6x + 3)\, dx$$

$$= x^3 - 3x^2 + 3x + k$$

where k is the constant of integration.

To evaluate k, we use the initial condition (2), which implies that $f(2) = 9$, or

$$9 = f(2) = 2^3 - 3(2)^2 + 3(2) + k$$

or $k = 7$. Hence, the required rule of definition of the function f is

$$f(x) = x^3 - 3x^2 + 3x + 7$$

3. Let $M(t)$ denote United Motors' market share at year t. Then

$$M(t) = \int f(t)\, dt$$

$$= \int (-0.01875t^2 + 0.15t - 1.2)\, dt$$

$$= -0.00625t^3 + 0.075t^2 - 1.2t + C$$

To determine the value of C, we use the initial condition $M(0) = 48.4$, obtaining $C = 48.4$. Therefore,

$$M(t) = -0.00625t^3 + 0.075t^2 - 1.2t + 48.4$$

In particular, United Motors' market share of new cars at the beginning of 1999 is given by

$$M(12) = -0.00625(12)^3 + 0.075(12)^2$$
$$-1.2(12) + 48.4 = 34$$

or 34%.

6.2 INTEGRATION BY SUBSTITUTION

In Section 6.1 we developed certain rules of integration that are closely related to the corresponding rules of differentiation in Chapters 3 and 5. In this section, we introduce a method of integration called the **method of substitution,** which is related to the Chain Rule for differentiating functions. When used in conjunction with the rules of integration developed earlier, the method of substitution is a powerful tool for integrating a large class of functions.

How the Method of Substitution Works

Consider the indefinite integral

$$\int 2(2x + 4)^5 \, dx \tag{3}$$

One way of evaluating this integral is to expand the expression $(2x + 4)^5$ and then integrate the resulting integrand term by term. As an alternative approach, let us see if we can simplify the integral by making a change of variable. Write

$$u = 2x + 4$$

with differential

$$du = 2 \, dx$$

If we substitute these quantities into equation (3), we obtain

$$\int 2(2x + 4)^5 \, dx = \int (2x + 4)^5 (2 \, dx) = \int u^5 \, du$$

↑
rewriting

$$\begin{cases} u = 2x + 4 \\ du = 2 \, dx \end{cases}$$

Now, the last integral involves a power function and is easily evaluated using Rule 2 of Section 6.1. Thus,

$$\int u^5 \, du = \frac{1}{6} u^6 + C$$

Therefore, using this result and replacing u by $u = 2x + 4$, we obtain

$$\int 2(2x + 4)^5 \, dx = \frac{1}{6} (2x + 4)^6 + C$$

We can verify that the foregoing result is indeed correct by computing

$$\frac{d}{dx} \left[\frac{1}{6} (2x + 4)^6 \right] = \frac{1}{6} \cdot 6(2x + 4)^5 (2) \quad \text{(Using the Chain Rule)}$$

$$= 2(2x + 4)^5$$

and observing that the last expression is just the integrand of (3).

The Method of Integration by Substitution

In order to see why the approach used in evaluating the integral in (3) is successful, write

$$f(x) = x^5 \qquad \text{and} \qquad g(x) = 2x + 4$$

Then $g'(x) = 2\, dx$. Furthermore, the integrand of (3) is just the composition of f and g. Thus,

$$(f \circ g)(x) = f(g(x))$$
$$= [g(x)]^5 = (2x + 4)^5$$

Therefore, (3) can be written as

$$\int f(g(x))g'(x)\, dx \tag{4}$$

Next, let us show that an integral having the form (4) can always be written as

$$\int f(u)\, du \tag{5}$$

Suppose F is an antiderivative of f. By the Chain Rule, we have

$$\frac{d}{dx}[F(g(x))] = F'(g(x))g'(x)$$

and therefore,

$$\int F'(g(x))g'(x)\, dx = F(g(x)) + C$$

Letting $F' = f$ and making the substitution $u = g(x)$, we have

$$\int f(g(x))g'(x)\, dx = F(u) + C = \int F'(u)\, du = \int f(u)\, du$$

as we wished to show. Thus, if the transformed integral is readily evaluated, as is the case with the integral (3), then the method of substitution will prove successful.

Before we look at more examples, let us summarize the steps involved in integration by substitution.

INTEGRATION BY SUBSTITUTION	
Step 1	Let $u = g(x)$, where $g(x)$ is part of the integrand, usually the "inside function" of the composite function $f(g(x))$.
Step 2	Compute $du = g'(x)\, dx$.
Step 3	Use the substitution $u = g(x)$ and $du = g'(x)\, dx$ to convert the *entire* integral into one involving *only u*.
Step 4	Evaluate the resulting integral.
Step 5	Replace u by $g(x)$ to obtain the final solution as a function of x.

REMARK   Sometimes we need to consider different choices of g for the substitution $u = g(x)$ in order to carry out step 3 and/or step 4.   o o o

EXAMPLE 1   Evaluate $\int 2x(x^2 + 3)^4 \, dx$.

Solution

Step 1   Observe that the integrand involves the composite function $(x^2 + 3)^4$ with "inside function" $g(x) = x^2 + 3$. So, we choose $u = x^2 + 3$.

Step 2   Compute $du = 2x \, dx$.

Step 3   Making the substitution $u = x^2 + 3$ and $du = 2x \, dx$, we obtain

$$\int 2x(x^2 + 3)^4 \, dx = \int (x^2 + 3)^4 (2x \, dx) = \int u^4 \, du$$

$$\uparrow$$
$$\text{rewriting}$$

an integral involving only the variable u.

Step 4   Evaluate

$$\int u^4 \, du = \frac{1}{5} u^5 + C$$

Step 5   Replacing u by $x^2 + 3$, we obtain

$$\int 2x(x^2 + 3)^4 \, dx = \frac{1}{5}(x^2 + 3)^5 + C$$ o o o

EXAMPLE 2   Evaluate $\int 3\sqrt{3x + 1} \, dx$.

Solution

Step 1   The integrand involves the composite function $\sqrt{3x + 1}$ with "inside function" $g(x) = 3x + 1$. So, let $u = 3x + 1$.

Step 2   Compute $du = 3 \, dx$.

Step 3   Making the substitution $u = 3x + 1$ and $du = 3 \, dx$, we obtain

$$\int 3\sqrt{3x + 1} \, dx = \int \sqrt{3x + 1}(3 \, dx) = \int \sqrt{u} \, du$$

an integral involving only the variable u.

Step 4   Evaluate

$$\int \sqrt{u} \, du = \int u^{1/2} \, du = \frac{2}{3} u^{3/2} + C$$

Step 5   Replacing u by $u = 3x + 1$, we obtain

$$\int 3\sqrt{3x + 1} \, dx = \frac{2}{3}(3x + 1)^{3/2} + C$$ o o o

EXAMPLE 3   Evaluate $\int x^2(x^3 + 1)^{3/2}\,dx$.

Solution

Step 1 The integrand contains the composite function $(x^3 + 1)^{3/2}$ with "inside function" $g(x) = x^3 + 1$. So, let $u = x^3 + 1$.

Step 2 Compute $du = 3x^2\,dx$.

Step 3 Making the substitution $u = x^3 + 1$ and $du = 3x^2\,dx$, or $x^2\,dx = \frac{1}{3}\,du$, we obtain

$$\int x^2(x^3 + 1)^{3/2}\,dx = \int (x^3 + 1)^{3/2}(x^2\,dx)$$

$$= \int u^{3/2}\left(\frac{1}{3}\,du\right) = \frac{1}{3}\int u^{3/2}\,du$$

an integral involving only the variable u.

Step 4 We evaluate

$$\frac{1}{3}\int u^{3/2}\,du = \frac{1}{3}\cdot\frac{2}{5}u^{5/2} + C = \frac{2}{15}u^{5/2} + C$$

Step 5 Replacing u by $x^3 + 1$, we obtain

$$\int x^2(x^3 + 1)^{3/2}\,dx = \frac{2}{15}(x^3 + 1)^{5/2} + C$$

○ ○ ○

Let $f(x) = x^2(x^3 + 1)^{3/2}$. Using the result of Example 3, we see that an antiderivative of f is $F(x) = \frac{2}{15}(x^3 + 1)^{5/2}$. However, in terms of u (where $u = x^3 + 1$), an antiderivative of f is $G(u) = \frac{2}{15}u^{5/2}$. Compute $F(2)$. Next, suppose we want to compute $F(2)$ using the function G instead. At what value of u should you evaluate $G(u)$ in order to obtain the desired result? Explain your answer.

In the remaining examples, we drop the practice of labeling the steps involved in evaluating each integral.

EXAMPLE 4   Evaluate $\int e^{-3x}\,dx$.

Solution Let $u = -3x$ so that $du = -3\,dx$, or $dx = -\frac{1}{3}\,du$. Then

$$\int e^{-3x}\,dx = \int e^u\left(-\frac{1}{3}\,du\right) = -\frac{1}{3}\int e^u\,du$$

$$= -\frac{1}{3}e^u + C = -\frac{1}{3}e^{-3x} + C$$

○ ○ ○

EXAMPLE 5   Evaluate $\int \frac{x}{3x^2 + 1} dx$.

Solution   Let $u = 3x^2 + 1$. Then $du = 6x\, dx$, or $x\, dx = \frac{1}{6} du$. Making the appropriate substitutions, we have

$$\int \frac{x}{3x^2 + 1} dx = \int \frac{1/6}{u} du$$

$$= \frac{1}{6} \int \frac{1}{u} du$$

$$= \frac{1}{6} \ln |u| + C$$

$$= \frac{1}{6} \ln(3x^2 + 1) + C \qquad \text{(Since } 3x^2 + 1 > 0\text{)}$$

○ ○ ○

EXAMPLE 6   Evaluate $\int \frac{(\ln x)^2}{2x} dx$.

Solution   Let $u = \ln x$.

Then
$$du = \frac{d}{dx} (\ln x)\, dx = \frac{1}{x} dx$$

$$\int \frac{(\ln x)^2}{2x} dx = \frac{1}{2} \int \frac{(\ln x)^2}{x} dx$$

Thus,
$$= \frac{1}{2} \int u^2\, du$$

$$= \frac{1}{6} u^3 + C$$

$$= \frac{1}{6} (\ln x)^3 + C$$

○ ○ ○

Suppose $\int f(u)\, du = F(u) + C$.

1. Show that $\int f(ax + b)\, dx = \frac{1}{a} F(ax + b) + C$.

2. How can you use this result to facilitate the evaluation of integrals such as $\int (2x + 3)^5\, dx$ and $\int e^{3x-2}\, dx$? Explain your answer.

Applications

Examples 7 and 8 show how the method of substitution can be used in practical situations.

EXAMPLE 7   In 1990, the head of the research and development department of the Soloron Corporation claimed that the cost of producing solar cell panels would drop at the rate of

$$\frac{58}{(3t + 2)^2} \qquad (0 \le t \le 10)$$

dollars per peak watt for the next t years, with $t = 0$ corresponding to the beginning of the year 1990. (A peak watt is the power produced at noon on a sunny day.) In 1990, the panels, which are used for photovoltaic power systems, cost $10 per peak watt. Find an expression giving the cost per peak watt of producing solar cell panels at the beginning of year t. What will the cost be at the beginning of 2000?

Solution   Let $C(t)$ denote the cost per peak watt for producing solar cell panels at the beginning of year t. Then

$$C'(t) = -\frac{58}{(3t + 2)^2}$$

Integrating, we find that

$$C(t) = \int \frac{-58}{(3t + 2)^2}\, dt$$

$$= -58 \int (3t + 2)^{-2}\, dt$$

Let $u = 3t + 2$ so that

$$du = 3\, dt \qquad \text{or} \qquad dt = \frac{1}{3}\, du$$

Then

$$C(t) = -58 \left(\frac{1}{3}\right) \int u^{-2}\, du$$

$$= -\frac{58}{3}(-1)u^{-1} + k$$

$$= \frac{58}{3(3t + 2)} + k$$

where k is an arbitrary constant. To determine the value of k, note that the cost per peak watt of producing solar cell panels at the beginning of 1990 ($t = 0$) was 10, or $C(0) = 10$. This gives

$$C(0) = \frac{58}{3(2)} + k = 10$$

or $k = \frac{1}{3}$. Therefore, the required expression is given by

$$C(t) = \frac{58}{3(3t + 2)} + \frac{1}{3}$$

$$= \frac{58 + (3t + 2)}{3(3t + 2)}$$

$$= \frac{t + 20}{3t + 2}$$

The cost per peak watt for producing solar cell panels at the beginning of 2000 is given by

$$C(10) = \frac{10 + 20}{3(10) + 2} \approx 0.94$$

or approximately $.94 per peak watt.

○ ○ ○

EXPLORING WITH TECHNOLOGY

Refer to Example 7.

a. Use a graphing utility to plot the graph of

$$C(t) = \frac{t + 20}{3t + 2}$$

using the viewing rectangle $[0, 10] \times [0, 5]$. Then use the numerical differentiation capability of the graphing utility to compute $C'(10)$.

b. Plot the graph of

$$C'(t) = -\frac{58}{(3t + 2)^2}$$

using the viewing rectangle $[0, 10] \times [-10, 0]$. Then use the evaluation capability of the graphing utility to find $C'(10)$. Is this value of $C'(10)$ the same as that obtained in part (a)? Explain your answer.

○   ○   ○

EXAMPLE 8   A study prepared by the marketing department of the Universal Instruments Company forecasts that, after its new line of Galaxy Home Computers is introduced into the market, sales will grow at the rate of

$$2000 - 1500e^{-0.05t} \qquad (0 \le t \le 60)$$

units per month. Find an expression that gives the total number of computers that will sell t months after they become available on the market. How many computers will Universal sell in the first year they are on the market?

Solution   Let $N(t)$ denote the total number of computers that may be expected to be sold t months after their introduction in the market. Then, the rate of growth of sales is given by $N'(t)$ units per month. Thus,

$$N'(t) = 2000 - 1500e^{-0.05t}$$

so that
$$N(t) = \int (2000 - 1500e^{-0.05t})\, dt$$
$$= \int 2000\, dt - 1500 \int e^{-0.05t}\, dt$$

Upon integrating the second integral by the method of substitution, we obtain

$$N(t) = 2000t + \frac{1500}{0.05} e^{-0.05t} + C \qquad \text{(Let } u = -0.05t,$$
$$\text{then } du = -0.05\, dt.)$$
$$= 2000t + 30{,}000e^{-0.05t} + C$$

To determine the value of C, note that the number of computers sold at the end of month 0 is nil, so $N(0) = 0$. This gives

$$N(0) = 30{,}000 + C = 0 \qquad \text{(Since } e^0 = 1)$$

or $C = -30{,}000$. Therefore, the required expression is given by

$$N(t) = 2000t + 30{,}000e^{-0.05t} - 30{,}000$$
$$= 2000t + 30{,}000(e^{-0.05t} - 1)$$

The number of computers that Universal can expect to sell in the first year is given by

$$N(12) = 2000(12) + 30{,}000(e^{-0.05(12)} - 1)$$
$$= 10{,}464$$

or 10,464 units.

○ ○ ○

SELF-CHECK EXERCISES 6.2

1. Evaluate $\int \sqrt{2x + 5}\, dx$.

2. Evaluate $\int \dfrac{x^2}{(2x^3 + 1)^{3/2}}\, dx$.

3. Evaluate $\int xe^{2x^2-1}\, dx$.

4. According to a joint study conducted by Oxnard's Environmental Management Department and a state government agency, the concentration of carbon monoxide in the air due to automobile exhaust is increasing at the rate given by

$$f(t) = \frac{8(0.1t + 1)}{300(0.2t^2 + 4t + 64)^{1/3}}$$

parts per million per year t. Currently, the concentration of carbon monoxide due to automobile exhaust is 0.16 part per million. Find an expression giving the concentration of carbon monoxide t years from now.

Solutions to Self-Check Exercises 6.2 can be found on page 453.

6.2 EXERCISES

In exercises 1–50, evaluate the indefinite integral.

1. $\int 4(4x + 3)^4 \, dx$

2. $\int 4x(2x^2 + 1)^7 \, dx$

3. $\int (x^3 - 2x)^2(3x^2 - 2) \, dx$

4. $\int (3x^2 - 2x + 1)(x^3 - x^2 + x)^4 \, dx$

5. $\int \dfrac{4x}{(2x^2 + 3)^3} \, dx$

6. $\int \dfrac{3x^2 + 2}{(x^3 + 2x)^2} \, dx$

7. $\int 3t^2 \sqrt{t^3 + 2} \, dt$

8. $\int 3t^2(t^3 + 2)^{3/2} \, dt$

9. $\int (x^2 - 1)^9 x \, dx$

10. $\int x^2(2x^3 + 3)^4 \, dx$

11. $\int \dfrac{x^4}{1 - x^5} \, dx$

12. $\int \dfrac{x^2}{\sqrt{x^3 - 1}} \, dx$

13. $\int \dfrac{2}{x - 2} \, dx$

14. $\int \dfrac{x^2}{x^3 - 3} \, dx$

15. $\int \dfrac{0.3x - 0.2}{0.3x^2 - 0.4x + 2} \, dx$

16. $\int \dfrac{2x^2 + 1}{0.2x^3 + 0.3x} \, dx$

17. $\int \dfrac{x}{3x^2 - 1} \, dx$

18. $\int \dfrac{x^2 - 1}{x^3 - 3x + 1} \, dx$

19. $\int e^{-2x} \, dx$

20. $\int e^{-0.02x} \, dx$

21. $\int e^{2-x} \, dx$

22. $\int e^{2t+3} \, dt$

23. $\int xe^{-x^2} \, dx$

24. $\int x^2 e^{x^3 - 1} \, dx$

25. $\int (e^x - e^{-x}) \, dx$

26. $\int (e^{2x} + e^{-3x}) \, dx$

27. $\int \dfrac{e^x}{1 + e^x} \, dx$

28. $\int \dfrac{e^{2x}}{1 + e^{2x}} \, dx$

29. $\int \dfrac{e^{\sqrt{x}}}{\sqrt{x}} \, dx$

30. $\int \dfrac{e^{-1/x}}{x^2} \, dx$

31. $\int \dfrac{e^{3x} + x^2}{(e^{3x} + x^3)^3} \, dx$

32. $\int \dfrac{e^x - e^{-x}}{(e^x + e^{-x})^{3/2}} \, dx$

33. $\int e^{2x}(e^{2x} + 1)^3 \, dx$

34. $\int e^{-x}(1 + e^{-x}) \, dx$

35. $\int \dfrac{\ln 5x}{x} \, dx$

36. $\int \dfrac{(\ln u)^3}{u} \, du$

37. $\int \dfrac{1}{x \ln x} \, dx$

38. $\int \dfrac{1}{x(\ln x)^2} \, dx$

39. $\int \dfrac{\sqrt{\ln x}}{x} \, dx$

40. $\int \dfrac{(\ln x)^{7/2}}{x} \, dx$

41. $\int \left(xe^{x^2} - \dfrac{x}{x^2 + 2} \right) dx$

42. $\int \left(xe^{-x^2} + \dfrac{e^x}{e^x + 3} \right) dx$

43. $\int \dfrac{x + 1}{\sqrt{x} - 1} \, dx$

[Hint: Let $u = \sqrt{x} - 1$.]

44. $\int \dfrac{e^{-u} - 1}{e^{-u} + u} \, du$

[Hint: Let $v = e^{-u} + u$.]

45. $\int x(x - 1)^5 \, dx$

[Hint: $u = x - 1$ implies $x = u + 1$.]

46. $\int \dfrac{t}{t + 1} \, dt$

$\left[\text{Hint: } \dfrac{t}{t + 1} = 1 - \dfrac{1}{t + 1}. \right]$

47. $\int \dfrac{1 - \sqrt{x}}{1 + \sqrt{x}} \, dx$

[Hint: Let $u = 1 + \sqrt{x}$.]

48. $\int \dfrac{1 + \sqrt{x}}{1 - \sqrt{x}} \, dx$

[Hint: Let $u = 1 - \sqrt{x}$.]

49. $\int v^2(1 - v)^6 \, dv$

[Hint: Let $u = 1 - v$.]

50. $\int x^3(x^2 + 1)^{3/2} \, dx$

[Hint: Let $u = x^2 + 1$.]

In exercises 51–54, find the function f given that the slope of the tangent line to the graph of f at any point (x, f(x)) is f'(x) and that the graph of f passes through the given point.

51. $f'(x) = 5(2x - 1)^4$; $(1, 3)$

52. $f'(x) = \dfrac{3x^2}{2\sqrt{x^3 - 1}}$; $(1, 1)$

53. $f'(x) = -2xe^{-x^2 + 1}$; $(1, 0)$

54. $f'(x) = 1 - \dfrac{2x}{x^2 + 1}$; $(0, 2)$

55. **Student Enrollment**   The registrar of Kellogg University estimates that the total student enrollment in the Continuing Education division will grow at the rate of

$$N'(t) = 2000(1 + 0.2t)^{-3/2}$$

students per year t years from now. If the current student enrollment is 1000, find an expression giving the total student enrollment t years from now. What will the student enrollment be 5 years from now?

56. TV Viewers: Newsmagazine Shows   The number of viewers of a weekly TV newsmagazine show, introduced in the 1993 season, has been increasing at the rate of

$$3\left(2 + \frac{1}{2}t\right)^{-1/3} \qquad (1 \le t \le 6)$$

million viewers per year in its tth year on the air. The number of viewers of the program during its first year on the air is given by $9(5/2)^{2/3}$ million. Find how many viewers are expected in the 1998 season.

57. Demand: Ladies' Boots   The rate of change of the unit price p (in dollars) of Apex ladies' boots is given by

$$p'(x) = \frac{-250x}{(16 + x^2)^{3/2}}$$

where x is the quantity demanded daily in units of a hundred. Find the demand function for these boots if the quantity demanded daily is 300 pairs ($x = 3$) when the unit price is $50 per pair.

58. Supply: Ladies' Boots   The rate of change of the unit price p (in dollars) of Apex ladies' boots is given by

$$p'(x) = \frac{240x}{(5 - x)^2}$$

where x is the number of pairs that the supplier will make available in the market daily when the unit price is $$p$ per pair. Find the supply equation for these boots if the quantity the supplier is willing to make available is 200 pairs daily ($x = 2$) when the unit price is $50 per pair.

59. Oil Spill   In calm waters, the oil spilling from the ruptured hull of a grounded tanker forms an oil slick that is circular in shape. If the radius r of the circle is increasing at the rate of

$$r'(t) = \frac{30}{\sqrt{2t + 4}}$$

feet per minute t minutes after the rupture occurs, find an expression for the radius at any time t. How large is the polluted area 16 minutes after the rupture occurred? [*Hint:* $r(0) = 0$.]

60. Life Expectancy of a Female   Suppose that in a certain country the life expectancy at birth of a female is chang-

ing at the rate of

$$g'(t) = \frac{5.45218}{(1 + 1.09t)^{0.9}}$$

years per year. Here t is measured in years, and $t = 0$ corresponds to the beginning of 1900. Find an expression $g(t)$ giving the life expectancy at birth (in years) of a female in that country if the life expectancy at the beginning of 1900 is 50.02 years. What is the life expectancy at birth of a female born in the year 2000 in that country?

61. Average Birth Height of Boys   Using data collected at Kaiser Hospital, pediatricians estimate that the average height of male children changes at the rate of

$$h'(t) = \frac{52.8706e^{-0.3277t}}{(1 + 2.449e^{-0.3277t})^2}$$

inches per year, where the child's height $h(t)$ is measured in inches and t, the child's age, is measured in years, with $t = 0$ corresponding to the age at birth. Find an expression $h(t)$ for the average height of a boy at age t if the height at birth of an average child is 19.4 inches. What is the height of an average 8-year-old boy?

62. Learning Curves   The average student enrolled in the 20-week Advanced Shorthand course at the American Institute of Stenography progresses according to the rule

$$N'(t) = 6e^{-0.05t} \qquad (0 \le t \le 20)$$

where $N'(t)$ measures the rate of change in the number of words per minute of dictation the student takes in shorthand after t weeks in the course. Assuming that the average student enrolled in this course begins with a dictation speed of 60 words per minute, find an expression $N(t)$ that gives the dictation speed of the student after t weeks in the course.

63. Sales: Loudspeakers   Two thousand pairs of Acrosonic model F loudspeaker systems were sold in the first year they appeared in the market. Since then, sales of these loudspeaker systems have been growing at the rate of

$$f'(t) = 2000(3 - 2e^{-t})$$

units per year, where t denotes the number of years these systems have been on the market. Determine the number of systems that were sold in the first 5 years after their introduction.

64. Amount of Glucose in the Bloodstream   Suppose that a patient is given a continuous intravenous infusion of glucose at a constant rate of r milligrams per minute. Then, the rate at which the amount of glucose in the bloodstream is changing at time t due to this infusion is given by

$$A'(t) = re^{-at}$$

milligrams per minute, where a is a positive constant associated with the rate at which excess glucose is eliminated from the bloodstream and is dependent on the patient's metabolism rate. Derive an expression for the amount of glucose in the bloodstream at time t.

[*Hint:* $A(0) = 0$.]

65. **Concentration of a Drug in an Organ** A drug is carried into an organ of volume V cm³ by a liquid that enters the organ at the rate of a cm³/sec and leaves it at the rate of b cm³/sec. The concentration of the drug in the liquid entering the organ is c g/cm³. If the concentration of the drug in the organ at time t is increasing at the rate of

$$x'(t) = \frac{1}{V}(ac - bx_0)e^{-bt/V}$$

g/cm³ per second, and the concentration of the drug in the organ initially is x_0 g/cm³, show that the concentration of the drug in the organ at time t is given by

$$x(t) = \frac{ac}{b} + \left(x_0 - \frac{ac}{b}\right)e^{-bt/V}$$

SOLUTIONS TO SELF–CHECK EXERCISES 6.2

1. Let $u = 2x + 5$. Then $du = 2\,dx$, or $dx = \frac{1}{2}\,du$. Making the appropriate substitutions, we have

$$\int \sqrt{2x + 5}\,dx = \int \sqrt{u}\left(\frac{1}{2}\,du\right) = \frac{1}{2}\int u^{1/2}\,du$$

$$= \frac{1}{2}\left(\frac{2}{3}\right)u^{3/2} + C$$

$$= \frac{1}{3}(2x + 5)^{3/2} + C$$

2. Let $u = 2x^3 + 1$, so that $du = 6x^2\,dx$, or $x^2\,dx = \frac{1}{6}\,du$. Making the appropriate substitutions, we have

$$\int \frac{x^2}{(2x^3 + 1)^{3/2}}\,dx = \int \frac{\left(\frac{1}{6}\right)du}{u^{3/2}} = \frac{1}{6}\int u^{-3/2}\,du$$

$$= \left(\frac{1}{6}\right)(-2)u^{-1/2} + C$$

$$= -\frac{1}{3}(2x^3 + 1)^{-1/2} + C$$

$$= -\frac{1}{3\sqrt{2x^3 + 1}} + C$$

3. Let $u = 2x^2 - 1$, so that $du = 4x\,dx$, or $x\,dx = \frac{1}{4}\,du$. Then

$$\int xe^{2x^2-1}\,dx = \frac{1}{4}\int e^u\,du$$

$$= \frac{1}{4}e^u + C$$

$$= \frac{1}{4}e^{2x^2-1} + C$$

4. Let $C(t)$ denote the concentration of carbon monoxide in the air due to automobile exhaust t years from now. Then

$$C'(t) = f(t) = \frac{8(0.1t + 1)}{300(0.2t^2 + 4t + 64)^{1/3}}$$

$$= \frac{8}{300}(0.1t + 1)(0.2t^2 + 4t + 64)^{-1/3}$$

Integrating, we find

$$C(t) = \int \frac{8}{300}(0.1t + 1)(0.2t^2 + 4t + 64)^{-1/3}\, dt$$

$$= \frac{8}{300}\int (0.1t + 1)(0.2t^2 + 4t + 64)^{-1/3}\, dt$$

Let $u = 0.2t^2 + 4t + 64$, so that $du = (0.4t + 4)\, dt = 4(0.1t + 1)\, dt$, or

$$(0.1t + 1)\, dt = \frac{1}{4}\, du$$

Then

$$C(t) = \frac{8}{300}\left(\frac{1}{4}\right)\int u^{-1/3}\, du$$

$$= \frac{1}{150}\left(\frac{3}{2}u^{2/3}\right) + k$$

$$= 0.01(0.2t^2 + 4t + 64)^{2/3} + k$$

where k is an arbitrary constant. To determine the value of k, we use the condition $C(0) = 0.16$, obtaining

$$C(0) = 0.16 = 0.01(64)^{2/3} + k$$

or

$$0.16 = 0.16 + k$$

and $k = 0$. Therefore,

$$C(t) = 0.01(0.2t^2 + 4t + 64)^{2/3}$$

6.3   AREA AND THE DEFINITE INTEGRAL

An Intuitive Look

Suppose that a certain state's annual rate of petroleum consumption over a 4-year period is constant and is given by the function

$$f(t) = 1.2 \qquad (0 \le t \le 4)$$

where t is measured in years and $f(t)$ in millions of barrels per year. Then the state's total petroleum consumption over the period of time in question is

$$(1.2)(4 - 0) \qquad \text{(Rate of consumption · time elapsed)}$$

Figure 6.5
The total petroleum consumption is given by the area of the rectangular region.

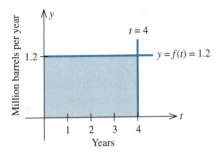

or 4.8 million barrels. If you examine the graph of f shown in Figure 6.5 you will see that this total is just the area of the rectangular region bounded above by the graph of f, below by the t-axis, and to the left and right by the vertical lines $t = 0$ (the y-axis) and $t = 4$, respectively.

Figure 6.6 shows the actual petroleum consumption of a certain New England state over a 4-year period from 1990 ($t = 0$) to 1994 ($t = 4$). Observe that the rate of consumption is not constant; that is, the function f is not a constant function. What is the state's total petroleum consumption over this 4-year period? It seems reasonable to conjecture that it is given by the "area" of the region bounded above by the graph of f, below by the t-axis, and to the left and right by the vertical lines $t = 0$ and $t = 4$, respectively.

Figure 6.6
The daily petroleum consumption is given by the "area" of the shaded region.

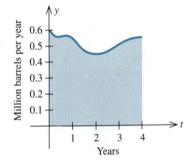

This example raises two questions:

1. What is the "area" of the region shown in Figure 6.6?

2. How do we compute this area?

The Area Problem

Figure 6.7
The area under the graph of f on $[a, b]$.

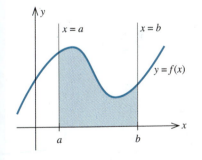

The preceding example touches on the second fundamental problem in calculus: Calculate the area of the region bounded by the graph of a nonnegative function f, the x-axis, and the vertical lines $x = a$ and $x = b$ (Figure 6.7). This area is called the **area under the graph of f** on the interval $[a, b]$, or from a to b.

Defining Area—Two Examples

Just as we used the slopes of secant lines (quantities that we could compute) to help us define the slope of the tangent line to a point on the graph of a function, we now adopt a parallel approach and use the areas of rectangles (quantities that we can compute) to help us define the area under the graph of a function. We begin by looking at a specific example.

EXAMPLE 1   Let $f(x) = x^2$ and consider the region R under the graph of f on the interval $[0, 1]$ (Figure 6.8a). In order to obtain an approximation of the area of R, let us construct four nonoverlapping rectangles as follows: Divide the interval $[0, 1]$ into four subintervals

$$\left[0, \frac{1}{4}\right], \quad \left[\frac{1}{4}, \frac{1}{2}\right], \quad \left[\frac{1}{2}, \frac{3}{4}\right], \quad \text{and} \quad \left[\frac{3}{4}, 1\right]$$

of equal length $1/4$. Next, construct four rectangles with these subintervals as bases and with heights given by the values of the function at the midpoints

$$\frac{1}{8}, \quad \frac{3}{8}, \quad \frac{5}{8}, \quad \text{and} \quad \frac{7}{8}$$

of each subinterval. Then each of these rectangles has width $1/4$ and height

$$f\left(\frac{1}{8}\right), \quad f\left(\frac{3}{8}\right), \quad f\left(\frac{5}{8}\right), \quad \text{and} \quad f\left(\frac{7}{8}\right)$$

respectively (Figure 6.8b).

Figure 6.8
The area of the region under the graph of f on [0, 1] in (a) is approximated by the sum of the areas of the four rectangles in (b).

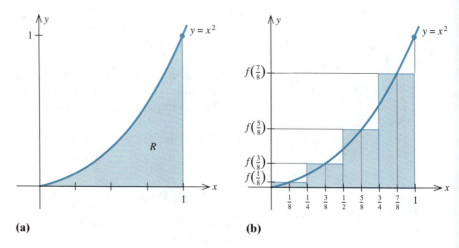

(a)   (b)

If we approximate the area A of S by the sum of the areas of the four rectangles, we obtain

$$A \approx \frac{1}{4}f\left(\frac{1}{8}\right) + \frac{1}{4}f\left(\frac{3}{8}\right) + \frac{1}{4}f\left(\frac{5}{8}\right) + \frac{1}{4}f\left(\frac{7}{8}\right)$$

$$= \frac{1}{4}\left[f\left(\frac{1}{8}\right) + f\left(\frac{3}{8}\right) + f\left(\frac{5}{8}\right) + f\left(\frac{7}{8}\right)\right]$$

$$= \frac{1}{4}\left[\left(\frac{1}{8}\right)^2 + \left(\frac{3}{8}\right)^2 + \left(\frac{5}{8}\right)^2 + \left(\frac{7}{8}\right)^2\right] \qquad \text{(Recall that } f(x) = x^2.\text{)}$$

$$= \frac{1}{4}\left(\frac{1}{64} + \frac{9}{64} + \frac{25}{64} + \frac{49}{64}\right) = \frac{21}{64}$$

or approximately 0.328125 square units.

Following the procedure of Example 1, we can obtain approximations of the area of the region R using any number n of rectangles ($n = 4$ in Example 1). Figure 6.9a shows the approximation of the area A of R using 8 rectangles ($n = 8$), and Figure 6.9b shows the approximation of the area A of R using 16 rectangles.

Figure 6.9
As n increases, the number of rectangles increases and the approximation improves.

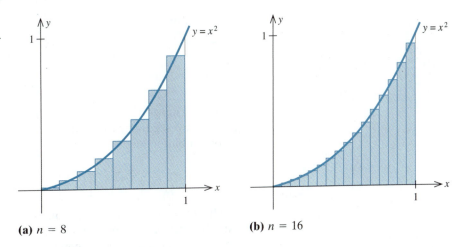

(a) $n = 8$ **(b)** $n = 16$

These figures suggest that the approximations seem to get better as n increases. This is borne out by the results given in Table 6.1, which were obtained using a computer.

Table 6.1

Number of Rectangles n	4	8	16	32	64	100	200
Approximation of A	0.328125	0.332031	0.333008	0.333252	0.333313	0.333325	0.333331

Our computations seem to suggest that the approximations approach the number 1/3 as n gets larger and larger. This result suggests that we *define* the area of the region under the graph of $f(x) = x^2$ on the interval $[0, 1]$ to be 1/3 square units.

In Example 1 we chose the *midpoint* of each subinterval as the point at which to evaluate $f(x)$ to obtain the height of the approximating rectangle. Let us consider another example, this time choosing the *left endpoint* of each subinterval.

EXAMPLE 2   Let R be the region under the graph of $f(x) = 16 - x^2$ on the interval $[1, 3]$. Find an approximation of the area A of R using four subintervals of $[1, 3]$ of equal length and picking the left endpoint of each subinterval to evaluate $f(x)$ to obtain the height of the approximating rectangle.

Figure 6.10
The area of R in (a) is approxi-
mated by the sum of the areas of the
four rectangles in (b).

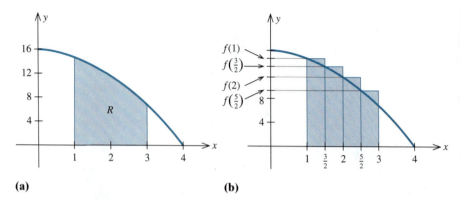

(a)   **(b)**

Solution The graph of f is sketched in Figure 6.10a. Since the length of
[1, 3] is 2, we see that the length of each subinterval is 2/4, or 1/2. Therefore,
the four subintervals are

$$\left[1, \frac{3}{2}\right], \quad \left[\frac{3}{2}, 2\right], \quad \left[2, \frac{5}{2}\right], \quad \text{and} \quad \left[\frac{5}{2}, 3\right]$$

The left endpoints of these subintervals are 1, 3/2, 2, and 5/2, respectively,
so the heights of the approximating rectangles are $f(1)$, $f(\frac{3}{2})$, $f(2)$, and $f(\frac{5}{2})$,
respectively (Figure 6.10b). Therefore, the required approximation is

$$A \approx \frac{1}{2}f(1) + \frac{1}{2}f\left(\frac{3}{2}\right) + \frac{1}{2}f(2) + \frac{1}{2}f\left(\frac{5}{2}\right)$$

$$= \frac{1}{2}\left[f(1) + f\left(\frac{3}{2}\right) + f(2) + f\left(\frac{5}{2}\right)\right]$$

$$= \frac{1}{2}\left\{[16 - (1)^2] + \left[16 - \left(\frac{3}{2}\right)^2\right]\right.$$

$$\left. + [16 - (2)^2] + \left[16 - \left(\frac{5}{2}\right)^2\right]\right\} \qquad \text{[Recall that } f(x) = 16 - x^2.\text{]}$$

$$= \frac{1}{2}\left(15 + \frac{55}{4} + 12 + \frac{39}{4}\right) = \frac{101}{4}$$

or approximately 25.25 square units.   ○ ○ ○

Table 6.2 shows the approximations of the area A of the region R of
Example 2 when n rectangles are used for the approximation and the heights of
the approximating rectangles are found by evaluating $f(x)$ at the left endpoints.
Once again we see that the approximations seem to approach a unique
number as n gets larger and larger—this time the number is $23\frac{1}{3}$. This result

Table 6.2

Number of Rectangles n	4	10	100	1,000	10,000	50,000	100,000
Approximation of A	25.2500	24.1200	23.4132	23.3413	23.3341	23.3335	23.3334

suggests that we *define* the area of the region under the graph of $f(x) = 16 - x^2$ on the interval $[1, 3]$ to be $23\frac{1}{3}$ square units.

Defining Area—The General Case

Examples 1 and 2 point the way to defining the area A under the graph of an arbitrary but continuous and nonnegative function f on an interval $[a, b]$ (Figure 6.11a).

Figure 6.11
The area of the region under the graph of f on $[a, b]$ in (a) is approximated by the sum of the areas of the n rectangles shown in (b).

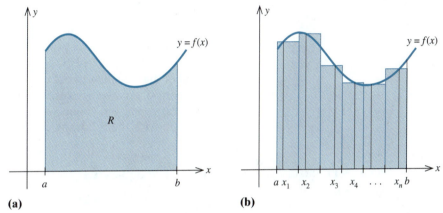

(a)   **(b)**

Divide the interval $[a, b]$ into n subintervals of equal length $\Delta x = (b - a)/n$. Next, pick n arbitrary points $x_1, x_2, \ldots, x_n$, called *representative points*, from the first, second, $\ldots$, and nth subintervals, respectively (Figure 6.11b). Then, approximating the area A of the region R by the n rectangles of width Δx and heights $f(x_1), f(x_2), \ldots, f(x_n)$, so that the areas of the rectangles are $f(x_1)\Delta x, f(x_2)\Delta x, \ldots, f(x_n)\Delta x$, we have

$$A \approx f(x_1)\Delta x + f(x_2)\Delta x + \cdots + f(x_n)\Delta x$$

The sum on the right-hand side of this expression is called a **Riemann sum** in honor of the German mathematician Bernhard Riemann (1826–1866). Now, as the earlier examples seem to suggest, the Riemann sum will approach a unique number as n becomes arbitrarily large.* We define this number to be the area A of the region R.

THE AREA UNDER THE GRAPH OF A FUNCTION

Let f be a nonnegative continuous function on $[a, b]$. Then the area of the region under the graph of f is

$$A = \lim_{n \to \infty} [f(x_1) + f(x_2) + \cdots + f(x_n)]\Delta x \qquad (6)$$

where $x_1, x_2, \ldots, x_n$ are arbitrary points in the n subintervals of $[a, b]$ of equal width $\Delta x = (b - a)/n$.

⦿ ⦿ ⦿

* Even though we chose the representative points to be the midpoints of the subintervals in Example 1 and the left endpoints in Example 2, it can be shown that each of the respective sums will always approach a unique number as n approaches infinity.

The Definite Integral

As we have just seen, the area under the graph of a continuous *nonnegative* function f on an interval $[a, b]$ is defined by the limit of the Riemann sum

$$\lim_{n \to \infty} [f(x_1)\Delta x + f(x_2)\Delta x + \cdots + f(x_n)\Delta x]$$

We now turn our attention to the study of limits of Riemann sums involving functions that are not necessarily nonnegative. Such limits arise in many applications of calculus.

For example, the calculation of the distance covered by a body traveling along a straight line involves evaluating a limit of this form. The computation of the total revenue realized by a company over a certain time period, the calculation of the total amount of electricity consumed in a typical home over a 24-hour period, the average concentration of a drug in a body over a certain interval of time, and the volume of a solid—all involve limits of this type.

We begin with the following definition.

DEFINITE INTEGRAL

Let f be defined on $[a, b]$. If

$$\lim_{n \to \infty} [f(x_1)\Delta x + f(x_2)\Delta x + \cdots + f(x_n)\Delta x]$$

exists for all choices of representative points $x_1, x_2, \ldots, x_n$ in the n subintervals of $[a, b]$ of equal width $\Delta x = (b - a)/n$, then this limit is called the **definite integral of f from a to b** and is denoted by $\int_a^b f(x)\, dx$. Thus,

$$\int_a^b f(x)\, dx = \lim_{n \to \infty} [f(x_1)\Delta x + f(x_2)\Delta x + \cdots + f(x_n)\Delta x] \qquad \textbf{(7)}$$

The number a is the **lower limit of integration** and the number b is the **upper limit of integration.**

REMARKS

1. If f is nonnegative, then the limit in (7) is the same as the limit in (6) and, therefore, the definite integral gives the area under the graph of f on $[a, b]$.

2. The limit in (7) is denoted by the integral sign $\int$ because, as we will see later, the definite integral and the antiderivative of a function f are related.

3. It is important to realize that the definite integral $\int_a^b f(x)\, dx$ is a number, whereas the indefinite integral $\int f(x)\, dx$ represents a family of functions (the antiderivatives of f).

4. If the limit in (7) exists, we say that f is **integrable** on the interval $[a, b]$.

∘ ∘ ∘

When Is a Function Integrable?

The following theorem, which we state without proof, guarantees that a continuous function is integrable.

INTEGRABILITY OF A FUNCTION	Let f be continuous on $[a, b]$. Then f is integrable on $[a, b]$; that is, the definite integral $\int_a^b f(x)\,dx$ exists.

Geometric Interpretation of the Definite Integral

If f is nonnegative and integrable on $[a, b]$, then we have the following geometric interpretation of the definite integral $\int_a^b f(x)\,dx$.

GEOMETRIC INTERPRETATION OF $\int_a^b f(x)\,dx$ FOR $f(x) \geq 0$ ON $[a, b]$	If f is nonnegative and continuous on $[a, b]$, then $$\int_a^b f(x)\,dx \qquad (8)$$ is equal to the area of the region under the graph of f on $[a, b]$ (see Figure 6.12).

Figure 6.12
If $f(x) \geq 0$ on $[a, b]$, then $\int_a^b f(x)\,dx$ = area under the graph of f on $[a, b]$.

Suppose f is nonpositive (that is, $f(x) \leq 0$) and continuous on $[a, b]$. Explain why the area of the region below the x-axis and above the graph of f is given by $-\int_a^b f(x)\,dx$.

Next, let's extend our geometric interpretation of the definite integral to include the case where f assumes both positive as well as negative values on $[a, b]$. Consider a typical Riemann sum of the function f,

$$f(x_1)\Delta x + f(x_2)\Delta x + \cdots + f(x_n)\Delta x$$

corresponding to a partition of $[a, b]$ into n subintervals of equal width $(b - a)/n$, where $x_1, x_2, \ldots, x_n$ are representative points in the subintervals. The sum consists of n terms in which a positive term corresponds to the area of a rectangle of height $f(x_k)$ [for some positive integer k] lying above the x-axis, and a negative term corresponds to the area of a rectangle of height $-f(x_k)$ lying below the x-axis. (See Figure 6.13, which depicts a situation with $n = 6$.)

As n gets larger and larger, the sums of the areas of the rectangles lying above the x-axis seem to give a better and better approximation of the area of the region lying above the x-axis (see Figure 6.14). Similarly, the sums of

Figure 6.13
The positive terms in the Riemann sum are associated with the areas of the rectangles that lie above the x-axis, and the negative terms are associated with the areas of those that lie below the x-axis.

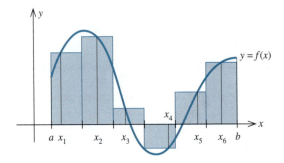

Figure 6.14
As n gets larger, the approximations get better. Here n = 12, and we are approximating with twice as many rectangles as in Figure 6.13.

the areas of those rectangles lying below the *x*-axis seem to give a better and better approximation of the area of the region lying below the *x*-axis.

These observations suggest the following geometric interpretation of the definite integral for an arbitrary continuous function on an interval $[a, b]$.

GEOMETRIC INTERPRETATION OF $\int_a^b f(x)\,dx$ ON $[a, b]$

If f is continuous on $[a, b]$, then

$$\int_a^b f(x)\,dx$$

is equal to the area of the region above $[a, b]$ minus the area of the region below $[a, b]$ (see Figure 6.15).

Figure 6.15
$\int_a^b f(x)\,dx = $ *Area of* $R_1 -$ *Area of* $R_2 +$ *Area of* R_3.

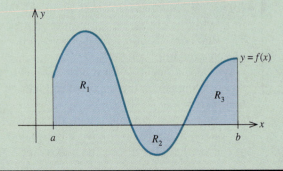

SELF-CHECK EXERCISE 6.3

Find an approximation of the area of the region R under the graph of $f(x) = 2x^2 + 1$ on the interval $[0, 3]$ using four subintervals of $[0, 3]$ of equal length and picking the midpoint of each subinterval as a representative point.

The Solution to Self-Check Exercise 6.3 can be found on page 465.

6.3 EXERCISES

In exercises 1 and 2, find an approximation of the area of the region R under the graph of f by computing the Riemann sum of f corresponding to the partition of the interval into the subintervals shown in the accompanying figures. In each case, use the midpoints of the subintervals as the representative points.

1.

2.

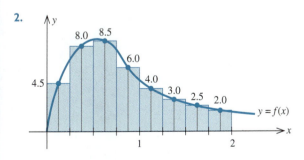

3. Let $f(x) = 3x$.
   a. Sketch the region R under the graph of f on the interval $[0, 2]$ and find its exact area using geometry.
   b. Use a Riemann sum with four subintervals of equal length ($n = 4$) to approximate the area of R. Choose the representative points to be the left endpoints of the subintervals.

c. Repeat (b) with eight subintervals of equal length ($n = 8$).
   d. Compare the approximations obtained in (b) and (c) with the exact area found in (a). Do the approximations improve with larger n?

4. Repeat exercise 3 choosing the representative points to be the right endpoints of the subintervals.

5. Let $f(x) = 4 - 2x$.
   a. Sketch the region R under the graph of f on the interval $[0, 2]$ and find its exact area using geometry.
   b. Use a Riemann sum with five subintervals of equal length ($n = 5$) to approximate the area of R. Choose the representative points to be the left endpoints of the subintervals.
   c. Repeat (b) with ten subintervals of equal length ($n = 10$).
   d. Compare the approximations obtained in (b) and (c) with the exact area found in (a). Do the approximations improve with larger n?

6. Repeat exercise 5 choosing the representative points to be the right endpoints of the subintervals.

7. Let $f(x) = x^2$ and compute the Riemann sum of f over the interval $[2, 4]$ using
   a. Two subintervals of equal length ($n = 2$)
   b. Five subintervals of equal length ($n = 5$)
   c. Ten subintervals of equal length ($n = 10$)
   In each case, choose the representative points to be the midpoints of the subintervals.
   d. Can you guess at the area of the region under the graph of f on the interval $[2, 4]$?

8. Repeat exercise 7 choosing the representative points to be the left endpoints of the subintervals.

9. Repeat exercise 7 choosing the representative points to be the right endpoints of the subintervals.

10. Let $f(x) = x^3$ and compute the Riemann sum of f over the interval $[0, 1]$ using
   a. Two subintervals of equal length ($n = 2$)
   b. Five subintervals of equal length ($n = 5$)
   c. Ten subintervals of equal length ($n = 10$)
   In each case, choose the representative points to be the midpoints of the subintervals.
   d. Can you guess at the area of the region under the graph of f on the interval $[0, 1]$?

11. Repeat exercise 10 choosing the representative points to be the left endpoints of the subintervals.

12. Repeat exercise 10 choosing the representative points to be the right endpoints of the subintervals.

In exercises 13–16, find an approximation of the area of the region R under the graph of the function f on the interval [a, b]. In each case, use n subintervals and choose the representative points as indicated.

13. $f(x) = x^2 + 1$; $[0, 2]$; $n = 5$; midpoints

14. $f(x) = 4 - x^2$; $[-1, 2]$; $n = 6$; left endpoints

15. $f(x) = \dfrac{1}{x}$; $[1, 3]$; $n = 4$; right endpoints

16. $f(x) = e^x$; $[0, 3]$; $n = 5$; midpoints

17. Real Estate Figure (a) shows a vacant lot with a 100-foot frontage in a development. To estimate its area, we introduce a coordinate system so that the x-axis coincides with the edge of the straight road forming the lower boundary of the property as shown in Figure (b). Then thinking of the upper boundary of the property as the graph of a continuous function f over the interval $[0, 100]$, we see that the problem is mathematically equivalent to that of finding the area under the graph of f on $[0, 100]$. To estimate the area of the lot using a Riemann sum, we divide the interval $[0, 100]$ into five equal subintervals of length 20 feet. Then, using surveyor's equipment, we measure the distance from the midpoint of each of these subintervals to the upper boundary of the property. These measurements give the values of $f(x)$ at $x = 10, 30, 50, 70$, and 90. What is the approximate area of the lot?

Road

(a)

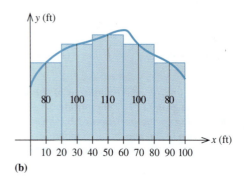

(b)

18. Real Estate Use the technique of exercise 17 to obtain an estimate of the area of the vacant lot shown in the accompanying figures.

Road

(a)

(b)

SOLUTION TO SELF-CHECK EXERCISE 6.3

The length of each subinterval is 3/4. Therefore, the four subintervals are

$$\left[0, \frac{3}{4}\right], \quad \left[\frac{3}{4}, \frac{3}{2}\right], \quad \left[\frac{3}{2}, \frac{9}{4}\right], \quad \text{and} \quad \left[\frac{9}{4}, 3\right]$$

The representative points are $\frac{3}{8}$, $\frac{9}{8}$, $\frac{15}{8}$, and $\frac{21}{8}$, respectively. Therefore, the required approximation is

$$
\begin{aligned}
A &= \frac{3}{4} f\left(\frac{3}{8}\right) + \frac{3}{4} f\left(\frac{9}{8}\right) + \frac{3}{4} f\left(\frac{15}{8}\right) + \frac{3}{4} f\left(\frac{21}{8}\right) \\
&= \frac{3}{4}\left[f\left(\frac{3}{8}\right) + f\left(\frac{9}{8}\right) + f\left(\frac{15}{8}\right) + f\left(\frac{21}{8}\right)\right] \\
&= \frac{3}{4}\left\{\left[2\left(\frac{3}{8}\right)^2 + 1\right] + \left[2\left(\frac{9}{8}\right)^2 + 1\right] + \left[2\left(\frac{15}{8}\right)^2 + 1\right] + \left[2\left(\frac{21}{8}\right)^2 + 1\right]\right\} \\
&= \frac{3}{4}\left(\frac{41}{32} + \frac{113}{32} + \frac{257}{32} + \frac{473}{32}\right) = \frac{663}{32}
\end{aligned}
$$

or approximately 20.72 square units.

6.4 THE FUNDAMENTAL THEOREM OF CALCULUS

The Fundamental Theorem of Calculus

In Section 6.3 we defined the definite integral of an arbitrary continuous function on an interval $[a, b]$ as a limit of Riemann sums. Calculating the value of a definite integral by actually taking the limit of such sums is tedious and in most cases impractical. It is important to realize that the numerical results we obtained in Examples 1 and 2 of Section 6.3 were *approximations* of the respective areas of the regions in question, even though these results enabled us to *conjecture* what the actual areas might be. Fortunately, there is a much better way of finding the exact value of a definite integral.

The following theorem shows how to evaluate the definite integral of a continuous function provided we can find an antiderivative of that function. Because of its importance in establishing the relationship between differentiation and integration, this theorem—discovered independently by Sir Isaac Newton (1642–1727) in England and Gottfried Wilhelm Leibniz (1646–1716) in Germany—is called the **Fundamental Theorem of Calculus.**

THE FUNDAMENTAL THEOREM OF CALCULUS

Let f be continuous on $[a, b]$. Then

$$\int_a^b f(x)\,dx = F(b) - F(a) \tag{9}$$

where F is any antiderivative of f; that is $F'(x) = f(x)$.

We will explain why this theorem is true at the end of this section.

When applying the Fundamental Theorem of Calculus, it is convenient to use the notation

$$F(x)\,\Big|_a^b = F(b) - F(a)$$

For example, using this notation, equation (9) is written

$$\int_a^b f(x)\,dx = F(x)\,\Big|_a^b = F(b) - F(a)$$

EXAMPLE 1 Let R be the region under the graph of $f(x) = x$ on the interval $[1, 3]$. Use the Fundamental Theorem of Calculus to find the area A of R and verify your result by elementary means.

Solution The region R is shown in Figure 6.16a. Since f is nonnegative on $[1, 3]$, the area of R is given by the definite integral of f from 1 to 3; that is

$$A = \int_1^3 x\,dx$$

Figure 6.16
The area of R can be computed in two different ways.

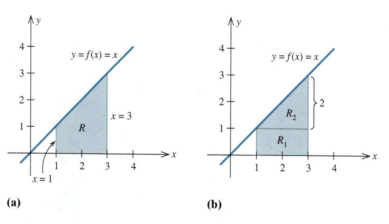

(a)   (b)

In order to evaluate the definite integral, observe that an antiderivative of $f(x) = x$ is $F(x) = \frac{1}{2}x^2 + C$, where C is an arbitrary constant. Therefore, by the Fundamental Theorem of Calculus, we have

$$A = \int_1^3 x\,dx = \frac{1}{2}x^2 + C\,\Big|_1^3$$

$$= \left(\frac{9}{2} + C\right) - \left(\frac{1}{2} + C\right) = 4 \text{ square units}$$

To verify this result by elementary means, observe that the area A is the area of the rectangle R_1 (width $\times$ height) plus the area of the triangle R_2 ($\frac{1}{2}$ base $\times$ height) (see Figure 6.16b); that is,

$$2(1) + \frac{1}{2}(2)(2) = 2 + 2 = 4$$

which agrees with the result obtained earlier.

Observe that in evaluating the definite integral in Example 1, the constant of integration "dropped out." This is true in general, for if $F(x) + C$ denotes an antiderivative of some function f, then

$$F(x) + C \Big|_a^b = [F(b) + C] - [F(a) + C]$$
$$= F(b) + C - F(a) - C$$
$$= F(b) - F(a)$$

With this fact in mind, we may, in all future computations involving the evaluations of a definite integral, drop the constant of integration from our calculations.

Finding the Area Under a Curve

Having seen how effective the Fundamental Theorem of Calculus is in helping us find the area of simple regions, we now use it to find the area of more complicated regions.

EXAMPLE 2   In Section 6.3, we conjectured that the area of the region R under the graph of $f(x) = x^2$ on the interval $[0, 1]$ was 1/3 square units. Use the Fundamental Theorem of Calculus to verify this conjecture.

Solution   The region R is reproduced in Figure 6.17. Observe that f is non-negative on $[0, 1]$, so the area of R is given by $A = \int_0^1 x^2 \, dx$. Since an antiderivative of $f(x) = x^2$ is $F(x) = \frac{1}{3}x^3$, we see, using the Fundamental Theorem

Figure 6.17
The area of R is $\int_0^1 x^2 \, dx = \frac{1}{3}$.

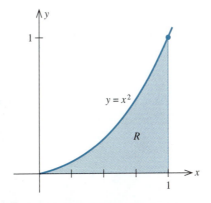

of Calculus, that

$$A = \int_0^1 x^2 \, dx = \frac{1}{3} x^3 \, \bigg|_0^1 = \frac{1}{3}(1) - \frac{1}{3}(0) = \frac{1}{3} \text{ square units}$$

as we wished to show.   ○ ○ ○

EXAMPLE 3 Find the area of the region R under the graph of $y = x^2 + 1$ from $x = -1$ to $x = 2$.

Solution The region R under consideration is shown in Figure 6.18. Using the Fundamental Theorem of Calculus, we find that the required area is

$$\int_{-1}^2 (x^2 + 1) \, dx = \left(\frac{1}{3} x^3 + x \right) \bigg|_{-1}^2$$

$$= \left[\frac{1}{3}(8) + 2 \right] - \left[\frac{1}{3}(-1)^3 + (-1) \right] = 6$$

or 6 square units.   ○ ○ ○

Evaluating Definite Integrals

In Examples 4 and 5, we use the rules of integration of Section 6.1 to help us evaluate the definite integrals.

EXAMPLE 4 Evaluate $\int_1^3 (3x^2 + e^x) \, dx$.

Solution

$$\int_1^3 (3x^2 + e^x) \, dx = x^3 + e^x \, \bigg|_1^3$$

$$= (27 + e^3) - (1 + e) = 26 + e^3 - e \qquad ○ ○ ○$$

EXAMPLE 5 Evaluate $\int_1^2 \left(\frac{1}{x} - \frac{1}{x^2} \right) dx$.

Solution

$$\int_1^2 \left(\frac{1}{x} - \frac{1}{x^2} \right) dx = \int_1^2 \left(\frac{1}{x} - x^{-2} \right) dx$$

$$= \ln |x| + \frac{1}{x} \bigg|_1^2$$

$$= \left(\ln 2 + \frac{1}{2} \right) - (\ln 1 + 1)$$

$$= \ln 2 - \frac{1}{2} \qquad \text{(Recall } \ln 1 = 0.)$$

○ ○ ○

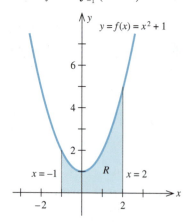

Figure 6.18
Area of $R = \int_{-1}^2 (x^2 + 1) \, dx$.

Consider the definite integral $\int_{-1}^{1} \frac{1}{x^2}\,dx$.

1. Show that a formal application of equation (9) leads to

$$\int_{-1}^{1} \frac{1}{x^2}\,dx = -\frac{1}{x}\Big|_{-1}^{1} = -1 - 1 = -2$$

2. Observe that $f(x) = \dfrac{1}{x^2}$ is positive at each value of x in $[-1, 1]$ where it is defined. Therefore, one might expect that the definite integral with integrand f has a positive value, if it exists.

3. Explain this apparent contradiction in the result (1) and the observation (2).

Applications

EXAMPLE 6 The management of Staedtler Office Equipment has determined that the daily marginal cost function associated with producing battery-operated pencil sharpeners is given by

$$C'(x) = 0.000006x^2 - 0.006x + 4$$

where $C'(x)$ is measured in dollars per unit and x denotes the number of units produced. Management has also determined that the daily fixed cost incurred in producing these pencil sharpeners is $100. Find Staedtler's daily total cost for producing (a) the first 500 units and (b) the 201st through 400th units.

Solution

a. Since $C'(x)$ is the marginal cost function, its antiderivative $C(x)$ is the total cost function. The daily fixed cost incurred in producing the pencil sharpeners is $C(0)$ dollars. Since the daily fixed cost is given as $100, we have $C(0) = 100$. We are required to find $C(500)$. Let us compute $C(500) - C(0)$, the net change in the total cost function $C(x)$ over the interval $[0, 500]$. Using the Fundamental Theorem of Calculus, we find

$$C(500) - C(0) = \int_{0}^{500} C'(x)\,dx$$

$$= \int_{0}^{500} (0.000006x^2 - 0.006x + 4)\,dx$$

$$= 0.000002x^3 - 0.003x^2 + 4x\,\Big|_{0}^{500}$$

$$= [0.000002(500)^3 - 0.003(500)^2 + 4(500)]$$
$$\quad - [0.000002(0)^3 - 0.003(0)^2 + 4(0)]$$

$$= 1500$$

Therefore $C(500) = 1500 + C(0) = 1500 + 100 = 1600$, so the total cost incurred daily by Staedtler in producing 500 pencil sharpeners is $1600.

b. The daily total cost incurred by Staedtler in producing the 201st through 400th units of battery-operated pencil sharpeners is given by

$$C(400) - C(200) = \int_{200}^{400} C'(x)\,dx$$

$$= \int_{200}^{400} (0.000006x^2 - 0.006x + 4)\,dx$$

$$= 0.000002x^3 - 0.003x^2 + 4x \Big|_{200}^{400}$$

$$= 552$$

or $552.   ◐ ◐ ◐

Since $C'(x)$ is nonnegative for x in the interval $(0, \infty)$, we have the following geometrical interpretation of the two definite integrals in Example 6: $\int_0^{500} C'(x)\,dx$ is the area of the region under the graph of the function C' from $x = 0$ to $x = 500$, shown in Figure 6.19a, and $\int_{200}^{400} C'(x)\,dx$ is the area of the region shown in Figure 6.19b from $x = 200$ to $x = 400$.

Figure 6.19

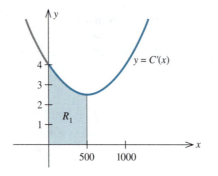

(a) Area of $R_1 = \int_0^{500} C'(x)\,dx$

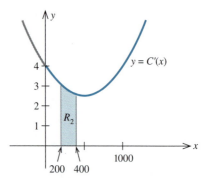

(b) Area of $R_2 = \int_{200}^{400} C'(x)\,dx$

EXAMPLE 7   An efficiency study conducted for the Elektra Electronics Company showed that the rate at which Space Commander walkie-talkies are assembled by the average worker t hours after starting work at 8 A.M. is given by the function

$$f(t) = -3t^2 + 12t + 15 \qquad (0 \le t \le 4)$$

Determine how many walkie-talkies can be assembled by the average worker in the first hour of the morning shift.

Solution Let $N(t)$ denote the number of walkie-talkies assembled by the average worker t hours after starting work in the morning shift. Then we have

$$N'(t) = f(t) = -3t^2 + 12t + 15$$

Therefore, the number of units assembled by the average worker in the first hour of the morning shift is

$$N(1) - N(0) = \int_0^1 N'(t)\, dt = \int_0^1 (-3t^2 + 12t + 15)\, dt$$

$$= -t^3 + 6t^2 + 15t \Big|_0^1 = -1 + 6 + 15$$

$$= 20, \quad \text{or} \quad 20 \text{ units}$$

○ ○ ○

EXPLORING WITH TECHNOLOGY

You can demonstrate graphically that $\int_0^x t\, dt = \frac{1}{2}x^2$ as follows:

1. Plot the graphs of $y1 = \int_0^x t\, dt$ and $y2 = \frac{1}{2}x^2$ on the same set of axes using the viewing rectangle $[-5, 5] \times [0, 10]$.

2. Compare the graphs of $y1$ and $y2$ and draw the desired conclusion.

○   ○   ○

EXAMPLE 8 A certain city's rate of electricity consumption is expected to grow exponentially with a growth constant of $k = 0.04$. If the present rate of consumption is 40 million kilowatt-hours (kwh) per year, what should the total production of electricity be over the next 3 years in order to meet the projected demand?

Solution If $R(t)$ denotes the expected rate of consumption of electricity t years from now, then

$$R(t) = 40e^{0.04t}$$

million kwh per year. Next, if $C(t)$ denotes the expected total consumption of electricity over a period of t years, then

$$C'(t) = R(t)$$

Therefore, the total consumption of electricity expected over the next 3 years is given by

$$\int_0^3 C'(t)\,dt = \int_0^3 40e^{0.04t}\,dt$$

$$= \frac{40}{0.04}e^{0.04t}\bigg|_0^3$$

$$= 1000(e^{0.12} - 1)$$

$$= 127.5, \quad \text{or} \quad 127.5 \text{ million kwh}$$

the amount that must be produced over the next 3 years in order to meet the demand.

● ● ●

The definite integral $\int_{-3}^3 \sqrt{9 - x^2}\,dx$ cannot be evaluated using the Fundamental Theorem of Calculus because the method of this section does not enable us to find an antiderivative of the integrand. But the integral can be evaluated by interpreting it as the area of a certain plane region. What is the region? And what is the value of the integral?

Validity of the Fundamental Theorem of Calculus

In order to demonstrate the plausibility of the Fundamental Theorem of Calculus for the case where f is nonnegative on an interval $[a, b]$, let us define an "area function" A as follows. Let $A(t)$ denote the area of the region R under the graph of $y = f(x)$ from $x = a$ to $x = t$, where $a \le t \le b$ (Figure 6.20).

If h is a small positive number, then $A(t + h)$ is the area of the region under the graph of $y = f(x)$ from $x = a$ to $x = t + h$. Therefore, the difference

$$A(t + h) - A(t)$$

is the area under the graph of $y = f(x)$ from $x = t$ to $x = t + h$ (Figure 6.21).

Figure 6.20
A(t) = Area under the graph of f from x = a to x = t.

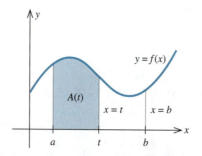

Figure 6.21
A(t + h) − A(t) = Area under the graph of f from x = t to x = t + h.

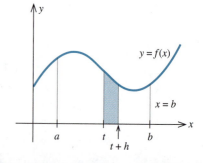

Figure 6.22
Area of rectangle = h · f(t).

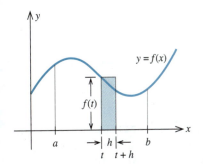

Now, the area of this last region can be approximated by the area of the rectangle of width h and height $f(t)$—that is, by the expression $h \cdot f(t)$ (Figure 6.22). Thus,

$$A(t + h) - A(t) \approx h \cdot f(t)$$

where the approximations improve as h is taken to be smaller and smaller. Dividing both sides of the foregoing relationship by h, we obtain

$$\frac{A(t + h) - A(t)}{h} \approx f(t)$$

Taking the limit as h approaches zero, we find, by the definition of the derivative, that the left-hand side is

$$\lim_{h \to 0} \frac{A(t + h) - A(t)}{h} = A'(t)$$

The right-hand side, which is independent of h, remains constant throughout the limiting process. Because the approximation becomes exact as h approaches zero, we find that

$$A'(t) = f(t)$$

Since the foregoing equation holds for all values of t in the interval $[a, b]$, we have shown that the *area function A* is an antiderivative of the function $f(x)$. By Theorem 1 of Section 6.1, we conclude that $A(x)$ must have the form

$$A(x) = F(x) + C$$

where F is any antiderivative of f and C is an arbitrary constant. To determine the value of C, observe that $A(a) = 0$. This condition implies that

$$A(a) = F(a) + C = 0$$

Figure 6.23
The area of R is given by A(b).

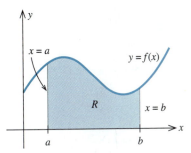

or $C = -F(a)$. Next, since the area of the region R is $A(b)$ (Figure 6.23), we see that the required area is

$$A(b) = F(b) + C$$
$$= F(b) - F(a)$$

Since the area of the region R is

$$\int_a^b f(x)\, dx$$

we have

$$\int_a^b f(x)\, dx = F(b) - F(a)$$

as we set out to show.

SELF–CHECK EXERCISES 6.4

1. Evaluate $\int_0^2 (x + e^x)\, dx$.

2. The daily marginal profit function associated with producing and selling Texa-Pep hot sauce is

$$P'(x) = -0.000006x^2 + 6$$

where x denotes the number of cases (each case contains 24 bottles) produced and sold daily and $P'(x)$ is measured in dollars per unit. The fixed cost is $400.
a. What is the total profit realizable from producing and selling 1000 cases of Texa-Pep per day?
b. What is the additional profit realizable if the production and sale of Texa-Pep are increased from 1000 to 1200 cases per day?

Solutions to Self-Check Exercises 6.4 can be found on page 478.

6.4 **EXERCISES**

In exercises 1–4, find the area of the region under the graph of the function f on the interval [a, b] using the Fundamental Theorem of Calculus. Then verify your result using geometry.

1. $f(x) = 2$; [1, 4]

2. $f(x) = 4$; [−1, 2]

3. $f(x) = 2x$; [1, 3]

4. $f(x) = -\frac{1}{4}x + 1$; [1, 4]

In exercises 5–16, find the area of the region under the graph of the function f on the interval [a, b].

5. $f(x) = 2x + 3$; [−1, 2]

6. $f(x) = 4x - 1$; [2, 4]

7. $f(x) = -x^2 + 4$; [−1, 2]

8. $f(x) = 4x - x^2$; [0, 4]

9. $f(x) = \frac{1}{x}$; [1, 2]

10. $f(x) = \frac{1}{x^2}$; [2, 4]

11. $f(x) = \sqrt{x}$; [1, 9]

12. $f(x) = x^3$; [1, 3]

13. $f(x) = 1 - \sqrt[3]{x}$; [−8, −1]

14. $f(x) = \frac{1}{\sqrt{x}}$; [1, 9]

15. $f(x) = e^x$; [0, 2]

16. $f(x) = e^x - x$; [1, 2]

In exercises 17–40, evaluate the definite integral.

17. $\int_2^4 3\, dx$

18. $\int_{-1}^2 -2\, dx$

19. $\int_1^3 (2x + 3)\, dx$

20. $\int_{-1}^0 (4 - x)\, dx$

21. $\int_{-1}^3 2x^2\, dx$

22. $\int_0^2 8x^3\, dx$

23. $\int_{-2}^2 (x^2 - 1)\, dx$

24. $\int_1^4 \sqrt{u}\, du$

25. $\int_1^8 4x^{1/3}\, dx$

26. $\int_1^4 2x^{-3/2}\, dx$

27. $\int_0^1 (x^3 - 2x^2 + 1)\, dx$

28. $\int_1^2 (t^5 - t^3 + 1)\, dt$

29. $\int_2^4 \frac{1}{x}\, dx$

30. $\int_1^3 \frac{2}{x}\, dx$

31. $\int_0^4 x(x^2 - 1)\, dx$

32. $\int_0^2 (x - 4)(x - 1)\, dx$

33. $\int_1^3 (t^2 - t)^2\, dt$

34. $\int_{-1}^1 (x^2 - 1)^2\, dx$

35. $\int_{-3}^{-1} \frac{1}{x^2}\, dx$

36. $\int_1^2 \frac{2}{x^3}\, dx$

37. $\int_1^4 \left(\sqrt{x} - \frac{1}{\sqrt{x}}\right) dx$

38. $\int_0^1 \sqrt{2x}(\sqrt{x} + \sqrt{2})\, dx$

39. $\int_1^4 \frac{3x^3 - 2x^2 + 4}{x^2}\, dx$

40. $\int_1^2 \left(1 + \frac{1}{u} + \frac{1}{u^2}\right) du$

41. Marginal Cost A division of Ditton Industries manufactures a deluxe toaster oven. Management has determined that the daily marginal cost function associated with producing these toaster ovens is given by

$$C'(x) = 0.0003x^2 - 0.12x + 20$$

where $C'(x)$ is measured in dollars per unit and x denotes the number of units produced. Management has also determined that the daily fixed cost incurred in the production is $800.
a. Find the total cost incurred by Ditton in producing the first 300 units of these toaster ovens per day.
b. What is the total cost incurred by Ditton in producing the 201st through 300th units per day?

42. Marginal Revenue The management of Ditton Industries has determined that the daily marginal revenue function associated with selling x units of their deluxe toaster ovens is given by

$$R'(x) = -0.1x + 40$$

where $R'(x)$ is measured in dollars per unit.
a. Find the daily total revenue realized from the sale of 200 units of the toaster oven.
b. Find the additional revenue realized when the production (and sales) level is increased from 200 to 300 units.

43. Marginal Profit Refer to exercise 41. The daily marginal profit function associated with the production and sales of the deluxe toaster ovens is known to be

$$P'(x) = -0.0003x^2 + 0.02x + 20$$

where x denotes the number of units manufactured and sold daily and $P'(x)$ is measured in dollars per unit.
a. Find the total profit realizable from the manufacture and sale of 200 units of the toaster ovens per day.
[*Hint:* $P(200) - P(0) = \int_0^{200} P'(x)\, dx$, $P(0) = -800$.]

b. What is the additional daily profit realizable if the production and sale of the toaster ovens are increased from 200 to 220 units per day?

44. Efficiency Studies Tempco Electronics, a division of Tempco Toys, Inc., manufactures an electronic football game. An efficiency study showed that the rate at which the games are assembled by the average worker t hours

after starting work at 8 A.M. is

$$-\frac{3}{2}t^2 + 6t + 20 \qquad (0 \le t \le 4)$$

units per hour.
a. Find the total number of games the average worker can be expected to assemble in the 4-hour morning shift.
b. How many units can the average worker be expected to assemble in the first hour of the morning shift? In the second hour of the morning shift?

45. Speedboat Racing In a recent pretrial run for the world water speed record, the velocity of the *Sea Falcon II* t seconds after firing the booster rocket was given by

$$v(t) = -t^2 + 20t + 440 \qquad (0 \le t \le 20)$$

feet per second. Find the distance covered by the boat over the 20-second period after the booster rocket was activated.
[*Hint:* The distance is given by $\int_0^{20} v(t)\, dt$.]

46. U.S. Census According to the U.S. Census Bureau, the number of Americans aged 45 to 54 (which stood at 25 million at the beginning of 1990) will grow at the rate of

$$R(t) = 0.00933t^3 + 0.019t^2 - 0.10833t + 1.3467$$

million people per year, t years from the beginning of 1990. How many Americans aged 45 to 54 will be added to the population between 1990 and the year 2000?
Source: U.S. Census Bureau

47. Air Purification To test air purifiers, the engineers run a purifier in a smoke-filled 10×20-foot room. While conducting a test for a certain brand of air purifier, it was determined that the amount of smoke in the room was decreasing at the rate of

$$R(t) = 0.00032t^4 - 0.01872t^3 + 0.3948t^2$$
$$- 3.83t + 17.63 \qquad (0 \le t \le 20)$$

percent of the (original) amount of the smoke per minute, t minutes after the start of the test. How much smoke was left in the room 5 minutes after the start of the test? Ten minutes after the start of the test?
Source: Consumer Reports

USING TECHNOLOGY

Finding Definite Integrals

Some graphing utilities have an operation for finding the definite integral of a function. If your graphing utility has this capability, use it to work through the example and exercises of this section.

EXAMPLE 1   Use the numerical integral operation of a graphing utility to evaluate

$$\int_{-1}^{2} \frac{2x + 4}{(x^2 + 1)^{3/2}} \, dx$$

Solution   Using the numerical integral operation of a graphing utility, we find

$$\int_{-1}^{2} \frac{2x + 4}{(x^2 + 1)^{3/2}} \, dx \approx 6.92592225992$$

EXERCISES

In exercises 1–4, use a graphing utility to find the area of the region under the graph of f on the interval [a, b]. Express your answer to four decimal places.

1. $f(x) = 0.002x^5 + 0.032x^4 - 0.2x^2 + 2;\ [-1.1, 2.2]$

2. $f(x) = x\sqrt{x^3 + 1};\ [1, 2]$

3. $f(x) = \sqrt{x}e^{-x};\ [0, 3]$

4. $f(x) = \dfrac{\ln x}{\sqrt{1 + x^2}};\ [1, 2]$

In exercises 5–10, use a graphing utility to evaluate the definite integral.

5. $\displaystyle\int_{-1.2}^{2.3} (0.2x^4 - 0.32x^3 + 1.2x - 1)\ dx$

6. $\displaystyle\int_{1}^{3} x(x^4 - 1)^{3.2}\ dx$

7. $\displaystyle\int_{0}^{2} \frac{3x^3 + 2x^2 + 1}{2x^2 + 3}\ dx$

8. $\displaystyle\int_{1}^{2} \frac{\sqrt{x} + 1}{2x^2 + 1}\ dx$

9. $\displaystyle\int_{0}^{2} \frac{e^x}{\sqrt{x^2 + 1}}\ dx$

10. $\displaystyle\int_{1}^{3} e^{-x}\ln(x^2 + 1)\ dx$

11. Rework exercise 46, Exercises 6.4.

12. Rework exercise 47, Exercises 6.4.

SOLUTIONS TO SELF-CHECK EXERCISES 6.4

I.
$$\int_0^2 (x + e^x)\, dx = \frac{1}{2}x^2 + e^x \Big|_0^2$$

$$= \left[\frac{1}{2}(2)^2 + e^2\right] - \left[\frac{1}{2}(0) + e^0\right]$$

$$= 2 + e^2 - 1$$

$$= e^2 + 1$$

2. a. We want $P(1000)$. But

$$P(1000) - P(0) = \int_0^{1000} P'(x)\, dx = \int_0^{1000} (-0.000006x^2 + 6)\, dx$$

$$= -0.000002x^3 + 6x \Big|_0^{1000}$$

$$= -0.000002(1000)^3 + 6(1000)$$

$$= 4000$$

So, $P(1000) = 4000 + P(0) = 4000 - 400$, or \$3600 per day $[P(0) = -C(0)]$.

b. The additional profit realizable is given by

$$\int_{1000}^{1200} P'(x)\, dx = -0.000002x^3 + 6x \Big|_{1000}^{1200}$$

$$= [-0.000002(1200)^3 + 6(1200)]$$

$$\quad - [-0.000002(1000)^3 + 6(1000)]$$

$$= 3744 - 4000$$

$$= -256$$

That is, the company sustains a loss of \$256 per day if production is increased to 1200 cases a day.

6.5 EVALUATING DEFINITE INTEGRALS

This section continues our discussion of the applications of the Fundamental Theorem of Calculus.

Properties of the Definite Integral

Before going on, we list the following useful properties of the definite integral, some of which parallel the rules of integration of Section 6.1.

| **PROPERTIES OF THE DEFINITE INTEGRAL** | Let f and g be integrable functions; then |

1. $\int_a^a f(x)\,dx = 0$

2. $\int_a^b f(x)\,dx = -\int_b^a f(x)\,dx$

3. $\int_a^b cf(x)\,dx = c\int_a^b f(x)\,dx \qquad (c, \text{a constant})$

4. $\int_a^b [f(x) \pm g(x)]\,dx = \int_a^b f(x)\,dx \pm \int_a^b g(x)\,dx$

5. $\int_a^b f(x)\,dx = \int_a^c f(x)\,dx + \int_c^b f(x)\,dx \qquad (a < c < b)$

Property 5 states that if c is a number lying between a and b so that the interval $[a, b]$ is divided into the intervals $[a, c]$ and $[c, b]$, then the integral of f over the interval $[a, b]$ may be expressed as the sum of the integral of f over the interval $[a, c]$ and the integral of f over the interval $[c, b]$.

Property 5 has the following geometrical interpretation when f is nonnegative. By definition

$$\int_a^b f(x)\,dx$$

is the area of the region under the graph of $y = f(x)$ from $x = a$ to $x = b$ (Figure 6.24). Similarly, we interpret the definite integrals

$$\int_a^c f(x)\,dx \qquad \text{and} \qquad \int_c^b f(x)\,dx$$

as the areas of the regions under the graph of $y = f(x)$ from $x = a$ to $x = c$ and from $x = c$ to $x = b$, respectively. Since the two regions do not overlap, we see that

$$\int_a^b f(x)\,dx = \int_a^c f(x)\,dx + \int_c^b f(x)\,dx$$

Figure 6.24
$\int_a^b f(x)\,dx = \int_a^c f(x)\,dx$
$\qquad + \int_c^b f(x)\,dx.$

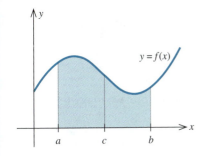

The Method of Substitution for Definite Integrals

Our first example shows two approaches generally used when evaluating a definite integral using the method of substitution.

EXAMPLE 1 Evaluate

$$\int_0^4 x\sqrt{9 + x^2}\,dx$$

Solution

Method 1 We first find the corresponding indefinite integral

$$I = \int x\sqrt{9 + x^2}\,dx$$

Make the substitution

$$u = 9 + x^2$$

so that

$$du = \frac{d}{dx}(9 + x^2)\,dx$$

$$= 2x\,dx$$

or

$$x\,dx = \frac{1}{2}\,du \qquad \text{(Dividing both sides by 2)}$$

Then

$$I = \int \frac{1}{2}\sqrt{u}\,du = \frac{1}{2}\int u^{1/2}\,du$$

$$= \frac{1}{3}u^{3/2} + C = \frac{1}{3}(9 + x^2)^{3/2} + C \qquad \begin{array}{l}\text{(Substituting}\\ 9 + x^2 \text{ for } u)\end{array}$$

Using this result, we now evaluate the given definite integral.

$$\int_0^4 x\sqrt{9 + x^2}\,dx = \frac{1}{3}(9 + x^2)^{3/2}\Big|_0^4$$

$$= \frac{1}{3}[(9 + 16)^{3/2} - 9^{3/2}]$$

$$= \frac{1}{3}(125 - 27) = \frac{98}{3} = 32\frac{2}{3}$$

Method 2 *Changing the Limits of Integration:* As before, we make the substitution

$$u = 9 + x^2 \qquad\qquad \textbf{(10)}$$

so that

$$du = 2x\,dx$$

or

$$x\,dx = \frac{1}{2}\,du$$

Next, observe that the given definite integral is evaluated *with respect to x* with the range of integration given by the interval $[0, 4]$. If we perform the integration *with respect to u* via the substitution (10), then we must adjust the range of integration to reflect the fact that the integration is being performed with respect to the new variable u. To determine the proper range of integration, note that when $x = 0$, equation (10) implies that

$$u = 9 + 0^2 = 9$$

which gives the required lower limit of integration with respect to u. Similarly, when $x = 4$,

$$u = 9 + 16 = 25$$

is the required upper limit of integration with respect to u. Thus, the range of integration when the integration is performed with respect to u is given by the interval $[9, 25]$. Therefore, we have

$$\int_0^4 x\sqrt{9 + x^2}\, dx = \int_9^{25} \frac{1}{2}\sqrt{u}\, du = \frac{1}{2}\int_9^{25} u^{1/2}\, du$$

$$= \frac{1}{3}u^{3/2}\bigg|_9^{25} = \frac{1}{3}(25^{3/2} - 9^{3/2})$$

$$= \frac{1}{3}(125 - 27) = \frac{98}{3} = 32\frac{2}{3}$$

which agrees with the result obtained using Method 1.   ◦ ◦ ◦

EXPLORING WITH TECHNOLOGY

Refer to Example 1. You can confirm the results obtained there by using a graphing utility as follows:

a. Use the numerical integration operation of the graphing utility to evaluate

$$\int_0^4 x\sqrt{9 + x^2}\, dx$$

b. Evaluate $\frac{1}{2}\int_9^{25}\sqrt{u}\, du$.

c. Conclude that $\int_0^4 x\sqrt{9 + x^2}\, dx = \frac{1}{2}\int_9^{25}\sqrt{u}\, du$.

◦ ◦ ◦

EXAMPLE 2 Evaluate

$$\int_0^2 xe^{2x^2}\, dx$$

Solution Let $u = 2x^2$ so that $du = 4x\, dx$, or $x\, dx = \frac{1}{4}\, du$. When $x = 0$, $u = 0$, and when $x = 2$, $u = 8$. This gives the lower and upper limits of integration with respect to u. Making the indicated substitutions, we find

$$\int_0^2 xe^{2x^2}\, dx = \int_0^8 \frac{1}{4}e^u\, du = \frac{1}{4}e^u\bigg|_0^8 = \frac{1}{4}(e^8 - 1)$$ ◦ ◦ ◦

EXAMPLE 3 Evaluate

$$\int_0^1 \frac{x^2}{x^3 + 1}\, dx$$

Solution Let $u = x^3 + 1$ so that $du = 3x^2\,dx$, or $x^2\,dx = \frac{1}{3}\,du$. When $x = 0$, $u = 1$, and when $x = 1$, $u = 2$. This gives the lower and upper limits of integration with respect to u. Making the indicated substitutions, we find

$$\int_0^1 \frac{x^2}{x^3 + 1}\,dx = \frac{1}{3}\int_1^2 \frac{du}{u} = \frac{1}{3}\ln|u|\Big|_1^2$$

$$= \frac{1}{3}(\ln 2 - \ln 1) = \frac{1}{3}\ln 2$$

Finding the Area Under a Curve

EXAMPLE 4 Find the area of the region R under the graph of $f(x) = e^{(1/2)x}$ from $x = -1$ to $x = 1$.

Solution The region R is shown in Figure 6.25. Its area is given by

$$A = \int_{-1}^1 e^{(1/2)x}\,dx$$

In order to evaluate this integral, we make the substitution

$$u = \frac{1}{2}x$$

so that $du = \frac{1}{2}\,dx$ or $dx = 2\,du$

When $x = -1$, $u = -\frac{1}{2}$, and when $x = 1$, $u = \frac{1}{2}$. Making the indicated substitutions, we obtain

$$A = \int_{-1}^1 e^{(1/2)x}\,dx = 2\int_{-1/2}^{1/2} e^u\,du$$

$$= 2e^u\Big|_{-1/2}^{1/2}$$

$$= 2(e^{1/2} - e^{-1/2})$$

or approximately 2.08 square units.

Figure 6.25
Area of $R = \int_{-1}^1 e^{(1/2)x}dx$.

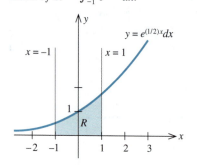

$y = e^{(1/2)x}dx$

$x = -1$ $x = 1$

R

Let f be a function defined piecewise by the rule

$$f(x) = \begin{cases} \sqrt{x} & \text{if } 0 \le x \le 1 \\ \dfrac{1}{x} & \text{if } 1 < x \le 2 \end{cases}$$

How would you use property 5 of definite integrals to find the area of the region under the graph of f on $[0, 2]$? What is the area?

Average Value of a Function

The *average value* of a function over an interval provides us with an application of the definite integral. Recall that the average value of a set of n numbers is the number

$$\frac{y_1 + y_2 + \cdots + y_n}{n}$$

Now, suppose that f is a continuous function defined on $[a, b]$. Let us divide the interval $[a, b]$ into n subintervals of equal length $(b - a)/n$. Choose points $x_1, x_2, \ldots, x_n$ in the first, second, $\ldots$, and nth subintervals, respectively. Then the average value of the numbers $f(x_1), f(x_2), \ldots, f(x_n)$, given by

$$\frac{f(x_1) + f(x_2) + \cdots + f(x_n)}{n}$$

is an approximation of the average of all the values of $f(x)$ on the interval $[a, b]$. This expression can be written in the form

$$\frac{(b - a)}{(b - a)} \left[f(x_1) \cdot \frac{1}{n} + f(x_2) \cdot \frac{1}{n} + \cdots + f(x_n) \cdot \frac{1}{n} \right]$$

$$= \frac{1}{b - a} \left[f(x_1) \cdot \frac{b - a}{n} + f(x_2) \cdot \frac{b - a}{n} + \cdots + f(x_n) \cdot \frac{b - a}{n} \right]$$

$$= \frac{1}{b - a} [f(x_1)\Delta x + f(x_2)\Delta x + \cdots + f(x_n)\Delta x] \qquad \textbf{(11)}$$

As n gets larger and larger, the expression (11) approximates the average value of $f(x)$ over $[a, b]$ with increasing accuracy. But the sum inside the brackets in (11) is a Riemann sum of the function f over $[a, b]$. In view of this, we have

$$\lim_{n \to \infty} \left[\frac{f(x_1) + f(x_2) + \cdots + f(x_n)}{n} \right]$$

$$= \frac{1}{b - a} \lim_{n \to \infty} [f(x_1)\Delta x + f(x_2)\Delta x + \cdots + f(x_n)\Delta x]$$

$$= \frac{1}{b - a} \int_a^b f(x)\, dx$$

This discussion motivates the following definition.

THE AVERAGE VALUE OF A FUNCTION

Suppose that f is integrable on $[a, b]$. Then the **average value of f** over $[a, b]$ is

$$\frac{1}{b - a} \int_a^b f(x)\, dx$$

EXAMPLE 5   Find the average value of the function $f(x) = \sqrt{x}$ over the interval $[0, 4]$.

Solution   The required average value is given by

$$\frac{1}{4 - 0}\int_0^4 \sqrt{x}\,dx = \frac{1}{4}\int_0^4 x^{1/2}\,dx$$

$$= \frac{1}{6}x^{3/2}\bigg|_0^4$$

$$= \frac{4}{3}$$

○ ○ ○

Applications

EXAMPLE 6   The interest rates charged by Madison Finance on auto loans for used cars over a certain 6-month period in 1996 are approximated by the function

$$r(t) = -\frac{1}{12}t^3 + \frac{7}{8}t^2 - 3t + 12 \qquad (0 \le t \le 6)$$

where t is measured in months and $r(t)$ is the annual percentage rate. What is the average rate on auto loans extended by Madison over the 6-month period?

Solution   The average rate over the 6-month period in question is given by

$$\frac{1}{6 - 0}\int_0^6 \left(-\frac{1}{12}t^3 + \frac{7}{8}t^2 - 3t + 12\right)dt$$

$$= \frac{1}{6}\left(-\frac{1}{48}t^4 + \frac{7}{24}t^3 - \frac{3}{2}t^2 + 12t\right)\bigg|_0^6$$

$$= \frac{1}{6}\left[-\frac{1}{48}(6^4) + \frac{7}{24}(6^3) - \frac{3}{2}(6^2) + 12(6)\right]$$

$$= 9$$

or 9% per year.

○ ○ ○

EXAMPLE 7   The amount of a certain drug in a patient's body t days after it has been administered is

$$C(t) = 5e^{-0.2t}$$

units. Determine the average amount of the drug present in the patient's body for the first 4 days after the drug has been administered.

Solution The average amount of the drug present in the patient's body for the first 4 days after it has been administered is given by

$$\frac{1}{4 - 0} \int_0^4 5e^{-0.2t}\, dt = \frac{5}{4} \int_0^4 e^{-0.2t}\, dt$$

$$= \frac{5}{4} \left[\left(-\frac{1}{0.2} \right) e^{-0.2t} \Big|_0^4 \right]$$

$$= \frac{5}{4} (-5e^{-0.8} + 5)$$

$$\approx 3.44$$

Figure 6.26
The average value of f over [a, b] is k.

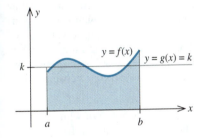

or approximately 3.44 units.

We now give a geometrical interpretation of the average value of a function f over an interval $[a, b]$. Suppose that $f(x)$ is nonnegative so that the definite integral

$$\int_a^b f(x)\, dx$$

gives the area under the graph of f from $x = a$ to $x = b$ (Figure 6.26). Observe that, in general, the "height" $f(x)$ varies from point to point. Can we replace $f(x)$ by a constant function $g(x) = k$ (which has constant height) such that the areas under each of the two functions f and g are the same? If so, since the area under the graph of g from $x = a$ to $x = b$ is $k(b - a)$, we have

$$k(b - a) = \int_a^b f(x)\, dx$$

or

$$k = \frac{1}{b - a} \int_a^b f(x)\, dx$$

so that k is the average value of f over $[a, b]$. Thus, the average value of a function f over an interval $[a, b]$ is the height of a rectangle with base of length $(b - a)$ that has the same area as that of the region under the graph of f from $x = a$ to $x = b$.

SELF-CHECK EXERCISES 6.5

1. Evaluate $\int_0^2 \sqrt{2x + 5}\, dx$.

2. Find the average value of the function $f(x) = 1 - x^2$ over the interval $[-1, 2]$.

3. The median price of a house in a southwestern state between January 1, 1991, and January 1, 1996, is approximated by the function

$$f(t) = t^3 - 7t^2 + 17t + 140 \qquad (0 \le t \le 5)$$

where $f(t)$ is measured in thousands of dollars and t is expressed in years ($t = 0$ corresponds to the beginning of 1991). Determine the average median price of a house over that time interval.

Solutions to Self-Check Exercises 6.5 can be found on page 488.

6.5 EXERCISES

In exercises 1–28, evaluate the given definite integral.

1. $\int_0^2 x(x^2 - 1)^3 \, dx$

2. $\int_0^1 x^2(2x^3 - 1)^4 \, dx$

3. $\int_0^1 x\sqrt{5x^2 + 4} \, dx$

4. $\int_1^3 x\sqrt{3x^2 - 2} \, dx$

5. $\int_0^2 x^2(x^3 + 1)^{3/2} \, dx$

6. $\int_1^5 (2x - 1)^{5/2} \, dx$

7. $\int_0^1 \frac{1}{\sqrt{2x + 1}} \, dx$

8. $\int_0^2 \frac{x}{\sqrt{x^2 + 5}} \, dx$

9. $\int_1^2 (2x - 1)^4 \, dx$

10. $\int_1^2 (2x + 4)(x^2 + 4x - 8)^3 \, dx$

11. $\int_{-1}^1 x^2(x^3 + 1)^4 \, dx$

12. $\int_{-1}^1 (x^3 + \frac{3}{4})(x^4 + 3x)^{-2} \, dx$

13. $\int_1^5 x\sqrt{x - 1} \, dx$

14. $\int_1^4 x\sqrt{x + 1} \, dx$

[*Hint:* Let $u = x + 1$.]

15. $\int_0^2 xe^{x^2} \, dx$

16. $\int_0^1 e^{-x} \, dx$

17. $\int_0^1 (e^{2x} + x^2 + 1) \, dx$

18. $\int_0^2 (e^t - e^{-t}) \, dt$

19. $\int_{-1}^1 xe^{x^2+1} \, dx$

20. $\int_0^4 \frac{e^{\sqrt{x}}}{\sqrt{x}} \, dx$

21. $\int_3^6 \frac{2}{x - 2} \, dx$

22. $\int_0^1 \frac{x}{1 + 2x^2} \, dx$

23. $\int_1^2 \frac{x^2 + 2x}{x^3 + 3x^2 - 1} \, dx$

24. $\int_0^1 \frac{e^x}{1 + e^x} \, dx$

25. $\int_1^2 \left(4e^{2u} - \frac{1}{u}\right) du$

26. $\int_1^2 \left(1 + \frac{1}{x} + e^x\right) dx$

27. $\int_1^2 \left(2e^{-4x} - \frac{1}{x^2}\right) dx$

28. $\int_1^2 \frac{\ln x}{x} \, dx$

In exercises 29–38, find the average value of the given function f over the indicated interval [a, b].

29. $f(x) = 2x + 3$; $[0, 2]$

30. $f(x) = 8 - x$; $[1, 4]$

31. $f(x) = 2x^2 - 3$; $[1, 3]$

32. $f(x) = 4 - x^2$; $[-2, 3]$

33. $f(x) = x^2 + 2x - 3$; $[-1, 2]$

34. $f(x) = x^3$; $[-1, 1]$

35. $f(x) = \sqrt{2x + 1}$; $[0, 4]$

36. $f(x) = e^{-x}$; $[0, 4]$

37. $f(x) = xe^{x^2}$; $[0, 2]$

38. $f(x) = \frac{1}{x + 1}$; $[0, 2]$

39. World Production of Coal   A study proposed in 1980 by researchers from the major producers and consumers of the world's coal concluded that coal could and must play an important role in fueling global economic growth over the next 20 years. The world production of coal in 1980 was 3.5 billion metric tons. If output were to increase at the rate of $3.5e^{0.05t}$ billion metric tons per year in year t ($t = 0$ corresponding to 1980), determine how much coal will be produced worldwide between 1980 and the end of the century.

40. Newton's Law of Cooling   A bottle of white wine at room temperature (68°F) is placed in a refrigerator at 4 P.M. Its temperature after t hours is changing at the rate of

$$-18e^{-0.6t}$$

degrees Fahrenheit per hour. By how many degrees will the temperature of the wine have dropped by 7 P.M.? What will the temperature of the wine be at 7 P.M.?

41. Net Investment Flow   The net investment flow (rate of capital formation) of the giant conglomerate LTF Incorporated is projected to be

$$t\sqrt{\frac{1}{2}t^2 + 1}$$

million dollars per year in year t. Find the accruement on the company's capital stock in the second year.

Hint: The amount is given by

$$\int_1^2 t\sqrt{\frac{1}{2}t^2 + 1} \, dt$$

42. Oil Production   Based on a preliminary report by a geological survey team, it is estimated that a newly discovered oil field can be expected to produce oil at the rate of

$$R(t) = \frac{600t^2}{t^3 + 32} + 5 \qquad (0 \le t \le 20)$$

thousand barrels per year, t years after production begins. Find the amount of oil that the field can be expected to yield during the first 5 years of production, assuming that the projection holds true.

43. Depreciation: Double Declining-Balance Method
Suppose that a tractor purchased at a price of $60,000 is to be depreciated by the *double declining-balance method* over a period of 10 years. It can be shown that the rate at which the book value will be decreasing is given by

$$R(t) = 13388.61e^{-0.22314t} \qquad (0 \le t \le 10)$$

dollars per year at year t. Find the amount by which the book value of the tractor will depreciate over the first 5 years of its life.

44. Velocity of a Car A car moves along a straight road in such a way that its velocity (in ft/sec) at any time t (in sec) is given by

$$v(t) = 3t\sqrt{16 - t^2} \qquad (0 \le t \le 4)$$

Find the distance traveled by the car in the 4 seconds from $t = 0$ to $t = 4$.

45. Average Temperature The temperature (in °F) in Boston over a 12-hour period on a certain December day was given by

$$T = -0.05t^3 + 0.4t^2 + 3.8t + 5.6 \qquad (0 \le t \le 12)$$

where t is measured in hours, with $t = 0$ corresponding to 6 A.M. Determine the average temperature on that day over the 12-hour period from 6 A.M. to 6 P.M.

46. Whale Population A group of marine biologists estimates that if certain conservation measures are implemented, the population of an endangered species of whale will be

$$N(t) = 3t^3 + 2t^2 - 10t + 600 \qquad (0 \le t \le 10)$$

where $N(t)$ denotes the population at the end of year t. Find the average population of the whales over the next 10 years.

47. Cable TV Subscribers The manager of the Tele-Star Cable Television Service estimates that the total number of subscribers to the service in a certain city t years from now will be

$$N(t) = -\frac{40,000}{\sqrt{1 + 0.2t}} + 50,000$$

Find the average number of cable television subscribers over the next 5 years if this prediction holds true.

48. Average Yearly Sales The sales of the Universal Instruments Company in the first t years of its operation are approximated by the function

$$S(t) = t\sqrt{0.2t^2 + 4}$$

where $S(t)$ is measured in millions of dollars. What were Universal's average yearly sales over its first 5 years of operation?

49. Concentration of a Drug in the Bloodstream The concentration of a certain drug in a patient's bloodstream t hours after injection is

$$C(t) = \frac{0.2t}{t^2 + 1}$$

milligrams per cubic centimeter. Determine the average concentration of the drug in the patient's bloodstream over the first 4 hours after the drug is injected.

50. Refer to exercise 44. Find the average velocity of the car over the time interval [0, 4].

51. Flow of Blood in an Artery According to a law discovered by nineteenth-century physician Jean Louis Marie Poiseuille, the velocity of blood (in cm/sec) r centimeters from the central axis of an artery is given by

$$v(r) = k(R^2 - r^2)$$

where k is a constant and R is the radius of the artery. Find the average velocity of blood along a radius of the artery (see accompanying figure).

[*Hint:* Evaluate $\dfrac{1}{R}\displaystyle\int_0^R v(r)\,dr$.]

Blood vessel

52. Prove property 1 of the definite integral.
[*Hint:* Let F be an antiderivative of f, and use the definition of the definite integral.]

53. Prove property 2 of the definite integral.
[*Hint:* See exercise 52.]

54. Verify by direct computation that

$$\int_1^3 x^2\,dx = -\int_3^1 x^2\,dx$$

55. Prove property 3 of the definite integral.
[*Hint:* See exercise 52.]

56. Verify by direct computation that

$$\int_1^9 2\sqrt{x}\, dx = 2\int_1^9 \sqrt{x}\, dx$$

57. Verify by direct computation that

$$\int_0^1 (1 + x - e^x)\, dx = \int_0^1 dx + \int_0^1 x\, dx - \int_0^1 e^x\, dx$$

What properties of the definite integral are demonstrated in this exercise?

58. Verify by direct computation that

$$\int_0^3 (1 + x^3)\, dx = \int_0^1 (1 + x^3)\, dx + \int_1^3 (1 + x^3)\, dx$$

What property of the definite integral is demonstrated here?

59. Verify by direct computation that

$$\int_0^3 (1 + x^3)\, dx = \int_0^1 (1 + x^3)\, dx$$
$$+ \int_1^2 (1 + x^3)\, dx + \int_2^3 (1 + x^3)\, dx$$

hence showing that property 5 may be extended.

SOLUTIONS TO SELF-CHECK EXERCISES 6.5

1. Let $u = 2x + 5$. Then $du = 2\, dx$, or $dx = \frac{1}{2}\, du$. Also, when $x = 0$, $u = 5$ and when $x = 2$, $u = 9$. Therefore,

$$\int_0^2 \sqrt{2x + 5}\, dx = \int_0^2 (2x + 5)^{1/2}\, dx$$

$$= \frac{1}{2} \int_5^9 u^{1/2}\, du$$

$$= \left(\frac{1}{2}\right)\left(\frac{2}{3} u^{3/2}\right)\Big|_5^9$$

$$= \frac{1}{3} [9^{3/2} - 5^{3/2}]$$

$$= \frac{1}{3} (27 - 5\sqrt{5})$$

2. The required average value is given by

$$\frac{1}{2 - (-1)} \int_{-1}^2 (1 - x^2)\, dx = \frac{1}{3} \int_{-1}^2 (1 - x^2)\, dx$$

$$= \frac{1}{3} \left(x - \frac{1}{3} x^3\right)\Big|_{-1}^2$$

$$= \frac{1}{3} \left[\left(2 - \frac{8}{3}\right) - \left(-1 + \frac{1}{3}\right)\right]$$

$$= 0$$

3. The average median price of a house over the stated time interval is given by

$$\frac{1}{5 - 0} \int_0^5 (t^3 - 7t^2 + 17t + 140)\, dt = \frac{1}{5} \left(\frac{1}{4} t^4 - \frac{7}{3} t^3 + \frac{17}{2} t^2 + 140t\right)\Big|_0^5$$

$$= \frac{1}{5} \left[\frac{1}{4}(5)^4 - \frac{7}{3}(5)^3 + \frac{17}{2}(5)^2 + 140(5)\right]$$

$$= 155.417$$

or $155,417.

6.6 AREA BETWEEN TWO CURVES

Suppose that a certain country's petroleum consumption is expected to grow at the rate of $f(t)$ million barrels per year, t years from now, for the next 5 years. Then the country's total petroleum consumption over the period of time in question is given by the area under the graph of f on the interval $[0, 5]$ (Figure 6.27).

Next, suppose that because of the implementation of certain energy-conservation measures, the rate of growth of petroleum consumption is expected to be $g(t)$ million barrels per year instead. Then the country's projected total petroleum consumption over the 5-year period is given by the area under the graph of g on the interval $[0, 5]$ (Figure 6.28).

Figure 6.27
At a rate of consumption $f(t)$ million barrels per year, the total petroleum consumption is given by the area of the region under the graph of f.

Figure 6.28
At a rate of consumption of $g(t)$ million barrels per year, the total petroleum consumption is given by the area of the region under the graph of g.

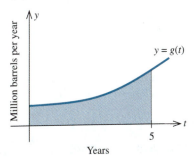

Therefore, the area of the shaded region S lying between the graphs of f and g on the interval $[0, 5]$ (Figure 6.29) gives the amount of petroleum that would be saved over the 5-year period because of the conservation measures.

Figure 6.29
The area of S gives the amount of petroleum that would be saved over the 5-year period.

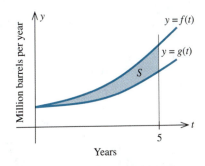

But the area of S is given by

Area under the graph of f on $[a, b]$ − Area under the graph of g on $[a, b]$

$$= \int_0^5 f(t)\, dt - \int_0^5 g(t)\, dt$$

$$= \int_0^5 [f(t) - g(t)]\, dt \qquad \text{(By property 4, Section 6.5)}$$

This example shows that some practical problems can be solved by finding the area of a region between two curves, which, in turn, can be found by evaluating an appropriate definite integral.

Finding the Area Between Two Curves

We now turn our attention to the general problem of finding the area of a plane region bounded both above and below by the graphs of functions. First, consider the situation in which the graph of one function lies above that of another. More specifically, let R be the region in the xy-plane (Figure 6.30) that is bounded above by the graph of a continuous function f, below by a continuous function g where $f(x) \geq g(x)$ on $[a, b]$, and to the left and right by the vertical lines $x = a$ and $x = b$, respectively. From the figure, we see that

$$\text{Area of } R = \text{Area under } f(x) - \text{Area under } g(x)$$

$$= \int_a^b f(x)\, dx - \int_a^b g(x)\, dx$$

$$= \int_a^b [f(x) - g(x)]\, dx$$

upon using property 4 of the definite integral.

Figure 6.30
Area of $R = \int_a^b [f(x) - g(x)]\, dx$.

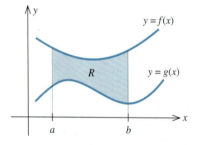

THE AREA BETWEEN TWO CURVES

Let f and g be continuous functions such that $f(x) \geq g(x)$ on the interval $[a, b]$. Then the area of the region bounded above by $y = f(x)$ and below by $y = g(x)$ on $[a, b]$ is given by

$$\int_a^b [f(x) - g(x)]\, dx \qquad (12)$$

Even though we assumed that both f and g were nonnegative in the derivation of (12), it may be shown that this equation is valid if f and g are not nonnegative (see exercise 48). Also, observe that if $g(x)$ is 0 for all x—that is, when the lower boundary of the region R is the x-axis—equation (12) gives the area of the region under the curve $y = f(x)$ from $x = a$ to $x = b$, as we would expect.

EXAMPLE 1   Find the area of the region bounded by the x-axis, the graph of $y = -x^2 + 4x - 8$, and the lines $x = -1$ and $x = 4$.

Solution   The region R under consideration is shown in Figure 6.31. We can view R as the region bounded above by the graph of $f(x) = 0$ (the x-axis) and below by the graph of $g(x) = -x^2 + 4x - 8$ on $[-1, 4]$. Therefore, the area of R is given by

$$\int_a^b [f(x) - g(x)]\, dx = \int_{-1}^4 [0 - (-x^2 + 4x - 8)]\, dx$$

$$= \int_{-1}^4 (x^2 - 4x + 8)\, dx$$

$$= \frac{1}{3}x^3 - 2x^2 + 8x \Big|_{-1}^4$$

$$= \left[\frac{1}{3}(64) - 2(16) + 8(4)\right] - \left[\frac{1}{3}(-1) - 2(1) + 8(-1)\right]$$

$$= 31\frac{2}{3}$$

Figure 6.31
Area of $R = -\int_{-1}^4 g(x)\, dx$.

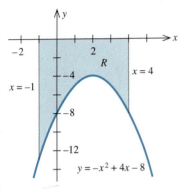

or $31\frac{2}{3}$ square units.   ◦ ◦ ◦

Figure 6.32
Area of $R = \int_1^2 [f(x) - g(x)]\, dx$.

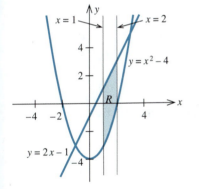

EXAMPLE 2   Find the area of the region R bounded by the graphs of

$$f(x) = 2x - 1 \quad \text{and} \quad g(x) = x^2 - 4$$

and the vertical lines $x = 1$ and $x = 2$.

Solution   We first sketch the graphs of the functions $f(x) = 2x - 1$ and $g(x) = x^2 - 4$, and the vertical lines $x = 1$ and $x = 2$, and identify the region R whose area is to be calculated (Figure 6.32).

Since the graph of f always lies above that of g for x in the interval $[1, 2]$, we see by equation (12) that the required area is given by

$$\int_1^2 [f(x) - g(x)]\, dx = \int_1^2 [(2x - 1) - (x^2 - 4)]\, dx$$

$$= \int_1^2 (-x^2 + 2x + 3)\, dx$$

$$= -\frac{1}{3}x^3 + x^2 + 3x \Big|_1^2$$

$$= \left(-\frac{8}{3} + 4 + 6\right) - \left(-\frac{1}{3} + 1 + 3\right) = \frac{11}{3}$$

Figure 6.33
Area of $R = \int_{-1}^3 [f(x) - g(x)]\, dx$.

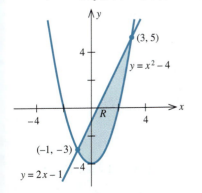

or $3\frac{2}{3}$ square units.   ◦ ◦ ◦

EXAMPLE 3   Find the area of the region R that is completely enclosed by the graphs of the functions

$$f(x) = 2x - 1 \quad \text{and} \quad g(x) = x^2 - 4$$

Solution   The region R is shown in Figure 6.33. First we find the points of intersection of the two curves. To do this, we solve the system that comprises the two equations $y = 2x - 1$ and $y = x^2 - 4$. Equating the two values of y gives

$$x^2 - 4 = 2x - 1$$

$$x^2 - 2x - 3 = 0$$

$$(x + 1)(x - 3) = 0$$

so $x = -1$ or $x = 3$. That is, the two curves intersect when $x = -1$ and $x = 3$.

Observe that we could also view the region R as the region bounded above by the graph of the function $f(x) = 2x - 1$, below by the graph of the function $g(x) = x^2 - 4$, and to the left and right by the vertical lines $x = -1$ and $x = 3$, respectively.

Next, since the graph of the function f always lies above that of the function g on $[-1, 3]$, we can use (12) to compute the desired area:

$$\int_a^b [f(x) - g(x)]\, dx = \int_{-1}^3 [(2x - 1) - (x^2 - 4)]\, dx$$

$$= \int_{-1}^3 (-x^2 + 2x + 3)\, dx$$

$$= -\frac{1}{3}x^3 + x^2 + 3x \Big|_{-1}^3$$

$$= (-9 + 9 + 9) - \left(\frac{1}{3} + 1 - 3\right) = \frac{32}{3}$$

or $10\frac{2}{3}$ square units.   ◦ ◦ ◦

EXAMPLE 4 Find the area of the region R bounded by the graphs of the functions

$$f(x) = x^2 - 2x - 1 \quad \text{and} \quad g(x) = -e^x - 1$$

and the vertical lines $x = -1$ and $x = 1$.

Figure 6.34
Area of $R = \int_{-1}^1 [f(x) - g(x)]\, dx$.

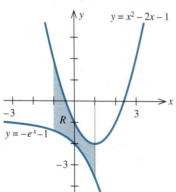

Solution The region R is shown in Figure 6.34. Since the graph of the function f always lies above that of the function g, the area of the region R is given by

$$\int_a^b [f(x) - g(x)]\, dx = \int_{-1}^1 [(x^2 - 2x - 1) - (-e^x - 1)]\, dx$$

$$= \int_{-1}^1 (x^2 - 2x + e^x)\, dx$$

$$= \frac{1}{3}x^3 - x^2 + e^x \Big|_{-1}^1$$

$$= \left(\frac{1}{3} - 1 + e\right) - \left(-\frac{1}{3} - 1 + e^{-1}\right)$$

$$= \frac{2}{3} + e - \frac{1}{e} \text{ square units}$$ ◦ ◦ ◦

Equation (12), which gives the area of the region between the curves $y = f(x)$ and $y = g(x)$ for $a \le x \le b$, is valid when the graph of the function *f* lies above that of the function *g* over the interval $[a, b]$. Example 5 shows how to use (12) to find the area of a region when the latter condition does not hold.

Figure 6.35
Area of R_1 = area of R_2.

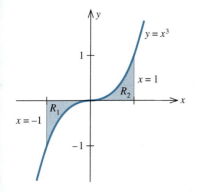

EXAMPLE 5   Find the area of the region bounded by the graph of the function $f(x) = x^3$, the *x*-axis, and the lines $x = -1$ and $x = 1$.

Solution   The region *R* under consideration can be thought of as comprising the two subregions R_1 and R_2, as shown in Figure 6.35.

Recall that the *x*-axis is represented by the function $g(x) = 0$. Since $g(x) \geq f(x)$ on $[-1, 0]$, we see that the area of R_1 is given by

$$\int_a^b [g(x) - f(x)]\, dx = \int_{-1}^0 (0 - x^3)\, dx = -\int_{-1}^0 x^3\, dx$$

$$= -\frac{1}{4}x^4 \Big|_{-1}^0 = 0 - \left(-\frac{1}{4}\right) = \frac{1}{4}$$

To find the area of R_2, we observe that $f(x) \geq g(x)$ on $[0, 1]$, so it is given by

$$\int_a^b [f(x) - g(x)]\, dx = \int_0^1 (x^3 - 0)\, dx = \int_0^1 x^3\, dx$$

$$= \frac{1}{4}x^4 \Big|_0^1 = \left(\frac{1}{4}\right) - 0 = \frac{1}{4}$$

Therefore, the area of *R* is $\frac{1}{4} + \frac{1}{4}$, or $\frac{1}{2}$, square units.

By making use of symmetry, we could have obtained the same result by computing

$$-2 \int_{-1}^0 x^3\, dx \qquad \text{or} \qquad 2 \int_0^1 x^3\, dx$$

as you may verify.   ◦ ◦ ◦

A function is *even* if it satisfies the condition $f(-x) = f(x)$, and it is *odd* if it satisfies the condition $f(-x) = -f(x)$. Show that the graph of an even function is symmetric with respect to the *y*-axis while the graph of an odd function is symmetric with respect to the origin. Explain why

$$\int_{-a}^a f(x)\, dx = 2 \int_0^a f(x)\, dx \quad \text{if } f \text{ is even}$$

and

$$\int_{-a}^a f(x)\, dx = 0 \qquad\qquad \text{if } f \text{ is odd}$$

EXAMPLE 6   Find the area of the region completely enclosed by the graphs of the functions

$$f(x) = x^3 - 3x + 3 \qquad \text{and} \qquad g(x) = x + 3$$

Figure 6.36
Area of R_1 + Area of R_2 =
$\int_{-2}^{0} [f(x) - g(x)]\, dx +$
$\int_{0}^{2} [g(x) - f(x)]\, dx.$

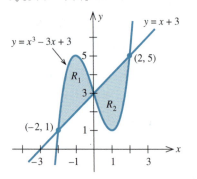

Solution   First sketch the graphs of $y = x^3 - 3x + 3$ and $y = x + 3$, and then identify the required region R. We can view the region R as being composed of the two subregions R_1 and R_2, as shown in Figure 6.36. By solving the equations $y = x + 3$ and $y = x^3 - 3x + 3$ simultaneously, we find the points of intersection of the two curves. Equating the two values of y, we have

$$x^3 - 3x + 3 = x + 3$$

$$x^3 - 4x = 0$$

or

$$x(x^2 - 4) = 0$$

$$x(x + 2)(x - 2) = 0 \quad \text{or} \quad x = 0, -2, \text{ or } 2$$

Hence, the points of intersection of the two curves are $(-2, 1)$, $(0, 3)$, and $(2, 5)$.

For $-2 \leq x \leq 0$, we see that the graph of the function f lies above that of the function g, so the area of the region R_1 is, by virtue of (12),

$$\int_{-2}^{0} [(x^3 - 3x + 3) - (x + 3)]\, dx = \int_{-2}^{0} (x^3 - 4x)\, dx$$

$$= \frac{1}{4}x^4 - 2x^2 \Big|_{-2}^{0}$$

$$= -(4 - 8)$$

$$= 4 \quad \text{or} \quad 4 \text{ square units}$$

For $0 \leq x \leq 2$, the graph of the function g lies above that of the function f, and the area of R_2 is given by

$$\int_{0}^{2} [(x + 3) - (x^3 - 3x + 3)]\, dx = \int_{0}^{2} (-x^3 + 4x)\, dx$$

$$= -\frac{1}{4}x^4 + 2x^2 \Big|_{0}^{2}$$

$$= -4 + 8$$

$$= 4 \quad \text{or} \quad 4 \text{ square units}$$

Therefore, the required area is the sum of the area of the two regions $R_1 + R_2$—that is, $4 + 4$, or 8 square units.   ○ ○ ○

Application

EXAMPLE 7   In a 1989 study for a developing country's Economic Development Board, government economists and energy experts concluded that if the Energy Conservation Bill were implemented in 1990, the country's oil consumption for the next 5 years would be expected to grow in accordance with the model

$$R(t) = 20e^{0.05t}$$

where t is measured in years ($t = 0$ corresponding to the year 1990) and $R(t)$ in millions of barrels per year. Without the government-imposed conservation measures, however, the expected rate of growth of oil consumption would be given by

$$R_1(t) = 20e^{0.08t}$$

millions of barrels per year. Using these models, determine how much oil would have been saved from 1990 through 1995 if the bill had been implemented.

Solution   Under the Energy Conservation Bill, the total amount of oil that would have been consumed between 1990 and 1995 is given by

$$\int_0^5 R(t)\, dt = \int_0^5 20e^{0.05t}\, dt \tag{13}$$

Without the bill, the total amount of oil that would have been consumed between 1990 and 1995 is given by

$$\int_0^5 R_1(t)\, dt = \int_0^5 20e^{0.08t}\, dt \tag{14}$$

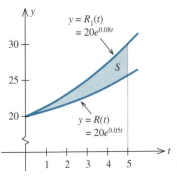

Figure 6.37
Area of $S = \int_0^5 [R_1(t) - R(t)]\, dt$.

Equation (13) may be interpreted as the area of the region under the curve $y = R(t)$ from $t = 0$ to $t = 5$. Similarly, we interpret (14) as the area of the region under the curve $y = R_1(t)$ from $t = 0$ to $t = 5$. Furthermore, note that the graph of $y = R_1(t) = 20e^{0.08t}$ always lies on or above the graph of $y = R(t) = 20e^{0.05t}$ ($t \geq 0$). Thus, the area of the shaded region S in Figure 6.37 shows the amount of oil that would have been saved from 1990 to 1995 if the Energy Conservation Bill had been implemented. But the area of the region S is given by

$$\int_0^5 [R_1(t) - R(t)]\, dt = \int_0^5 [20e^{0.08t} - 20^{0.05t}]\, dt$$

$$= 20 \int_0^5 (e^{0.08t} - e^{0.05t})\, dt$$

$$= 20 \left(\frac{e^{0.08t}}{0.08} - \frac{e^{0.05t}}{0.05} \right) \Bigg|_0^5$$

$$= 20 \left[\left(\frac{e^{0.4}}{0.08} - \frac{e^{0.25}}{0.05} \right) - \left(\frac{1}{0.08} - \frac{1}{0.05} \right) \right]$$

$$\approx 9.3 \quad \text{or} \quad \text{approximately 9.3 square units}$$

Thus, the amount of oil that would have been saved is 9.3 million barrels.

SELF-CHECK EXERCISES 6.6

1. Find the area of the region bounded by the graphs of $f(x) = x^2 + 2$ and $g(x) = 1 - x$ and the vertical lines $x = 0$ and $x = 1$.

2. Find the area of the region completely enclosed by the graphs of $f(x) = -x^2 + 6x + 5$ and $g(x) = x^2 + 5$.

 3. The management of the Kane Corporation, which operates a chain of hotels, expects its profits to grow at the rate of $1 + t^{2/3}$ million dollars per year t years from now. However, with renovations and improvements of existing hotels and proposed acquisitions of new hotels, Kane's profits are expected to grow at the rate of $t - 2\sqrt{t} + 4$ million dollars per year in the next decade. What additional profits are expected over the next 10 years if the group implements the proposed plans?

Solutions to Self-Check Exercises 6.6 can be found on page 500.

EXERCISES

In exercises 1–8, find the area of the shaded region.

1.

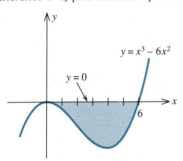

$y = x^3 - 6x^2$

$y = 0$

2.

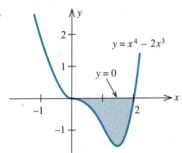

$y = x^4 - 2x^3$

$y = 0$

3.

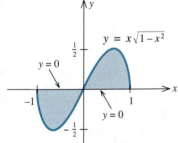

$y = x\sqrt{1 - x^2}$

$y = 0$

$y = 0$

4.

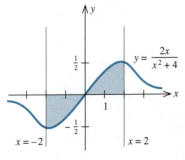

$y = \dfrac{2x}{x^2 + 4}$

$x = -2$ $x = 2$

5.

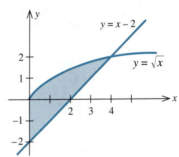

$y = x - 2\sqrt{x}$

6.

$y = x - 2$

$y = \sqrt{x}$

7.

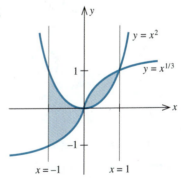

$y = x^2$

$y = x^{1/3}$

$x = -1$ $x = 1$

8.

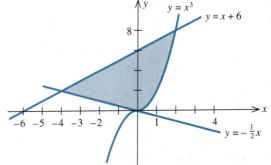

$y = x^3$

$y = x + 6$

$y = -\dfrac{1}{2}x$

In exercises 9–16, sketch the graph and find the area of the region bounded below by the graph of each of the given functions and above by the x-axis from x = a to x = b.

9. $f(x) = -x^2$; $a = -1$, $b = 2$

10. $f(x) = x^2 - 4$; $a = -2$, $b = 2$

11. $f(x) = x^2 - 5x + 4$; $a = 1$, $b = 3$

12. $f(x) = x^3$; $a = -1$, $b = 0$

13. $f(x) = -1 - \sqrt{x}$; $a = 0$, $b = 9$

14. $f(x) = \frac{1}{2}x - \sqrt{x}$; $a = 0$, $b = 4$

15. $f(x) = -e^{(1/2)x}$; $a = -2$, $b = 4$

16. $f(x) = -xe^{-x^2}$; $a = 0$, $b = 1$

In exercises 17–26, sketch the graphs of the functions f and g and find the area of the region enclosed by these graphs and the vertical lines x = a and x = b.

17. $f(x) = x^2 + 3$, $g(x) = 1$; $a = 1$, $b = 3$

18. $f(x) = x + 2$, $g(x) = x^2 - 4$; $a = -1$, $b = 2$

19. $f(x) = -x^2 + 2x + 3$, $g(x) = -x + 3$; $a = 0$, $b = 2$

20. $f(x) = 9 - x^2$, $g(x) = 2x + 3$; $a = -1$, $b = 1$

21. $f(x) = x^2 + 1$, $g(x) = \frac{1}{3}x^3$; $a = -1$, $b = 2$

22. $f(x) = \sqrt{x}$, $g(x) = -\frac{1}{2}x - 1$; $a = 1$, $b = 4$

23. $f(x) = \dfrac{1}{x}$, $g(x) = 2x - 1$; $a = 1$, $b = 4$

24. $f(x) = x^2$, $g(x) = \dfrac{1}{x^2}$; $a = 1$, $b = 3$

25. $f(x) = e^x$, $g(x) = \dfrac{1}{x}$; $a = 1$, $b = 2$

26. $f(x) = x$, $g(x) = e^{2x}$; $a = 1$, $b = 3$

In exercises 27–34, sketch the graph and find the area of the region bounded by the graph of the function f and the lines y = 0, x = a, and x = b.

27. $f(x) = x$; $a = -1$, $b = 2$

28. $f(x) = x^2 - 2x$; $a = -1$, $b = 1$

29. $f(x) = -x^2 + 4x - 3$; $a = -1$, $b = 2$

30. $f(x) = x^3 - x^2$; $a = -1$, $b = 1$

31. $f(x) = x^3 - 4x^2 + 3x$; $a = 0$, $b = 2$

32. $f(x) = 4x^{1/3} + x^{4/3}$; $a = -1$, $b = 8$

33. $f(x) = e^x - 1$; $a = -1$, $b = 3$

34. $f(x) = xe^{x^2}$; $a = 0$, $b = 2$

In exercises 35–40, sketch the graph and find the area of the region completely enclosed by the graphs of the given functions f and g.

35. $f(x) = x + 2$ and $g(x) = x^2 - 4$

36. $f(x) = -x^2 + 4x$ and $g(x) = 2x - 3$

37. $f(x) = x^2$ and $g(x) = x^3$

38. $f(x) = x^3 - 6x^2 + 9x$ and $g(x) = x^2 - 3x$

39. $f(x) = \sqrt{x}$ and $g(x) = x^2$

40. $f(x) = 2x$ and $g(x) = x\sqrt{x + 1}$

41. Effect of Advertising on Revenue   In the accompanying figure, the function f gives the rate of change of Odyssey Travel's revenue with respect to the amount x it spends on advertising with their current advertising agency. By engaging the services of a different advertising agency, it is expected that Odyssey's revenue will grow at the rate given by the function g. Give an interpretation of the area A of the region S, and find an expression for A in terms of a definite integral involving f and g.

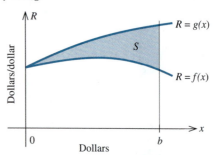

42. Pulse Rate During Exercise   In the accompanying figure, the function f gives the rate of increase of an individual's pulse rate when he walked a prescribed course on a treadmill 6 months ago. The function g gives the rate of increase of his pulse rate when he recently walked the same prescribed course. Give an interpretation of the area A of the region S and find an expression for A in terms of a definite integral involving f and g.

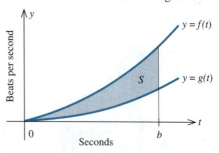

43. Air Purification   In order to study the effectiveness of air purifiers in removing smoke, engineers run each purifier in a smoke-filled 10×20-foot room. In the accompanying figure, the function f gives the rate of change of the smoke level per minute, t minutes after the start of the test, when a Brand A purifier is used. The function g gives the rate of change of the smoke level per minute when a Brand B purifier is used.
a. Give an interpretation of the area of the region S.
b. Find an expression for the area of S in terms of a definite integral involving f and g.

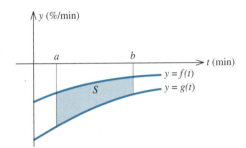

44. Turbo-Charged Engine vs. Standard Engine   In tests conducted by Auto Test Magazine on two identical models of the Phoenix Elite—one equipped with a standard engine and the other with a turbo-charger—it was found that the acceleration of the former is given by

$$a = f(t) = 4 + 0.8t \qquad (0 \le t \le 12)$$

feet per second per second, t seconds after starting from rest at full throttle, whereas the acceleration of the latter is given by

$$a = g(t) = 4 + 1.2t + 0.03t^2 \qquad (0 \le t \le 12)$$

feet per second per second. How much faster is the turbo-charged model moving than the model with the standard engine, at the end of a 10-second test run at full throttle?

45. Alternative Energy Sources   Because of the increasingly important role played by coal as a viable alternative energy source, the production of coal has been growing at the rate of

$$3.5e^{0.05t}$$

billion metric tons per year t years from 1980 (which corresponds to $t = 0$). Had it not been for the energy crisis, the rate of production of coal since 1980 might have been only

$$3.5e^{0.01t}$$

billion metric tons per year t years from 1980. Determine how much additional coal will be produced between 1980 and the end of the century as an alternate energy source.

46. Effect of TV Advertising on Car Sales   Carl Williams, the new proprietor of Carl Williams Auto Sales, estimates that with extensive television advertising, car sales over the next several years could be increasing at the rate of

$$5e^{0.3t}$$

thousand cars per year t years from now, instead of at the current rate of

$$(5 + 0.5t^{3/2})$$

thousand cars per year t years from now. Find how many more cars Mr. Williams expects to sell over the next 5 years by implementing his advertising plans.

47. Population Growth   In an endeavor to curb population growth in a Southeast Asian island state, the government has decided to launch an extensive propaganda campaign. Without curbs, the government expects the rate of population growth to have been

$$60e^{0.02t}$$

thousand people per year t years from now, over the next 5 years. However, successful implementation of the proposed campaign is expected to result in a population growth rate of

$$-t^2 + 60$$

thousand people per year t years from now, over the next 5 years. Assuming that the campaign is mounted, how many fewer people will there be in that country 5 years from now than there would have been if no curbs had been imposed?

48. Show that the area of a region R bounded above by the graph of a function f and below by the graph of a function g from $x = a$ to $x = b$ is given by

$$\int_a^b [f(x) - g(x)] \, dx$$

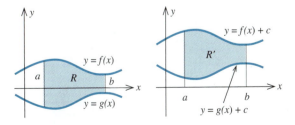

Hint: The validity of the formula was verified earlier for the case when both f and g were nonnegative. Now let f and g be two functions such that $f(x) \geq g(x)$ for $a \leq x \leq b$. Then there exists some nonnegative constant c such that the curves $y = f(x) + c$ and $y = g(x) + c$ are translated in the y-direction in such a way that the region R' has the same area as the region R (see the figures on page 499). Show that the area of R' is given by

$$\int_a^b \{[f(x) + c] - [g(x) + c]\} \, dx = \int_a^b [f(x) - g(x)] \, dx$$

SOLUTIONS TO SELF-CHECK EXERCISES 6.6

1. The region in question is shown in the accompanying figure. Since the graph of the function f lies above that of the function g for $0 \leq x \leq 1$, we see that the

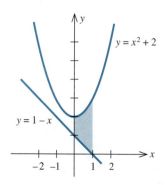

required area is given by

$$\int_0^1 [(x^2 + 2) - (1 - x)] \, dx = \int_0^1 (x^2 + x + 1) \, dx$$

$$= \frac{1}{3}x^3 + \frac{1}{2}x^2 + x \Big|_0^1$$

$$= \frac{1}{3} + \frac{1}{2} + 1$$

$$= \frac{11}{6}$$

or $\frac{11}{6}$ square units.

2. The region in question is shown in the accompanying figure. To find the points of intersection of the two curves, we solve the equation

$$-x^2 + 6x + 5 = x^2 + 5$$

$$2x^2 - 6x = 0$$

$$2x(x - 3) = 0$$

giving $x = 0$ or $x = 3$. Therefore, the points of intersection are $(0, 5)$ and $(3, 14)$.

Since the graph of f always lies above that of g for $0 \leq x \leq 3$, we see that the

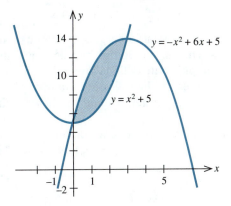

required area is given by

$$\int_0^3 [(-x^2 + 6x + 5) - (x^2 + 5)]\, dx = \int_0^3 (-2x^2 + 6x)\, dx$$

$$= -\frac{2}{3}x^3 + 3x^2 \Big|_0^3$$

$$= -18 + 27$$

$$= 9$$

or 9 square units.

3. The additional profits realizable over the next 10 years are given by

$$\int_0^{10} [(t - 2\sqrt{t} + 4) - (1 + t^{2/3})]\, dt$$

$$= \int_0^{10} (t - 2t^{1/2} + 3 - t^{2/3})\, dt$$

$$= \frac{1}{2}t^2 - \frac{4}{3}t^{3/2} + 3t - \frac{3}{5}t^{5/3} \Big|_0^{10}$$

$$= \frac{1}{2}(10)^2 - \frac{4}{3}(10)^{3/2} + 3(10) - \frac{3}{5}(10)^{5/3}$$

$$\approx 9.99$$

or approximately \$10 million.

USING TECHNOLOGY

Finding the Area Between Two Curves

The numerical integral operation can also be used to find the area between two curves. We do this by using the numerical integral operation to evaluate an appropriate definite integral or the sum (difference) of appropriate definite integrals. In the following example, the intersection operation is also used to advantage to help us find the limits of integration.

EXAMPLE 1 Use a graphing utility to find the area of the region R that is completely enclosed by the graphs of the functions

$$f(x) = 2x^3 - 8x^2 + 4x - 3 \quad \text{and} \quad g(x) = 3x^2 + 10x - 11$$

Solution The graphs of f and g in the viewing rectangle $[-3, 4] \times [-20, 5]$ are shown in Figure T.1. Using the intersection operation of a graphing utility,

Figure T1
The region R is completely enclosed by the graphs of f and g.

we find the x-coordinates of the points of intersection of the two graphs to be approximately -1.04 and 0.65, respectively. Since the graph of f lies above that of g on the interval $[-1.04, 0.65]$, we see that the area of R is given by

$$A = \int_{-1.04}^{0.65} [(2x^3 - 8x^2 + 4x - 3) - (3x^2 + 10x - 11)]\, dx$$

$$= \int_{-1.04}^{0.65} (2x^3 - 11x^2 - 6x + 8)\, dx$$

Using the numerical integral function of a graphing utility, we find $A \approx 9.87$, and so the area of R is approximately 9.87 square units. ○ ○ ○

EXERCISES

In exercises 1–6, use a graphing utility to (a) plot the graphs of the functions f and g and (b) find the area of the region enclosed by these graphs and the vertical lines $x = a$ and $x = b$. Express your answers accurate to four decimal places.

1. $f(x) = x^3(x - 5)^4$, $g(x) = 0$; $a = 1$, $b = 3$

2. $f(x) = x - \sqrt{1 - x^2}$, $g(x) = 0$; $a = -\frac{1}{2}$, $b = \frac{1}{2}$

3. $f(x) = x^{1/3}(x + 1)^{1/2}$, $g(x) = x^{-1}$; $a = 1.2$, $b = 2$

4. $f(x) = 2$, $g(x) = \ln(1 + x^2)$; $a = -1$, $b = 1$

5. $f(x) = \sqrt{x}$, $g(x) = \dfrac{x^2 - 3}{x^2 + 1}$; $a = 0$, $b = 3$

6. $f(x) = \dfrac{4}{x^2 + 1}$, $g(x) = x^4$; $a = -1$, $b = 1$

In exercises 7–12, use a graphing utility to (a) plot the graphs of the functions f and g and (b) find the area of the region totally enclosed by the graphs of these functions.

7. $f(x) = 2x^3 - 8x^2 + 4x - 3$ and $g(x) = -3x^2 + 10x - 10$

8. $f(x) = x^4 - 2x^2 + 2$ and $g(x) = 4 - 2x^2$

9. $f(x) = 2x^3 - 3x^2 + x + 5$ and $g(x) = e^{2x} - 3$

10. $f(x) = \frac{1}{2}x^2 - 3$ and $g(x) = \ln x$

11. $f(x) = xe^{-x}$ and $g(x) = x - 2\sqrt{x}$

12. $f(x) = e^{-x^2}$ and $g(x) = x^4$

6.7 APPLICATIONS OF THE DEFINITE INTEGRAL TO BUSINESS AND ECONOMICS

In this section, we consider several applications of the definite integral in the fields of business and economics.

Consumers' and Producers' Surplus

We begin by deriving a formula for computing the consumers' surplus. Suppose $p = D(x)$ is the demand function that relates the unit price p of a commodity to the quantity x demanded of it. Furthermore, suppose that a fixed unit market price has been established for the commodity and that corresponding to this unit price, the quantity demanded is $\bar{x}$ units (Figure 6.38). Then those consumers who would be willing to pay a unit price higher than $\bar{p}$ for the commodity would, in effect, experience a savings. This difference between what the consumers *would* be willing to pay for $\bar{x}$ units of the commodity and what they *actually* pay for them is called the **consumers' surplus**.

To derive a formula for computing the consumers' surplus, divide the interval $[0, \bar{x}]$ into n subintervals, each of length $\Delta x = \bar{x}/n$, and denote the right endpoints of these subintervals by $x_1, x_2, \ldots, x_n = \bar{x}$ (Figure 6.39).

We observe in Figure 6.39 that there are consumers who would pay a unit price of at least $D(x_1)$ dollars for the first Δx units of the commodity instead of the market price of $\bar{p}$ dollars per unit. The savings to these consumers is approximated by

$$D(x_1)\Delta x - \bar{p}\Delta x = [D(x_1) - \bar{p}]\Delta x$$

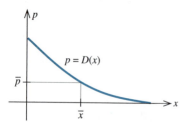

Figure 6.38
D(x) is a demand function.

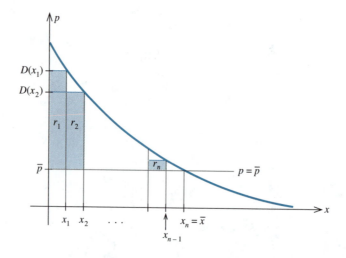

Figure 6.39
Approximating consumers' surplus by the sum of the rectangles r_1, $r_2, \ldots, r_n$.

which is the area of the rectangle r_1. Pursuing the same line of reasoning, we find that the savings to the consumers who would be willing to pay a unit price of at least $D(x_2)$ dollars for the next Δx units (from x_1 through x_2) of the commodity, instead of the market price of $\bar{p}$ dollars per unit, is approximated by

$$D(x_2)\Delta x - \bar{p}\Delta x = [D(x_2) - \bar{p}]\Delta x$$

Continuing, we approximate the total savings to the consumers in purchasing $\bar{x}$ units of the commodity by the sum

$$[D(x_1) - \bar{p}]\Delta x + [D(x_2) - \bar{p}]\Delta x + \cdots + [D(x_n) - \bar{p}]\Delta x$$
$$= [D(x_1) + D(x_2) + \cdots + D(x_n)]\Delta x - \underbrace{[\bar{p}\Delta x + \bar{p}\Delta x + \cdots + \bar{p}\Delta x]}_{n \text{ terms}}$$
$$= [D(x_1) + D(x_2) + \cdots + D(x_n)]\Delta x - n\bar{p}\Delta x$$
$$= [D(x_1) + D(x_2) + \cdots + D(x_n)]\Delta x - \bar{p}\bar{x}$$

Now, the first term in the last expression is the Riemann sum of the demand function $p = D(x)$ over the interval $[0, \bar{x}]$ with representative points $x_1, x_2, \ldots, x_n$. Letting n approach infinity, we obtain the following formula for the consumers' surplus CS.

CONSUMERS' SURPLUS	The **consumers' surplus** is given by $$CS = \int_0^{\bar{x}} D(x)\, dx - \bar{p}\bar{x} \qquad (15)$$ where D is the demand function, $\bar{p}$ is the unit market price, and $\bar{x}$ is the quantity sold.

The consumers' surplus is given by the area of the region bounded above by the demand curve $p = D(x)$ and below by the straight line $p = \bar{p}$ from $x = 0$ to $x = \bar{x}$ (Figure 6.40). We can also see this if we rewrite equation (15) in the form

$$\int_0^{\bar{x}} [D(x) - \bar{p}]\, dx$$

and interpret the result geometrically.

Analogously, we can derive a formula for computing the producers' surplus. Suppose that $p = S(x)$ is the supply equation that relates the unit price p of a certain commodity to the quantity x that the supplier will make available in the market at that price.

Again, suppose that a fixed market price $\bar{p}$ has been established for the commodity and that, corresponding to this unit price, a quantity of $\bar{x}$ units

Figure 6.40
Consumers' surplus.

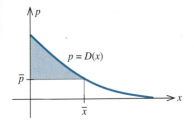

will be made available in the market by the supplier (Figure 6.41). Then the suppliers who would be willing to make the commodity available at a lower price stand to gain from the fact that the market price is set as such. The difference between what the suppliers actually receive and what they would be willing to receive is called the **producers' surplus.** Proceeding in a manner similar to the derivation of the equation for computing the consumers' surplus, we find that the producers' surplus *PS* is defined as follows:

THE PRODUCERS' **SURPLUS**	The **producers' surplus** is given by $$PS = \bar{p}\bar{x} - \int_0^{\bar{x}} S(x)\, dx \qquad (16)$$ where $S(x)$ is the supply function, $\bar{p}$ is the unit market price, and $\bar{x}$ is the quantity supplied.

Geometrically, the producers' surplus is given by the area of the region bounded above by the straight line $p = \bar{p}$ and below by the supply curve $p = S(x)$ from $x = 0$ to $x = \bar{x}$ (Figure 6.42).

Figure 6.41
S(x) is a supply function.

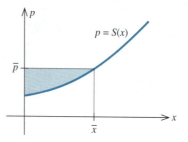

Figure 6.42
Producers' surplus.

We can also show that the last statement is true by converting equation (16) to the form

$$\int_0^{\bar{x}} [\bar{p} - S(x)]\, dx$$

and interpreting the definite integral geometrically.

EXAMPLE 1   The demand function for a certain make of 10-speed bicycle is given by

$$p = D(x) = -0.001x^2 + 250$$

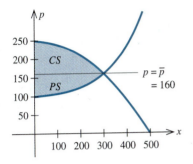

Figure 6.43
Consumers' surplus and producers' surplus when market price = equilibrium price.

where p is the unit price in dollars and x is the quantity demanded in units of a thousand. The supply function for these bicycles is given by

$$p = S(x) = 0.0006x^2 + 0.02x + 100$$

where p stands for the unit price in dollars and x stands for the number of bicycles that the supplier will put on the market, in units of a thousand. Determine the consumers' surplus and the producers' surplus if the market price of a bicycle is set at the equilibrium price.

Solution   Recall that the equilibrium price is the unit price of the commodity when market equilibrium occurs. We determine the equilibrium price by solving for the point of intersection of the demand curve and the supply curve (Figure 6.43). To solve the system of equations

$$p = -0.001x^2 + 250$$

and

$$p = 0.0006x^2 + 0.02x + 100$$

we simply substitute the first equation into the second, obtaining

$$0.0006x^2 + 0.02x + 100 = -0.001x^2 + 250$$
$$0.0016x^2 + 0.02x - 150 = 0$$
$$16x^2 + 200x - 1{,}500{,}000 = 0$$
$$2x^2 + 25x - 187{,}500 = 0$$

Factoring this last equation, we obtain

$$(2x + 625)(x - 300) = 0$$

Thus, $x = -625/2$ or $x = 300$. The first number lies outside the interval of interest, so we are left with the solution $x = 300$, with a corresponding value of

$$p = -0.001(300)^2 + 250 = 160$$

Thus, the equilibrium point is (300, 160); that is, the equilibrium quantity is 300,000 and the equilibrium price is $160. Setting the market price at $160 per unit and using formula (15) with $\bar{p} = 160$ and $\bar{x} = 300$, we find that the consumers' surplus is given by

$$CS = \int_0^{300} (-0.001x^2 + 250)\, dx - (160)(300)$$

$$= \left(-\frac{1}{3{,}000}x^3 + 250x\right)\Big|_0^{300} - 48{,}000$$

$$= -\frac{300^3}{3{,}000} + (250)(300) - 48{,}000$$

$$= 18{,}000$$

or $18,000,000.

(Recall that x is measured in units of a thousand.) Next, using (16), we find that the producers' surplus is given by

$$PS = (160)(300) - \int_0^{300} (0.0006x^2 + 0.02x + 100)\, dx$$

$$= 48{,}000 - (0.0002x^3 + 0.01x^2 + 100x)\Big|_0^{300}$$

$$= 48{,}000 - [(0.0002)(300)^3 + (0.01)(300)^2 + 100(300)]$$

$$= 11{,}700$$

or $11,700,000.

● ● ●

The Future and Present Value of an Income Stream

To introduce the notion of the future and the present value of an income stream, suppose that a firm generates a stream of income over a period of time—the revenue generated by a large chain of retail stores over a 5-year period, for example. As the income is realized, it is reinvested and earns interest at a fixed rate. The *accumulated future income stream* over the 5-year period is the amount of money the firm ends up with at the end of that period.

The definite integral can be used to determine this **accumulated, or total, future income stream** over a period of time. The total future value of an income stream gives us a way to measure the value of such a stream. To find the total future value of an income stream, suppose that

$$R(t) = \text{Rate of income generation at any time } t \qquad \text{(Dollars per year)}$$

$$r = \text{Interest rate compounded continuously}$$

and $T = \text{Term}$ (In years)

Let us divide the time interval $[0, T]$ into n subintervals of equal length $\Delta t = T/n$ and denote the right endpoints of these intervals by $t_1, t_2, \ldots, t_n = T$, as shown in Figure 6.44.

If R is a continuous function on $[0, T]$, then $R(t)$ will not differ by much from $R(t_1)$ in the subinterval $[0, t_1]$ provided that the subinterval is small (which is true if n is large). Therefore, the income generated over the time interval $[0, t_1]$ is approximately

$$R(t_1)\Delta t \qquad \text{(Constant rate of income · Length of time)}$$

dollars. The future value of this amount, T years from now, calculated as if it were earned at time t_1, is

$$[R(t_1)\Delta t]e^{r(T-t_1)} \qquad \text{[Equation (10), Section 5.3]}$$

dollars. Similarly, the income generated over the time interval $[t_1, t_2]$ is approximately $P(t_2)\Delta t$ dollars and has a future value, T years from now, of approximately

$$[R(t_2)\Delta t]e^{r(T-t_2)}$$

Figure 6.44
The time interval $[0, T]$ is partitioned into n subintervals.

$0 \quad t_1 \quad t_2 \quad t_3 \cdots t_{n-1} \quad t_n = \mathrm{T}$

dollars. Therefore, the sum of the future values of the income stream generated over the time interval $[0, T]$ is approximately

$$R(t_1)e^{r(T-t_1)}\Delta t + R(t_2)e^{r(T-t_2)}\Delta t + \cdots + R(t_n)e^{r(T-t_n)}\Delta t$$
$$= e^{rT}[R(t_1)e^{-rt_1}\Delta t + R(t_2)e^{-rt_2}\Delta t + \cdots + R(t_n)e^{-rt_n}\Delta t]$$

dollars. But this sum is just the Riemann sum of the function $e^{rT}R(t)e^{-rt}$ over the interval $[0, T]$ with representative points $t_1, t_2, \ldots, t_n$. Letting n approach infinity, we obtain the following result.

ACCUMULATED OR TOTAL FUTURE VALUE OF AN INCOME STREAM

The **accumulated,** or **total, future value** after T years of an income stream of $R(t)$ dollars per year, earning interest at the rate of r per year compounded continuously, is given by

$$A = e^{rT} \int_0^T R(t)e^{-rt}\, dt \tag{17}$$

EXAMPLE 2 Crystal Car Wash recently bought an automatic car-washing machine that is expected to generate $40,000 in revenue per year, t years from now, for the next 5 years. If the income is reinvested in a business earning interest at the rate of 12% per year compounded continuously, find the total accumulated value of this income stream at the end of 5 years.

Solution We are required to find the total future value of the given income stream after 5 years. Using equation (17) with $R(t) = 40,000$, $r = 0.12$, and $T = 5$, we see that the required value is given by

$$e^{0.12(5)} \int_0^5 40,000e^{-0.12t}\, dt$$

$$= e^{0.6}\left[-\frac{40,000}{0.12}e^{-0.12t}\right]\Big|_0^5 \qquad \text{(Integrate using the substitution } u = -0.12t.\text{)}$$

$$= -\frac{40,000e^{0.6}}{0.12}(e^{-0.6} - 1) \approx 274,039.60$$

or approximately $274,040.

           ○ ○ ○

Another way of measuring the value of an income stream is by considering its present value. The **present value** of an income stream of $R(t)$ dollars per year over a term of T years, earning interest at the rate of r per year compounded continuously, is the principal P that will yield the same accumulated value as the income stream itself when P is invested today for a period of T years at the same rate of interest. In other words,

$$Pe^{rT} = e^{rT} \int_0^T R(t)e^{-rt}\, dt$$

Dividing both sides of the equation by e^{rT} gives the following result.

PRESENT VALUE OF AN INCOME STREAM

The **present value of an income stream** of $R(t)$ dollars per year, earning interest at the rate of r per year compounded continuously, is given by

$$PV = \int_0^T R(t)e^{-rt}\, dt \qquad \textbf{(18)}$$

EXAMPLE 3   The owner of a local cinema is considering two alternative plans for renovating and improving the theater. Plan A calls for an immediate cash outlay of $250,000, whereas plan B requires an immediate cash outlay of $180,000. It has been estimated that adopting plan A would result in a net income stream generated at the rate of

$$f(t) = 630,000$$

dollars per year, whereas adopting plan B would result in a net income stream generated at the rate of

$$g(t) = 580,000$$

dollars per year for the next 3 years. If the prevailing interest rate for the next 5 years were 10% per year, which plan would generate a higher net income by the end of 3 years?

Solution   Since the initial outlay is $250,000, we find, using equation (18) with $R(t) = 630,000$, $r = 0.1$, and $T = 3$, that the present value of the net income under plan A is given by

$$\int_0^3 630,000e^{-0.1t}\, dt - 250,000$$

$$= \frac{630,000}{-0.1} e^{-0.1t} \Big|_0^3 - 250,000 \qquad \text{(Integrate using the substitution } u = -0.1t.)$$

$$= -6,300,000e^{-0.3} + 6,300,000 - 250,000$$

$$\approx 1,382,845$$

or approximately $1,382,845.

To find the present value of the net income under plan B, we use (18) with $R(t) = 580,000$, $r = 0.1$, and $T = 3$, obtaining

$$\int_0^3 580,000e^{-0.1t}\, dt - 180,000$$

dollars. Proceeding as in the previous computation, we see that the required value is $1,323,254 (exercise 8).

Comparing the present value of each plan, we conclude that plan A would generate a higher net income by the end of 3 years.   ○ ○ ○

REMARK   The function R in Example 3 is a constant function. If R is not a constant function, then we may need more sophisticated techniques of integration to evaluate the integral in (18). Exercises 7.1 and 7.2 in Chapter 7 contain problems of this type.   ○ ○ ○

The Amount and Present Value of an Annuity

An annuity is a sequence of payments made at regular time intervals. The time period in which these payments are made is called the *term* of the annuity. Although the payments need not be equal in size, they are equal in many important applications, and we will assume that they are equal in our discussion. Examples of annuities are regular deposits to a savings account, monthly home mortgage payments, and monthly insurance payments.

The **amount of an annuity** is the sum of the payments plus the interest earned. A formula for computing the amount of an annuity A can be derived with the help of (17). Let

$$P = \text{Size of each payment in the annuity}$$
$$r = \text{Interest rate compounded continuously}$$
$$T = \text{Term of the annuity (in years)}$$
and
$$m = \text{Number of payments per year}$$

The payments into the annuity constitute a constant income stream of $R(t) = mP$ dollars per year. With this value of $R(t)$, (17) yields

$$A = e^{rT} \int_0^T R(t)e^{-rt}\, dt = e^{rT} \int_0^T mPe^{-rt}\, dt$$

$$= mPe^{rT} \left[-\frac{e^{-rt}}{r} \right]\Bigg|_0^T$$

$$= mPe^{rT} \left[-\frac{e^{-rT}}{r} + \frac{1}{r} \right] = \frac{mP}{r}(e^{rT} - 1)$$

This leads us to the following formula.

AMOUNT OF AN ANNUITY

The **amount of an annuity** is

$$A = \frac{mP}{r}(e^{rT} - 1) \tag{19}$$

where P, r, T, and m are as defined earlier.

EXAMPLE 4   On January 1, 1985, Mr. Chapman deposited $2000 into an Individual Retirement Account (IRA) paying interest at the rate of 10% per year compounded continuously. Assuming that he deposits $2000 annually into the account, how much will he have in his IRA at the beginning of the year 2001?

Solution   We use (19), with $P = 2000$, $r = 0.1$, $T = 16$, and $m = 1$, obtaining

$$A = \frac{2000}{0.1}(e^{1.6} - 1)$$

$$\approx 79{,}060.65$$

Thus, Mr. Chapman will have approximately $79,061 in his account at the beginning of the year 2001.

○ ○ ○

EXPLORING WITH TECHNOLOGY

Refer to Example 4. Suppose Mr. Chapman wishes to know how much he will have in his IRA at any time in the future, not just at the beginning of 2001, as you were asked to compute in the example.

a. Using formula (17) and the relevant data from Example 4, show that the required amount at any time x (x measured in years, $x > 0$) is given by

$$A = f(x) = 20{,}000(e^{0.1x} - 1)$$

b. Use a graphing utility to plot the graph of f using the viewing rectangle $[0, 30] \times [2000, 400{,}000]$.

c. Using **ZOOM** and **TRACE**, or using the function evaluation capability of your graphing utility, use the result of part (b) to verify the result obtained in Example 4. Comment on the advantage of the mathematical model found in part (a).

○   ○   ○

Using (18), we can derive the following formula for the *present value* of an annuity.

PRESENT VALUE OF AN ANNUITY

The **present value of an annuity** is given by

$$PV = \frac{mP}{r}(1 - e^{-rT}) \qquad \textbf{(20)}$$

where P, r, T, and m are as defined earlier.

EXAMPLE 5   Mr. Carson, the proprietor of a hardware store, wants to establish a fund from which he will withdraw $1000 per month for the next 10 years. If the fund earns interest at the rate of 9% per year compounded continuously, how much money does he need to establish the fund?

Solution We want to find the present value of an annuity with $P = 1000$, $r = 0.09$, $T = 10$, and $m = 12$. Using equation (20), we find

$$PV = \frac{12,000}{0.09}(1 - e^{-(0.09)(10)})$$

$$\approx 79,124.04$$

Thus, Mr. Carson needs approximately $79,124 to establish the fund. ₒ ₒ ₒ

Lorentz Curves and Income Distributions

One method used by economists to study the distribution of income in a society is based on the **Lorentz curve,** named after American statistician M. D. Lorentz. To describe the Lorentz curve, let $f(x)$ denote the proportion of the total income received by the poorest $100x\%$ of the population for $0 \leq x \leq 1$. Using this terminology, $f(0.3) = 0.1$ simply states that the lowest 30% of the income recipients receive 10% of the total income.

The function f has the following properties:

1. The domain of f is $[0, 1]$.

2. The range of f is $[0, 1]$.

3. $f(0) = 0$ and $f(1) = 1$.

4. $f(x) \leq x$ for every x in $[0, 1]$.

5. f is increasing on $[0, 1]$.

The first two properties follow from the fact that both x and $f(x)$ are fractions of a whole. Property 3 is a statement that 0% of the income recipients receive 0% of the total income and 100% of the income recipients receive 100% of the total income. Property 4 follows from the fact that the lowest $100x\%$ of the income recipients cannot receive more than $100x\%$ of the total income. A typical Lorentz curve is shown in Figure 6.45.

EXAMPLE 6 A developing country's income distribution is described by the function

$$f(x) = \frac{19}{20}x^2 + \frac{1}{20}x$$

a. Sketch the Lorentz curve for the given function.

b. Compute $f(0.2)$ and $f(0.8)$ and interpret your results.

Solution

a. The Lorentz curve is shown in Figure 6.46.

b.
$$f(0.2) = \frac{19}{20}(0.2)^2 + \frac{1}{20}(0.2) = 0.048$$

so that the lowest 20% of the people receive 4.8% of the total income.

$$f(0.8) = \frac{19}{20}(0.8)^2 + \frac{1}{20}(0.8) = 0.648$$

so that the lowest 80% of the people receive 64.8% of the total income.

ₒ ₒ ₒ

Figure 6.45
A Lorentz curve.

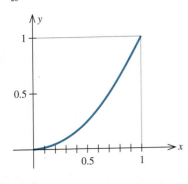

Figure 6.46
The Lorentz curve $f(x) = \frac{19}{20}x^2 + \frac{1}{20}x$.

Figure 6.47
The closer the Lorentz curve is to the line, the more equitable the income distribution.

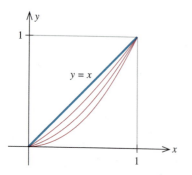

Next, let us consider the Lorentz curve described by the function $y = f(x) = x$. Since exactly $100x\%$ of the total income is received by the lowest $100x\%$ of income recipients, the line $y = x$ is called the **line of complete equality.** For example, 10% of the total income is received by the lowest 10% of income recipients, 20% of the total income is received by the lowest 20% of income recipients, and so on. Now, it is evident that the closer a Lorentz curve is to this line, the more equitable the income distribution is among the income recipients. But the proximity of a Lorentz curve to the line of complete equality is reflected by the area between the Lorentz curve and the line $y = x$ (Figure 6.47). The closer the curve is to the line, the smaller the enclosed area.

This observation suggests that we may define a number, called the **coefficient of inequality,** or **Gini Index,** of a Lorentz curve, as the ratio of the area between the line of complete equality and the Lorentz curve to the area under the line of complete equality. Since the area under the line of complete equality is $\frac{1}{2}$, we see that the coefficient of inequality is given by the following formula.

COEFFICIENT OF INEQUALITY OF A LORENTZ CURVE

The **coefficient of inequality** of a Lorentz curve is

$$L = 2 \int_0^1 [x - f(x)]\, dx \qquad \text{(21)}$$

The coefficient of inequality is a number between 0 and 1. For example, a coefficient of zero implies that the income distribution is perfectly uniform.

EXAMPLE 7   In a study conducted by a certain country's Economic Development Board with regard to the income distribution of certain segments of the country's work force, it was found that the Lorentz curves for the distribution of income of medical doctors and of movie actors and actresses are described by the functions

$$f(x) = \frac{14}{15}x^2 + \frac{1}{15}x \qquad \text{and} \qquad g(x) = \frac{5}{8}x^4 + \frac{3}{8}x$$

respectively. Compute the coefficient of inequality for each Lorentz curve. Which profession has a more equitable income distribution?

Solution   The required coefficients of inequality are, respectively,

$$L_1 = 2 \int_0^1 \left[x - \left(\frac{14}{15}x^2 + \frac{1}{15}x \right) \right] dx = 2 \int_0^1 \left(\frac{14}{15}x - \frac{14}{15}x^2 \right) dx$$

$$= \frac{28}{15} \int_0^1 (x - x^2)\, dx = \frac{28}{15} \left(\frac{1}{2}x^2 - \frac{1}{3}x^3 \right) \Big|_0^1$$

$$= \frac{14}{45} \approx 0.311$$

and $L_2 = 2 \int_0^1 \left[x - \left(\frac{5}{8} x^4 + \frac{3}{8} x \right) \right] dx = 2 \int_0^1 \left(\frac{5}{8} x - \frac{5}{8} x^4 \right) dx$

$$= \frac{5}{4} \int_0^1 (x - x^4) \, dx = \frac{5}{4} \left(\frac{1}{2} x^2 - \frac{1}{5} x^5 \right) \Big|_0^1$$

$$= \frac{15}{40} \approx 0.375$$

We conclude that in this country the incomes of medical doctors are more evenly distributed than the incomes of movie actors and actresses.   ০ ০ ০

SELF-CHECK EXERCISE 6.7

  The demand function for a certain make of exercise bicycle that is sold exclusively through cable television is

$$p = d(x) = \sqrt{9 - 0.02x}$$

where p is the unit price in hundreds of dollars and x is the quantity demanded per week. The corresponding supply function is given by

$$p = s(x) = \sqrt{1 + 0.02x}$$

where p has the same meaning as before and x is the number of exercise bicycles the supplier will make available at price p. Determine the consumers' surplus and the producers' surplus if the unit price is set at the equilibrium price.

The solution to Self-Check Exercise 6.7 can be found on page 517.

6.7 EXERCISES

A calculator is recommended for this exercise set.

1. **Consumers' Surplus**   The demand function for a certain make of cartridge typewriter ribbon is given by

$$p = -0.01x^2 - 0.1x + 6$$

where p is the unit price in dollars and x is the quantity demanded each week, measured in units of a thousand. Determine the consumers' surplus if the market price is set at \$4 per cartridge.

2. **Consumers' Surplus**   The demand function for a certain brand of compact disc is given by

$$p = -0.01x^2 - 0.2x + 8$$

where p is the wholesale unit price in dollars and x is the quantity demanded each week, measured in units

of a thousand. Determine the consumers' surplus if the wholesale market price is set at \$5 per disc.

3. **Consumers' Surplus**   It is known that the quantity demanded of a certain make of portable hair dryer is x hundred units per week, and the corresponding wholesale unit price is

$$p = \sqrt{225 - 5x}$$

dollars. Determine the consumers' surplus if the wholesale market price is set at \$10 per unit.

4. **Producers' Surplus**   The supplier of the portable hair dryers in exercise 3 will make x hundred units of hair dryers available in the market when the wholesale unit

price is

$$p = \sqrt{36 + 1.8x}$$

dollars. Determine the producers' surplus if the whole-sale market price is set at $9 per unit.

5. Producers' Surplus   The supply function for the compact discs of exercise 2 is given by

$$p = 0.01x^2 + 0.1x + 3$$

where p is the unit wholesale price in dollars and x stands for the quantity that will be made available in the market by the supplier, measured in units of a thousand. Determine the producers' surplus if the wholesale market price is set at the equilibrium price.

6. Consumers' and Producers' Surplus   The management of the Titan Tire Company has determined that the quantity demanded x of their Super Titan tires per week is related to the unit price p by the relation

$$p = 144 - x^2$$

where p is measured in dollars and x is measured in units of a thousand. Titan will make x units of the tires available in the market if the unit price is

$$p = 48 + \frac{1}{2}x^2$$

dollars. Determine the consumers' surplus and the producers' surplus when the market unit price is set at the equilibrium price.

7. Consumers' and Producers' Surplus   The quantity demanded x (in units of a hundred) of the Mikado miniature cameras per week is related to the unit price p (in dollars) by

$$p = -0.2x^2 + 80$$

and the quantity x (in units of a hundred) that the supplier is willing to make available in the market is related to the unit price p (in dollars) by

$$p = 0.1x^2 + x + 40$$

If the market price is set at the equilibrium price, find the consumers' surplus and the producers' surplus.

8. Refer to Example 3, page 510. Verify that

$$\int_0^3 580,000e^{-0.1t}\, dt - 180,000 \approx 1,323,254$$

9. Present Value of an Investment   Suppose that an investment is expected to generate income at the rate of

$$R(t) = 200,000$$

dollars per year for the next 5 years. Find the present value of this investment if the prevailing interest rate is 8% per year compounded continuously.

10. Franchises   Ms. Gordon purchased a 15-year franchise for a computer outlet store that is expected to generate income at the rate of

$$R(t) = 400,000$$

dollars per year. If the prevailing interest rate is 10% per year compounded continuously, find the present value of the franchise.

11. The Amount of an Annuity   Find the amount of an annuity if $250 a month is paid into it for a period of 20 years earning interest at the rate of 8% per year compounded continuously.

12. The Amount of an Annuity   Find the amount of an annuity if $400 a month is paid into it for a period of 20 years earning interest at the rate of 8% per year compounded continuously.

13. The Amount of an Annuity   Mr. Jorgenson deposits $150 per month in a savings account paying 8% per year compounded continuously. Estimate the amount that will be in his account after 15 years.

14. Custodial Accounts   The Armstrongs wish to establish a custodial account to finance their children's education. If they deposit $200 monthly for 10 years in a savings account paying 9% per year compounded continuously, find how much their savings account will be worth at the end of this period.

15. IRA Accounts   Refer to Example 4, page 511. Suppose that Mr. Chapman makes his IRA payment on April 1, 1985, and annually thereafter. If interest is paid at the same initial rate, approximately how much will Mr. Chapman have in his account at the beginning of the year 2001?

16. Present Value of an Annuity   Estimate the present value of an annuity if payments are $800 monthly for 12 years and the account earns interest at the rate of 10% per year compounded continuously.

17. Present Value of an Annuity   Estimate the present value of an annuity if payments are $1200 monthly for 15 years and the account earns interest at the rate of 10% per year compounded continuously.

18. Lottery Payments   A state lottery commission pays the winner of the "Million Dollar" lottery 20 annual installments of $50,000 each. If the prevailing interest rate is 8% per year compounded continuously, find the present value of the winning ticket.

19. **Reverse Annuity Mortgages** Mr. Sinclair wishes to supplement his retirement income by $300 per month for the next 10 years. He plans to obtain a reverse annuity mortgage (RAM) on his home to meet this need. Estimate the amount of the mortgage he will require if the prevailing interest rate is 12% per year compounded continuously.

20. **Reverse Annuity Mortgage** Refer to exercise 19. Miss Santos wishes to supplement her retirement income by $400 per month for the next 15 years by obtaining a RAM. Estimate the amount of the mortgage she will require if the prevailing interest rate is 9% per year compounded continuously.

21. **Lorentz Curves** A certain country's income distribution is described by the function

$$f(x) = \frac{15}{16}x^2 + \frac{1}{16}x$$

a. Sketch the Lorentz curve for this function.
b. Compute $f(0.4)$ and $f(0.9)$ and interpret your results.

22. **Lorentz Curves** In a study conducted by a certain country's Economic Development Board, it was found that the Lorentz curve for the distribution of income of college teachers was described by the function

$$f(x) = \frac{13}{14}x^2 + \frac{1}{14}x$$

and that of lawyers by the function

$$g(x) = \frac{9}{11}x^4 + \frac{2}{11}x$$

a. Compute the coefficient of inequality for each Lorentz curve.
b. Which profession has a more equitable income distribution?

23. **Lorentz Curves** A certain country's income distribution is described by the function

$$f(x) = \frac{14}{15}x^2 + \frac{1}{15}x$$

a. Sketch the Lorentz curve for this function.
b. Compute $f(0.3)$ and $f(0.7)$.

24. **Lorentz Curves** In a study conducted by a certain country's Economic Development Board, it was found that the Lorentz curve for the distribution of income of stockbrokers was described by the function

$$f(x) = \frac{11}{12}x^2 + \frac{1}{12}x$$

and that of high school teachers by the function

$$g(x) = \frac{5}{6}x^2 + \frac{1}{6}x$$

a. Compute the coefficient of inequality for each Lorentz curve.
b. Which profession has a more equitable income distribution?

SOLUTION TO SELF-CHECK EXERCISE 6.7

We find the equilibrium price and equilibrium quantity by solving the system of equations

$$p = \sqrt{9 - 0.02x}$$

and
$$p = \sqrt{1 + 0.02x}$$

simultaneously. Substituting the first equation into the second, we have

$$\sqrt{9 - 0.02x} = \sqrt{1 + 0.02x}$$

Squaring both sides of the equation then leads to

$$9 - 0.02x = 1 + 0.02x$$

giving
$$x = 200$$

USING TECHNOLOGY

Consumers' Surplus and Producers' Surplus

1. Resolve Example 1, Section 6.7, using a graphing utility.

[*Hint:* Use the intersection operation to find the equilibrium quantity and the equilibrium price. Use the numerical integration operation to evaluate the definite integral.]

2. Resolve exercise 7, Section 6.7, using a graphing utility.

[*Hint:* See exercise 1.]

3. The demand function for a certain brand of travel alarm clocks is given by

$$p = -0.01x^2 - 0.3x + 10$$

where p is the wholesale unit price in dollars and x is the quantity demanded each month, measured in units of a thousand. The supply function for this brand of clocks is given by

$$p = -0.01x^2 + 0.2x + 4$$

where p has the same meaning as before and x is the quantity, in thousands, the supplier will make available in the marketplace per month. Determine the consumers' surplus and the producers' surplus when the market unit price is set at the equilibrium price.

4. The quantity demanded of a certain make of compact disc organizer is x thousand units per week, and the corresponding wholesale unit price is

$$p = \sqrt{400 - 8x}$$

dollars. The supplier of the organizers will make x thousand units available in the market when the unit wholesale price is

$$p = 0.02x^2 + 0.04x + 5$$

dollars. Determine the consumers' surplus and the producers' surplus when the market unit price is set at the equilibrium price.

Therefore,
$$p = \sqrt{9 - 0.02(200)}$$
$$= \sqrt{5} \approx 2.24$$

and the equilibrium price is $224 and the equilibrium quantity is 200. The consumers' surplus is given by

$$CS = \int_0^{200} \sqrt{9 - 0.02x}\, dx - (2.24)(200)$$

$$= \int_0^{200} (9 - 0.02x)^{1/2}\, dx - 448 \qquad \text{(Integrating by substitution)}$$

$$= -\frac{1}{0.02}\left(\frac{2}{3}\right)(9 - 0.02x)^{3/2} \Big|_0^{200} - 448$$

$$= -\frac{1}{0.03}(5^{3/2} - 9^{3/2}) - 448$$

$$\approx 79.32$$

or approximately $7932.
    Next, the producers' surplus is given by

$$PS = (2.24)(200) - \int_0^{200} \sqrt{1 + 0.02x}\, dx$$

$$= 448 - \int_0^{200} (1 + 0.02x)^{1/2}\, dx$$

$$= 448 - \frac{1}{0.02}\left(\frac{2}{3}\right)(1 + 0.02x)^{3/2} \Big|_0^{200}$$

$$= 448 - \frac{1}{0.03}(5^{3/2} - 1)$$

$$\approx 108.66$$

or approximately $10,866.

6.8 VOLUMES OF SOLIDS OF REVOLUTION

Finding the Volume of a Solid of Revolution

In this section we will use the notion of a Riemann sum to help us develop a formula for computing the volume of a solid that results when a region is revolved about an axis. The solid is called a **solid of revolution.** To find the volume of a solid of revolution, suppose that the plane region under the curve defined by a nonnegative continuous function $y = f(x)$ between $x = a$ and $x = b$ is revolved about the x-axis (Figure 6.48).

Figure 6.48

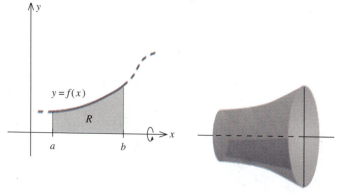

(a) Region R under the curve

(b) Solid obtained by revolving R about the x-axis

To derive a formula for finding the volume of the solid of revolution, we divide the interval $[a, b]$ into n subintervals, each of equal length, by means of the points $x_0 = a, x_1, x_2, \ldots, x_n = b$, so that the length of each subinterval is given by $\Delta x = (b - a)/n$. Also, let $p_1, p_2, \ldots, p_n$ be points in the subintervals $[x_0, x_1], [x_1, x_2], \ldots, [x_{n-1}, x_n]$, respectively (Figure 6.49).

Figure 6.49

The region R is revolved about the x-axis.

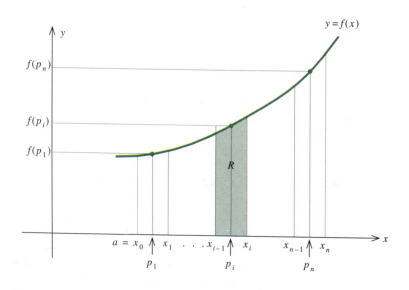

Let us concentrate on that part of the solid of revolution swept out by revolving the region under the curve between $x = x_{i-1}$ and $x = x_i$ about the x-axis. The volume ΔV_i of this object, shown in Figure 6.50a, may be approximated by the volume of the disk shown in Figure 6.50b, of radius $f(p_i)$ and width Δx, obtained by revolving the rectangular region of height $f(p_i)$ and width Δx about the x-axis (Figure 6.50c).

Figure 6.50

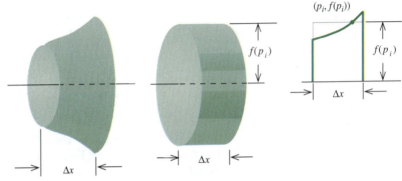

(a) Volume ΔV_i of object   (b) Volume of disk   (c) Approximating rectangle

Figure 6.51
The solid of revolution is approximated by n disks.

Since the volume of a disk (cylinder) is equal to $\pi r^2 h$, we see that

$$\Delta V_i \approx \pi [f(p_i)]^2 \, \Delta x$$

This analysis suggests that the volume of the solid of revolution may be approximated by the sum of the volumes of n suitable disks—namely,

$$\pi [f(p_1)]^2 \, \Delta x + \pi [f(p_2)]^2 \, \Delta x + \cdots + \pi [f(p_n)]^2 \, \Delta x$$

(Figure 6.51). This last expression is the Riemann sum of the function $g(x) = \pi [f(x)]^2$ over the interval $[a, b]$. Letting n approach infinity, we obtain the following formula for the volume of a solid of revolution.

VOLUME OF A SOLID OF REVOLUTION

The volume V of the solid of revolution obtained by revolving the region below the graph of $y = f(x)$ from $x = a$ to $x = b$ about the x-axis is

$$V = \pi \int_a^b [f(x)]^2 \, dx \qquad (22)$$

EXAMPLE 1   Find the volume of the solid of revolution obtained by revolving the region under the curve $y = f(x) = e^{-x}$ from $x = 0$ to $x = 1$ about the x-axis.

Figure 6.52

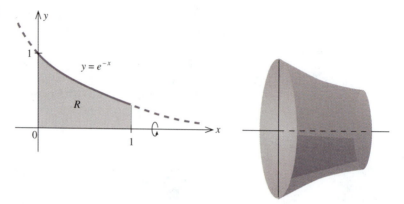

(a) The region under the curve
$y = e^{-x}$ from $x = 0$ to $x = 1$

(b) The solid of revolution
obtained when R is revolved
about the x-axis

Solution   The region under consideration, together with the resulting solid
of revolution, is shown in Figure 6.52. Using formula (22), we find that the
required volume is given by

$$\pi \int_0^1 (e^{-x})^2 \, dx = \pi \int_0^1 e^{-2x} \, dx$$

$$= -\frac{\pi}{2} e^{-2x} \Big|_0^1 = \frac{\pi}{2}(1 - e^{-2})$$

cubic units.

○ ○ ○

EXAMPLE 2   Derive a formula for the volume of a right circular cone of
radius r and height h.

Solution   The cone is shown in Figure 6.53a. It is the solid of revolution
obtained by revolving the region R of Figure 6.53b about the x-axis. The

Figure 6.53

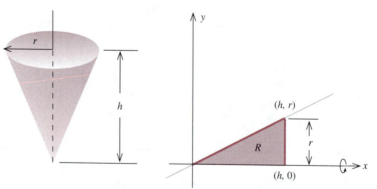

(a) A right circular cone
of radius r and height h

(b) R is the region under the straight
line $y = \dfrac{r}{h} x$ from $x = 0$ to $x = h$.

region R is the region under the straight line $y = f(x) = (r/h)x$ between $x = 0$ and $x = h$. Using formula (22), we see that the required volume is given by

$$V = \pi \int_0^h \left[\left(\frac{r}{h}\right)x \right]^2 dx = \frac{\pi r^2}{h^2} \int_0^h x^2 \, dx$$

$$= \left(\frac{\pi r^2}{3h^2}\right)x^3 \Big|_0^h = \frac{1}{3}\pi r^2 h$$

cubic units.   ○ ○ ○

Next, let us consider the solid of revolution obtained by revolving the region R bounded above by the graph of the nonnegative function $f(x)$ and below by the graph of the nonnegative function $g(x)$, from $x = a$ to $x = b$, about the x-axis (Figure 6.54).

Figure 6.54

(a) R is the region bounded by the curves $y = f(x)$ and $y = g(x)$ from $x = a$ to $x = b$.

(b) The solid of revolution obtained by revolving R about the x-axis

Figure 6.55

R is the region bounded by the curves $y = e^x$ and $y = x^2 + \frac{1}{2}$ from $x = 0$ to $x = 1$.

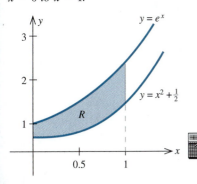

To derive a formula for computing the volume of such a solid of revolution, observe that the required volume is the volume of the solid of revolution obtained by revolving the region under the curve $y = f(x)$ from $x = a$ to $x = b$ about the x-axis *minus* the volume of the solid of revolution obtained by revolving the region under the curve $y = g(x)$ from $x = a$ to $x = b$ about the x-axis. Thus, the required volume is given by

$$V = \pi \int_a^b [f(x)]^2 \, dx - \pi \int_a^b [g(x)]^2 \, dx$$

or

$$V = \pi \int_a^b \{[f(x)]^2 - [g(x)]^2\} \, dx$$

(23)

EXAMPLE 3 Find the volume of the solid of revolution obtained by revolving the region bounded by the curves $y = e^x$ and $y = x^2 + \frac{1}{2}$ from $x = 0$ to $x = 1$ about the x-axis (Figure 6.55).

Solution Using formula (23) with $f(x) = e^x$ and $g(x) = x^2 + \frac{1}{2}$, we see that the required volume is given by

$$V = \pi \int_0^1 \left\{ [e^x]^2 - \left[x^2 + \frac{1}{2} \right]^2 \right\} dx$$

$$= \pi \int_0^1 \left(e^{2x} - x^4 - x^2 - \frac{1}{4} \right) dx$$

$$= \pi \left(\frac{1}{2} e^{2x} - \frac{1}{5} x^5 - \frac{1}{3} x^3 - \frac{1}{4} x \right) \Big|_0^1$$

$$= \pi \left[\left(\frac{1}{2} e^2 - \frac{1}{5} - \frac{1}{3} - \frac{1}{4} \right) - \frac{1}{2} \right]$$

$$= \frac{\pi}{60} (30e^2 - 77)$$

or approximately 7.57 cubic units.

EXAMPLE 4 Find the solid of revolution obtained by revolving the region bounded by the curves $y = f(x) = x^2$ and $y = g(x) = x^3$ about the x-axis.

Solution We first find the point(s) of intersection of the two curves by solving the system of simultaneous equations

$$\left. \begin{array}{l} y = x^2 \\ y = x^3 \end{array} \right\}$$

We have

$$x^3 = x^2$$

$$x^3 - x^2 = 0$$

$$x^2(x - 1) = 0$$

so that $x = 0$ or $x = 1$, and the points of intersection are $(0, 0)$ and $(1, 1)$ (Figure 6.56).

Next, observe that $g(x) \leq f(x)$ for all values of x between $x = 0$ and $x = 1$, so that using formula (23), we find the required volume to be

$$V = \pi \int_0^1 [(x^2)^2 - (x^3)^2] \, dx$$

$$= \pi \int_0^1 (x^4 - x^6) \, dx$$

$$= \pi \left(\frac{1}{5} x^5 - \frac{1}{7} x^7 \right) \Big|_0^1$$

$$= \pi \left(\frac{1}{5} - \frac{1}{7} \right)$$

$$= \frac{2\pi}{35}$$

cubic units.

Figure 6.56

R is the region bounded by the curves $y = x^2$ and $y = x^3$.

Application

EXAMPLE 5 The external fuel tank for a space shuttle has a shape that may be obtained by revolving the region under the curve

$$f(x) = \begin{cases} 4\sqrt{10} & \text{if } -120 \le x \le 10 \\ \frac{1}{5}\sqrt{x}(30 - x) & \text{if } 10 < x \le 30 \end{cases}$$

from $x = -120$ to $x = 30$ (all measurements are given in feet) about the x-axis (Figure 6.57). The tank carries liquid hydrogen for fueling the shuttle's three main engines. Estimate the capacity of the tank (231 cubic inches = 1 gallon).

Figure 6.57

The solid of revolution obtained by revolving the region under the curve y = f(x) about the x-axis.

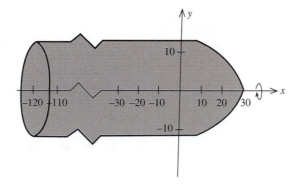

Solution The volume of the tank is given by

$$V = \pi \int_{-120}^{30} [f(x)]^2 \, dx$$

$$= \pi \int_{-120}^{10} (4\sqrt{10})^2 \, dx + \pi \int_{10}^{30} \left[\frac{1}{5}\sqrt{x}(30 - x) \right]^2 dx$$

$$= 160\pi \int_{-120}^{10} dx + \frac{\pi}{25} \int_{10}^{30} x(30 - x)^2 \, dx$$

$$= 160\pi x \Big|_{-120}^{10} + \frac{\pi}{25} \int_{10}^{30} (900x - 60x^2 + x^3) \, dx$$

$$= 160\pi(130) + \frac{\pi}{25} \left(450x^2 - 20x^3 + \frac{1}{4}x^4 \right) \Big|_{10}^{30}$$

$$= 20{,}800\pi + \frac{\pi}{25} \left\{ \left[450(30)^2 - 20(30)^3 + \frac{1}{4}(30)^4 \right] \right.$$

$$\left. - \left[450(10)^2 - 20(10)^3 + \frac{1}{4}(10^4) \right] \right\}$$

$$= 20{,}800\pi + 1600\pi = 22{,}400\pi$$

or approximately 70,372 cubic feet. Therefore, its capacity is approximately $(70{,}372)(12^3)/231$, or approximately 526,419 gallons.   ❍ ❍ ❍

SELF–CHECK EXERCISE 6.8

Find the volume of the solid of revolution obtained by revolving the region bounded by the graphs of $f(x) = 5 - x^2$ and $g(x) = 1$ about the x-axis.

The solution to Self-Check Exercise 6.8 can be found on page 527.

6.8 EXERCISES

In exercises 1–10, find the volume of the solid of revolution obtained by revolving the given region under the curve $y = f(x)$ from $x = a$ to $x = b$ about the x-axis.

1. $y = 3x$; $a = 0$, $b = 1$

2. $y = x + 1$; $a = 0$, $b = 2$

3. $y = \sqrt{x}$; $a = 1$, $b = 4$

4. $y = \sqrt{x + 1}$; $a = 0$, $b = 3$

5. $y = \sqrt{1 + x^2}$; $a = 0$, $b = 1$

6. $y = \sqrt{4 - x^2}$; $a = -2$, $b = 2$

7. $y = 1 - x^2$; $a = -1$, $b = 1$

8. $y = 2x - x^2$; $a = 0$, $b = 2$

9. $y = e^x$; $a = 0$, $b = 1$

10. $y = e^{-3x}$; $a = 0$, $b = 1$

In exercises 11–18, find the volume of the solid of revolution obtained by revolving the given region bounded above by the curve $y = f(x)$ and below by the curve $y = g(x)$ from $x = a$ to $x = b$ about the x-axis.

11. $f(x) = x$ and $g(x) = x^2$; $a = 0$, $b = 1$

12. $f(x) = 1$ and $g(x) = \sqrt{x}$; $a = 0$, $b = 1$

13. $f(x) = 4 - x^2$ and $g(x) = 3$; $a = -1$, $b = 1$

14. $f(x) = 8$ and $g(x) = x^{3/2}$; $a = 0$, $b = 4$

15. $f(x) = \sqrt{16 - x^2}$ and $g(x) = x$; $a = 0$, $b = 2\sqrt{2}$

16. $f(x) = e^x$ and $g(x) = \dfrac{1}{x}$; $a = 1$, $b = 2$

17. $f(x) = e^x$ and $g(x) = e^{-x}$; $a = 0$, $b = 1$

18. $f(x) = e^x$ and $g(x) = \sqrt{x}$; $a = 1$, $b = 2$

In exercises 19–26, find the volume of the solid of revolution obtained by revolving the region bounded by the graphs of the given functions about the x-axis.

19. $y = x$ and $y = \sqrt{x}$

20. $y = x^{1/3}$ and $y = x^2$

21. $y = \dfrac{1}{2}x + 3$ and $y = x^2$

22. $y = 2x$ and $y = x\sqrt{x + 1}$

23. $y = x^2$ and $y = 4 - x^2$

24. $y = x^2$ and $y = x(4 - x)$

25. $y = \dfrac{1}{x}$, $y = x$,   and $y = 2x$

[*Hint:* You will need to evaluate two integrals.]

26. $y = \sqrt{x - 1}$ and $y = (x - 1)^2$

27. By computing the volume of the solid of revolution obtained by revolving the region under the semicircle $y = \sqrt{r^2 - x^2}$ from $x = -r$ to $x = r$ about the x-axis, show that the volume of a sphere of radius r is $\frac{4}{3}\pi r^3$ cubic units.

28. **Volume of a Football**   Find the volume of the prolate spheroid (a solid of revolution in the shape of a football) obtained by revolving the region under the graph of the function $y = \frac{2}{5}\sqrt{25 - x^2}$ from $x = -5$ to $x = 5$ about the x-axis.

29. Capacity of a Man-made Lake   A man-made lake is approximately circular and has a cross section that is a region bounded above by the x-axis and below by the graph of $y = 10\left[\left(\dfrac{x}{100}\right)^2 - 1\right]$ (see the accompanying figure). Find the approximate capacity of the lake. [*Hint:* Find the volume of the solid of revolution obtained by revolving the shaded region about the y-axis. Thus, $V = \pi \displaystyle\int_{-10}^{0} [f(y)]^2 \, dy$, where $f(y)$ is obtained by solving the given equation for x in terms of y.]

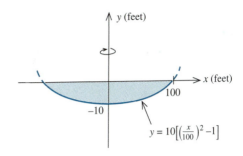

SOLUTION TO SELF-CHECK EXERCISE 6.8

To find the points of intersection of the two curves, we solve the equation $5 - x^2 = 1$, giving $x = \pm 2$ (see the figure). Using formula (23), we find the required volume

$$V = \pi \int_{-2}^{2} [(5 - x^2)^2 - 1^2] \, dx$$

$$= 2\pi \int_{0}^{2} (25 - 10x^2 + x^4 - 1) \, dx \qquad \text{(By symmetry)}$$

$$= 2\pi \int_{0}^{2} (x^4 - 10x^2 + 24) \, dx$$

$$= 2\pi \left(\frac{1}{5}x^5 - \frac{10}{3}x^3 + 24x\right)\Big|_{0}^{2}$$

$$= 2\pi \left(\frac{32}{5} - \frac{80}{3} + 48\right)$$

$$= \frac{832\pi}{15} \text{ cubic units}$$

CHAPTER 6 SUMMARY OF PRINCIPAL FORMULAS AND TERMS

Formulas

1. Indefinite integral of a constant

$$\int k\,du = ku + C$$

2. Power Rule

$$\int u^n\,du = \frac{u^{n+1}}{n+1} + C$$

3. Constant Multiple Rule

$$\int kf(u)\,du = k\int f(u)\,du$$
$(k, \text{ a constant})$

4. Sum Rule

$$\int [f(u) \pm g(u)]\,du$$
$$= \int f(u)\,du \pm \int g(u)\,du$$

5. Indefinite integral of the exponential function

$$\int e^u\,du = e^u + C$$

6. Indefinite integral of $f(u) = \dfrac{1}{u}$

$$\int \frac{du}{u} = \ln|u| + C$$

7. Method of substitution

$$\int f'(g(x))g'(x)\,dx = \int f'(u)\,du$$

8. Fundamental Theorem of Calculus

$$\int_a^b f(x)\,dx$$
$$= F(b) - F(a), F'(x) = f(x)$$

9. Area between two curves

$$\int_a^b [f(x) - g(x)]\,dx, f(x) \geq g(x)$$

10. Definite integral as the limit of a sum

$$\int_a^b f(x)\,dx = \lim_{n\to\infty} S_n,$$
where S_n is a Riemann sum

11. Average value of f over $[a, b]$

$$\frac{1}{b-a}\int_a^b f(x)\,dx$$

12. Consumers' surplus

$$CS = \int_0^{\bar{x}} D(x)\,dx - \bar{p}\bar{x}$$

13. Producers' surplus

$$PS = \bar{p}\bar{x} - \int_0^{\bar{x}} S(x)\,dx$$

14. Accumulated (future) value of an income stream

$$A = e^{rT}\int_0^T R(t)e^{-rt}\,dt$$

15. Present value of an income stream

$$PV = \int_0^T R(t)e^{-rt}\,dt$$

16. Present value of an annuity

$$PV = \frac{mP}{r}(1 - e^{-rT})$$

17. Amount of an annuity $\qquad$ $A = \dfrac{mP}{r}(e^{rT} - 1)$

18. Coefficient of inequality of a Lorentz curve $\qquad$ $L = 2\displaystyle\int_0^1 [x - f(x)]\,dx$

19. Volume of a solid of revolution $\qquad$ $V = \pi\displaystyle\int_a^b [f(x)]^2\,dx$

Terms

Antiderivative

Integration

Indefinite integral

Integrand

Constant of integration

Riemann sum

Definite integral

Limits of integration

Lorentz curve

Solid of revolution

CHAPTER 6 REVIEW EXERCISES

In exercises 1–20, evaluate each indefinite integral.

1. $\displaystyle\int (x^3 + 2x^2 - x)\,dx$

2. $\displaystyle\int (\tfrac{1}{3}x^3 - 2x^2 + 8)\,dx$

3. $\displaystyle\int \left(x^4 - 2x^3 + \dfrac{1}{x^2}\right) dx$

4. $\displaystyle\int (x^{1/3} - \sqrt{x} + 4)\,dx$

5. $\displaystyle\int x(2x^2 + x^{1/2})\,dx$

6. $\displaystyle\int (x^2 + 1)(\sqrt{x} - 1)\,dx$

7. $\displaystyle\int \left(x^2 - x + \dfrac{2}{x} + 5\right) dx$

8. $\displaystyle\int \sqrt{2x + 1}\,dx$

9. $\displaystyle\int (3x - 1)(3x^2 - 2x + 1)^{1/3}\,dx$

10. $\displaystyle\int x^2(x^3 + 2)^{10}\,dx$

11. $\displaystyle\int \dfrac{x - 1}{x^2 - 2x + 5}\,dx$

12. $\displaystyle\int 2e^{-2x}\,dx$

13. $\displaystyle\int (x + \tfrac{1}{2})e^{x^2 + x + 1}\,dx$

14. $\displaystyle\int \dfrac{e^{-x} - 1}{(e^{-x} + x)^2}\,dx$

15. $\displaystyle\int \dfrac{(\ln x)^5}{x}\,dx$

16. $\displaystyle\int \dfrac{\ln x^2}{x}\,dx$

17. $\displaystyle\int x^3(x^2 + 1)^{10}\,dx$

18. $\displaystyle\int x\sqrt{x + 1}\,dx$

19. $\displaystyle\int \dfrac{x}{\sqrt{x - 2}}\,dx$

20. $\displaystyle\int \dfrac{3x}{\sqrt{x + 1}}\,dx$

In exercises 21–32, evaluate each definite integral.

21. $\displaystyle\int_0^1 (2x^3 - 3x^2 + 1)\,dx$

22. $\displaystyle\int_0^2 (4x^3 - 9x^2 + 2x - 1)\,dx$

23. $\displaystyle\int_1^4 (\sqrt{x} + x^{-3/2})\,dx$

24. $\displaystyle\int_0^1 20x(2x^2 + 1)^4\,dx$

25. $\displaystyle\int_{-1}^0 12(x^2 - 2x)(x^3 - 3x^2 + 1)^3\,dx$

26. $\displaystyle\int_4^7 x\sqrt{x - 3}\,dx$

27. $\displaystyle\int_0^2 \dfrac{x}{x^2 + 1}\,dx$

28. $\displaystyle\int_0^1 \dfrac{dx}{(5 - 2x)^2}$

29. $\displaystyle\int_0^2 \dfrac{4x}{\sqrt{1 + 2x^2}}\,dx$

30. $\displaystyle\int_0^2 xe^{(-1/2)x^2}\,dx$

31. $\displaystyle\int_{-1}^0 \dfrac{e^{-x}}{(1 + e^{-x})^2}\,dx$

32. $\displaystyle\int_1^e \dfrac{\ln x}{x}\,dx$

In exercises 33–36, find the function f given that the slope of the tangent line to the graph at any point (x, f(x)) is f'(x) and that the graph of f passes through the given point.

33. $f'(x) = 3x^2 - 4x + 1$; $(1, 1)$

34. $f'(x) = \dfrac{x}{\sqrt{x^2 + 1}}; (0, 1)$

35. $f'(x) = 1 - e^{-x}; (0, 2)$

36. $f'(x) = \dfrac{\ln x}{x}; (1, -2)$

37. Let $f(x) = -2x^2 + 1$, and compute the Riemann sum of f over the interval $[1, 2]$ by partitioning the interval into five subintervals of the same length ($n = 5$), where the points p_i ($1 \le i \le 5$) are taken to be the *right* endpoints of the respective subintervals.

38. The management of the National Electric Corporation has determined that the daily marginal cost function associated with producing their automatic drip coffeemakers is given by

$$C'(x) = 0.00003x^2 - 0.03x + 20$$

where $C'(x)$ is measured in dollars per unit and x denotes the number of units produced. Management has also determined that the daily fixed cost incurred in producing these coffeemakers is $500. What is the total cost incurred by National in producing the first 400 coffeemakers per day?

39. Refer to exercise 38. Management has also determined that the daily marginal revenue function associated with producing and selling their coffeemakers is given by

$$R'(x) = -0.03x + 60$$

where x denotes the number of units produced and sold and $R'(x)$ is measured in dollars per unit.
a. Determine the revenue function $R(x)$ associated with producing and selling these coffeemakers.
b. What is the demand equation relating the wholesale unit price to the quantity of coffeemakers demanded?

40. The Franklin National Life Insurance Company purchased a new computer for $200,000. If the rate at which the computer's resale value changes is given by the function

$$V'(t) = 3800(t - 10)$$

where t is the length of time since the purchase date and $V'(t)$ is measured in dollars per year, find an expression $V(t)$ that gives the resale value of the computer after t years. How much would the computer cost after 6 years?

41. The marketing department of the Vista Vision Corporation forecasts that sales of their new line of projection television systems will grow at the rate of

$$3000 - 2000e^{-0.04t} \qquad (0 \le t \le 24)$$

units per month once they are introduced into the market. Find an expression giving the total number of units of the projection television systems that Vista may expect to sell t months from the time they are put on the market. How many units of the television systems can Vista expect to sell during the first year?

42. Due to the increasing cost of fuel, the manager of the City Transit Authority estimates that the number of commuters using the city subway system will increase at the rate of

$$3000(1 + 0.4t)^{-1/2} \qquad (0 \le t \le 36)$$

per month t months from now. If 100,000 commuters are currently using the system, find an expression giving the total number of commuters who will be using the subway t months from now. How many commuters will be using the subway six months from now?

43. The management of a division of Ditton Industries has determined that the daily marginal cost function associated with producing their hot-air corn poppers is given by

$$C'(x) = 0.00003x^2 - 0.03x + 10$$

where $C'(x)$ is measured in dollars per unit and x denotes the number of units manufactured. Management has also determined that the daily fixed cost incurred in producing these corn poppers is $600. Find the total cost incurred by Ditton in producing the first 500 corn poppers.

44. In 1980 the world produced 3.5 billion metric tons of coal. If output increased at the rate of

$$3.5e^{0.04t}$$

billion metric tons per year in year t ($t = 0$ corresponds to 1980), determine how much coal was produced worldwide between 1980 and the end of 1985.

45. Find the area of the region under the curve $y = 3x^2 + 2x + 1$ from $x = -1$ to $x = 2$.

46. Find the area of the region under the curve $y = e^{2x}$ from $x = 0$ to $x = 2$.

47. Find the area of the region bounded by the graph of the function $y = 1/x^2$, the x-axis, and the lines $x = 1$ and $x = 3$.

48. Find the area of the region bounded by the curve $y = -x^2 - x + 2$ and the x-axis.

49. Find the area of the region bounded by the graphs of the functions $f(x) = e^x$ and $g(x) = x$ and the vertical lines $x = 0$ and $x = 2$.

50. Find the area of the region that is completely enclosed by the graphs of $f(x) = x^4$ and $g(x) = x$.

51. Find the area of the region between the curve $y = x(x - 1)(x - 2)$ and the x-axis.

52. Based on current production techniques, the rate of oil production from a certain oil well t years from now is estimated to be

$$R_1(t) = 100e^{0.05t}$$

thousand barrels per year. Based on a new production technique, however, it is estimated that the rate of oil production from that oil well t years from now will be

$$R_2(t) = 100e^{0.08t}$$

thousand barrels per year. Determine how much additional oil will be produced over the next 10 years if the new technique is adopted.

53. Find the average value of the function

$$f(x) = \frac{x}{\sqrt{x^2 + 16}}$$

over the interval $[0, 3]$.

54. The demand function for a brand of blank videocassettes is given by

$$p = -0.01x^2 - 0.2x + 23$$

where p is the wholesale unit price in dollars and x is the quantity demanded each week, measured in units of a thousand. Determine the consumers' surplus if the wholesale unit price is $8 per cassette.

55. The quantity demanded x (in units of a hundred) of the Sportsman 5×7 tents, per week, is related to the unit price p (in dollars) by the equation

$$p = -0.1x^2 - x + 40$$

The quantity x (in units of a hundred) that the supplier is willing to make available in the market is related to

the unit price by

$$p = 0.1x^2 + 2x + 20$$

If the market price is set at the equilibrium price, find the consumers' surplus and the producers' surplus.

56. Mr. Cunningham plans to deposit $4000 per year in his Keogh Retirement Account. If interest is compounded continuously at the rate of 8% per year, how much will he have in his retirement account after 20 years?

57. Miss Parker sold her house under an installment contract whereby the buyer gave her a down payment of $9000 and agreed to make monthly payments of $925 per month for 30 years. If the prevailing interest rate is 12% per year compounded continuously, find the present value of the purchase price of the house.

58. Ms. Carlson purchased a 10-year franchise for a health spa that is expected to generate income at the rate of

$$P(t) = 80,000$$

dollars per year. If the prevailing interest rate is 10% per year compounded continuously, find the present value of the franchise.

59. A certain country's income distribution is described by the function

$$f(x) = \frac{17}{18}x^2 + \frac{1}{18}x$$

a. Sketch the Lorentz curve for this function.
b. Compute $f(0.3)$ and $f(0.6)$ and interpret your results.
c. Compute the coefficient of inequality for this Lorentz curve.

60. The population of a certain Sunbelt city, currently 80,000, is expected to grow exponentially in the next 5 years with a growth constant of 0.05. If the prediction comes true, what will the average population of the city be over the next 5 years?

61. Find the volume of the solid of revolution obtained by revolving the region under the curve $f(x) = 1/x$ from $x = 1$ to $x = 3$ about the x-axis.

62. Find the volume of the solid of revolution obtained by revolving the region bounded above by the curve $f(x) = \sqrt{x}$ and below by the curve $g(x) = x^2$ about the x-axis.

In addition to the basic rules of integration developed in Chapter 6, there are more sophisticated techniques for finding the antiderivatives of functions. We begin this chapter by looking at the method of integration by parts. We then look at a technique of integration that involves using tables of integrals that have been compiled for this purpose. We also look at numerical methods of integration, which enable us to obtain approximate solutions to definite integrals, especially those whose exact value cannot be found otherwise. More specifically, we study the Trapezoidal Rule and Simpson's Rule. Numerical integration methods are especially useful when the integrand is known at only discrete points. Finally, we learn how to evaluate integrals in which the intervals of integration are unbounded. Such integrals, called *improper integrals*, play an important role in the study of probability.

What is the area of the oil spill caused by a grounded tanker? In Example 5, page 558, you will see how to determine the area of the oil spill.

7

ADDITIONAL TOPICS IN INTEGRATION

7.1 INTEGRATION BY PARTS

The Method of Integration by Parts

Integration by parts is another technique of integration that, like the method of substitution discussed in Chapter 6, is based on a corresponding rule of differentiation. In this case, the rule of differentiation is the Product Rule, which asserts that if f and g are differentiable functions, then

$$\frac{d}{dx}[f(x)g(x)] = f(x)g'(x) + g(x)f'(x) \tag{1}$$

If we integrate both sides of equation (1) with respect to x, we obtain

$$\int \frac{d}{dx}f(x)g(x)\,dx = \int f(x)g'(x)\,dx + \int g(x)f'(x)\,dx$$

or

$$f(x)g(x) = \int f(x)g'(x)\,dx + \int g(x)f'(x)\,dx$$

This last equation, which may be written in the form

$$\int f(x)g'(x)\,dx = f(x)g(x) - \int g(x)f'(x)\,dx \tag{2}$$

is called the formula for **integration by parts.** This formula is useful since it enables us to express one indefinite integral in terms of another that may be easier to evaluate. Formula (2) may be simplified by letting

$$u = f(x) \qquad dv = g'(x)\,dx$$
$$du = f'(x)\,dx \qquad v = g(x)$$

giving the following version of the formula for integration by parts.

INTEGRATION BY PARTS FORMULA

$$\int u\,dv = uv - \int v\,du \tag{3}$$

EXAMPLE 1   Evaluate $\int xe^x\,dx$.

Solution   No method of integration developed thus far enables us to evaluate the given indefinite integral in its present form. Therefore, we attempt to write it in terms of an indefinite integral that is easier to evaluate. Let us use the integration by parts formula (3) by letting

$$u = x \quad \text{and} \quad dv = e^x\,dx$$

so that

$$du = dx \quad \text{and} \quad v = e^x$$

Therefore,

$$\int xe^x\, dx = \int u\, dv$$

$$= uv - \int v\, du$$

$$= xe^x - \int e^x\, dx$$

$$= xe^x - e^x + C$$

$$= (x - 1)e^x + C$$

The success of the method of integration by parts depends on the proper choice of u and dv. For example, if we had chosen

$$u = e^x \quad \text{and} \quad dv = x\, dx$$

in the last example, then

$$du = e^x\, dx \quad \text{and} \quad v = \frac{1}{2}x^2$$

so (3) would have yielded

$$\int xe^x\, dx = \int u\, dv$$

$$= uv - \int v\, du$$

$$= \frac{1}{2}x^2e^x - \int \frac{1}{2}x^2e^x\, dx$$

Since the indefinite integral on the right-hand side of this equation is not readily evaluated (it is, in fact, more complicated than the original integral!), choosing u and dv as shown has not helped us evaluate the given indefinite integral.

In general, we can use the following guidelines.

GUIDELINES FOR CHOOSING u AND dv

Choose u and dv so that

1. du is simpler than u.
2. dv is easy to integrate.

EXAMPLE 2 Evaluate $\int x \ln x\, dx$.

Solution Letting

$$u = \ln x \quad \text{and} \quad dv = x\, dx$$

we have

$$du = \frac{1}{x}\, dx \quad \text{and} \quad v = \frac{1}{2}x^2$$

Therefore,

$$\int x \ln x \, dx = \int u \, dv = uv - \int v \, du$$

$$= \frac{1}{2}x^2 \ln x - \int \frac{1}{2}x^2 \cdot \left(\frac{1}{x}\right) dx$$

$$= \frac{1}{2}x^2 \ln x - \frac{1}{2}\int x \, dx$$

$$= \frac{1}{2}x^2 \ln x - \frac{1}{4}x^2 + C$$

$$= \frac{1}{4}x^2(2 \ln x - 1) + C$$

○ ○ ○

EXAMPLE 3 Evaluate $\int \dfrac{xe^x}{(x+1)^2} \, dx$.

Solution Let

$$u = xe^x \quad \text{and} \quad dv = \frac{1}{(x+1)^2} \, dx$$

Then

$$du = (xe^x + e^x) \, dx = e^x(x+1) \, dx \quad \text{and} \quad v = -\frac{1}{x+1}$$

Therefore, $\int \dfrac{xe^x}{(x+1)^2} \, dx = \int u \, dv = uv - \int v \, du$

$$= xe^x \left(\frac{-1}{x+1}\right) - \int \left(-\frac{1}{x+1}\right) e^x(x+1) \, dx$$

$$= -\frac{xe^x}{x+1} + \int e^x \, dx$$

$$= -\frac{xe^x}{x+1} + e^x + C$$

$$= \frac{e^x}{x+1} + C$$

○ ○ ○

The next example shows that repeated applications of the technique of integration by parts is sometimes required to evaluate an integral.

EXAMPLE 4 Evaluate $\int x^2 e^x \, dx$.

Solution Let

$$u = x^2 \qquad \text{and} \quad dv = e^x \, dx$$

so that

$$du = 2x \, dx \quad \text{and} \qquad v = e^x$$

Therefore,

$$\int x^2 e^x \, dx = \int u \, dv = uv - \int v \, du$$

$$= x^2 e^x - \int e^x(2x) \, dx$$

$$= x^2 e^x - 2 \int xe^x \, dx$$

To complete the solution of the problem, we need to evaluate the integral

$$\int xe^x \, dx$$

But this integral may be found using integration by parts. In fact, you will recognize that this integral is precisely that of Example 1. Using the results obtained there, we now find

$$\int x^2 e^x \, dx = x^2 e^x - 2[(x-1)e^x] + C = e^x(x^2 - 2x + 2) + C \qquad \circ \; \circ \; \circ$$

---

a. Use the method of integration by parts to derive the formula

$$\int x^n e^{ax} \, dx = \frac{1}{a} x^n e^{ax} - \frac{n}{a} \int x^{n-1} e^{ax} \, dx$$

where n is a positive integer and a is a real number.

b. Use the formula of part (a) to evaluate

$$\int x^3 e^x \, dx$$

[*Hint:* You may find the results of Example 4 helpful.]

Application

EXAMPLE 5   The estimated rate at which oil will be produced from a certain oil well t years after production has begun is given by

$$R(t) = 100te^{-0.1t}$$

thousand barrels per year. Find an expression that describes the total production of oil at the end of year t.

Solution   Let $T(t)$ denote the total production of oil from the well at the end of year t ($t \geq 0$). Then the rate of oil production will be given by $T'(t)$ thousand barrels per year. Thus,

$$T'(t) = R(t) = 100te^{-0.1t}$$

so

$$T(t) = \int 100te^{-0.1t} \, dt$$

$$= 100 \int te^{-0.1t} \, dt$$

We use the technique of integration by parts to evaluate this integral. Let

$$u = t \quad \text{and} \quad dv = e^{-0.1t} \, dt$$

so that

$$du = dt \quad \text{and} \quad v = -\frac{1}{0.1} e^{-0.1t} = -10e^{-0.1t}$$

Therefore,
$$T(t) = 100\left[-10te^{-0.1t} + 10\int e^{-0.1t}\,dt\right]$$
$$= 100[-10te^{-0.1t} - 100e^{-0.1t}] + C$$
$$= -1000e^{-0.1t}(t + 10) + C$$

To determine the value of C, note that the total quantity of oil produced at the end of year 0 is nil, so $T(0) = 0$. This gives

$$T(0) = -1000(10) + C = 0$$

or
$$C = 10,000$$

Thus, the required production function is given by

$$T(t) = -1000e^{-0.1t}(t + 10) + 10,000$$

○ ○ ○

EXPLORING WITH TECHNOLOGY

Refer to Example 5.
a. Use a graphing utility to plot the graph of

$$T(t) = -1000e^{-0.1t}(t + 10) + 10,000$$

in the viewing rectangle $[0, 100] \times [0, 12{,}000]$. Use **TRACE** to see what happens when t is very large.
b. Verify the result of part (a) by evaluating $\lim_{t\to\infty} T(t)$. Interpret your results.

○ ○ ○

SELF-CHECK EXERCISES 7.1

1. Evaluate $\int x^2 \ln x \, dx$.

2. Since the inauguration of Ryan's Express at the beginning of 1998, the number of passengers (in millions) flying on this commuter airline has been growing at the rate of

$$R(t) = 0.1 + 0.2te^{-0.4t}$$

passengers per year ($t = 0$ corresponds to the beginning of 1998). Assuming that this trend continues through 2002, determine how many passengers will have flown on Ryan's Express by that time.

Solutions to Self-Check Exercises 7.1 can be found on page 540.

7.1 | EXERCISES

In exercises 1–26, evaluate each indefinite integral.

1. $\int xe^{2x}\,dx$

2. $\int xe^{-x}\,dx$

3. $\int xe^{x/4}\,dx$

4. $\int 6xe^{3x}\,dx$

5. $\int (e^x - x)^2\,dx$

6. $\int (e^{-x} + x)^2\,dx$

7. $\int (x + 1)e^x\,dx$

8. $\int (x - 3)e^{3x}\,dx$

9. $\int x(x + 1)^{-3/2}\,dx$

10. $\int x(x + 4)^{-2}\,dx$

11. $\int x\sqrt{x - 5}\,dx$

12. $\int \dfrac{x}{\sqrt{2x + 3}}\,dx$

13. $\int x \ln 2x\,dx$

14. $\int x^2 \ln 2x\,dx$

15. $\int x^3 \ln x\,dx$

16. $\int \sqrt{x} \ln x\,dx$

17. $\int \sqrt{x} \ln \sqrt{x}\,dx$

18. $\int \dfrac{\ln x}{\sqrt{x}}\,dx$

19. $\int \dfrac{\ln x}{x^2}\,dx$

20. $\int \dfrac{\ln x}{x^3}\,dx$

21. $\int \ln x\,dx$

[*Hint:* Let $u = \ln x$ and $dv = dx$.]

22. $\int \ln(x + 1)\,dx$

23. $\int x^2 e^{-x}\,dx$

[*Hint:* Integrate by parts twice.]

24. $\int e^{-\sqrt{x}}\,dx$

[*Hint:* First make the substitution $u = \sqrt{x}$; then integrate by parts.]

25. $\int x(\ln x)^2\,dx$

[*Hint:* Integrate by parts twice.]

26. $\int x \ln(x + 1)\,dx$

[*Hint:* First make the substitution $u = x + 1$; then integrate by parts.]

In exercises 27–32, evaluate each definite integral by using the method of integration by parts.

27. $\int_0^{\ln 2} xe^x\,dx$

28. $\int_0^2 xe^{-x}\,dx$

29. $\int_1^4 \ln x\,dx$

30. $\int_1^2 x \ln x\,dx$

31. $\int_0^2 xe^{2x}\,dx$

32. $\int_0^1 x^2 e^{-x}\,dx$

33. Find the function f given that the slope of the tangent line to the graph of f at any point $(x, f(x))$ is xe^{-2x} and that the graph passes through the point $(0, 3)$.

34. Find the function f given that the slope of the tangent line to the graph of f at any point $(x, f(x))$ is $x\sqrt{x + 1}$ and that the graph passes through the point $(3, 6)$.

35. Find the area of the region under the graph of $f(x) = \ln x$ from $x = 1$ to $x = 5$.

36. Find the area of the region under the graph of $f(x) = xe^{-x}$ from $x = 0$ to $x = 3$.

37. Velocity of a Dragster   The velocity of a dragster t seconds after leaving the starting line is

$$100te^{-0.2t}$$

feet per second. What is the distance covered by the dragster in the first 10 seconds of its run?

38. Production of Steam Coal   In keeping with the projected increase in worldwide demand for steam coal, the boiler-firing fuel used for generating electricity, the management of Consolidated Mining has decided to step up its mining operations. Plans call for increasing the yearly production of steam coal by

$$2te^{-0.05t}$$

million metric tons per year for the next 20 years. The current yearly production is 20 million metric tons. Find a function that describes Consolidated's total production of steam coal at the end of t years. How much coal will Consolidated have produced over the next 20 years if this plan is carried out?

39. Concentration of a Drug in the Bloodstream   The concentration of a certain drug in a patient's bloodstream t hours after it has been administered is given by $C(t) = 3te^{-t/3}$ mg/ml. Find the average concentration of the drug in the patient's bloodstream over the first 12 hours after administration.

40. Alcohol-Related Traffic Accidents As a result of increasingly stiff laws aimed at reducing the number of alcohol-related traffic accidents in a certain state, preliminary data indicate that the number of such accidents has been changing at the rate of

$$R(t) = -10 - te^{0.1t}$$

accidents per month t months after the laws took effect. There were 982 alcohol-related accidents in the year before the enactment of the laws. Determine how many alcohol-related accidents were expected during the first year the laws were in effect.

41. Record Sales Sales of the latest recording by Brittania, a British rock group, are currently $2te^{-0.1t}$ units per week (each unit representing 10,000 albums), where t denotes the number of weeks since the recording's release. Find an expression that gives the total number of units of record albums sold as a function of t.

42. Rate of Return on an Investment Suppose that an investment is expected to generate income at the rate of

$$P(t) = 30,000 + 800t$$

dollars per year for the next 5 years. Find the present value of this investment if the prevailing interest rate is 8% per year compounded continuously.

[*Hint:* Use formula (18), Section 6.7 (page 510).]

43. Present Value of a Franchise Ms. Hart purchased a 15-year franchise for a computer outlet store that is expected to generate income at the rate of

$$P(t) = 50,000 + 3000t$$

dollars per year. If the prevailing interest rate is 10% per year compounded continuously, find the present value of the franchise.

[*Hint:* Use formula (18), Section 6.7 (page 510).]

44. Growth of HMOs The membership of the Cambridge Community Health Plan, Inc. (a health maintenance organization), is projected to grow at the rate of $9\sqrt{t+1} \ln \sqrt{t+1}$ thousand people per year, t years from now. If the HMO's current membership is 50,000, what will the membership be 5 years from now?

SOLUTIONS TO SELF-CHECK EXERCISES 7.1

1. Let $u = \ln x$ and $dv = x^2 \, dx$, so that $du = \dfrac{1}{x} \, dx$ and $v = \dfrac{1}{3}x^3$. Then

$$\int x^2 \ln x \, dx = \int u \, dv = uv - \int v \, du$$

$$= \frac{1}{3}x^3 \ln x - \int \frac{1}{3}x^2 \, dx$$

$$= \frac{1}{3}x^3 \ln x - \frac{1}{9}x^3 + C$$

$$= \frac{1}{9}x^3 (3 \ln x - 1) + C$$

2. Let $N(t)$ denote the total number of passengers who will have flown on Ryan's Express by the end of year t. Then $N'(t) = R(t)$, so that

$$N(t) = \int R(t) \, dt$$

$$= \int (0.1 + 0.2te^{-0.4t}) \, dt$$

$$= \int 0.1 \, dt + 0.2 \int te^{-0.4t} \, dt$$

We now use the technique of integration by parts on the second integral. Letting $u = t$ and $dv = e^{-0.4t}\, dt$, we have

$$du = dt \quad \text{and} \quad v = -\frac{1}{0.4}e^{-0.4t} = -2.5e^{-0.4t}$$

Therefore,

$$N(t) = 0.1t + 0.2\left[-2.5te^{-0.4t} + 2.5\int e^{-0.4t}\, dt \right]$$

$$= 0.1t - 0.5te^{-0.4t} - \frac{0.5}{0.4}e^{-0.4t} + C$$

$$= 0.1t - 0.5(t + 2.5)e^{-0.4t} + C$$

To determine the value of C, note that $N(0) = 0$, which gives

$$N(0) = -0.5(2.5) + C = 0$$

or

$$C = 1.25$$

Therefore,

$$N(t) = 0.1t - 0.5(t + 2.5)e^{-0.4t} + 1.25$$

The number of passengers who will have flown on Ryan's Express by the end of 2002 is given by

$$N(5) = 0.1(5) - 0.5(5 + 2.5)e^{-0.4(5)} + 1.25$$

$$= 1.242493$$

that is, 1,242,493 passengers.

7.2 INTEGRATION USING TABLES OF INTEGRALS

A Table of Integrals

We have studied several techniques for finding an antiderivative of a function. However, useful as they are, these techniques are not always applicable. There are, of course, numerous other methods for finding an antiderivative of a function. Extensive lists of integration formulas have been compiled based on these methods.

   A small sample of the integration formulas that can be found in many mathematical handbooks is given in the following table of integrals. The formulas are grouped according to the basic form of the integrand. Note that it may be necessary to modify the integrand of the integral to be evaluated in order to use one of these formulas.

Table of integrals

Forms Involving $a + bu$

1. $\displaystyle\int \frac{u\,du}{a + bu} = \frac{1}{b^2}\left[a + bu - a\ln|a + bu|\right] + C$

2. $\displaystyle\int \frac{u^2\,du}{a + bu} = \frac{1}{2b^3}\left[(a + bu)^2 - 4a(a + bu) + 2a^2\ln|a + bu|\right] + C$

3. $\displaystyle\int \frac{u\,du}{(a + bu)^2} = \frac{1}{b^2}\left[\frac{a}{a + bu} + \ln|a + bu|\right] + C$

4. $\displaystyle\int u\sqrt{a + bu}\,du = \frac{2}{15b^2}(3bu - 2a)(a + bu)^{3/2} + C$

5. $\displaystyle\int \frac{u\,du}{\sqrt{a + bu}} = \frac{2}{3b^2}(bu - 2a)\sqrt{a + bu} + C$

6. $\displaystyle\int \frac{du}{u\sqrt{a + bu}} = \frac{1}{\sqrt{a}}\ln\left|\frac{\sqrt{a + bu} - \sqrt{a}}{\sqrt{a + bu} + \sqrt{a}}\right| + C \qquad \text{if } a > 0$

Forms Involving $\sqrt{a^2 + u^2}$

7. $\displaystyle\int \sqrt{a^2 + u^2}\,du = \frac{u}{2}\sqrt{a^2 + u^2} + \frac{a^2}{2}\ln\left|u + \sqrt{a^2 + u^2}\right| + C$

8. $\displaystyle\int u^2\sqrt{a^2 + u^2}\,du = \frac{u}{8}(a^2 + 2u^2)\sqrt{a^2 + u^2} - \frac{a^4}{8}\ln\left|u + \sqrt{a^2 + u^2}\right| + C$

9. $\displaystyle\int \frac{du}{\sqrt{a^2 + u^2}} = \ln\left|u + \sqrt{a^2 + u^2}\right| + C$

10. $\displaystyle\int \frac{du}{u\sqrt{a^2 + u^2}} = -\frac{1}{a}\ln\left|\frac{\sqrt{a^2 + u^2} + a}{u}\right| + C$

11. $\displaystyle\int \frac{du}{u^2\sqrt{a^2 + u^2}} = -\frac{\sqrt{a^2 + u^2}}{a^2 u} + C$

12. $\displaystyle\int \frac{du}{(a^2 + u^2)^{3/2}} = \frac{u}{a^2\sqrt{a^2 + u^2}} + C$

Forms Involving $\sqrt{u^2 - a^2}$

13. $\displaystyle\int \sqrt{u^2 - a^2}\,du = \frac{u}{2}\sqrt{u^2 - a^2} - \frac{a^2}{2}\ln\left|u + \sqrt{u^2 - a^2}\right| + C$

14. $\displaystyle\int u^2\sqrt{u^2 - a^2}\,du = \frac{u}{8}(2u^2 - a^2)\sqrt{u^2 - a^2} - \frac{a^4}{8}\ln\left|u + \sqrt{u^2 - a^2}\right| + C$

15. $\displaystyle\int \frac{\sqrt{u^2 - a^2}}{u^2}\,du = -\frac{\sqrt{u^2 - a^2}}{u} + \ln\left|u + \sqrt{u^2 - a^2}\right| + C$

16. $\displaystyle\int \frac{du}{\sqrt{u^2 - a^2}} = \ln\left|u + \sqrt{u^2 - a^2}\right| + C$

17. $\displaystyle\int \frac{du}{u^2 \sqrt{u^2 - a^2}} = \frac{\sqrt{u^2 - a^2}}{a^2 u} + C$

18. $\displaystyle\int \frac{du}{(u^2 - a^2)^{3/2}} = -\frac{u}{a^2 \sqrt{u^2 - a^2}} + C$

Forms Involving $\sqrt{a^2 - u^2}$

19. $\displaystyle\int \frac{\sqrt{a^2 - u^2}}{u} \, du = \sqrt{a^2 - u^2} - a \ln \left| \frac{a + \sqrt{a^2 - u^2}}{u} \right| + C$

20. $\displaystyle\int \frac{du}{u \sqrt{a^2 - u^2}} = -\frac{1}{a} \ln \left| \frac{a + \sqrt{a^2 - u^2}}{u} \right| + C$

21. $\displaystyle\int \frac{du}{u^2 \sqrt{a^2 - u^2}} = -\frac{\sqrt{a^2 - u^2}}{a^2 u} + C$

22. $\displaystyle\int \frac{du}{(a^2 - u^2)^{3/2}} = \frac{u}{a^2 \sqrt{a^2 - u^2}} + C$

Forms Involving e^{au} and $\ln u$

23. $\displaystyle\int u e^{au} \, du = \frac{1}{a^2} (au - 1) e^{au} + C$

24. $\displaystyle\int u^n e^{au} \, du = \frac{1}{a} u^n e^{au} - \frac{n}{a} \int u^{n-1} e^{au} \, du$

25. $\displaystyle\int \frac{du}{1 + be^{au}} = u - \frac{1}{a} \ln (1 + be^{au}) + C$

26. $\displaystyle\int \ln u \, du = u \ln u - u + C$

27. $\displaystyle\int u^n \ln u \, du = \frac{u^{n+1}}{(n+1)^2} [(n+1) \ln u - 1] + C \qquad (n \neq -1)$

28. $\displaystyle\int \frac{du}{u \ln u} = \ln |\ln u| + C$

29. $\displaystyle\int (\ln u)^n \, du = u(\ln u)^n - n \int (\ln u)^{n-1} \, du$

Using a Table of Integrals

We now consider several examples that illustrate how the table of integrals can be used to evaluate an integral.

EXAMPLE 1 Use the table of integrals to evaluate $\displaystyle\int \frac{2x \, dx}{\sqrt{3 + x}}$.

Solution We first write

$$\int \frac{2x \, dx}{\sqrt{3 + x}} = 2 \int \frac{x \, dx}{\sqrt{3 + x}}$$

Since $\sqrt{3 + x}$ is of the form $\sqrt{a + bu}$, with $a = 3$, $b = 1$, and $u = x$, we use formula 5:

$$\int \frac{u\,du}{\sqrt{a + bu}} = \frac{2}{3b^2}(bu - 2a)\sqrt{a + bu} + C$$

All the formulas given in the table of integrals can be verified by direct computation. Describe a method you would use and apply it to verify a formula of your choice.

obtaining

$$2\int \frac{x}{\sqrt{3 + x}}\,dx = 2\left[\frac{2}{3(1)}(x - 6)\sqrt{3 + x}\right] + C$$

$$= \frac{4}{3}(x - 6)\sqrt{3 + x} + C$$

EXAMPLE 2   Use the table of integrals to evaluate $\int x^2 \sqrt{3 + x^2}\,dx$.

Solution   Observe that if we write 3 as $(\sqrt{3})^2$, then $3 + x^2$ has the form $\sqrt{a^2 + u^2}$, with $a = \sqrt{3}$ and $u = x$. Using formula 8,

$$\int u^2 \sqrt{a^2 + u^2}\,du = \frac{u}{8}(a^2 + 2u^2)\sqrt{a^2 + u^2} - \frac{a^4}{8}\ln|u + \sqrt{a^2 + u^2}| + C$$

we obtain

$$\int x^2 \sqrt{3 + x^2}\,dx = \frac{x}{8}(3 + 2x^2)\sqrt{3 + x^2} - \frac{9}{8}\ln|x + \sqrt{3 + x^2}| + C$$

EXAMPLE 3   Use the table of integrals to evaluate

$$\int_3^4 \frac{dx}{x^2\sqrt{50 - 2x^2}}$$

Solution   We first evaluate the indefinite integral

$$I = \int \frac{dx}{x^2\sqrt{50 - 2x^2}}$$

Observe that $\sqrt{50 - 2x^2} = \sqrt{2(25 - x^2)} = \sqrt{2}\sqrt{25 - x^2}$, so we can write

$$I = \frac{1}{\sqrt{2}}\int \frac{dx}{x^2\sqrt{25 - x^2}} = \frac{\sqrt{2}}{2}\int \frac{dx}{x^2\sqrt{25 - x^2}}\qquad\left(\frac{1}{\sqrt{2}} = \frac{\sqrt{2}}{2}\right)$$

Next, using formula 21,

$$\int \frac{du}{u^2\sqrt{a^2 - u^2}} = -\frac{\sqrt{a^2 - u^2}}{a^2 u} + C$$

with $a = 5$ and $u = x$, we find

$$I = \frac{\sqrt{2}}{2}\left[-\frac{\sqrt{25 - x^2}}{25x} \right]$$

$$= -\left(\frac{\sqrt{2}}{50}\right)\frac{\sqrt{25 - x^2}}{x}$$

Finally, using this result, we obtain

$$\int_3^4 \frac{dx}{x^2 \sqrt{50 - 2x^2}} = -\frac{\sqrt{2}}{50}\frac{\sqrt{25 - x^2}}{x}\bigg|_3^4$$

$$= -\frac{\sqrt{2}}{50}\frac{\sqrt{25 - 16}}{4} - \left(-\frac{\sqrt{2}}{50}\frac{\sqrt{25 - 9}}{3} \right)$$

$$= -\frac{3\sqrt{2}}{200} + \frac{2\sqrt{2}}{75} = \frac{7\sqrt{2}}{600}$$

◗ ◗ ◗

The formulas given in the table of integrals were derived using various techniques, including the method of substitution and the method of integration by parts studied earlier. For example, formula 1,

$$\int \frac{u \, du}{a + bu} = \frac{1}{b^2}[a + bu - a \ln |a + bu|] + C$$

can be derived using the method of substitution. Show how this is done.

As illustrated in the next example, we may need to apply a formula more than once in order to evaluate an integral.

EXAMPLE 4   Use the table of integrals to evaluate $\int x^2 e^{(-1/2)x} \, dx$.

Solution   Scanning the table of integrals for a formula involving e^{ax} in the integrand, we are led to formula 24,

$$\int u^n e^{au} \, du = \frac{1}{a}u^n e^{au} - \frac{n}{a}\int u^{n-1}e^{au} \, du$$

With $n = 2$, $a = -\frac{1}{2}$, and $u = x$, we have

$$\int x^2 e^{(-1/2)x} \, dx = \left(\frac{1}{-1/2}\right)x^2 e^{(-1/2)x} - \frac{2}{(-1/2)}\int xe^{(-1/2)x} \, dx$$

$$= -2x^2 e^{(-1/2)x} + 4\int xe^{(-1/2)x} \, dx$$

If we use formula 24 once again, with $n = 1$, $a = -\frac{1}{2}$, and $u = x$, to evaluate the integral on the right, we obtain

$$\int x^2 e^{(-1/2)x}\, dx = -2x^2 e^{(-1/2)x} + 4\left[\left(\frac{1}{-1/2}\right) xe^{(-1/2)x} - \frac{1}{(-1/2)}\int e^{(-1/2)x}\, dx\right]$$

$$= -2x^2 e^{(-1/2)x} + 4\left[-2xe^{(-1/2)x} + 2 \cdot \frac{1}{(-1/2)} e^{(-1/2)x}\right] + C$$

$$= -e^{(-1/2)x}(2x^2 + 8x + 16) + C \qquad \text{o o o}$$

 EXAMPLE 5   A study prepared for the National Association of Realtors estimated that the mortgage rate over the next t months will be

$$r(t) = \frac{8t + 100}{t + 10} \qquad (0 \le t \le 24)$$

percent per year. If the prediction holds true, what will the average mortgage rate be over the next 12 months?

Solution   The average mortgage rate over the next 12 months will be given by

$$A = \frac{1}{12 - 0} \int_0^{12} \frac{8t + 100}{t + 10}\, dt$$

$$= \frac{1}{12}\left[\int_0^{12} \frac{8t}{t + 10}\, dt + \int_0^{12} \frac{100}{t + 10}\, dt\right]$$

$$= \frac{8}{12} \int_0^{12} \frac{t}{t + 10}\, dt + \frac{100}{12} \int_0^{12} \frac{1}{t + 10}\, dt$$

Using formula 1,

$$\int \frac{u\, du}{a + bu} = \frac{1}{b^2}[a + bu - a \ln |a + bu|] + C \qquad (a = 10, b = 1, u = t)$$

to evaluate the first integral, we have

$$A = \left(\frac{2}{3}\right)[10 + t - 10 \ln (10 + t)]\Big|_0^{12} + \left(\frac{25}{3}\right) \ln (10 + t)\Big|_0^{12}$$

$$= \left(\frac{2}{3}\right)[(22 - 10 \ln 22) - (10 - 10 \ln 10)] + \left(\frac{25}{3}\right)[\ln 22 - \ln 10]$$

$$\approx 9.31$$

or approximately 9.31% per year. $\qquad$ o o o

SELF-CHECK EXERCISES 7.2

1. Use the table of integrals to evaluate

$$\int_0^2 \frac{dx}{(5 - x^2)^{3/2}}$$

 2. During a flu epidemic, the number of children in the Easton Middle School who contracted influenza t days after the outbreak began was given by

$$N(t) = \frac{200}{1 + 9e^{-0.8t}}$$

Determine the average number of children who contracted the flu in the first 10 days of the epidemic.

Solutions to Self-Check Exercises 7.2 can be found on page 549.

7.2 EXERCISES

In exercises 1–32, use the table of integrals in this section to evaluate each of the given integrals.

1. $\displaystyle \int \frac{2x}{2 + 3x}\,dx$

2. $\displaystyle \int \frac{x}{(1 + 2x)^2}\,dx$

3. $\displaystyle \int \frac{3x^2}{2 + 4x}\,dx$

4. $\displaystyle \int \frac{x^2}{3 + x}\,dx$

5. $\displaystyle \int x^2 \sqrt{9 + 4x^2}\,dx$

6. $\displaystyle \int x^2 \sqrt{4 + x^2}\,dx$

7. $\displaystyle \int \frac{dx}{x\sqrt{1 + 4x}}$

8. $\displaystyle \int_0^2 \frac{x + 1}{\sqrt{2 + 3x}}\,dx$

9. $\displaystyle \int_0^2 \frac{dx}{\sqrt{9 + 4x^2}}$

10. $\displaystyle \int \frac{dx}{x\sqrt{4 + 8x^2}}$

11. $\displaystyle \int \frac{dx}{(9 - x^2)^{3/2}}$

12. $\displaystyle \int \frac{dx}{(2 - x^2)^{3/2}}$

13. $\displaystyle \int x^2 \sqrt{x^2 - 4}\,dx$

14. $\displaystyle \int_3^5 \frac{dx}{x^2 \sqrt{x^2 - 9}}$

15. $\displaystyle \int \frac{\sqrt{4 - x^2}}{x}\,dx$

16. $\displaystyle \int_0^1 \frac{dx}{(4 - x^2)^{3/2}}$

17. $\displaystyle \int xe^{2x}\,dx$

18. $\displaystyle \int \frac{dx}{1 + e^{-x}}$

19. $\displaystyle \int \frac{dx}{(x + 1)\ln(1 + x)}$

[*Hint:* First use the substitution $u = x + 1$.]

20. $\displaystyle \int \frac{x}{(x^2 + 1)\ln(x^2 + 1)}\,dx$

[*Hint:* First use the substitution $u = x^2 + 1$.]

21. $\displaystyle \int \frac{e^{2x}}{(1 + 3e^x)^2}\,dx$

22. $\displaystyle \int \frac{e^{2x}}{\sqrt{1 + 3e^x}}\,dx$

23. $\displaystyle \int \frac{3e^x}{1 + e^{(1/2)x}}\,dx$

24. $\displaystyle \int \frac{dx}{1 - 2e^{-x}}$

25. $\displaystyle \int \frac{\ln x}{x(2 + 3\ln x)}\,dx$

26. $\displaystyle \int_1^e (\ln x)^2\,dx$

27. $\displaystyle \int_0^1 x^2 e^x\,dx$

28. $\displaystyle \int x^3 e^{2x}\,dx$

29. $\displaystyle \int x^2 \ln x\,dx$

30. $\displaystyle \int x^3 \ln x\,dx$

31. $\displaystyle \int (\ln x)^3\,dx$

32. $\displaystyle \int (\ln x)^4\,dx$

33. Consumers' Surplus Refer to Section 6.7. The demand function for Apex ladies' boots is

$$p = \frac{250}{\sqrt{16 + x^2}}$$

where p is the wholesale unit price in dollars and x is the quantity demanded daily, in units of a hundred. Find the consumers' surplus if the wholesale price is set at $50 per pair.

34. Producers' Surplus   Refer to Section 6.7. The supplier of Apex ladies' boots will make x hundred pairs of the boots available in the market daily when the wholesale unit price is

$$p = \frac{30x}{5 - x}$$

dollars. Find the producers' surplus if the wholesale price is set at $50 per pair.

35. Amusement Park Attendance   The management of Astro World ("The Amusement Park of the Future") estimates that the number of visitors (in thousands) entering the amusement park t hours after opening time at 9 A.M. is given by

$$R(t) = \frac{60}{(2 + t^2)^{3/2}}$$

per hour. Determine the number of visitors admitted by noon.

36. Voter Registration   The number of voters in a certain district of a city is expected to grow at the rate of

$$R(t) = \frac{3000}{\sqrt{4 + t^2}}$$

people per year t years from now. If the number of voters at present is 20,000, find how many voters will be in the district 5 years from now.

37. Growth of Fruit Flies   Based on data collected during an experiment, a biologist found that the number of fruit flies (*Drosophila*) with a limited food supply could be approximated by the exponential model

$$N(t) = \frac{1000}{1 + 24e^{-0.02t}}$$

where t denotes the number of days since the beginning of the experiment. Find the average number of fruit flies in the colony in the first 10 days of the experiment; in the first 20 days.

38. VCR Ownership   According to estimates by Paul Kroger Associates, the percentage of households that own videocassette recorders (VCRs) is given by

$$P(t) = \frac{68}{1 + 21.67e^{-0.62t}} \qquad (0 \le t \le 12)$$

where t is measured in years, with $t = 0$ corresponding to the beginning of 1981. Find the average percentage of households owning VCRs from the beginning of 1981 to the beginning of 1993.
Source: Paul Kroger Associates

39. Recycling Programs   The commissioner of the City of Newton Department of Public Works estimates that the number of people in the city who have been recycling their magazines in year t following the introduction of the recycling program at the beginning of 1990 is

$$N(t) = \frac{100,000}{2 + 3e^{-0.2t}}$$

Find the average number of people who have recycled their magazines during the first 5 years since the program was introduced.

40. Franchises   Ms. Lane purchased a 10-year franchise for a fast-food restaurant that is expected to generate income at the rate of $R(t) = 250,000 + 2000t^2$ dollars per year, t years from now. If the prevailing interest rate is 10% per year compounded continuously, find the present value of the franchise.

[*Hint:* Use formula (18), Section 6.7.]

41. Accumulated Value of an Income Stream   The revenue of Virtual Reality, a video-game arcade, is generated at the rate of $R(t) = 20,000t$ dollars. If the revenue is invested t years from now in a business earning interest at the rate of 15% per year compounded continuously, find the accumulated value of this stream of income at the end of 5 years.

[*Hint:* Use formula (18), Section 6.7.]

42. Lorentz Curves   In a study conducted by a certain country's Economic Development Board regarding the income distribution of certain segments of the country's work force, it was found that the Lorentz curve for the distribution of income of college professors is described by the function

$$g(x) = \frac{1}{3} x \sqrt{1 + 8x}$$

Compute the coefficient of inequality of the Lorentz curve.

[*Hint:* Use formula (21), Section 6.7.]

SOLUTIONS TO SELF-CHECK EXERCISES 7.2

1. Using formula 22, page 543, with $a^2 = 5$, we see that

$$\int_0^2 \frac{dx}{(5 - x^2)^{3/2}} = \frac{x}{5\sqrt{5 - x^2}}\bigg|_0^2$$

$$= \frac{2}{5\sqrt{5 - 4}}$$

$$= \frac{2}{5}$$

2. The average number of children who contracted the flu in the first 10 days of the epidemic is given by

$$A = \frac{1}{10}\int_0^{10} \frac{200}{1 + 9e^{-0.8t}}\, dt = 20\int_0^{10} \frac{dt}{1 + 9e^{-0.8t}}$$

$$= 20\left[t + \frac{1}{0.8}\ln(1 + 9e^{-0.8t})\right]\bigg|_0^{10} \qquad \text{(Formula 25, } a = -0.8,$$
$$\qquad\qquad\qquad b = 9, u = t)$$

$$= 20\left[10 + \frac{1}{0.8}\ln(1 + 9e^{-8})\right] - 20\left(\frac{1}{0.8}\right)\ln 10$$

$$\approx 200.07537 - 57.56463$$

$$\approx 143$$

or 143 students.

7.3 NUMERICAL INTEGRATION

Approximating Definite Integrals

One method of measuring cardiac output is to inject 5 to 10 mg of a dye into a vein leading to the heart. After making its way through the lungs, the dye returns to the heart and is pumped into the aorta, where its concentration is measured at equal time intervals. The graph of the function c in Figure 7.1 shows the concentration of dye in a person's aorta, measured at 2-second intervals after 5 mg of dye have been injected. The person's cardiac output (measured in l/min) is computed using the formula

$$R = \frac{60D}{\int_0^{28} c(t)\, dt} \tag{4}$$

where D is the quantity of dye injected (see exercise 40, page 565).

Now, in order to use formula (4), we need to evaluate the definite integral

$$\int_0^{28} c(t)\, dt$$

Figure 7.1
The function c gives the concentra-
tion of a dye measured at the aorta.
The graph is constructed by drawing
a smooth curve through a set of dis-
crete points.

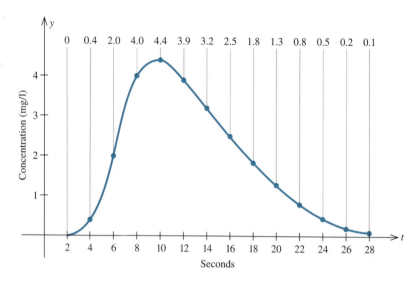

But we do not have the algebraic rule defining the integrand c for all values of t in $[0, 28]$. In fact, we are given its values only at a set of discrete points in that interval. In situations such as this, the Fundamental Theorem of Calculus proves useless because we cannot find an antiderivative of c. (We will complete the solution to this problem in Example 4.)

Other situations also arise in which an integrable function has an antiderivative that cannot be found in terms of elementary functions (functions that can be expressed as a finite combination of algebraic, exponential, logarithmic, and trigonometric functions). Examples of such functions are

$$f(x) = e^{x^2}, \quad g(x) = x^{-1/2}e^x, \quad \text{and} \quad h(x) = \frac{1}{\ln x}$$

Riemann sums provide us with a good approximation of a definite integral, provided the number of subintervals in the partitions is large enough. But there are better techniques and formulas, called *quadrature formulas,* that give a more efficient way of computing approximate values of definite integrals. In this section, we look at two rather simple but effective ways of approximating definite integrals.

The Trapezoidal Rule

We assume that $f(x) \geq 0$ on $[a, b]$ in order to simplify the derivation of the Trapezoidal Rule, but the result is valid without this restriction. We begin by subdividing the interval $[a, b]$ into n subintervals of equal length Δx, by means of the $(n + 1)$ points $x_0 = a, x_1, x_2, \ldots, x_n = b$, where n is a positive integer (Figure 7.2).

Then the length of each subinterval is given by

$$\Delta x = \frac{b - a}{n}$$

Figure 7.2
The area under the curve is equal to the sum of the n subregions R_1, R_2, . . . , R_n.

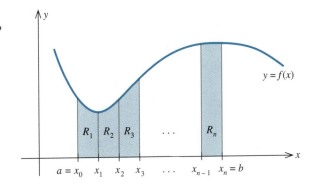

Furthermore, as we saw earlier, we may view the definite integral

$$\int_a^b f(x)\, dx$$

as the area of the region R under the curve $y = f(x)$ between $x = a$ and $x = b$. This area is given by the sum of the areas of the n nonoverlapping subregions R_1, R_2, . . . , R_n, such that R_1 represents the region under the curve $y = f(x)$ from $x = x_0$ to $x = x_1$, and so on.

Figure 7.3
The area of R_1 is approximated by the area of the trapezoid.

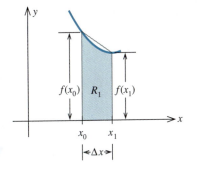

The basis for the Trapezoidal Rule lies in the approximation of each of the regions R_1, R_2, . . . , R_n by a suitable trapezoid. This often leads to a much better approximation than one obtained by means of rectangles (a Riemann sum).

Let us consider the subregion R_1, shown magnified for the sake of clarity in Figure 7.3. Observe that the area of the region R_1 may be approximated by the trapezoid of width Δx whose parallel sides are of lengths $f(x_0)$ and $f(x_1)$. The area of the trapezoid is given by

$$\left[\frac{f(x_0) + f(x_1)}{2} \right] \Delta x \qquad \text{(Average of the lengths of the parallel sides times the width)}$$

square units. Similarly, the area of the region R_2 may be approximated by the trapezoid of width Δx and sides of lengths $f(x_1)$ and $f(x_2)$. The area of the trapezoid is given by

$$\left[\frac{f(x_1) + f(x_2)}{2} \right] \Delta x$$

Similarly, we see that the area of the last (nth) approximating trapezoid is given by

$$\left[\frac{f(x_{n-1}) + f(x_n)}{2} \right] \Delta x$$

Then the area of the region R is approximated by the sum of the areas of the n trapezoids, that is,

$$\left[\frac{f(x_0) + f(x_1)}{2}\right] \Delta x + \left[\frac{f(x_1) + f(x_2)}{2}\right] \Delta x + \cdots + \left[\frac{f(x_{n-1}) + f(x_n)}{2}\right] \Delta x$$

$$= \frac{\Delta x}{2} [f(x_0) + f(x_1) + f(x_1) + f(x_2) + \cdots + f(x_{n-1}) + f(x_n)]$$

$$= \frac{\Delta x}{2} [f(x_0) + 2f(x_1) + 2f(x_2) + \cdots + 2f(x_{n-1}) + f(x_n)]$$

Since the area of the region R is given by the value of the definite integral we wished to approximate, we are led to the following approximation formula, which is called the **Trapezoidal Rule.**

TRAPEZOIDAL RULE

$$\int_a^b f(x)\, dx \approx \frac{\Delta x}{2} [f(x_0) + 2f(x_1) + 2f(x_2) \\ + \cdots + 2f(x_{n-1}) + f(x_n)]$$

(5)

where $\Delta x = \dfrac{b - a}{n}$.

The approximation generally improves with larger values of n.

 EXAMPLE 1   Approximate the value of

$$\int_1^2 \frac{1}{x}\, dx$$

using the Trapezoidal Rule with $n = 10$. Compare this result with the exact value of the integral.

Solution   Here $a = 1$, $b = 2$, and $n = 10$, so

$$\Delta x = \frac{b - a}{n} = \frac{1}{10} = 0.1$$

and

$$x_0 = 1, \quad x_1 = 1.1, \quad x_2 = 1.2, \quad x_3 = 1.3, \quad \ldots, \quad x_9 = 1.9, \quad x_{10} = 2$$

The Trapezoidal Rule yields

$$\int_1^2 \frac{1}{x}\, dx \approx \frac{0.1}{2}\left[1 + 2\left(\frac{1}{1.1}\right) + 2\left(\frac{1}{1.2}\right) + 2\left(\frac{1}{1.3}\right) + \cdots + 2\left(\frac{1}{1.9}\right) + \frac{1}{2}\right]$$

$$\approx 0.693771$$

In this case, we can easily compute the actual value of the definite integral under consideration. In fact,

$$\int_1^2 \frac{1}{x}\,dx = \ln x \Big|_1^2 = \ln 2 - \ln 1 = \ln 2$$

$$\approx 0.693147$$

Thus, the Trapezoidal Rule with $n = 10$ yields a result with an error of 0.000624, to six decimal places.   ◦ ◦ ◦

 EXAMPLE 2   The demand function for a certain brand of perfume is given by

$$p = D(x) = \sqrt{10{,}000 - 0.01x^2}$$

where p is the unit price in dollars and x is the quantity demanded each week, measured in ounces. Find the consumers' surplus if the market price is set at $60 per ounce.

Solution   When $p = 60$, we have

$$\sqrt{10{,}000 - 0.01x^2} = 60$$
$$10{,}000 - 0.01x^2 = 3600$$
$$x^2 = 640{,}000$$

or $x = 800$ since x must be nonnegative. Next, using the consumers' surplus formula (page 505) with $\bar{p} = 60$ and $\bar{x} = 800$, we see that the consumers' surplus is given by

$$CS = \int_0^{800} \sqrt{10{,}000 - 0.01x^2}\,dx - (60)(800)$$

It is not easy to evaluate this definite integral by finding an antiderivative of the integrand. Instead, let us use the Trapezoidal Rule with $n = 10$.
With $a = 0$ and $b = 800$, we find that

$$\Delta x = \frac{b - a}{n} = \frac{800}{10} = 80$$

and

$$x_0 = 0, \quad x_1 = 80, \quad x_2 = 160, \quad x_3 = 240, \quad \ldots, \quad x_9 = 720, \quad x_{10} = 800$$

so

$$\int_0^{800} \sqrt{10{,}000 - 0.01x^2}\,dx$$

$$\approx \frac{80}{2}[100 + 2\sqrt{10{,}000 - (0.01)(80)^2}$$

$$+ 2\sqrt{10{,}000 - (0.01)(160)^2} + \cdots + 2\sqrt{10{,}000 - (0.01)(720)^2}$$

$$+ \sqrt{10{,}000 - (0.01)(800)^2}]$$

$$= 40[100 + 199.3590 + 197.4234 + 194.1546 + 189.4835$$
$$+ 183.3030 + 175.4537 + 165.6985$$
$$+ 153.6750 + 138.7948 + 60]$$
$$\approx 70{,}293.82$$

Therefore, the consumers' surplus is approximately $70{,}294 - 48{,}000$, or $22,294.

Explain how you would approximate the value of $\int_0^2 f(x)\, dx$ using the Trapezoidal Rule with $n = 10$, where

$$f(x) = \begin{cases} \sqrt{1 + x^2} & \text{if } 0 \le x \le 1 \\[2mm] \dfrac{2}{\sqrt{1 + x^2}} & \text{if } 1 < x \le 2 \end{cases}$$

and find the value.

Simpson's Rule

Before stating Simpson's Rule, let us review the two rules we have used in approximating a definite integral. Let f be a continuous nonnegative function defined on the interval $[a, b]$. Suppose that the interval $[a, b]$ is partitioned by means of the $n + 1$ equally spaced points $x_0 = a, x_1, x_2, \ldots, x_n = b$, where n is a positive integer, so that the length of each subinterval is $\Delta x = (b - a)/n$ (Figure 7.4).

Figure 7.4
The area under the curve is equal to the sum of the n subregions $R_1, R_2, \ldots, R_n$.

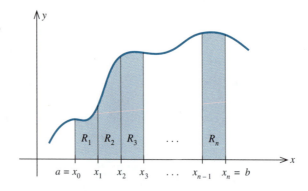

Let us concentrate on the portion of the graph of $y = f(x)$ defined on the interval $[x_0, x_2]$. In using a Riemann sum to approximate the definite integral, we are, in effect, approximating the function $f(x)$ on $[x_0, x_1]$ by the *constant* function $y = f(p_1)$, where p_1 is chosen to be a point in $[x_0, x_1]$; the function $f(x)$ on $[x_1, x_2]$ by the constant function $y = f(p_2)$, where p_2 lies in $[x_1, x_2]$; and so on. Using a Riemann sum, we see that the area of the region under the curve $y = f(x)$ between $x = a$ and $x = b$ is approximated by the area under the approximating "step" function (Figure 7.5a).

Figure 7.5

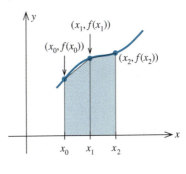

(a) The area under the curve is approximated by the area of the rectangles.

(b) The area under the curve is approximated by the area of the trapezoids.

When we use the Trapezoidal Rule, we are, in effect, approximating the function $f(x)$ on the interval $[x_0, x_1]$ by a *linear* function through the two points $(x_0, f(x_0))$ and $(x_1, f(x_1))$; the function $f(x)$ on $[x_1, x_2]$ by a *linear* function through the two points $(x_1, f(x_1))$ and $(x_2, f(x_2))$; and so on. Thus, the Trapezoidal Rule simply approximates the actual area of the region under the curve $y = f(x)$ from $x = a$ to $x = b$ by the area under the approximating polygonal curve (Figure 7.5b).

A natural extension of the preceding idea is to approximate portions of the graph of $y = f(x)$ by means of portions of the graphs of second-degree polynomials (parts of parabolas). It can be shown that, given any three noncollinear points, there is a unique parabola that passes through the given points. Choose the points $(x_0, f(x_0))$, $(x_1, f(x_1))$, and $(x_2, f(x_2))$ corresponding to the first three points of the partition. Then we can approximate the function $f(x)$ on $[x_0, x_2]$ by means of a quadratic function whose graph contains these three points (Figure 7.6).

Although we will not do so here, it can be shown that the area under the parabola between $x = x_0$ and $x = x_2$ is given by

$$\frac{\Delta x}{3}[f(x_0) + 4f(x_1) + f(x_2)]$$

square units. Repeating this argument on the interval $[x_2, x_4]$, we see that the area under the curve between $x = x_2$ and $x = x_4$ is approximated by the area under the parabola between x_2 and x_4—that is, by

$$\frac{\Delta x}{3}[f(x_2) + 4f(x_3) + f(x_4)]$$

square units. Proceeding, we conclude that if n is even (Why?), then the area under the curve $y = f(x)$ from $x = a$ to $x = b$ may be approximated by the

Figure 7.6

Simpson's Rule approximates the area under the curve by the area under the parabola.

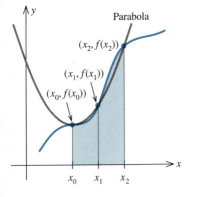

sum of the areas under the $n/2$ approximating parabolas—that is,

$$\frac{\Delta x}{3}[f(x_0) + 4f(x_1) + f(x_2)] + \frac{\Delta x}{3}[f(x_2) + 4f(x_3) + f(x_4)] + \cdots$$

$$+ \frac{\Delta x}{3}[f(x_{n-2}) + 4f(x_{n-1}) + f(x_n)]$$

$$= \frac{\Delta x}{3}[f(x_0) + 4f(x_1) + f(x_2) + f(x_2) + 4f(x_3) + f(x_4) + \cdots$$

$$+ f(x_{n-2}) + 4f(x_{n-1}) + f(x_n)]$$

$$= \frac{\Delta x}{3}[f(x_0) + 4f(x_1) + 2f(x_2) + 4f(x_3)$$

$$+ 2f(x_4) + \cdots + 4f(x_{n-1}) + f(x_n)]$$

The preceding is the derivation of the approximation formula known as **Simpson's Rule.**

SIMPSON'S RULE

$$\int_a^b f(x)\,dx \approx \frac{\Delta x}{3}[f(x_0) + 4f(x_1) + 2f(x_2) + 4f(x_3) + 2f(x_4)$$

$$+ \cdots + 4f(x_{n-1}) + f(x_n)] \tag{6}$$

where $\Delta x = \dfrac{b-a}{n}$ and n is even.

In using this rule, remember that n must be even.

  EXAMPLE 3   Find an approximation of

$$\int_1^2 \frac{1}{x}\,dx$$

using Simpson's Rule with $n = 10$. Compare this result with that of Example 1 and also with the exact value of the integral.

Solution   We have $a = 1$, $b = 2$, $f(x) = 1/x$, and $n = 10$, so

$$\Delta x = \frac{b-a}{n} = \frac{1}{10} = 0.1$$

and

$$x_0 = 1, \quad x_1 = 1.1, \quad x_2 = 1.2, \quad x_3 = 1.3, \quad \ldots, \quad x_9 = 1.9, \quad x_{10} = 2$$

Simpson's Rule yields

$$\int_1^2 \frac{1}{x}\,dx \approx \frac{0.1}{3}\left[f(1) + 4f(1.1) + 2f(1.2) + \cdots + 4f(1.9) + f(2)\right]$$

$$= \frac{0.1}{3}\left[1 + 4\left(\frac{1}{1.1}\right) + 2\left(\frac{1}{1.2}\right) + 4\left(\frac{1}{1.3}\right) + 2\left(\frac{1}{1.4}\right) + 4\left(\frac{1}{1.5}\right)\right.$$

$$\left. + 2\left(\frac{1}{1.6}\right) + 4\left(\frac{1}{1.7}\right) + 2\left(\frac{1}{1.8}\right) + 4\left(\frac{1}{1.9}\right) + \frac{1}{2}\right]$$

$$\approx 0.693150$$

The Trapezoidal Rule with $n = 10$ yielded an approximation of 0.693771, which is 0.000624 off the value of ln $2 \approx 0.693147$ to six decimal places. Simpson's Rule yields an approximation with an error of 0.000003, a definite improvement over the Trapezoidal Rule.   ◐ ◐ ◐

EXAMPLE 4   Solve the problem posed at the beginning of this section. Recall that we wished to find a person's cardiac output by using the formula

$$R = \frac{60D}{\displaystyle\int_0^{28} c(t)\,dt}$$

where D (the quantity of dye injected) is equal to 5 mg and the function c has the graph shown in Figure 7.7. Use Simpson's Rule with $n = 14$ to estimate the value of the integral.

Figure 7.7
The function c gives the concentration of a dye measured at the aorta. The graph is constructed by drawing a smooth curve through a set of discrete points.

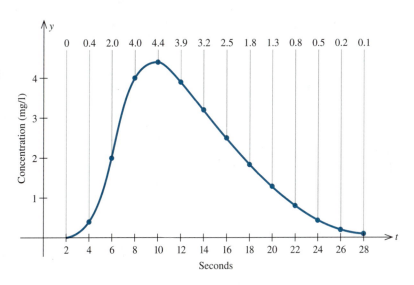

Solution   Using Simpson's Rule with $n = 14$ and $\Delta t = 2$ so that

$$t_0 = 0, \quad t_1 = 2, \quad t_2 = 4, \quad t_3 = 6, \quad \ldots, \quad t_{14} = 28$$

we obtain

$$\int_0^{28} c(t)\,dt \approx \frac{2}{3}[c(0) + 4c(2) + 2c(4) + 4c(6) + \cdots$$

$$+ 4c(26) + c(28)]$$

$$= \frac{2}{3}[0 + 4(0) + 2(0.4) + 4(2.0) + 2(4.0)$$

$$+ 4(4.4) + 2(3.9) + 4(3.2) + 2(2.5) + 4(1.8)$$

$$+ 2(1.3) + 4(0.8) + 2(0.5) + 4(0.2) + 0.1]$$

$$\approx 49.9$$

Therefore, the person's cardiac output is

$$R \approx \frac{60(5)}{49.9} \approx 6.0$$

or 6.0 l/min.   ○ ○ ○

EXAMPLE 5 An oil spill off the coastline was caused by a ruptured tank in a grounded oil tanker. From aerial photographs, the Coast Guard was able to obtain the dimensions of the oil spill (Figure 7.8). Using Simpson's Rule with $n = 10$, estimate the area of the oil spill.

Figure 7.8
Simpson's Rule can be used to calculate the area of the oil spill.

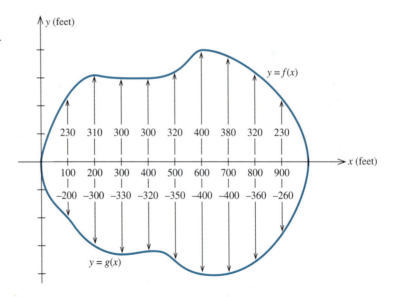

Solution We may think of the area affected by the oil spill as the area of the plane region bounded above by the graph of the function $f(x)$ and below by the graph of the function $g(x)$ between $x = 0$ and $x = 1000$ (Figure 7.8). Then, the required area is given by

$$A = \int_0^{1000} [f(x) - g(x)]\,dx$$

Using Simpson's Rule with $n = 10$ and $\Delta x = 100$ so that

$$x_0 = 0, \quad x_1 = 100, \quad x_2 = 200, \quad \ldots, \quad x_{10} = 1000$$

we have

$$
\begin{aligned}
A &= \int_0^{1000} [f(x) - g(x)]\, dx \\
&\approx \frac{\Delta x}{3} \{[f(x_0) - g(x_0)] + 4[f(x_1) - g(x_1)] + 2[f(x_2) - g(x_2)] \\
&\quad + \cdots + 4[f(x_9) - g(x_9)] + [f(x_{10}) - g(x_{10})]\} \\
&= \frac{100}{3} \{[0 - 0] + 4[230 - (-200)] + 2[310 - (-300)] \\
&\quad + 4[300 - (-330)] + 2[300 - (-320)] + 4[320 - (-350)] \\
&\quad + 2[400 - (-400)] + 4[380 - (-400)] + 2[320 - (-360)] \\
&\quad + 4[230 - (-260)] + [0 - 0]\} \\
&= \frac{100}{3} [0 + 4(430) + 2(610) + 4(630) + 2(620) + 4(670) \\
&\quad + 2(800) + 4(780) + 2(680) + 4(490) + 0] \\
&= \frac{100}{3} (17{,}420) \\
&\approx 580{,}667
\end{aligned}
$$

or approximately 580,667 square feet.   ◦ ◦ ◦

Explain how you would approximate the value of $\int_0^2 f(x)\, dx$ using Simpson's Rule with $n = 10$, where

$$
f(x) = \begin{cases} \sqrt{1 + x^2} & \text{if } 0 \le x \le 1 \\ \dfrac{2}{\sqrt{1 + x^2}} & \text{if } 1 < x \le 2 \end{cases}
$$

and find the value.

Error Analysis

The following results give the bounds on the errors incurred when the Trapezoidal Rule and Simpson's Rule are used to approximate a definite integral (proof omitted).

ERRORS IN THE TRAPEZOIDAL AND SIMPSON APPROXIMATIONS

Suppose that the definite integral

$$\int_a^b f(x)\, dx$$

is approximated with n subintervals.

1. The *maximum* error incurred in using the Trapezoidal Rule is

$$\frac{M(b-a)^3}{12n^2} \tag{7}$$

where M is a number such that $|f''(x)| \le M$ for all x in $[a, b]$.

2. The *maximum* error incurred in using Simpson's Rule is

$$\frac{M(b-a)^5}{180n^4} \tag{8}$$

where M is a number such that $|f^{(4)}(x)| \le M$ for all x in $[a, b]$.

REMARK   In many instances, the actual error is less than the upper error bounds given.   ◦ ◦ ◦

EXAMPLE 6   Find bounds on the errors incurred when

$$\int_1^2 \frac{1}{x}\, dx$$

is approximated using (a) the Trapezoidal Rule and (b) Simpson's Rule with $n = 10$. Compare these with the actual errors found in Examples 1 and 3.

Solution

a. Here $a = 1$, $b = 2$, and $f(x) = 1/x$. Next, to find a value for M, we compute

$$f'(x) = -\frac{1}{x^2} \quad \text{and} \quad f''(x) = \frac{2}{x^3}$$

Since $f''(x)$ is positive and decreasing on $(1, 2)$ (Why?), it attains its maximum value of 2 at $x = 1$, the left endpoint of the interval. Therefore, if we take $M = 2$, then $|f''(x)| \le 2$. Using (7), we see that the maximum error incurred is

$$\frac{2(2-1)^3}{12(10)^2} = \frac{2}{1200} = 0.0016667$$

The actual error found in Example 1, 0.000624, is much less than the upper bound just found.

b. We compute

$$f'''(x) = \frac{-6}{x^4} \quad \text{and} \quad f^{(4)}(x) = \frac{24}{x^5}$$

Since $f^{(4)}(x)$ is positive and decreasing on $(1, 2)$ (just look at $f^{(5)}$ to verify this fact), it attains its maximum at the left endpoint of $[1, 2]$. Now,

$$f^{(4)}(1) = 24$$

and so we may take $M = 24$. Using (8), we obtain the maximum error of

$$\frac{24(2 - 1)^5}{180(10)^4} = 0.0000133$$

The actual error is 0.000003 (see Example 3).   ◦ ◦ ◦

SELF-CHECK EXERCISES 7.3

1. Use the Trapezoidal Rule and Simpson's Rule with $n = 8$ to approximate the value of the definite integral

$$\int_0^2 \frac{1}{\sqrt{1 + x^2}} \, dx$$

2. The graph in the accompanying figure shows the consumption of petroleum in the United States in quadrillion BTU, from 1976 to 1990. Using Simpson's Rule with $n = 14$, estimate the average consumption during the 14-year period.

Source: The World Almanac

Solutions to Self-Check Exercises 7.3 can be found on page 565.

7.3 EXERCISES

A calculator is recommended for this exercise set. In exercises 1–14, use the Trapezoidal Rule and Simpson's Rule to approximate the value of each definite integral. Compare your result with the exact value of the integral.

1. $\int_0^2 x^2\,dx; n = 6$

2. $\int_1^3 (x^2 - 1)\,dx; n = 4$

3. $\int_0^1 x^3\,dx; n = 4$

4. $\int_1^2 x^3\,dx; n = 6$

5. $\int_1^2 \frac{1}{x}\,dx; n = 4$

6. $\int_1^2 \frac{1}{x}\,dx; n = 8$

7. $\int_1^2 \frac{1}{x^2}\,dx; n = 4$

8. $\int_0^1 \frac{1}{1 + x}\,dx; n = 4$

9. $\int_0^4 \sqrt{x}\,dx; n = 8$

10. $\int_0^2 x\sqrt{2x^2 + 1}\,dx; n = 6$

11. $\int_0^1 e^{-x}\,dx; n = 6$

12. $\int_0^1 xe^{-x^2}\,dx; n = 6$

13. $\int_1^2 \ln x\,dx; n = 4$

14. $\int_0^1 x\ln(x^2 + 1)\,dx; n = 8$

In exercises 15–22, use the Trapezoidal Rule and Simpson's Rule to approximate the value of each definite integral.

15. $\int_0^1 \sqrt{1 + x^3}\,dx; n = 4$

16. $\int_0^2 x\sqrt{1 + x^3}\,dx; n = 4$

17. $\int_0^2 \frac{1}{\sqrt{x^3 + 1}}\,dx; n = 4$

18. $\int_0^1 \sqrt{1 - x^2}\,dx; n = 4$

19. $\int_0^2 e^{-x^2}\,dx; n = 4$

20. $\int_0^1 e^{x^2}\,dx; n = 6$

21. $\int_1^2 x^{-1/2}e^x\,dx; n = 4$

22. $\int_2^4 \frac{dx}{\ln x}; n = 6$

In exercises 23–28, find a bound on the error in approximating the given definite integral using (a) the Trapezoidal Rule and (b) Simpson's Rule with n intervals.

23. $\int_{-1}^2 x^5\,dx; n = 10$

24. $\int_0^1 e^{-x}\,dx; n = 8$

25. $\int_1^3 \frac{1}{x}\,dx; n = 10$

26. $\int_1^3 \frac{1}{x^2}\,dx; n = 8$

27. $\int_0^2 \frac{1}{\sqrt{1 + x}}\,dx; n = 8$

28. $\int_1^3 \ln x\,dx; n = 10$

29. Trial Run of an Attack Submarine In a submerged trial run of an attack submarine, a reading of the sub's velocity was made every quarter hour, as shown in the accompanying table. Use the Trapezoidal Rule to esti-mate the distance traveled by the submarine during the 2-hour period.

Time t (hours)	0	$\frac{1}{4}$	$\frac{1}{2}$	$\frac{3}{4}$
Velocity $V(t)$ (mph)	19.5	24.3	34.2	40.5

Time t (hours)	1	$\frac{5}{4}$	$\frac{3}{2}$	$\frac{7}{4}$	2
Velocity $V(t)$ (mph)	38.4	26.2	18	16	8

30. Real Estate Cooper Realty is considering develop-ment of a time-sharing condominium resort complex along the oceanfront property illustrated in the accom-panying graph. In order to obtain an estimate of the area of this property, measurements of the distances from the edge of a straight road, which defines one boundary of the property, to the corresponding points on the shoreline are made at 100-foot intervals. Using Simpson's Rule with $n = 10$, estimate the area of the oceanfront property.

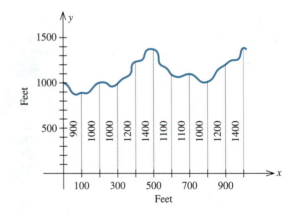

31. Fuel Consumption of Domestic Cars Thanks to smaller and more fuel-efficient models, American car-makers have doubled their average fuel economy over a 13-year period from 1974 to 1987. The graph depicted in the figure gives the average fuel consumption in miles per gallon (mpg) of domestic-built cars over the period under consideration ($t = 0$ corresponds to the beginning of 1974). Use the Trapezoidal Rule to estimate the average fuel consumption of the domestic car built during this period.

[*Hint:* Approximate the integral $\frac{1}{13} \int_0^{13} f(t)\, dt$.]

32. Average Temperature The graph in the accompanying figure shows the daily mean temperatures recorded during one September in Cameron Highlands. Using (a) the Trapezoidal Rule and (b) Simpson's Rule with $n = 10$, estimate the average temperature during that month.

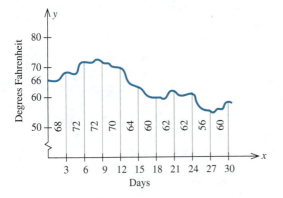

33. Consumers' Surplus Refer to Section 6.7. The demand equation for the Sicard wristwatch is given by

$$p = \frac{50}{0.01x^2 + 1} \qquad (0 \le x \le 20)$$

where x (measured in units of a thousand) is the quantity demanded per week and p is the unit price in dollars. Use (a) the Trapezoidal Rule and (b) Simpson's Rule (take $n = 8$) to estimate the consumers' surplus if the market price is $25 per watch.

34. Producers' Surplus Refer to Section 6.7. The supply function for the audio compact disc manufactured by the Herald Record Company is given by

$$p = S(x) = \sqrt{0.01x^2 + 0.11x + 38}$$

where p is the unit wholesale price in dollars and x stands for the quantity that will be made available in the market by the supplier, measured in units of a thousand. Use (a) the Trapezoidal Rule and (b) Simpson's Rule (take $n = 8$) to estimate the producers' surplus if the wholesale price is $8 per disc.

35. Air Pollution The amount of nitrogen dioxide, a brown gas that impairs breathing, present in the atmosphere on a certain May day in the city of Long Beach has been approximated by

$$A(t) = \frac{136}{1 + 0.25(t - 4.5)^2} + 28 \qquad (0 \le t \le 11)$$

where $A(t)$ is measured in pollutant standard index (PSI), t is measured in hours, and $t = 0$ corresponds to 7 A.M. Use the Trapezoidal Rule with $n = 10$ to estimate the average PSI between 7 A.M. and noon.
[*Hint:* $\frac{1}{5} \int_0^5 A(t)\, dt$.]

36. Growth of Service Industries It has been estimated that service industries, which currently make up 30% of the nonfarm work force in a certain country, will continue to grow at the rate of

$$R(t) = 5e^{1/(t+1)}$$

percent per decade t decades from now. Estimate the percentage of the nonfarm work force in the service industries one decade from now.
[*Hint:* (a) Show that the desired answer is given by $30 + \int_0^1 5e^{1/(t+1)}\, dt$ and (b) use Simpson's Rule with $n = 10$ to approximate the definite integral.]

37. Length of Infants at Birth Medical records of infants delivered at the Kaiser Memorial Hospital show that the percentage of infants whose length at birth is between 19 and 21 inches is given by

$$P = 100 \int_{19}^{21} \frac{1}{2.6\sqrt{2\pi}}\, e^{-1/2[(x-20)/2.6]^2}\, dx$$

Use Simpson's Rule with $n = 10$ to estimate P.

In 1992 Chesebro was appointed professor of medicine at Harvard Medical School, after having conducted cardiovascular research at the Mayo Clinic for a number of years. In addition, he was made Associate Director for Research in the Cardiac Unit at Massachusetts General Hospital, one of Harvard's teaching affiliates.

James H. Chesebro, M.D.

Institution: Harvard Medical School, Massachusetts General Hospital

For over 20 years, James Chesebro has worked as an investigative cardiologist, diagnosing and treating patients suffering from heart and blood-vessel diseases. He specializes in researching ways to prevent blood clots from narrowing the heart's arteries. Even when patients have coronary bypass operations to correct problems, stresses Chesebro, "the piece of vein used to detour around blocked arteries is prone to plug up with clots."

Throughout his career, Chesebro has investigated the body's clotting mechanisms, as well as substances that may prevent, or even dissolve, clots. A study conducted in 1982 showed that bypass patients taking dipyridamole and aspirin were nearly three times less likely to develop clots in vein grafts than patients given placebos.

Recently, Chesebro has been working with a substance called hirudin derived from leech saliva. Hirudin has proved quite beneficial in blocking arterial blood-clot formations.

Determining the correct medication and dosage involves intense quantitative research. Colleagues from other branches of science such as nuclear medicine, molecular biology, and biochemistry play key roles in the research process. Chesebro and his colleagues have to understand how the heart's physiology and pharmacological variables such as distribution rates, concentration levels, elimination rates, and biological effects of particular substances dictate the choice and dosage of medication.

In considering whether a particular medication will bind effectively with specific body cells, researchers must determine bond tightness, speed, and length of time to reach optimal concentration. Researchers rely on complicated equations such as "integrating the area under a curve to find the amount of medication in the body at a given time," notes Chesebro. Without calculus, these equations can't be solved.

Whether physicians rely on medication or surgical procedures such as balloon angioplasty to open clogged passages, other factors play a major role in the outcome. Cholesterol and blood sugar levels, patient age, blood pressure, the geometry of the blockage, and the velocity and turbulence of blood flow all warrant consideration. According to Chesebro, "linear modeling can be used to predict the contribution of patient variables and local blood-vessel variables in the outcome of opening blocked arteries."

The bottom line? Calculus has been a key contributor to Chesebro's success in preventing disabling arterial blood clots.

38. Tread Lives of Tires Under normal driving conditions the percentage of Super Titan radial tires expected to have a useful tread life of between 30,000 and 40,000 miles is given by

$$P = 100 \int_{30,000}^{40,000} \frac{1}{2000\sqrt{2\pi}} e^{-1/2[(x-40,000)/2000]^2} \, dx$$

Use Simpson's Rule with $n = 10$ to estimate P.

39. Measuring Cardiac Output Eight milligrams of a dye are injected into a vein leading to an individual's heart. The concentration of the dye in the aorta measured at 2-second intervals is shown in the accompanying table. Use Simpson's Rule and the formula of Example 4 to estimate the person's cardiac output.

t	0	2	4	6	8	10	12
$C(t)$	0	0	2.8	6.1	9.7	7.6	4.8

t	14	16	18	20	22	24
$C(t)$	3.7	1.9	0.8	0.3	0.1	0

40. Derive the formula

$$R = \frac{60D}{\int_0^T C(t) \, dt}$$

for calculating the cardiac output of a person in l/min. Here $C(t)$ is the concentration of dye in the aorta (in mg/l) at time t (in seconds) for t in $[0, T]$, and D is the amount of dye (in mg) injected into a vein leading to the heart.

[*Hint:* Partition the interval $[0, T]$ into n subintervals of equal length Δt. The amount of dye that flows past the measuring point in the aorta during the time interval $[0, \Delta t]$ is approximately $C(t_1)(R\Delta t)/60$ (concentration times volume). Therefore, the total amount of dye measured at the aorta is

$$\frac{[C(t_1)R\Delta t + C(t_2)R\Delta t + \cdots + C(t_n)R\Delta t]}{60} = D$$

Take the limit of the Riemann sum to obtain

$$R = \frac{60D}{\int_0^T C(t) \, dt}$$

SOLUTIONS TO SELF-CHECK EXERCISES 7.3

1. We have $x = 0$, $b = 2$, and $n = 8$, so

$$\Delta x = \frac{b-a}{n} = \frac{2}{8} = 0.25$$

and $x_0 = 0$, $x_1 = 0.25$, $x_2 = 0.50$, $x_3 = 0.75, \ldots, x_7 = 1.75$, and $x_8 = 2$. The Trapezoidal Rule gives

$$\int_0^2 \frac{1}{\sqrt{1+x^2}} \, dx$$

$$\approx \frac{0.25}{2}\left[1 + \frac{2}{\sqrt{1+(0.25)^2}} + \frac{2}{\sqrt{1+(0.5)^2}} + \cdots \right.$$

$$\left. + \frac{2}{\sqrt{1+(1.75)^2}} + \frac{1}{\sqrt{5}}\right]$$

$$\approx 0.125(1 + 1.9403 + 1.7889 + 1.6000 + 1.4142 + 1.2494$$

$$+ 1.1094 + 0.9923 + 0.4472)$$

$$\approx 1.4427$$

Using Simpson's Rule with $n = 8$ gives

$$\int_0^2 \frac{1}{\sqrt{1 + x^2}} \, dx$$

$$\approx \frac{0.25}{3} \left[1 + \frac{4}{\sqrt{1 + (0.25)^2}} + \frac{2}{\sqrt{1 + (0.5)^2}} + \frac{4}{\sqrt{1 + (0.75)^2}} \right.$$

$$\left. + \cdots + \frac{4}{\sqrt{1 + (1.75)^2}} + \frac{1}{\sqrt{5}} \right]$$

$$\approx \frac{0.25}{3} (1 + 3.8806 + 1.7889 + 3.2000 + 1.4142 + 2.4988 + 1.1094$$

$$+ 1.9846 + 0.4472)$$

$$\approx 1.4436$$

2. The average consumption of petroleum during the 14-year period is given by

$$\frac{1}{14} \int_0^{14} f(x) \, dx$$

where f is the function describing the given graph. Using Simpson's Rule with $a = 0$, $b = 14$, and $n = 14$ so that $\Delta x = 1$ and

$$x_0 = 0, \quad x_1 = 1, \quad x_2 = 2, \quad \ldots, \quad x_{14} = 14$$

we have

$$\frac{1}{14} \int_0^{14} f(x) \, dx$$

$$\approx \left(\frac{1}{14} \right) \left(\frac{1}{3} \right) [f(x_0) + 4f(x_1) + 2f(x_2) + 4f(x_3) + \cdots + 4f(x_{13}) + f(x_{14})]$$

$$= \frac{1}{42} [35 + 4(36.6) + 2(37.6) + 4(36.6) + 2(34.2)$$

$$+ 4(32.0) + 2(30.2) + 4(30.2) + 2(31.0) + 4(31.0)$$

$$+ 2(32.2) + 4(32.9) + 2(34.2) + 4(34.2) + 33.6]$$

$$\approx 33.4$$

or approximately 33.4 quadrillion BTU per year.

7.4 IMPROPER INTEGRALS

Improper Integrals

All of the definite integrals we have encountered have had finite intervals of integration. In many applications, however, we are concerned with integrals that have unbounded intervals of integration. Such integrals are called **improper integrals**.

Figure 7.9
The area of the unbounded region R can be approximated by a definite integral.

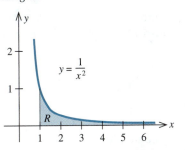

Figure 7.10
Area of shaded region $= \int_1^b \frac{1}{x^2} dx.$

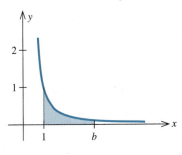

To lead us to the definition of an improper integral of a function f over an infinite interval, consider the problem of finding the area of the region R under the curve $y = f(x) = 1/x^2$ and to the right of the vertical line $x = 1$, as shown in Figure 7.9. Because the interval over which the integration must be performed is unbounded, the method of integration presented previously cannot be applied directly in solving this problem. However, we can approximate the region R by the definite integral

$$\int_1^b \frac{1}{x^2} dx \tag{9}$$

which gives the area of the region under the curve $y = f(x) = 1/x^2$ from $x = 1$ to $x = b$ (Figure 7.10). You can see that the approximation of the region R by the definite integral (9) improves as the upper limit of integration, b, becomes larger and larger. Figure 7.11 illustrates the situation for $b = 2, 3,$ and 4.

Figure 7.11
As b increases, the approximation of R by the definite integral improves.

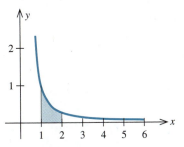

(a) Area of region under the graph of f on $[1, 2]$.

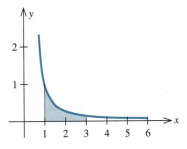

(b) Area of region under the graph of f on $[1, 3]$.

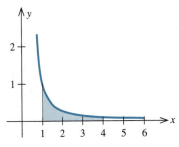

(c) Area of region under the graph of f on $[1, 4]$.

This observation suggests that if we define a function $I(b)$ by

$$I(b) = \int_1^b \frac{1}{x^2}\, dx \tag{10}$$

then we can find the area of the required region R by evaluating the limit of $I(b)$ as b tends to infinity; that is, the area of R is given by

$$\lim_{b \to \infty} I(b) = \lim_{b \to \infty} \int_1^b \frac{1}{x^2}\, dx \tag{11}$$

EXAMPLE 1

a. Evaluate the definite integral $I(b)$ in equation (10).

b. Compute $I(b)$ for $b = 10, 100, 1000,$ and $10,000$.

c. Evaluate the limit in equation (11).

d. Interpret the results of (b) and (c).

Solution

a. $I(b) = \int_1^b \frac{1}{x^2}\, dx = -\frac{1}{x}\Big|_1^b = -\frac{1}{b} + 1$

b. From the result of (a),

$$I(b) = 1 - \frac{1}{b}$$

Therefore,

$$I(10) = 1 - \frac{1}{10} = 0.9$$

$$I(100) = 1 - \frac{1}{100} = 0.99$$

$$I(1000) = 1 - \frac{1}{1000} = 0.999$$

$$I(10,000) = 1 - \frac{1}{10,000} = 0.9999$$

c. Once again, using the result of (a), we find

$$\lim_{b \to \infty} I(b) = \lim_{b \to \infty} \int_1^b \frac{1}{x^2}\, dx$$

$$= \lim_{b \to \infty} \left(1 - \frac{1}{b} \right)$$

$$= 1$$

d. The result of (c) tells us that the area of the region R is 1 square unit. The results of the computations performed in (b) reinforce our expectation that $I(b)$ should approach 1, the area of the region R, as b approaches infinity.

         ○ ○ ○

The preceding discussion and the results of Example 1 suggest that we define the improper integral of a continuous function f over the unbounded interval $[a, \infty)$ as follows.

IMPROPER INTEGRAL OF f OVER $[a, \infty)$

Let f be a continuous function on the unbounded interval $[a, \infty)$. Then the improper integral of f over $[a, \infty)$ is defined by

$$\int_a^\infty f(x)\,dx = \lim_{b \to \infty} \int_a^b f(x)\,dx \tag{12}$$

if the limit exists.

If the limit exists, the improper integral is said to be **convergent.** An improper integral for which the limit in equation (12) fails to exist is said to be **divergent.**

EXAMPLE 2   Evaluate $\displaystyle\int_2^\infty \frac{1}{x}\,dx$ if it converges.

Solution
$$\int_2^\infty \frac{1}{x}\,dx = \lim_{b \to \infty} \int_2^b \frac{1}{x}\,dx$$

$$= \lim_{b \to \infty} \ln x \,\Big|_2^b$$

$$= \lim_{b \to \infty} (\ln b - \ln 2)$$

Since $\ln b \to \infty$ as $b \to \infty$, the limit does not exist and we conclude that the given improper integral is divergent.

         ○ ○ ○

a. Suppose f is continuous and nonnegative on $[0, \infty)$. Furthermore, suppose $\lim_{x \to \infty} f(x) = L$, where L is a positive number. What can you say about the convergence of the improper integral $\int_0^\infty f(x)\,dx$? Explain your answer and illustrate with an example.

b. Suppose f is continuous and nonnegative on $[0, \infty)$ and satisfies the condition $\lim_{x \to \infty} f(x) = 0$. What can you say about $\int_0^\infty f(x)\,dx$? Explain and illustrate your answer with examples.

EXAMPLE 3 Find the area of the region R under the curve $y = e^{-x/2}$ for $x \geq 0$.

Solution The region R is shown in Figure 7.12. Taking $b > 0$, we compute the area of the region under the curve $y = e^{-x/2}$ from $x = 0$ to $x = b$; namely,

$$I(b) = \int_0^b e^{-x/2} \, dx = -2e^{-x/2} \Big|_0^b = -2e^{-b/2} + 2$$

Then the area of the region R is given by

$$\lim_{b \to \infty} I(b) = \lim_{b \to \infty} (2 - 2e^{-b/2}) = 2 - 2 \lim_{b \to \infty} \frac{1}{e^{b/2}}$$

$$= 2$$

or 2 square units.   ○ ○ ○

Figure 7.12

Area of $R = \displaystyle\int_0^\infty e^{-x/2} \, dx$.

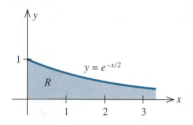

EXPLORING WITH TECHNOLOGY

You can see how fast the improper integral in Example 3 converges, as follows:

a. Use a graphing utility to plot the graph of $I(b) = 2 - 2e^{-b/2}$ in the viewing window $[0, 50] \times [0, 3]$.

b. Use **TRACE** to follow the values of y for increasing values of x starting at the origin.

○   ○   ○

The improper integral defined in equation (12) has an interval of integration that is unbounded on the right. Improper integrals with intervals of integration that are unbounded on the left also arise in practice and are defined in a similar manner.

IMPROPER INTEGRAL OF f OVER $(-\infty, b]$

Let f be a continuous function on the unbounded interval $(-\infty, b]$. Then the improper integral of f over $(-\infty, b]$ is defined by

$$\int_{-\infty}^b f(x) \, dx = \lim_{a \to -\infty} \int_a^b f(x) \, dx \tag{13}$$

if the limit exists.

In this case, the improper integral is said to be *convergent*. Otherwise, the improper integral is said to be *divergent*.

EXAMPLE 4 Find the area of the region R bounded above by the x-axis, below by the curve $y = -e^{2x}$, and on the right by the vertical line $x = 1$.

Figure 7.13

Area of $R = -\int_{-\infty}^{1} -e^{2x}\, dx$.

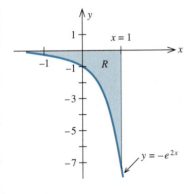

Solution The region R is shown in Figure 7.13. Taking $a < 1$, compute the area of the region bounded above by the x-axis $(y = 0)$ and below by the curve $y = -e^{2x}$ from $x = a$ to $x = 1$; namely,

$$I(a) = \int_a^1 [0 - (-e^{2x})]\, dx = \int_a^1 e^{2x}\, dx$$

$$= \frac{1}{2} e^{2x} \Big|_a^1 = \frac{1}{2} e^2 - \frac{1}{2} e^{2a}$$

Then the area of the required region is given by

$$\lim_{a \to -\infty} I(a) = \lim_{a \to -\infty} \left(\frac{1}{2} e^2 - \frac{1}{2} e^{2a} \right)$$

$$= \frac{1}{2} e^2 - \frac{1}{2} \lim_{a \to -\infty} e^{2a}$$

$$= \frac{1}{2} e^2$$ ο ο ο

Another improper integral found in practical applications involves the integration of a function f over the unbounded interval $(-\infty, \infty)$.

IMPROPER INTEGRAL OF f OVER $(-\infty, \infty)$

Let f be a continuous function over the unbounded interval $(-\infty, \infty)$. Let c be any real number, and suppose both the improper integrals

$$\int_{-\infty}^{c} f(x)\, dx \quad \text{and} \quad \int_{c}^{\infty} f(x)\, dx$$

are convergent. Then the improper integral of f over $(-\infty, \infty)$ is defined by

$$\int_{-\infty}^{\infty} f(x)\, dx = \int_{-\infty}^{c} f(x)\, dx + \int_{c}^{\infty} f(x)\, dx \qquad \textbf{(14)}$$

In this case, we say that the improper integral on the left in equation (14) is convergent. If either one of the two improper integrals on the right in (14) is divergent, then the improper integral on the left is not defined.

REMARK   Usually we choose $c = 0$.   ο ο ο

EXAMPLE 5 Evaluate the improper integral

$$\int_{-\infty}^{\infty} xe^{-x^2}\, dx$$

and give a geometrical interpretation of the results.

Solution Take the point c in equation (14) to be $c = 0$. Let us first evaluate

$$\int_{-\infty}^{0} xe^{-x^2} \, dx = \lim_{a \to -\infty} \int_{a}^{0} xe^{-x^2} \, dx$$

$$= \lim_{a \to -\infty} -\frac{1}{2} e^{-x^2} \Big|_{a}^{0}$$

$$= \lim_{a \to -\infty} \left[-\frac{1}{2} + \frac{1}{2} e^{-a^2} \right] = -\frac{1}{2}$$

Next, we evaluate

$$\int_{0}^{\infty} xe^{-x^2} \, dx = \lim_{b \to \infty} \int_{0}^{b} xe^{-x^2} \, dx$$

$$= \lim_{b \to \infty} -\frac{1}{2} e^{-x^2} \Big|_{0}^{b}$$

$$= \lim_{b \to \infty} \left[-\frac{1}{2} e^{-b^2} + \frac{1}{2} \right] = \frac{1}{2}$$

Therefore,

$$\int_{-\infty}^{\infty} xe^{-x^2} \, dx = \int_{-\infty}^{0} xe^{-x^2} \, dx + \int_{0}^{\infty} xe^{-x^2} \, dx$$

$$= -\frac{1}{2} + \frac{1}{2}$$

$$= 0$$

The graph of $y = xe^{-x^2}$ is sketched in Figure 7.14. A glance at the figure tells us that the improper integral

$$\int_{-\infty}^{0} xe^{-x^2} \, dx$$

gives the negative of the area of the region R_1, bounded above by the x-axis, below by the curve $y = xe^{-x^2}$, and on the right by the y-axis ($x = 0$). However, the improper integral

$$\int_{0}^{\infty} xe^{-x^2} \, dx$$

gives the area of the region R_2 under the curve $y = xe^{-x^2}$ for $x \geq 0$. Since the graph of f is symmetrical with respect to the origin, the area of R_1 is equal to the area of R_2. In other words,

$$\int_{-\infty}^{0} xe^{-x^2} \, dx = -\int_{0}^{\infty} xe^{-x^2} \, dx$$

Therefore,

$$\int_{-\infty}^{\infty} xe^{-x^2} \, dx = \int_{-\infty}^{0} xe^{-x^2} \, dx + \int_{0}^{\infty} xe^{-x^2} \, dx$$

$$= -\int_{0}^{\infty} xe^{-x^2} \, dx + \int_{0}^{\infty} xe^{-x^2} \, dx$$

$$= 0$$

as was shown earlier.

Figure 7.14

$$\int_{-\infty}^{\infty} xe^{-x^2} \, dx =$$

$$\int_{-\infty}^{0} xe^{-x^2} \, dx + \int_{0}^{\infty} xe^{-x^2} \, dx.$$

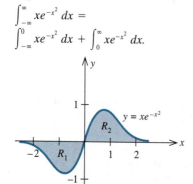

● ● ●

Perpetuities

Recall from Section 6.7 that the present value of an annuity is given by

$$PV \approx mP \int_0^T e^{-rt}\, dt = \frac{mP}{r}(1 - e^{-rT}) \qquad \textbf{(15)}$$

Now, if the payments of an annuity are allowed to continue indefinitely, we have what is called a **perpetuity.** The present value of a perpetuity may be approximated by the improper integral

$$PV \approx mP \int_0^\infty e^{-rt}\, dt$$

obtained from formula (15) by allowing the term of the annuity, T, to approach infinity. Thus

$$mP \int_0^\infty e^{-rt}\, dt = \lim_{b \to \infty} mP \int_0^b e^{-rt}\, dt$$

$$= mP \lim_{b \to \infty} \int_0^b e^{-rt}\, dt$$

$$= mP \lim_{b \to \infty} \left[-\frac{1}{r} e^{-rt} \Big|_0^b \right]$$

$$= mP \lim_{b \to \infty} \left(-\frac{1}{r} e^{-rb} + \frac{1}{r} \right) = \frac{mP}{r}$$

THE PRESENT VALUE OF A PERPETUITY

The **present value PV of a perpetuity** is given by

$$PV = \frac{mP}{r} \qquad \textbf{(16)}$$

where m is the number of payments per year, P is the size of each payment, and r is the interest rate (compounded continuously).

EXAMPLE 6   The Robinson family wishes to create a scholarship fund at a college. If a scholarship in the amount of $5000 is awarded annually beginning one year from now, find the amount of the endowment they are required to make now. Assume that this fund will earn interest at a rate of 8% per year compounded continuously.

Solution   The amount of the endowment, A, is given by the present value of a perpetuity, with $m = 1$, $P = 5000$, and $r = 0.08$. Using formula (16), we find

$$A = \frac{(1)(5000)}{0.08}$$

$$= 62{,}500$$

or $62,500.

The improper integral also plays an important role in the study of probability theory, as we will see in Section 7.5.

SELF–CHECK EXERCISES 7.4

1. Evaluate $\int_{-\infty}^{\infty} \frac{x^3}{(1+x^4)^{3/2}} dx$.

2. Suppose that an income stream is expected to continue indefinitely. Then the present value of such a stream can be calculated from the formula for the present value of an income stream by letting T approach infinity. Thus, the required present value is given by

$$PV = \int_0^{\infty} P(t)e^{-rt}\, dt$$

Suppose that Marcia has an oil well in her backyard that generates a stream of income given by

$$P(t) = 20e^{-0.02t}$$

where $P(t)$ is in thousands of dollars per year and t is the time in years from the present. Assuming that the prevailing interest rate in the foreseeable future is 10% per year compounded continuously, what is the present value of the income stream?

Solutions to Self-Check Exercises 7.4 can be found on page 576.

7.4 EXERCISES

In exercises 1–10, find the area of the region under the given curve $y = f(x)$ over the indicated interval.

1. $f(x) = \dfrac{2}{x^2}; x \geq 3$

2. $f(x) = \dfrac{2}{x^3}; x \geq 2$

3. $f(x) = \dfrac{1}{(x-2)^2}; x \geq 3$

4. $f(x) = \dfrac{2}{(x+1)^3}; x \geq 0$

5. $f(x) = \dfrac{1}{x^{3/2}}; x \geq 1$

6. $f(x) = \dfrac{3}{x^{5/2}}; x \geq 4$

7. $f(x) = \dfrac{1}{(x+1)^{5/2}}; x \geq 0$

8. $f(x) = \dfrac{1}{(1-x)^{3/2}}; x \leq 0$

9. $f(x) = e^{2x}; x \leq 2$

10. $f(x) = xe^{-x^2}; x \geq 0$

11. Find the area of the region bounded by the x-axis and the graph of the function

$$f(x) = \frac{x}{(1+x^2)^2}$$

12. Find the area of the region bounded by the x-axis and the graph of the function

$$f(x) = \frac{e^x}{(1+e^x)^2}$$

13. Consider the improper integral

$$\int_0^{\infty} \sqrt{x}\, dx$$

a. Evaluate $I(b) = \int_0^b \sqrt{x}\, dx$.

b. Show that

$$\lim_{b \to \infty} I(b) = \infty$$

thus proving that the given improper integral is divergent.

14. Consider the improper integral

$$\int_1^\infty x^{-2/3}\, dx$$

a. Evaluate $I(b) = \int_1^b x^{-2/3}\, dx$.

b. Show that

$$\lim_{b\to\infty} I(b) = \infty$$

thus proving that the given improper integral is divergent.

In exercises 15–40, evaluate each improper integral whenever it is convergent.

15. $\int_1^\infty \dfrac{3}{x^4}\, dx$

16. $\int_1^\infty \dfrac{1}{x^3}\, dx$

17. $\int_4^\infty \dfrac{2}{x^{3/2}}\, dx$

18. $\int_1^\infty \dfrac{1}{\sqrt{x}}\, dx$

19. $\int_1^\infty \dfrac{4}{x}\, dx$

20. $\int_2^\infty \dfrac{3}{x}\, dx$

21. $\int_{-\infty}^0 \dfrac{1}{(x-2)^3}\, dx$

22. $\int_2^\infty \dfrac{1}{(x+1)^2}\, dx$

23. $\int_1^\infty \dfrac{1}{(2x-1)^{3/2}}\, dx$

24. $\int_{-\infty}^0 \dfrac{1}{(4-x)^{3/2}}\, dx$

25. $\int_0^\infty e^{-x}\, dx$

26. $\int_0^\infty e^{-x/2}\, dx$

27. $\int_{-\infty}^0 e^{2x}\, dx$

28. $\int_{-\infty}^0 e^{3x}\, dx$

29. $\int_1^\infty \dfrac{e^{\sqrt{x}}}{\sqrt{x}}\, dx$

30. $\int_1^\infty \dfrac{e^{-\sqrt{x}}}{\sqrt{x}}\, dx$

31. $\int_{-\infty}^0 xe^x\, dx$

32. $\int_0^\infty xe^{-2x}\, dx$

33. $\int_{-\infty}^\infty x\, dx$

34. $\int_{-\infty}^\infty x^3\, dx$

35. $\int_{-\infty}^\infty x^3(1 + x^4)^{-2}\, dx$

36. $\int_{-\infty}^\infty x(x^2 + 4)^{-3/2}\, dx$

37. $\int_{-\infty}^\infty xe^{1-x^2}\, dx$

38. $\int_{-\infty}^\infty \left(x - \dfrac{1}{2}\right)e^{-x^2+x-1}\, dx$

39. $\int_{-\infty}^\infty \dfrac{e^{-x}}{1 + e^{-x}}\, dx$

40. $\int_{-\infty}^\infty \dfrac{xe^{-x^2}}{1 + e^{-x^2}}\, dx$

41. The Amount of an Endowment   A university alumni group wishes to provide an annual scholarship in the amount of \$1500 beginning next year. If the scholarship fund will earn interest at the rate of 8% per year compounded continuously, find the amount of the endowment the alumni are required to make now.

42. The Amount of an Endowment   Mr. Thompson wishes to establish a fund to provide a university medical center with an annual research grant of \$50,000 beginning next year. If the fund will earn interest at the rate of 9% per year compounded continuously, find the amount of the endowment he is required to make now.

43. Perpetual Net Income Streams   The present value of a perpetual stream of income that flows continually at the rate of $P(t)$ dollars per year is given by the formula

$$PV \approx \int_0^\infty P(t)e^{-rt}\, dt$$

where r is the rate at which interest is compounded continuously. Using this formula, find the present value of a perpetual net income stream that is generated at the rate of

$$P(t) = 10{,}000 + 4000t$$

dollars per year.

$$\left[\text{Hint: } \lim_{b\to 0} \dfrac{b}{e^{rb}} = 0.\right]$$

44. Establishing a Trust Fund   Mrs. Wilkinson wants to establish a trust fund that will provide her children and heirs with a perpetual annuity in the amount of

$$P(t) = 20 + t$$

thousand dollars per year beginning next year. If the trust fund will earn interest at the rate of 10% per year compounded continuously, find the amount that she must place in the trust fund now.

[*Hint:* Use the formula given in exercise 43.]

45. Capital Value   The capital value (present sale value) CV of property that can be rented on a perpetual basis for R dollars annually is approximated by the formula

$$CV \approx \int_0^\infty Re^{-it}\, dt$$

where i is the prevailing continuous interest rate.
a. Show that $CV = R/i$.
b. Find the capital value of property that can be rented at \$10,000 annually when the prevailing continuous interest rate is 12% per year.

SOLUTIONS TO SELF-CHECK EXERCISES 7.4

1. Write

$$\int_{-\infty}^{\infty} \frac{x^3}{(1+x^4)^{3/2}}\, dx = \int_{-\infty}^{0} \frac{x^3}{(1+x^4)^{3/2}}\, dx + \int_{0}^{\infty} \frac{x^3}{(1+x^4)^{3/2}}\, dx$$

Now,

$$\int_{-\infty}^{0} \frac{x^3}{(1+x^4)^{3/2}}\, dx$$

$$= \lim_{a \to -\infty} \int_{a}^{0} x^3 (1+x^4)^{-3/2}\, dx$$

$$= \lim_{a \to -\infty} \frac{1}{4}(-2)(1+x^4)^{-1/2}\Big|_{a}^{0} \qquad \text{(Integrating by substitution)}$$

$$= -\frac{1}{2}\lim_{a \to -\infty}\left[1 - \frac{1}{(1+a^4)^{1/2}}\right]$$

$$= -\frac{1}{2}$$

Similarly, you can show that

$$\int_{0}^{\infty} \frac{x^3}{(1+x^4)^{3/2}}\, dx = \frac{1}{2}$$

Therefore,

$$\int_{-\infty}^{\infty} \frac{x^3}{(1+x^4)^{3/2}}\, dx = -\frac{1}{2} + \frac{1}{2} = 0$$

2. The required present value is given by

$$PV = \int_{0}^{\infty} 20 e^{-0.02t} e^{-0.10t}\, dt$$

$$= 20 \int_{0}^{\infty} e^{-0.12t}\, dt$$

$$= 20 \lim_{b \to \infty} \int_{0}^{b} e^{-0.12t}\, dt$$

$$= -\frac{20}{0.12} \lim_{b \to \infty} e^{-0.12t}\Big|_{0}^{b}$$

$$= -\frac{500}{3} \lim_{b \to \infty} (e^{-0.12b} - 1)$$

$$= \frac{500}{3}$$

or approximately \$166,667.

CHAPTER 7 SUMMARY OF PRINCIPAL FORMULAS AND TERMS

Formulas

1. Integration by parts

$$\int u \, dv = uv - \int v \, du$$

2. Improper integral of f over $[a, \infty)$

$$\int_a^\infty f(x) \, dx = \lim_{b \to \infty} \int_a^b f(x) \, dx$$

3. Improper integral of f over $(-\infty, b]$

$$\int_{-\infty}^b f(x) \, dx = \lim_{a \to -\infty} \int_a^b f(x) \, dx$$

4. Improper integral of f over $(-\infty, \infty)$

$$\int_{-\infty}^\infty f(x) \, dx = \int_{-\infty}^c f(x) \, dx + \int_c^\infty f(x) \, dx$$

5. Present value of a perpetuity

$$PV = \frac{mP}{r}$$

6. Trapezoidal Rule

$$\int_a^b f(x) \, dx \approx \frac{\Delta x}{2} [f(x_0) + 2f(x_1) + 2f(x_2) + \cdots + 2f(x_{n-1}) + f(x_n)]$$

where $\Delta x = \dfrac{b - a}{n}$

7. Simpson's Rule

$$\int_a^b f(x) \, dx \approx \frac{\Delta x}{3} [f(x_0) + 4f(x_1) + 2f(x_2) + 4f(x_3) + 2f(x_4) + \cdots + 4f(x_{n-1}) + f(x_n)]$$

where $\Delta x = \dfrac{b - a}{n}$

8. Maximum error for Trapezoidal Rule

$$\frac{M(b - a)^3}{12n^2}, \text{ where } |f''(x)| \le M$$
$$(a \le x \le b)$$

9. Maximum error for Simpson's Rule

$$\frac{M(b - a)^5}{180n^4}, \text{ where } |f^{(4)}(x)| \le M$$
$$(a \le x \le b)$$

Terms

Improper integral

Convergent integral

Divergent integral

Perpetuity

CHAPTER 7 REVIEW EXERCISES

In exercises 1–6, evaluate each integral.

1. $\displaystyle\int 2xe^{-x}\,dx$

2. $\displaystyle\int xe^{4x}\,dx$

3. $\displaystyle\int \ln 5x\,dx$

4. $\displaystyle\int_{1}^{4}\ln 2x\,dx$

5. $\displaystyle\int_{0}^{1} xe^{-2x}\,dx$

6. $\displaystyle\int_{0}^{2} xe^{2x}\,dx$

7. Find the function f given that the slope of the tangent line to the graph of f at any point $(x, f(x))$ is

$$f'(x) = \frac{\ln x}{\sqrt{x}}$$

and that the graph of f passes through the point $(1, -2)$.

8. Find the function f given that the slope of the tangent line to the graph of f at any point $(x, f(x))$ is

$$f'(x) = \frac{e^x}{1 + e^x}$$

and that the graph of f passes through the point $(0, 0)$.

In exercises 9–14, use the table of integrals in Section 7.2 to evaluate each of the given integrals.

9. $\displaystyle\int \frac{x^2\,dx}{(3 + 2x)^2}$

10. $\displaystyle\int \frac{2x}{\sqrt{2x + 3}}\,dx$

11. $\displaystyle\int x^2 e^{4x}\,dx$

12. $\displaystyle\int \frac{dx}{(x^2 - 25)^{3/2}}$

13. $\displaystyle\int \frac{dx}{x^2\sqrt{x^2 - 4}}$

14. $\displaystyle\int 8x^3 \ln 2x\,dx$

In exercises 15–20, evaluate each improper integral whenever it is convergent.

15. $\displaystyle\int_{0}^{\infty} e^{-2x}\,dx$

16. $\displaystyle\int_{-\infty}^{0} e^{3x}\,dx$

17. $\displaystyle\int_{3}^{\infty} \frac{2}{x}\,dx$

18. $\displaystyle\int_{2}^{\infty} \frac{1}{(x + 2)^{3/2}}\,dx$

19. $\displaystyle\int_{2}^{\infty} \frac{dx}{(1 + 2x)^2}$

20. $\displaystyle\int_{1}^{\infty} 3e^{1-x}\,dx$

In exercises 21–24, use the Trapezoidal Rule and Simpson's Rule to approximate the value of the definite integral.

21. $\displaystyle\int_{1}^{3} \frac{dx}{1 + \sqrt{x}}; n = 4$

22. $\displaystyle\int_{0}^{1} e^{x^2}\,dx; n = 4$

23. $\displaystyle\int_{-1}^{1} \sqrt{1 + x^4}\,dx; n = 4$

24. $\displaystyle\int_{1}^{3} \frac{e^x}{x}\,dx; n = 4$

25. The supply equation for the GTC Slim-Phone is given by

$$p = 2\sqrt{25 + x^2}$$

where p is the unit price in dollars and x is the quantity demanded per month in units of ten thousand. Find the producers' surplus if the market price is \$26. Use the table of integrals in Section 7.2 to evaluate the definite integral.

26. The sales of the Starr Communication Company's newest video cartridge game, Laser Beams, are currently

$$te^{-0.05t}$$

units per month (each unit representing 1000 cartridges), where t denotes the number of months since the release of the cartridge. Find an expression that gives the total number of units of video cartridge games sold as a function of t. How many video games will be sold by the end of the first year?

27. The demand equation for a computer software program is given by

$$p = 2\sqrt{325 - x^2}$$

where p is the unit price in dollars and x is the quantity demanded per month in units of a thousand. Find the consumers' surplus if the market price is \$30. Evaluate the definite integral using Simpson's Rule with $n = 10$.

28. From aerial photographs, the Coast Guard was able to determine the dimensions of an oil spill along an embankment on a coastline, as shown in the accompanying figure. Using (a) the Trapezoidal Rule and (b) Simpson's Rule with $n = 10$, estimate the area of the oil spill.

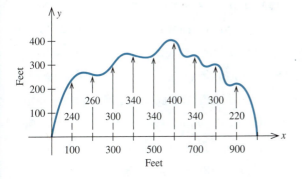

29. Mr. Lindsey wishes to establish a memorial fund at the Newtown Hospital in the amount of $10,000 per year beginning next year. If the fund earns interest at a rate of 9% per year compounded continuously, find the amount of endowment that he is required to make now.

Up to now we have dealt with functions involving one variable. In many real-life situations, however, we encounter quantities that depend on two or more quantities. For example, the Consumer Price Index (CPI) compiled every month by the Bureau of Labor Statistics depends on the price of more than 95,000 consumer items from gas to groceries. In order to study such relationships, we need the notion of a function of several variables, and we study those first in this chapter. Next, generalizing the concept of the derivative of a function of one variable, we are led to the idea of the *partial derivatives* of a function of two or more variables. Partial derivatives enable us to study the rate of change of a function with respect to one variable while holding all the other variables constant. We next learn how to find the maximum and/or minimum values of a function of several variables. For example, we learn how a manufacturer can maximize profits by producing the right amount of each product. As another example of the application of optimization theory, we learn how to find an equation of the straight line that "best" fits a set of data points scattered about a straight line.

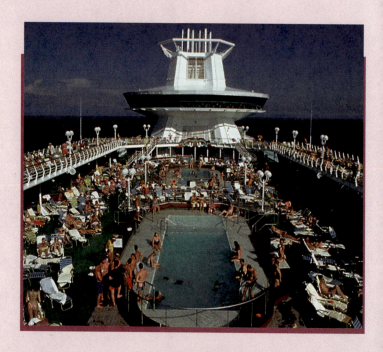

The operators of the Viking Princess, *a luxury cruise liner, are thinking about adding another swimming pool to the ship. The chief engineer has suggested that an area in the form of an ellipse, located in the rear of the promenade deck, would be a suitable site. Subject to this constraint, what are the dimensions of the largest pool that can be built? See Example 5, page 640, to see how to solve this problem.*

8

CALCULUS OF SEVERAL VARIABLES

8.1 FUNCTIONS OF SEVERAL VARIABLES

Up to now, our study of calculus has been restricted to functions of one variable. In many practical situations, however, the formulation of a problem results in a mathematical model that involves a function of two or more variables. For example, suppose Ace Novelty Company determines that the profits are $6, $5, and $4 for three types of souvenirs it produces. Let x, y, and z denote the number of type-A, type-B, and type-C souvenirs to be made. Then the company's profit is given by

$$P = 6x + 5y + 4z$$

and P is a function of the three variables x, y, and z.

Functions of Two Variables

Although this chapter deals with real-valued functions of several variables, most of our definitions and results are stated in terms of a function of two variables. One reason for adopting this approach, as you will soon see, is that there is a geometrical interpretation for this special case, which serves as an important visual aid. We can then draw upon the experience gained from studying the two-variable case to help us understand the concepts and results connected with the more general case, which, by and large, is just a simple extension of the lower-dimensional case.

A FUNCTION OF TWO VARIABLES	A **real-valued function of two variables,** f, consists of
	1. a set A of ordered pairs of real numbers (x, y) called the **domain** of the function and
	2. a rule that associates with each ordered pair in the domain of f one and only one real number, denoted by $z = f(x, y)$.

The variables x and y are called **independent variables,** and the variable z, which is dependent on the values of x and y, is referred to as a **dependent variable.**

As in the case of a real-valued function of one real variable, the number $z = f(x, y)$ is called the **value of f** at the point (x, y). And, unless specified, the domain of the function f will be taken to be the largest possible set for which the rule defining f is meaningful.

EXAMPLE 1   Let f be the function defined by

$$f(x, y) = x + xy + y^2 + 2$$

Compute $f(0, 0)$, $f(1, 2)$, and $f(2, 1)$.

Solution   We have

$$f(0, 0) = 0 + (0)(0) + 0^2 + 2 = 2$$
$$f(1, 2) = 1 + (1)(2) + 2^2 + 2 = 9$$

and $$f(2, 1) = 2 + (2)(1) + 1^2 + 2 = 7$$ ○ ○ ○

The domain of a function of two variables $f(x, y)$ is a set of ordered pairs of real numbers and may therefore be viewed as a subset of the xy-plane.

EXAMPLE 2   Find the domain of each of the following functions.

a. $f(x, y) = x^2 + y^2$ **b.** $g(x, y) = \dfrac{2}{x - y}$

c. $h(x, y) = \sqrt{1 - x^2 - y^2}$

Solution

a. $f(x, y)$ is defined for all real values of x and y, so the domain of the function f is the set of all points (x, y) in the xy-plane.

b. $g(x, y)$ is defined for all $x \neq y$, so the domain of the function g is the set of all points in the xy-plane except those lying on the line $y = x$ (Figure 8.1a).

c. We require that $1 - x^2 - y^2 \geq 0$ or $x^2 + y^2 \leq 1$, which is just the set of all points (x, y) lying on and inside the circle of radius 1 with center at the origin (Figure 8.1b).

Figure 8.1

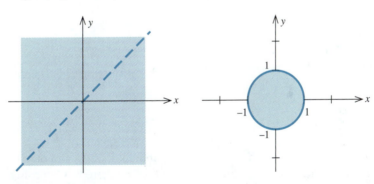

(a) Domain of g.   **(b)** Domain of h.   ○ ○ ○

Applications

EXAMPLE 3   The Acrosonic Company manufactures a bookshelf loud-speaker system that may be bought fully assembled or in a kit. The demand equations that relate the unit prices, p and q, to the quantities demanded weekly, x and y, of the assembled and kit versions of the loud-speaker systems are given by

$$p = 300 - \frac{1}{4}x - \frac{1}{8}y \quad \text{and} \quad q = 240 - \frac{1}{8}x - \frac{3}{8}y$$

a. What is the weekly total revenue function $R(x, y)$?

b. What is the domain of the function R?

Solution

a. The weekly revenue realizable from the sale of x units of the assembled speaker systems at p dollars per unit is given by xp dollars. Similarly, the weekly revenue realizable from the sale of y units of the kits at q dollars per unit is given by yq dollars. Therefore, the weekly total revenue function R is given by

$$R(x, y) = xp + yq$$

$$= x\left(300 - \frac{1}{4}x - \frac{1}{8}y\right) + y\left(240 - \frac{1}{8}x - \frac{3}{8}y\right)$$

$$= -\frac{1}{4}x^2 - \frac{3}{8}y^2 - \frac{1}{4}xy + 300x + 240y$$

Figure 8.2
The domain of $R(x, y)$.

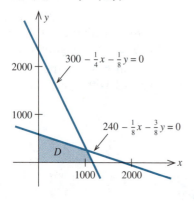

b. To find the domain of the function R, let us observe that the quantities x, y, p, and q must be nonnegative. This observation leads to the following system of linear inequalities:

$$300 - \frac{1}{4}x - \frac{1}{8}y \geq 0$$

$$240 - \frac{1}{8}x - \frac{3}{8}y \geq 0$$

$$x \geq 0$$

$$y \geq 0$$

The domain of the function R is sketched in Figure 8.2.   ○ ○ ○

Suppose the total profit of a two-product company is given by $P(x, y)$, where x denotes the number of units of the first product produced and sold, and y denotes the number of units of the second product produced and sold. Fix $x = a$, where a is a positive number so that (a, y) is in the domain of P. Describe and give an economic interpretation of the function $f(y) = P(a, y)$. Next, fix $y = b$, where b is a positive number so that (x, b) is in the domain of P. Describe and give an economic interpretation of the function $g(x) = P(x, b)$.

EXAMPLE 4   The monthly payment that amortizes a loan of A dollars in t years when the interest rate is r per year is given by

$$P = f(A, r, t) = \frac{Ar}{12\left[1 - \left(1 + \dfrac{r}{12}\right)^{-12t}\right]}$$

Find the monthly payment for a home mortgage of $90,000 to be amortized over 30 years when the interest rate is 10% per year.

Solution Letting $A = 90,000$, $r = 0.1$, and $t = 30$, we find the required monthly payment to be

$$P = f(90,000, 0.1, 30) = \frac{90,000(0.1)}{12\left[1 - \left(1 + \dfrac{0.1}{12}\right)^{-360}\right]}$$

$$\approx 789.81$$

or approximately $789.81.

Figure 8.3
The three-dimensional Cartesian co-ordinate system.

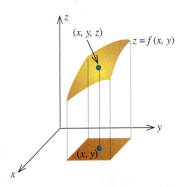

Graphs of Functions of Two Variables

In order to graph a function of two variables, we need a three-dimensional coordinate system. This is readily constructed by adding a third axis to the plane Cartesian coordinate system in such a way that the three resulting axes are mutually perpendicular and intersect at O. Observe that, by construction, the zeros of the three number scales coincide at the origin of the **three-dimensional Cartesian coordinate system** (Figure 8.3).

Figure 8.4

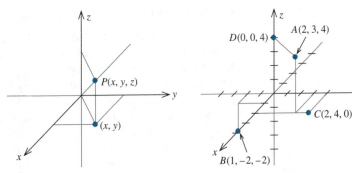

(a) A point in three-dimensional space.

(b) Some sample points in three-dimensional space.

Figure 8.5
The graph of a function in three-dimensional space.

A point in three-dimensional space can now be represented uniquely in this coordinate system by an **ordered triple** of numbers (x, y, z), and, conversely, every ordered triple of real numbers (x, y, z) represents a point in three-dimensional space (Figure 8.4a). For example, the points $A(2, 3, 4)$, $B(1, -2, -2)$, $C(2, 4, 0)$, and $D(0, 0, 4)$ are shown in Figure 8.4b.

Now, if $f(x, y)$ is a function of two variables x and y, the domain of f is a subset of the xy-plane. Let $z = f(x, y)$ so that there is one and only one point $(x, y, z) \equiv (x, y, f(x, y))$ associated with each point (x, y) in the domain of f. The totality of all such points makes up the **graph** of the function f and is, except for certain degenerate cases, a surface in three-dimensional space (Figure 8.5).

In interpreting the graph of a function $f(x, y)$, one often thinks of the value $z = f(x, y)$ of the function at the point (x, y) as the "height" of the point (x, y, z) on the graph of f. If $f(x, y) > 0$, then the point (x, y, z) is

$f(x, y)$ units above the xy-plane; if $f(x, y) < 0$, then the point (x, y, z) is $|f(x, y)|$ units below the xy-plane.

In general, it is quite difficult to draw the graph of a function of two variables, but techniques have been developed that enable us to generate such graphs with minimum effort using a computer. Figure 8.6 shows the computer-generated graphs of two functions.

Figure 8.6
Two computer-generated graphs of functions of two variables.

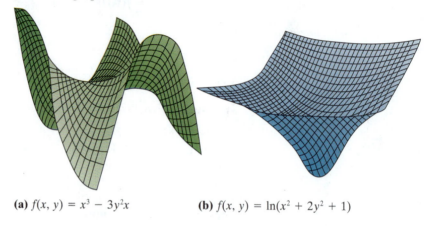

(a) $f(x, y) = x^3 - 3y^2x$ **(b)** $f(x, y) = \ln(x^2 + 2y^2 + 1)$

Level Curves

As mentioned earlier, the graph of a function of two variables is often difficult to sketch, and we will not develop a systematic procedure for sketching it. Instead, we describe a method that is used in constructing topographical maps. This method is relatively easy to apply and conveys sufficient information to enable one to obtain a feel for the graph of the function.

Suppose that $f(x, y)$ is a function of two variables x and y, with a graph as shown in Figure 8.7. If c is some value of the function f, then the equation $f(x, y) = c$ describes a curve lying on the plane $z = c$ called the **trace** of the graph of f in the plane $z = c$. If this trace is projected onto the xy-plane, the

Figure 8.7
The graph of the function $z = f(x, y)$ and its intersection with the plane $z = c$.

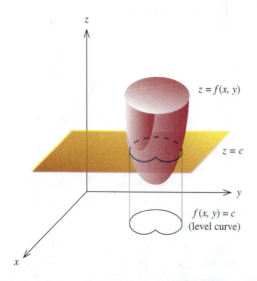

resulting curve in the xy-plane is called a **level curve.** By drawing the level curves corresponding to several admissible values of c, we obtain a **contour map.** Observe that, by construction, every point on a particular level curve corresponds to a point on the surface $z = f(x, y)$ that is a certain fixed distance from the xy-plane. Thus, by elevating or depressing the level curves that make up the contour map in one's mind, it is possible to obtain a feel for the general shape of the surface represented by the function f. Figure 8.8a shows a part of a mountain range with one peak; Figure 8.8b is the associated contour map.

Figure 8.8

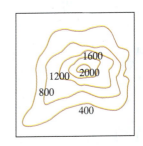

(a) A peak on a mountain range.

(b) A contour map for the mountain peak.

EXAMPLE 5   Sketch a contour map for the function $f(x, y) = x^2 + y^2$.

Solution   The level curves are the graphs of the equation $x^2 + y^2 = c$ for nonnegative numbers c. Taking $c = 0, 1, 4, 9$, and 16, for example, we obtain

$$c = \ 0 : x^2 + y^2 = 0$$
$$c = \ 1 : x^2 + y^2 = 1$$
$$c = \ 4 : x^2 + y^2 = 4 = 2^2$$
$$c = \ 9 : x^2 + y^2 = 9 = 3^2$$
$$c = 16 : x^2 + y^2 = 16 = 4^2$$

Figure 8.9

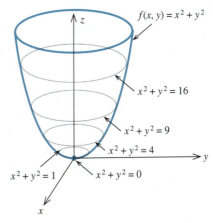

(a) Level curves of $f(x, y) = x^2 + y^2$.

(b) The graph of $f(x, y) = x^2 + y^2$.

The five level curves are concentric circles with center at the origin and radius given by $r = 0, 1, 2, 3$, and 4, respectively (Figure 8.9a). A sketch of the graph of $f(x, y) = x^2 + y^2$ is included for your reference in Figure 8.9b.   ○ ○ ○

Figure 8.10
Level curves for $f(x, y) = 2x^2 - y$.

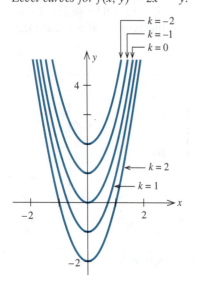

EXAMPLE 6   Sketch the level curves for the function $f(x, y) = 2x^2 - y$ corresponding to $z = -2, -1, 0, 1$, and 2.

Solution   The level curves are the graphs of the equation $2x^2 - y = k$ or $y = 2x^2 - k$ for $k = -2, -1, 0, 1$, and 2. The required level curves are shown in Figure 8.10.   ○ ○ ○

Level curves of functions of two variables are found in many practical applications. For example, if $f(x, y)$ denotes the temperature at a location within the continental United States with longitude x and latitude y at a certain time of day, then the temperature at the point (x, y) is given by the "height" of the surface, represented by $z = f(x, y)$. In this situation, the level curve $f(x, y) = k$ is a curve superimposed on a map of the United States connecting points having the same temperature at a given time (Figure 8.11). These level curves are called **isotherms.**

Similarly, if $f(x, y)$ gives the barometric pressure at the location (x, y), then the level curves of the function f are called **isobars,** lines connecting points having the same barometric pressure at a given time.

As a final example, suppose that $P(x, y, z)$ is a function of three variables x, y, and z giving the profit realized when x, y, and z units of three products A, B, and C, respectively, are produced and sold. Then the equation $P(x, y, z) = k$, where k is a constant, represents a surface in three-dimensional space called a **level surface** of P. In this situation, the level surface represented by $P(x, y, z) = k$ represents the product mix that results in a profit of exactly k dollars. Such a level surface is called an **isoprofit surface.**

Figure 8.11
*Isotherms: curves connecting points
that have the same temperature.*

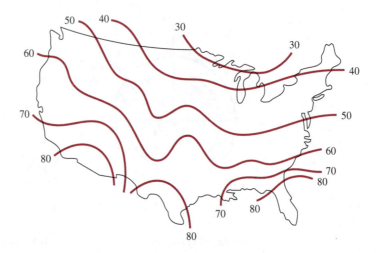

○ ○ ○

SELF-CHECK EXERCISES 8.1

1. Let $f(x, y) = x^2 - 3xy + \sqrt{x + y}$. Compute $f(1, 3)$ and $f(-1, 1)$. Is the point $(-1, 0)$ in the domain of f?

2. Find the domain of $f(x, y) = \dfrac{1}{x} + \dfrac{1}{x - y} - e^{x+y}$.

 3. The Odyssey Travel Agency has a monthly advertising budget of $20,000. Odyssey's management estimates that if they spend x dollars on newspaper advertising and y dollars on television advertising, then the monthly revenue will be

$$f(x, y) = 30x^{1/4}y^{3/4}$$

dollars. What will the monthly revenue be if Odyssey spends $5000 per month on newspaper ads and $15,000 per month on television ads? If Odyssey spends $4000 per month on newspaper ads and $16,000 per month on television ads?

Solutions to Self-Check Exercises 8.1 can be found on page 592.

8.1 EXERCISES

1. Let $f(x, y) = 2x + 3y - 4$. Compute $f(0, 0)$, $f(1, 0)$, $f(0, 1)$, $f(1, 2)$, and $f(2, -1)$.

2. Let $g(x, y) = 2x^2 - y^2$. Compute $g(1, 2)$, $g(2, 1)$, $g(1, 1)$, $g(-1, 1)$, and $g(2, -1)$.

3. Let $f(x, y) = x^2 + 2xy - x + 3$. Compute $f(1, 2)$, $f(2, 1)$, $f(-1, 2)$, and $f(2, -1)$.

4. Let $h(x, y) = (x + y)/(x - y)$. Compute $h(0, 1)$, $h(-1, 1)$, $h(2, 1)$, and $h(\pi, -\pi)$.

5. Let $g(s, t) = 3s\sqrt{t} + t\sqrt{s} + 2$. Compute $g(1, 2)$, $g(2, 1)$, $g(0, 4)$, and $g(4, 9)$.

6. Let $f(x, y) = xye^{x^2+y^2}$. Compute $f(0, 0)$, $f(0, 1)$, $f(1, 1)$, and $f(-1, -1)$.

7. Let $h(s, t) = s \ln t - t \ln s$. Compute $h(1, e)$, $h(e, 1)$, and $h(e, e)$.

8. Let $f(u, v) = (u^2 + v^2)e^{uv^2}$. Compute $f(0, 1)$, $f(-1, -1)$, $f(a, b)$, and $f(b, a)$.

9. Let $g(r, s, t) = re^{s/t}$. Compute $g(1, 1, 1)$, $g(1, 0, 1)$, and $g(-1, -1, -1)$.

10. Let $g(u, v, w) = (ue^{vw} + ve^{uw} + we^{uv})/(u^2 + v^2 + w^2)$. Compute $g(1, 2, 3)$ and $g(3, 2, 1)$.

In exercises 11–18, find the domain of the function.

11. $f(x, y) = 2x + 3y$

12. $g(x, y, z) = x^2 + y^2 + z^2$

13. $h(u, v) = \dfrac{uv}{u - v}$

14. $f(s, t) = \sqrt{s^2 + t^2}$

15. $g(r, s) = \sqrt{rs}$

16. $f(x, y) = e^{-xy}$

17. $h(x, y) = \ln(x + y - 5)$

18. $h(u, v) = \sqrt{4 - u^2 - v^2}$

In exercises 19–24, sketch the level curves of the function corresponding to the given values of z.

19. $f(x, y) = 2x + 3y$; $z = -2, -1, 0, 1, 2$

20. $f(x, y) = -x^2 + y$; $z = -2, -1, 0, 1, 2$

21. $f(x, y) = 2x^2 + y$; $z = -2, -1, 0, 1, 2$

22. $f(x, y) = xy$; $z = -4, -2, 2, 4$

23. $f(x, y) = \sqrt{16 - x^2 - y^2}$; $z = 0, 1, 2, 3, 4$

24. $f(x, y) = e^x - y$; $z = -2, -1, 0, 1, 2$

25. The volume of a cylindrical tank of radius r and height h is given by

$$V = f(r, h) = \pi r^2 h$$

Find the volume of a cylindrical tank of radius 1.5 ft and height 4 ft.

26. IQs The IQ (intelligence quotient) of a person whose mental age is m years and whose chronological age is c years is defined as

$$f(m, c) = \frac{100m}{c}$$

What is the IQ of a nine-year-old child who has a mental age of 13.5 years?

27. Poiseuille's Law Poiseuille's Law states that the resistance R, measured in dynes, of blood flowing in a blood vessel of length l and radius r (both in centimeters) is given by

$$R = f(l, r) = \frac{kl}{r^4}$$

where k is the viscosity of blood (in dyne-sec/cm²). What is the resistance, in terms of k, of blood flowing through an arteriole 4 cm long and of radius 0.1 cm?

28. Revenue Functions The Country Workshop manufactures both finished and unfinished furniture for the home. The estimated quantities demanded each week of its rolltop desks in the finished and unfinished versions are x and y units when the corresponding unit prices are

$$p = 200 - \frac{1}{5}x - \frac{1}{10}y$$

$$q = 160 - \frac{1}{10}x - \frac{1}{4}y$$

dollars, respectively.
a. What is the weekly total revenue function $R(x, y)$?
b. Find the domain of the function R.

29. For the total revenue function $R(x, y)$ of exercise 28, compute $R(100, 60)$ and $R(60, 100)$. Interpret your results.

30. Revenue Functions Weston Publishing Company publishes a deluxe edition and a standard edition of

its English language dictionary. Weston's management estimates that the number of deluxe editions demanded is x copies per day, and the number of standard editions demanded is y copies per day when the unit prices are

$$p = 20 - 0.005x - 0.001y$$

$$q = 15 - 0.001x - 0.003y$$

dollars, respectively.
a. Find the daily total revenue function $R(x, y)$.
b. Find the domain of the function R.

31. For the total revenue function $R(x, y)$ of exercise 30, compute $R(300, 200)$ and $R(200, 300)$. Interpret your results.

 32. Volume of a Gas The volume of a certain mass of gas is related to its pressure and temperature by the formula

$$V = \frac{30.9T}{P}$$

where the volume V is measured in liters, the temperature T is measured in degrees Kelvin (obtained by adding 273° to the Celsius temperature), and the pressure P is measured in millimeters of mercury pressure.
a. Find the domain of the function V.
b. Calculate the volume of the gas at standard temperature and pressure—that is, when $T = 273$ K and $P = 760$ mm of mercury.

 33. Surface Area of a Human Body An empirical formula by E. F. Dubois relates the surface area S of a human body (in square meters) to its weight W in kilograms and its height H in centimeters. The formula

$$S = 0.007184W^{0.425}H^{0.725}$$

is used by physiologists in metabolism studies.
a. Find the domain of the function S.
b. What is the surface area of a human body that weighs 70 kilograms and has a height of 178 centimeters?

 34. Arson for Profit A study of arson for profit was conducted by a team of paid civilian experts and police detectives appointed by the mayor of a large city. It was found that the number of suspicious fires in that city in 1992 was very closely related to the concentration of tenants in the city's public housing and to the level of reinvestment in the area in conventional mortgages by the ten largest banks. In fact, the number of fires was closely approximated by the formula

$$N(x, y) = \frac{100(1000 + 0.03x^2y)^{1/2}}{(5 + 0.2y)^2} \qquad \begin{array}{l} (0 \le x \le 150; \\ 5 \le y \le 35) \end{array}$$

where x denotes the number of persons per census tract and y denotes the level of reinvestment in the area in cents per dollar deposited. Using this formula, estimate the total number of suspicious fires in the districts of the city where the concentration of public housing tenants was 100 per census tract and the level of reinvestment was 20 cents per dollar deposited.

35. Continuously Compounded Interest If a principal of P dollars is deposited in an account earning interest at the rate of r per year compounded continuously, then the accumulated amount at the end of t years is given by

$$A = f(P, r, t) = Pe^{rt}$$

dollars. Find the accumulated amount at the end of 3 years if a sum of $10,000 is deposited in an account earning interest at the rate of 10% per year.

36. Home Mortgages The monthly payment that amortizes a loan of A dollars in t years when the interest rate is r per year is given by

$$P = f(A, r, t) = \frac{Ar}{12\left[1 - \left(1 + \dfrac{r}{12}\right)^{-12t}\right]}$$

a. Find the monthly payment for a home mortgage of $100,000 that will be amortized over 30 years with an interest rate of 8% per year. With an interest rate of 10% per year.
b. Find the monthly payment for a home mortgage of $100,000 that will be amortized over 20 years with an interest rate of 8% per year.

37. Home Mortgages Suppose that a home buyer secures a bank loan of A dollars to purchase a house. If the interest rate charged is r per year and the loan is to be amortized in t years, then the principal repayment at

the end of i months is given by

$$B = f(A, r, t, i)$$

$$= A\left[\frac{\left(1 + \dfrac{r}{12}\right)^i - 1}{\left(1 + \dfrac{r}{12}\right)^{12t} - 1}\right] \qquad (0 \le i \le 12t)$$

Suppose the Blakelys borrow a sum of $80,000 from a bank to help finance the purchase of a house and the bank charges interest at a rate of 9% per year. If the Blakelys agree to repay the loan in equal installments over 30 years, how much will they owe the bank after the 60th payment (5 years)? After the 240th payment (20 years)?

38. Force Generated by a Centrifuge A centrifuge is a machine designed for the specific purpose of subjecting materials to a sustained centrifugal force. The actual amount of centrifugal force, F, expressed in dynes (1 gram of force = 980 dynes) is given by

$$F = f(M, S, R) = \frac{\pi^2 S^2 M R}{900}$$

where S is in revolutions per minute (rpm), M is in grams, and R is in centimeters. Show that an object revolving at the rate of 600 rpm in a circle with radius 10 cm generates a centrifugal force that is approximately 40 times gravity.

39. Wilson Lot Size Formula The Wilson lot size formula in economics states that the optimal quantity Q of goods for a store to order is given by

$$Q = f(C, N, h) = \sqrt{\frac{2CN}{h}}$$

where C is the cost of placing an order, N is the number of items the store sells per week, and h is the weekly holding cost for each item. Find the most economical quantity of 10-speed bicycles to order if it costs the store $20 to place an order and $5 to hold a bicycle for a week and the store expects to sell 40 bicycles a week.

SOLUTIONS TO SELF-CHECK EXERCISES 8.1

1. $f(1, 3) = 1^2 - 3(1)(3) + \sqrt{1 + 3} = -6$
$f(-1, 1) = (-1)^2 - 3(-1)(1) + \sqrt{-1 + 1} = 4$

The point $(-1, 0)$ is not in the domain of f because the term $\sqrt{x + y}$ is not defined when $x = -1$ and $y = 0$. In fact, the domain of f consists of all real values of x and y that satisfy the inequality $x + y \geq 0$, the shaded half-plane shown in the accompanying figure.

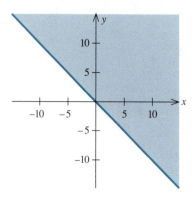

2. Since division by zero is not permitted, we see that $x \neq 0$ and $x - y \neq 0$. Therefore, the domain of f is the set of all points in the xy-plane not containing the y-axis $(x = 0)$ and the straight line $x = y$.

3. If Odyssey spends $5000 per month on newspaper ads $(x = 5000)$ and $15,000 per month on television ads $(y = 15,000)$, then its monthly revenue will be given by

$$f(5000, 15,000) = 30(5000)^{1/4}(15,000)^{3/4}$$
$$\approx 341,926.06$$

or approximately $341,926. If the agency spends $4000 per month on newspaper ads and $16,000 per month on television ads, then its monthly revenue will be given by

$$f(4000, 16,000) = 30(4000)^{1/4}(16,000)^{3/4}$$
$$\approx 339,411.26$$

or approximately $339,411.

8.2 PARTIAL DERIVATIVES

Partial Derivatives

For a function $f(x)$ of one variable x, there is no ambiguity when we speak about the rate of change of $f(x)$ with respect to x since x must be constrained to move along the x-axis. The situation becomes more complicated, however,

Figure 8.12
We can approach a point in the plane from infinitely many directions.

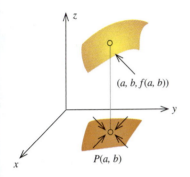

when we study the rate of change of a function of two or more variables. For example, the domain D of a function of two variables $f(x, y)$ is a subset of the plane (Figure 8.12), so if $P(a, b)$ is any point in the domain of f, there are infinitely many directions from which one can approach the point P. We may therefore ask for the rate of change of f at P along any of these directions.

However, we will not deal with this general problem. Instead, we will restrict ourselves to studying the rate of change of the function $f(x, y)$ at a point $P(a, b)$ in each of two *preferred directions*—namely, the direction parallel to the x-axis and the direction parallel to the y-axis. Let $y = b$, where b is a constant, so that $f(x, b)$ is a function of the one variable x. Since the equation $z = f(x, y)$ is the equation of a surface, the equation $z = f(x, b)$ is the equation of the curve C on the surface formed by the intersection of the surface and the plane $y = b$ (Figure 8.13).

Figure 8.13
The curve C is formed by the intersection of the plane $y = b$ with the surface $z = f(x, y)$.

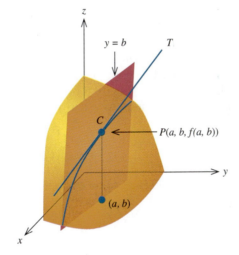

Because $f(x, b)$ is a function of one variable x, we may compute the derivative of f with respect to x at $x = a$. This derivative, obtained by keeping the variable y fixed and differentiating the resulting function $f(x, y)$ with respect to x, is called the **first partial derivative of f with respect to x** at (a, b), written

$$\frac{\partial z}{\partial x}(a, b) \quad \text{or} \quad \frac{\partial f}{\partial x}(a, b) \quad \text{or} \quad f_x(a, b)$$

Thus, $\dfrac{\partial z}{\partial x}(a, b) = \dfrac{\partial f}{\partial x}(a, b) = f_x(a, b) = \lim\limits_{h \to 0} \dfrac{f(a + h, b) - f(a, b)}{h}$

provided the limit exists. The first partial derivative of f with respect to x at (a, b) measures both the slope of the tangent line T to the curve C and the rate of change of the function f in the x-direction when $x = a$ and $y = b$. We also write

$$\left.\frac{\partial f}{\partial x}\right|_{(a,b)} \equiv f_x(a, b)$$

Similarly, we define the **first partial derivative of f with respect to y** at (a, b), written

$$\frac{\partial z}{\partial y}(a, b) \quad \text{or} \quad \frac{\partial f}{\partial y}(a, b) \quad \text{or} \quad f_y(a, b)$$

as the derivative obtained by keeping the variable x fixed and differentiating the resulting function $f(x, y)$ with respect to y. That is,

$$\frac{\partial z}{\partial y}(a, b) = \frac{\partial f}{\partial y}(a, b) = f_y(a, b)$$

$$= \lim_{k \to 0} \frac{f(a, b + k) - f(a, b)}{k}$$

if the limit exists. The first partial derivative of f with respect to y at (a, b) measures both the slope of the tangent line T to the curve C, obtained by holding x constant (Figure 8.14), and the rate of change of the function f in the y-direction when $x = a$ and $y = b$. We write

$$\left.\frac{\partial f}{\partial y}\right|_{(a,b)} \equiv f_y(a, b)$$

Before looking at some examples, let us summarize these definitions.

Figure 8.14
The first partial derivative of f with respect to y at (a, b) measures the slope of the tangent line T to the curve C with x held constant.

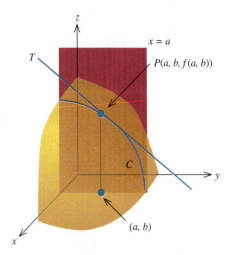

FIRST PARTIAL DERIVATIVES OF $f(x, y)$

Suppose that $f(x, y)$ is a function of the two variables x and y. Then the first partial derivative of f with respect to x at the point (x, y) is

$$\frac{\partial f}{\partial x} = \lim_{h \to 0} \frac{f(x + h, y) - f(x, y)}{h}$$

provided the limit exists. The first partial derivative of f with respect to y at the point (x, y) is

$$\frac{\partial f}{\partial y} = \lim_{k \to 0} \frac{f(x, y + k) - f(x, y)}{k}$$

provided the limit exists.

EXAMPLE I   Find the partial derivatives $\partial f/\partial x$ and $\partial f/\partial y$ of the function

$$f(x, y) = x^2 - xy^2 + y^3$$

What is the rate of change of the function f in the x-direction at the point $(1, 2)$? What is the rate of change of the function f in the y-direction at the point $(1, 2)$?

Solution   To compute $\partial f/\partial x$, think of the variable y as a constant and differentiate the resulting function of x with respect to x. Let us write

$$f(x, y) = x^2 - xy^2 + y^3$$

where the variable y to be treated as a constant is shown in color. Then

$$\frac{\partial f}{\partial x} = 2x - y^2$$

To compute $\partial f/\partial y$, think of the variable x as being fixed—that is, as a constant—and differentiate the resulting function of y with respect to y. In this case,

$$f(x, y) = x^2 - xy^2 + y^3$$

so that

$$\frac{\partial f}{\partial y} = -2xy + 3y^2$$

The rate of change of the function f in the x-direction at the point $(1, 2)$ is given by

$$f_x(1, 2) = \frac{\partial f}{\partial x}\bigg|_{(1, 2)} = 2(1) - 2^2 = -2$$

That is, f decreases 2 units for each unit increase in the x-direction, y being kept constant ($y = 2$). The rate of change of the function f in the y-direction at the point $(1, 2)$ is given by

$$f_y(1, 2) = \frac{\partial f}{\partial y}\bigg|_{(1,2)} = -2(1)(2) + 3(2)^2 = 8$$

That is, f increases 8 units for each unit increase in the y-direction, x being kept constant $(x = 1)$.

○ ○ ○

Refer to the Group Discussion question on page 584. Suppose the management of the company has decided that the projected sales of the first product is a units. Describe how you might help management decide how many units of the second product the company should produce and sell in order to maximize the company's total profit. Justify your method to management. Suppose, however, management feels that b units of the second product can be manufactured and sold. How would you help management decide how many units of the first product to manufacture in order to maximize the company's total profit?

EXAMPLE 2   Compute the first partial derivatives of each function.

a. $f(x, y) = \dfrac{xy}{x^2 + y^2}$ **b.** $g(s, t) = (s^2 - st + t^2)^5$

c. $h(u, v) = e^{u^2 - v^2}$ **d.** $f(x, y) = \ln(x^2 + 2y^2)$

Solution

a. To compute $\partial f/\partial x$, think of the variable y as a constant. Thus,

$$f(x, y) = \frac{xy}{x^2 + y^2}$$

so that, upon using the Quotient Rule, we have

$$\frac{\partial f}{\partial x} = \frac{(x^2 + y^2)y - xy(2x)}{(x^2 + y^2)^2}$$

$$= \frac{y(y^2 - x^2)}{(x^2 + y^2)^2}$$

with simplification and factorization. To compute $\partial f/\partial y$, think of the variable x as a constant. Thus,

$$f(x, y) = \frac{xy}{x^2 + y^2}$$

so that, upon using the Quotient Rule once again, we obtain

$$\frac{\partial f}{\partial y} = \frac{(x^2 + y^2)x - xy(2y)}{(x^2 + y^2)^2}$$

$$= \frac{x(x^2 - y^2)}{(x^2 + y^2)^2}$$

b. To compute $\partial g / \partial s$, we treat the variable t as if it were a constant. Thus,

$$g(s, t) = (s^2 - st + t^2)^5$$

Using the General Power Rule, we find

$$\frac{\partial g}{\partial s} = 5(s^2 - st + t^2)^4 \cdot (2s - t)$$

$$= 5(2s - t)(s^2 - st + t^2)^4$$

To compute $\partial g / \partial t$, we treat the variable s as if it were a constant. Thus,

$$g(s, t) = (s^2 - st + t^2)^5$$

$$\frac{\partial g}{\partial t} = 5(s^2 - st + t^2)^4(-s + 2t)$$

$$= 5(2t - s)(s^2 - st + t^2)^4$$

c. To compute $\partial h / \partial u$, think of the variable v as a constant. Thus,

$$h(u, v) = e^{u^2 - v^2}$$

Using the Chain Rule for exponential functions, we have

$$\frac{\partial h}{\partial u} = e^{u^2 - v^2} \cdot 2u$$

$$= 2u e^{u^2 - v^2}$$

Next, we treat the variable u as if it were a constant,

$$h(u, v) = e^{u^2 - v^2}$$

and we obtain

$$\frac{\partial h}{\partial v} = e^{u^2 - v^2} \cdot (-2v)$$

$$= -2v e^{u^2 - v^2}$$

d. To compute $\partial f / \partial x$, think of the variable y as a constant. Thus,

$$f(x, y) = \ln(x^2 + 2y^2)$$

so that the Chain Rule for logarithmic functions gives

$$\frac{\partial f}{\partial x} = \frac{2x}{x^2 + 2y^2}$$

Next, treating the variable x as if it were a constant, we find

$$f(x, y) = \ln(x^2 + 2y^2)$$

$$\frac{\partial f}{\partial y} = \frac{4y}{x^2 + 2y^2}$$

○ ○ ○

a. Let (a, b) be a point in the domain of $f(x, y)$. Put $g(x) = f(x, b)$, and suppose g is differentiable at $x = a$. Explain why you can find $f_x(a, b)$ by computing $g'(a)$. How would you go about calculating $f_y(a, b)$ using a similar technique? Give a geometric interpretation of these processes.

b. Let $f(x, y) = x^2y^3 - 3x^2y + 2$. Use the method of part (a) to find $f_x(1, 2)$ and $f_y(1, 2)$.

To compute the partial derivative of a function of several variables with respect to one variable, say x, we think of the other variables as if they were constants and differentiate the resulting function with respect to x.

EXAMPLE 3   Compute the first partial derivatives of the function

$$w = f(x, y, z) = xyz - xe^{yz} + x \ln y$$

Solution   Here we have a function of three variables, x, y, and z, and we are required to compute

$$\frac{\partial f}{\partial x}, \quad \frac{\partial f}{\partial y}, \quad \text{and} \quad \frac{\partial f}{\partial z}$$

To compute f_x, we think of the other two variables, y and z, as fixed, and we differentiate the resulting function of x with respect to x, thereby obtaining

$$f_x = yz - e^{yz} + \ln y$$

To compute f_y, we think of the other two variables, x and z, as constants, and we differentiate the resulting function of y with respect to y. We then obtain

$$f_y = xz - xze^{yz} + \frac{x}{y}$$

Finally, to compute f_z, we treat the variables x and y as constants and differentiate the function f with respect to z, obtaining

$$f_z = xy - xye^{yz} \qquad \circ \ \circ \ \circ$$

EXPLORING WITH TECHNOLOGY

Refer to the preceding Group Discussion question. Let

$$f(x, y) = \frac{e^{\sqrt{xy}}}{(1 + xy^2)^{3/2}}$$

a. Compute $g(x) = f(x, 1)$ and use a graphing utility to plot the graph of g in the viewing rectangle $[0, 2] \times [0, 2]$.
b. Use the differentiation operation of your graphing utility to find $g'(1)$ and hence $f_x(1, 1)$.
c. Compute $h(y) = f(1, y)$ and use a graphing utility to plot the graph of h in the viewing rectangle $[0, 2] \times [0, 2]$.
d. Use the differentiation operation of your graphing utility to find $h'(1)$ and hence $f_y(1, 1)$.

$$\circ \quad \circ \quad \circ$$

The Cobb-Douglas Production Function

For an economic interpretation of the first partial derivatives of a function of two variables, let us turn our attention to the function

$$f(x, y) = ax^b y^{1-b} \tag{1}$$

where a and b are positive constants with $0 < b < 1$. This function is called the **Cobb-Douglas production function.** Here x stands for the amount of money expended for labor, y stands for the cost of capital equipment (buildings, machinery, and other tools of production), and the function f measures the output of the finished product (in suitable units) and is called, accordingly, the **production function.**

The partial derivative f_x is called the **marginal productivity of labor.** It measures the rate of change of production with respect to the amount of money expended for labor, with the level of capital expenditure held constant. Similarly, the partial derivative f_y, called the **marginal productivity of capital,** measures the rate of change of production with respect to the amount expended on capital, with the level of labor expenditure held fixed.

EXAMPLE 4   A certain country's production in the early years following World War II is described by the function

$$f(x, y) = 30x^{2/3} y^{1/3}$$

units, when x units of labor and y units of capital were used.

a. Compute f_x and f_y.

b. What are the marginal productivity of labor and the marginal productivity of capital when the amounts expended on labor and capital are 125 units and 27 units, respectively?

c. Should the government have encouraged capital investment rather than increasing expenditure on labor to increase the country's productivity?

Solution

a.
$$f_x = 30 \cdot \frac{2}{3} x^{-1/3} y^{1/3} = 20 \left(\frac{y}{x}\right)^{1/3}$$

$$f_y = 30x^{2/3} \cdot \frac{1}{3} y^{-2/3} = 10 \left(\frac{x}{y}\right)^{2/3}$$

b. The required marginal productivity of labor is given by

$$f_x(125, 27) = 20 \left(\frac{27}{125}\right)^{1/3} = 20 \left(\frac{3}{5}\right)$$

or 12 units per unit increase in labor expenditure (capital expenditure is held constant at 27 units). The required marginal productivity of capital is given by

$$f_y(125, 27) = 10 \left(\frac{125}{27}\right)^{2/3} = 10 \left(\frac{25}{9}\right)$$

or $27\frac{7}{9}$ units per unit increase in capital expenditure (labor outlay is held constant at 125 units).

c. From the results of (b), we see that a unit increase in capital expenditure resulted in a much faster increase in productivity than a unit increase in labor expenditure would have. Therefore, the government should have encouraged increased spending on capital rather than on labor during the early years of reconstruction.   ◦ ◦ ◦

Substitute and Complementary Commodities

For another application of the first partial derivatives of a function of two variables in the field of economics, let us consider the relative demands of two commodities. We say that the two commodities are **substitute** (competitive) **commodities** if a decrease in the demand for one results in an increase in the demand for the other. Examples of competitive commodities are coffee and tea. Conversely, two commodities are referred to as **complementary commodities** if a decrease in the demand for one results in a decrease in the demand for the other as well. Examples of complementary commodities are automobiles and tires.

We now derive a criterion for determining whether two commodities A and B are substitute or complementary. Suppose that the demand equations that relate the quantities demanded, x and y, to the unit prices, p and q, of the two commodities are given by

$$x = f(p, q) \quad \text{and} \quad y = g(p, q)$$

Let us consider the partial derivative $\partial f/\partial p$. Since f is the demand function for commodity A, we see that, for fixed q, f is typically a decreasing function of p—that is, $\partial f/\partial p < 0$. Now, if the two commodities were substitute commodities, then the quantity demanded of commodity B would increase with respect to p—that is, $\partial g/\partial p > 0$. A similar argument with p fixed shows that if A and B are substitute commodities, then $\partial f/\partial q > 0$. Thus, the two commodities A and B are substitute commodities if

$$\frac{\partial f}{\partial q} > 0 \quad \text{and} \quad \frac{\partial g}{\partial p} > 0$$

Similarly, A and B are complementary commodities if

$$\frac{\partial f}{\partial q} < 0 \quad \text{and} \quad \frac{\partial g}{\partial p} < 0$$

SUBSTITUTE AND COMPLEMENTARY COMMODITIES

Two commodities A and B are **substitute commodities** if

$$\frac{\partial f}{\partial q} > 0 \quad \text{and} \quad \frac{\partial g}{\partial p} > 0 \tag{2}$$

Two commodities A and B are **complementary commodities** if

$$\frac{\partial f}{\partial q} < 0 \quad \text{and} \quad \frac{\partial g}{\partial p} < 0 \tag{3}$$

EXAMPLE 5 Suppose that the daily demand for butter is given by

$$x = f(p,q) = \frac{3q}{1+p^2}$$

and the daily demand for margarine is given by

$$y = g(p,q) = \frac{2p}{1+\sqrt{q}} \quad (p>0, q>0)$$

where p and q denote the prices per pound (in dollars) of butter and margarine, respectively, and x and y are measured in millions of pounds. Determine whether these two commodities are substitute, complementary, or neither.

Solution We compute

$$\frac{\partial f}{\partial q} = \frac{3}{1+p^2}$$

and

$$\frac{\partial g}{\partial p} = \frac{2}{1+\sqrt{q}}$$

Since

$$\frac{\partial f}{\partial q} > 0 \quad \text{and} \quad \frac{\partial g}{\partial p} > 0$$

for all values of $p > 0$ and $q > 0$, we conclude that butter and margarine are substitute commodities.

Second-Order Partial Derivatives

The first partial derivatives $f_x(x, y)$ and $f_y(x, y)$ of a function $f(x, y)$ of the two variables x and y are also functions of x and y. As such, we may differentiate each of the functions f_x and f_y to obtain the **second-order partial derivatives of f.** (See Figure 8.15.) Thus, differentiating the function f_x with respect to x leads to the second partial derivative

$$f_{xx} \equiv \frac{\partial^2 f}{\partial x^2} = \frac{\partial}{\partial x}(f_x)$$

Figure 8.15
A schematic showing the four sec-ond-order partial derivatives of f.

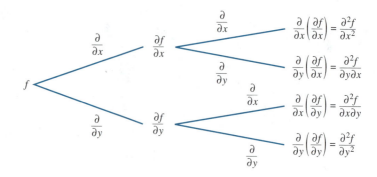

However, differentiation of f_x with respect to y leads to the second partial derivative

$$f_{xy} \equiv \frac{\partial^2 f}{\partial y \partial x} = \frac{\partial}{\partial y}(f_x)$$

Similarly, differentiation of the function f_y with respect to x and with respect to y leads to

$$f_{yx} \equiv \frac{\partial^2 f}{\partial x \partial y} = \frac{\partial}{\partial x}(f_y)$$

and

$$f_{yy} \equiv \frac{\partial^2 f}{\partial y^2} = \frac{\partial}{\partial y}(f_y)$$

respectively. Note that, in general, it is not true that $f_{xy} = f_{yx}$. However, in most practical applications, f_{xy} and f_{yx} are equal.

EXAMPLE 6   Find the second-order partial derivatives of the function

$$f(x, y) = x^3 - 3x^2y + 3xy^2 + y^2$$

Solution   The first partial derivatives of f are

$$f_x = \frac{\partial}{\partial x}(x^3 - 3x^2y + 3xy^2 + y^2)$$
$$= 3x^2 - 6xy + 3y^2$$
$$f_y = \frac{\partial}{\partial y}(x^3 - 3x^2y + 3xy^2 + y^2)$$
$$= -3x^2 + 6xy + 2y$$

Therefore,
$$f_{xx} = \frac{\partial}{\partial x}(f_x) = \frac{\partial}{\partial x}(3x^2 - 6xy + 3y^2)$$
$$= 6x - 6y = 6(x - y)$$

$$f_{xy} = \frac{\partial}{\partial y}(f_x) = \frac{\partial}{\partial y}(3x^2 - 6xy + 3y^2)$$

$$= -6x + 6y = 6(y - x)$$

$$f_{yx} = \frac{\partial}{\partial x}(f_y) = \frac{\partial}{\partial x}(-3x^2 + 6xy + 2y)$$

$$= -6x + 6y = 6(y - x)$$

$$f_{yy} = \frac{\partial}{\partial y}(f_y) = \frac{\partial}{\partial y}(-3x^2 + 6xy + 2y)$$

$$= 6x + 2$$

○ ○ ○

EXAMPLE 7   Find the second-order partial derivatives of the function

$$f(x, y) = e^{xy^2}$$

Solution   We have

$$f_x = \frac{\partial}{\partial x}(e^{xy^2})$$

$$= y^2 e^{xy^2}$$

and

$$f_y = \frac{\partial}{\partial y}(e^{xy^2})$$

$$= 2xye^{xy^2}$$

so the required second-order partial derivatives of f are

$$f_{xx} = \frac{\partial}{\partial x}(f_x) = \frac{\partial}{\partial x}(y^2 e^{xy^2})$$

$$= y^4 e^{xy^2}$$

$$f_{xy} = \frac{\partial}{\partial y}(f_x) = \frac{\partial}{\partial y}(y^2 e^{xy^2})$$

$$= 2ye^{xy^2} + 2xy^3 e^{xy^2}$$

$$= 2ye^{xy^2}(1 + xy^2)$$

$$f_{yx} = \frac{\partial}{\partial x}(f_y) = \frac{\partial}{\partial x}(2xye^{xy^2})$$

$$= 2ye^{xy^2} + 2xy^3 e^{xy^2}$$

$$= 2ye^{xy^2}(1 + xy^2)$$

$$f_{yy} = \frac{\partial}{\partial y}(f_y) = \frac{\partial}{\partial y}(2xye^{xy^2})$$

$$= 2xe^{xy^2} + (2xy)(2xy)e^{xy^2}$$

$$= 2xe^{xy^2}(1 + 2xy^2)$$

○ ○ ○

SELF–CHECK EXERCISES 8.2

1. Compute the first partial derivatives of $f(x, y) = x^3 - 2xy^2 + y^2 - 8$.

2. Find the first partial derivatives of $f(x, y) = x \ln y + ye^x - x^2$ at $(0, 1)$ and interpret your results.

3. Find the second-order partial derivatives of the function of Self-Check exercise 1.

4. A certain country's production is described by the function

$$f(x, y) = 60x^{1/3}y^{2/3}$$

when x units of labor and y units of capital are used.

a. What are the marginal productivity of labor and the marginal productivity of capital when the amounts expended on labor and capital are 125 units and 8 units, respectively?

b. Should the government encourage capital investment rather than increased expenditure on labor at this time in order to increase the country's productivity?

Solutions to Self-Check Exercises 8.2 can be found on page 608.

8.2 EXERCISES

In exercises 1–22, find the first partial derivatives of each function.

1. $f(x, y) = 2x + 3y + 5$

2. $f(x, y) = 2xy$

3. $g(x, y) = 2x^2 + 4y + 1$

4. $f(x, y) = 1 + x^2 + y^2$

5. $f(x, y) = \dfrac{2y}{x^2}$

6. $f(x, y) = \dfrac{x}{1 + y}$

7. $g(u, v) = \dfrac{u - v}{u + v}$

8. $f(x, y) = \dfrac{x^2 - y^2}{x^2 + y^2}$

9. $f(s, t) = (s^2 - st + t^2)^3$

10. $g(s, t) = s^2t + st^{-3}$

11. $f(x, y) = (x^2 + y^2)^{2/3}$

12. $f(x, y) = x\sqrt{1 + y^2}$

13. $f(x, y) = e^{xy+1}$

14. $f(x, y) = (e^x + e^y)^5$

15. $f(x, y) = x \ln y + y \ln x$

16. $f(x, y) = x^2e^{y^2}$

17. $g(u, v) = e^u \ln v$

18. $f(x, y) = \dfrac{e^{xy}}{x + y}$

19. $f(x, y, z) = xyz + xy^2 + yz^2 + zx^2$

20. $g(u, v, w) = \dfrac{2uvw}{u^2 + v^2 + w^2}$

21. $h(r, s, t) = e^{rst}$

22. $f(x, y, z) = xe^{y/z}$

In exercises 23–32, evaluate the first partial derivatives of the function at the given point.

23. $f(x, y) = x^2y + xy^2$; $(1, 2)$

24. $f(x, y) = x^2 + xy + y^2 + 2x - y$; $(-1, 2)$

25. $f(x, y) = x\sqrt{y} + y^2$; (2, 1)

26. $g(x, y) = \sqrt{x^2 + y^2}$; (3, 4)

27. $f(x, y) = \dfrac{x}{y}$; (1, 2)

28. $f(x, y) = \dfrac{x + y}{x - y}$; (1, −2)

29. $f(x, y) = e^{xy}$: (1, 1)

30. $f(x, y) = e^x \ln y$; (0, e)

31. $f(x, y, z) = x^2yz^3$; (1, 0, 2)

32. $f(x, y, z) = x^2y^2 + z^2$; (1, 1, 2)

In exercises 33–40, find the second-order partial derivatives of the given function. In each case, show that the mixed partial derivatives f_{xy} and f_{yx} are equal.

33. $f(x, y) = x^2y + xy^3$

34. $f(x, y) = x^3 + x^2y + x + 4$

35. $f(x, y) = x^2 − 2xy + 2y^2 + x − 2y$

36. $f(x, y) = x^3 + x^2y^2 + y^3 + x + y$

37. $f(x, y) = \sqrt{x^2 + y^2}$

38. $f(x, y) = x\sqrt{y} + y\sqrt{x}$

39. $f(x, y) = e^{-x/y}$

40. $f(x, y) = \ln(1 + x^2y^2)$

41. **Productivity of a Country** The productivity of a South American country is given by the function

$$f(x, y) = 20x^{3/4}y^{1/4}$$

when x units of labor and y units of capital are used.
a. What are the marginal productivity of labor and the marginal productivity of capital when the amounts expended on labor and capital are 256 units and 16 units, respectively?
b. Should the government encourage capital investment rather than increased expenditure on labor at this time in order to increase the country's productivity?

42. **Productivity of a Country** The productivity of a country in Western Europe is given by the function

$$f(x, y) = 40x^{4/5}y^{1/5}$$

when x units of labor and y units of capital are used.
a. What are the marginal productivity of labor and the marginal productivity of capital when the amounts ex-

pended on labor and capital are 32 units and 243 units, respectively?
b. Should the government encourage capital investment rather than increased expenditure on labor at this time in order to increase the country's productivity?

43. **Land Prices** The rectangular region R shown in the accompanying figure represents a city's financial district. The price of land within the district is approximated by the function

$$p(x, y) = 200 − 10\left(x − \frac{1}{2}\right)^2 − 15(y − 1)^2$$

where $p(x, y)$ is the price of land at the point (x, y) in dollars per square foot and x and y are measured in miles. Compute $\dfrac{\partial p}{\partial x}$ (0, 1) and $\dfrac{\partial p}{\partial y}$ (0, 1) and interpret your results.

44. **Complementary and Substitute Commodities** In a survey conducted by *Home Entertainment* magazine, it was determined that the demand equation for videocassette recorders (VCRs) is given by

$$x = f(p, q) = 10,000 − 10p + 0.2q^2$$

and the demand equation for videodisc players is given by

$$y = g(p, q) = 5000 + 0.8p^2 − 20q$$

where p and q denote the unit prices (in dollars) for the cassette recorders and disc players, respectively, and x

Using Technology

Finding Partial Derivatives at a Given Point

Suppose $f(x, y)$ is a function of two variables and we wish to compute

$$f_x(a, b) = \frac{\partial f}{\partial x}\bigg|_{(a, b)}$$

Recall that in computing $\partial f / \partial x$, we think of y as being fixed. But in this situation, we are evaluating $\partial f / \partial x$ at (a, b). Therefore, we set y equal to b. Doing this leads to the function g of one variable, x, defined by

$$g(x) = f(x, b)$$

It follows from the definition of the partial derivative that

$$f_x(a, b) = g'(a)$$

Thus, the value of the partial derivative $\partial f / \partial x$ at a given point (a, b) can be found by evaluating the derivative of a function of one variable. In particular, the latter can be found by using the numerical derivative operation of a graphing utility. We find $f_y(a, b)$ in a similar manner.

EXAMPLE 1 Let $f(x, y) = (1 + xy^2)^{3/2}e^{x^2y}$. Find (a) $f_x(1, 2)$ and (b) $f_y(1, 2)$.

Solution

a. Define $g(x) = f(x, 2) = (1 + 4x)^{3/2}e^{2x^2}$. Using the numerical derivative operation to find $g'(1)$, we obtain

$$f_x(1, 2) = g'(1) \approx 429.5832249$$

b. Define $h(y) = f(1, y) = (1 + y^2)^{3/2}e^y$. Using the numerical derivative operation to find $h'(2)$, we obtain

$$f_y(1, 2) = h'(2) \approx 181.74674899 \qquad \circ \; \circ \; \circ$$

EXERCISES

In exercises 1–6, use a graphing utility to compute $\partial f/\partial x$ and $\partial f/\partial y$ at the given point.

1. $f(x, y) = \sqrt{x}(2 + xy^2)^{1/3}$; $(1, 2)$

2. $f(x, y) = \sqrt{xy}(1 + 2xy)^{2/3}$; $(1, 4)$

3. $f(x, y) = \dfrac{x + y^2}{1 + x^2 y}$; $(1, 2)$

4. $f(x, y) = \dfrac{xy^2}{(\sqrt{x} + \sqrt{y})^2}$; $(4, 1)$

5. $f(x, y) = e^{-xy^2}(x + y)^{1/3}$; $(1, 1)$

6. $f(x, y) = \dfrac{\ln(\sqrt{x} + y^2)}{x^2 + y^2}$; $(4, 1)$

and y denote the number of cassette recorders and disc players demanded per week. Determine whether these two products are substitute, complementary, or neither.

45. Complementary and Substitute Commodities   In the survey mentioned in exercise 44, it was determined that the demand equation for VCRs is given by

$$x = f(p, q) = 10,000 - 10p - e^{0.5q}$$

and the demand equation for blank tapes for the VCRs is given by

$$y = g(p, q) = 50,000 - 4000q - 10p$$

where p and q denote the unit prices, respectively, and x and y denote the number of VCRs and the number of blank VCR tapes demanded per week. Determine whether these two products are substitute, complementary, or neither.

46. Complementary and Substitute Commodities   Refer to exercise 28, Exercises 8.1. Show that the finished and unfinished home furniture manufactured by the Country Workshop are substitute commodities.

[*Hint:* Solve the system of equations for x and y in terms of p and q.]

47. Revenue Functions   The total weekly revenue (in dollars) of the Country Workshop associated with manufacturing and selling rolltop desks is given by the function

$$R(x, y) = -0.2x^2 - 0.25y^2 - 0.2xy + 200x + 160y$$

where x denotes the number of finished units and y denotes the number of unfinished units manufactured and sold per week. Compute $\partial R/\partial x$ and $\partial R/\partial y$ when $x = 300$ and $y = 250$. Interpret your results.

48. Profit Functions   The monthly profit (in dollars) of the Bond and Barker Department Store depends on the level of inventory x (in thousands of dollars) and the floor space y (in thousands of square feet) available for display of the merchandise, as given by the equation

$$P(x, y) = -0.02x^2 - 15y^2 + xy$$
$$+ 39x + 25y - 20,000$$

Compute $\partial P/\partial x$ and $\partial P/\partial y$ when $x = 4000$ and $y = 150$. Interpret your results. Repeat with $x = 5000$ and $y = 150$.

49. Volume of a Gas   The volume V (in liters) of a certain mass of gas is related to its pressure P (in millimeters of mercury) and its temperature T (in degrees Kelvin) by the law

$$V = \frac{30.9T}{P}$$

Compute $\partial V/\partial T$ and $\partial V/\partial P$ when $T = 300$ and $P = 800$. Interpret your results.

50. Surface Area of a Human Body   The formula

$$S = 0.007184W^{0.425}H^{0.725}$$

gives the surface area S of a human body (in square meters) in terms of its weight W in kilograms and its height H in centimeters. Compute $\partial S/\partial W$ and $\partial S/\partial H$ when $W = 70$ kg and $H = 180$ cm. Interpret your results.

SOLUTIONS TO SELF-CHECK EXERCISES 8.2

1.

$$f_x = \frac{\partial f}{\partial x} = 3x^2 - 2y^2$$

$$f_y = \frac{\partial f}{\partial y} = -2x(2y) + 2y$$

$$= 2y(1 - 2x)$$

2.

$$f_x = \ln y + ye^x - 2x; \quad f_y = \frac{x}{y} + e^x$$

In particular, $f_x(0, 1) = \ln 1 + 1e^0 - 2(0) = 1$

$$f_y(0, 1) = \frac{0}{1} + e^0 = 1$$

The results tell us that at the point $(0, 1)$, $f(x, y)$ increases 1 unit for each unit increase in the x-direction, y being kept constant; $f(x, y)$ also increases 1 unit for each unit increase in the y-direction, x being kept constant.

3. From the results of Self-Check exercise 1,

$$f_x = 3x^2 - 2y^2$$

Therefore,

$$f_{xx} = \frac{\partial}{\partial x}(3x^2 - 2y^2) = 6x$$

and $f_{xy} = \frac{\partial}{\partial y}(3x^2 - 2y^2) = -4y$

Also, from the results of Self-Check exercise 1,

$$f_y = 2y(1 - 2x)$$

and so

$$f_{yx} = \frac{\partial}{\partial x}[2y(1 - 2x)] = -4y$$

and $f_{yy} = \frac{\partial}{\partial y}[2y(1 - 2x)] = 2(1 - 2x)$

4. a. The marginal productivity of labor when the amounts expended on labor and capital are x and y units, respectively, is given by

$$f_x(x, y) = 60\left(\frac{1}{3}x^{-2/3}\right)y^{2/3} = 20\left(\frac{y}{x}\right)^{2/3}$$

In particular, the required marginal productivity of labor is given by

$$f_x(125, 8) = 20\left(\frac{8}{125}\right)^{2/3} = 20\left(\frac{4}{25}\right)$$

or 3.2 units per unit increase in labor expenditure, capital expenditure being held constant at 8 units. Next, we compute

$$f_y(x, y) = 60x^{1/3}\left(\frac{2}{3}y^{-1/3}\right) = 40\left(\frac{x}{y}\right)^{1/3}$$

and deduce that the required marginal productivity of capital is given by

$$f_y(125, 8) = 40\left(\frac{125}{8}\right)^{1/3} = 40\left(\frac{5}{2}\right)$$

or 100 units per unit increase in capital expenditure, labor expenditure being held constant at 125 units.

b. The results of (a) tell us that the government should encourage increased spending on capital rather than on labor.

8.3 MAXIMA AND MINIMA OF FUNCTIONS OF SEVERAL VARIABLES

Maxima and Minima

In Chapter 4 we saw how solving a practical problem formulated in terms of a function f of one variable often centers on determining an extreme value of f with respect to that variable. For example, we can find a firm's production level x that will yield a maximum profit by finding the absolute maximum value of the profit function $P(x)$ with respect to x. Conversely, to find a manufacturer's production level x that will result in a minimal cost of operation, we find the absolute minimum value of the cost function $C(x)$ with respect to x.

The notion of an extreme value of a function plays an equally important role in the case of a function of several variables. As with a function of one variable, it is important to distinguish between the concept of a relative maximum (relative minimum) of f and that of an absolute maximum (absolute minimum) of f. More specifically, a function $f(x, y)$ of two variables has a **relative maximum** at a point (a, b) in the domain of f if $f(x, y) \le f(a, b)$ for all points (x, y) that are sufficiently close to the point (a, b). Similarly, f has a **relative minimum** at (a, b) if $f(x, y) \ge f(a, b)$ for all points (x, y) that are sufficiently close to the point (a, b). Geometrically, this means that no point on the graph of f corresponding to points (x, y) lying sufficiently close to the point (a, b) can be above (or below) the point $(a, b, f(a, b))$. Figure 8.16

Figure 8.16
f has a relative maximum at (a, b) and a relative minimum at (c, d).

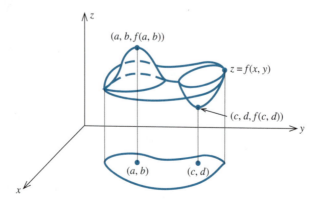

depicts the graph of a function f of two variables that has a relative maximum at the point (a, b) and a relative minimum at the point (c, d).

A function $f(x, y)$ of two variables has an **absolute maximum** at a point (a, b) if

$$f(x, y) \leq f(a, b)$$

for *all* points (x, y) in the domain of f. Similarly, a function $f(x, y)$ has an **absolute minimum** at a point (a, b) if

$$f(x, y) \geq f(a, b)$$

for *all* points (x, y) in the domain of f. Geometrically, the absolute maximum and the absolute minimum of a function f are, respectively, the highest and the lowest points on the graph of f.

Just as in the case of a function of one variable, a relative extremum (relative maximum or relative minimum) may or may not be an absolute extremum. However, in order to simplify matters, we will assume that whenever an absolute extremum exists, it will occur at a point where f has a relative extremum.

Just as the first and second derivatives play an important role in determining the relative extrema of a function of one variable, the first and second partial derivatives are powerful tools for locating and classifying the relative extrema of functions of several variables.

Suppose now that a differentiable function $f(x, y)$ of two variables has a relative maximum (relative minimum) at a point (a, b) in the domain of f. From Figure 8.17 it is clear that at the point (a, b), the slope of the "tangent

Figure 8.17

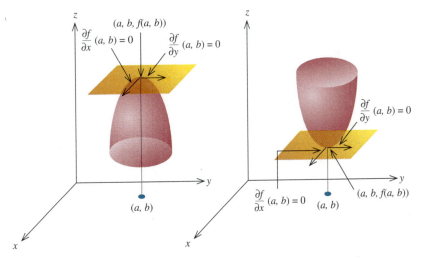

(a) f has a relative maximum at (a, b).

(b) f has a relative minimum at (a, b).

lines" to the surface in any direction must be zero. In particular, this implies that both

$$\frac{\partial f}{\partial x}(a, b) \quad \text{and} \quad \frac{\partial f}{\partial y}(a, b)$$

must be zero.

The point (a, b) is called a **critical point** of the function f. In case we are tempted to conclude that a critical point of a function f must automatically be a relative extremum of f, let us consider the graph of the function f in Figure 8.18. Here we have both

$$\frac{\partial f}{\partial x}(a, b) = 0 \quad \text{and} \quad \frac{\partial f}{\partial y}(a, b) = 0$$

Figure 8.18
The point $(a, b, f(a, b))$ is called a saddle point.

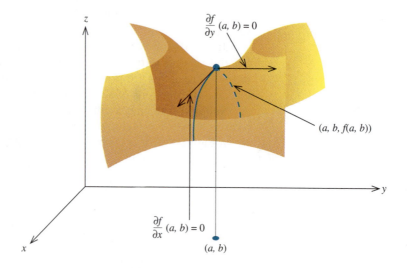

The point $(a, b, f(a, b))$ is neither a relative maximum nor a relative minimum of the function f since there are points nearby that are higher and others that are lower than it. Such a point is called a **saddle point.** Thus, as in the case of a function of one variable, we may conclude that a critical point of a function of two (or more) variables is only a candidate for a relative extremum of f.

To determine the nature of a critical point of a function $f(x, y)$ of two variables, we use the second partial derivatives of f. The resulting test, which helps us classify these points, is called the **Second Derivative Test** and is incorporated into the following procedure for finding and classifying the relative extrema of f.

DETERMINING RELATIVE EXTREMA

1. Find the critical points of $f(x, y)$ by solving the system of simultaneous equations

$$f_x = 0$$

$$f_y = 0$$

2. The Second Derivative Test: Let

$$D(x, y) = f_{xx}f_{yy} - f_{xy}^2$$

Then

a. $D(a, b) > 0$ and $f_{xx}(a, b) < 0$ implies that $f(x, y)$ has a **relative maximum** at the point (a, b).

b. $D(a, b) > 0$ and $f_{xx}(a, b) > 0$ implies that $f(x, y)$ has a **relative minimum** at the point (a, b).

c. $D(a, b) < 0$ implies that $f(x, y)$ has neither a relative maximum nor a relative minimum at the point (a, b).

d. $D(a, b) = 0$ implies that the test is inconclusive, so some other technique must be used to solve the problem.

EXAMPLE 1   Find the relative extrema of the function

$$f(x, y) = x^2 + y^2$$

Solution   We have

$$f_x = 2x$$

$$f_y = 2y$$

To find the critical point(s) of f, we set $f_x = 0$ and $f_y = 0$ and solve the resulting system of simultaneous equations

$$2x = 0$$

$$2y = 0$$

obtaining $x = 0$, $y = 0$, or $(0, 0)$, as the sole critical point of f. Next, we apply the Second Derivative Test to determine the nature of the critical point $(0, 0)$. We compute

$$f_{xx} = 2, \quad f_{xy} = 0, \quad f_{yy} = 2$$

and

$$D(x, y) = f_{xx}f_{yy} - f_{xy}^2 = (2)(2) - 0 = 4$$

In particular, $D(0, 0) = 4$. Since $D(0, 0) > 0$ and $f_{xx}(0, 0) = 2 > 0$, we conclude that $f(x, y)$ has a relative minimum at the point $(0, 0)$. The relative minimum value, 0, also happens to be the absolute minimum of f. The graph of the function f, shown in Figure 8.19, confirms these results.   ◦ ◦ ◦

Figure 8.19
The graph of $f(x, y) = x^2 + y^2$.

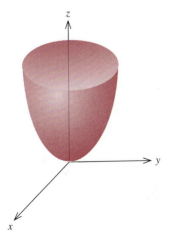

EXAMPLE 2 Find the relative extrema of the function

$$f(x, y) = 3x^2 - 4xy + 4y^2 - 4x + 8y + 4$$

Solution We have

$$f_x = 6x - 4y - 4$$
$$f_y = -4x + 8y + 8$$

To find the critical points of f, we set $f_x = 0$ and $f_y = 0$ and solve the resulting system of simultaneous equations

$$6x - 4y = 4$$
$$-4x + 8y = -8$$

Multiplying the first equation by 2 and the second equation by 3, we obtain the equivalent system

$$12x - 8y = 8$$
$$-12x + 24y = -24$$

Adding the two equations gives $16y = -16$, or $y = -1$. We substitute this value for y into either equation in the system to get $x = 0$. Thus, the only critical point of f is $(0, -1)$. Next, we apply the Second Derivative Test to determine whether the point $(0, -1)$ gives rise to a relative extremum of f. We compute

$$f_{xx} = 6, \quad f_{xy} = -4, \quad f_{yy} = 8$$

and

$$D(x, y) = f_{xx}f_{yy} - f_{xy}^2 = (6)(8) - (-4)^2 = 32$$

Since $D(0, -1) = 32 > 0$ and $f_{xx}(0, -1) = 6 > 0$, we conclude that $f(x, y)$ has a relative minimum at $(0, -1)$. The value of $f(x, y)$ at the point $(0, -1)$ is given by

$$f(0, -1) = 3(0)^2 - 4(0)(-1) + 4(-1)^2 - 4(0) + 8(-1) + 4 = 0$$

Suppose $f(x, y)$ has a relative extremum (relative maximum or relative minimum) at a point (a, b). Let $g(x) = f(x, b)$ and $h(y) = f(a, y)$. Assuming that f and g are differentiable, explain why $g'(a) = 0$ and $h'(b) = 0$. Explain why these results are equivalent to the conditions $f_x(a, b) = 0$ and $f_y(a, b) = 0$.

EXAMPLE 3 Find the relative extrema of the function

$$f(x, y) = 4y^3 + x^2 - 12y^2 - 36y + 2$$

Solution To find the critical points of f, we set $f_x = 0$ and $f_y = 0$ simultaneously, obtaining

$$f_x = 2x = 0$$
$$f_y = 12y^2 - 24y - 36 = 0$$

The first equation implies that $x = 0$. The second equation implies that

$$y^2 - 2y - 3 = 0$$
$$(y + 1)(y - 3) = 0$$

a. Refer to the Second Derivative Test. Can the condition $f_{xx}(a, b) < 0$ in part 2a be replaced by the condition $f_{yy}(a, b) < 0$? Explain your answer. How about the condition $f_{xx}(a, b) > 0$ in part 2b?

b. Let $f(x, y) = x^4 + y^4$. Show that $(0, 0)$ is a critical point of f and that $D(0, 0) = 0$. Explain why f has a relative (in fact, an absolute) minimum at $(0, 0)$. Does this contradict the Second Derivative Test? Explain your answer.

that is, $y = -1$ or 3. Therefore, there are two critical points of the function f—namely, $(0, -1)$ and $(0, 3)$. Next, we apply the Second Derivative Test to determine the nature of each of the two critical points. We compute

$$f_{xx} = 2, \quad f_{xy} = 0, \quad f_{yy} = 24y - 24 = 24(y - 1)$$

Therefore,

$$D(x, y) = f_{xx}f_{yy} - f_{xy}^2 = 48(y - 1)$$

For the point $(0, -1)$,

$$D(0, -1) = 48(-1 - 1) = -96 < 0$$

Since $D(0, -1) < 0$, we conclude that the point $(0, -1)$ gives a saddle point of f. For the point $(0, 3)$,

$$D(0, 3) = 48(3 - 1) = 96 > 0$$

Since $D(0, 3) > 0$ and $f_{xx}(0, 3) > 0$, we conclude that the function f has a relative minimum at the point $(0, 3)$. Furthermore, since

$$f(0, 3) = 4(3)^3 + (0)^2 - 12(3)^2 - 36(3) + 2$$
$$= -106$$

we see that the relative minimum value of f is -106.   ○ ○ ○

Applications

As in a practical optimization problem involving a function of one variable, the solution to an optimization problem involving a function of several variables calls for finding the *absolute* extremum of the function. Determining the absolute extremum of a function of several variables is more difficult than merely finding the relative extrema of the function. However, in many situations, the absolute extremum of a function actually coincides with the largest relative extremum of the function that occurs in the interior of its domain. We assume that the problems considered here belong to this category. Furthermore, the existence of the absolute extremum (solution) of a practical problem is often deduced from the geometric or physical nature of the problem.

EXAMPLE 4   The total weekly revenue (in dollars) that Acrosonic Company realizes in producing and selling its bookshelf loudspeaker systems is given by

$$R(x, y) = -\frac{1}{4}x^2 - \frac{3}{8}y^2 - \frac{1}{4}xy + 300x + 240y$$

where x denotes the number of fully assembled units and y denotes the number of kits produced and sold per week. The total weekly cost attributable to the production of these loudspeakers is

$$C(x, y) = 180x + 140y + 5000$$

dollars, where x and y have the same meaning as before. Determine how many assembled units and how many kits Acrosonic should produce per week to maximize its profit.

Solution   The contribution to Acrosonic's weekly profit from the production and sale of the bookshelf loudspeaker systems is given by

$$P(x, y) = R(x, y) - C(x, y)$$

$$= \left(-\frac{1}{4} x^2 - \frac{3}{8} y^2 - \frac{1}{4} xy + 300x + 240y \right) - (180x + 140y + 5000)$$

$$= -\frac{1}{4} x^2 - \frac{3}{8} y^2 - \frac{1}{4} xy + 120x + 100y - 5000$$

To find the relative maximum of the profit function $P(x, y)$, we first locate the critical point(s) of P. Setting $P_x(x, y)$ and $P_y(x, y)$ equal to zero, we obtain

$$P_x = -\frac{1}{2} x - \frac{1}{4} y + 120 = 0$$

$$P_y = -\frac{3}{4} y - \frac{1}{4} x + 100 = 0$$

Solving the first of these equations for y yields

$$y = -2x + 480$$

which, upon substitution into the second equation, yields

$$-\frac{3}{4}(-2x + 480) - \frac{1}{4} x + 100 = 0$$

$$6x - 1440 - x + 400 = 0$$

or $$x = 208$$

We substitute this value of x into the equation $y = -2x + 480$ to get

$$y = 64$$

Therefore, the function P has the sole critical point $(208, 64)$. To show that the point $(208, 64)$ is a solution to our problem, we use the Second Derivative Test. We compute

$$P_{xx} = -\frac{1}{2}, \quad P_{xy} = -\frac{1}{4}, \quad P_{yy} = -\frac{3}{4}$$

so

$$D(x, y) = \left(-\frac{1}{2} \right)\left(-\frac{3}{4} \right) - \left(-\frac{1}{4} \right)^2 = \frac{3}{8} - \frac{1}{16} = \frac{5}{16}$$

In particular, $D(208, 64) = 5/16 > 0$.

Since $D(208, 64) > 0$ and $P_{xx}(208, 64) < 0$, the point $(208, 64)$ yields a relative maximum of P. This relative maximum is also the absolute maximum of P. We conclude that Acrosonic can maximize its weekly profit by manufacturing 208 assembled units and 64 kits of their bookshelf loudspeaker systems. The maximum weekly profit realizable from the production and sale of these loudspeaker systems is given by

$$P(208, 64) = -\frac{1}{4}(208)^2 - \frac{3}{8}(64)^2 - \frac{1}{4}(208)(64)$$

$$+ 120(208) + 100(64) - 5000$$

$$= 10{,}680$$

or $10,680.

EXAMPLE 5 A television relay station will serve towns A, B, and C, whose relative locations are shown in Figure 8.20. Determine a site for the location of the station if the sum of the squares of the distances from each town to the site is minimized.

Solution Suppose the required site is located at the point $P(x, y)$. With the aid of the distance formula, we find that the square of the distance from town A to the site is

$$(x - 30)^2 + (y - 20)^2$$

The respective distances from towns B and C to the site are found in a similar manner, so that the sum of the squares of the distances from each town to the site is given by

$$f(x, y) = (x - 30)^2 + (y - 20)^2 + (x + 20)^2$$
$$+ (y - 10)^2 + (x - 10)^2 + (y + 10)^2$$

To find the relative minimum of $f(x, y)$, we first find the critical point(s) of f. Using the Chain Rule to find $f_x(x, y)$ and $f_y(x, y)$ and setting each equal to zero, we obtain

$$f_x = 2(x - 30) + 2(x + 20) + 2(x - 10) = 6x - 40 = 0$$
$$f_y = 2(y - 20) + 2(y - 10) + 2(y + 10) = 6y - 40 = 0$$

from which we deduce that $(20/3, 20/3)$ is the sole critical point of f. Since

$$f_{xx} = 6, \quad f_{xy} = 0, \quad f_{yy} = 6$$

we have

$$D(x, y) = f_{xx}f_{yy} - f_{xy}^2 = (6)(6) - 0 = 36$$

Since $D(20/3, 20/3) > 0$ and $f_{xx}(20/3, 20/3) > 0$, we conclude that the point $(20/3, 20/3)$ yields a relative minimum of f. Thus, the required site has coordinates $x = 20/3$ and $y = 20/3$.

Figure 8.20
Locating a site for a television relay station.

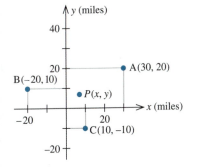

SELF-CHECK EXERCISES 8.3

1. Let $f(x, y) = 2x^2 + 3y^2 - 4xy + 4x - 2y + 3$.
   a. Find the critical point of f.
   b. Use the Second Derivative Test to classify the nature of the critical point.
   c. Find the relative extremum of f if it exists.

2. Robertson Controls Company manufactures two models of setback thermostats: a standard mechanical thermostat and a deluxe electronic thermostat. Robertson's monthly revenue (in hundreds of dollars) is

$$R(x, y) = -\frac{1}{8}x^2 - \frac{1}{2}y^2 - \frac{1}{4}xy + 20x + 60y$$

where x (in units of a hundred) denotes the number of mechanical thermostats manufactured and y (in units of a hundred) denotes the number of electronic thermostats manufactured per month. The total monthly cost incurred in producing these thermostats is

$$C(x, y) = 7x + 20y + 280$$

hundred dollars. Find how many thermostats of each model Robertson should manufacture per month in order to maximize its profits. What is the maximum profit?

Solutions to Self-Check Exercises 8.3 can be found on page 620.

8.3 EXERCISES

In exercises 1–20, find the critical point(s) of the function. Then use the Second Derivative Test to classify the nature of each point, if possible. Finally, determine the relative extrema of the function.

1. $f(x, y) = 1 - 2x^2 - 3y^2$

2. $f(x, y) = x^2 - xy + y^2 + 1$

3. $f(x, y) = x^2 - y^2 - 2x + 4y + 1$

4. $f(x, y) = 2x^2 + y^2 - 4x + 6y + 3$

5. $f(x, y) = x^2 + 2xy + 2y^2 - 4x + 8y - 1$

6. $f(x, y) = x^2 - 4xy + 2y^2 + 4x + 8y - 1$

7. $f(x, y) = 2x^3 + y^2 - 9x^2 - 4y + 12x - 2$

8. $f(x, y) = 2x^3 + y^2 - 6x^2 - 4y + 12x - 2$

9. $f(x, y) = x^3 + y^2 - 2xy + 7x - 8y + 4$

10. $f(x, y) = 2y^3 - 3y^2 - 12y + 2x^2 - 6x + 2$

11. $f(x, y) = x^3 - 3xy + y^3 - 2$

12. $f(x, y) = x^3 - 2xy + y^2 + 5$

13. $f(x, y) = xy + \frac{4}{x} + \frac{2}{y}$

14. $f(x, y) = \frac{x}{y^2} + xy$

15. $f(x, y) = x^2 - e^{y^2}$

16. $f(x, y) = e^{x^2 - y^2}$

17. $f(x, y) = e^{x^2 + y^2}$

18. $f(x, y) = e^{xy}$

19. $f(x, y) = \ln(1 + x^2 + y^2)$

20. $f(x, y) = xy + \ln x + 2y^2$

21. Maximizing Profit   The total weekly revenue (in dollars) the Country Workshop realized in manufacturing and selling its rolltop desks is given by

$$R(x, y) = -0.2x^2 - 0.25y^2 - 0.2xy + 200x + 160y$$

where x denotes the number of finished units and y denotes the number of unfinished units manufactured and sold per week. The total weekly cost attributable to the manufacture of these desks is given by

$$C(x, y) = 100x + 70y + 4000$$

dollars. Determine how many finished units and how many unfinished units the company should manufacture per week in order to maximize its profit. What is the maximum profit realizable?

22. **Maximizing Profit** The total daily revenue (in dollars) that Weston Publishing Company realizes in publishing and selling its English language dictionaries is given by

$$R(x, y) = -0.005x^2 - 0.003y^2 - 0.002xy$$
$$+ 20x + 15y$$

where x denotes the number of deluxe copies and y denotes the number of standard copies published and sold daily. The total daily cost of publishing these dictionaries is given by

$$C(x, y) = 6x + 3y + 200$$

dollars. Determine how many deluxe copies and how many standard copies Weston should publish per day to maximize its profits. What is the maximum profit realizable?

23. **Maximum Price** The rectangular region R shown in the accompanying figure represents the financial district of a city. The price of land within the district is approximated by the function

$$p(x, y) = 200 - 10\left(x - \frac{1}{2}\right)^2 - 15(y - 1)^2$$

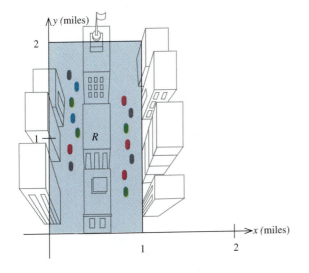

where $p(x, y)$ is the price of land at the point (x, y) in dollars per square foot and x and y are measured in miles. At what point in the financial district is the price of land highest?

24. **Maximizing Profit** C & G Imports, Inc., imports two brands of white wine, one from Germany and the other from Italy. The German wine costs $4 a bottle, and the Italian wine can be obtained for $3 a bottle. It has been estimated that if the German wine retails at p dollars per bottle and the Italian wine is sold for q dollars per bottle, then

$$2000 - 150p + 100q$$

bottles of the German wine and

$$1000 + 80p - 120q$$

bottles of the Italian wine will be sold per week. Determine the unit price for each brand that will allow C & G to realize the largest possible weekly profit.

25. **Determining the Optimal Site** An auxiliary electric power station will serve three communities, A, B, and C, whose relative locations are shown in the accompanying figure. Determine where the power station should be located if the sum of the squares of the distances from each community to the site is minimized.

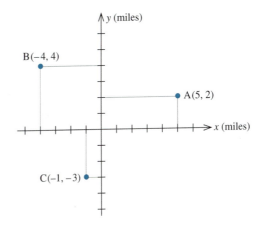

26. **Packaging** An open rectangular box having a volume of 108 cubic inches is to be constructed from a tin sheet. Find the dimensions of such a box if the amount of material used in its construction is to be minimal. [*Hint:* Let the dimensions of the box be x'' by y'' by z''. Then

$$xyz = 108$$

and the amount of material used is given by

$$S = xy + 2yz + 2xz$$

Show that

$$S = f(x, y) = xy + \frac{216}{x} + \frac{216}{y}$$

Minimize $f(x, y)$.]

27. Packaging Postal regulations specify that the combined length and girth of a parcel sent by parcel post may not exceed 108 inches. Find the dimensions of the rectangular package that would have the greatest possible volume under these regulations.

[*Hint:* Let the dimensions of the box be x'' by y'' by z'' (see the figure). Then

$$2x + 2z + y = 108$$

and the volume

$$V = xyz$$

Show that

$$V = f(x, z) = 108xz - 2x^2z - 2xz^2$$

Maximize $f(x, z)$.]

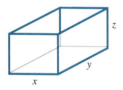

28. Minimizing Heating and Cooling Costs A building in the shape of a rectangular box is to have a volume of 12,000 cubic feet (see the figure). It is estimated that the annual heating and cooling costs will be $2 per square foot for the top, $4 per square foot for the front and back, and $3 per square foot for the sides. Find the dimensions of the building that will result in a minimal annual heating and cooling cost. What is the minimal annual heating and cooling cost?

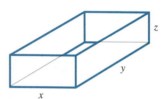

29. Packaging An open box having a volume of 48 cubic inches is to be constructed. If the box is to include a partition that is parallel to a side of the box, as shown in the figure, and the amount of material used is to be minimal, what should the dimensions of the box be?

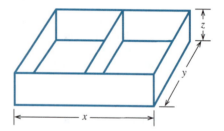

SOLUTIONS TO SELF-CHECK EXERCISES 8.3

I. a. To find the critical point(s) of f, we solve the system of equations

$$f_x = 4x - 4y + 4 = 0$$
$$f_y = -4x + 6y - 2 = 0$$

obtaining $x = -2$ and $y = -1$. Thus, the only critical point of f is $(-2, -1)$.
b. We have $f_{xx} = 4$, $f_{xy} = -4$, and $f_{yy} = 6$, so

$$D(x, y) = f_{xx}f_{yy} - f_{xy}^2$$
$$= (4)(6) - (-4)^2 = 8$$

Since $D(-2, -1) > 0$ and $f_{xx}(-2, -1) > 0$, we conclude that f has a relative minimum at the point $(-2, -1)$.

c. The relative minimum value of $f(x, y)$ at the point $(-2, -1)$ is

$$f(-2, -1) = 2(-2)^2 + 3(-1)^2 - 4(-2)(-1) + 4(-2) - 2(-1) + 3$$
$$= 0$$

2. Robertson's monthly profit is

$$P(x, y) = R(x, y) - C(x, y)$$
$$= \left(-\frac{1}{8}x^2 - \frac{1}{2}y^2 - \frac{1}{4}xy + 20x + 60y\right) - (7x + 20y + 280)$$
$$= -\frac{1}{8}x^2 - \frac{1}{2}y^2 - \frac{1}{4}xy + 13x + 40y - 280$$

The critical point of P is found by solving the system

$$P_x = -\frac{1}{4}x - \frac{1}{4}y + 13 = 0$$

$$P_y = -\frac{1}{4}x - y + 40 = 0$$

giving $x = 16$ and $y = 36$. Thus $(16, 36)$ is the critical point of P. Next,

$$P_{xx} = -\frac{1}{4}, \quad P_{xy} = -\frac{1}{4}, \quad P_{yy} = -1$$

and

$$D(x, y) = f_{xx}f_{yy} - f_{xy}^2$$
$$= \left(-\frac{1}{4}\right)(-1) - \left(-\frac{1}{4}\right)^2 = \frac{3}{16}$$

Since $D(16, 36) > 0$ and $P_{xx}(16, 36) < 0$, the point $(16, 36)$ yields a relative maximum of P. We conclude that the monthly profit is maximized by manufacturing 1600 mechanical and 3600 electronic setback thermostats per month. The maximum monthly profit realizable is

$$P(16, 36) = -\frac{1}{8}(16)^2 - \frac{1}{2}(36)^2 - \frac{1}{4}(16)(36) + 13(16) + 40(36) - 280$$

$$= 544$$

or $54,400.

8.4 THE METHOD OF LEAST SQUARES

The Method of Least Squares

In Section 1.4, Example 10, we saw how a linear equation can be used to approximate the sales trend for a local sporting goods store. As we saw there, one use of a **trend line** is to predict a store's future sales. Recall that we

Figure 8.21

A scatter diagram.

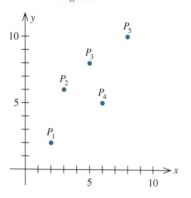

obtained the line by requiring that it pass through two data points, the rationale being that such a line seems to *fit* the data reasonably well.

In this section we describe a general method, known as the **method of least squares,** for determining a straight line that, in some sense, *best* fits a set of data points when the points are scattered about a straight line. In order to illustrate the principle behind the method of least squares, suppose, for simplicity, that we are given five data points,

$$P_1(x_1, y_1) \quad P_2(x_2, y_2), \quad P_3(x_3, y_3), \quad P_4(x_4, y_4), \quad P_5(x_5, y_5)$$

that describe the relationship between the two variables x and y. By plotting these data points, we obtain a graph called a **scatter diagram** (Figure 8.21).

If we try to fit a straight line to these data points, the line will miss the first, second, third, fourth, and fifth data points by the amounts $d_1, d_2, d_3, d_4,$ and d_5, respectively (Figure 8.22).

The **principle of least squares** states that the straight line L that fits the data points best is the one chosen by requiring that the sum of the squares of $d_1, d_2, \ldots, d_5$, that is,

$$d_1^2 + d_2^2 + d_3^2 + d_4^2 + d_5^2$$

Figure 8.22

The approximating line misses each point by the amounts $d_1, d_2, \ldots, d_5$.

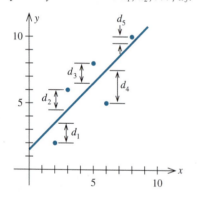

be made as small as possible. If we think of the amount d_1 as the error made when the value y_1 is approximated by the corresponding value of y lying on the straight line L, and d_2 as the error made when the value y_2 is approximated by the corresponding value of y, and so on, then it can be seen that the least-squares criterion calls for minimizing the sum of the squares of the errors. The line L obtained in this manner is called the **least-squares line,** or **regression line.**

To find a method for computing the regression line L, suppose L has the representation $y = f(x) = mx + b$, where m and b are to be determined. Observe that

$$\begin{aligned}
&d_1^2 + d_2^2 + d_3^2 + d_4^2 + d_5^2 \\
&= [f(x_1) - y_1]^2 + [f(x_2) - y_2]^2 + [f(x_3) - y_3]^2 \\
&\quad + [f(x_4) - y_4]^2 + [f(x_5) - y_5]^2 \\
&= (mx_1 + b - y_1)^2 + (mx_2 + b - y_2)^2 + (mx_3 + b - y_3)^2 \\
&\quad + (mx_4 + b - y_4)^2 + (mx_5 + b - y_5)^2
\end{aligned}$$

and may be viewed as a function of the two variables m and b. Thus, the least-squares criterion is equivalent to minimizing the function

$$\begin{aligned}
f(m, b) &= (mx_1 + b - y_1)^2 + (mx_2 + b - y_2)^2 + (mx_3 + b - y_3)^2 \\
&\quad + (mx_4 + b - y_4)^2 + (mx_5 + b - y_5)^2
\end{aligned}$$

with respect to m and b. Using the Chain Rule, we compute

$$\begin{aligned}
\frac{\partial f}{\partial m} &= 2(mx_1 + b - y_1)x_1 + 2(mx_2 + b - y_2)x_2 + 2(mx_3 + b - y_3)x_3 \\
&\quad + 2(mx_4 + b - y_4)x_4 + 2(mx_5 + b - y_5)x_5 \\
&= 2[mx_1^2 + bx_1 - x_1y_1 + mx_2^2 + bx_2 - x_2y_2 + mx_3^2 + bx_3 - x_3y_3 \\
&\quad + mx_4^2 + bx_4 - x_4y_4 + mx_5^2 + bx_5 - x_5y_5]
\end{aligned}$$

$$= 2[(x_1^2 + x_2^2 + x_3^2 + x_4^2 + x_5^2)m + (x_1 + x_2 + x_3 + x_4 + x_5)b$$
$$- (x_1 y_1 + x_2 y_2 + x_3 y_3 + x_4 y_4 + x_5 y_5)]$$

and

$$\frac{\partial f}{\partial b} = 2(mx_1 + b - y_1) + 2(mx_2 + b - y_2) + 2(mx_3 + b - y_3)$$
$$+ 2(mx_4 + b - y_4) + 2(mx_5 + b - y_5)$$
$$= 2[(x_1 + x_2 + x_3 + x_4 + x_5)m + 5b - (y_1 + y_2 + y_3 + y_4 + y_5)]$$

Setting

$$\frac{\partial f}{\partial m} = 0 \quad \text{and} \quad \frac{\partial f}{\partial b} = 0$$

gives

$$(x_1^2 + x_2^2 + x_3^2 + x_4^2 + x_5^2)m + (x_1 + x_2 + x_3 + x_4 + x_5)b$$
$$= x_1 y_1 + x_2 y_2 + x_3 y_3 + x_4 y_4 + x_5 y_5$$

and

$$(x_1 + x_2 + x_3 + x_4 + x_5)m + 5b = y_1 + y_2 + y_3 + y_4 + y_5$$

Solving these two simultaneous equations for m and b then leads to an equation $y = mx + b$ of a straight line.

Before looking at an example, we state a more general result whose derivation is identical to the special case involving the five data points just discussed.

THE METHOD OF LEAST SQUARES

Suppose we are given n data points

$$P_1(x_1, y_1), \quad P_2(x_2, y_2), \quad P_3(x_3, y_3), \quad \ldots, \quad P_n(x_n, y_n)$$

Then the least-squares (regression) line for the data is given by the linear equation

$$y = f(x) = mx + b$$

where the constants m and b satisfy the equations

$$(x_1^2 + x_2^2 + x_3^2 + \cdots + x_n^2)m + (x_1 + x_2 + x_3 + \cdots + x_n)b$$
$$= x_1 y_1 + x_2 y_2 + x_3 y_3 + \cdots + x_n y_n \tag{4}$$

and

$$(x_1 + x_2 + x_3 + \cdots + x_n)m + nb$$
$$= y_1 + y_2 + y_3 + \cdots + y_n \tag{5}$$

simultaneously. Equations (4) and (5) are called **normal equations.**

EXAMPLE 1   Find an equation of the least-squares line for the data

$$P_1(1, 1), \quad P_2(2, 3), \quad P_3(3, 4), \quad P_4(4, 3), \quad P_5(5, 6)$$

Solution   Here we have $n = 5$ and

$$x_1 = 1, \quad x_2 = 2, \quad x_3 = 3, \quad x_4 = 4, \quad x_5 = 5$$
$$y_1 = 1, \quad y_2 = 3, \quad y_3 = 4, \quad y_4 = 3, \quad y_5 = 6$$

so equation (4) becomes

$$(1 + 4 + 9 + 16 + 25)m + (1 + 2 + 3 + 4 + 5)b$$
$$= 1 + 6 + 12 + 12 + 30$$

or $55m + 15b = 61$ **(6)**

and equation (5) becomes

$$(1 + 2 + 3 + 4 + 5)m + 5b = 1 + 3 + 4 + 3 + 6$$

or $15m + 5b = 17$ **(7)**

Solving equation (7) for b gives

$$b = -3m + \frac{17}{5}$$ **(8)**

which, upon substitution into equation (6), gives

$$15\left(-3m + \frac{17}{5}\right) + 55m = 61$$

$$-45m + 51 + 55m = 61$$

$$10m = 10$$

or $m = 1$

Substituting this value of m into equation (8) gives

$$b = -3 + \frac{17}{5}$$

$$= \frac{2}{5} = 0.4$$

Therefore, the required least-squares line is

$$y = x + 0.4$$

The scatter diagram and the regression line are shown in Figure 8.23.

○ ○ ○

Figure 8.23
The scatter diagram and the least-squares line $y = x + 0.4$.

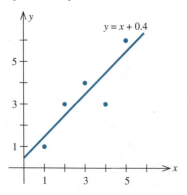

Refer to the data in Example 1.
a. Show that the sum of the errors, $d_1 + d_2 + \cdots + d_5$, where

$$d_1 = f(x_1) - y_1, \quad d_2 = f(x_2) - y_2, \quad \ldots, \quad d_5 = f(x_5) - y_5$$

is equal to zero.

b. Find the sum of the squares of the errors, $d_1^2 + d_2^2 + \cdots + d_5^2$.

c. In the method of least squares, explain why we consider the sum of the squares of the errors rather than just the sum of the errors.

d. The square root of the average of the sum of the squares of the errors is called the *root-mean-square (rms) error*. Find the rms error for the problem at hand. [*Hint:* The rms error is given by $\sqrt{D/5}$, where D is the quantity computed in part (b).]

e. Explain why minimizing the rms error is equivalent to minimizing the sum of the squares of the errors. (Recall that we used this result when we derived the normal equations.)

Applications

EXAMPLE 2 The proprietor of Leisure Travel Service compiled the following data relating the firm's annual profit to its annual advertising expenditure (both measured in thousands of dollars).

Annual Advertising Expenditure (x)	12	14	17	21	26	30
Annual Profit (y)	60	70	90	100	100	120

a. Determine an equation of the least-squares line for these data.

b. Draw a scatter diagram and the least-squares line for these data.

c. Use the result obtained in (a) to predict Leisure Travel's annual profit if the annual advertising budget is $20,000.

Solution

a. The calculations required for obtaining the normal equations are summarized as follows:

	x	y	x^2	xy
	12	60	144	720
	14	70	196	980
	17	90	289	1,530
	21	100	441	2,100
	26	100	676	2,600
	30	120	900	3,600
Sum	120	540	2646	11,530

The normal equations are

$$6b + 120m = 540 \tag{9}$$

$$120b + 2646m = 11{,}530 \tag{10}$$

Solving equation (9) for b gives

$$b = -20m + 90 \tag{11}$$

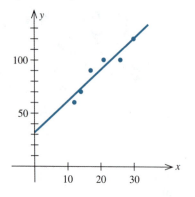

which, upon substitution into equation (10), gives

$$120(-20m + 90) + 2646m = 11,530$$
$$-2400m + 10,800 + 2646m = 11,530$$
$$246m = 730$$

or

$$m = 2.97$$

Substituting this value of m into equation (11) gives

$$b = -20(2.97) + 90 = 30.6$$

Therefore, the required least-squares line is given by

$$y = f(x) = 2.97x + 30.6$$

b. The scatter diagram and the least-squares line are shown in Figure 8.24.

c. Leisure Travel's predicted annual profit corresponding to an annual budget of $20,000 is given by

$$f(20) = 2.97(20) + 30.6 = 90$$

or $90,000.

EXAMPLE 3   A market research study conducted for Century Communications Company provided the following data on the projected monthly sales x (in thousands) of Century's videocassette version of a box-office hit adventure movie with a proposed wholesale unit price of p dollars.

p	38	36	34.5	30	28.5
x	2.2	5.4	7.0	11.5	14.6

a. Find the demand equation if the demand curve is the least-squares line for these data.

b. Suppose the total monthly cost function associated with producing and distributing the videocassette movies is given by

$$C(x) = 4x + 25$$

where x denotes the number of units (in thousands) produced and sold and $C(x)$ is in thousands of dollars. Determine the unit wholesale price that will maximize Century's monthly profit.

Solution

a. The calculations required for obtaining the normal equations may be summarized as follows:

	x	p	x^2	xp
	2.2	38	4.84	83.6
	5.4	36	29.16	194.4
	7.0	34.5	49	241.5
	11.5	30	132.25	345
	14.6	28.5	213.16	416.1
Sum	40.7	167	428.41	1280.6

The normal equations are

$$5b + 40.7m = 167$$
$$40.7b + 428.41m = 1280.6$$

Solving this system of linear equations simultaneously, we find that

$$m = -0.81 \quad \text{and} \quad b = 39.99$$

Therefore, the required least-squares line is given by

$$p = f(x) = -0.81x + 39.99$$

which is the required demand equation, provided $0 \le x \le 49.37$.

b. The total revenue function in this case is given by

$$R(x) = xp = -0.81x^2 + 39.99x$$

and since the total cost function is

$$C(x) = 4x + 25$$

we see that the profit function is

$$P(x) = -0.81x^2 + 39.99x - (4x + 25)$$
$$= -0.81x^2 + 35.99x - 25$$

To find the absolute minimum of $P(x)$ over the closed interval $[0, 49.37]$, we compute

$$P'(x) = -1.62x + 35.99$$

Since $P'(x) = 0$, we find $x = 22.22$ as the only critical point of P. Finally, from the table

x	0	22.22	49.37
$P(x)$	-25	374.78	-222.47

we see that the optimal wholesale price is $22.22 dollars per videocassette.

∘ ∘ ∘

SELF-CHECK EXERCISES 8.4

1. Find an equation of the least-squares line for the data

$$P_1(0, 3), \quad P_2(2, 6.5), \quad P_3(4, 10), \quad P_4(6, 16), \quad P_5(7, 16.5)$$

2. The following data obtained from the U.S. Department of Commerce give the percentage of people over age 65 who have high school diplomas.

Year (x)	0	6	11	16	22	26
Percent with Diploma (y)	19	25	30	35	44	48

Here $x = 0$ corresponds to the beginning of the year 1959.
a. Find an equation of the least-squares line for the given data.
b. Assuming that this trend continues, what percentage of people over 65 will have high school diplomas at the beginning of the year 2001 ($x = 42$)?
Source: U.S. Department of Commerce

Solutions to Self-Check Exercises 8.4 can be found on page 632.

8.4 EXERCISES

A calculator is recommended for this exercise set. In exercises 1–6, (a) find the equation of the least-squares line for the given data and (b) draw a scatter diagram for the given data and graph the least-squares line.

1.

x	1	2	3	4
y	4	6	8	11

2.

x	1	3	5	7	9
y	9	8	6	3	2

3.

x	1	2	3	4	4	6
y	4.5	5	3	2	3.5	1

4.

x	1	1	2	3	4	4	5
y	2	3	3	3.5	3.5	4	5

5. $P_1(1, 3)$, $P_2(2, 5)$, $P_3(3, 5)$, $P_4(4, 7)$, $P_5(5, 8)$

6. $P_1(1, 8)$, $P_2(2, 6)$, $P_3(5, 6)$, $P_4(7, 4)$, $P_5(10, 1)$

7. College Admissions The following data were compiled by the admissions office at Faber College during the past 5 years. The data relate the number of college brochures and follow-up letters (x) sent to a preselected list of high school juniors who had taken the PSAT and the number of completed applications (y) received from these students (both measured in units of 1000).

x	4	4.5	5	5.5	6
y	0.5	0.6	0.8	0.9	1.2

a. Determine the equation of the least-squares line for these data.
b. Draw a scatter diagram and the least-squares line for these data.
c. Use the result obtained in (a) to predict the number of completed applications that might be expected if 6400 brochures and follow-up letters are sent out during the next year.

8. Net Sales The management of Kaldor, Inc., a manufacturer of electric motors, submitted the following data in the annual report to its stockholders. The table shows the net sales (in millions of dollars) during the 5 years that have elapsed since the new management team took over. (The first year the firm operated under the new management corresponds to the time period $x = 1$, and the four subsequent years correspond to $x = 2, 3, 4, 5$.)

Year (x)	1	2	3	4	5
Net Sales (y)	426	437	460	473	477

a. Determine the equation of the least-squares line for these data.
b. Draw a scatter diagram and the least-squares line for these data.
c. Use the result obtained in (a) to predict the net sales for the upcoming year.

9. SAT Verbal Scores The following data were compiled by the superintendent of schools in a large metropolitan area. The table shows the average SAT verbal scores of high school seniors during the 5 years since the district implemented the "back-to-basics" program.

Year (x)	1	2	3	4	5
Average Score (y)	436	438	428	430	426

a. Determine the equation of the least-squares line for these data.

b. Draw a scatter diagram and the least-squares line for these data.

c. Use the result obtained in (a) to predict the average SAT verbal score of high school seniors 2 years from now ($x = 7$).

10. **Auto Operating Costs** The following figures were compiled by Clarke, Kingsley, and Company, a consulting firm that specializes in auto operating costs, relating the annual mileage (in thousands of miles) that an average new compact car is driven to the cost per mile (in cents) of operating the car.

Annual Mileage (x)	5	10	15	20	25	30
Cost per Mile (y)	50.3	34.8	30.1	27.4	25.6	23.5

a. Determine an equation of the least-squares line for these data.

b. Draw a scatter diagram and the least-squares line for these data.

c. Use the result obtained in (a) to estimate the cost per mile of operating a new company car if it is driven 8000 miles during the first year of ownership.
Source: Clarke, Kingsley, and Company

11. **Size of Average Farm** The size of the average farm in the United States has been growing steadily over the years. The following data, obtained from the U.S. Department of Agriculture, give the size of the average farm y (in acres) from 1940 through 1991. (Here $x = 0$ corresponds to the beginning of the year 1940.)

Year (x)	0	10	20	30	40	51
Size of Farm (y)	168	213	297	374	427	467

a. Find the equation of the least-squares line for these data.

b. Use the result of (a) to estimate the size of the average farm in the year 2000.
Source: The World Almanac

12. **Welfare Costs** According to the Massachusetts Department of Welfare, the spending (in billions of dollars) by Medicaid, the national health-care plan for the poor, over the 5-year period from 1988 to 1992 is summarized

in the following table. (Here $x = 0$ represents the beginning of the year 1988.)

Year (x)	0	1	2	3	4
Expenditure (y)	1.550	1.662	1.786	1.888	2.009

a. Find an equation of the least-squares line for these data.

b. Use the result of (a) to estimate Medicaid spending for the year 1996, assuming the trend continued.
Source: Massachusetts Department of Welfare

13. **Mass Transit Subsidies** The following table gives the projected state subsidies (in millions of dollars) to the Massachusetts Bay Transit Authority (MBTA) over a 5-year period.

Year (x)	1	2	3	4	5
Subsidy (y)	20	24	26	28	32

a. Find an equation of the least-squares line for these data.

b. Use the result of (a) to estimate the state subsidy to the MBTA for the eighth year ($x = 8$).
Source: Massachusetts Bay Transit Authority

14. **Social Security Wage Base** The Social Security (FICA) wage base (in thousands of dollars) from 1993 to 1998 is given in the following table:

Year	1993	1994	1995	1996	1997	1998
Wage Base (y)	57.6	60.6	61.2	62.7	65.4	68.4

a. Find an equation of the least-squares line for these data. (Let $x = 1$ represent the year 1993.)

b. Use your result of (a) to estimate the FICA wage base in the year 2002.
Source: The World Almanac

15. **Production of All-Aluminum Cans** Steel has been playing a decreasing role in the manufacture of beverage cans in the United States. According to the Can Manufacturing Institute, the use of bimetallic cans has been dwindling whereas the use of all-aluminum cans has been growing steadily. The accompanying table gives the production of all-aluminum cans over the period from 1975 through 1989.

Using Technology

Finding an Equation of a Least-Squares Line

A graphing utility is especially useful in calculating an equation of the least-squares line for a set of data. We simply enter the given data in the form of lists into the calculator and then use the linear regression function to obtain the coefficients of the required equation.

EXAMPLE 1 Find an equation of the least-squares line for the following data:

x	1.1	2.3	3.2	4.6	5.8	6.7	8
y	-5.8	-5.1	-4.8	-4.4	-3.7	-3.2	-2.5

Solution   First we enter the data as

$$x_1 = 1.1, \quad y_1 = -5.8, \quad x_2 = 2.3, \quad y_2 = -5.1, \quad x_3 = 3.2, \quad y_3 = -4.8, \quad x_4 = 4.6,$$

$$y_4 = -4.4, \quad x_5 = 5.8, \quad y_5 = -3.7, \quad x_6 = 6.7, \quad y_6 = -3.2, \quad x_7 = 8, \quad y_7 = -2.5$$

Then, using the linear regression function from the statistics menu, we find

$$a = -6.29996900666, \quad b = 0.460560979389, \quad \text{corr.} = .994488871079, \quad n = 7$$

Therefore, an equation of the least-squares line ($y = a + bx$) is

$$y = -6.3333 + 0.46x$$

The correlation coefficient of .99449 attests to the excellent fit of the regression line. (The closer to 1 it is, the better the correlation.)   ○ ○ ○

EXAMPLE 2 According to Pacific Gas and Electric, the nation's largest utility company, the demand for electricity from 1990 through the year 2000 is as follows:

t	0	2	4	6	8	10
y	333	917	1500	2117	2667	3292

Here $t = 0$ corresponds to 1990, and y gives the amount of electricity demanded in year t measured in megawatts. Find an equation of the least-squares line for these data.

Source: Pacific Gas and Electric

Solution   First, we enter the data as

$$x_1 = 0, \quad y_1 = 333, \quad x_2 = 2, \quad y_2 = 917, \quad x_3 = 4, \quad y_3 = 1500$$

$$x_4 = 6, \quad y_4 = 2117, \quad x_5 = 8, \quad y_5 = 2667, \quad x_6 = 10, \quad y_6 = 3292$$

Then, using the linear regression function from the statistics menu, we find

$$a = 328.476190476 \quad \text{and} \quad b = 295.171428571$$

Therefore, an equation of the least-squares line is

$$y = 328 + 295t$$

○ ○ ○

EXERCISES

In exercises 1–4, find an equation of the least-squares line for the data.

1.

x	2.1	3.4	4.7	5.6	6.8	7.2
y	8.8	12.1	14.8	16.9	19.8	21.1

2.

x	1.1	2.4	3.2	4.7	5.6	7.2
y	−0.5	1.2	2.4	4.4	5.7	8.1

3.

x	−2.1	−1.1	0.1	1.4	2.5	4.2	5.1
y	6.2	4.7	3.5	1.9	0.4	−1.4	−2.5

4.

x	−1.12	0.1	1.24	2.76	4.21	6.82
y	7.61	4.9	2.74	−0.47	−3.51	−8.94

5. Waste Generation   According to data from the Council on Environmental Quality, the amount of waste (in millions of tons per year) generated in the United States from 1960 to 1990 was:

Year	1960	1965	1970	1975
Amount (y)	81	100	120	124

Year	1980	1985	1990
Amount (y)	140	152	164

a. Find an equation of the least-squares line for these data. (Let x be in units of 5, and let $x = 1$ represent 1960.)
b. Use the result of (a) to estimate the amount of waste generated in the year 2000, assuming the trend continues.
Source: Council on Environmental Quality

6. Median Price of Homes   According to data from the Association of Realtors, the median price (in thousands of dollars) of existing homes in a certain metropolitan area from 1982 to 1992 was:

Year	1982	1983	1984	1985	1986	1987
Price (y)	66.4	69.8	72.8	76.0	79.6	83.1

Year	1988	1989	1990	1991	1992
Price (y)	86.3	89.5	92.3	96.0	99.5

a. Find an equation of the least-squares line for these data. (Let $x = 1$ represent 1982.)
b. Use the result of (a) to estimate the median price of a house in 1997, assuming the trend continued.
Source: Association of Realtors

Year	1975	1977	1979	1981
No. of Cans (in billions)	16.7	26	33.3	48.3

Year	1983	1985	1987	1989
No. of Cans (in billions)	57	65.8	74.2	83.3

a. Find an equation of the least-squares line for these data. (Let $x = 1$ represent 1975.)
b. Use the result of (a) to estimate the number of cans produced in 1993, assuming the trend continued.
Source: Can Manufacturing Institute

16. Health-Care Spending   The following data, compiled by the Organization for Economic Cooperation and Development (OECD) in 1990, gives the per capita gross domestic product (GDP) and the corresponding per capita spending on health care for selected countries.

Country	Turkey	Spain	Netherlands
GDP (per capita in thousands of dollars)	4.25	10	14
Health-care Spending (per capita in dollars)	178	667	1194

Country	Sweden	Switzerland	Canada
GDP (per capita in thousands of dollars)	15.5	17.8	19.5
Health-care Spending (per capita in dollars)	1500	1388	1640

a. Letting x denote a country's GDP (in thousands of dollars per capita) and y denote the per capita health-care spending (in dollars), find an equation of the least-squares line for these data giving the typical relationship between GDP and health-care spending for the selected countries.
b. The per capita GDP of the United States is $20,000. If the health-care spending of the United States were in line with that of these sample OECD countries, what would it be? [*Note:* The actual per capita health-care spending of the United States in 1990 was $2444.]
Source: Organization for Economic Cooperation and Development

SOLUTIONS TO SELF-CHECK EXERCISES 8.4

1. We first construct the table:

	x	y	x^2	xy
	0	3	0	0
	2	6.5	4	13
	4	10	16	40
	6	16	36	96
	7	16.5	49	115.5
Sum	19	52	105	264.5

The normal equations are

$$5b + 19m = 52$$

$$19b + 105m = 264.5$$

Solving the first equation for b gives

$$b = -3.8m + 10.4$$

which, upon substitution into the second equation, gives

$$19(-3.8m + 10.4) + 105m = 264.5$$

$$-72.2m + 197.6 + 105m = 264.5$$

$$32.8m = 66.9$$

or

$$m = 2.04$$

Substituting this value of m into the expression for b found earlier gives

$$b = -3.8(2.04) + 10.4 = 2.65$$

Therefore, the required least-squares line has the equation

$$y = 2.04x + 2.65$$

2. a. The calculations required for obtaining the normal equations may be summarized as follows:

x	y	x^2	xy
0	19	0	0
6	25	36	150
11	30	121	330
16	35	256	560
22	44	484	968
26	48	676	1248
Sum 81	201	1573	3256

The normal equations are

$$6b + 81m = 201$$

$$81b + 1573m = 3256$$

Solving this system of linear equations simultaneously, we find

$$m = 1.13 \quad \text{and} \quad b = 18.23$$

Therefore, the required least-squares line has the equation

$$y = f(x) = 1.13x + 18.23$$

b. The percentage of people over age 65 who will have high school diplomas at the beginning of 2001 is given by

$$f(42) = 1.13(42) + 18.23$$
$$= 65.69$$

or approximately 65.7%.

8.5 CONSTRAINED MAXIMA AND MINIMA AND THE METHOD OF LAGRANGE MULTIPLIERS

Constrained Relative Extrema

In Section 8.3 we studied the problem of determining the relative extremum of a function $f(x, y)$ without placing any restrictions on the independent variables x and y except, of course, that the point (x, y) lie in the domain of f. Such a relative extremum of a function f is referred to as an **unconstrained relative extremum of f.** However, in many practical optimization problems, we must maximize or minimize a function in which the independent variables are subjected to certain further constraints.

In this section we discuss a powerful method for determining the relative extrema of a function $f(x, y)$ whose independent variables x and y are required to satisfy one or more constraints of the form $g(x, y) = 0$. Such a relative extremum of a function f is called a **constrained relative extremum of f.** We can see the difference between an unconstrained extremum of a function $f(x, y)$ of two variables and a constrained extremum of f, where the independent variables x and y are subjected to a constraint of the form $g(x, y) = 0$, by considering the geometry of the two cases. Figure 8.25a depicts the graph

Figure 8.25

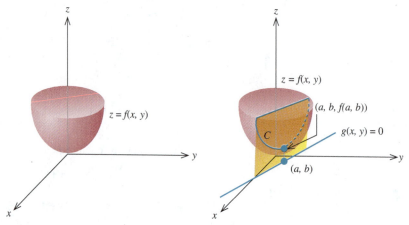

(a) $f(x, y)$ has an unconstrained relative extremum at $(0, 0)$.

(b) $f(x, y)$ has a constrained relative extremum at $(a, b, f(a, b))$.

of a function $f(x, y)$ that has an unconstrained relative minimum at the point $(0, 0)$. However, when the independent variables x and y are subjected to an equality constraint of the form $g(x, y) = 0$, the points (x, y, z) that satisfy both $z = f(x, y)$ and the constraint equation $g(x, y) = 0$ lie on a curve C. Therefore, the constrained relative minimum of f must also lie on C (Figure 8.25b).

Our first example involves an equality constraint $g(x, y) = 0$ in which we solve for the variable y explicitly in terms of x. In this case, we may apply the technique used in Chapter 4 to find the relative extrema of a function of one variable.

EXAMPLE I   Find the relative minimum of the function

$$f(x, y) = 2x^2 + y^2$$

subject to the constraint $g(x, y) = x + y - 1 = 0$.

Solution   Solving the constraint equation for y explicitly in terms of x, we obtain $y = -x + 1$. Substituting this value of y into the function $f(x, y) = 2x^2 + y^2$ results in a function of x:

$$h(x) = 2x^2 + (-x + 1)^2 = 3x^2 - 2x + 1$$

The function h describes the curve C lying on the graph of f on which the constrained relative minimum of f occurs. To find this point, use the technique developed in Chapter 4 to determine the relative extrema of a function of one variable:

$$h'(x) = 6x - 2 = 2(3x - 1)$$

Setting $h'(x) = 0$ gives $x = \frac{1}{3}$ as the sole critical point of the function h. Next, we find

$$h''(x) = 6$$

and, in particular,

$$h''\left(\frac{1}{3}\right) = 6 > 0$$

Therefore, by the Second Derivative Test, the point $x = \frac{1}{3}$ gives rise to a relative minimum of h. Substitute this value of x into the constraint equation $x + y - 1 = 0$ to get $y = \frac{2}{3}$. Thus, the point $(\frac{1}{3}, \frac{2}{3})$ gives rise to the required constrained relative minimum of f. Since

$$f\left(\frac{1}{3}, \frac{2}{3}\right) = 2\left(\frac{1}{3}\right)^2 + \left(\frac{2}{3}\right)^2 = \frac{2}{3}$$

the required constrained relative minimum value of f is $\frac{2}{3}$ at the point $(\frac{1}{3}, \frac{2}{3})$. It may be shown that $\frac{2}{3}$ is, in fact, a constrained absolute minimum value of f (Figure 8.26).

Figure 8.26
f has a constrained absolute minimum of $\frac{2}{3}$ at $(\frac{1}{3}, \frac{2}{3})$.

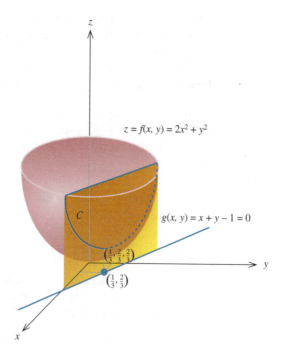

The Method of Lagrange Multipliers

The major drawback of the technique used in Example 1 is that it relies on our ability to solve the constraint equation $g(x, y) = 0$ for y explicitly in terms of x. This is not always an easy task. Moreover, even when we can solve the constraint equation $g(x, y) = 0$ for y explicitly in terms of x, the resulting function of one variable that is to be optimized may turn out to be unnecessarily complicated. Fortunately, an easier method exists. This method, called the **Method of Lagrange Multipliers** (Joseph Lagrange, 1736–1813), follows:

THE METHOD OF LAGRANGE MULTIPLIERS

To find the relative extremum of the function $f(x, y)$ subject to the constraint $g(x, y) = 0$ (assuming that these extreme values exist),

1. Form an auxiliary function

$$F(x, y, \lambda) = f(x, y) + \lambda g(x, y)$$

called the Lagrangian function (the variable λ is called the Lagrange multiplier).

2. Solve the system that comprises the equations

$$F_x = 0, \quad F_y = 0, \quad \text{and} \quad F_\lambda = 0$$

for all values of x, y, and λ.

3. Evaluate f at each of the points (x, y) found in step 2. The largest (smallest) of these values is the maximum (minimum) value of f.

Let us re-solve Example 1 using the Method of Lagrange Multipliers.

EXAMPLE 2   Using the Method of Lagrange Multipliers, find the relative minimum of the function

$$f(x, y) = 2x^2 + y^2$$

subject to the constraint $x + y = 1$.

Solution   Write the constraint equation $x + y = 1$ in the form $g(x, y) = x + y - 1 = 0$. Then form the Lagrangian function

$$F(x, y, \lambda) = f(x, y) + \lambda g(x, y)$$
$$= 2x^2 + y^2 + \lambda(x + y - 1)$$

To find the critical point(s) of the function F, solve the system that comprises the equations

$$F_x = 4x + \lambda = 0$$
$$F_y = 2y + \lambda = 0$$
$$F_\lambda = x + y - 1 = 0$$

Solving the first and second equations in this system for x and y in terms of λ, we obtain

$$x = -\frac{1}{4}\lambda \quad \text{and} \quad y = -\frac{1}{2}\lambda$$

which, upon substitution into the third equation, yields

$$-\frac{1}{4}\lambda - \frac{1}{2}\lambda - 1 = 0 \quad \text{or} \quad \lambda = -\frac{4}{3}$$

Therefore, $x = \frac{1}{3}$ and $y = \frac{2}{3}$, and $(\frac{1}{3}, \frac{2}{3})$ affords a constrained minimum of the function f, in agreement with the result obtained earlier.   ⊙ ⊙ ⊙

The Method of Lagrange Multipliers may be used to solve a problem involving a function of three or more variables, as illustrated in the next example.

EXAMPLE 3   Use the Method of Lagrange Multipliers to find the minimum of the function

$$f(x, y, z) = 2xy + 6yz + 8xz$$

subject to the constraint

$$xyz = 12,000$$

(*Note:* The existence of the minimum is suggested by the geometry of the problem.)

Solution   Write the constraint equation $xyz = 12,000$ in the form $g(x, y, z) = xyz - 12,000$. Then the Lagrangian function is

$$F(x, y, z, \lambda) = f(x, y, z) + \lambda g(x, y, z)$$
$$= 2xy + 6yz + 8xz + \lambda(xyz - 12,000)$$

To find the critical point(s) of the function F, we solve the system that comprises the equations

$$F_x = 2y + 8z + \lambda yz = 0$$
$$F_y = 2x + 6z + \lambda xz = 0$$
$$F_z = 6y + 8x + \lambda xy = 0$$
$$F_\lambda = xyz - 12{,}000 = 0$$

Solving the first three equations of the system for λ in terms of x, y, and z, we have

$$\lambda = -\frac{2y + 8z}{yz}$$

$$\lambda = -\frac{2x + 6z}{xz}$$

$$\lambda = -\frac{6y + 8x}{xy}$$

Equating the first two expressions for λ leads to

$$\frac{2y + 8z}{yz} = \frac{2x + 6z}{xz}$$

or

$$2xy + 8xz = 2xy + 6yz$$

$$x = \frac{3}{4}y$$

Next, equating the second and third expressions for λ in the same system yields

$$\frac{2x + 6z}{xz} = \frac{6y + 8x}{xy}$$

$$2xy + 6yz = 6yz + 8xz$$

and

$$z = \frac{1}{4}y$$

Finally, substituting these values of x and z into the equation $xyz - 12{,}000 = 0$, the fourth equation of the first system of equations, we have

$$\left(\frac{3}{4}y\right)(y)\left(\frac{1}{4}y\right) - 12{,}000 = 0$$

$$y^3 = \frac{(12{,}000)(4)(4)}{3} = 64{,}000$$

or $y = 40$. The corresponding values of x and z are given by $x = \frac{3}{4}(40) = 30$ and $z = \frac{1}{4}(40) = 10$. Therefore, we see that the point $(30, 40, 10)$ gives the constrained minimum of f. The minimum value is

$$f(30, 40, 10) = 2(30)(40) + 6(40)(10) + 8(30)(10) = 7200$$

Refer to Example 3.

a. By solving the constraint equation $xyz = 12{,}000$ for z in terms of x and y and then substituting this value of z into the function

$$f(x, y, z) = 2xy + 6yz + 8xz$$

show that the problem of finding the constrained minimum of f is equivalent to finding the unconstrained minimum of the function

$$g(x, y) = 2xy + \frac{72{,}000}{x} + \frac{96{,}000}{y}$$

b. Use the technique of Section 8.3 to find the minimum of g and hence the constrained minimum of f.

Applications

EXAMPLE 4   Refer to Example 3, Section 8.1. The total weekly profit (in dollars) that Acrosonic Company realized in producing and selling its bookshelf loudspeaker systems is given by the profit function

$$P(x, y) = -\frac{1}{4}x^2 - \frac{3}{8}y^2 - \frac{1}{4}xy + 120x + 100y - 5000$$

where x denotes the number of fully assembled units and y denotes the number of kits produced and sold per week. Acrosonic's management decides that production of these loudspeaker systems should be restricted to a total of exactly 230 units per week. Under this condition, how many fully assembled units and how many kits should be produced per week to maximize Acrosonic's weekly profit?

Solution   The problem is equivalent to maximizing the function

$$P(x, y) = -\frac{1}{4}x^2 - \frac{3}{8}y^2 - \frac{1}{4}xy + 120x + 100y - 5000$$

subject to the constraint

$$g(x, y) = x + y - 230 = 0$$

The Lagrangian function is

$$F(x, y, \lambda) = P(x, y) + \lambda g(x, y)$$
$$= -\frac{1}{4}x^2 - \frac{3}{8}y^2 - \frac{1}{4}xy + 120x + 100y$$
$$- 5000 + \lambda(x + y - 230)$$

To find the critical point(s) of F, solve the following system of equations:

$$F_x = -\frac{1}{2}x - \frac{1}{4}y + 120 + \lambda = 0$$

$$F_y = -\frac{3}{4}y - \frac{1}{4}x + 100 + \lambda = 0$$

$$F_\lambda = x + y - 230 = 0$$

Solving the first equation of this system for λ, we obtain

$$\lambda = \frac{1}{2}x + \frac{1}{4}y - 120$$

which, upon substitution into the second equation, yields

$$-\frac{3}{4}y - \frac{1}{4}x + 100 + \frac{1}{2}x + \frac{1}{4}y - 120 = 0$$

or

$$-\frac{1}{2}y + \frac{1}{4}x - 20 = 0$$

Solving the last equation for y gives

$$y = \frac{1}{2}x - 40$$

When we substitute this value of y into the third equation of the system, we have

$$x + \frac{1}{2}x - 40 - 230 = 0$$

or

$$x = 180$$

The corresponding value of y is $(\frac{1}{2})(180) - 40$, or 50. Thus, the required constrained relative maximum of P occurs at the point (180, 50). Again, we can show that the point (180, 50) in fact yields a constrained absolute maximum for P. Thus, Acrosonic's profit is maximized by producing 180 assembled and 50 kit versions of their bookshelf loudspeaker systems. The maximum weekly profit realizable is given by

$$P(180, 50) = -\frac{1}{4}(180)^2 - \frac{3}{8}(50)^2 - \frac{1}{4}(180)(50)$$

$$+ 120(180) + 100(50) - 5000$$

$$= 10{,}312.5$$

or $10,312.50.   • • •

Figure 8.27

A rectangular pool will be built in the elliptical poolside area.

Figure 8.28

We want to find the largest rectangle that can be inscribed in the ellipse described by $x^2 + 4y^2 = 3600$.

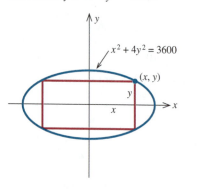

EXAMPLE 5   The operators of the *Viking Princess*, a luxury cruise liner, are contemplating the addition of another swimming pool to the ship. The chief engineer has suggested that an area in the form of an ellipse located in the rear of the promenade deck would be suitable for this purpose. This location would provide a poolside area with sufficient space for passenger movement and placement of deck chairs (Figure 8.27). It has been determined that the ellipse may be described by the equation $x^2 + 4y^2 = 3600$, where x and y are measured in feet. *Viking's* operators would like to know the dimensions of the rectangular pool with the largest possible area that would meet these requirements.

Solution   In order to solve this problem, we need to find the rectangle inscribed in the ellipse with equation $x^2 + 4y^2 = 3600$ and having the largest area. Letting the sides of the rectangle be $2x$ and $2y$ feet, we see that the area of the rectangle is $A = 4xy$ (Figure 8.28). Furthermore, the point (x, y) must be constrained to move along the ellipse so that it satisfies the equation $x^2 + 4y^2 = 3600$. Thus, the problem is equivalent to maximizing the function

$$f(x, y) = 4xy$$

subject to the constraint $g(x, y) = x^2 + 4y^2 - 3600 = 0$. The Lagrangian function is

$$F(x, y, \lambda) = f(x, y) + \lambda g(x, y)$$
$$= 4xy + \lambda(x^2 + 4y^2 - 3600)$$

To find the critical point(s) of F, we solve the following system of equations:

$$F_x = 4y + 2\lambda x = 0$$
$$F_y = 4x + 8\lambda y = 0$$
$$F_\lambda = x^2 + 4y^2 - 3600 = 0$$

Solving the first equation of this system for λ, we obtain

$$\lambda = -\frac{2y}{x}$$

which, upon substitution into the second equation, yields

$$4x + 8\left(-\frac{2y}{x}\right)(y) = 0 \quad \text{or} \quad x^2 - 4y^2 = 0$$

that is, $x = \pm 2y$. Substituting these values of x into the third equation of the system, we have

$$4y^2 + 4y^2 - 3600 = 0$$

or, upon solving, $y = \pm\sqrt{450} = \pm 15\sqrt{2}$. The corresponding values of x are $\pm 30\sqrt{2}$. Because both x and y must be nonnegative, we have $x = 30\sqrt{2}$ and $y = 15\sqrt{2}$. Thus, the dimensions of the pool with maximum area are $30\sqrt{2}$ ft by $60\sqrt{2}$ ft, or approximately 42 ft × 85 ft. ○ ○ ○

EXAMPLE 6 Suppose that x units of labor and y units of capital are required to produce

$$f(x, y) = 100x^{3/4}y^{1/4}$$

units of a certain product (recall that this is a Cobb-Douglas production function). If each unit of labor costs \$200, each unit of capital costs \$300, and a total of \$60,000 is available for production, determine how many units of labor and how many units of capital should be used in order to maximize production.

Solution The total cost of x units of labor at \$200 per unit and y units of capital at \$300 per unit is equal to $200x + 300y$ dollars. But \$60,000 is budgeted for production, so $200x + 300y = 60,000$, which we rewrite as

$$g(x, y) = 200x + 300y - 60,000 = 0$$

To maximize $f(x, y) = 100x^{3/4}y^{1/4}$ subject to the constraint $g(x, y) = 0$, we form the Lagrangian function

$$F(x, y, \lambda) = f(x, y) + \lambda g(x, y)$$
$$= 100x^{3/4}y^{1/4} + \lambda(200x + 300y - 60,000)$$

To find the critical point(s) of F, we solve the system that comprises the equations

$$F_x = 75x^{-1/4}y^{1/4} + 200\lambda = 0$$
$$F_y = 25x^{3/4}y^{-3/4} + 300\lambda = 0$$
$$F_\lambda = 200x + 300y - 60{,}000 = 0$$

Solving the first equation for λ, we have

$$\lambda = -\frac{75x^{-1/4}y^{1/4}}{200} = -\frac{3}{8}\left(\frac{y}{x}\right)^{1/4}$$

which, when substituted into the second equation, yields

$$25\left(\frac{x}{y}\right)^{3/4} + 300\left(-\frac{3}{8}\right)\left(\frac{y}{x}\right)^{1/4} = 0$$

Multiplying the last equation by $(x/y)^{1/4}$ then gives

$$25\left(\frac{x}{y}\right) - \frac{900}{8} = 0$$

or

$$x = \left(\frac{900}{8}\right)\left(\frac{1}{25}\right)y = \frac{9}{2}y$$

Substituting this value of x into the third equation of the first system of equations, we have

$$200\left(\frac{9}{2}y\right) + 300y - 60{,}000 = 0$$

from which we deduce that $y = 50$. Hence, $x = 225$. Thus, maximum production is achieved when 225 units of labor and 50 units of capital are used.   ο ο ο

When used in the context of Example 6, the negative of the Lagrange multiplier λ is called the **marginal productivity of money.** That is, if one additional dollar is available for production, then approximately $-\lambda$ units of a product can be produced. Here

$$\lambda = -\frac{3}{8}\left(\frac{y}{x}\right)^{1/4} = -\frac{3}{8}\left(\frac{50}{225}\right)^{1/4} \approx -0.257$$

so, in this case, the marginal productivity of money is 0.257. For example, if $65{,}000$ is available for production instead of the originally budgeted figure of $60{,}000$, then the maximum production may be boosted from the original

$$f(225, 50) = 100(225)^{3/4}(50)^{1/4}$$

or 15,448 units, to

$$15{,}448 + 5000(0.257)$$

or 16,733 units.

SELF-CHECK EXERCISES 8.5

1. Use the Method of Lagrange Multipliers to find the relative maximum of the function

$$f(x, y) = -2x^2 - y^2$$

subject to the constraint $3x + 4y = 12$.

2. The total monthly profit of Robertson Controls Company in manufacturing and selling x hundred of its standard mechanical setback thermostats and y hundred of its deluxe electronic setback thermostats per month is given by the total profit function

$$P(x, y) = -\frac{1}{8}x^2 - \frac{1}{2}y^2 - \frac{1}{4}xy + 13x + 40y - 280$$

where P is in hundreds of dollars. If the production of setback thermostats is to be restricted to a total of exactly 4000 per month, how many of each model should Robertson manufacture in order to maximize its monthly profits? What is the maximum monthly profit?

Solutions to Self-Check Exercises 8.5 can be found on page 645.

8.5 EXERCISES

In exercises 1–16, use the Method of Lagrange Multipliers to optimize the given function subject to the given constraint.

1. Minimize the function $f(x, y) = x^2 + 3y^2$ subject to the constraint $x + y - 1 = 0$.

2. Minimize the function $f(x, y) = x^2 + y^2 - xy$ subject to the constraint $x + 2y - 14 = 0$.

3. Maximize the function $f(x, y) = 2x + 3y - x^2 - y^2$ subject to the constraint $x + 2y = 9$.

4. Maximize the function $f(x, y) = 16 - x^2 - y^2$ subject to the constraint $x + y - 6 = 0$.

5. Minimize the function $f(x, y) = x^2 + 4y^2$ subject to the constraint $xy = 1$.

6. Minimize the function $f(x, y) = xy$ subject to the constraint $x^2 + 4y^2 = 4$.

7. Maximize the function $f(x, y) = x + 5y - 2xy - x^2 - 2y^2$ subject to the constraint $2x + y = 4$.

8. Maximize the function $f(x, y) = xy$ subject to the constraint $2x + 3y - 6 = 0$.

9. Maximize the function $f(x, y) = xy^2$ subject to the constraint $9x^2 + y^2 = 9$.

10. Minimize the function $f(x, y) = \sqrt{y^2 - x^2}$ subject to the constraint $x + 2y - 5 = 0$.

11. Find the maximum and minimum values of the function $f(x, y) = xy$ subject to the constraint $x^2 + y^2 = 16$.

12. Find the maximum and minimum values of the function $f(x, y) = e^{xy}$ subject to the constraint $x^2 + y^2 = 8$.

13. Find the maximum and minimum values of the function $f(x, y) = xy^2$ subject to the constraint $x^2 + y^2 = 1$.

14. Maximize the function $f(x, y, z) = xyz$ subject to the constraint $2x + 2y + z = 84$.

15. Minimize the function $f(x, y, z) = x^2 + y^2 + z^2$ subject to the constraint $3x + 2y + z = 6$.

16. Find the maximum value of the function $f(x, y, z) = x + 2y - 3z$ subject to the constraint $z = 4x^2 + y^2$.

17. Maximizing Profit The total weekly profit (in dollars) realized by the Country Workshop in manufacturing and selling its rolltop desks is given by the profit function

$$P(x, y) = -0.2x^2 - 0.25y^2 - 0.2xy$$
$$+ 100x + 90y - 4000$$

where x stands for the number of finished units and y denotes the number of unfinished units manufactured and sold per week. The company's management decides to restrict the manufacture of these desks to a total of exactly 200 units per week. How many finished and how many unfinished units should be manufactured per week to maximize the company's weekly profit?

18. Maximizing Profit The total daily profit (in dollars) realized by Weston Publishing Company in publishing and selling its dictionaries is given by the profit function

$$P(x, y) = -0.005x^2 - 0.003y^2 - 0.002xy$$
$$+ 14x + 12y - 200$$

where x stands for the number of deluxe editions and y denotes the number of standard editions sold daily. Weston's management decides that publication of these dictionaries should be restricted to a total of exactly 400 copies per day. How many deluxe copies and how many standard copies should be published per day to maximize Weston's daily profit?

19. Minimizing Construction Costs The management of UNICO Department Store decides to enclose an 800-ft² area outside their building to display potted plants. The enclosed area will be a rectangle, one side of which is provided by the external walls of the store. Two sides of the enclosure will be made of pine board, and the fourth side will be made of galvanized steel fencing material. If the pine board fencing costs $6 per running foot and the steel fencing costs $3 per running foot, determine the dimensions of the enclosure that will cost the least to erect.

20. Minimizing Container Costs The Betty Moore Company requires that its corned beef hash containers have a capacity of 64 cubic inches, be right circular cylinders, and be made of a tin alloy. Find the radius and height of the least expensive container that can be made if the metal for the side and bottom costs 4 cents per square inch and the metal for the pull-off lid costs 2 cents per square inch.

Hint: Let the radius and height of the container be r and h inches, respectively. Then the volume of the container is $\pi r^2 h = 64$, and the cost is given by

$$C(r, h) = 8\pi rh + 6\pi r^2$$

21. Minimizing Construction Costs An open rectangular box is to be constructed from material that costs $3 per square foot for the bottom and $1 per square foot for its sides. Find the dimensions of the box of greatest volume that can be constructed for $36.

22. Minimizing Construction Costs A closed rectangular box having a volume of 4 cubic feet is to be constructed. If the material for the sides costs $1 per square foot and the material for the top and bottom costs $1.50 per square foot, find the dimensions of the box that can be constructed with minimum cost.

23. Maximizing Sales The Ross-Simons Company has a monthly advertising budget of $60,000. Their marketing department estimates that if they spend x dollars on newspaper advertising and y dollars on television advertising, then the monthly sales will be given by

$$z = f(x, y) = 90x^{1/4}y^{3/4}$$

dollars. Determine how much money Ross-Simons should spend on newspaper ads and on television ads per month to maximize its monthly sales.

24. Maximizing Production John Mills, the proprietor of Mills Engine Company, a manufacturer of model airplane engines, finds that it takes x units of labor and y units of capital to produce

$$f(x, y) = 100x^{3/4}y^{1/4}$$

units of the product. If a unit of labor costs $100, a unit of capital costs $200, and $200,000 is budgeted for production, determine how many units should be expended on labor and how many units should be expended on capital in order to maximize production.

25. Use the Method of Lagrange Multipliers to solve exercise 28, Exercises 8.3.

Solutions to Self-Check Exercises 8.5

1. Write the constraint equation in the form $g(x, y) = 3x + 4y - 12 = 0$. Then the Lagrangian function is

$$F(x, y, \lambda) = -2x^2 - y^2 + \lambda(3x + 4y - 12)$$

To find the critical point(s) of F, we solve the system

$$F_x = -4x + 3\lambda = 0$$
$$F_y = -2y + 4\lambda = 0$$
$$F_\lambda = 3x + 4y - 12 = 0$$

Solving the first two equations for x and y in terms of λ, we find $x = \frac{3}{4}\lambda$ and $y = 2\lambda$. Substituting these values of x and y into the third equation of the system yields

$$3\left(\frac{3}{4}\lambda\right) + 4(2\lambda) - 12 = 0$$

or $\lambda = 48/41$. Therefore, $x = (3/4)(48/41) = 36/41$ and $y = 2(48/41) = 96/41$, and we see that the point $(36/41, 96/41)$ gives the constrained maximum of f. The maximum value is

$$f\left(\frac{36}{41}, \frac{96}{41}\right) = -2\left(\frac{36}{41}\right)^2 - \left(\frac{96}{41}\right)^2$$
$$= -\frac{11{,}808}{1{,}681} = -\frac{288}{41}$$

2. We want to maximize

$$P(x, y) = -\frac{1}{8}x^2 - \frac{1}{2}y^2 - \frac{1}{4}xy + 13x + 40y - 280$$

subject to the constraint

$$g(x, y) = x + y - 40 = 0$$

The Lagrangian function is

$$F(x, y, \lambda) = P(x, y) + \lambda g(x, y)$$
$$= -\frac{1}{8}x^2 - \frac{1}{2}y^2 - \frac{1}{4}xy + 13x$$
$$+ 40y - 280 + \lambda(x + y - 40)$$

To find the critical points of F, we solve the following system of equations:

$$F_x = -\frac{1}{4}x - \frac{1}{4}y + 13 + \lambda = 0$$

$$F_y = -\frac{1}{4}x - y + 40 + \lambda = 0$$

$$F_\lambda = x + y - 40 = 0$$

Subtracting the first equation from the second gives

$$-\frac{3}{4}y + 27 = 0 \quad \text{or} \quad y = 36$$

Substituting this value of y into the third equation yields $x = 4$. Therefore, in order to maximize its monthly profits, Robertson should manufacture 400 standard and 3600 deluxe thermostats. The maximum monthly profit is given by

$$P(4, 36) = -\frac{1}{8}(4)^2 - \frac{1}{2}(36)^2 - \frac{1}{4}(4)(36)$$
$$+ 13(4) + 40(36) - 280$$
$$= 526$$

or $52,600.

8.6 TOTAL DIFFERENTIALS

Differentials of Two or More Variables

In Section 3.7 we defined the differential of a function of one variable $y = f(x)$ to be

$$dy = f'(x)\, dx$$

In particular, we showed that if the actual change $dx = \Delta x$ in the independent variable x is small, then the differential dy provides us with an approximation of the actual change Δy in the dependent variable y; that is,

$$\Delta y = f(x + \Delta x) - f(x)$$
$$\approx f'(x)\, dx$$

The concept of the differential extends readily to a function of two or more variables.

TOTAL DIFFERENTIAL

Let $z = f(x, y)$ define a differentiable function of x and y.

1. The **differentials** of the independent variables x and y are $dx = \Delta x$ and $dy = \Delta y$.

2. The **differential** of the dependent variable z is

$$dz = \frac{\partial f}{\partial x}\, dx + \frac{\partial f}{\partial y}\, dy \qquad (12)$$

Thus, analogous to the one-variable case, the total differential of z is a linear function of dx and dy. Furthermore, it provides us with an approximation of the exact change in z,

$$\Delta z = f(x + \Delta x, y + \Delta y) - f(x, y)$$

corresponding to a net change Δx in x from x to $x + \Delta x$ and a net change Δy in y from y to $y + \Delta y$; that is,

$$\Delta z \approx dz = \frac{\partial f}{\partial x}(x, y)\, dx + \frac{\partial f}{\partial y}(x, y)\, dy \tag{13}$$

provided $\Delta x = dx$ and $\Delta y = dy$ are sufficiently small.

EXAMPLE 1   Let $z = 2x^2y + y^3$.

a. Find the differential dz of z.

b. Find the approximate change in z when x changes from $x = 1$ to $x = 1.01$ and y changes from $y = 2$ to $y = 1.98$.

c. Find the actual change in z when x changes from $x = 1$ to $x = 1.01$ and y changes from $y = 2$ to $y = 1.98$. Compare the result with that obtained in (b).

Solution

a. Let $f(x, y) = 2x^2y + y^3$. Then the required differential is

$$dz = \frac{\partial f}{\partial x}\, dx + \frac{\partial f}{\partial y}\, dy = 4xy\, dx + (2x^2 + 3y^2)\, dy$$

b. Here $x = 1$, $y = 2$, and $dx = 1.01 - 1 = 0.01$ and $dy = 1.98 - 2 = -0.02$. Therefore,

$$\Delta z \approx dz = 4(1)(2)(0.01) + [2(1) + 3(4)](-0.02) = -0.20$$

c. The actual change in z is given by

$$\begin{aligned}
\Delta z &= f(1.01, 1.98) - f(1, 2) \\
&= [2(1.01)^2(1.98) + (1.98)^3] - [2(1)^2(2) + (2)^3] \\
&\approx 11.801988 - 12 \\
&= -0.1980
\end{aligned}$$

We see that $\Delta z \approx dz$, as expected.   ○ ○ ○

Applications

EXAMPLE 2   The weekly total revenue of Acrosonic Company resulting from the production and sales of x fully assembled bookshelf loudspeaker systems and y kit versions of the same loudspeaker system is

$$R(x, y) = -\frac{1}{4}x^2 - \frac{3}{8}y^2 - \frac{1}{4}xy + 300x + 240y$$

dollars. Determine the approximate change in Acrosonic's weekly total revenue when the level of production is increased from 200 assembled units and 60 kits per week to 206 assembled units and 64 kits per week.

Solution   The approximate change in the weekly total revenue is given by the total differential R at $x = 200$ and $y = 60$, $dx = 206 - 200 = 6$ and $dy = 64 - 60 = 4$; that is, by

$$dR = \frac{\partial R}{\partial x}\,dx + \frac{\partial R}{\partial y}\,dy\,\Big|_{\substack{x=200,\,y=60 \\ dx=6,\,dy=4}}$$

$$= \left(-\frac{1}{2}x - \frac{1}{4}y + 300 \right)\Big|_{(200,\,60)} \cdot (6)$$

$$+ \left(-\frac{3}{4}y - \frac{1}{4}x + 240 \right)\Big|_{(200,\,60)} \cdot (4)$$

$$= (-100 - 15 + 300)6 + (-45 - 50 + 240)4$$

$$= 1690$$

or $1690.

○ ○ ○

EXAMPLE 3   The production for a certain country in the early years following World War II is described by the function

$$f(x, y) = 30x^{2/3}y^{1/3}$$

units, when x units of labor and y units of capital were utilized. Find the approximate change in output if the amount expended on labor had been decreased from 125 units to 123 units and the amount expended on capital had been increased from 27 to 29 units. Is your result as expected given the result of Example 4(c), Section 8.2?

Solution   The approximate change in output is given by the total differential of f at $x = 125$, $y = 27$, $dx = 123 - 125 = -2$, and $dy = 29 - 27 = 2$; that is, by

$$df = \frac{\partial f}{\partial x}\,dx + \frac{\partial f}{\partial y}\,dy\,\Big|_{\substack{x=125,\,y=27 \\ dx=-2,\,dy=2}}$$

$$= 20x^{-1/3}y^{1/3}\,\Big|_{(125,\,27)} \cdot (-2) + 10x^{2/3}y^{-2/3}\,\Big|_{(125,\,27)} \cdot (2)$$

$$= 20\left(\frac{27}{125} \right)^{1/3}(-2) + 10\left(\frac{125}{27} \right)^{2/3}(2)$$

$$= -20\left(\frac{3}{5} \right)(2) + 10\left(\frac{25}{9} \right)(2) = \frac{284}{9}$$

or $31\frac{5}{9}$ units. This result is fully compatible with the result of Example 4, where the recommendation was to encourage increased spending on capital rather than on labor.

○ ○ ○

If f is a function of the three variables x, y, and z, then the total differential of $w = f(x, y, z)$ is defined to be

$$dw = \frac{\partial f}{\partial x} dx + \frac{\partial f}{\partial y} dy + \frac{\partial f}{\partial z} dz$$

where $dx = \Delta x$, $dy = \Delta y$, and $dz = \Delta z$ are the actual changes in the independent variables x, y, and z as x changes from $x = a$ to $x = a + \Delta x$, y changes from $y = b$ to $y = b + \Delta y$, and z changes from $z = c$ to $z = c + \Delta z$, respectively.

○ ○ ○

EXAMPLE 4   Find the maximum percentage error in calculating the volume of a rectangular box if an error of at most 1% is made in measuring the length, width, and height of the box.

Solution   Let x, y, and z denote the length, width, and height, respectively, of the rectangular box. Then the volume of the box is given by $V = f(x, y, z) = xyz$ cubic units. Now suppose the true dimensions of the rectangular box are a, b, and c units, respectively. Since the error committed in measuring the length, width, and height of the box is at most 1%, we have

$$|\Delta x| = |x - a| \leq 0.01a$$
$$|\Delta y| = |y - b| \leq 0.01b$$
$$|\Delta z| = |z - c| \leq 0.01c$$

Therefore, the maximum error in calculating the volume of the box is

$$|\Delta V| \approx |dV| = \left| \frac{\partial f}{\partial x} dx + \frac{\partial f}{\partial y} dy + \frac{\partial f}{\partial z} dz \right|_{x=a, y=b, z=c}$$

$$= \left| yz\, dx + xz\, dy + xy\, dz \right|_{x=a, y=b, z=c}$$

$$= |bc\, dx + ac\, dy + ab\, dz|$$

$$\leq bc|dx| + ac|dy| + ab|dz|$$

$$\leq bc(0.01a) + ac(0.01b) + ab(0.01c)$$

$$= (0.03)abc$$

Since the actual volume of the box is abc cubic units, we see that the maximum percentage error in calculating its volume is

$$\frac{|\Delta V|}{V|_{(a,b,c)}} \approx \frac{(0.03)abc}{abc} = 0.03$$

that is, approximately 3%.

○ ○ ○

Refer to Example 4, where we found the maximum percentage error in calculating the volume of the rectangular box to be *approximately* 3%. What is the precise maximum percentage error?

SELF-CHECK EXERCISE 8.6

Let f be a function defined by $z = f(x, y) = 3xy^2 - 4y$. Find the total differential of f at $(-1, 3)$. Then find the approximate change in z when x changes from $x = -1$ to $x = -0.98$ and y changes from $y = 3$ to $y = 3.01$.

The solution to Self-Check Exercise 8.6 can be found on page 651.

8.6 EXERCISES

In exercises 1–18, find the total differential of the function.

1. $f(x, y) = x^2 + 2y$

2. $f(x, y) = 2x^2 + 3y^2$

3. $f(x, y) = 2x^2 - 3xy + 4x$

4. $f(x, y) = xy^3 - x^2y^2$

5. $f(x, y) = \sqrt{x^2 + y^2}$

6. $f(x, y) = (x + 3y^2)^{1/3}$

7. $f(x, y) = \dfrac{5y}{x - y}$

8. $f(x, y) = \dfrac{x + y}{x - y}$

9. $f(x, y) = 2x^5 - ye^{-3x}$

10. $f(x, y) = xye^{x+y}$

11. $f(x, y) = x^2e^y + y \ln x$

12. $f(x, y) = \ln(x^2 + y^2)$

13. $f(x, y, z) = xy^2z^3$

14. $f(x, y, z) = x\sqrt{y} + y\sqrt{z}$

15. $f(x, y, z) = \dfrac{x}{y + z}$

16. $f(x, y, z) = \dfrac{x + y}{y + z}$

17. $f(x, y, z) = xyz + xe^{yz}$

18. $f(x, y, z) = \sqrt{e^x + e^y + ze^{xy}}$

In exercises 19–30, find the approximate change in z when the point (x, y) changes from (x₀, y₀) to (x₁, y₁).

19. $f(x, y) = 4x^2 - xy$; from $(1, 2)$ to $(1.01, 2.02)$

20. $f(x, y) = 2x^2 - 2x^3y^2 - y^3$; from $(-1, 2)$ to $(-0.98, 2.01)$

21. $f(x, y) = x^{2/3}y^{1/2}$; from $(8, 9)$ to $(7.97, 9.03)$

22. $f(x, y) = \sqrt{x^2 + y^2}$; from $(1, 3)$ to $(1.03, 3.03)$

23. $f(x, y) = \dfrac{x}{x - y}$; from $(-3, -2)$ to $(-3.02, -1.98)$

24. $f(x, y) = \dfrac{x - y}{x + y}$; from $(-3, -2)$ to $(-3.02, -1.98)$

25. $f(x, y) = 2xe^{-y}$; from $(4, 0)$ to $(4.03, 0.03)$

26. $f(x, y) = \sqrt{x}e^y$; from $(1, 1)$ to $(1.01, 0.98)$

27. $f(x, y) = xe^{xy} - y^2$; from $(-1, 0)$ to $(-0.97, 0.03)$

28. $f(x, y) = xe^{-y} + ye^{-x}$; from $(1, 1)$ to $(1.01, 0.90)$

29. $f(x, y) = x \ln x + y \ln x$; from $(2, 3)$ to $(1.98, 2.89)$

30. $f(x, y) = \ln(xy)^{1/2}$; from $(5, 10)$ to $(5.05, 9.95)$

31. **Effect of Inventory and Floor Space on Profit**   The monthly profit (in dollars) of Bond and Barker Department Store depends on the level of inventory x (in thousands of dollars) and the floor space y (in thousands of square feet) available for display of the merchandise, as given by the equation

$$P(x, y) = -0.02x^2 - 15y^2 + xy + 39x + 25y - 20,000$$

Currently, the level of inventory is \$4,000,000 ($x = 4000$) and the floor space is 150,000 square feet ($y = 150$). Find the anticipated change in monthly profit if management increases the level of inventory by \$500,000 and decreases the floor space for display of merchandise by 10,000 square feet.

32. **Effect of Production on Profit**   The Country Workshop's total weekly profit (in dollars) realized in manufacturing and selling its rolltop desks is given by

$$P(x, y) = -0.2x^2 - 0.25y^2 - 0.2xy + 100x + 90y - 4000$$

where x stands for the number of finished units and y denotes the number of unfinished units manufactured and sold per week. Currently, the weekly output is 190 finished and 105 unfinished units. Determine the approximate change in the total weekly profit if the sole proprietor of the Country Workshop decides to increase the number of finished units to 200 per week and decrease the number of unfinished units to 100 per week.

33. **Revenue of a Travel Agency** The Odyssey Travel Agency's monthly revenue (in thousands of dollars) depends on the amount of money x (in thousands) spent on advertising per month and the number of agents y in its employ in accordance with the rule

$$R(x, y) = -x^2 - 0.5y^2 + xy + 8x + 3y + 20$$

Currently, the amount of money spent on advertising is $10,000 per month and there are 15 agents in the agency's employ. Estimate the change in revenue resulting from an increase of $1000 per month in advertising expenditure and a decrease of one agent.

34. **Effect of Capital and Labor on Productivity** The productivity of a South American country is given by the function

$$f(x, y) = 20x^{3/4}y^{1/4}$$

when x units of labor and y units of capital are utilized. Find the approximate change in output if the amount expended on labor is decreased from 256 to 254 units and the amount expended on capital is increased from 16 to 18 units.

35. **Error in Calculating the Volume of a Cylinder** The radius and height of a right circular cylinder are measured with a possible error of 0.1 cm. Approximate the maximum error in calculating the volume of the cylinder if the measured dimensions $r = 8$ cm and $h = 20$ cm are used.

36. **Error in Calculating the Surface Area of a Human** The formula

$$S = 0.007184W^{0.425}H^{0.725}$$

gives the surface area S of a human body (in square meters) in terms of its weight W in kilograms and its height H in centimeters. If an error of 1% is made in measuring the weight of a person and an error of 2% is made in measuring the height, what is the percentage error in the measurement of the person's surface area?

37. **Error in Calculating Total Resistance** The total resistance R of three resistors with resistance R_1, R_2, and R_3, connected in parallel, is given by the relationship

$$\frac{1}{R} = \frac{1}{R_1} + \frac{1}{R_2} + \frac{1}{R_3}$$

If R_1, R_2, and R_3 are measured at 100, 200, and 300 ohms, respectively, with a maximum error of 1% in each measurement, find the approximate maximum error in the calculated value of R.

38. **Error in Measuring Arterial Blood Flow** The flow of blood through an arteriole in cubic centimeters per second is given by

$$V = \frac{\pi p r^4}{8kl}$$

where l (in cm) is the length of the arteriole, r (in cm) is its radius, p (in dyne/cm^2) is the difference in pressure between the two ends of the arteriole, and k is the viscosity of blood (in dyne-sec/cm^2). Find the approximate percentage change in the flow of blood if an error of 2% is made in measuring the length of the arteriole and an error of 1% is made in measuring its radius. Assume that p and k are constant.

SOLUTION TO SELF-CHECK EXERCISE 8.6

We find

$$\frac{\partial f}{\partial x} = 3y^2 \quad \text{and} \quad \frac{\partial f}{\partial y} = 6xy - 4$$

so that

$$\frac{\partial f}{\partial x}(-1, 3) = 3(3)^2 = 27$$

and

$$\frac{\partial f}{\partial y}(-1, 3) = 6(-1)(3) - 4 = -22$$

Therefore, the total differential is

$$dz = \frac{\partial f}{\partial x}(-1, 3)\, dx + \frac{\partial f}{\partial y}(-1, 3)\, dy$$

$$= 27\, dx - 22\, dy$$

Now $dx = -0.98 - (-1) = 0.02$ and $dy = 3.01 - 3 = 0.01$, so that the approximate change in z is

$$dz = 27(0.02) - 22(0.01) = 0.32$$

8.7 DOUBLE INTEGRALS

A Geometric Interpretation of the Double Integral

In order to introduce the notion of the integral of a function of two variables, let us first recall the definition of the definite integral of a continuous function of one variable $y = f(x)$ over the interval $[a, b]$. We first divide the interval $[a, b]$ into n subintervals, each of equal length, by the points $x_0 = a < x_1 < x_2 < \cdots < x_n = b$ and define the Riemann sum by

$$S_n = f(p_1)h + f(p_2)h + \cdots + f(p_n)h$$

where $h = (b - a)/n$ and p_i is an arbitrary point in the interval $[x_{i-1}, x_i]$. The definite integral of f over $[a, b]$ is defined as the limit of the Riemann sum S_n as n tends to infinity, whenever it exists. Furthermore, recall that when f is a nonnegative continuous function on $[a, b]$, then the ith term of the Riemann sum, $f(p_i)h$, is an approximation (by the area of a rectangle) of the area under that part of the graph of $y = f(x)$ between $x = x_{i-1}$ and $x = x_i$, so that the Riemann sum S_n provides us with an approximation of the area under the curve $y = f(x)$ from $x = a$ to $x = b$. The integral

$$\int_a^b f(x)\, dx = \lim_{n \to \infty} S_n$$

gives the *actual* area under the curve from $x = a$ to $x = b$.

Now suppose that $f(x, y)$ is a continuous function of two variables defined over a region R. For simplicity, we assume for the moment that R is a rectangular region in the plane (Figure 8.29). Let us construct a Riemann sum for this function over the rectangle R by following a procedure that parallels the case for a function of one variable over an interval I. We begin by observing that the analogue of a *partition* in the two-dimensional case is a rectangular **grid** comprising mn rectangles, each of length h and width k, as a result of partitioning the side of the rectangle R of length $(b - a)$ into m segments and the side of length $(d - c)$ into n segments. By construction

$$h = \frac{b - a}{m} \quad \text{and} \quad k = \frac{d - c}{n}$$

A sample grid with $m = 5$ and $n = 4$ is shown in Figure 8.30.

Let us label the rectangles $R_1, R_2, R_3, \ldots, R_{mn}$. If (x_i, y_i) is *any* point in $R_i\ (1 \leq i \leq mn)$, then the **Riemann sum of** $f(x, y)$ **over the region R** is defined as

$$S(m, n) = f(x_1, y_1)hk + f(x_2, y_2)hk + \cdots + f(x_{mn}, y_{mn})hk$$

Figure 8.29

$f(x, y)$ is a function defined over a rectangular region R.

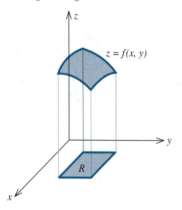

Figure 8.30

Grid with $m = 5$ and $n = 4$.

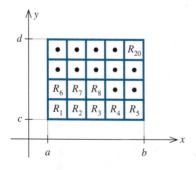

If the limit of $S(m, n)$ exists as both m and n tend to infinity, we call this limit the value of the **double integral of $f(x, y)$ over the region R** and denote it by

$$\iint_R f(x, y) \, dA$$

If $f(x, y)$ is a nonnegative function, then the number $f(x_i, y_i)hk$ is the volume of a prism with base R_i and height $f(x_i, y_i)$ that provides us with an approximation of the volume of the solid under the surface $z = f(x, y)$ and bounded below by the rectangular region R_i (Figure 8.31).

Figure 8.31

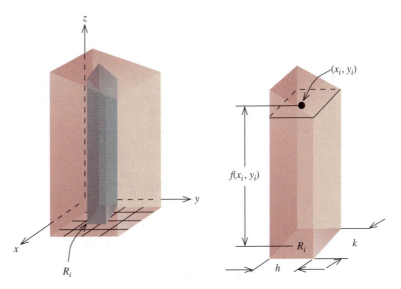

(a) Approximating the volume of a solid with rectangular prisms

(b) A prism with base R_i and height $f(x_i, y_i)$

Therefore, the Riemann sum $S(m, n)$ gives us an approximation of the volume of the solid bounded above by the surface $z = f(x, y)$ and below by the plane region R. As both m and n tend to infinity, the Riemann sum $S(m, n)$ approaches the *actual* volume under the solid.

Evaluating a Double Integral over a Rectangular Region

Let us turn our attention to the evaluation of the double integral

$$\iint_R f(x, y) \, dA$$

where R is the rectangular region shown in Figure 8.30. As in the case of the definite integral of a function of one variable, it turns out that the double integral can be evaluated without our having to first find an appropriate Riemann sum and then taking the limit of that sum. Instead, as we will now see, the technique calls for evaluating two single integrals—the so-called **iterated integrals**—in succession, using a process that might be called "antipartial differentiation." The technique is described in the following result, which we state without proof.

Let R be the rectangle defined by the inequalities $a \le x \le b$ and $c \le y \le d$ (Figure 8.30). Then

$$\iint_R f(x, y)\, dA = \int_c^d \left[\int_a^b f(x, y)\, dx \right] dy \tag{14}$$

where the iterated integrals on the right-hand side are evaluated as follows. We first compute the integral

$$\int_a^b f(x, y)\, dx$$

by treating y as if it were a constant and integrating the resulting function of x with respect to x (the "dx" reminds us that we are integrating with respect to x). In this manner we obtain a value for the integral that may contain the variable y. Thus,

$$\int_a^b f(x, y)\, dx = g(y)$$

for some function g. Substituting this value into equation (14) gives

$$\int_c^d g(y)\, dy$$

which may be integrated in the usual manner.

EXAMPLE 1   Evaluate $\iint_R f(x, y)\, dA$, where $f(x, y) = x + 2y$ and R is the rectangle defined by $1 \le x \le 4$ and $1 \le y \le 2$.

Solution   Using equation (14), we find

$$\iint_R f(x, y)\, dA = \int_1^2 \left[\int_1^4 (x + 2y)\, dx \right] dy$$

To compute

$$\int_1^4 (x + 2y)\, dx$$

we treat y as if it were a constant (remember that the "dx" reminds us that we are integrating with respect to x). We obtain

$$\int_1^4 (x + 2y)\, dx = \frac{1}{2}x^2 + 2xy \Big|_{x=1}^{x=4}$$

$$= \left[\frac{1}{2}(16) + 2(4)y \right] - \left[\frac{1}{2}(1) + 2(1)y \right]$$

$$= \frac{15}{2} + 6y$$

Thus,

$$\iint_R f(x, y)\, dA = \int_1^2 \left(\frac{15}{2} + 6y\right) dy = \left(\frac{15}{2}y + 3y^2\right)\Big|_1^2$$

$$= (15 + 12) - \left(\frac{15}{2} + 3\right) = 16\frac{1}{2}$$

o o o

Evaluating a Double Integral over a Plane Region

Up to now, we have assumed that the region over which a double integral is to be evaluated is rectangular. In fact, however, it is possible to compute the double integral of functions over rather arbitrary regions. The next theorem, which we state without proof, expands the number of types of regions over which we may integrate.

THEOREM 1

a. Suppose $g_1(x)$ and $g_2(x)$ are continuous functions on $[a, b]$ and the region R is defined by $R = \{(x, y) \mid g_1(x) \leq y \leq g_2(x); a \leq x \leq b\}$. Then

$$\iint_R f(x, y)\, dA = \int_a^b \left[\int_{g_1(x)}^{g_2(x)} f(x, y)\, dy\right] dx \qquad (15)$$

(Figure 8.32a).

b. Suppose $h_1(y)$ and $h_2(y)$ are continuous functions on $[c, d]$ and the region R is defined by $R = \{(x, y) \mid h_1(y) \leq x \leq h_2(y); c \leq y \leq d\}$. Then

$$\iint_R f(x, y)\, dA = \int_c^d \left[\int_{h_1(y)}^{h_2(y)} f(x, y)\, dx\right] dy \qquad (16)$$

(Figure 8.32b).

Figure 8.32

(a)   (b)

REMARKS

1. Observe that in equation (15) the lower and upper limits of integration with respect to y are given by $y = g_1(x)$ and $y = g_2(x)$. This is to be expected since, for a fixed value of x lying between $x = a$ and $x = b$, y runs between the lower curve defined by $y = g_1(x)$ and the upper curve defined by $y = g_2(x)$ [see Figure 8.32a]. Observe, too, that in the special case when $g_1(x) = c$ and $g_2(x) = d$, the region R is rectangular, and (15) reduces to (14).

2. For a fixed value of y, x runs between $x = h_1(y)$ and $x = h_2(y)$, giving the indicated limits of integration with respect to x in equation (16) [see Figure 8.32b].

3. Note that the two curves in Figure 8.32b are not graphs of functions of x (use the vertical-line test) but are graphs of functions of y. It is this observation that justifies the approach leading to equation (16).　　　　○ ○ ○

We now look at a few examples.

EXAMPLE 2　Evaluate $\iint_R f(x, y)\, dA$ given that $f(x, y) = x^2 + y^2$ and R is the region bounded by the graphs of $g_1(x) = x$ and $g_2(x) = 2x$ for $0 \le x \le 2$.

Figure 8.33
R is the region bounded by $g_1(x) = x$ and $g_2(x) = 2x$ for $0 \le x \le 2$.

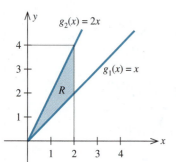

Solution　The region under consideration is shown in Figure 8.33. Using equation (15), we find

$$\iint_R f(x, y)\, dA = \int_0^2 \left[\int_x^{2x} (x^2 + y^2)\, dy \right] dx$$

$$= \int_0^2 \left[\left(x^2 y + \frac{1}{3} y^3 \right) \Big|_x^{2x} \right] dx$$

$$= \int_0^2 \left[\left(2x^3 + \frac{8}{3} x^3 \right) - \left(x^3 + \frac{1}{3} x^3 \right) \right] dx$$

$$= \int_0^2 \frac{10}{3} x^3\, dx = \frac{5}{6} x^4 \Big|_0^2 = 13\frac{1}{3}$$　　○ ○ ○

EXAMPLE 3　Evaluate $\iint_R f(x, y)\, dA$, where $f(x, y) = xe^y$ and R is the plane region bounded by the graphs of $y = x^2$ and $y = x$.

Figure 8.34
R is the region bounded by $y = x^2$ and $y = x$.

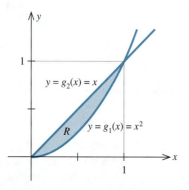

Solution　The region in question is shown in Figure 8.34. The point of intersection of the two curves is found by solving the equation $x^2 = x$, giving $x = 0$ and $x = 1$. Using equation (15), we find

$$\iint_R f(x, y)\, dA = \int_0^1 \left[\int_{x^2}^x xe^y\, dy \right] dx = \int_0^1 \left[xe^y \Big|_{x^2}^x \right] dx$$

$$= \int_0^1 (xe^x - xe^{x^2})\, dx = \int_0^1 xe^x\, dx - \int_0^1 xe^{x^2}\, dx$$

and integrating the first integral on the right-hand side by parts,

$$= \left[(x - 1)e^x - \frac{1}{2} e^{x^2} \right] \Big|_0^1$$

$$= -\frac{1}{2} e - \left(-1 - \frac{1}{2} \right) = \frac{1}{2}(3 - e)$$　　○ ○ ○

Refer to Example 3.

a. You can also view the region R as an example of the region shown in Figure 8.32b. Doing so, find the functions h_1 and h_2 and the numbers c and d.

b. Find an expression for $\iint_R f(x, y)\, dA$ in terms of iterated integrals using formula (16).

c. Evaluate the iterated integrals of part (b) and hence verify the result of Example 3. (*Hint:* The table of integrals in Section 7.2 may be helpful.)

d. Does viewing the region R in two different ways make a difference?

The next example not only illustrates the use of equation (16) but also shows that it may be the only viable way to evaluate the given double integral.

EXAMPLE 4   Evaluate

$$\iint_R xe^{y^2}\, dA$$

where R is the plane region bounded by the y-axis, $x = 0$, the horizontal line $y = 4$, and the graph of $y = x^2$.

Figure 8.35
R is the region bounded by the y-axis, $x = 0$, $y = 4$, and $y = x^2$.

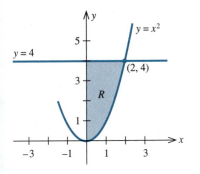

Solution   The region R is shown in Figure 8.35. The point of intersection of the line $y = 4$ and the graph of $y = x^2$ is found by solving the equation $x^2 = 4$, giving $x = 2$ and the required point $(2, 4)$. Using equation (15) with $y = g_1(x) = x^2$ and $y = g_2(x) = 4$ leads to

$$\iint_R xe^{y^2}\, dA = \int_0^2 \left[\int_{x^2}^4 xe^{y^2}\, dy \right] dx$$

Now evaluation of the integral

$$\int_{x^2}^4 xe^{y^2}\, dy = x \int_{x^2}^4 e^{y^2}\, dy$$

calls for finding the antiderivative of the integrand e^{y^2} in terms of elementary functions, a task that, as was pointed out in Section 7.3, cannot be done. Let us begin afresh and attempt to make use of equation (16).

Since the equation $y = x^2$ is equivalent to the equation $x = \sqrt{y}$, which clearly expresses x as a function of y, we may write, with $x = h_1(y) = 0$ and $h_2(y) = \sqrt{y}$,

$$\iint_R xe^{y^2}\, dA = \int_0^4 \left[\int_0^{\sqrt{y}} xe^{y^2}\, dx \right] dy = \int_0^4 \left[\frac{1}{2}x^2 e^{y^2} \Big|_0^{\sqrt{y}} \right] dy$$

$$= \int_0^4 \frac{1}{2} ye^{y^2}\, dy = \frac{1}{4} e^{y^2} \Big|_0^4 = \frac{1}{4}\left(e^{16} - 1\right)$$

SELF-CHECK EXERCISE 8.7

Evaluate $\displaystyle\iint_R (x + y)\, dA$, where R is the region bounded by the graphs of $g_1(x) = x$ and $g_2(x) = x^{1/3}$.

The solution to Self-Check Exercise 8.7 can be found on page 659.

8.7 EXERCISES

In exercises 1–25, evaluate the double integral
$\displaystyle\iint_R f(x, y)\, dA$ *for the given function $f(x, y)$ and the region R.*

1. $f(x, y) = y + 2x$; R is the rectangle defined by $1 \le x \le 2$ and $0 \le y \le 1$.

2. $f(x, y) = x + 2y$; R is the rectangle defined by $-1 \le x \le 2$ and $0 \le y \le 2$.

3. $f(x, y) = xy^2$; R is the rectangle defined by $-1 \le x \le 1$ and $0 \le y \le 1$.

4. $f(x, y) = 12xy^2 + 8y^3$; R is the rectangle defined by $0 \le x \le 1$ and $0 \le y \le 2$.

5. $f(x, y) = x/y$; R is the rectangle defined by $-1 \le x \le 2$ and $1 \le y \le e^3$.

6. $f(x, y) = \dfrac{xy}{1 + y^2}$; R is the rectangle defined by $-2 \le x \le 2$ and $0 \le y \le 1$.

7. $f(x, y) = 4xe^{2x^2 + y}$; R is the rectangle defined by $0 \le x \le 1$ and $-2 \le y \le 0$.

8. $f(x, y) = \dfrac{y}{x^2} e^{y/x}$; R is the rectangle defined by $1 \le x \le 2$ and $0 \le y \le 1$.

9. $f(x, y) = \ln y$; R is the rectangle defined by $0 \le x \le 1$ and $1 \le y \le e$.

10. $f(x, y) = \dfrac{\ln y}{x}$; R is the rectangle defined by $1 \le x \le e^2$ and $1 \le y \le e$.

11. $f(x, y) = x + 2y$; R is bounded by $x = 0$, $x = 1$, $y = 0$, and $y = x$.

12. $f(x, y) = xy$; R is bounded by $x = 0$, $x = 1$, $y = 0$, and $y = x$.

13. $f(x, y) = 2x + 4y$; R is bounded by $x = 1$, $x = 3$, $y = 0$, and $y = x + 1$.

14. $f(x, y) = 2 - y$; R is bounded by $x = -1$, $x = 1 - y$, $y = 0$, and $y = 2$.

15. $f(x, y) = x + y$; R is bounded by $x = 0$, $x = \sqrt{y}$, $y = 0$, and $y = 4$.

16. $f(x, y) = x^2 y^2$; R is bounded by $x = 0$, $x = 1$, $y = x^2$, and $y = x^3$.

17. $f(x, y) = y$; R is bounded by $x = 0$, $x = \sqrt{4 - y^2}$, $y = 0$, and $y = 2$.

18. $f(x, y) = \dfrac{y}{x^3 + 2}$; R is bounded by $x = 0$, $x = 1$, $y = 0$, and $y = x$.

19. $f(x, y) = 2xe^y$; R is bounded by $x = 0$, $x = 1$, $y = 0$, and $y = x$.

20. $f(x, y) = 2x$; R is bounded by $x = e^{2y}$, $x = y$, $y = 0$, and $y = 1$.

21. $f(x, y) = ye^x$; R is bounded by $y = \sqrt{x}$ and $y = x$.

22. $f(x, y) = xe^{-y^2}$; R is bounded by $x = 0$, $y = x^2$, and $y = 4$.

23. $f(x, y) = e^{y^2}$; R is bounded by $x = 0$, $x = 1$, $y = 2x$, and $y = 2$.

24. $f(x, y) = y$; R is bounded by $x = 1$, $x = e$, $y = 0$, and $y = \ln x$.

25. $f(x, y) = ye^{x^3}$; R is bounded by $x = y/2$, $x = 1$, $y = 0$, and $y = 2$.

SOLUTION TO SELF–CHECK EXERCISE 8.7

The region R is shown in the figure. The points of intersection of the two curves are found by solving the equation $x = x^{1/3}$, giving $x = 0$ and $x = 1$. Using equation (15), we find

$$
\iint_R (x + y)\, dA = \int_0^1 \left[\int_x^{x^{1/3}} (x + y)\, dy \right] dx
$$

$$
= \int_0^1 \left[xy + \frac{1}{2} y^2 \Big|_x^{x^{1/3}} \right] dx
$$

$$
= \int_0^1 \left[\left(x^{4/3} + \frac{1}{2} x^{2/3} \right) - \left(x^2 + \frac{1}{2} x^2 \right) \right] dx
$$

$$
= \int_0^1 \left(x^{4/3} + \frac{1}{2} x^{2/3} - \frac{3}{2} x^2 \right) dx
$$

$$
= \frac{3}{7} x^{7/3} + \frac{3}{10} x^{5/3} - \frac{1}{2} x^3 \Big|_0^1
$$

$$
= \frac{3}{7} + \frac{3}{10} - \frac{1}{2} = \frac{8}{35}
$$

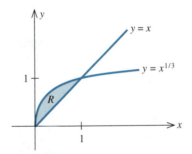

8.8 APPLICATIONS OF DOUBLE INTEGRALS

In this section we will give some sample applications involving the double integral.

Finding the Volume of a Solid by Double Integrals

As we saw in the last section, the double integral

$$
\iint_R f(x, y)\, dA
$$

gives the volume of the solid bounded by the graph of $f(x, y)$ over the region R.

THE VOLUME OF A SOLID UNDER A SURFACE	Let R be a region in the xy-plane, and let f be continuous and nonnegative on R. Then the volume of the solid bounded above by the surface $z = f(x, y)$ and below by R is given by $$V = \iint_R f(x, y)\, dA$$

Figure 8.36
The plane region R defined by $y = \sqrt{1 - x^2}$, $0 \le x \le 1$.

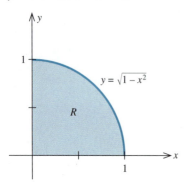

EXAMPLE I   Find the volume of the solid bounded above by the plane $z = f(x, y) = y$ and below by the plane region R defined by $y = \sqrt{1 - x^2}$, $0 \le x \le 1$.

Solution   The region R is sketched in Figure 8.36. Observe that $f(x, y) = y \ge 0$ for $y \in R$. Therefore, the required volume is given by

$$\iint_R y\, dA = \int_0^1 \left[\int_0^{\sqrt{1-x^2}} y\, dy \right] dx = \int_0^1 \left[\frac{1}{2} y^2 \Big|_0^{\sqrt{1-x^2}} \right] dx$$

$$= \int_0^1 \frac{1}{2}(1 - x^2)\, dx = \frac{1}{2}\left(x - \frac{1}{3} x^3 \right)\Big|_0^1 = \frac{1}{3}$$

or $\frac{1}{3}$ cubic unit. The solid is shown in Figure 8.37. Note that it is not necessary to make a sketch of the solid in order to compute its volume.

Figure 8.37
The solid bounded above by the plane $z = y$ and below by the plane region defined by $y = \sqrt{1 - x^2}$, $0 \le x \le 1$.

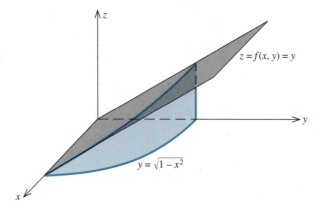

Population of a City

Suppose that the plane region R represents a certain district of a city and $f(x, y)$ gives the population density (the number of people per square mile) at any point (x, y) in R. Enclose the set R by a rectangle and construct a grid for it in the usual manner. In any rectangular region of the grid that has no point

Figure 8.38

The rectangular region R representing a certain district of a city is enclosed by a rectangular grid.

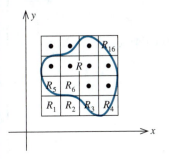

in common with R, set $f(x_i, y_i)hk = 0$ (see Figure 8.38). Then, corresponding to any grid covering the set R, the general term of the Riemann sum $f(x_i, y_i)hk$ (population density times area) gives the number of people who live in that part of the city corresponding to the rectangular region R_i. Therefore, the Riemann sum gives an approximation of the number of people living in the district represented by R and, in the limit, the double integral

$$\iint_R f(x, y)\, dA$$

gives the actual number of people living in the district under consideration.

EXAMPLE 2 The population density of a certain city is described by the function

$$f(x, y) = 10{,}000 e^{-0.2|x| - 0.1|y|}$$

where the origin $(0, 0)$ gives the location of the city hall. What is the population inside the rectangular area described by

$$R = \{(x, y) \mid -10 \le x \le 10;\ -5 \le y \le 5\}$$

if x and y are in miles? (See Figure 8.39.)

Solution By symmetry, it suffices to compute the population in the first quadrant. (Why?) Then, upon observing that in this quadrant

$$f(x, y) = 10{,}000 e^{-0.2x - 0.1y} = 10{,}000 e^{-0.2x} e^{-0.1y}$$

we see that the population in R is given by

$$\iint_R f(x, y)\, dA = 4 \int_0^{10} \left[\int_0^5 10{,}000 e^{-0.2x} e^{-0.1y}\, dy \right] dx$$

$$= 4 \int_0^{10} \left[-100{,}000 e^{-0.2x} e^{-0.1y} \Big|_0^5 \right] dx$$

$$= 400{,}000(1 - e^{-0.5}) \int_0^{10} e^{-0.2x}\, dx$$

$$= 2{,}000{,}000(1 - e^{-0.5})(1 - e^{-2})$$

or approximately 680,438 people.   ○ ○ ○

Figure 8.39

The rectangular region R represents a certain district of a city.

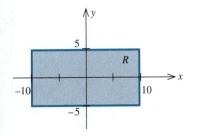

Average Value of a Function

In Section 6.5 we showed that the average value of a continuous function $f(x)$ over an interval $[a, b]$ is given by

$$\frac{1}{b - a} \int_a^b f(x)\, dx$$

a. Consider the improper double integral $\iint\limits_D f(x, y) \, dA$ of the continuous function f of

two variables defined over the plane region

$$D = \{(x, y) \mid 0 \leq x < \infty \text{ and } 0 \leq y < \infty\}$$

Using the definition of improper integrals of functions of one variable (Section 7.4), explain why it makes sense to define

$$\iint\limits_D f(x, y) \, dA = \lim_{N \to \infty} \int_0^N \left[\lim_{M \to \infty} \int_0^M f(x, y) \, dx \right] dy$$

$$= \lim_{M \to \infty} \int_0^M \left[\lim_{N \to \infty} \int_0^N f(x, y) \, dy \right] dx$$

provided the limits exist.

b. Refer to Example 2. Assuming that the population density of the city is described by

$$f(x, y) = 10,000 e^{-0.2|x| - 0.1|y|}$$

for $-\infty < x < \infty$ and $-\infty < y < \infty$, show that the population outside the rectangular region

$$R = \{(x, y) \mid -10 \leq x \leq 10; \ -5 \leq y \leq 5\}$$

of Example 2 is given by

$$4 \iint\limits_D f(x, y) \, dx \, dy - 680,438$$

(recall that 680,438 is the approximate population inside R).

c. Use the results of parts (a) and (b) to determine the population of the city outside the rectangular area R.

That is, the average value of a function over $[a, b]$ is the integral of f over $[a, b]$ divided by the length of the interval. An analogous result holds for a function of two variables $f(x, y)$ over a plane region R. To see this, enclose R by a rectangle and construct a rectangular grid. Let (x_i, y_i) be any point in the rectangle R_i of area hk. Now the average value of the mn numbers $f(x_1, y_1), f(x_2, y_2), \ldots, f(x_{mn}, y_{mn})$ is given by

$$\frac{f(x_1, y_1) + f(x_2, y_2) + \cdots + f(x_{mn}, y_{mn})}{mn}$$

which can also be written as

$$\frac{hk}{hk} \left[\frac{f(x_1, y_1) + f(x_2, y_2) + \cdots + f(x_{mn}, y_{mn})}{mn} \right]$$

$$= \frac{1}{(mn)hk} [f(x_1, y_1) + f(x_2, y_2) + \cdots + f(x_{mn}, y_{mn})] hk$$

Now the area of R is approximated by the sum of the mn rectangles (*omitting those having no points in common with R*), each of area hk. Note that this is the denominator of the previous expression. Therefore, taking the limit as m and n both tend to infinity, we obtain the following formula for the *average value of $f(x, y)$ over R*.

AVERAGE VALUE OF $f(x, y)$ OVER A REGION R

If f is integrable over the plane region R, then its average value over R is given by

$$\frac{\displaystyle\iint_R f(x, y)\, dA}{\text{Area of } R}$$

or

$$\frac{\displaystyle\iint_R f(x, y)\, dA}{\displaystyle\iint_R dA} \tag{17}$$

REMARK   If we let $f(x, y) = 1$ for all (x, y) in R, then

$$\iint_R f(x, y)\, dA = \iint_R dA = \text{Area of } R$$

○ ○ ○

EXAMPLE 3   Find the average value of the function $f(x, y) = xy$ over the plane region defined by $y = e^x$, $0 \le x \le 1$.

Solution   The region R is shown in Figure 8.40. The area of the region R is given by

$$\int_0^1 \left[\int_0^{e^x} dy \right] dx = \int_0^1 \left[y \Big|_0^{e^x} \right] dx$$
$$= \int_0^1 e^x\, dx$$
$$= e^x \Big|_0^1$$
$$= e - 1$$

square units. We would obtain the same result had we viewed the area of this region as the area of the region under the curve $y = e^x$ from $x = 0$ to $x = 1$.

Figure 8.40
The plane region R defined by $y = e^x$, $0 \le x \le 1$.

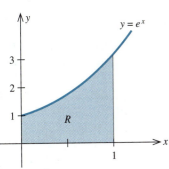

Next, we compute

$$\iint_R f(x, y)\, dA = \int_0^1 \left[\int_0^{e^x} xy\, dy \right] dx$$

$$= \int_0^1 \left[\frac{1}{2} xy^2 \Big|_0^{e^x} \right] dx$$

$$= \int_0^1 \frac{1}{2} xe^{2x}\, dx$$

$$= \frac{1}{4} xe^{2x} - \frac{1}{8} e^{2x} \Big|_0^1 \qquad \text{(Integrating by parts)}$$

$$= \left(\frac{1}{4} e^2 - \frac{1}{8} e^2 \right) + \frac{1}{8}$$

$$= \frac{1}{8} (e^2 + 1)$$

square units. Therefore, the required average value is given by

$$\frac{\displaystyle\iint_R f(x, y)\, dA}{\displaystyle\iint_R dA} = \frac{\frac{1}{8}(e^2 + 1)}{e - 1} = \frac{e^2 + 1}{8(e - 1)}$$

○ ○ ○

 EXAMPLE 4   (Refer to Example 2.) The population density of a certain city (number of people per square mile) is described by the function

$$f(x, y) = 10{,}000e^{-0.2|x| - 0.1|y|}$$

where the origin gives the location of the city hall. What is the average population density inside the rectangular area described by

$$R = \{(x, y) \mid -10 \le x \le 10;\ -5 \le y \le 5\}$$

where x and y are measured in miles?

Solution   From the results of Example 2, we know that

$$\iint_R f(x, y)\, dA \approx 680{,}438$$

From Figure 8.39, we see that the area of the plane rectangular region R is $(20)(10)$, or 200, square miles. Therefore, the average population inside R is

$$\frac{\displaystyle\iint_R f(x, y)\, dA}{\displaystyle\iint_R dA} = \frac{680{,}438}{200} = 3402.19$$

or approximately 3402 people per square mile.

○ ○ ○

SELF-CHECK EXERCISE 8.8

The population density of a coastal town located on an island is described by the function

$$f(x, y) = \frac{5000xe^y}{1 + 2x^2} \qquad (0 \le x \le 4, -2 \le y \le 0)$$

where x and y are measured in miles (see the figure). What is the population inside the rectangular area defined by $R = \{(x, y) \mid 0 \le x \le 4, -2 \le y \le 0\}$? What is the average population density in the area?

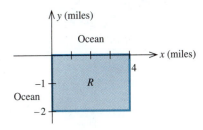

The solution to Self-Check Exercise 8.8 can be found on page 668.

8.8 EXERCISES

In exercises 1–8, use a double integral to find the volume of the solid shown in the figure.

1.

2.

3.

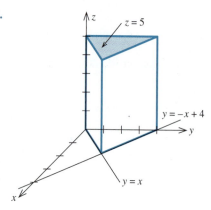

$z = 5$

$y = -x + 4$

$y = x$

4.

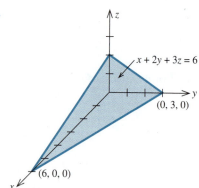

$x + 2y + 3z = 6$

$(0, 3, 0)$

$(6, 0, 0)$

5.

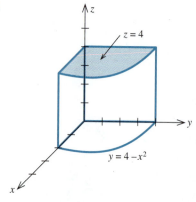

$z = 4$

$y = 4 - x^2$

6.

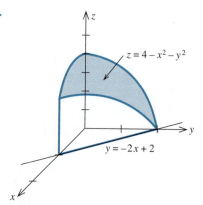

$z = 4 - x^2 - y^2$

$y = -2x + 2$

7.

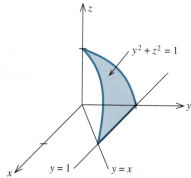

$y^2 + z^2 = 1$

$y = 1$

$y = x$

8.

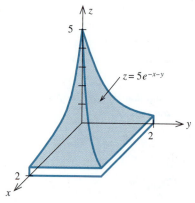

5

$z = 5e^{-x-y}$

2

2

In exercises 9–16, find the volume of the solid bounded above by the surface $z = f(x, y)$ and below by the plane region R.

9. $f(x, y) = 4 - 2x - y$; $R = \{(x, y) \mid 0 \le x \le 1$ and $0 \le y \le 2\}$

10. $f(x, y) = 2x + y$; R is the triangle bounded by $y = 2x$, $y = 0$, and $x = 2$.

11. $f(x, y) = x^2 + y^2$; R is the rectangle with vertices $(0, 0)$, $(1, 0)$, $(1, 2)$, and $(0, 2)$.

12. $f(x, y) = e^{x+2y}$; R is the triangle with vertices $(0, 0)$, $(1, 0)$, and $(0, 1)$.

13. $f(x, y) = 2xe^y$; R is the triangle bounded by $y = x$, $y = 2$, and $x = 0$.

14. $f(x, y) = \dfrac{2y}{1 + x^2}$; R is the region bounded by $y = \sqrt{x}$, $y = 0$, and $x = 4$.

15. $f(x, y) = 2x^2y$; R is the region bounded by the graphs of $y = x$ and $y = x^2$.

16. $f(x, y) = x$; R is the region in the first quadrant bounded by the semicircle $y = \sqrt{16 - x^2}$, the x-axis, and the y-axis.

In exercises 17–22, find the average value of the given function f(x, y) over the plane region R.

17. $f(x, y) = 6x^2y^3$; $R = \{(x, y) \mid 0 < x \le 2 \text{ and } 0 \le y \le 3\}$

18. $f(x, y) = x + 2y$; R is the triangle with vertices $(0, 0)$, $(1, 0)$, and $(1, 1)$.

19. $f(x, y) = xy$; R is the triangle bounded by $y = x$, $y = 2 - x$, and $y = 0$.

20. $f(x, y) = e^{-x^2}$; R is the triangle with vertices $(0, 0)$, $(1, 0)$, and $(1, 1)$.

21. $f(x, y) = xe^y$; R is the triangle with vertices $(0, 0)$, $(1, 0)$, and $(1, 1)$.

22. $f(x, y) = \ln x$; R is the region bounded by the graphs of $y = 2x$ and $y = 0$ from $x = 1$ to $x = 3$.   [*Hint:* Use integration by parts.]

23. **Population Density** The population density of a coastal town is described by the function

$$f(x, y) = \frac{10{,}000e^y}{1 + 0.5|x|} \qquad (-10 \le x \le 10, -4 \le y \le 0)$$

where x and y are measured in miles (see the accompanying figure).

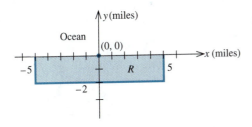

What is the population inside the rectangular area described by

$$R = \{(x, y) \mid -5 \le x \le 5, -2 \le y \le 0\}$$

24. **Average Population Density** Refer to exercise 23. Find the average population density inside the rectangular area R.

25. **Average Profit** The Country Workshop's total weekly profit (in dollars) realized in manufacturing and selling its rolltop desks is given by the profit function

$$P(x, y) = -0.2x^2 - 0.25y^2 - 0.2xy + 100x + 90y - 4000$$

where x stands for the number of finished units and y stands for the number of unfinished units manufactured and sold per week. Find the average weekly profit if the number of finished units manufactured and sold varies between 180 and 200 and the number of unfinished units varies between 100 and 120 per week.

26. **Average Price of Land** The rectangular region R shown in the accompanying figure represents a city's financial district. The price of land within the district is approximated by the function

$$p(x, y) = 200 - 10\left(x - \frac{1}{2}\right)^2 - 15(y - 1)^2$$

where $p(x, y)$ is the price of land at the point (x, y) in dollars per square foot and x and y are measured in miles. What is the average price of land per square foot in the district?

SOLUTION TO SELF-CHECK EXERCISE 8.8

The population in R is given by

$$\iint_R f(x, y)\, dA = \int_0^4 \left[\int_{-2}^0 \frac{5000xe^y}{1 + 2x^2}\, dy \right] dx$$

$$= \int_0^4 \left[\frac{5000xe^y}{1 + 2x^2} \Big|_{-2}^0 \right] dx$$

$$= 5000(1 - e^{-2}) \int_0^4 \frac{x}{1 + 2x^2}\, dx$$

$$= 5000(1 - e^{-2}) \left[\frac{1}{4} \ln(1 + 2x^2) \Big|_0^4 \right]$$

$$= 5000(1 - e^{-2}) \left(\frac{1}{4} \right) \ln 33$$

or approximately 3779 people. The average population density inside R is

$$\frac{\displaystyle\iint_R f(x, y)\, dA}{\displaystyle\iint_R dA} = \frac{3779}{(2)(4)}$$

or approximately 472 people per square mile.

CHAPTER 8 SUMMARY OF PRINCIPAL TERMS

Terms

Three-dimensional Cartesian coordinate system

Level curve

Function of two variables:
Domain

First partial derivative

Second partial derivative

Relative maximum

Relative minimum

Absolute maximum

Absolute minimum

Critical point

Saddle point

Second Derivative Test

Constrained relative extremum

Total differential

Riemann sum

Double integral

Cobb-Douglas production function

Marginal productivity of labor

Marginal productivity of capital

Substitute commodities

Complementary commodities

Method of least squares

Least-squares line

Regression line

Normal equations

Method of Lagrange Multipliers

Volume of a solid under a surface

CHAPTER 8 REVIEW EXERCISES

1. Let $f(x, y) = xy/(x^2 + y^2)$. Compute $f(0, 1), f(1, 0)$, and $f(1, 1)$. Does $f(0, 0)$ exist?

2. Let $h(x, y, z) = xye^z + (x/y)$. Compute $h(1, 1, 0)$, $h(-1, 1, 1)$, and $h(1, -1, 1)$.

3. Find the domain of the function

$$f(u, v) = \frac{\sqrt{u}}{u - v}$$

4. Find the domain of the function

$$f(x, y, z) = \frac{xy\sqrt{z}}{(1 - x)(1 - y)(1 - z)}$$

In exercises 5 and 6, sketch the level curves of the given function corresponding to the given values of z.

5. $z = f(x, y) = y - x^2$; $z = -2, -1, 0, 1, 2$

6. $z = f(x, y) = e^{xy}$; $z = 1, 2, 3$

In exercises 7–12, compute the first partial derivatives of the given functions.

7. $f(x, y) = x^2y^3 + 3xy^2 + \dfrac{x}{y}$

8. $f(u, v) = \sqrt{uv^2 - 2u}$ 9. $g(x, y) = \dfrac{xy}{x^2 + y^2}$

10. $h(x, y) = (2xy + 3y^2)^5$

11. $f(x, y) = (x^2 + y^2)e^{x^2 + y^2}$

12. $f(x, y) = \ln\left(1 + \dfrac{x^2}{y^2}\right)$

In exercises 13–16, compute the second-order partial derivatives of the given functions.

13. $f(x, y) = x^3 - 2x^2y + y^2 + x - 2y$

14. $f(x, y) = (2x^2 + 3y^2)^3$

15. $g(x, y) = e^{x^2 + y^2}$ 16. $h(s, t) = \ln\left(\dfrac{s}{t}\right)$

17. Let $f(x, y, z) = x^3y^2z + xy^2z + 3xy - 4z$. Compute $f_x(1, 1, 0)$, $f_y(1, 1, 0)$, and $f_z(1, 1, 0)$, and interpret your results.

In exercises 18 and 19, find the total differential of each function at the given point.

18. $f(x, y) = (x^2 + y^4)^{3/2}$; (3, 2)

19. $f(x, y) = xe^{x-y} + x \ln y$; (1, 1)

In exercises 20 and 21, find the approximate change in z when the point (x, y) changes from (x_0, y_0) to (x_1, y_1).

20. $f(x, y) = 2x^2y^3 + 3y^2x^2 - 2xy$; from $(1, -1)$ to $(1.02, -0.98)$

21. $f(x, y) = 4x^{3/4}y^{1/4}$; from $(16, 81)$ to $(17, 80)$

In exercises 22–26, find the critical point(s) of the given functions. Then use the Second Derivative Test to classify the nature of each of these points, if possible. Finally, determine the relative extrema of each function.

22. $f(x, y) = 2x^2 + y^2 - 8x - 6y + 4$

23. $f(x, y) = x^2 + 3xy + y^2 - 10x - 20y + 12$

24. $f(x, y) = x^3 + y^2 - 4xy + 17x - 10y + 8$

25. $f(x, y) = e^{2x^2 + y^2}$

26. $f(x, y) = \ln(x^2 + y^2 - 2x - 2y + 4)$

27. Find the equation of the least-squares line for these data points:

$$(-1, 0), \quad (1, 3), \quad (3, 4), \quad (5, 3), \quad (7, 5)$$

In exercises 28–31, use the Method of Lagrange Multipliers to optimize the given functions subject to the given constraints.

28. Maximize the function $f(x, y) = -3x^2 - y^2 + 2xy$ subject to the constraint $2x + y = 4$.

29. Minimize the function $f(x, y) = 2x^2 + 3y^2 - 6xy + 4x - 9y + 10$ subject to the constraint $x + y = 1$.

30. Find the maximum and minimum values of the function $f(x, y) = 2x - 3y + 1$ subject to the constraint $2x^2 + 3y^2 - 125 = 0$.

31. Find the maximum and minimum values of the function $f(x, y) = e^{x-y}$ subject to the constraint $x^2 + y^2 = 1$.

In exercises 32–35, evaluate the double integrals.

32. $f(x, y) = 3x - 2y$; R is the rectangle defined by $2 \leq x \leq 4$ and $-1 \leq y \leq 2$.

33. $f(x, y) = e^{-x-2y}$; R is the rectangle defined by $0 \leq x \leq 2$ and $0 \leq y \leq 1$.

34. $f(x, y) = 2x^2y$; R is bounded by $x = 0$, $x = 1$, $y = x^2$, and $y = x^3$.

35. $f(x, y) = y/x$; R is bounded by $x = 1$, $x = 2$, $y = 1$, and $y = x$.

In exercises 36 and 37, find the volume of the solid bounded above by the surface $z = f(x, y)$ and below by the plane region R.

36. $f(x, y) = 4x^2 + y^2$; $R = \{0 \leq x \leq 2 \text{ and } 0 \leq y \leq 1\}$

37. $f(x, y) = x + y$; R is the region bounded by $y = x^2$, $y = 4x$, and $y = 4$.

38. Find the average value of the function

$$f(x, y) = xy + 1$$

over the plane region R bounded by $y = x^2$ and $y = 2x$.

39. A division of Ditton Industries makes a 16-speed and a 10-speed electric blender. The company's management estimates that x units of the 16-speed model and y units of the 10-speed model are demanded daily when the unit prices are

$$p = 80 - 0.02x - 0.1y$$
and $$q = 60 - 0.1x - 0.05y$$

dollars, respectively.
a. Find the daily total revenue function $R(x, y)$.
b. Find the domain of the function R.
c. Compute $R(100, 300)$ and interpret your result.

40. In a survey conducted by *Home Entertainment* magazine, it was determined that the demand equation for compact disc (CD) players is given by

$$x = f(p, q) = 900 - 9p - e^{0.4q}$$

whereas the demand equation for compact audio discs is given by

$$y = g(p, q) = 20,000 - 3000q - 4p$$

where p and q denote the unit prices (in dollars) for the CD players and audio discs, respectively, and x and y denote the number of CD players and audio discs demanded per week. Determine whether these two products are substitute, complementary, or neither.

41. The total daily profit function (in dollars) of Weston Publishing Company realized in publishing and selling its English language dictionaries is given by

$$P(x, y) = -0.0005x^2 - 0.003y^2 - 0.002xy$$
$$+ 14x + 12y - 200$$

where x denotes the number of deluxe copies and y denotes the number of standard copies published and sold daily. Currently the number of deluxe and standard copies of the dictionaries published and sold daily are 1000 and 1700, respectively. Determine the approximate daily change in the total daily profit if the number of deluxe copies is increased to 1050 and the number of standard copies is decreased to 1650 per day.

42. Odyssey Travel Agency's monthly revenue depends on the amount of money x (in thousands of dollars) spent on advertising per month and the number of agents y in its employ in accordance with the rule

$$R(x, y) = -x^2 - 0.5y^2 + xy + 8x + 3y + 20$$

Determine the amount of money the agency should spend per month and the number of agents it should employ in order to maximize its monthly revenue.

43. The following data were compiled by the Bureau of Television Advertising in a large metropolitan area, giving the average daily TV viewing time per household in that area over the years 1988 to 1996.

Year	1988	1990
Daily Viewing Time (y)	6 hr 9 min	6 hr 30 min

Year	1992	1994	1996
Daily Viewing Time (y)	6 hr 36 min	7 hr	7 hr 16 min

a. Find the least-squares line for these data. (Let $x = 1$ represent the year 1988.)
b. Estimate the average daily TV viewing time per household in the year 2000.

44. The owner of the Rancho Grande wants to enclose a rectangular piece of grazing land along the straight portion of a river and then subdivide it using a fence running parallel to the sides. No fencing is required along the river. If the material for the sides costs $3 per running yard and the material for the divider costs $2 per running yard, what will the dimensions of a 303,750-yard pasture be if the cost of fencing material is kept to a minimum?

45. The production of Q units of a commodity is related to the amount of labor x and the amount of capital y (in suitable units) expended by the equation

$$Q = f(x, y) = x^{3/4}y^{1/4}$$

If an expenditure of 100 units is available for production, how should it be apportioned between labor and capital so that Q is maximized?

[*Hint:* Use the Method of Lagrange Multipliers to maximize the function Q subject to the constraint $x + y = 100$.]

Many applied problems can be formulated as an equation involving the derivative, or differential, of the function to be determined. These equations are called *differential equations*. In this chapter we show how differential equations are used to solve problems involving the growth of an amount of money earning interest compounded continuously, the growth of a population of bacteria, the decay of radioactive material, and the rate at which a person learns a new subject, just to name a few.

If the amount of fertilizer used to cultivate land on a wheat farm is increased, will the crop yield increase substantially? Researchers at a midwestern university found that a new experimental fertilizer increased the wheat yield of the land at the university's experimental field station. In Example 2, page 690, you will see how they used a differential equation to estimate the crop yield.

DIFFERENTIAL EQUATIONS

• • •

9.1 DIFFERENTIAL EQUATIONS

Models Involving Differential Equations

We first encountered differential equations in Section 6.1. Recall that a **differential equation** is an equation that involves an unknown function and its derivative(s). Here are some examples of differential equations:

$$\frac{dy}{dx} = xe^x, \qquad \frac{dy}{dx} + 2y = x^2, \qquad \frac{d^2y}{dt^2} + \left(\frac{dy}{dt}\right)^3 + ty - 8 = 0$$

Differential equations appear in practically every branch of applied mathematics, and the study of such equations remains one of the most active areas in mathematics. As you will see in the next few examples, models involving differential equations often arise from the mathematical formulation of practical problems.

Unrestricted Growth Models The unrestricted growth model was first discussed in Chapter 5. There we saw that the size of a population at any time t, $Q(t)$, increases at a rate that is proportional to $Q(t)$ itself. Thus,

$$\frac{dQ}{dt} = kQ \qquad\qquad (1)$$

where k is a constant of proportionality. This is a differential equation involving the unknown function Q and its derivative Q'.

Restricted Growth Models In many applications the quantity $Q(t)$ does not exhibit unrestricted growth but approaches some definite upper bound. The learning curves and logistic functions we discussed in Chapter 5 are examples of restricted growth models. Let us derive the mathematical models that lead to these functions.

Suppose $Q(t)$ does not exceed some number C, called the *carrying capacity of the environment*. Furthermore, suppose the rate of growth of this quantity is *proportional* to the difference between its upper bound and its current size. The resulting differential equation is

$$\frac{dQ}{dt} = k(C - Q) \qquad\qquad (2)$$

where k is a constant of proportionality. Observe that if the initial population is small relative to C, then the rate of growth of Q is relatively large. But as $Q(t)$ approaches C, the difference $C - Q(t)$ approaches zero, and so does the rate of growth of Q. In Section 9.3 you will see that the solution of the differential equation (2) is a function that describes a learning curve (Figure 9.1).

Next, let us consider a restricted growth model in which the rate of growth of a quantity $Q(t)$ is *jointly proportional* to its current size and the difference between its upper bound and its current size; that is,

$$\frac{dQ}{dt} = kQ(C - Q) \qquad\qquad (3)$$

Figure 9.1

$Q(t)$, the solution of a differential equation, describes a learning curve.

where k is a constant of proportionality. Observe that when $Q(t)$ is small relative to C, the rate of growth of Q is approximately proportional to Q. But as $Q(t)$ approaches C, the growth rate slows down to zero. Next, if $Q > C$, then $dQ/dt < 0$ and the quantity is decreasing with time, with the decay rate slowing down as Q approaches C. We will show later that the solution of the differential equation (3) is just the logistic function we discussed in Chapter 5. Its graph is shown in Figure 9.2.

Stimulus Response   In the quantitative theory of psychology, one model that describes the relationship between a stimulus S and the resulting response R is the Weber-Fechner Law. This law asserts that the rate of change of a reaction R is inversely proportional to the stimulus S. Mathematically, this law may be expressed as

$$\frac{dR}{dS} = \frac{k}{S} \tag{4}$$

where k is a constant of proportionality. Furthermore, suppose that the threshold level, the lowest level of stimulation at which sensation is detected, is S_0. Then we have the condition $R = 0$ when $S = S_0$; that is, $R(S_0) = 0$. The graph of R versus S is shown in Figure 9.3.

Mixture Problems   Our next example is a typical mixture problem. Suppose that a tank initially contains 10 gallons of pure water. Brine containing 3 pounds of salt per gallon flows into the tank at a rate of 2 gallons per minute, and the well-stirred mixture flows out of the tank at the same rate. How much salt is in the tank at any given time?

Let us formulate this problem mathematically. Suppose $A(t)$ denotes the amount of salt in the tank at any time t. Then the derivative dA/dt, the rate of change of the amount of salt at any time t, must satisfy the condition

$$\frac{dA}{dt} = (\text{Rate of salt flowing in}) - (\text{Rate of salt flowing out})$$

(Figure 9.4). But the rate at which salt flows into the tank is given by

$$(2 \text{ gal/min})(3 \text{ lb/gal}) \qquad [(\text{Rate of flow}) \times (\text{Concentration})]$$

or 6 lb/min. Since the rate at which the solution leaves the tank is the same as the rate at which the brine is poured into it, that tank contains 10 gallons of the mixture at any time t. Since the salt content at any time t is A pounds, the concentration of the mixture is $(A/10)$ lb/gal. Therefore, the rate at which salt flows out of the tank is given by

$$(2 \text{ gal/min}) \left(\frac{A}{10} \text{ lb/gal} \right)$$

or $(A/5)$ lb/min. Therefore, we are led to the differential equation

$$\frac{dA}{dt} = 6 - \frac{A}{5} \tag{5}$$

An additional condition arises from the fact that initially there is no salt in the solution. This condition may be expressed mathematically as $A = 0$ when $t = 0$ or, more concisely, $A(0) = 0$.

We will solve each of the differential equations we have introduced here in Section 9.3.

Solutions of Differential Equations

Suppose we are given a differential equation involving the derivative(s) of a function y. Recall that a **solution** to a differential equation is any function $f(x)$ that satisfies the differential equation. Thus, $y = f(x)$ is a solution of the differential equation provided that the replacement of y and its derivative(s) by the function $f(x)$ and its corresponding derivatives reduces the given differential equation to an identity for all values of x.

EXAMPLE 1   Show that the function $f(x) = e^{-x} + x - 1$ is a solution of the differential equation

$$y' + y = x$$

Solution   Let

$$y = f(x) = e^{-x} + x - 1$$

so that

$$y' = f'(x) = -e^{-x} + 1$$

Substituting this last equation into the left side of the given equation yields

$$\overbrace{(-e^{-x} + 1)}^{y'} + \overbrace{(e^{-x} + x - 1)}^{y} = -e^{-x} + 1 + e^{-x} + x - 1 = x$$

which is equal to the right side of the given equation for all values of x. Therefore, $f(x) = e^{-x} + x - 1$ is a solution of the given differential equation.

○ ○ ○

In the preceding example, we verified that $y = e^{-x} + x - 1$ is a solution of the differential equation $y' + y = x$. This is by no means the only solution of the differential equation, as the next example shows.

EXAMPLE 2   Show that any function of the form $f(x) = ce^{-x} + x - 1$, where c is a constant, is a solution of the differential equation

$$y' + y = x$$

Solution   Let

$$y = f(x) = ce^{-x} + x - 1$$

so that

$$y' = f'(x) = -ce^{-x} + 1$$

Substituting the last equation into the left side of the given differential equation yields

$$\overbrace{-ce^{-x} + 1}^{y'} + \overbrace{ce^{-x} + x - 1}^{y} = x$$

and we have verified the assertion.

○ ○ ○

It can be shown that *every* solution of the differential equation $y' + y = x$ must have the form $y = ce^{-x} + x - 1$, where c is a constant; therefore, this is the **general solution** of the differential equation $y' + y = x$. Figure 9.5 shows a family of solutions of this differential equation for selected values of c.

Figure 9.5
Some solutions of $y' + y = x$.

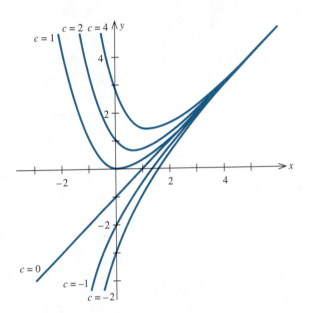

Recall that a solution obtained by assigning a specific value to the constant c is called a **particular solution** of the differential equation. For example, the particular solution $y = e^{-x} + x - 1$ of Example 1 is obtained from the general solution by taking $c = 1$. In practice, a particular solution of a differential equation is obtained from the general solution of the differential equation by requiring that the solution and/or its derivative(s) satisfy certain conditions at one or more values of x.

EXAMPLE 3   Use the results of Example 2 to find the particular solution of the equation $y' + y = x$ that satisfies the condition $y(0) = 0$; that is, $f(0) = 0$, where f denotes the solution.

Solution   From the results of Example 2, we see that the general solution of the given differential equation is given by

$$y = f(x) = ce^{-x} + x - 1$$

Using the given condition, we see that

$$f(0) = ce^0 + 0 - 1 = c - 1 = 0 \quad \text{or} \quad c = 1$$

Therefore, the required particular solution is $y = e^{-x} + x - 1$.   ⊙ ⊙ ⊙

Consider the differential equation $\dfrac{dy}{dx} = F(x, y)$, and suppose $y = f(x)$ is a solution of the differential equation.

a. If (a, b) is a point in the domain of F, explain why $F(a, b)$ gives the slope of f at $x = a$.

b. For the differential equation $\dfrac{dy}{dx} = \dfrac{x}{y}$, compute $F(x, y)$ for selected integral values of x and y. (For example, try $x = \pm 2, \pm1, 0, \pm1, \pm2$ and $y = \pm2, \pm1, \pm2, \pm3$.) Verify that if you draw a lineal element (a tiny line segment) having slope $F(x, y)$ through each point (x, y), you obtain a *direction field* similar to the one shown in the figure.

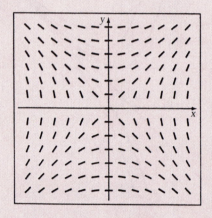

c. The direction field associated with the differential equation hints at the solution curves for the differential equation. Sketch a few solution curves for the differential equation.

REMARK   You will be asked to verify your answer to part (c) in the next section.

SELF-CHECK EXERCISES 9.1

1. Consider the differential equation

$$xy' + 2y = 4x^2$$

a. Show that $y = x^2 + (c/x^2)$ is the general solution of the differential equation.
b. Find the particular solution of the differential equation that satisfies $y(1) = 4$.

2. The population of a certain species grows at a rate directly proportional to the square root of its size. If the initial population is N_0, find the population at any time t. Formulate but do not solve the problem.

Solutions to Self-Check Exercises 9.1 can be found on page 680.

9.1 EXERCISES

In exercises 1–12, verify that y is a solution of the given differential equation.

1. $y = x^2$; $xy' + y = 3x^2$

2. $y = e^x$; $y' - y = 0$

3. $y = \frac{1}{2} + ce^{-x^2}$, c any constant; $y' + 2xy = x$

4. $y = Ce^{kx}$, C any constant; $dy/dx = ky$

5. $y = e^{-2x}$; $y'' + y' - 2y = 0$

6. $y = C_1e^x + C_2e^{2x}$; $y'' - 3y' + 2y = 0$

7. $y = C_1e^{-2x} + C_2xe^{-2x}$; $y'' + 4y' + 4y = 0$

8. $y = C_1 + C_2x^{1/3}$; $3xy'' + 2y' = 0$

9. $y = \dfrac{C_1}{x} + C_2\dfrac{\ln x}{x}$; $x^2y'' + 3xy' + y = 0$

10. $y = C_1e^x + C_2xe^x + C_3x^2e^x$; $y''' - 3y'' + 3y' - y = 0$

11. $y = C - Ae^{-kt}$, A and C constants; $dy/dt = k(C - y)$

12. $y = \dfrac{C}{1 + Ae^{-Ckt}}$, A and C constants; $dy/dt = ky(C - y)$

In exercises 13–18, verify that y is a general solution of the given differential equation. Then find a particular solution of the differential equation that satisfies the side condition.

13. $y = Cx^2 - 2x$; $y' - 2(y/x) = 2$; $y(1) = 10$

14. $y = Ce^{-x^2}$; $y' = -2xy$; $y(0) = y_0$

15. $y = C/x$; $y' + (1/x)y = 0$; $y(1) = 1$

16. $y = Ce^{2x} - 2x - 1$; $y' - 2y - 4x = 0$; $y(0) = 3$

17. $y = \dfrac{Ce^x}{x} + \dfrac{1}{2}xe^x$; $y' + \left(\dfrac{1-x}{x}\right)y = e^x$; $y(1) = -\dfrac{1}{2}e$

18. $y = C_1x^3 + C_2x^2$; $x^2y'' - 4xy' + 6y = 0$; $y(2) = 0$ and $y'(2) = 4$

19. **Radioactive Decay** A radioactive substance decays at a rate directly proportional to the amount present. If

the substance is present in the amount of Q_0 grams initially $(t = 0)$, find the amount present at some later time. Formulate the problem in terms of a differential equation with a side condition. Do not solve it.

20. **Supply and Demand** Let $S(t)$ denote the supply of a certain commodity as a function of time t. Suppose that the rate of change of the supply is proportional to the difference between the demand $D(t)$ and the supply. Find a differential equation that describes this situation.

21. **Net Investment** The management of a company has decided that the level of investment should not exceed C dollars. Furthermore, management has decided that the rate of net investment (the rate of change of the total capital invested) should be proportional to the difference between C and the total capital invested. Formulate but do not solve the problem in terms of a differential equation.

22. **Lambert's Law of Absorption** Lambert's law of absorption states that the percentage of incident light, L, absorbed in passing through a thin layer of material, x, is proportional to the thickness of the material. If, for a certain material, x_0 inches of the material reduces the light to half its intensity, how much additional material is needed to reduce the intensity to a quarter of its initial value? Formulate but do not solve the problem in terms of a differential equation with a side condition.

23. **Concentration of a Drug in the Bloodstream** The rate at which the concentration of a drug in the bloodstream decreases is proportional to the concentration at any time t. Initially, the concentration of the drug in the bloodstream is C_0 gm/ml. What is the concentration of the drug in the bloodstream at any time t? Formulate but do not solve the problem in terms of a differential equation with a side condition.

24. **Amount of Glucose in the Bloodstream** Suppose that glucose is infused into the bloodstream at a constant rate of C grams per minute and, at the same time, the glucose is converted and removed from the bloodstream at a rate proportional to the amount of glucose present. Show that the amount of glucose $A(t)$ present in the bloodstream at any time t is governed by the differential

equation

$$A' = C - kA$$

where k is a constant.

25. **Newton's Law of Cooling** Newton's law of cooling states that the temperature of a body drops at a rate that is proportional to the difference between the temperature y of the body and the constant temperature C of the surrounding medium (assume that the temperature of the body is initially greater than C). Show that Newton's law of cooling may be expressed as the differential equation

$$\frac{dy}{dt} = -k(y - C), \qquad y(0) = y_0$$

where y_0 denotes the temperature of the body before immersion in the medium.

26. **Fisk's Law** Suppose that a cell of volume V cc is surrounded by a homogeneous chemical solution of concentration C gm/cc. Let y denote the concentration of the solute inside the cell at any time t, and suppose that, initially, the concentration is y_0. Fisk's law, named after the German physiologist Adolf Fisk (1829–1901), states that the rate of change of the concentration of solute inside the cell at any time t is proportional to the difference between the concentration of the solute outside the cell and the concentration inside the cell and inversely proportional to the volume of the cell. Show that Fisk's law may be expressed as the differential equation

$$\frac{dy}{dt} = \frac{k}{V}(C - y), \qquad y(0) = y_0$$

where k is a constant. (*Note:* The constant of proportionality, k, depends on the area and permeability of the cell membrane.)

27. **Allometric Laws** Suppose that $x(t)$ denotes the weight of an animal's organ at time t and $g(t)$ denotes the size of another organ in the same animal at the same time t. An allometric law (allometry is the study of the relative growth of a part in relation to an entire organism) states that the relative growth rate of one organ, $(dx/dt)/x$, is proportional to the relative growth rate of the other, $(dy/dt)/y$. Show that this allometric law may be stated in terms of the differential equation

$$\frac{1}{x}\frac{dx}{dt} = k\frac{1}{y}\frac{dy}{dt}$$

where k is a constant.

28. **Gompertz Growth Curve** Suppose that a quantity $Q(t)$ does not exceed some number C; that is, $Q(t) \le C$ for all t. Suppose further that the rate of growth of $Q(t)$ is jointly poportional to its current size and the difference between its upper bound and the natural logarithm of its current size. What is the size of the quantity $Q(t)$ at any time t? Show that the mathematical formulation of this problem leads to the differential equation

$$\frac{dQ}{dt} = kQ(C - \ln Q)$$

with the side condition $Q(0) = Q_0$, where Q_0 denotes the size of the quantity present initially. The graph of $Q(t)$ is called the **Gompertz growth curve.** This model, like the ones leading to the learning curve and the logistice curve, describes restricted growth.

SOLUTIONS TO SELF-CHECK EXERCISES 9.1

I. a. We compute

$$y' = 2x - \frac{2c}{x^3}$$

Substituting this into the left side of the given differential equation gives

$$x\left(2x - \frac{2c}{x^3}\right) + 2\left(x^2 + \frac{c}{x^2}\right) = 2x^2 - \frac{2c}{x^2} + 2x^2 + \frac{2c}{x^2} = 4x^2$$

which equals the expression on the right side of the differential equation, and this verifies the assertion.

b. Using the given condition, we have

$$4 = 1^2 + \frac{c}{1^2} \quad \text{or} \quad c = 3$$

and the required particular solution is

$$y = x^2 + \frac{3}{x^2}$$

2. Let N denote the size of the population at any time t. Then the required differential equation is

$$\frac{dN}{dt} = kN^{1/2}$$

and the initial condition is $N(0) = N_0$.

9.2 SEPARATION OF VARIABLES

The Method of Separation of Variables

Differential equations are classified according to their basic form. A compelling reason for this categorization is that different methods are used to solve different types of equations.

A differential equation may be classified by the order of its derivative. A differential equation is of **order n** if the highest derivative of the unknown function appearing in the equation is of order n. For example, the differential equations

$$y' = xe^x \quad \text{and} \quad y' + 2y = x^2$$

are first-order equations, whereas the differential equation

$$\frac{d^2y}{dt^2} + \left(\frac{dy}{dt}\right)^3 + ty - 8 = 0$$

is of order two. For the remainder of this chapter we will restrict our study to first-order differential equations.

In this section we will describe a method for solving an important class of first-order differential equations that can be written in the form

$$\frac{dy}{dx} = f(x)g(y)$$

where $f(x)$ is a function of x only and $g(y)$ is a function of y only. Such differential equations are said to be **separable** because the variables can be separated. Equations (1) through (5) are first-order separable differential equations. As another example, the equation

$$\frac{dQ}{dt} = kQ(C - Q)$$

has the form $dQ/dt = f(t)g(Q)$, where $f(t) = k$ and $g(Q) = Q(C - Q)$, and so is separable. On the other hand, the differential equation

$$\frac{dy}{dx} = xy^2 + 2$$

is *not* separable.

Separable first-order equations can be solved using the *method of separation of variables*.

THE METHOD OF SEPARATION OF VARIABLES

Suppose that we are given a first-order separable differential equation in the form

$$\frac{dy}{dx} = f(x)g(y) \tag{6}$$

Step I Write equation (6) in the form

$$\frac{dy}{g(y)} = f(x)\,dx \tag{7}$$

When written in this form, the variables in (7) are said to be *separated*.

Step 2 Integrate each side of equation (7) with respect to the appropriate variable.

We will justify this method at the end of this section.

Solving Separable Differential Equations

EXAMPLE I   Find the general solution of the first-order differential equation

$$y' = \frac{xy}{x^2 + 1}$$

Solution

Step I   Observe that the given differential equation has the form

$$\frac{dy}{dx} = \left(\frac{x}{x^2 + 1}\right) y = f(x)g(y)$$

where $f(x) = x/(x^2 + 1)$ and $g(y) = y$, and is therefore separable. Separating the variables, we obtain

$$\frac{dy}{y} = \left(\frac{x}{x^2 + 1}\right) dx$$

Step 2 Integrating each side of the last equation with respect to the appropriate variable, we have

$$\int \frac{dy}{y} = \int \frac{x}{x^2 + 1}\, dx$$

or

$$\ln |y| + C_1 = \frac{1}{2}\, \ln(x^2 + 1) + C_2$$

$$\ln |y| = \frac{1}{2}\, \ln(x^2 + 1) + C_2 - C_1$$

where C_1 and C_2 are arbitrary constants of integration. Letting C denote the constant such that $C_2 - C_1 = \ln |C|$, we have

$$\ln |y| = \frac{1}{2}\, \ln(x^2 + 1) + \ln |C|$$

$$= \ln \sqrt{x^2 + 1} + \ln |C|$$

$$= \ln |C \sqrt{x^2 + 1}| \qquad (\ln A + \ln B = \ln AB)$$

so the general solution is

$$y = C\sqrt{x^2 + 1} \qqu\qquad \text{◑ ◑ ◑}$$

EXPLORING WITH TECHNOLOGY

Refer to Example 1, where it was shown that the general solution of the given differential equation is $y = C \sqrt{x^2 + 1}$. Use a graphing utility to plot the graphs of the members of this family of solutions corresponding to $C = -3, -2, -1, 0, 1, 2,$ and 3. Use the standard viewing rectangle.

◑ ◑ ◑

EXAMPLE 2 Find the particular solution of the differential equation

$$ye^x + (y^2 - 1)y' = 0$$

that satisfies the condition $y(0) = 1$.

Solution

Step 1 Writing the given differential equation in the form

$$ye^x + (y^2 - 1)\frac{dy}{dx} = 0 \quad \text{or} \quad (y^2 - 1)\frac{dy}{dx} = -ye^x$$

and separating the variables, we obtain

$$\frac{y^2 - 1}{y}\, dy = -e^x\, dx$$

Step 2   Integrating each side of this equation with respect to the appropriate variable, we have

$$\int \frac{y^2 - 1}{y} \, dy = -\int e^x \, dx$$

$$\int \left(y - \frac{1}{y} \right) dy = -\int e^x \, dx$$

$$\frac{1}{2} y^2 - \ln |y| = -e^x + C_1$$

$$y^2 - \ln y^2 = -2e^x + C \qquad (C = 2C_1)$$

Using the condition $y(0) = 1$, we have

$$1 - \ln 1 = -2 + C \quad \text{or} \quad C = 3$$

Therefore, the required solution is given by

$$y^2 - \ln y^2 = -2e^x + 3$$

$\circ \ \circ \ \circ$

Example 2 is an initial value problem. In general, an **initial value problem** is a differential equation with enough conditions specified at a point to determine a particular (unique) solution.* Also observe that the solution of Example 2 appeared as an implicit equation involving x and y. This often happens when we solve separable differential equations.

EXAMPLE 3   Find an equation f given that (1) the slope of the tangent line to the graph of f at any point $P(x, y)$ is given by the expression $-x/(2y)$ and (2) the graph of f passes through the point $P(1, 2)$.

Solution   The slope of the tangent line to the graph of f at any point $P(x, y)$ is given by the derivative

$$y' = \frac{dy}{dx} = -\frac{x}{2y}$$

which is a separable first-order differential equation. Separating the variables, we obtain

$$2y \, dy = -x \, dx$$

$\circ \ \circ \ \circ$

* One condition is sufficient for a first-order differential equation.

Figure 9.6
The graph of $x^2 + 2y^2 = 9$.

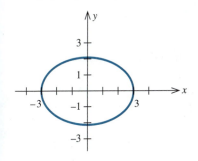

which, upon integration, yields

$$y^2 = -\frac{1}{2}x^2 + C_1$$

or
$$x^2 + 2y^2 = C \qquad (C = 2C_1)$$

where C is an arbitrary constant.

To evaluate C, we use the second condition, which implies that when $x = 1$, $y = 2$. This gives

$$1^2 + 2(2^2) = C \quad \text{or} \quad C = 9$$

Hence, the required equation is

$$x^2 + 2y^2 = 9$$

The graph of the equation f appears in Figure 9.6.

Refer to the Group Discussion Problem on page 678 in Section 9.1. Use the Method of Separation of Variables to solve the differential equation $\dfrac{dy}{dx} = \dfrac{x}{y}$ and thus verify that your solution to (c) in the problem is indeed correct.

Justification of the Method of Separation of Variables

To justify the method of separation of variables, let us consider the separable equation (6) in its general form

$$\frac{dy}{dx} = f(x)g(y)$$

If $g(y) \neq 0$, we may rewrite the equation in the form

$$\frac{1}{g(y)}\frac{dy}{dx} - f(x) = 0$$

Now, suppose that G is an antiderivative of $1/g$ and F is an antiderivative of f. Using the Chain Rule, we see that

$$\frac{d}{dx}[G(y) - F(x)] = G'(y)\frac{dy}{dx} - F'(x) = \frac{1}{g(y)}\frac{dy}{dx} - f(x)$$

Therefore,

$$\frac{d}{dx}[G(y) - F(x)] = 0$$

and so
$$G(y) - F(x) = C \qquad (C, \text{a constant})$$

But the last equation is equivalent to

$$G(y) = F(x) + C$$

or

$$\int \frac{dy}{g(y)} = \int f(x)\, dx$$

which is precisely the result of step 2 in the method of separation of variables.

SELF-CHECK EXERCISE 9.2

Find the solution of the differential equation $y' = 2x^2 y + 2x^2$ that satisfies the equation $y(0) = 0$.

The solution to Self-Check Exercise 9.2 can be found on page 687.

9.2 EXERCISES

In exercises 1–16, find the general solution of the first-order differential equations by separating variables.

1. $y' = \dfrac{x+1}{y^2}$

2. $y' = \dfrac{x^2}{y}$

3. $y' = \dfrac{e^x}{y^2}$

4. $y' = -\dfrac{x}{y}$

5. $y' = 2y$

6. $y' = 2(y-1)$

7. $y' = xy^2$

8. $y' = \dfrac{2y}{x+1}$

9. $y' = -2(3y+4)$

10. $y' = \dfrac{2y+3}{x^2}$

11. $y' = \dfrac{x^2+1}{3y^2}$

12. $y' = \dfrac{xe^x}{2y}$

13. $y' = \sqrt{\dfrac{y}{x}}$

14. $y' = \dfrac{xy^2}{\sqrt{1+x^2}}$

15. $y' = \dfrac{y\ln x}{x}$

16. $y' = \dfrac{(x-4)y^4}{x^3(y^2-3)}$

In exercises 17–28, find the solution of the initial value problem.

17. $y' = \dfrac{2x}{y}; y(1) = -2$

18. $y' = xe^{-y}; y(0) = 1$

19. $y' = 2 - y; y(0) = 3$

20. $y' = y/x; y(1) = 1$

21. $y' = 3xy - 2x; y(0) = 1$

22. $y' = xe^{x^2}y; y(0) = 1$

23. $y' = \dfrac{xy}{x^2+1}; y(0) = 1$

24. $y' = x^2 y^{-1/2}; y(1) = 1$

25. $y' = xye^x; y(1) = 1$

26. $y' = 2xe^{-y}; y(0) = 1$

27. $y' = 3x^2 e^{-y}; y(0) = 1$

28. $y' = \dfrac{y^2}{x-2}; y(3) = 1$

29. Find an equation defining a function f given that (1) the slope of the tangent line to the graph of f at any point $P(x, y)$ is given by the expression $dy/dx = (3x^2)/(2y)$ and (2) the graph of f passes through the point $(1, 3)$.

30. Find a function f given that (1) the slope of the tangent line to the graph of f at any point $P(x, y)$ is given by the expression $dy/dx = 3xy$ and (2) the graph of f passes through the point $(0, 2)$.

31. Exponential Decay Use separation of variables to solve the differential equation

$$\frac{dQ}{dt} = -kQ, \qquad Q(0) = Q_0$$

where k and Q_0 are positive constants, describing exponential decay.

32. Fisk's Law Refer to exercise 26, Section 9.1. Use separation of variables to solve the differential equation

$$\frac{dy}{dt} = \frac{k}{V}(C - y), \qquad y(0) = y_0$$

where k, V, C, and y_0 are constants with $C - y > 0$. Find $\lim_{t \to \infty} y$ and interpret your result.

33. **Concentration of Glucose in the Bloodstream** Refer to exercise 24, Section 9.1. Use separation of variables to solve the differential equation $A' = C - kA$, where C and k are positive constants.

    Hint: Rewrite the given differential equation in the form

    $$\frac{dA}{dt} = k\left(\frac{C}{k} - A\right)$$

34. **Allometric Laws** Refer to exercise 27, Section 9.1. Use separation of variables to solve the differential equation

    $$\frac{1}{x}\frac{dx}{dt} = k\frac{1}{y}\frac{dy}{dt}$$

    where k is a constant.

35. **Supply and Demand** Assume that the rate of change of the supply of a commodity is proportional to the difference between the demand and the supply so that

    $$\frac{dS}{dt} = k(D - S)$$

    where k is a constant of proportionality. Suppose that D is constant and $S(0) = S_0$. Find a formula for $S(t)$.

36. **Supply and Demand** Assume that the rate of change of the unit price of a commodity is proportional to the difference between the demand and the supply so that

    $$\frac{dp}{dt} = k(D - S)$$

    where k is a constant of proportionality. Suppose that $D = 50 - 2p$, $S = 5 + 3p$, and $p(0) = 4$. Find a formula for $p(t)$.

SOLUTION TO SELF-CHECK EXERCISE 9.2

Writing the differential equation in the form

$$\frac{dy}{dx} = 2x^2(y + 1)$$

and separating variables, we obtain

$$\frac{dy}{y + 1} = 2x^2\, dx$$

Integrating each side of the last equation with respect to the appropriate variable, we have

$$\int \frac{dy}{y + 1} = \int 2x^2\, dx$$

or

$$\ln|y + 1| = \frac{2}{3}x^3 + C$$

Using the initial condition $y(0) = 0$, we have

$$\ln 1 = C \quad \text{or} \quad C = 0$$

Therefore,

$$\ln|y + 1| = \frac{2}{3}x^3$$

$$y + 1 = e^{(2/3)x^3}$$

or

$$y = e^{(2/3)x^3} - 1$$

9.3 APPLICATIONS OF SEPARABLE DIFFERENTIAL EQUATIONS

In this section we will look at some applications of first-order separable differential equations. We begin by reexamining some of the applications discussed in Section 9.1.

Unrestricted Growth Models

The differential equation describing an unrestricted growth model is given by

$$\frac{dQ}{dt} = kQ$$

where $Q(t)$ represents the size of a certain population at any time t, and k is a positive constant. Separating the variables in this differential equation, we have

$$\frac{dQ}{Q} = k \, dt$$

which, upon integration, yields

$$\int \frac{dQ}{Q} = \int k \, dt$$

$$\ln |Q| = kt + C_1$$

or $$Q = e^{kt+C_1} = Ce^{kt}$$

where $C = e^{C_1}$ is an arbitrary positive constant. Thus, we may write the solution as

$$Q(t) = Ce^{kt}$$

Figure 9.7
An unrestricted growth model.

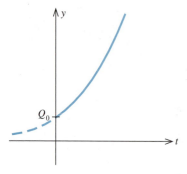

Observe that if the quantity present initially is denoted by Q_0, then $Q(0) = Q_0$. Applying this condition yields the equation

$$Ce^0 = Q_0 \quad \text{or} \quad C = Q_0$$

Therefore, the model for unrestricted exponential growth with initial population Q_0 is given by

$$Q(t) = Q_0 e^{kt} \tag{8}$$

(Figure 9.7).

EXAMPLE 1   Under ideal laboratory conditions, the rate of growth of bacteria in a culture is proportional to the size of the culture at any time t. Suppose that 10,000 bacteria are present initially in a culture and 60,000 are present 2 hours later. How many bacteria will there be in the culture at the end of 4 hours?

Solution   Let $Q(t)$ denote the number of bacteria present in the culture at time t. Then

$$\frac{dQ}{dt} = kQ$$

where k is a constant of proportionality. Solving this separable first-order differential equation, we obtain

$$Q(t) = Q_0 e^{kt} \qquad \text{[Equation (8)]}$$

where Q_0 denotes the initial bacteria population. Since $Q_0 = 10{,}000$, we have

$$Q(t) = 10{,}000 e^{kt}$$

Next, the condition that 60,000 bacteria are present 2 hours later translates into $Q(2) = 60{,}000$, or

$$60{,}000 = 10{,}000 e^{2k}$$
$$e^{2k} = 6$$
$$e^{k} = 6^{1/2}$$

Thus, the number of bacteria present at any time t is given by

$$Q(t) = 10{,}000 e^{kt} = 10{,}000 (e^{k})^{t}$$
$$= (10{,}000) 6^{(t/2)}$$

In particular, the number of bacteria present in the culture at the end of 4 hours is given by

$$Q(4) = 10{,}000 (6^{4/2})$$
$$= 360{,}000 \qquad \circ \ \circ \ \circ$$

Restricted Growth Models

From Section 9.1 we see that a differential equation describing a restricted growth model is given by

$$\frac{dQ}{dt} = k(C - Q) \tag{9}$$

where both k and C are positive constants. To solve this separable first-order differential equation, we first separate the variables, obtaining

$$\frac{dQ}{C - Q} = k \, dt$$

Integrating each side with respect to the appropriate variable yields

$$\int \frac{dQ}{C - Q} = \int k \, dt$$
$$-\ln |C - Q| = kt + d \qquad (d, \text{ an arbitrary constant})$$
$$\ln |C - Q| = -kt - d \tag{10}$$
$$C - Q = e^{-kt-d} = e^{-kt} e^{-d}$$

or $\qquad\qquad Q(t) = C - A e^{-kt}$

where we have denoted the constant e^{-d} by A. This is the equation of the learning curve studied in Chapter 5 (Figure 9.8).

Figure 9.8
A restricted exponential growth model.

EXAMPLE 2   In an experiment conducted by researchers of the Agriculture Department of a midwestern university, it was found that the maximum yield of wheat in the university's experimental field station was 150 bushels per acre. Furthermore, the researchers discovered that the rate at which the yield of wheat increased was governed by the differential equation

$$\frac{dQ}{dx} = k(150 - Q)$$

where $Q(x)$ denotes the yield in bushels per acre and x is the amount in pounds of an experimental fertilizer used per acre of land. Data obtained in the experiment indicated that 10 pounds of fertilizer per acre of land would result in a yield of 80 bushels of wheat per acre, whereas 20 pounds of fertilizer per acre of land would result in a yield of 120 bushels of wheat per acre. Determine the yield if 30 pounds of fertilizer were used per acre.

Solution   The given differential equation has the same form as equation (9) with $C = 150$. Solving it directly or using the result obtained in the solution of equation (9), we see that the yield per acre is given by

$$Q(x) = 150 - Ae^{-kx}$$

The first of the given conditions implies that $Q(10) = 80$; that is,

$$150 - Ae^{-10k} = 80$$

or $A = 70e^{10k}$. Therefore,

$$Q(x) = 150 - 70e^{10k}e^{-kx}$$
$$= 150 - 70e^{-k(x-10)}$$

The second of the given conditions implies that $Q(20) = 120$, or

$$150 - 70e^{-k(20-10)} = 120$$
$$70e^{-10k} = 30$$
$$e^{-10k} = \frac{3}{7}$$

Taking the logarithm of each side of the equation, we find

$$\ln e^{-10k} = \ln\left(\frac{3}{7}\right)$$
$$-10k = \ln 3 - \ln 7 \approx -0.8473$$

or $$k \approx 0.085$$

Therefore,

$$Q(x) = 150 - 70e^{-0.085(x-10)}$$

Figure 9.9

$Q(x)$ is a function relating crop yield to the amount of fertilizer used.

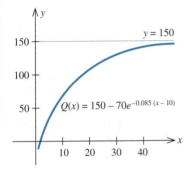

$Q(x) = 150 - 70e^{-0.085\,(x - 10)}$

In particular, when $x = 30$, we have

$$Q(30) = 150 - 70e^{-0.085(20)}$$
$$= 150 - 70e^{-1.7}$$
$$\approx 137$$

so the yield would be 137 bushels per acre if 30 pounds of fertilizer were used per acre. The graph of $Q(x)$ is shown in Figure 9.9. Note that in the solution of this problem, it was not necessary to find the specific value of k. (Why?)

❍ ❍ ❍

Next, let us consider a differential equation describing another type of restricted growth:

$$\frac{dQ}{dt} = kQ(C - Q)$$

where k and C are positive constants. Separating variables leads to

$$\frac{dQ}{Q(C - Q)} = k\, dt$$

Integrating each side of this equation with respect to the appropriate variable, we have

$$\int \frac{1}{Q(C - Q)}\, dQ = \int k\, dt$$

As it stands, the integrand on the left side of this equation is not in a form that can be easily integrated. However, observe that

$$\frac{1}{Q(C - Q)} = \frac{1}{C}\left[\frac{1}{Q} + \frac{1}{C - Q}\right]$$

as you may verify by adding the terms between the brackets on the right-hand side. Making use of this identity, we have

$$\int \frac{1}{C}\left[\frac{1}{Q} + \frac{1}{C - Q}\right] dQ = \int k\, dt$$

$$\int \frac{dQ}{Q} + \int \frac{dQ}{C - Q} = Ck \int dt$$

$$\ln |Q| - \ln |C - Q| = Ckt + b \qquad (b, \text{an arbitrary constant})$$

$$\ln \left|\frac{Q}{C - Q}\right| = Ckt + b$$

$$\frac{Q}{C - Q} = e^{Ckt+b} = e^b e^{Ckt} = Be^{Ckt} \qquad (B = e^b)$$

$$Q = CBe^{Ckt} - QBe^{Ckt}$$

$$(1 + Be^{Ckt})Q = CBe^{Ckt}$$

and

$$Q = \frac{CBe^{Ckt}}{1 + Be^{Ckt}}$$

or

$$Q(t) = \frac{C}{1 + Ae^{-Ckt}} \qquad \left(A = \frac{1}{B}\right) \qquad \textbf{(11)}$$

(see Figure 9.2). In its final form, this function is equivalent to the logistic function encountered in Chapter 5.

EXAMPLE 3   During a flu epidemic, 5% of the 5000 army personnel stationed at Fort MacArthur had contracted influenza at time $t = 0$. Furthermore, the rate at which they were contracting influenza was jointly proportional to the number of personnel who had already contracted the disease and the noninfected population. If 20% of the personnel had contracted the flu by the tenth day, find the number of personnel who had contracted the flu by the thirteenth day.

Solution   Let $Q(t)$ denote the number of army personnel who had contracted the flu after t days. Then

$$\frac{dQ}{dt} = kQ(5000 - Q)$$

We may solve this separable differential equation directly, or we may appeal to the result for the more general problem obtained earlier. Opting for the latter and noting that $C = 5000$, we find that, using equation (11),

$$Q(t) = \frac{5000}{1 + Ae^{-5000kt}}$$

The condition that 5% of the population had contracted influenza at time $t = 0$ implies that

$$Q(0) = \frac{5000}{1 + A} = 250$$

from which we see that $A = 19$. Therefore,

$$Q(t) = \frac{5000}{1 + 19e^{-5000kt}}$$

Next, the condition that 20% of the population had contracted influenza by the tenth day implies that

$$Q(10) = \frac{5000}{1 + 19e^{-50,000k}} = 1000$$

or

$$1 + 19e^{-50,000k} = 5$$

$$e^{-50,000k} = \frac{4}{19}$$

$$-50,000k = \ln 4 - \ln 19$$

and
$$k = -\frac{1}{50,000}(\ln 4 - \ln 19)$$

$$\approx 0.0000312$$

Therefore,

$$Q(t) = \frac{5000}{1 + 19e^{-0.156t}}$$

In particular, the number of army personnel who had contracted the flu by the thirteenth day is given by

$$Q(13) = \frac{5000}{1 + 19e^{-0.156(13)}} = \frac{5000}{1 + 19e^{-2.028}} \approx 1428$$

or approximately 1428, or 29%. The graph of $Q(t)$ is shown in Figure 9.10.

Figure 9.10
An epidemic model.

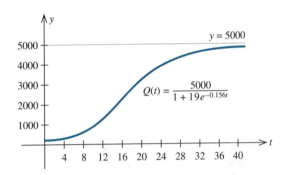

Consider the model for restricted growth described by the differential equation $dQ/dt = kQ(C - Q)$ with the solution given in equation (11).

a. Show that the rate of growth of Q is greatest at $t = (\ln A)/kC$. [*Hint:* Use both the differential equation and equation (11).]

b. Refer to Example 3. At what time is the number of influenza cases increasing at the greatest rate?

EXAMPLE 4 Derive the Weber-Fechner Law describing the relationship between a stimulus S and the resulting response R by solving the differential equation (4) subject to the condition $R = 0$ when $S = S_0$, where S_0 is the threshold level.

Solution The differential equation under consideration is

$$\frac{dR}{dS} = \frac{k}{S}$$

and is separable. Separating the variables, we have

$$dR = k\frac{dS}{S}$$

which, upon integration, yields

$$\int dR = k \int \frac{dS}{S}$$
$$R = k \ln S + C$$

where C is an arbitrary constant. Using the condition $R = 0$ when $S = S_0$ gives

$$0 = k \ln S_0 + C$$

or

$$C = -k \ln S_0$$

Substituting this value of C in the expression for R leads to

$$R = k \ln S - k \ln S_0$$

or

$$R = k \ln \frac{S}{S_0}$$

the required relationship between R and S. The graph of R is shown in Figure 9.11.

○ ○ ○

Figure 9.11
The Weber-Fechner Law.

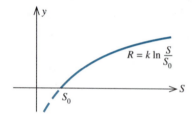

$R = k \ln \dfrac{S}{S_0}$

EXAMPLE 5   Suppose that a tank initially contains 10 gallons of pure water. Brine containing 3 pounds of salt per gallon flows into the tank at a rate of 2 gallons per minute, and the well-stirred mixture flows out of the tank at the same rate. How much salt is present at the end of 10 minutes? How much salt is present in the long run?

Solution   The problem was formulated mathematically on page 675, and we were led to the differential equation

$$\frac{dA}{dt} = 6 - \frac{A}{5}$$

subject to the condition $A(0) = 0$. The differential equation

$$\frac{dA}{dt} = 6 - \frac{A}{5} = \frac{30 - A}{5}$$

is separable. Separating the variables and integrating, we obtain

$$\int \frac{dA}{30 - A} = \int \frac{1}{5} dt$$

$$-\ln |30 - A| = \frac{1}{5}t + b \qquad (b, \text{ a constant})$$

$$\ln |30 - A| = -\frac{1}{5}t - b$$

$$30 - A = e^{-b} e^{-t/5}$$

and

$$A = 30 - Ce^{-t/5} \qquad (C = e^{-b})$$

Figure 9.12
The solution of the differential equation $\dfrac{dA}{dt} = 6 - \dfrac{A}{5}$.

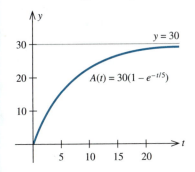

The condition $A(0) = 0$ implies that

$$0 = 30 - C$$

giving $C = 30$, and so

$$A(t) = 30(1 - e^{-t/5})$$

The amount of salt present after 10 minutes is given by

$$A(10) = 30(1 - e^{-2}) \approx 25.94$$

or 25.94 pounds. The amount of salt present in the long run is given by

$$\lim_{t \to \infty} A(t) = \lim_{t \to \infty} 30(1 - e^{-t/5}) = 30$$

or 30 pounds (Figure 9.12).   ⊙ ⊙ ⊙

SELF-CHECK EXERCISE 9.3

Newton's law of cooling states that the temperature of an object drops at a rate that is proportional to the difference in the temperature between the object and that of the surrounding medium. Suppose that an apple pie is taken out of the oven at a temperature of 200°F and placed on the counter in a room where the temperature is 70°F. If the temperature of the apple pie is 150°F after 5 minutes, find its temperature $y(t)$ as a function of time t.

The solution to Self-Check Exercise 9.3 can be found on page 697.

9.3 EXERCISES

A calculator is recommended for these exercises.

1. Chemical Decomposition   The rate of decomposition of a certain chemical substance is directly proportional to the amount present at any time t. If y_0 grams of the chemical are present at time $t = 0$, find an expression for the amount present at any time t.

2. Growth of Bacteria   Under ideal laboratory conditions, the rate of growth of bacteria in a culture is proportional to the size of the culture at any time t. Suppose that 2000 bacteria are present initially in the culture and 5000 are present 1 hour later. How many bacteria will be in the culture at the end of 2 hours?

3. World Population Growth   The world population at the beginning of 1980 was 4.5 billion. Assuming that the population continues to grow at its present rate of approximately 2% per year, find a function $Q(t)$ that expresses the world population (in billions) as a function of time t (in years). What will the world population be at the beginning of the year 2005?

4. Population Growth   The population of a certain community is increasing at a rate directly proportional to the population at any time t. In the last 3 years the population has doubled. How long will it take for the population to triple?

5. Lambert's Law of Absorption   According to Lambert's law of absorption, the percentage of incident light, L, absorbed in passing through a thin layer of material, x, is proportional to the thickness of the material. For a certain material, if one-half inch of the material reduces the light to half of its intensity, how much additional material is needed to reduce the intensity to one-fourth of its initial value?

6. **Savings Accounts**   An amount of money deposited in a savings account grows at a rate proportional to the amount present. (It can be shown that an amount of money grows in this manner if it earns interest compounded continuously.) Suppose that $10,000 is deposited in a fixed account earning interest at the rate of 10% per year compounded continuously.
   a. What is the accumulated amount after 5 years?
   b. How long does it take for the original deposit to double in value?

7. **Chemical Reactions**   In a certain chemical reaction, a substance is converted into another substance at a rate proportional to the square of the amount of the first substance present at any time t. Initially ($t = 0$) 50 grams of the first substance was present; 1 hour later, only 10 grams of it remained. Find an expression that gives the amount of the first substance present at any time t. What is the amount present after 2 hours?

8. **Newton's Law of Cooling**   Newton's law of cooling states that the rate at which the temperature of an object changes is directly proportional to the difference in temperature between the object and that of the surrounding medium. Suppose that a horseshoe heated to a temperature of 100°C is immersed in a large tank of water at a (constant) temperature at 30°C at time $t = 0$. Three minutes later the temperature of the horseshoe is reduced to 70°C. Derive an expression that gives the temperature of the horseshoe at any time t. What is the temperature of the horseshoe 5 minutes after it has been immersed in the water?

9. **Newton's Law of Cooling**   Newton's law of cooling states that the rate at which the temperature of an object changes is directly proportional to the difference in temperature between the object and that of the surrounding medium. Suppose that a cup of coffee is prepared with boiling water (212°F) and left to cool on the counter in a room where the temperature is 72°F. If the temperature of the coffee is 140°F after 2 minutes, determine when the coffee will be cool enough to drink (say, 110°F).

10. **Learning Curves**   The American Stenographic Institute finds that the average student taking Elementary Shorthand will progress at a rate given by

$$\frac{dQ}{dt} = k(80 - Q)$$

in a 20-week course, where $Q(t)$ measures the number of words of dictation a student can take per minute after t weeks in the course. If the average student can take 50 words of dictation per minute after 10 weeks in the course, how many words per minute can the average student take after completing the course?

11. **Training Personnel**   The personnel manager of the Gibraltar Insurance Company estimates that the number of insurance claims an experienced clerk can process in a day is 40. Furthermore, the rate at which a clerk can process insurance claims during the tth week of training is proportional to the difference between the maximum number possible (40) and the number he or she can process in the tth week. If the number of claims the average trainee can process after two weeks on the job is 10 per day, determine how many claims the average trainee can process after six weeks on the job.

12. **Effect of Immigration on Population Growth**   Suppose that a country's population at any time t grows in accordance with the rule

$$\frac{dP}{dt} = kP + I$$

where P denotes the population at any time t, k is a positive constant reflecting the natural growth rate of the population, and I is a constant giving the (constant) rate of immigration into the country. If the total population of the country at time $t = 0$ is P_0, find an expression for the population at any time t.

13. **Effect of Immigration on Population Growth**   Refer to exercise 12. The population of the United States in the year 1980 ($t = 0$) was 226.5 million. Suppose that the natural growth rate is 0.8% annually ($k = 0.008$) and net immigration is allowed at the rate of half a million people per year ($I = 0.5$) until the end of the century. What will the U.S. population be in the year 2005?

14. **Sinking Funds**   The proprietor of the Carson Hardware Store has decided to set up a sinking fund for the purpose of purchasing a computer 2 years from now. It is expected that the purchase will involve a sum of $30,000. The fund grows at the rate of

$$\frac{dA}{dt} = rA + P$$

where A denotes the size of the fund at any time t, r is the annual interest rate earned by the fund compounded continuously, and P is the amount (in dollars) paid into the fund by the proprietor per year (assume this is done on a frequent basis in small deposits over the year so that it is essentially continuous). If the fund earns 10% interest per year compounded continuously, determine the size of the yearly investment the proprietor should pay into the fund.

15. Spread of a Rumor   The rate at which a rumor spreads through an Alpine village of 400 residents is jointly proportional to the number of residents who have heard it and the number who have not. Initially, 10 residents heard the rumor, but two days later this number increased to 80. Find the number of people who will have heard the rumor after a week.

16. Growth of a Fruit Fly Colony   A biologist has determined that the maximum number of fruit flies that can be sustained in a carefully controlled environment (with a limited supply of space and food) is 400. Suppose that the rate at which the population of the colony increases obeys the rule

$$\frac{dQ}{dt} = kQ(C - Q)$$

where C is the carrying capacity (400) and Q denotes the number of fruit flies in the colony at any time t. If the initial population of fruit flies in the experiment is 10 and it grows to 45 after ten days, determine the population of the colony of fruit flies on the twentieth day.

17. Gompertz Growth Curves   Refer to exercise 28, Section 9.1. Consider the differential equation

$$\frac{dQ}{dt} = kQ(C - \ln Q)$$

with the side condition $Q(0) = Q_0$. The solution $Q(t)$ describes restricted growth and has a graph known as the Gompertz curve. Using separation of variables, solve this differential equation.

18. Chemical Reaction Rates   Two chemical solutions, one containing N molecules of chemical A and another containing M molecules of chemical B, are mixed together at time $t = 0$. The molecules from the two chemicals combine to form another chemical solution containing y (AB) molecules. The rate at which the AB molecules are formed, dy/dt, is called the **reaction rate** and is jointly proportional to $(N - y)$ and $(M - y)$. Thus,

$$\frac{dy}{dt} = k(N - y)(M - y)$$

where k is a constant (we assume the temperature of the chemical mixture remains constant during the interaction). Solve this differential equation with the side condition $y(0) = 0$ assuming that $N - y > 0$ and $M - y > 0$.

Hint: Use the identity

$$\frac{1}{(N - y)(M - y)} = \frac{1}{M - N}\left(\frac{1}{N - y} - \frac{1}{M - y}\right)$$

19. Mixture Problems   A tank initially contains 20 gallons of pure water. Brine containing 2 pounds of salt per gallon flows into the tank at a rate of 3 gallons per minute, and the well-stirred mixture flows out of the tank at the same rate. How much salt is present in the tank at any time? How much salt is present at the end of 20 minutes? How much salt is present in the long run?

20. Mixture Problems   A tank initially contains 50 gallons of brine, in which 10 pounds of salt is dissolved. Brine containing 2 pounds of dissolved salt per gallon flows into the tank at the rate of 2 gallons per minute, and the well-stirred mixture flows out of the tank at the same rate. How much salt is present in the tank at the end of 10 minutes?

SOLUTION TO SELF-CHECK EXERCISE 9.3

We are required to solve the initial value problem

$$\frac{dy}{dt} = -k(y - 70) \qquad (k, \text{ a constant of proportionality})$$

$$y(0) = 200$$

To solve the differential equation, we separate variables and integrate, obtaining

$$\int \frac{dy}{y - 70} = \int -k \, dt$$

$$\ln |y - 70| = -kt + d \qquad (d, \text{an arbitrary constant})$$

$$y - 70 = e^{-kt+d} = Ae^{-kt} \qquad (A = e^d)$$

or
$$y = 70 + Ae^{-kt}$$

Using the initial condition $y(0) = 200$, we find

$$200 = 70 + A$$

or $A = 130$. Therefore,

$$y = 70 + 130e^{-kt}$$

To determine the value of k, we use the fact that $y(5) = 150$, obtaining

$$150 = 70 + 130e^{-5k}$$

$$130e^{-5k} = 80$$

$$e^{-5k} = \frac{80}{130}$$

$$-5k = \ln \frac{80}{130}$$

$$k = -\frac{1}{5} \ln \frac{80}{130}$$

$$= 0.097$$

Therefore, the required expression is

$$y(t) = 70 + 130e^{-0.097t}$$

9.4 APPROXIMATE SOLUTIONS OF DIFFERENTIAL EQUATIONS

Euler's Method

As in the case of definite integrals, there are many differential equations whose exact solutions cannot be found using any of the available methods. In such cases, we must once again resort to approximate solutions to the problems at hand.

Many numerical methods have been developed for efficient computation of approximate solutions to differential equations. In this section, we will look at a method for solving the problem

$$\frac{dy}{dx} = F(x, y), \qquad y(x_0) = y_0 \tag{12}$$

Euler's Method, named after its discoverer, Leonhard Euler (1707–1783), illustrates quite clearly the idea behind the method for finding the approximate solution of equation (12). Basically, the technique calls for approximating the actual solution $y = f(x)$ at certain selected values of x. The values of f between two adjacent values of x are then found by linear interpolation. This situation is depicted geometrically in Figure 9.13. Thus, in Euler's Method, the actual solution curve of the differential equation is approximated by a suitable polygonal curve. We now describe the method in greater detail.

Figure 9.13
Using Euler's Method, the actual solution curve of the differential equation is approximated by a polygonal curve.

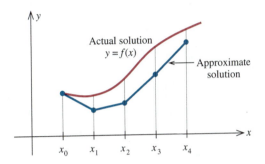

Let h be a small positive number, and let $x_n = x_0 + nh$ ($n = 0, 1, 2, 3, \ldots$). Then

$$x_1 = x_0 + h, \quad x_2 = x_0 + 2h, \quad x_3 = x_0 + 3h, \quad \ldots$$

Thus, the points $x_0, x_1, x_2, x_3, \ldots$ are spaced evenly apart and the distance between any two adjacent points is h units.

We begin by finding an approximation $y = \tilde{f}_1(x)$ to the actual solution $y = f(x)$ on the subinterval $[x_0, x_1]$. Observe that the *initial* condition $y(x_0) = y_0$ of (12) tells us that the point (x_0, y_0) lies on the solution curve. Now, Euler's Method calls for approximating the part of the graph of f on the interval $[x_0, x_1]$ by a straight line segment that is tangent to the graph of f at the point (x_0, y_0). To find this approximating linear function $y = \tilde{f}_1(x)$, recall that the differential equation $y' = F(x, y)$ gives the slope of the tangent line to the graph of $y = f(x)$ at any point (x, y) lying on the graph. In particular, the slope of the required straight line segment is equal to $F(x_0, y_0)$. Therefore, using the point-slope form of the equation of a line, we see that the equation of the straight line segment is

$$y - y_0 = F(x_0, y_0)(x - x_0)$$

or
$$y = y_0 + F(x_0, y_0)(x - x_0)$$

Thus, the required approximation to the actual solution $y = f(x)$ on the interval $[x_0, x_1]$ is given by the linear function

$$\tilde{f}_1(x) = y_0 + F(x_0, y_0)(x - x_0) \tag{13}$$

This situation is depicted graphically in Figure 9.14.

Figure 9.14
$\tilde{f}_1(x) = y_0 + F(x_0, y_0)(x - x_0)$ is the
equation of the straight line used to
approximate $f(x)$ on $[x_0, x_1]$.

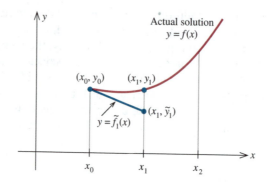

Next, to find an approximation $y = \tilde{f}_2(x)$ to the actual solution $y = f(x)$ on the subinterval $[x_1, x_2]$, observe that at $x = x_1$, the true value $y_1 = f(x)$ is approximated by the number

$$\tilde{y}_1 = \tilde{f}_1(x_1) = y_0 + F(x_0, y_0)(x_1 - x_0)$$
$$= y_0 + F(x_0, y_0)h \qquad (\text{Since } x_1 - x_0 = h)$$

According to Euler's Method, the approximating function in this interval has a graph that is just a line segment beginning at the point $(x_1, \tilde{y}_1)$ and having an appropriate slope. Ideally, this slope should be the slope of the tangent line to the graph of $y = f(x)$ at the point (x_1, y_1). But since this point is supposedly unknown to us, a practical alternative is to use the point $(x_1, \tilde{y}_1)$ in lieu of the point (x_1, y_1) to compute an approximation of the desired slope—namely, $F(x_1, \tilde{y}_1)$. Once again, using the point-slope form of the equation of a line, we find that the equation of the required tangent line is

$$y - \tilde{y}_1 = F(x_1, \tilde{y}_1)(x - x_1)$$

or

$$y = \tilde{y}_1 + F(x_1, \tilde{y}_1)(x - x_1)$$

so the required approximating linear function on the interval $[x_1, x_2]$ is given by

$$\tilde{f}_2(x) = \tilde{y}_1 + F(x_1, \tilde{y}_1)(x - x_1) \tag{14}$$

(Figure 9.15).

Figure 9.15
$\tilde{f}_2(x) = \tilde{y}_1 + F(x_1, \tilde{y}_1)(x - x_1)$ is the
equation of the straight line used to
approximate $f(x)$ on $[x_1, x_2]$.

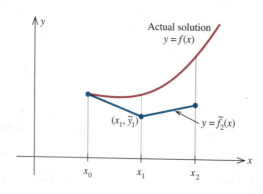

Proceeding, we find that the actual value $y_2 = f(x_2)$ is approximated by the number

$$\tilde{y}_2 = \tilde{f}_2(x_2) = \tilde{y}_1 + F(x_1, \tilde{y}_1)(x_2 - x_1)$$
$$= \tilde{y}_1 + F(x_1, \tilde{y}_1)h$$

and the approximating function on the interval $[x_2, x_3]$ is given by

$$\tilde{f}_3(x) = \tilde{y}_2 + F(x_2, \tilde{y}_2)(x - x_2)$$

and so on.

In many practical applications we are interested only in finding an approximation of the solution of the differential equation (12) at some specific value of x, say $x = b$. In such cases, it is not important to compute the approximating functions $\tilde{f}_1, \tilde{f}_2, \tilde{f}_3, \ldots$. It suffices to compute the numbers $\tilde{y}_1, \tilde{y}_2, \tilde{y}_3, \ldots, \tilde{y}_n$ at the points $x_1, x_2, x_3, \ldots, x_n$ with h chosen to be

$$h = \frac{b - x_0}{n}$$

where n is a sufficiently large positive integer. In this case, we see that

$$x_1 = x_0 + h, \quad x_2 = x_0 + 2h, \quad \ldots, \quad x_n = x_0 + nh = x_0 + (b - x_0) = b$$

so that $\tilde{y}_n = \tilde{f}_n(x_n) = \tilde{f}_n(b)$ does in fact yield the required approximation. We may summarize the procedure as follows:

EULER'S METHOD

Suppose we are given the differential equation

$$\frac{dy}{dx} = F(x, y)$$

subject to the initial condition $y(x_0) = y_0$ and we wish to find an approximation of $y(b)$, where b is a number greater than x_0. Compute

$$h = \frac{b - x_0}{n}$$

$$x_1 = x_0 + h, \quad x_2 = x_0 + 2h, \quad x_3 = x_0 + 3h, \quad \ldots, \quad x_n = x_0 + nh = b$$

and

$$\tilde{y}_0 = y_0$$
$$\tilde{y}_1 = \tilde{y}_0 + hF(x_0, \tilde{y}_0)$$
$$\tilde{y}_2 = \tilde{y}_1 + hF(x_1, \tilde{y}_1)$$
$$\vdots$$
$$\tilde{y}_n = \tilde{y}_{n-1} + hF(x_{n-1}, \tilde{y}_{n-1})$$

Then $\tilde{y}_n$ gives an approximation of the true value $y(b)$ of the solution to the initial value problem at $x = b$.

Solving Differential Equations with Euler's Method

 EXAMPLE 1 Use Euler's Method with $n = 8$ to obtain an approximation of the initial value problem

$$y' = x - y, \qquad y(0) = 1$$

when $x = 2$.

Solution Here $x_0 = 0$ and $b = 2$, so that, taking $n = 8$, we find

$$h = \frac{2 - 0}{8} = \frac{1}{4}$$

and

$$x_0 = 0, \quad x_1 = \frac{1}{4}, \quad x_2 = \frac{1}{2}, \quad x_3 = \frac{3}{4}, \quad x_4 = 1,$$

$$x_5 = \frac{5}{4}, \quad x_6 = \frac{3}{2}, \quad x_7 = \frac{7}{4}, \quad x_8 = b = 2$$

Also,

$$F(x, y) = x - y \quad \text{and} \quad y_0 = y(0) = 1$$

Therefore, the approximations of the actual solution at the points $x_0, x_1, x_2, \ldots, x_n = b$ are

$$\tilde{y}_0 = y_0 = 1$$

$$\tilde{y}_1 = \tilde{y}_0 + hF(x_0, \tilde{y}_0) = 1 + \frac{1}{4}(0 - 1) = \frac{3}{4}$$

$$\tilde{y}_2 = \tilde{y}_1 + hF(x_1, \tilde{y}_1) = \frac{3}{4} + \frac{1}{4}\left(\frac{1}{4} - \frac{3}{4}\right) = \frac{5}{8}$$

$$\tilde{y}_3 = \tilde{y}_2 + hF(x_2, \tilde{y}_2) = \frac{5}{8} + \frac{1}{4}\left(\frac{1}{2} - \frac{5}{8}\right) = \frac{19}{32}$$

$$\tilde{y}_4 = \tilde{y}_3 + hF(x_3, \tilde{y}_3) = \frac{19}{32} + \frac{1}{4}\left(\frac{3}{4} - \frac{19}{32}\right) = \frac{81}{128}$$

$$\tilde{y}_5 = \tilde{y}_4 + hF(x_4, \tilde{y}_4) = \frac{81}{128} + \frac{1}{4}\left(1 - \frac{81}{128}\right) = \frac{371}{512}$$

$$\tilde{y}_6 = \tilde{y}_5 + hF(x_5, \tilde{y}_5) = \frac{371}{512} + \frac{1}{4}\left(\frac{5}{4} - \frac{371}{512}\right) = \frac{1753}{2048}$$

$$\tilde{y}_7 = \tilde{y}_6 + hF(x_6, \tilde{y}_6) = \frac{1753}{2048} + \frac{1}{4}\left(\frac{3}{2} - \frac{1753}{2048}\right) = \frac{8331}{8192}$$

$$\tilde{y}_8 = \tilde{y}_7 + hF(x_7, \tilde{y}_7) = \frac{8331}{8192} + \frac{1}{4}\left(\frac{7}{4} - \frac{8331}{8192}\right) = \frac{39,329}{32,768}$$

Thus, the approximate value of $y(2)$ is

$$\frac{39,329}{32,768} \approx 1.2002$$

EXAMPLE 2 Use Euler's Method with (a) $n = 5$ and (b) $n = 10$ to approximate the solution of the initial value problem

$$y' = -2xy^2, \qquad y(0) = 1$$

on the interval $[0, 0.5]$. Find the actual solution of the initial value problem. Finally, sketch the graphs of the approximate solutions and the actual solution for $0 \le x \le 0.5$ on the same set of axes.

Solution

a. Here $x_0 = 0$ and $b = 0.5$. Taking $n = 5$, we find

$$h = \frac{0.5 - 0}{5} = 0.1$$

and $x_0 = 0$, $x_1 = 0.1$, $x_2 = 0.2$, $x_3 = 0.3$, $x_4 = 0.4$, and $x_5 = b = 0.5$. Also,

$$F(x, y) = -2xy^2 \quad \text{and} \quad y_0 = y(0) = 1$$

Therefore,

$$\tilde{y}_0 = y_0 = 1$$
$$\tilde{y}_1 = \tilde{y}_0 + hF(x_0, \tilde{y}_0) = 1 + 0.1(-2)(0)(1)^2 = 1$$
$$\tilde{y}_2 = \tilde{y}_1 + hF(x_1, \tilde{y}_1) = 1 + 0.1(-2)(0.1)(1)^2 = 0.98$$
$$\tilde{y}_3 = \tilde{y}_2 + hF(x_2, \tilde{y}_2) = 0.98 + 0.1(-2)(0.2)(0.98)^2 = 0.9416$$
$$\tilde{y}_4 = \tilde{y}_3 + hF(x_3, \tilde{y}_3) = 0.9416 + 0.1(-2)(0.3)(0.9416)^2 = 0.8884$$
$$\tilde{y}_5 = \tilde{y}_4 + hF(x_4, \tilde{y}_4) = 0.8884 + 0.1(-2)(0.4)(0.8884)^2 = 0.8253$$

b. Here $x_0 = 0$ and $b = 0.5$. Taking $n = 10$, we find

$$h = \frac{0.5 - 0}{10}$$

and $x_0 = 0$, $x_1 = 0.05$, $x_2 = 0.10$, ..., $x_9 = 0.45$, and $x_{10} = 0.5 = b$. Proceeding as in (a), we obtain the approximate solutions listed in the following table:

x	0.00	0.05	0.10	0.15	0.20	0.25
$\tilde{y}_n$	1.0000	1.0000	0.9950	0.9851	0.9705	0.9517

x	0.30	0.35	0.40	0.45	0.50
$\tilde{y}_n$	0.9291	0.9032	0.8746	0.8440	0.8119

To obtain the actual solution of the differential equation, we separate variables, obtaining

$$\frac{dy}{y^2} = -2x \, dx$$

Integrating each side of the last equation with respect to the appropriate variables, we have

$$\int \frac{dy}{y^2} = -\int 2x \, dx$$

or
$$-\frac{1}{y} + C_1 = -x^2 + C_2$$

$$\frac{1}{y} = x^2 + C \qquad (C = C_1 - C_2)$$

and
$$y = \frac{1}{x^2 + C}$$

Using the condition $y(0) = 1$, we have

$$1 = \frac{1}{0 + C} \quad \text{or} \quad C = 1$$

Therefore, the required solution is given by

$$y = \frac{1}{x^2 + 1}$$

The graphs of the approximate solutions and the actual solution are sketched in Figure 9.16.

Figure 9.16
The approximate solutions and the actual solution to an initial value problem.

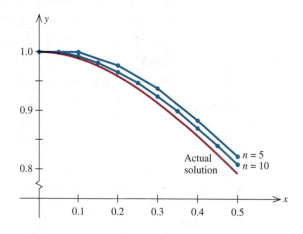

SELF–CHECK EXERCISE 9.4

Use Euler's Method with $n = 5$ to obtain an approximation of the initial value problem

$$y' = 2x + y, \qquad y(0) = 1$$

when $x = 1$.

The solution to Self-Check Exercise 9.4 can be found on page 705.

9.4 EXERCISES

A calculator is recommended for this exercise set.
In exercises 1–10, use Euler's Method with (a) n = 4 and (b) n = 6 to obtain an approximation of each initial value problem when x = b.

1. $y' = x + y$, $y(0) = 1$; $b = 1$

2. $y' = x - 2y$, $y(0) = 1$; $b = 2$

3. $y' = 2x - y + 1$, $y(0) = 2$; $b = 2$

4. $y' = 2xy$, $y(0) = 1$; $b = 0.5$

5. $y' = -2xy^2$, $y(0) = 1$; $b = 0.5$

6. $y' = x^2 + y^2$, $y(0) = 1$; $b = 1.5$

7. $y' = \sqrt{x + y}$, $y(1) = 1$; $b = 1.5$

8. $y' = (1 + x^2)^{-1}$, $y(0) = 0$; $b = 1$

9. $y' = x/y$, $y(0) = 1$; $b = 1$

10. $y' = xy^{1/3}$, $y(0) = 1$; $b = 1$

In exercises 11–15, use Euler's Method with n = 5 to obtain an approximate solution to each initial value problem over the indicated interval.

11. $y' = \dfrac{1}{2}xy$, $y(0) = 1$; $0 \le x \le 1$

12. $y' = x^2y$, $y(0) = 2$; $0 \le x \le 0.6$

13. $y' = 2x - y + 1$, $y(0) = 2$; $0 \le x \le 1$

14. $y' = x + y^2$, $y(0) = 0$; $0 \le x \le 0.5$

15. $y' = x^2 + y$, $y(0) = 1$; $0 \le x \le 0.5$

16. **Growth of Service Industries** It has been estimated that service industries, which currently make up 30% of the nonfarm work force in a certain country, will continue to grow at the rate of

$$R(t) = 5e^{1/(t+1)}$$

percent per decade t decades from now. Estimate the percentage of the nonfarm work force in the service industries one decade from now.
[*Hint:* (a) Show that the desired answer is $P(1)$, where P is the solution of the initial value problem

$$P' = 5e^{1/(t+1)}, \qquad P(0) = 30$$

(b) Use Euler's Method with $n = 10$ to approximate the solution.]

SOLUTION TO SELF-CHECK EXERCISE 9.4

Here $x_0 = 0$ and $b = 1$, so that, taking $n = 5$, we find

$$h = \frac{1 - 0}{5} = \frac{1}{5}$$

and

$$x_0 = 0, \quad x_1 = \frac{1}{5}, \quad x_2 = \frac{2}{5}, \quad x_3 = \frac{3}{5}, \quad x_4 = \frac{4}{5}, \quad x_5 = b = 1$$

Also,

$$F(x, y) = 2x + y \quad \text{and} \quad y_0 = y(0) = 1$$

Therefore, the approximations of the actual solution at the points $x_0, x_1, x_2, \ldots,$ $x_5 = 1$ are

$$\tilde{y}_0 = y_0 = 1$$

$$\tilde{y}_1 = \tilde{y}_0 + hF(x_0, \tilde{y}_0) = 1 + \frac{1}{5}(0 + 1) = \frac{6}{5}$$

$$\tilde{y}_2 = \tilde{y}_1 + hF(x_1, \tilde{y}_1) = \frac{6}{5} + \frac{1}{5}\left(\frac{2}{5} + \frac{6}{5}\right) = \frac{38}{25}$$

$$\tilde{y}_3 = \tilde{y}_2 + hF(x_2, \tilde{y}_2) = \frac{38}{25} + \frac{1}{5}\left(\frac{4}{5} + \frac{38}{25}\right) = \frac{248}{125}$$

$$\tilde{y}_4 = \tilde{y}_3 + hF(x_3, \tilde{y}_3) = \frac{248}{125} + \frac{1}{5}\left(\frac{6}{5} + \frac{248}{125}\right) = \frac{1638}{625}$$

$$\tilde{y}_5 = \tilde{y}_4 + hF(x_4, \tilde{y}_4) = \frac{1638}{625} + \frac{1}{5}\left(\frac{8}{5} + \frac{1638}{625}\right) = \frac{10{,}828}{3125}$$

Thus, the approximate value of $y(1)$ is

$$\frac{10{,}828}{3125} \approx 3.4650$$

CHAPTER 9 SUMMARY OF PRINCIPAL TERMS

Terms

Differential equation

General solution of a
   differential equation

Particular solution of a
   differential equation

First-order differential
   equation

Separable differential equation

Method of separation of variables

Initial value problem

Euler's Method

CHAPTER 9 REVIEW EXERCISES

In exercises 1–3, verify that y is a solution of the differential equation.

1. $y = C_1 e^{2x} + C_2 e^{-3x}$; $y'' + y' - 6y = 0$

2. $y = 2e^{2x} + 3x - 2$; $y'' - y' - 2y = -6x + 1$

3. $y = Cx^{-4/3}$; $4xy^3\, dx + 3x^2 y^2\, dy = 0$

In exercises 4 and 5, verify that y is a general solution of the given differential equation and find a particular solution of the differential equation satisfying the side condition.

4. $y = \dfrac{1}{x^2 - C}$; $\dfrac{dy}{dx} = -2xy^2$; $y(0) = 1$

5. $y = (9x + C)^{-1/3}$; $\dfrac{dy}{dx} = -3y^4$; $y(0) = \dfrac{1}{2}$

In exercises 6–11, solve the differential equation.

6. $y' = \dfrac{x^3 + 1}{y^2}$

7. $\dfrac{dy}{dt} = 2(4 - y)$

8. $y' = \dfrac{y \ln x}{x}$

9. $y' = 3x^2 y^2 + y^2$; $y(0) = -2$

10. $y' = x^2(1 - y)$; $y(0) = -2$

11. $\dfrac{dy}{dx} = -\dfrac{3}{2} x^2 y$; $y(0) = 3$

12. Find a function f given that (1) the slope of the tangent line to the graph of f at any point $P(x, y)$ is given by the expression

$$y' = -\frac{4xy}{x^2 + 1}$$

and (2) the graph of f passes through the point $(1, 1)$.

In exercises 13–16, use Euler's Method with (a) $n = 4$ and (b) $n = 6$ to obtain an approximation of each of the following initial value problems when $x = b$.

13. $y' = x + y^2$, $y(0) = 0$; $b = 1$

14. $y' = x^2 + 2y^2$; $y(0) = 0$; $b = 1$

15. $y' = 1 + 2xy^2$, $y(0) = 0$; $b = 1$

16. $y' = e^x + y^2$, $y(0) = 0$; $b = 1$

In exercises 17 and 18, use Euler's Method with $n = 5$ to obtain an approximate solution to each initial value problem over the indicated interval.

17. $y' = 2xy$, $y(0) = 1$; $0 \le x \le 1$

18. $y' = x^2 + y^2$, $y(0) = 1$; $0 \le x \le 1$

19. The resale value of a certain machine decreases at a rate proportional to the machine's purchase price. The machine was purchased at $50,000 and was worth $32,000 two years later.
    a. Find an expression for the resale value of the machine at any time t.
    b. Find the value of the machine after 5 years.

20. The Brentano-Stevens Law, which describes the rate of change of a response R to a stimulus S, is given by

$$\frac{dR}{dS} = k \cdot \frac{R}{S}$$

where k is a positive constant. Solve this differential equation.

21. HAL Corporation invests P dollars per year (assume this is done on a frequent basis in small deposits over the year so that it is essentially continuous) into a fund earning interest at the rate of $r\%$ per year compounded continuously. Then the size of the fund, A, grows at a rate given by

$$\frac{dA}{dt} = rA + P$$

Suppose that $A = 0$ when $t = 0$. Determine the size of the fund after t years. What is the size of the fund after

5 years if $P = 50,000$ (dollars) and $r = 12$ (percent per year)?

22. Suppose that the demand, D, for a certain commodity is constant and that the rate of change in the supply, S, over time t is proportional to the difference between the demand and the supply so that

$$\frac{dS}{dt} = k(D - S)$$

Find an expression for the supply at any time t if the supply at $t = 0$ is S_0.

23. Mrs. Benson placed an 11-lb rib roast, which had been standing at room temperature (68°F), into a 350°F oven at 4 P.M. At 6 P.M. the temperature of the roast was 118°F. At what time would the temperature of the roast have been 150°F (medium rare)?

   [*Hint*: Use Newton's law of cooling (or heating).]

24. The American Stenographic Institute finds that the average student taking the Advanced Shorthand course will progress at a rate given by

$$\frac{dQ}{dt} = k(120 - Q)$$

in a 20-week course, where $Q(t)$ measures the number of words of dictation the student can take per minute after t weeks in the course. [Assume $Q(0) = 60$.] If the average student can take 90 words of dictation per minute after ten weeks in the course, how many words per minute can the average student take after completing the course?

25. A rumor to the effect that a rental increase was imminent was first heard by four residents of the Chatham West Condominium Complex. The rumor spread through the complex of 200 single-family dwellings at a rate jointly proportional to the number of families who had heard it and the number who had not. Two days later, the number of families who had heard the rumor had increased to 40. Find how many families had heard the rumor after five days.

26. A tank initially contains 40 gallons of pure water. Brine containing 3 pounds of salt per gallon flows into the tank at a rate of 4 gallons per minute, and the well-stirred mixture flows out of the tank at the same rate. How much salt is in the tank at any time t? How much salt is in the tank in the long run?

The systematic study of probability began in the seventeenth century, when certain aristocrats wanted to discover superior strategies to use in the gaming rooms of Europe. Some of the best mathematicians of the period were engaged in this pursuit. Since then, applications of probability have evolved in virtually every sphere of human endeavor that contains an element of uncertainty.

In this chapter we will take a look at the role of calculus in the study of *probability* involving a *continuous random variable*. We will see how probability can be used to find the average life span of a certain brand of color television tube, the average waiting time for patients in a health clinic, and the percentage of a current Mediterranean population who have serum cholesterol levels between 160 and 180 mg/dl—to name but a few applications.

Where do we go from here? Some of the top 10% of this senior class will further their education at one of the campuses of the state university system. In Example 6, page 744, we will determine the minimum grade point average a senior needs to be eligible for admission to one of the state universities.

PROBABILITY AND CALCULUS

10.1 PROBABILITY DISTRIBUTIONS OF RANDOM VARIABLES

Probability

We begin by mentioning some elementary notations important in the study of probability. For our purpose, an **experiment** is an activity with observable results called **outcomes,** or **sample points.** The totality of all outcomes is the **sample space** of the experiment. A subset of the sample space is called an **event** of the experiment.

Now, given an event associated with an experiment, our primary objective is to determine the likelihood that this event will occur. This likelihood, or **probability of the event,** is a number between 0 and 1 and may be viewed as the proportionate number of times that the event will occur if the experiment associated with the event is repeated indefinitely under independent and similar conditions. By way of example, let us consider the simple experiment of tossing an unbiased coin and observing whether it lands "heads" (H) or "tails" (T). Since the coin is *unbiased*, we see that the probability of each outcome is $\frac{1}{2}$, abbreviated

$$P(\text{H}) = \frac{1}{2} \quad \text{and} \quad P(\text{T}) = \frac{1}{2}$$

Discrete Random Variables

In many situations, it is desirable to assign numerical values to the outcomes of an experiment. For example, suppose that an experiment consists of casting a die and observing the face that lands up. If we let X denote the outcome of the experiment, then X assumes one of the values 1, 2, 3, 4, 5, or 6. Because the values assumed by X depend on the outcomes of a chance experiment, the outcome X is referred to as a **random variable.** In this case, the random variable X is also said to be **finite discrete** since it can assume only a finite number of integer values.

The function P that associates with each value of a random variable its probability of occurrence is called a *probability function*. It is *discrete* since its domain consists of a finite set. In general, a discrete probability function is defined as follows:

DISCRETE PROBABILITY FUNCTION

A **discrete probability function** P with domain $\{x_1, x_2, \ldots, x_n\}$ satisfies these conditions:

1. $0 \leq P(x_i) \leq 1$
2. $P(x_1) + P(x_2) + \cdots + P(x_n) = 1$

REMARK   These conditions imply that the probability assigned to an outcome must be nonnegative and less than or equal to 1 and that the sum of all probabilities must be 1.   ○ ○ ○

Histograms

A discrete probability function or distribution may be exhibited graphically by means of a **histogram.** To construct a histogram of a particular probability distribution, first locate the values of the random variable on a number line. Then, above each such number, erect a rectangle with width 1 and height equal to the probability associated with that value of the random variable.

For example, consider the data in Table 10.1, the number of cars observed waiting in line at 2-minute intervals between 3 and 5 P.M. on a certain Friday at the drive-in teller of the Westwood Savings Bank and the corresponding frequency of occurrence. If we divide each number on the right of the table by 60 (the sum of these numbers), then we obtain the respective probabilities associated with the random variable X, when X assumes the values 0, 1, 2, . . . , 8. For example,

$$P(X = 0) = \frac{2}{60} \approx 0.03$$

$$P(X = 1) = \frac{9}{60} = 0.15 \quad \text{and so on}$$

The resulting probability distribution is shown in Table 10.2.

The histogram associated with this probability distribution is shown in Figure 10.1.

Table 10.1

Number of Cars	Frequency of Occurrence
0	2
1	9
2	16
3	12
4	8
5	6
6	4
7	2
8	1

Table 10.2
Probability distribution for the random variable X.

x	$P(X = x)$
0	0.03
1	0.15
2	0.27
3	0.20
4	0.13
5	0.10
6	0.07
7	0.03
8	0.02

Figure 10.1

Probability distribution of the number of cars waiting in line.

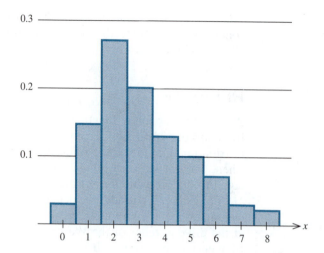

Observe that the area of a rectangle in a histogram is associated with a value of a random variable X and that this area gives precisely the probability associated with that value of X. This follows since each rectangle, by construction, has width 1 and height corresponding to the probability associated with the value of the random variable.

Another consequence arising from the method of construction of a histogram is that the probability associated with more than one value of the random variable X is given by the sum of the areas of the rectangles associated with those values of X. For example, the probability that three or four cars are in line is given by

$$P(X = 3) + P(X = 4)$$

which may be obtained from the histogram by adding the areas of the rectangles associated with the values 3 and 4 of the random variable X. Thus, the required probability is

$$P(X = 3) + P(X = 4) = 0.20 + 0.13 = 0.33$$

Continuous Random Variables

A random variable x that can assume any value in an interval is called a **continuous random variable.** Examples of continuous random variables are the life span of a light bulb, the length of a telephone call, the length of an infant at birth, the daily amount of rainfall in Boston, and the life span of a certain plant species. For the remainder of this chapter, we will be interested primarily in continuous random variables.

Consider an experiment in which the associated random variable x has the interval $[a, b]$ as its sample space. Then an event of the experiment is any subset of $[a, b]$. For example, if x denotes the life span of a light bulb, then the sample space associated with the experiment is $[0, \infty)$, and the event that a light bulb selected at random has a life span between 500 and 600 hours inclusive is described by the interval [500, 600] or, equivalently, by the inequality $500 \le x \le 600$. The probability that the light bulb will have a life span of between 500 and 600 hours is denoted by $P(500 \le x \le 600)$.

In general, we will be interested in computing $P(a \le x \le b)$, the probability that a random variable x assumes a value in the interval $a \le x \le b$. This computation is based on the notion of a probability density function, which we now introduce.

PROBABILITY DENSITY FUNCTION

A **probability density function** of a random variable x is a nonnegative function f having the following properties:

1. The total area of the region under the graph of f is equal to 1 (see Figure 10.2a).

2. The probability that an observed value of the random variable x lies in the interval $[a, b]$ is given by

$$P(a \le x \le b) = \int_a^b f(x)\, dx$$

(see Figure 10.2b).

Figure 10.2

(a) Area of $R = 1$

(b) $P(a \le x \le b)$ is the probability that an outcome of an experiment will lie between a and b.

A few comments are in order. First, a probability density function of a random variable x may be constructed using methods that range from theoretical considerations of the problem on the one extreme to an interpretation of data associated with the experiment on the other. Second, property 1 states that the probability that a continuous random variable takes on a value lying in its range is 1, a certainty, which is expected. Third, property 2 states that the probability that the random variable x assumes a value in an interval $a \le x \le b$ is given by the area of the region between the graph of f and the x-axis from $x = a$ to $x = b$. Because the area under one point of the graph of f is equal to zero, we see immediately that $P(a \le x \le b) = P(a < x \le b) = P(a \le x < b) = P(a < x < b)$.

EXAMPLE 1   Show that each of the following functions satisfies the non-negativity condition and property 1 of probability density functions:

a. $f(x) = \dfrac{2}{27} x(x - 1) \qquad (1 \le x \le 4)$

b. $f(x) = \dfrac{1}{3} e^{(-1/3)x} \qquad (0 \le x < \infty)$

Solution

a. Since the factors x and $(x - 1)$ are both nonnegative, we see that $f(x) \ge 0$ on $[1, 4]$. Next, we compute

$$\int_1^4 \frac{2}{27} x(x - 1) \, dx = \frac{2}{27} \int_1^4 (x^2 - x) \, dx$$

$$= \frac{2}{27} \left(\frac{1}{3} x^3 - \frac{1}{2} x^2 \right) \Big|_1^4$$

$$= \frac{2}{27} \left[\left(\frac{64}{3} - 8 \right) - \left(\frac{1}{3} - \frac{1}{2} \right) \right]$$

$$= \frac{2}{27} \left(\frac{27}{2} \right)$$

$$= 1$$

showing that property 1 of probability density functions holds as well.

b. First, $f(x) = \frac{1}{3} e^{(-1/3)x} \ge 0$ for all values of x in $[0, \infty)$. Next,

$$\int_0^\infty \frac{1}{3} e^{(-1/3)x} \, dx = \lim_{b \to \infty} \int_0^b \frac{1}{3} e^{(-1/3)x} \, dx$$

$$= \lim_{b \to \infty} -e^{(-1/3)x} \Big|_0^b$$

$$= \lim_{b \to \infty} \left(-e^{(-1/3)b} + 1 \right)$$

$$= 1$$

so the area under the graph of $f(x) = \frac{1}{3} e^{(-1/3)x}$ is equal to 1, as we set out to show.

    ◐ ◐ ◐

EXAMPLE 2

a. Determine the value of the constant k such that the function $f(x) = kx^2$ is a probability density function on the interval $[0, 5]$.

b. If x is a continuous random variable with the probability density function given in (a), compute the probability that x will assume a value between $x = 1$ and $x = 2$.

c. Find the probability that x will assume a value at $x = 3$.

Solution

a. We compute

$$\int_0^5 kx^2\, dx = k \int_0^5 x^2\, dx$$

$$= \frac{k}{3} x^3 \Big|_0^5$$

$$= \frac{125}{3} k$$

Since this value must be equal to 1, we find that $k = 3/125$.

b. The required probability is given by

$$P(1 \le x \le 2) = \int_1^2 f(x)\, dx$$

$$= \int_1^2 \frac{3}{125} x^2\, dx$$

$$= \frac{1}{125} x^3 \Big|_1^2$$

$$= \frac{1}{125} (8 - 1)$$

$$= \frac{7}{125}$$

Figure 10.3
$P(1 \le x \le 2)$ *for the probability density function* $y = \dfrac{3}{125} x^2$.

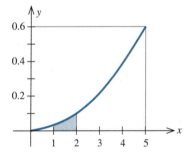

The graph of the probability density function f and the area corresponding to the probability $P(1 \le x \le 2)$ are shown in Figure 10.3.

c. The required probability is given by

$$P(x = 3) = \int_3^3 f(x)\, dx$$

$$= \int_3^3 \frac{3}{125} x^2\, dx = 0$$

      ○ ○ ○

REMARK   Observe that, in general, the probability that x will assume a value at a point $x = a$ is zero since $P(x = a) = \int_a^a f(x)\, dx = 0$ by property 1 of the definite integral (page 479).

      ○ ○ ○

EXAMPLE 3   TKK Products Corporation manufactures a 200-watt electric light bulb. Laboratory tests show that the life spans of these light bulbs have a distribution described by the probability density function

$$f(x) = 0.001 e^{-0.001x}$$

Determine the probability that a light bulb will have each of these life spans:

a. 500 hours or less
b. More than 500 hours
c. More than 1000 hours but less than 1500 hours

Solution   Let x denote the life span of a light bulb.

a. The probability that a light bulb will have a life span of 500 hours or less is given by

$$P(0 \le x \le 500) = \int_0^{500} 0.001e^{-0.001x}\,dx$$

$$= -e^{-0.001x}\Big|_0^{500}$$

$$= -e^{-0.5} + 1$$

$$\approx 0.3935$$

b. The probability that a light bulb will have a life span of more than 500 hours is given by

$$P(x > 500) = \int_{500}^{\infty} 0.001e^{-0.001x}\,dx$$

$$= \lim_{b \to \infty} \int_{500}^{b} 0.001e^{-0.001x}\,dx$$

$$= \lim_{b \to \infty} -e^{-0.001x}\Big|_{500}^{b}$$

$$= \lim_{b \to \infty} \left(-e^{-0.001b} + e^{-0.5} \right)$$

$$= e^{-0.5} \approx 0.6065$$

This result may also be obtained by observing that

$$P(x > 500) = 1 - P(x \le 500)$$

$$= 1 - 0.3935 \qquad \text{[Using the result from (a)]}$$

$$\approx 0.6065$$

c. The probability that a light bulb will have a life span of more than 1000 hours but less than 1500 hours is given by

$$P(1000 < x < 1500) = \int_{1000}^{1500} 0.001e^{-0.001x}\,dx$$

$$= -e^{-0.001x}\Big|_{1000}^{1500}$$

$$= -e^{-1.5} + e^{-1}$$

$$\approx -0.2231 + 0.3679$$

$$= 0.1448 \qquad \qquad \circ \; \circ \; \circ$$

The probability density function of Example 3 has the form

$$f(x) = ke^{-kx}$$

where $x \ge 0$ and k is a positive constant. Its graph is shown in Figure 10.4. This probability function is called an **exponential density function,** and the random variable associated with it is said to be **exponentially distributed.** Exponential random variables are used to represent the life span of electronic components, the duration of telephone calls, the waiting time in a doctor's

Figure 10.4
The area under the graph of the exponential density function is equal to 1.

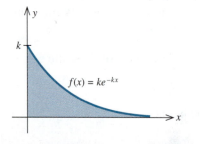

$f(x) = ke^{-kx}$

office, and the time between successive flight arrivals and departures in an airport, to mention but a few applications.

SELF-CHECK EXERCISES 10.1

1. Determine the value of the constant k such that the function $f(x) = k(4x - x^2)$ is a probability density function on the interval $[0, 4]$.

2. Suppose that x is a continuous random variable with the probability density function of Exercise 1. Find the probability that x will assume a value between $x = 1$ and $x = 3$.

Solutions to Self-Check Exercises 10.1 can be found on page 721.

10.1 EXERCISES

In exercises 1–10, show that the given function is a probability density function on the given interval.

1. $f(x) = \dfrac{1}{3}$; $(3 \le x \le 6)$

2. $f(x) = \dfrac{1}{\pi}$; $(\pi \le x \le 2\pi)$

3. $f(x) = \dfrac{2}{32} x$; $(2 \le x \le 6)$

4. $f(x) = \dfrac{3}{8} x^2$; $(0 \le x \le 2)$

5. $f(x) = \dfrac{2}{9} (3x - x^2)$; $(0 \le x \le 3)$

6. $f(x) = \dfrac{3}{32} (x - 1)(5 - x)$; $(1 \le x \le 5)$

7. $f(x) = \dfrac{12 - x}{72}$; $(0 \le x \le 12)$

8. $f(x) = 20(x^3 - x^4)$; $(0 \le x \le 1)$

9. $f(x) = \dfrac{8}{7x^2}$; $(1 \le x \le 8)$

10. $f(x) = \dfrac{3}{14} \sqrt{x}$; $(1 \le x \le 4)$

In exercises 11–18, find the value of the constant k so that the given function is a probability density function in the given interval.

11. $f(x) = k$; $[1, 4]$

12. $f(x) = kx$; $[0, 4]$

13. $f(x) = k(4 - x)$; $[0, 4]$

14. $f(x) = kx^3$; $[0, 1]$

15. $f(x) = k\sqrt{x}$; $[0, 4]$

16. $f(x) = \dfrac{k}{x}$; $[1, 5]$

17. $f(x) = \dfrac{k}{x^3}$; $[1, \infty)$

18. $f(x) = ke^{-x/2}$; $[0, \infty)$

In exercises 19–24, f is a probability density function defined on the given interval. Find the indicated probabilities.

19. $f(x) = \dfrac{1}{12} x$; $[1, 5]$

   a. $P(2 \le x \le 4)$ b. $P(1 \le x \le 4)$
   c. $P(x \ge 2)$ d. $P(x = 2)$

20. $f(x) = \dfrac{1}{9} x^2$; $[0, 3]$

   a. $P(1 \le x \le 2)$ b. $P(1 < x \le 3)$
   c. $P(x \le 2)$ d. $P(x = 1)$

21. $f(x) = \dfrac{3}{32} (4 - x^2)$; $[-2, 2]$

   a. $P(-1 \le x \le 1)$ b. $P(x \le 0)$
   c. $P(x > -1)$ d. $P(x = 0)$

22. $f(x) = \dfrac{3}{16} \sqrt{x}$; $[0, 4]$

   a. $P(1 < x < 3)$ b. $P(x \le 2)$
   c. $P(x = 2)$ d. $P(x \ge 1)$

23. $f(x) = \dfrac{1}{4\sqrt{x}}$; $[1, 9]$

   a. $P(x \ge 4)$ b. $P(1 \le x < 8)$
   c. $P(x = 3)$ d. $P(x \le 4)$

24. $f(x) = \dfrac{1}{2} e^{-x/2}$; $[0, \infty)$

   a. $P(x \le 4)$ b. $P(1 < x < 2)$
   c. $P(x = 50)$ d. $P(x \ge 2)$

25. Life Span of a Plant The life span of a certain plant species (in days) is described by the probability density function

$$f(x) = \frac{1}{100} e^{-x/100}$$

a. Find the probability that a plant of this species will live for 100 days or less.
b. Find the probability that a plant of this species will live longer than 120 days.
c. Find the probability that a plant of this species will live longer than 60 days but less than 140 days.

26. Waiting Time at a Health Clinic The average waiting time in minutes for patients arriving at the Newtown Health Clinic between 1 and 4 P.M. on a weekday is an exponentially distributed random variable x with associated probability density function $f(x) = \frac{1}{15} e^{-(1/15)x}$.
a. What is the probability that a patient arriving at the clinic between 1 and 4 P.M. will have to wait longer than 15 minutes?
b. What is the probability that a patient arriving at the clinic between 1 and 4 P.M. will have to wait between 10 and 12 minutes?

27. Reliability of Robots The National Welding Company uses industrial robots in some of its assembly line operations. Management has determined that the lengths of time in hours between breakdowns are exponentially distributed with probability density function $f(t) = 0.001e^{-0.001t}$.
a. What is the probability that a robot selected at random will break down after between 600 and 800 hours of use?
b. What is the probability that a robot will break down after 1200 hours of use?

28. Waiting Time at an Expressway Tollbooth Suppose that the time intervals in seconds between arrivals of successive cars at an expressway tollbooth during rush hour are exponentially distributed with associated probability density function $f(t) = \frac{1}{8} e^{-(1/8)t}$. Find the probability that the average time interval between arrivals of successive cars is more than 8 seconds.

29. Mail-Order Phone Calls A study conducted by Uni-Mart, a mail-order department store, reveals that the time intervals in minutes between incoming telephone calls on its toll-free 800 line between 10 A.M. and 2 P.M. are exponentially distributed with probability density function $f(t) = \frac{1}{30} e^{-(1/30)t}$. What is the probability that the time interval between successive calls is more than 2 minutes?

30. Reliability of a Microprocessor The microprocessors manufactured by United Motor Works Company, which are used in automobiles to regulate fuel consumption, are guaranteed against defects for 20,000 miles of use. Tests conducted in the laboratory under simulated driving conditions reveal that the distances driven in miles before the microprocessors break down are exponentially distributed with probability density function $f(x) = 0.00001e^{-0.00001x}$. What is the probability that a microprocessor selected at random will fail during the warranty period?

31. Probability of Snowfall The amount of snowfall (in feet) in a remote region of Alaska in the month of January is a continuous random variable with probability density function $f(x) = \frac{2}{9}x(3 - x), 0 \le x \le 3$. Find the probability that the amount of snowfall will be between 1 and 2 feet; more than 1 foot.

32. Number of Chips in Chocolate Chip Cookies The number of chocolate chips in each cookie of a certain brand has a distribution described by the probability density function

$$f(x) = \frac{1}{36}(6x - x^2), \qquad 0 \le x \le 6$$

Find the probability that the number of chocolate chips in a randomly chosen cookie is fewer than two.

33. Life Expectancy of Color Television Tubes The life expectancy (in years) of a certain brand of color television tube is a continuous random variable with probability density function

$$f(t) = 9(9 + t^2)^{-3/2}, \qquad 0 \le t < \infty$$

Find the probability that a randomly chosen television tube will last more than 4 years.
[*Hint:* Integrate using a table of integrals.]

USING TECHNOLOGY

GRAPHING A HISTOGRAM

A graphing calculator can be used to plot the histogram for a given set of data, as illustrated by the example.

EXAMPLE 1 A survey of 90,000 households conducted in 1995 revealed the following percentage of women who wear each given shoe size:

Shoe Size	<5	5–5½	6–6½	7–7½	8–8½	9–9½	10–10½	>10½
Percentage of Women	1	5	15	27	29	14	7	2

Source: Footwear Market Insights survey

a. Plot a histogram for the given data.
b. What percentage of women in the survey wear size $7-7\frac{1}{2}$ or $8-8\frac{1}{2}$ shoes?

Solution **a.** Let X denote the random variable taking on the values 1 through 8, where 1 corresponds to a shoe size less than 5, 2 corresponds to a shoe size of $5-5\frac{1}{2}$, and so on. Entering the values of X as $x_1 = 1$, $x_2 = 2$, ..., $x_8 = 8$, and the corresponding values of Y as $y_1 = 1$, $y_2 = 5$, ..., $y_8 = 2$, and then using the Draw function from the Statistics menu, we draw the histogram shown in Figure T1.

Figure T1
The histogram for the given data using the viewing rectangle $[0, 9] \times [0, 35]$.

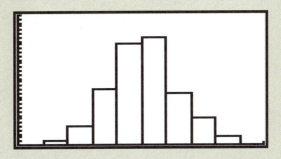

b. The probability that a woman participating in the survey wears size $7-7\frac{1}{2}$ or $8-8\frac{1}{2}$ shoes is given by

$$P(X = 4) + P(X = 5) = 0.27 + 0.29 = 0.56$$

which tells us that 56% of the women wear either size $7-7\frac{1}{2}$ or size $8-8\frac{1}{2}$ shoes.

◦ ◦ ◦

EXERCISES

1. Graph the histogram associated with the data given in Table 10.2 page 711. Compare your graph with that given in Figure 10.1, page 712.

2. Distribution of Families by Size   A survey was conducted by the Public Housing Authority in a certain community among 1000 families to determine the distribution of families by size. The results follow.

Family Size	2	3	4	5
Frequency of Occurrence	350	200	245	125

Family Size	6	7	8
Frequency of Occurrence	66	10	4

a. Find the probability distribution of the random variable X, where X denotes the number of persons in a randomly chosen family.
b. Graph the histogram corresponding to the probability distribution found in (a).

3. Waiting Lines   The accompanying data were obtained in a study conducted by the manager of Sav-More Supermarket. In this study the number of customers waiting in line at the express checkout at the beginning of each 3-minute interval between 9 A.M. and 12 noon on Saturday was observed.

Number of Customers	0	1	2	3	4
Frequency of Occurrence	1	4	2	7	14

Number of Customers	5	6	7	8	9	10
Frequency of Occurrence	8	10	6	3	4	1

a. Find the probability distribution of the random variable X, where X denotes the number of customers observed waiting in line.
b. Graph the histogram representing this probability distribution.

4. Television Pilots   After the private screening of a new television pilot, audience members were asked to rate the new show on a scale of 1 to 10 (10 being the highest rating). From a group of 140 people, these responses were obtained:

Rating	1	2	3	4	5
Frequency of Occurrence	1	4	3	11	23

Rating	6	7	8	9	10
Frequency of Occurrence	21	28	29	16	4

Let the random variable X denote the rating given to the show by a randomly chosen audience member. Find the probability distribution and graph the histogram associated with these data.

SOLUTIONS TO SELF-CHECK EXERCISES 10.1

1. We compute

$$\int_0^4 k(4x - x^2)\, dx = k\left(2x^2 - \frac{1}{3}x^3\right)\Big|_0^4$$

$$= k\left(32 - \frac{64}{3}\right)$$

$$= \frac{32}{3}k$$

Since this value must be equal to 1, we have $k = \frac{3}{32}$.

2. The required probability is given by

$$P(1 \le x \le 3) = \int_1^3 f(x)\, dx$$

$$= \int_1^3 \frac{3}{32}(4x - x^2)\, dx$$

$$= \frac{3}{32}\left(2x^2 - \frac{1}{3}x^3\right)\Big|_1^3$$

$$= \frac{3}{32}\left[(18 - 9) - \left(2 - \frac{1}{3}\right)\right] = \frac{11}{16}$$

10.2 EXPECTED VALUE AND STANDARD DEVIATION

Expected Value

The average value of a set of numbers is a familiar notion to most people. For example, to compute the average of the four numbers 12, 16, 23, and 37, we simply add these numbers and divide the resulting sum by 4, giving the average as

$$\frac{12 + 16 + 23 + 37}{4} = \frac{88}{4}$$

or 22. In general, we have the following definition:

AVERAGE OR MEAN

The average, or mean, of the n numbers $x_1, x_2, \ldots, x_n$ is $\bar{x}$ (read "x bar"), where

$$\bar{x} = \frac{x_1 + x_2 + \cdots + x_n}{n}$$

EXAMPLE 1   The number of cars observed waiting in line at the beginning of each 2-minute interval between 3 and 5 P.M. on a certain Friday at the drive-in teller of Westwood Savings Bank and the corresponding frequency of occurrence are shown in Table 10.3. Find the average number of cars observed waiting in line at the beginning of each 2-minute interval during the 2-hour period.

Table 10.3

Number of Cars	0	1	2	3	4	5	6	7	8
Frequency of Occurrence	2	9	16	12	8	6	4	2	1

Solution   Observe from Table 10.3 that the number 0 (of cars) occurs twice, the number 1 occurs nine times, and so on. There are altogether

$$2 + 9 + 16 + 12 + 8 + 6 + 4 + 2 + 1 = 60$$

numbers to be averaged. Therefore, the required average is given by

$$\frac{0 \cdot 2 + 1 \cdot 9 + 2 \cdot 16 + 3 \cdot 12 + 4 \cdot 8 + 5 \cdot 6 + 6 \cdot 4 + 7 \cdot 2 + 8 \cdot 1}{60} \approx 3.1 \quad \textbf{(1)}$$

or approximately 3.1 cars.   ○ ○ ○

Let us reconsider the expression in equation (1) that gives the average of the frequency distribution shown in Table 10.3. Dividing each term by the denominator, we can rewrite the expression in the form

$$0 \cdot \left(\frac{2}{60}\right) + 1 \cdot \left(\frac{9}{60}\right) + 2 \cdot \left(\frac{16}{60}\right) + 3 \cdot \left(\frac{12}{60}\right) + 4 \cdot \left(\frac{8}{60}\right) + 5 \cdot \left(\frac{6}{60}\right)$$

$$+ 6 \cdot \left(\frac{4}{60}\right) + 7 \cdot \left(\frac{2}{60}\right) + 8 \cdot \left(\frac{1}{60}\right)$$

Observe that each term in the sum is a product of two factors; the first factor is the value assumed by the random variable X, where X denotes the number of cars observed waiting in line, and the second factor is just the probability associated with that value of the random variable. This observation suggests the following general method for calculating the *expected value* (that is, the average, or mean) of a random variable X that assumes a finite number of values from the knowledge of its probability distribution:

EXPECTED VALUE OF A RANDOM VARIABLE X   Let X denote a random variable that assumes the values $x_1, x_2, \ldots, x_n$ with associated probabilities $p_1, p_2, \ldots, p_n$, respectively. Then the **expected value** of X, $E(X)$, is given by

$$E(X) = x_1 p_1 + x_2 p_2 + \cdots + x_n p_n \quad \textbf{(2)}$$

REMARK   The numbers $x_1, x_2, \ldots, x_n$ may be positive, zero, or negative. For example, such a number will be positive if it represents a profit and negative if it represents a loss.   ○ ○ ○

EXAMPLE 2   Refer to Example 1. Let the random variable X denote the number of cars observed waiting in line. Use the probability distribution for the random variable X (Table 10.4) to solve Example 1 again.

Table 10.4
Probability distribution for the random variable X.

x	0	1	2	3	4	5	6	7	8
$P(X = x)$	0.03	0.15	0.27	0.20	0.13	0.10	0.07	0.03	0.02

Solution   Let X denote the number of cars observed waiting in line. Then the average number of cars observed waiting in line is given by the expected value of X—that is, by

$$E(X) = 0 \cdot (0.03) + 1 \cdot (0.15) + 2 \cdot (0.27) + 3 \cdot (0.20) + 4 \cdot (0.13)$$
$$+ 5 \cdot (0.10) + 6 \cdot (0.07) + 7 \cdot (0.03) + 8 \cdot (0.02)$$
$$= 3.1$$

or 3.1 cars, which agrees with the earlier result.   ○ ○ ○

The expected value of a random variable X is a measure of the central tendency of the probability distribution associated with X. In respected trials of an experiment with a random variable X, the average of the observed values of X gets closer and closer to the expected value of X as the number of trials gets larger and larger. Geometrically, the expected value of a random variable X has the following simple interpretation: If a laminate is made of the histogram of a probability distribution associated with a discrete random variable X, then the expected value of X corresponds to the point on the base of the laminate at which the latter will balance perfectly when the point is directly over a fulcrum (Figure 10.5).

Figure 10.5
Expected value of a random variable X.

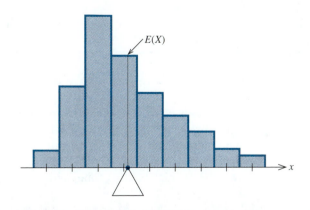

Now suppose that x is a continuous random variable and f is the probability density function associated with it. For simplicity, let us first assume that $a \leq x \leq b$. Divide the interval $[a, b]$ into n subintervals of equal length $\Delta x = (b - a)/n$ by means of the $(n + 1)$ points $x_0 = a, x_1, x_2, \ldots, x_n = b$ (Figure 10.6). To find an approximation of the average value, or expected value, of

Figure 10.6

Approximating the expected value of a random variable x on [a, b] by a Riemann sum.

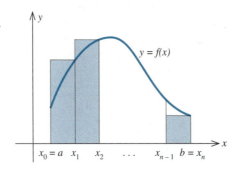

x on the interval $[a, b]$, let us treat x as if it were a discrete random variable that takes on the values $x_1, x_2, \ldots, x_n$ with probabilities $p_1, p_2, \ldots, p_n$. Then

$$E(x) \approx x_1 p_1 + x_2 p_2 + \cdots + x_n p_n$$

But p_1 is the probability that x is in the interval $[x_0, x_1]$, and this is just the area under the graph of f from $x = x_0$ to $x = x_1$, which may be approximated by $f(x_1) \Delta x$. The probabilities $p_2, \ldots, p_n$ may be approximated in a similar manner. Thus,

$$E(x) \approx x_1 f(x_1) \Delta x + x_2 f(x_2) \Delta x + \cdots + x_n f(x_n) \Delta x$$

which is seen to be the Riemann sum of the function $g(x) = xf(x)$ over the interval $[a, b]$. Letting n approach infinity, we obtain the following formula:

EXPECTED VALUE OF A CONTINUOUS RANDOM VARIABLE	Suppose that the function f defined on the interval $[a, b]$ is the probability density function associated with a continuous random variable x. Thus the **expected value** of x is

$$E(x) = \int_a^b xf(x)\, dx \qquad (3)$$

If either $a = -\infty$ or $b = \infty$, then the integral in (3) becomes an improper integral.

The expected value of a random variable is used in many practical applications. For example, if x represents the life span of a certain brand of electronic component, then the expected value of x gives the average life span of these components. If x measures the waiting time in a doctor's office, then $E(x)$ gives the average waiting time, and so on.

EXAMPLE 3 Show that if a continuous random variable x is exponentially distributed with the probability density function

$$f(x) = ke^{-kx}$$

then the expected value of x, $E(x)$, is equal to $1/k$. Using this result, determine the average life span of a 200-watt electric light bulb manufactured by TKK Products Corporation of Example 3, page 715.

Solution We compute

$$E(x) = \int_0^\infty xf(x)\,dx$$

$$= \int_0^\infty kxe^{-kx}\,dx$$

$$= k \lim_{b\to\infty} \int_0^b xe^{-kx}\,dx$$

Integrating by parts with

$$u = x \quad \text{and} \quad dv = e^{-kx}\,dx$$

so that

$$du = dx \quad \text{and} \quad v = -\frac{1}{k}e^{-kx}$$

we have

$$E(x) = k \lim_{b\to\infty} \left[-\frac{1}{k}xe^{-kx}\Big|_0^b + \frac{1}{k}\int_0^b e^{-kx}\,dx \right]$$

$$= k \lim_{b\to\infty} \left[-\left(\frac{1}{k}\right)be^{-kb} - \frac{1}{k^2}e^{-kx}\Big|_0^b \right]$$

$$= k \lim_{b\to\infty} \left[-\left(\frac{1}{k}\right)be^{-kb} - \frac{1}{k^2}e^{-kb} + \frac{1}{k^2} \right]$$

$$= -\lim_{b\to\infty} \frac{b}{e^{kb}} - \frac{1}{k}\lim_{b\to\infty} \frac{1}{e^{kb}} + \frac{1}{k}\lim_{b\to\infty} 1$$

Now, by taking a sequence of values of b that approaches infinity, for example, $b = 10, 100, 1000, 10{,}000, \ldots$, we see that, for a fixed k,

$$\lim_{b\to\infty} \frac{b}{e^{kb}} = 0$$

Therefore,

$$E(x) = \frac{1}{k}$$

as we set out to show. From our previous work we know that $k = 0.001$. So the average life span of the TKK light bulb is $1/(0.001) = 1000$ hours. ○ ○ ○

Before considering another example, let us summarize the result obtained in Example 3.

THE AVERAGE VALUE OF AN EXPONENTIAL DENSITY FUNCTION

If a continuous random variable x is exponentially distributed with probability density function

$$f(x) = ke^{-kx}$$

then the **expected value** of x is given by

$$E = \frac{1}{k}$$

EXAMPLE 4 On a typical Monday morning, the time between successive arrivals of planes at Jackson International Airport is an exponentially distributed random variable x with expected value of 10 (minutes).

a. Find the probability density function associated with x.

b. What is the probability that between 6 and 8 minutes will elapse between successive arrivals of planes?

c. What is the probability that the time between successive arrivals will be more than 15 minutes?

Solution

a. Since x is exponentially distributed, the associated probability density function has the form $f(x) = ke^{-kx}$. Next, since the expected value of x is 10, we see that

$$E(x) = \frac{1}{k} = 10$$

or

$$k = \frac{1}{10} = 0.1$$

so the required probability density function is

$$f(x) = 0.1e^{-0.1x}$$

b. The probability that between 6 and 8 minutes will elapse between successive arrivals is given by

$$P(6 \le x \le 8) = \int_6^8 0.1e^{-0.1x}\,dx = -e^{-0.1x}\Big|_6^8$$
$$= -e^{-0.8} + e^{-0.6}$$
$$\approx 0.100$$

c. The probability that the time between successive arrivals will be more than 15 minutes is given by

$$P(x > 15) = \int_{15}^{\infty} 0.1e^{-0.1x}\,dx$$
$$= \lim_{b \to \infty} \int_{15}^{b} 0.1e^{-0.1x}\,dx$$
$$= \lim_{b \to \infty} \left[-e^{-0.1x}\Big|_{15}^{b} \right]$$
$$= \lim_{b \to \infty} (-e^{-0.1b} + e^{-1.5}) = e^{-1.5}$$
$$\approx 0.22$$

Variance

The mean, or expected value, of a random variable enables us to express an important property of the probability distribution associated with the random variable in terms of a single number. But knowing the location, or central tendency, of a probability distribution alone is usually not enough to give a reasonably accurate picture of the probability distribution. Consider, for example, the two probability distributions whose histograms appear in Figure 10.7. Both distributions have the same expected value, or mean, $\mu = 4$ (the Greek letter μ is read "mu"). Note that the probability distribution with the histogram shown in Figure 10.7a is closely concentrated about its mean μ, whereas the one with the histogram shown in Figure 10.7b is widely dispersed, or spread, about its mean.

Figure 10.7
The histograms of two probability distributions.

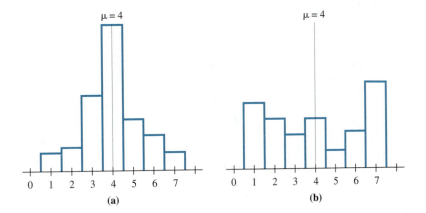

(a)   (b)

As another example, suppose Mr. David Horowitz, host of the television show "The Consumer Advocate," decides to demonstrate the accuracy of the weights of two popular brands of potato chips. Ten packages of potato chips of each brand are selected at random and weighed carefully. The weights in ounces are listed in Table 10.5.

Table 10.5

Brand A	16.1	16	15.8	16	15.9	16.1	15.9	16	16	16.2
Brand B	16.3	15.7	15.8	16.2	15.9	16.1	15.7	16.2	16	16.1

We will verify in Example 7 that the mean weights for each of the two brands is 16 ounces. However, a cursory examination of the data now will show that the weights of the brand B packages exhibit much greater dispersion about the mean than those of brand A.

One measure of the degree of dispersion, or spread, of a probability distribution about its mean is the variance of the random variable associated with the probability distribution. A probability distribution with a small spread about its mean will have a small variance, whereas one with a larger spread will have a larger variance. Thus, the variance of the random variable associated with the probability distribution whose histogram appears in Figure 10.7a is smaller than the variance of the random variable associated with the probability distribution whose histogram is shown in Figure 10.7b. Also, as we will see in Example 7, the variance of the random variable associated with the weights of the brand A potato chips is smaller than that of the random variable associated with the weights of the brand B potato chips.

VARIANCE OF A RANDOM VARIABLE X

Suppose a random variable has the probability distribution

x	x_1	x_2	x_3	$\cdots$	x_n
$P(X = x)$	p_1	p_2	p_3	$\cdots$	p_n

and expected value

$$E(X) = \mu$$

Then the **variance** of the random variable X is

$$\text{Var}(X) = p_1(x_1 - \mu)^2 + p_2(x_2 - \mu)^2 + \cdots + p_n(x_n - \mu)^2 \qquad (4)$$

Let's look a little closer at formula (4). First, note that the numbers

$$x_1 - \mu, x_2 - \mu, \ldots, x_n - \mu \qquad (5)$$

measure the **deviations** of $x_1, x_2, \ldots, x_n$ from μ, respectively. Thus, the numbers

$$(x_1 - \mu)^2, (x_2 - \mu)^2, \ldots, (x_n - \mu)^2 \qquad (6)$$

measure the squares of the deviations of $x_1, x_2, \ldots, x_n$ from μ, respectively. Next, by multiplying each of the numbers in (6) by the probability associated with each value of the random variable X, we weight the numbers accordingly so that their sum is a measure of the variance of X about its mean. An attempt to define the variance of a random variable about its mean in a similar manner using the deviations rather than their squares would not be fruitful. Some of the deviations may be positive whereas others may be negative, and (because of cancellations) the sum will not give a satisfactory measure of the variance of the random variable.

EXAMPLE 5 Find the variance of the random variable X and of the random variable Y whose probability distributions are given in Table 10.6. These are the probability distributions associated with the histograms shown in Figures 10.7a and b.

Table 10.6

x	$P(X = x)$	y	$P(Y = y)$
1	0.05	1	0.2
2	0.075	2	0.15
3	0.2	3	0.1
4	0.375	4	0.15
5	0.15	5	0.05
6	0.1	6	0.1
7	0.05	7	0.25

Solution   The mean of the random variable X is given by

$$\mu_X = (1)(0.05) + (2)(0.075) + (3)(0.2) + (4)(0.375) + (5)(0.15)$$
$$+ (6)(0.1) + (7)(0.05)$$
$$= 4$$

Therefore, using formula (4) and the data from the probability distribution of X, we find that the variance of X is given by

$$\text{Var}\,(X) = (0.05)(1 - 4)^2 + (0.075)(2 - 4)^2 + (0.2)(3 - 4)^2$$
$$+ (0.375)(4 - 4)^2 + (0.15)(5 - 4)^2$$
$$+ (0.1)(6 - 4)^2 + (0.05)(7 - 4)^2$$
$$= 1.95$$

Next, we find that the mean of the random variable Y is given by

$$\mu_Y = (1)(0.2) + (2)(0.15) + (3)(0.1) + (4)(0.15) + (5)(0.05)$$
$$+ (6)(0.1) + (7)(0.25)$$
$$= 4$$

and so the variance of Y is given by

$$\text{Var}\,(Y) = (0.2)(1 - 4)^2 + (0.15)(2 - 4)^2 + (0.1)(3 - 4)^2$$
$$+ (0.15)(4 - 4)^2 + (0.05)(5 - 4)^2$$
$$+ (0.1)(6 - 4)^2 + (0.25)(7 - 4)^2$$
$$= 5.2$$

Note that Var (X) is smaller than Var (Y), which confirms the earlier observations about the spread, or dispersion, of the probability distribution of X and Y, respectively.   ◦ ◦ ◦

Standard Deviation

Because formula (4), which gives the variance of the random variable X, involves the squares of the deviations, the unit of measurement of Var (X) is the square of the unit of measurement of the values of X. For example, if the values assumed by the random variable X are measured in units of a gram, then Var (X) will be measured in units involving the *square* of a gram. To remedy this situation, one normally works with the square root of Var (X) rather than Var (X) itself. The former is called the *standard deviation of X*.

STANDARD DEVIATION OF A RANDOM VARIABLE X

The **standard deviation** of a random variable X, σ (pronounced "sigma"), is defined by

$$\sigma = \sqrt{\text{Var}(X)} \tag{7}$$
$$= \sqrt{p_1(x_1 - \mu)^2 + p_2(x_2 - \mu)^2 + \cdots + p_n(x_n - \mu)^2}$$

where $x_1, x_2, \ldots, x_n$ denote the values assumed by the random variable X and $p_1 = P(X = x_1)$, $p_2 = P(X = x_2)$, ..., $p_n = P(X = x_n)$.

EXAMPLE 6   Find the standard deviations of the random variables X and Y of Example 5.

Solution   From the results of Example 5, we have Var $(X) = 1.95$ and Var $(Y) = 5.2$. Taking their respective square roots, we have

$$\sigma_X = \sqrt{1.95}$$
$$\approx 1.40$$

and

$$\sigma_Y = \sqrt{5.2}$$
$$\approx 2.28$$

● ● ●

EXAMPLE 7   Let X and Y denote the random variables whose values are the weights of the brand A and brand B potato chips, respectively (see Table 10.5). Compute the means and standard deviations of X and Y and interpret your results.

Solution   The probability distributions of X and Y may be computed from the data in Table 10.7.

Table 10.7

x	Relative Frequency of Occurrence	$P(X = x)$	y	Relative Frequency of Occurrence	$P(Y = y)$
15.8	1	0.1	15.7	2	0.2
15.9	2	0.2	15.8	1	0.1
16.0	4	0.4	15.9	1	0.1
16.1	2	0.2	16.0	1	0.1
16.2	1	0.1	16.1	2	0.2
			16.2	2	0.2
			16.3	1	0.1

The means of X and Y are given by

$$\mu_X = (0.1)(15.8) + (0.2)(15.9) + (0.4)(16.0) + (0.2)(16.1)$$
$$+ (0.1)(16.2)$$
$$= 16$$
$$\mu_Y = (0.2)(15.7) + (0.1)(15.8) + (0.1)(15.9) + (0.1)(16.0)$$
$$+ (0.2)(16.1) + (0.2)(16.2) + (0.1)(16.3)$$
$$= 16$$

Therefore,

$$\text{Var}\,(X) = (0.1)(15.8 - 16)^2 + (0.2)(15.9 - 16)^2 + (0.4)(16 - 16)^2$$
$$+ (0.2)(16.1 - 16)^2 + (0.1)(16.2 - 16)^2$$
$$= 0.012$$

and

$$\text{Var}\,(Y) = (0.2)(15.7 - 16)^2 + (0.1)(15.8 - 16)^2 + (0.1)(15.9 - 16)^2$$
$$+ (0.1)(16 - 16)^2 + (0.2)(16.1 - 16)^2 + (0.2)(16.2 - 16)^2$$
$$+ (0.1)(16.3 - 16)^2$$
$$= 0.042$$

so the required standard deviations are

$$\sigma_X = \sqrt{\text{Var}\,(X)}$$
$$= \sqrt{0.012}$$
$$\approx 0.11$$

and

$$\sigma_Y = \sqrt{\text{Var}\,(Y)}$$
$$= \sqrt{0.042}$$
$$\approx 0.20$$

The mean of X and that of Y are both equal to 16. Therefore, the average weight of a package of potato chips of either brand is 16 ounces. However, the standard deviation of Y is greater than that of X. This tells us that the weights of the packages of brand B potato chips are more widely dispersed about the common mean of 16 than are those of brand A.   ◐ ◐ ◐

A useful alternative formula for the variance is

$$\sigma^2 = E(X^2) - \mu^2$$

where $E(X)$ is the expected value of X.

a. Establish the validity of the formula.
b. Use the formula to verify the calculations in Example 7.

Now suppose that x is a *continuous* random variable. Using an argument similar to that used earlier when we extended the result for the expected value from the discrete to the continuous case, we have the following definitions:

VARIANCE AND STANDARD DEVIATION OF A CONTINUOUS RANDOM VARIABLE

Let x be a continuous random variable with probability density function $f(x)$ on $[a, b]$. Then the variance of x is

$$\text{Var}(x) = \int_a^b (x - \mu)^2 f(x)\, dx \tag{8}$$

and the standard deviation of x is

$$\sigma = \sqrt{\text{Var}(x)} \tag{9}$$

EXAMPLE 8   Find the expected value, variance, and standard deviation of the random variable x associated with the probability density function

$$f(x) = \frac{32}{15x^3}$$

on $[1, 4]$.

Solution   Using (3), we find the mean of x:

$$\mu = \int_a^b xf(x)\, dx = \int_1^4 x \cdot \frac{32}{15x^3}\, dx$$

$$= \frac{32}{15} \int_1^4 x^{-2}\, dx = \frac{32}{15}\left[-\frac{1}{x}\right]_1^4$$

$$= \frac{32}{15}\left(-\frac{1}{4} + 1\right) = \frac{8}{5}$$

Next, using (8), we find

$$\text{Var}(x) = \int_a^b (x - \mu)^2 f(x)\, dx = \int_1^4 \left(x - \frac{8}{5}\right)^2 \cdot \frac{32}{15x^3}\, dx$$

$$= \frac{32}{15} \int_1^4 \left(x^2 - \frac{16}{5}x + \frac{64}{25}\right)\frac{1}{x^3}\, dx$$

$$= \frac{32}{15} \int_1^4 \left(\frac{1}{x} - \frac{16}{5}x^{-2} + \frac{64}{25}x^{-3}\right) dx$$

$$= \frac{32}{15}\left[\ln x + \frac{16}{5x} - \frac{32}{25x^2}\right]_1^4$$

$$= \frac{32}{15}\left[\left(\ln 4 + \frac{16}{5 \cdot 4} - \frac{32}{25 \cdot 16}\right) - \left(\ln 1 + \frac{16}{5} - \frac{32}{25}\right)\right]$$

$$= \frac{32}{15}\left(\ln 4 + \frac{4}{5} - \frac{2}{25} - \frac{16}{5} + \frac{32}{25}\right)$$

$$= \frac{32}{15}\left(\ln 4 - \frac{6}{5}\right) \approx 0.40$$

Finally, using (9), we find the required standard deviation to be

$$\sigma = \sqrt{\text{Var}(x)} \approx 0.63$$

● ● ●

Alternative Formula for Variance Using (8) to calculate the variance of a continuous random variable can be rather tedious. The following formula often makes this task easier:

$$\text{Var}(x) = \int_a^b x^2 f(x)\, dx - \mu^2 \qquad (10)$$

This equation follows from these computations:

$$
\begin{aligned}
\text{Var}(x) &= \int_a^b (x - \mu)^2 f(x)\, dx \\
&= \int_a^b (x^2 - 2\mu x + \mu^2) f(x)\, dx \\
&= \int_a^b x^2 f(x)\, dx - 2\mu \int_a^b x f(x)\, dx + \mu^2 \int_a^b f(x)\, dx \\
&= \int_a^b x^2 f(x)\, dx - 2\mu \cdot \mu + \mu^2 \qquad \left[\text{Since } \int_a^b x f(x)\, dx = \mu \ \text{ and } \ \int_a^b f(x)\, dx = 1\right] \\
&= \int_a^b x^2 f(x)\, dx - \mu^2
\end{aligned}
$$

EXAMPLE 9 Use formula (10) to calculate the variance of the random variable of Example 8.

Solution Using (10), we have

$$
\begin{aligned}
\text{Var}(x) &= \int_a^b x^2 f(x)\, dx - \mu^2 = \int_1^4 x^2 \cdot \frac{32}{15x^3}\, dx - \left(\frac{8}{5}\right)^2 \\
&= \frac{32}{15} \int_1^4 \frac{1}{x}\, dx - \frac{64}{25} = \frac{32}{15} \ln x \Big|_1^4 - \frac{64}{25} \\
&= \frac{32}{15} \ln 4 - \frac{64}{25} = \frac{32}{15}\left(\ln 4 - \frac{6}{5}\right)
\end{aligned}
$$

as obtained earlier.

● ● ●

10.2 EXERCISES

In exercises 1–14, find the mean, variance, and standard deviation of the random variable x associated with the given probability density function over the given interval.

1. $f(x) = \frac{1}{3}; [3, 6]$

2. $f(x) = \frac{1}{4}; [2, 6]$

3. $f(x) = \frac{3}{125} x^2; [0, 5]$

4. $f(x) = \frac{3}{8} x^2; [0, 2]$

5. $f(x) = \frac{3}{32}(x - 1)(5 - x); [1, 5]$

6. $f(x) = 20(x^3 - x^4); [0, 1]$

7. $f(x) = \frac{8}{7x^2}; [1, 8]$

8. $f(x) = \frac{4}{3x^2}; [1, 4]$

9. $f(x) = \frac{3}{14}\sqrt{x}; [1, 4]$

10. $f(x) = \frac{5}{2} x^{3/2}; [0, 1]$

11. $f(x) = \dfrac{3}{x^4}; [1, \infty)$

12. $f(x) = 3.5x^{-4.5}; [1, \infty)$

13. $f(x) = \dfrac{1}{4}e^{-x/4}; [0, \infty)$ [*Hint:* $\lim\limits_{x \to \infty} x^n e^{kx} = 0, k > 0$]

14. $f(x) = \dfrac{1}{9}xe^{-x/3}; [0, \infty)$ [*Hint:* $\lim\limits_{x \to \infty} x^n e^{kx} = 0, k > 0$]

15. **Life Span of a Plant** The life span of a certain plant species (in days) is described by the probability density function

$$f(x) = \frac{1}{100}e^{-x/100}$$

If a plant of this species is selected at random, how long can the plant be expected to live?

16. **Number of Chips in Chocolate Chip Cookies** The number of chocolate chips in a certain brand of cookies has a distribution described by the probability density function

$$f(x) = \frac{1}{36}(6x - x^2), \qquad 0 \le x \le 6$$

Find the expected number of chips in a cookie selected at random.

17. **Shopping Habits** The amount of time t (in minutes) a shopper spends browsing in the magazine section of a supermarket is a continuous random variable with probability density function

$$f(t) = \frac{2}{25}t, \qquad 0 \le t \le 5$$

How much time is a shopper chosen at random expected to spend in the magazine section?

18. **Reaction Time of a Motorist** The amount of time t (in seconds) it takes a motorist to react to a road emergency is a continuous random variable with probability density function

$$f(t) = \frac{9}{4t^3}, \qquad 1 \le t \le 3$$

What is the expected reaction time for a motorist chosen at random?

19. **Expected Snowfall** The amount of snowfall in feet in a remote region of Alaska in the month of January is

a continuous random variable with probability density function

$$f(x) = \frac{2}{9}x(3 - x), \qquad 0 \le x \le 3$$

Find the amount of snowfall one can expect in any given month of January in Alaska.

20. **Gas Station Sales** The amount of gas (in thousands of gallons) Al's Gas Station sells on a typical Monday is a continuous random variable with probability density function

$$f(x) = 4(x - 2)^3, \qquad 2 \le x \le 3$$

How much gas can the gas station expect to sell each Monday?

21. **Demand for Butter** The quantity demanded x (in thousands of pounds) of a certain brand of butter per week is a continuous random variable with probability density function

$$f(x) = \frac{6}{125}x(5 - x), \qquad 0 \le x \le 5$$

What is the expected demand for this brand of butter per week?

22. **Life Expectancy of Color Television Tubes** The life expectancy (in years) of a certain brand of color television tube is a continuous random variable with probability density function

$$f(t) = 9(9 + t^2)^{-3/2}, \qquad 0 \le t < \infty$$

How long is one of these color television tubes expected to last?

In exercises 23–28, find the median of the random variable x with the given probability density function defined on the given interval I. The median of x is defined to be the number m such that $P(x \le m) = \frac{1}{2}$. Observe that half of the x-values lie below m and the other half lie above m.

23. $f(x) = \dfrac{1}{6}; [2, 8]$

24. $f(x) = \dfrac{2}{15}x; [1, 4]$

25. $f(x) = \dfrac{3}{16}\sqrt{x}; [0, 4]$

26. $f(x) = \dfrac{1}{6\sqrt{x}}; [1, 16]$

27. $f(x) = \dfrac{3}{x^2}; [1, \infty)$

28. $f(x) = \dfrac{1}{2}e^{-x/2}; [0, \infty)$

USING TECHNOLOGY

FINDING THE MEAN AND STANDARD DEVIATION

The calculation of the mean and standard deviation of a random variable is facilitated by the use of a graphing utility.

EXAMPLE 1 A survey conducted in 1995 of the Fortune 1000 companies revealed the following age distribution of company directors.

Age	20–25	25–30	30–35	35–40	40–45	45–50	50–55	55–60
Number of Directors	1	6	28	104	277	607	1142	1413

Age		60–65	65–70	70–75	75–80	80–85	85–90
Number of Directors		1424	494	159	62	31	5

Source: Directorship

a. Plot a histogram for the given data.
b. Find the mean age and the standard deviation of the company directors.

Solution   **a.** Let X denote the random variable taking on the values 1 through 14, where 1 corresponds to the age bracket 20–25, 2 corresponds to the age bracket 25–30, and so on. Entering the values of X as $x_1 = 1$, $x_2 = 2$, ..., $x_{14} = 14$, and the corresponding values of Y as $y_1 = 1$, $y_2 = 6$, ..., $y_{14} = 5$, and then using the Draw function from the Statistics menu, we draw the histogram shown in Figure T1.

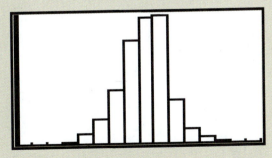

Figure T1
The histogram for the given data using the viewing rectangle $[0, 16] \times [0, 1500]$.

b. Using the appropriate function from the Statistics menu, we find that $\bar{x} = 8.4376$ and $\sigma x = 1.8463$; that is, the mean of X is $\mu \approx 8.4$ and the standard deviation is $\sigma \approx 1.8$. Thus, the average age of the directors is approximately 60 years old.

735

EXERCISES

1. a. Graph the histogram associated with the random variable X in Example 5, page 729.
b. Find the mean and the standard deviation for these data.

2. a. Graph the histogram associated with the random variable Y in Example 5, page 729.
b. Find the mean and the standard deviation for these data.

3. Driving Age Requirements The minimum age requirement for a regular driver's license differs from state to state. The frequency distribution for this age requirement in the 50 states is given in the table.

Minimum Age	15	16	17	18	19	21
Frequency of Occurrence	1	15	4	28	1	1

a. Describe a random variable X that is associated with these data.
b. Find the probability distribution for the random variable X.
c. Graph the histogram associated with these data.
d. Find the mean and the standard deviation for these data.

4. The distribution of the number of chocolate chips in a cookie is shown in the following table.

Number of Chocolate Chips (x)	0	1	2
$P(X = x)$	0.01	0.03	0.05

Number of Chocolate Chips (x)	3	4	5
$P(X = x)$	0.11	0.13	0.24

Number of Chocolate Chips (x)	6	7	8
$P(X = x)$	0.22	0.16	0.05

a. Describe a random variable X that is associated with these data.
b. Find the probability distribution for the random variable X.
c. Graph the histogram associated with these data.
d. Find the mean and the standard deviation for these data.

5. A sugar refiner uses a machine to pack sugar in 5-lb cartons. In order to check the machine's accuracy, cartons are selected at random and weighed. The results follow:

4.98	5.02	4.96	4.97	5.03
4.96	4.98	5.01	5.02	5.06
4.97	5.04	5.04	5.01	4.99
4.98	5.04	5.01	5.03	5.05
4.96	4.97	5.02	5.04	4.97
5.03	5.01	5.00	5.01	4.98

a. Describe a random variable X that is associated with these data.
b. Find the probability distribution for the random variable X.
c. Find the mean and the standard deviation for these data.

6. The scores of 25 students in a mathematics examination follow:

90	85	74	92	68	94	66
87	85	70	72	68	73	72
69	66	58	70	74	88	90
98	71	75	68			

a. Describe a random variable X that is associated with these data.
b. Find the probability distribution for the random variable X.
c. Find the mean and the standard deviation for these data.

10.3 NORMAL DISTRIBUTIONS

Normal Distributions

In Section 10.2 we saw the useful role played by exponential density functions in many applications. In this section we will look at yet another class of continuous probability distributions known as **normal distributions.** The normal distribution is without doubt the most important of all the probability distributions. Many phenomena, such as the heights of people in a given population, the weights of newborn infants, the IQs of college students, and the actual weights of 16-ounce packages of cereals, have probability distributions that are normal. The normal distribution also provides us with an accurate approximation to the distributions of many random variables associated with random sampling problems.

The general **normal probability density function** with mean μ and standard deviation σ is defined to be

$$f(x) = \frac{e^{-(1/2)[(x-\mu)/\sigma]^2}}{\sigma\sqrt{2\pi}}, \qquad -\infty < x < \infty$$

The graph of f, which is bell shaped, is called a **normal curve** (Figure 10.8).

Figure 10.8
A normal curve.

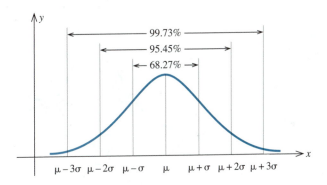

The normal curve (and therefore the corresponding normal distribution) is completely determined by its mean μ and standard deviation σ. In fact, the normal curve has the following characteristics, described in terms of these two parameters:*

1. The curve has a peak at $x = \mu$.

2. The curve is symmetrical with respect to the vertical line $x = \mu$.

∘ ∘ ∘

* The probability density function associated with this normal curve is given by

$$y = \frac{1}{\sigma\sqrt{2\pi}} e^{-(1/2)[(x-\mu)/\sigma]^2}$$

3. The curve always lies above the x-axis but approaches the x-axis as x extends indefinitely in either direction.

4. The area under the curve is 1.

5. For any normal curve, 68.27% of the area under the curve lies within 1 standard deviation of the mean (that is, between $\mu - \sigma$ and $\mu + \sigma$), 95.45% of the area lies within 2 standard deviations of the mean, and 99.73% of the area lies within 3 standard deviations of the mean.

Figure 10.9 shows two normal curves with different means μ_1 and μ_2 but the same deviation. Figure 10.10 shows two normal curves with the same mean but different standard deviations σ_1 and σ_2. (Which number is smaller?)

Figure 10.9

Two normal curves that have the same standard deviation but different means.

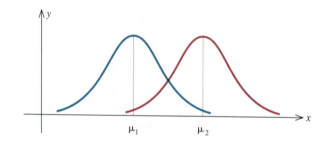

Figure 10.10

Two normal curves that have the same mean but different standard deviations.

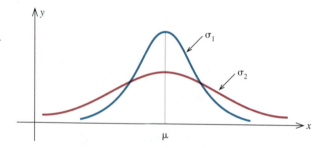

In general, the mean μ of a normal distribution determines where the center of the curve is located, whereas the standard deviation σ of a normal distribution determines the sharpness (or flatness) of the curve.

As this discussion reveals, there are infinitely many normal curves corresponding to different choices of the parameters μ and σ, which characterize such curves. Fortunately, any normal curve may be transformed into any other normal curve (as we will see later on), so in the study of normal curves it suffices to single out one such particular curve for special attention. The normal curve with mean $\mu = 0$ and standard deviation $\sigma = 1$ is called the **standard normal curve.** The corresponding distribution is called the **standard normal distribution.** The random variable itself is called the **standard normal variable** and is commonly denoted by Z.

EXPLORING WITH TECHNOLOGY

Consider the probability density function

$$f(x) = \frac{1}{\sqrt{2\pi}} e^{-x^2/2}$$

which is the formula given in the footnote on page 737 with $\mu = 0$ and $\sigma = 1$.
a. Use a graphing utility to plot the graph of f using the viewing rectangle $[-4, 4] \times [0, 0.5]$.
b. Use the numerical integration function of a graphing utility to find the area of the region under the graph of f on the intervals $[-1, 1]$, $[-2, 2]$, and $[-3, 3]$ and thus verify property 5 of normal distributions for the special case where $\mu = 0$ and $\sigma = 1$.

Computations of Probabilities Associated with Normal Distributions

Areas under the standard normal curve have been extensively computed and tabulated. Appendix Table 4 gives the areas of the regions under the standard normal curve to the left of the number z; these areas correspond, of course, to probabilities of the form $P(Z < z)$ or $P(Z \leq z)$. The next several examples illustrate the use of this table in computations involving the probabilities associated with the standard normal variable.

EXAMPLE I Let Z be the standard normal variable. By first making a sketch of the appropriate region under the standard normal curve, find these values

a. $P(Z < 1.24)$ **b.** $P(Z > 0.5)$ **c.** $P(0.24 < Z < 1.48)$
d. $P(-1.65 < Z < 2.02)$

Figure 10.11
$P(Z < 1.24)$

Solution

a. The region under the standard normal curve associated with the probability $P(Z < 1.24)$ is shown in Figure 10.11. To find the area of the required region using Appendix Table 4, we first locate the number 1.2 in the column and the number 0.04 in the row, both headed by z, and read off the number 0.8925 appearing in the body of the table. Thus,

$$P(Z < 1.24) = 0.8925$$

0   1.24

b. The region under the standard normal curve associated with the probability $P(Z > 0.5)$ is shown in Figure 10.12a. Observe, however, that the required area is, by virtue of the symmetry of the standard normal curve, equal to the shaded area shown in Figure 10.12b. Thus,

$$P(Z > 0.5) = P(Z < -0.5)$$
$$= 0.3085$$

Figure 10.12

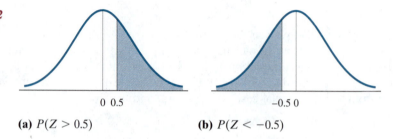

(a) $P(Z > 0.5)$ **(b)** $P(Z < -0.5)$

Figure 10.13
$P(0.24 < Z < 1.48)$

c. The probability $P(0.24 < Z < 1.48)$ is equal to the shaded area shown in Figure 10.13. This area is obtained by subtracting the area under the curve to the left of $z = 0.24$ from the area under the curve to the left of $z = 1.48$; that is,

$$P(0.24 < Z < 1.48) = P(Z < 1.48) - P(Z < 0.24)$$
$$= 0.9306 - 0.5948$$
$$= 0.3358$$

d. The probability $P(-1.65 < Z < 2.02)$ is given by the shaded area in Figure 10.14. We have

$$P(-1.65 < Z < 2.02) = P(Z < 2.02) - P(Z < -1.65)$$
$$= 0.9783 - 0.0495$$
$$= 0.9288$$ ○ ○ ○

Figure 10.14
$P(-1.65 < Z < 2.02)$

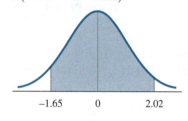

EXAMPLE 2   Let Z be the standard normal variable. Find the value of z if z satisfies

a. $P(Z < z) = 0.9474$ **b.** $P(Z > z) = 0.9115$
c. $P(-z < Z < z) = 0.7888$

Figure 10.15
$P(Z < z) = 0.9474$

Solution

a. Refer to Figure 10.15. We want the value of z such that the area of the region under the standard normal curve and to the left of $Z = z$ is 0.9474. Locating the number 0.9474 in Appendix Table 4 and reading back, we find that $z = 1.62$.
b. Since $P(Z > z)$ or, equivalently, the area of the region to the right of z is greater than 0.5, z must be negative (Figure 10.16). Therefore, $-z$ is positive. Furthermore, the area of the region to the right of z is the same as

Figure 10.16
$P(Z > z) = 0.9115$

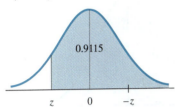

0.9115

z 0 $-z$

Figure 10.17
$P(-z < Z < z) = 0.7888$

0.7888

$-z$ 0 z

the area of the region to the left of $-z$:

$$P(Z > z) = P(Z < -z)$$
$$= 0.9115$$

Looking up the table, we find $-z = 1.35$, so $z = -1.35$.

c. The region associated with $P(-z < Z < z)$ is shown in Figure 10.17. Observe that by symmetry the area of this region is just double that of the area of the region between $Z = 0$ and $Z = z$; that is,

$$P(-z < Z < z) = 2P(0 < Z < z)$$

or

$$P(0 < Z < z) = P(Z < z) - \frac{1}{2}$$

(see Figure 10.18). Therefore,

$$\frac{1}{2}P(-z < Z < z) = P(Z < z) - \frac{1}{2}$$

or, solving for $P(Z < z)$,

$$P(Z < z) = \frac{1}{2} + \frac{1}{2}P(-z < Z < z)$$

$$= \frac{1}{2}(1 + 0.7888)$$

$$= 0.8944$$

Consulting the table, we find $z = 1.25$.   ◦ ◦ ◦

Figure 10.18

$P(0 < Z < z)$

$=$

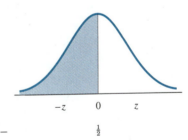

$-z$ 0 z

$P(Z < z)$

$-$

$-z$ 0 z

$\frac{1}{2}$

We now turn our attention to the computation of probabilities associated with normal distributions whose means and standard deviations are not necessarily equal to 0 and 1, respectively. As mentioned earlier, any normal curve may be transformed into the standard normal curve. In particular, it may be shown that if X is a normal random variable with mean μ and standard deviation σ, then it can be transformed into the standard normal random variable Z by means of the substitution

$$Z = \frac{X - \mu}{\sigma}$$

(see Figure 10.19).

Figure 10.19

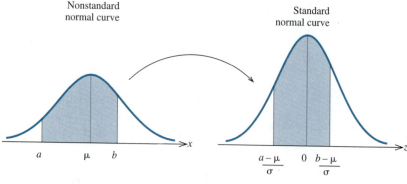

Nonstandard normal curve

Standard normal curve

Area under the
curve between a and b
$=$
Area under the curve
between $\dfrac{a-\mu}{\sigma}$ and $\dfrac{b-\mu}{\sigma}$

The area of the region under the normal curve (with random variable X) between $x = a$ and $x = b$ is *equal* to the area of the region under the standard normal curve between $z = (a - \mu)/\sigma$ and $z = (b - \mu)/\sigma$. In terms of probabilities associated with these distributions, we have

$$P(a < X < b) = P\left(\frac{a-\mu}{\sigma} < Z < \frac{b-\mu}{\sigma}\right) \tag{11}$$

Similarly, we have

$$P(X < b) = P\left(Z < \frac{b-\mu}{\sigma}\right) \tag{12}$$

and

$$P(X > a) = P\left(Z > \frac{a-\mu}{\sigma}\right) \tag{13}$$

Thus, with the help of (11)–(13), computations of probabilities associated with any normal distribution may be reduced to computations of areas of regions under the standard normal curve.

EXAMPLE 3   Suppose X is a normal random variable with $\mu = 100$ and $\sigma = 20$. Find these values:

a. $P(X < 120)$ **b.** $P(X > 70)$ **c.** $P(75 < X < 110)$

Solution

a. Using (12) with $\mu = 100$, $\sigma = 20$, and $b = 120$, we have

$$P(X < 120) = P\left(Z < \frac{120 - 100}{20}\right)$$

$$= P(Z < 1) = 0.8413 \qquad \text{(Using the table of values of } Z\text{)}$$

b. Using (13) with $\mu = 100$, $\sigma = 20$, and $a = 70$, we have

$$P(X > 70) = P\left(Z > \frac{70 - 100}{20}\right)$$

$$= P(Z > -1.5)$$

$$= P(Z < 1.5) = 0.9332 \qquad \text{(Using the table of values of } Z\text{)}$$

Figure 10.20

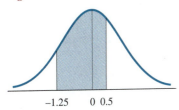

−1.25 0 0.5

c. Using (11) with $\mu = 100$, $\sigma = 20$, $a = 75$, and $b = 110$, we have

$$P(75 < X < 110)$$

$$= P\left(\frac{75 - 100}{20} < Z < \frac{110 - 100}{20}\right)$$

$$= P(-1.25 < Z < 0.5)$$

$$= P(Z < 0.5) - P(Z < -1.25) \quad \text{(see Figure 10.20)}$$

$$= 0.6915 - 0.1056 = 0.5859 \quad \text{(Using the table of values of } Z)$$

⁖ ⁖ ⁖

Applications Involving Normal Random Variables

EXAMPLE 4 The medical records of infants delivered at Kaiser Memorial Hospital show that the infants' birth weights in pounds are normally distributed with a mean of 7.4 and a standard deviation of 1.2. Find the probability that an infant selected at random from among those delivered at the hospital weighed more than 9.2 pounds at birth.

Solution Let X be the normal random variable denoting the birth weights of infants delivered at the hospital. Then the probability that an infant selected at random has a birth weight of more than 9.2 pounds is given by $P(X > 9.2)$. To compute $P(X > 9.2)$, we use formula (13) with $\mu = 7.4$, $\sigma = 1.2$, and $a = 9.2$. We find

$$P(X > 9.2) = P\left(Z > \frac{9.2 - 7.4}{1.2}\right) \quad \left[P(X > a) = P\left(Z > \frac{a - \mu}{\sigma}\right)\right]$$

$$= P(Z > 1.5)$$

$$= P(Z < -1.5)$$

$$= 0.0668$$

Thus, the probability that an infant delivered at the hospital weighed more than 9.2 pounds is 0.0668.

⁖ ⁖ ⁖

EXAMPLE 5 Idaho Natural Produce Corporation ships potatoes to its distributors in bags with a mean weight of 50 pounds and a standard deviation of 0.5 pound. If a bag of potatoes is selected at random from a shipment, what are the probabilities of these events?

a. The bag weighs more than 51 pounds.
b. It weighs less than 48 pounds.
c. It weighs between 49 and 51 pounds.

Solution Let X denote the weight of a bag of potatoes packed by the company. Then the mean and standard deviation of X are $\mu = 50$ and $\sigma = 0.5$, respectively.

a. The probability that a bag selected at random weighs more than 51 pounds is given by

$$P(X > 51) = P\left(Z > \frac{51 - 50}{0.5}\right) \qquad \left[P(X > a) = P\left(Z > \frac{a - \mu}{\sigma}\right)\right]$$

$$= P(Z > 2)$$
$$= P(Z < -2)$$
$$= 0.0228$$

b. The probability that a bag selected at random weighs less than 48 pounds is given by

$$P(X < 48) = P\left(Z < \frac{48 - 50}{0.5}\right) \qquad \left[P(X < b) = P\left(Z < \frac{b - \mu}{\sigma}\right)\right]$$

$$= P(Z < -4)$$
$$= 0$$

c. The probability that a bag selected at random weighs between 49 and 51 pounds is given by

$$P(49 < X < 51) = P\left(\frac{49 - 50}{0.5} < Z < \frac{51 - 50}{0.5}\right) \qquad \left[P(a < X < b) = P\left(\frac{a - \mu}{\sigma} < Z < \frac{b - \mu}{\sigma}\right)\right]$$

$$= P(-2 < Z < 2)$$
$$= P(Z < 2) - P(Z < -2)$$
$$= 0.9772 - 0.0228$$
$$= 0.9544$$

EXAMPLE 6   The grade-point average of the senior class of Jefferson High School is normally distributed with a mean of 2.7 and a standard deviation of 0.4 point. If a senior in the top 10% of his or her class is eligible for admission to any of the nine campuses of the state university system, what is the minimum grade-point average that a senior should have to ensure eligibility for admission to the state university system?

Solution   Let X denote the grade-point average of a randomly selected senior at Jefferson High School, and let x denote the minimum grade-point average to ensure his or her eligibility for admission to the university. Since only the top 10% are eligible for admission, x must satisfy the equation

$$P(X \geq x) = 0.1$$

Using formula (13) with $\mu = 2.7$ and $\sigma = 0.4$, we find

$$P(X \geq x) = P\left(Z \geq \frac{x - 2.7}{0.4}\right) = 0.1 \qquad \left[P(X > a) = P\left(Z > \frac{a - \mu}{\sigma}\right)\right]$$

But, this is equivalent to the equation

$$P\left(Z \leq \frac{x - 2.7}{0.4}\right) = 0.9 \qquad \text{(Why?)}$$

Consulting Appendix Table 4, we find

$$\frac{x - 2.7}{0.4} = 1.28$$

Upon solving for x, we obtain

$$x = (1.28)(0.4) + 2.7$$
$$\approx 3.2$$

Thus, to ensure eligibility for admission to one of the nine campuses of the state university system, a senior at Jefferson High School should have a minimum 3.2 grade-point average.   ⊙ ⊙ ⊙

SELF-CHECK EXERCISE 10.3

1. Let Z be a standard normal variable.
   a. Find the value of $P(-1.2 < Z < 2.1)$ by first sketching the appropriate region under the standard normal curve.
   b. Find the value of z if z satisfies

   $$P(-z < Z < z) = 0.8764$$

2. Let X be a normal random variable with $\mu = 80$ and $\sigma = 10$. Find these values:
   a. $P(X < 100)$
   b. $P(X > 60)$
   c. $P(70 < X < 90)$

3. The serum cholesterol levels (in mg/dl) in a current Mediterranean population are found to be normally distributed with a mean of 160 and a standard deviation of 50. Scientists at the National Heart, Lung, and Blood Institute consider this pattern ideal for a minimal risk of heart attacks. Find the percentage of the population who have blood cholesterol levels between 160 and 180 mg/dl.

Solutions to Self-Check Exercises 10.3 can be found on page 748.

10.3 EXERCISES

In exercises 1–6, find the value of the probability of the standard normal variable Z corresponding to the shaded area under the standard normal curve.

1. $P(Z < 1.45)$

1.45

2. $P(Z > 1.11)$

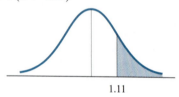

1.11

3. $P(Z < -1.75)$

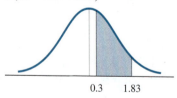

-1.75

4. $P(0.3 < Z < 1.83)$

0.3   1.83

5. $P(-1.32 < Z < 1.74)$

-1.32   1.74

6. $P(-2.35 < Z < -0.51)$

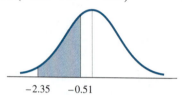

-2.35   -0.51

In exercises 7–14, (a) sketch the area under the standard normal curve corresponding to the given probability and (b) find the value of the probability of the standard normal variable Z corresponding to this area.

7. $P(Z < 1.37)$ **8.** $P(Z > 2.24)$

9. $P(Z < -0.65)$ **10.** $P(0.45 < Z < 1.75)$

11. $P(Z > -1.25)$

12. $P(-1.48 < Z < 1.54)$

13. $P(0.68 < Z < 2.02)$

14. $P(-1.41 < Z < -0.24)$

15. Let Z be the standard normal variable. Find the values of z if z satisfies
 a. $P(Z < z) = 0.8907$
 b. $P(Z < z) = 0.2090$

16. Let Z be the standard normal variable. Find the values of z if z satisfies
 a. $P(Z > z) = 0.9678$
 b. $P(-z < Z < z) = 0.8354$

17. Let Z be the standard normal variable. Find the values of z if z satisfies
 a. $P(Z > -z) = 0.9713$
 b. $P(Z < -z) = 0.9713$

18. Suppose X is a normal random variable with $\mu = 380$ and $\sigma = 20$. Find these values:
 a. $P(X < 405)$ **b.** $P(400 < X < 430)$
 c. $P(X > 400)$

19. Suppose X is a normal random variable with $\mu = 50$ and $\sigma = 5$. Find these values:
 a. $P(X < 60)$ **b.** $P(X > 43)$
 c. $P(46 < X < 58)$

20. Suppose X is a normal random variable with $\mu = 500$ and $\sigma = 75$. Find these values:

   a. $P(X < 750)$ **b.** $P(X > 350)$
   c. $P(400 < X < 600)$

21. Medical Records The medical records of infants delivered at Kaiser Memorial Hospital show that the infants' lengths at birth (in inches) are normally distibuted with a mean of 20 and a standard deviation of 2.6. Find the probability that an infant selected at random from among those delivered at the hospital measures

   a. more than 22 inches
   b. less than 18 inches
   c. between 19 and 21 inches

22. Factory Workers' Wages According to the data released by a city's Chamber of Commerce, the weekly wages of factory workers are normally distributed with a mean of $400 and a standard deviation of $50. Find the probability that a worker selected at random from the city has a weekly wage of

   a. less than $300
   b. more than $460
   c. between $350 and $450

23. Product Reliability TKK Products Corporation manufactures electric light bulbs in the 50-, 60-, 75-, and 100-watt range. Laboratory tests show that the lives of these light bulbs are normally distributed with a mean of 750 hours and a standard deviation of 75 hours. What is the probability that a TKK light bulb selected at random will burn

   a. for more than 900 hours?
   b. for less than 600 hours?
   c. between 750 and 900 hours?
   d. between 600 and 800 hours?

24. Education On the average, a student takes 100 words per minute midway through an advanced shorthand course at the American Institute of Stenography. Assuming that the students' dictation speeds are normally distributed and that the standard deviation is 20 words per minute, find the probability that a student randomly selected from the course could take dictation at a speed of

   a. more than 120 words per minute
   b. between 80 and 120 words per minute
   c. less than 80 words per minute

25. IQs The IQs of students at Wilson Elementary School were measured recently and found to be normally distributed with a mean of 100 and a standard deviation of 15. What is the probability that a student selected at random will have an IQ of

   a. 140 or higher?   **b.** 120 or higher?
   c. between 100 and 120?
   d. 90 or less?

26. Product Reliability The tread lives of Super Titan radial tires under normal driving conditions are normally distributed with a mean of 40,000 miles and a standard deviation of 2000 miles. What is the probability that a tire selected at random will have a tread life of more than 35,000 miles? If four new tires are installed on a car and they experience even wear, determine the probability that all four tires still have useful tread lives after 35,000 miles of driving.

27. Female Factory Workers' Wages According to data released by a city's Chamber of Commerce, the weekly wages (in dollars) of female factory workers are normally distributed with a mean of 375 and a standard deviation of 50. Find the probability that a female factory worker selected at random from the city has a weekly wage of $350 to $450.

28. Civil Service Exams To be eligible for further consideration, applicants for certain Civil Service positions must first pass a written qualifying examination on which a score of 70 or more must be obtained. In a recent examination, it was found that the scores were normally distributed with a mean of 60 points and a standard deviation of 10 points. Determine the percentage of applicants who passed the written qualifying examination.

29. Warranties The general manager of the Service Department of MCA Television Company has estimated that the time that elapses between the dates of purchase and the dates on which the 19-inch sets manufactured by the company first require service is normally distributed with a mean of 22 months and a standard deviation of 4 months. If MCA gives a 1-year warranty on parts and labor for these sets, determine what percentage of sets manufactured and sold may require service before the warranty period runs out.

30. Grade Distributions The scores on an economics examination are normally distributed with a mean of 72 and a standard deviation of 16. If the instructor assigns a grade of A to 10% of the class, what is the lowest score a student may have and still get an A?

31. Grade Distributions The scores on a sociology examination are normally distributed with a mean of 70 and a standard deviation of 10. If the instructor assigns A's to 15%, B's to 25%, C's to 40%, D's to 15%, and F's to 5% of the class, find the cutoff points for these grades.

Solution to Self-Check Exercise 10.3

1. a. The probability $P(-1.2 < Z < 2.1)$ is given by the shaded area in the accompanying figure. We have

$$P(-1.2 < Z < 2.1) = P(Z < 2.1) - P(Z < -1.2)$$
$$= 0.9821 - 0.1151$$
$$= 0.867$$

b. The region associated with $P(-z < Z < z)$ is shown in the accompanying figure. Observe that we have the following relationship:

$$P(Z < z) = \frac{1}{2}[1 + P(-z < Z < z)]$$

(see Example 2c). With $P(-z < Z < z) = 0.8764$, we find

$$P(Z < z) = \frac{1}{2}(1 + 0.8764)$$
$$= 0.9382$$

Consulting the table, we find $z = 1.54$.

2. Using the transformations (11), (12), and (13) and the table of values of Z, we have

a. $P(X < 100) = P\left(Z < \dfrac{100 - 80}{10}\right)$
$$= P(Z < 2)$$
$$= 0.9772$$

b. $P(X > 60) = P\left(Z > \dfrac{60 - 80}{10}\right)$

$\qquad = P(Z > -2)$

$\qquad = P(Z < 2)$

$\qquad = 0.9772$

c. $P(70 < X < 90) = P\left(\dfrac{70 - 80}{10} < Z < \dfrac{90 - 80}{10}\right)$

$\qquad = P(-1 < Z < 1)$

$\qquad = P(Z < 1) - P(Z < -1)$

$\qquad = 0.8413 - 0.1587$

$\qquad = 0.6826$

3. Let X be the normal random variable denoting the serum cholesterol levels (in mg/dl) in the current Mediterranean population under consideration. Thus, the percentage of the population having blood cholesterol levels between 160 and 180 mg/dl is given by $P(160 < X < 180)$. To compute $P(160 < X < 180)$, we use (11), with $\mu = 160$, $\sigma = 50$, $a = 160$, and $b = 180$. We find

$$P(160 < X < 180) = P\left(\dfrac{160 - 160}{50} < Z < \dfrac{180 - 160}{50}\right)$$

$$= P(0 < Z < 0.4)$$

$$P(Z < 0.4) - P(Z < 0)$$

$$= 0.6554 - 0.5000$$

$$= 0.1554$$

so approximately 15.5% of the population has blood cholesterol levels between 160 and 180 mg/dl.

CHAPTER 10 SUMMARY OF PRINCIPAL FORMULAS AND TERMS

Formulas

1. Expected value $\qquad E(x) = x_1 p_1 + x_2 p_2 + \cdots + x_n p_n$

2. Expected value of a continuous random variable $\qquad E(x) = \displaystyle\int_a^b x f(x)\, dx$

3. Exponential density function $\qquad f(x) = k e^{-kx}$

4. Expected value of an exponential density function $\qquad E(x) = \dfrac{1}{k}$

5. Variance of a continuous random variable $\qquad \text{Var}(x) = \displaystyle\int_a^b (x - \mu)^2 f(x)\, dx$

6. Standard deviation of a continuous random variable $\qquad \sigma = \sqrt{\text{Var}(x)}$

Terms

Experiment	Probability density
Outcomes (sample points)	function
Sample space	Average (mean)
Event	Expected value
Probability of an event	Variance of a random
Random variable	variable
Discrete probability	Standard deviation of a
function	random variable
Histogram	Normal distribution
Continuous random	Standard normal
variable	random variable

CHAPTER 10 REVIEW EXERCISES

In exercises 1–4, show that the function is a probability density function on the given interval.

1. $f(x) = \dfrac{1}{28}(2x + 3); [0, 4]$

2. $f(x) = \dfrac{3}{16}\sqrt{x}; [0, 4]$

3. $f(x) = \dfrac{1}{4}; [7, 11]$ **4.** $f(x) = \dfrac{4}{x^5}; [1, \infty)$

In exercises 5–8, find the value of the constant k so that the function is a probability density function in the given interval.

5. $f(x) = kx^2; [0, 9]$

6. $f(x) = \dfrac{k}{\sqrt{x}}; [1, 16]$

7. $f(x) = \dfrac{k}{x^2}; [1, 3]$ **8.** $f(x) = \dfrac{k}{x^{2.5}}; [1, \infty)$

In exercises 9–12, f is a probability density function defined on the given interval. Find the indicated probabilities.

9. $f(x) = \dfrac{2}{21}x; [2, 5]$

   a. $P(x \leq 4)$ **b.** $P(x = 4)$
   c. $P(3 \leq x \leq 4)$

10. $f(x) = \dfrac{1}{4}; [1, 5]$

   a. $P(2 \leq x \leq 4)$ **b.** $P(x \leq 3)$
   c. $P(x \geq 2)$

11. $f(x) = \dfrac{3}{16}\sqrt{x}; [0, 4]$

   a. $P(1 \leq x \leq 3)$ **b.** $P(x \leq 3)$
   c. $P(x = 2)$

12. $f(x) = \dfrac{1}{x^2}; [1, \infty)$

   a. $P(x \leq 10)$ **b.** $P(2 \leq x \leq 4)$
   c. $P(x \geq 2)$

In exercises 13–16, find the mean, variance, and standard deviation of the random variable x associated with the probability density function f over the given interval.

13. $f(x) = \dfrac{1}{5}; [2, 7]$

14. $f(x) = \dfrac{1}{28}(2x + 3); [0, 4]$

15. $f(x) = \dfrac{1}{4}(3x^2 + 1); [-1, 1]$

16. $f(x) = \dfrac{4}{x^5}; [1, \infty)$

In exercises 17–20, Z is the standard normal variable. Find the given probability.

17. $P(Z < 2.24)$ **18.** $P(Z > -1.24)$

19. $P(0.24 \leq Z \leq 1.28)$ **20.** $P(-1.37 \leq Z \leq 1.37)$

21. Suppose X is a normal random variable with $\mu = 80$ and $\sigma = 8$. Find these values:

a. $P(X \le 84)$ **b.** $P(X \ge 70)$

c. $P(75 \le X \le 85)$

22. Suppose X is a normal random variable with $\mu = 45$ and $\sigma = 3$. Find these values:

a. $P(X \le 50)$ **b.** $P(X \ge 40)$

c. $P(40 \le X \le 50)$

23. Records at Centerville Hospital indicate that the length of time in days that a maternity patient stays in the hospital has a probability density function given by

$$P(t) = \frac{1}{4} e^{-(1/4)t}$$

a. What is the probability that a woman entering the maternity wing will be there longer than six days?

b. What is the probability that a woman entering the maternity wing will be there less than two days?

c. What is the average length of time that a woman entering the maternity wing stays in the hospital?

24. The life span (in years) of a certain make of car battery is an exponentially distributed random variable with an expected value of 5. Find the probability that the life span of a battery is (a) less than 4 years, (b) more than 6 years, and (c) between 2 and 4 years.

In this chapter we show how certain functions can be represented by a *power series*. A power series involves infinitely many terms, but when truncated, it is just a polynomial. By approximating a function with a *Taylor polynomial*, we are often able to obtain approximate solutions to problems that we cannot otherwise solve.

We also look at a method, called the Newton-Raphson Method, for finding the zeros of a function. For example, Newton's Method can be used to find the critical points of a function, which, as you may recall, are candidates for the solution of the optimization problems considered in Chapter 4.

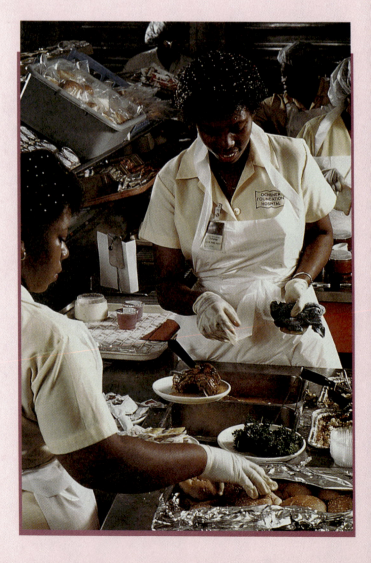

What percentage of the nonfarm work force will be in the service industries one decade from now? In Example 4, page 760, you will see how a Taylor polynomial can be used to help answer this question.

TAYLOR POLYNOMIALS AND INFINITE SERIES

11.1 TAYLOR POLYNOMIALS

As we saw earlier, it is not always possible to obtain an exact solution to a problem, and in such cases we have to settle for an approximate solution. In this section we will show how a function may be approximated near a given point by a polynomial. Polynomials, as we have seen time and again, are easy to work with; for example, they are easy to evaluate, differentiate, and integrate. Thus, by using polynomials rather than working with the original function itself, we can often obtain approximate solutions to a problem that we might otherwise not be able to solve.

Taylor Polynomials

Suppose we are given a differentiable function f and a point $x = a$ in the domain of f. Then the polynomial of degree 0 that best approximates f *near* $x = a$ is the constant polynomial

$$P_0(x) = f(a)$$

which coincides with f at $x = a$ (Figure 11.1a).

Figure 11.1

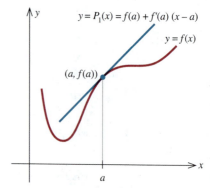

(a) $P_0(x) = f(a)$ is a zero-degree polynomial that approximates f near $x = a$.

(b) $P_1(x) = f(a) + f'(a)(x - a)$ is a first-degree polynomial that approximates f near $x = a$.

Now, unless f itself is a constant function, it is possible in most cases to obtain a better approximation of f *near* $x = a$ by using a polynomial function of degree 1. Recall that the linear function

$$L(x) = f(a) + f'(a)(x - a)$$

is just an equation of the tangent line to the graph of the function f at the point $(a, f(a))$ (Figure 11.1b). As such, the value of the function L coincides with the value of f at $x = a$, and its slope coincides with the slope of f at $x = a$; that is,

$$L(a) = f(a) \quad \text{and} \quad L'(a) = f'(a)$$

Let us write

$$L(x) = P_1(x)$$

Then
$$P_1(x) = f(a) + f'(a)(x - a)$$

is the required polynomial approximation (of degree 1) of f near $x = a$.

This discussion suggests that yet a better approximation to f at $x = a$ may be found by using a polynomial of degree 2, $P_2(x)$, and requiring that its value, slope, and concavity coincide with those of f at $x = a$. In other words, $P_2(x)$ should satisfy the three conditions

$$P_2(a) = f(a), \quad P_2'(a) = f'(a), \quad \text{and} \quad P_2''(a) = f''(a)$$

The third condition ensures that the graph of the polynomial bends in the right way, at least near $x = a$. Pursuing this line of reasoning, we are led to the search for a polynomial of degree n in $x - a$,

$$P_n(x) = a_0 + a_1(x - a) + a_2(x - a)^2$$
$$+ a_3(x - a)^3 + \cdots + a_n(x - a)^n$$

(where $a_0, a_1, \ldots, a_n$ are constants), that satisfies the conditions

$$P_n(a) = f(a), \quad P_n'(a) = f'(a), \quad P_n''(a) = f''(a), \quad \ldots, \quad P_n^{(n)}(a) = f^{(n)}(a) \quad \textbf{(1)}$$

To determine the required polynomial, we compute

$$P_n'(x) = a_1 + 2a_2(x - a) + 3a_3(x - a)^2 + \cdots + na_n(x - a)^{n-1}$$

$$P_n''(x) = 2a_2 + 3 \cdot 2a_3(x - a) + \cdots + n(n - 1)a_n(x - a)^{n-2}$$

$$P_n'''(x) = 3 \cdot 2a_3 + 4 \cdot 3 \cdot 2a_4(x - a) + \cdots + n(n - 1)(n - 2)a_n(x - a)^{n-3}$$

$$\vdots$$

$$P_n^{(n)}(x) = n(n - 1)(n - 2) \cdots (1)a_n$$

Setting $x = a$ in each of the expressions for $P_n(x), P_n'(x), P_n''(x), \ldots, P_n^{(n)}(x)$ in succession and using the conditions in (1), we find

$$P_n(a) = a_0 = f(a)$$

$$P_n'(a) = a_1 = f'(a)$$

$$P_n''(a) = 2a_2 = f''(a)$$

$$P_n'''(a) = 3 \cdot 2a_3 = f'''(a)$$

$$\vdots$$

$$P_n^{(n)}(a) = n(n - 1)(n - 2) \cdots (1)a_n = f^{(n)}(a)$$

from which we deduce that

$$a_0 = f(a), \quad a_1 = f'(a), \quad a_2 = \frac{1}{2}f''(a), \quad a_3 = \frac{1}{3 \cdot 2}f'''(a), \quad \ldots,$$

$$a_n = \frac{1}{n(n - 1)(n - 2) \cdots (1)} f^{(n)}(a)$$

Let us introduce the expression $n!$ (read "n factorial"), defined by

$$n! = n(n-1)(n-2)(n-3) \cdots 3 \cdot 2 \cdot 1 \quad \text{for } n \geq 1$$
$$0! = 1$$

Thus,

$$1! = 1 \qquad\qquad 4! = 4 \cdot 3 \cdot 2 \cdot 1 = 24$$
$$2! = 2 \cdot 1 = 2 \qquad 5! = 5 \cdot 4 \cdot 3 \cdot 2 \cdot 1 = 120$$
$$3! = 3 \cdot 2 \cdot 1 = 6$$

and so on. Using this notation, we may write the coefficients of $P_n(x)$ as

$$a_0 = f(a), \quad a_1 = f'(a), \quad a_2 = \frac{1}{2!} f''(a), \quad a_3 = \frac{1}{3!} f'''(a), \quad \ldots, \quad a_n = \frac{1}{n!} f^{(n)}(a)$$

so that the required polynomial is

$$P_n(x) = f(a) + f'(a)(x-a) + \frac{f''(a)}{2!}(x-a)^2 + \cdots$$
$$+ \frac{f^{(n)}(a)}{n!}(x-a)^n$$

THE nTH TAYLOR POLYNOMIAL	Suppose that the function f and its first n derivatives are defined at $x = a$. Then the nth Taylor polynomial of f at $x = a$ is the polynomial $$P_n(x) = f(a) + f'(a)(x-a)$$ $$+ \frac{f''(a)}{2!}(x-a)^2 + \cdots + \frac{f^{(n)}(a)}{n!}(x-a)^n \qquad (2)$$ which coincides with $f(x), f'(x), \ldots, f^{(n)}(x)$ at $x = a$; that is, $$P_n(a) = f(a), \quad P_n'(a) = f'(a), \quad \ldots, \quad P_n^{(n)}(a) = f^{(n)}(a)$$

Using a Taylor Polynomial to Approximate a Function

In many instances, the Taylor polynomial $P_n(x)$ provides us with a good approximation of $f(x)$ near $x = a$.

EXAMPLE I   Find the first four Taylor polynomials of $f(x) = e^x$ at $x = 0$, and sketch the graph of each polynomial superimposed upon the graph of $f(x) = e^x$.

Solution   Here $a = 0$, and since

$$f(x) = f'(x) = f''(x) = f'''(x) = f^{(4)}(x) = e^x$$

we find

$$f(0) = f'(0) = f''(0) = f'''(0) = f^{(4)}(0) = 1$$

Using formula (2) with $n = 1, 2, 3,$ and 4 in succession, we find that the first four Taylor polynomials are

$$P_1(x) = f(0) + f'(0)(x - 0) = 1 + x$$

$$P_2(x) = f(0) + f'(0)(x - 0) + \frac{f''(0)}{2!}(x - 0)^2 = 1 + x + \frac{1}{2}x^2$$

$$P_3(x) = f(0) + f'(0)(x - 0) + \frac{f''(0)}{2!}(x - 0)^2 + \frac{f'''(0)}{3!}(x - 0)^3$$

$$= 1 + x + \frac{1}{2}x^2 + \frac{1}{6}x^3$$

and $$P_4(x) = f(0) + f'(0)(x - 0) + \frac{f''(0)}{2!}(x - 0)^2 + \frac{f'''(0)}{3!}(x - 0)^3$$

$$+ \frac{f^{(4)}(0)}{4!}(x - 0)^4 = 1 + x + \frac{1}{2}x^2 + \frac{1}{6}x^3 + \frac{1}{24}x^4$$

The graphs of these polynomials are shown in Figure 11.2. Observe that the approximation of $f(x)$ near $x = 0$ improves as the degree of the approximating Taylor polynomial increases.

Figure 11.2
The graphs of the first four Taylor polynomials of $f(x) = e^x$ at $x = 0$ superimposed upon the graph of $f(x) = e^x$.

EXPLORING WITH TECHNOLOGY

Let $f(x) = xe^{-x}$.
a. Find the first four Taylor polynomials of f at $x = 0$.
b. Use a graphing utility to plot the graphs of f, P_1, P_2, P_3, and P_4 on the same set of axes in the viewing rectangle $[-0.5, 1] \times [-0.5, 0.5]$.
c. Comment on the approximation of f by the polynomial P_n near $x = 0$ for $n = 1, 2, 3,$ and 4. What happens to the approximation if x is "far" from the origin?

○ ○ ○

EXAMPLE 2 Find the nth Taylor polynomial of $f(x) = 1/(1 - x)$ at $x = 0$. Compute $P_5(0.1)$ and compare this result with $f(0.1)$.

Solution Here $a = 0$. Next, we find

$$f(x) = \frac{1}{1-x} = (1-x)^{-1}$$

$$f'(x) = -(1-x)^{-2}(-1) = (1-x)^{-2}$$

$$f''(x) = -2(1-x)^{-3}(-1) = 2(1-x)^{-3}$$

$$f'''(x) = 3 \cdot 2(1-x)^{-4}$$

$$f^{(4)}(x) = 4 \cdot 3 \cdot 2(1-x)^{-5}$$

$$\vdots$$

$$f^{(n)}(x) = n(n-1)(n-2) \cdots 1(1-x)^{-n-1}$$

so that

$$f(0) = 1, \quad f'(0) = 1, \quad f''(0) = 2, \quad f'''(0) = 3!, \quad f^{(4)}(0) = 4!, \quad \ldots, \quad f^{(n)}(0) = n!$$

Therefore, the required nth Taylor polynomial is

$$P_n(x) = 1 + \frac{1}{1}x + \frac{2}{2!}x^2 + \frac{3!}{3!}x^3 + \frac{4!}{4!}x^4 + \cdots + \frac{n!}{n!}x^n$$

$$= 1 + x + x^2 + x^3 + x^4 + \cdots + x^n$$

In particular,

$$P_5(x) = 1 + x + x^2 + x^3 + x^4 + x^5$$

so that

$$P_5(0.1) = 1 + 0.1 + (0.1)^2 + (0.1)^3 + (0.1)^4 + (0.1)^5$$
$$= 1.11111$$

On the other hand, the actual value of $f(x)$ at $x = 0.1$ is

$$f(0.1) = \frac{1}{1 - 0.1} = \frac{1}{0.9} = 1.111\ldots$$

Thus, the error incurred in approximating $f(0.1)$ by $P_5(0.1)$ is $0.00000111.\ldots$

○ ○ ○

Applications

EXAMPLE 3   Obtain the second Taylor polynomial of $f(x) = \sqrt{x}$ at $x = 25$ and use it to approximate the value of $\sqrt{26.5}$.

Solution   Here $a = 25$ and $n = 2$. Since

$$f(x) = x^{1/2}, \quad f'(x) = \frac{1}{2}x^{-1/2} = \frac{1}{2\sqrt{x}}, \quad \text{and} \quad f''(x) = -\frac{1}{4}x^{-3/2} = -\frac{1}{4x^{3/2}}$$

we find that

$$f(25) = 5, \quad f'(25) = \frac{1}{10}, \quad \text{and} \quad f''(25) = -\frac{1}{500}$$

Using formula (2), we obtain the required polynomial

$$P_2(x) = f(25) + f'(25)(x - 25) + \frac{f''(25)}{2!}(x - 25)^2$$

$$= 5 + \frac{1}{10}(x - 25) - \frac{1}{1000}(x - 25)^2$$

Next, using $P_2(x)$ as an approximation of $f(x)$ near $x = 25$, we find

$$\sqrt{26.5} = f(26.5)$$
$$\approx P_2(26.5)$$
$$= 5 + \frac{1}{10}(26.5 - 25) - \frac{1}{1000}(26.5 - 25)^2$$
$$\approx 5.14775$$

The exact value of $\sqrt{26.5}$, rounded off to five decimal places, is 5.14782. Thus, the error incurred in the approximation is 0.00007.

○ ○ ○

Suppose you want to find an approximation to the value of $\sqrt[3]{26.5}$ using a second Taylor polynomial of some function f expanded about $x = a$.

a. How would you choose f and a?

b. Obtain an approximation of $\sqrt[3]{26.5}$, and compare your result with that obtained using a calculator.

The next example shows how a Taylor polynomial can be used to approximate an integral that involves an integrand whose antiderivative cannot be expressed as an elementary function.

 EXAMPLE 4   It has been estimated that service industries, which currently make up 30% of the nonfarm work force in a certain country, will continue to grow at the rate of

$$R(t) = 5e^{1/(t+1)}$$

percent per decade, t decades from now. Estimate the percentage of the nonfarm work force in the service industries one decade from now.

Solution   The percentage of the nonfarm work force in the service industries t decades from now will be given by

$$P(t) = \int 5e^{1/(t+1)} \, dt, \qquad P(0) = 30$$

This integral cannot be expressed in terms of an elementary function. To obtain an approximate solution to the problem at hand, let us first make the substitution

$$u = \frac{1}{t + 1}$$

so that

$$t + 1 = \frac{1}{u} \quad \text{and} \quad t = \frac{1}{u} - 1$$

giving

$$dt = -\frac{1}{u^2} \, du$$

The integral becomes

$$\tilde{P}(u) = 5 \int e^u \left(-\frac{du}{u^2} \right) = -5 \int \frac{e^u}{u^2} \, du$$

Next, let us approximate e^u at $u = 0$ by a fourth-degree Taylor polynomial. Proceeding as in Example 1, we find that

$$e^u \approx 1 + u + \frac{u^2}{2!} + \frac{u^3}{3!} + \frac{u^4}{4!}$$

Thus,

$$\tilde{P}(u) \approx -5 \int \frac{1}{u^2} \left(1 + u + \frac{u^2}{2} + \frac{u^3}{6} + \frac{u^4}{24} \right) du$$

$$= -5 \int \left(\frac{1}{u^2} + \frac{1}{u} + \frac{1}{2} + \frac{u}{6} + \frac{u^2}{24} \right) du$$

$$= -5 \left(-\frac{1}{u} + \ln u + \frac{1}{2}u + \frac{u^2}{12} + \frac{u^3}{72} \right) + C$$

Therefore,

$$P(t) \approx -5\left[-(t+1) + \ln\left(\frac{1}{t+1}\right) + \frac{1}{2(t+1)} + \frac{1}{12(t+1)^2} + \frac{1}{72(t+1)^3}\right] + C$$

Using the condition $P(0) = 30$, we find

$$P(0) \approx -5\left(-1 + \ln 1 + \frac{1}{2} + \frac{1}{12} + \frac{1}{72}\right) + C = 30$$

or $C \approx 27.99$. So

$$P(t) \approx -5\left[-(t+1) + \ln\left(\frac{1}{t+1}\right) + \frac{1}{2(t+1)} + \frac{1}{12(t+1)^2} + \frac{1}{72(t+1)^3}\right]$$
$$+ 27.99$$

In particular, the percentage of the nonfarm work force in the service industries one decade from now will be given by

$$P(1) \approx -5\left[-2 + \ln\left(\frac{1}{2}\right) + \frac{1}{4} + \frac{1}{48} + \frac{1}{576}\right] + 27.99$$
$$\approx 40.09$$

or approximately 40.1%.

Errors in Taylor Polynomial Approximations

The **error** incurred in approximating $f(x)$ by its Taylor polynomial $P_n(x)$ at $x = a$ is

$$R_n(x) = f(x) - P_n(x)$$

A formula for $R_n(x)$ can be derived, but we will not give it here since from the point of view of applications, we are more interested in a bound for $|R_n(x)|$. A bound on $|R_n(x)|$ provides us with a bound on the error incurred when a Taylor polynomial of f is used to approximate $f(x)$. We have the following result, which we state without proof.

ERROR BOUND FOR TAYLOR POLYNOMIAL APPROXIMATIONS	Suppose that the function f and its first n derivatives are continuous on the interval $[c, d]$ containing the point $x = a$, and its $(n+1)$st derivative $f^{(n+1)}(x)$ exists for every x in (c, d). Then for all x in $[c, d]$, $$\|R_n(x)\| \leq \frac{M}{(n+1)!}\|x - a\|^{n+1} \qquad (3)$$ where M is any number such that $\|f^{(n+1)}(t)\| \leq M$ for all t between a and x.

EXAMPLE 5   Refer to Example 2. Use formula (3) to obtain an upper bound for the error incurred in approximating $f(0.1)$ by $P_5(0.1)$, where

$$f(x) = \frac{1}{1-x}$$

and $P_5(x)$ is the fifth Taylor polynomial of f at $x = 0$. Compare this result with the actual error incurred in the approximation.

Solution   First, we need to find a number M such that

$$|f^{(6)}(t)| \le M$$

for all t lying in the interval $[0, 0.1]$. From the solution of Example 2, we find that

$$f^{(6)}(t) = (6)(5)(4)(3)(2)(1)(1-t)^{-6-1}$$
$$= \frac{720}{(1-t)^7}$$

Observe that

$$f^{(6)}(t) = \frac{720}{(1-t)^7}$$

is increasing on the interval $[0, 0.1]$. [Just look at $f^{(7)}(t)$.] Consequently, its maximum is attained at the right endpoint $t = 0.1$. So we may take

$$M = \frac{720}{(1 - 0.1)^7} \approx 1505.3411$$

Next, with $a = 0$, $x = 0.1$, and $M = 1505.3411$, formula (3) gives

$$|R_5(0.1)| \le \frac{1505.3411}{(5+1)!}(0.1)^{5+1}$$
$$\approx 0.0000021$$

Finally, we refer once again to Example 2 to see that the actual error incurred in approximating $f(0.1)$ by $P_5(0.1)$ was $0.00000111 \ldots$, and this is less than the error bound 0.0000021 just obtained.   ⊙ ⊙ ⊙

Refer to the Group Discussion question on page 759. Obtain an upper bound for the error incurred in approximating $\sqrt[5]{26.5}$ using the Taylor polynomial you have chosen.

EXAMPLE 6 Use a fourth Taylor polynomial to obtain an approximation of $e^{0.5}$. Find a bound for the error in the approximation.

Solution The point $x = 0.5$ is near $x = 0$, so we will use the fourth Taylor polynomial $P_4(x)$ at $x = 0$ to approximate $e^{0.5}$. From Example 1, we see that

$$P_4(x) = 1 + x + \frac{1}{2}x^2 + \frac{1}{6}x^3 + \frac{1}{24}x^4$$

Therefore,

$$e^{0.5} \approx P_4(0.5) = 1 + 0.5 + \frac{1}{2}(0.5)^2 + \frac{1}{6}(0.5)^3 + \frac{1}{24}(0.5)^4$$

$$\approx 1.6484$$

To obtain an error bound for this approximation, we need to obtain an upper bound M for $|f^{(5)}(t)|$ for all t lying in the interval $[0, 0.5]$ (remember, $n = 4$). Now

$$f^{(5)}(t) = e^t$$

and since $e^t > 0$ for all values of t, $f^{(5)}(t) = |e^t| = e^t$. Next, we observe that e^t is an increasing function on $(-\infty, \infty)$ and in particular on $[0, 0.5]$. Therefore, e^t attains an absolute maximum at $x = 0.5$, the right endpoint of the interval in question; that is,

$$|f^{(5)}(t)| = |e^t| = e^t \leq e^{0.5} \quad \text{on } [0, 0.5]$$

Now, $e < 3$, so certainly

$$|f^{(5)}(t)| \leq e^{0.5} < 3^{0.5} = \sqrt{3}$$
$$< 2$$

Thus, we may take $M = 2$. Finally, with $a = 0$, $x = 0.5$, and this value of M, formula (3) gives

$$|R_4(0.5)| \leq \frac{2}{(4+1)!}|0.5 - 0|^{4+1}$$

$$= \frac{2}{5!}(0.5)^5$$

$$\approx 0.00052$$

So our result, $e^{0.5} \approx 1.6484$, is guaranteed to be accurate to within 0.0005.

● ● ●

REMARKS

1. Using a calculator, we see that $e^{0.5} \approx 1.6487$, rounded to four decimal places. Therefore, the error incurred in approximating this number by $P_4(0.5)$ is $(1.6487 - 1.6484)$, or 0.0003, which is less than the upper error bound of 0.0005 obtained.

2. The error bound obtained in Example 6 is not the best possible. This stems partially from our choice of M. In practical applications, it is not essential that the error bound be the best possible as long as it is sharp enough.

● ● ●

SELF-CHECK EXERCISES 11.1

1. Let $f(x) = \dfrac{1}{1 + x}$.

    a. Find the nth Taylor polynomial of f at $x = 0$.

    b. Use the third Taylor polynomial of f at $x = 0$ to obtain an approximation of $f(0.1)$. Compare this result with $f(0.1)$.

2. Refer to Exercise 1. Find a bound on the error in the approximation of $f(0.1)$.

Solutions to Self-Check Exercises 11.1 can be found on page 766.

11.1 EXERCISES

In exercises 1–10, find the first three Taylor polynomials of the given function at the indicated point.

1. $f(x) = e^{-x}$; $x = 0$

2. $f(x) = e^{2x}$; $x = 0$

3. $f(x) = \dfrac{1}{x + 1}$; $x = 0$

4. $f(x) = \dfrac{1}{1 - x}$; $x = 2$

5. $f(x) = \dfrac{1}{x}$; $x = 1$

6. $f(x) = \dfrac{1}{x + 2}$; $x = 1$

7. $f(x) = \sqrt{1 - x}$; $x = 0$

8. $f(x) = \sqrt{x}$; $x = 4$

9. $f(x) = \ln(1 - x)$; $x = 0$

10. $f(x) = xe^x$; $x = 0$

In exercises 11–20, find the nth Taylor polynomial of the given function at the indicated point.

11. $f(x) = x^4$ at $x = 2$, $n = 2$

12. $f(x) = x^5$ at $x = -1$, $n = 3$

13. $f(x) = \ln x$ at $x = 1$, $n = 4$

14. $f(x) = 1/x$ at $x = 2$, $n = 4$

15. $f(x) = e^x$ at $x = 1$, $n = 4$

16. $f(x) = e^{2x}$ at $x = 1$, $n = 4$

17. $f(x) = \sqrt{1 - x}$ at $x = 0$, $n = 3$

18. $f(x) = \sqrt{x + 1}$ at $x = 0$, $n = 3$

19. $f(x) = \dfrac{1}{2x + 3}$ at $x = 0$, $n = 3$

20. $f(x) = x^2 e^x$ at $x = 0$, $n = 2$

21. Find the nth Taylor polynomial of $f(x) = \dfrac{1}{1 + x}$ at $x = 0$. Compare $P_4(0.1)$ and compare this result with $f(0.1)$.

22. Find the third Taylor polynomial of $f(x) = \sqrt{3x + 1}$ at $x = 0$. Compute $P_3(0.2)$ and compare this result with $f(0.2)$.

23. Find the fourth Taylor polynomial of $f(x) = e^{-x/2}$ at $x = 0$ and use it to estimate $e^{-0.1}$.

24. Find the third Taylor polynomial of $f(x) = \ln x$ at $x = 1$ and use it to estimate $\ln 1.1$.

25. Find the second Taylor polynomial of $f(x) = \sqrt{x}$ at $x = 16$ and use it to estimate the value of $\sqrt{15.6}$.

26. Find the second Taylor polynomial of $f(x) = x^{1/3}$ at $x = 8$ and use it to estimate the value of $\sqrt[3]{8.1}$.

27. Find the third Taylor polynomial of $f(x) = \ln(x + 1)$ at $x = 0$ and use it to estimate the value of

$$\int_0^{1/2} \ln(x + 1)\, dx$$

Compare your result with the exact value found using integration by parts.

28. Use the fourth Taylor polynomial of $f(x) = \ln(1 + x)$ to approximate

$$\int_{0.1}^{0.2} \frac{\ln(x + 1)}{x}\, dx$$

29. Use the second Taylor polynomial of $f(x) = \sqrt{x}$ at $x = 4$ to approximate $\sqrt{4.06}$, and find a bound for the error in the approximation.

30. Use the third Taylor polynomial of $f(x) = e^{-x}$ at $x = 0$ to approximate $e^{-0.2}$, and find a bound for the error in the approximation.

31. Use the third Taylor polynomial of $f(x) = \dfrac{1}{1 - x}$ at $x = 0$ to approximate $f(0.2)$. Find a bound for the error in the approximation and compare your results to the exact value of $f(0.2)$.

32. Use the second Taylor polynomial of $f(x) = \sqrt{x+1}$ at $x = 0$ to approximate $f(0.1)$, and find a bound for the error in the approximation.

33. Use the second Taylor polynomial of $f(x) = \ln x$ at $x = 1$ to approximate $\ln 1.1$, and find a bound for the error in the approximation.

34. Use the second Taylor polynomial of $f(x) = \ln x$ at $x = 1$ to approximate $\ln 0.9$, and find a bound for the error in the approximation.

35. Let $f(x) = e^{-x}$.
  a. Find a bound in the error incurred in approximating $f(x)$ by the third Taylor polynomial, $P_3(x)$, of f at $x = 0$ in the interval $[0, 1]$.
  b. Estimate $\int_0^1 e^{-x}\, dx$ by computing $\int_0^1 P_3(x)\, dx$.
  c. What is the bound on the error in the approximation in (b)?
  [*Hint:* Compute $\int_0^1 B\, dx$, where B is a bound on $|R_3(x)|$ for x in $[0, 1]$.]
  d. What is the actual error?

36. a. Find a bound on the error incurred in approximating $f(x) = \ln x$ by the fourth Taylor polynomial, $P_4(x)$, of f at $x = 1$ in the interval $[1, 1.5]$.
  b. Approximate the area under the graph of f from $x = 1$ to $x = 1.5$ by computing
  $$\int_1^{1.5} P_4(x)\, dx$$
  c. What is the bound on the error in the approximation in (b)?
  [*Hint:* Compute $\int_1^{1.5} B\, dx$, where B is a bound on $|R_4(x)|$ for x in $[1, 1.5]$.]
  d. What is the actual error?
  [*Hint:* Use integration by parts to evaluate $\int_1^{1.5} \ln x\, dx$.]

37. Use the fourth Taylor polynomial at $x = 0$ to obtain an approximation of the area under the graph of $f(x) = e^{-x^2/2}$ from $x = 0$ to $x = 0.5$.
  [*Hint:* Use the substitution $u = -x^2/2$. See Example 4.]

38. Use the fourth Taylor polynomial at $x = 0$ to obtain an approximation of the area under the graph of $f(x) = \ln(1 + x^2)$ from $x = -\frac{1}{2}$ to $x = \frac{1}{2}$.
  [*Hint:* First find the Taylor polynomial of $f(u) = \ln(1 + u)$; then use the substitution $u = x^2$.]

39. Growth of Service Industries It has been estimated that service industries, which currently make up 30% of the nonfarm work force in a certain country, will continue to grow at the rate of
$$R(t) = 6e^{1/(2t+1)}$$
percent per decade t decades from now. Estimate the percentage of the nonfarm work force in service industries two decades from now.

40. Vacation Trends The Travel Data Center of a European country has estimated that the percentage of vacations traditionally taken during the months of July, August, and September will continue to decline according to the rule
$$R(t) = -2e^{1/(t+1)}$$
percent per decade ($t = 0$ corresponds to the year 1992) as more and more vacationers switch to fall and winter vacations. There were 37% summer vacationers in that country in 1992. Estimate the percentage of vacations that will be taken in the summer months in the year 2002.

41. Continuing Education Enrollment The registrar of Kellogg University estimates that the total student enrollment in the Continuing Education division will be given by
$$N(t) = -\frac{20{,}000}{\sqrt{1 + 0.2t}} + 21{,}000$$
where $N(t)$ denotes the number of students enrolled in the division t years from now. Using the second Taylor polynomial of N at $t = 0$, find an approximation of the average enrollment at Kellogg University between $t = 0$ and $t = 2$.

42. Concentration of Carbon Monoxide in the Air According to a joint study conducted by Oxnard's Environmental Management Department and a state government agency, the concentration of carbon monoxide in the air due to automobile exhaust t years from now is given by
$$C(t) = 0.01(0.2t^2 + 4t + 64)^{2/3}$$
parts per million. Use the second Taylor polynomial of C at $t = 0$ to obtain an approximation of the average level of concentration of carbon monoxide in the air between $t = 0$ and $t = 2$.

SOLUTIONS TO SELF-CHECK EXERCISES 11.1

1. a. We rewrite $f(x)$ as

$$f(x) = (1 + x)^{-1}$$

and compute

$$f'(x) = -(1 + x)^{-2}$$
$$f''(x) = (-1)(-2)(1 + x)^{-3} = 2(1 + x)^{-3}$$
$$f'''(x) = 2(-3)(1 + x)^{-4} = -3 \cdot 2(1 + x)^{-4}$$
$$f^{(4)}(x) = -3 \cdot 2(-4)(1 + x)^{-5} = 4 \cdot 3 \cdot 2(1 + x)^{-5}$$
$$\vdots$$
$$f^{(n)}(x) = (-1)^n n! (1 + x)^{-(n+1)}$$

Evaluating $f(x)$ and its derivatives at $x = 0$ gives

$$f(0) = 1, \quad f'(0) = -1, \quad f''(0) = 2!, \quad f'''(0) = -3!,$$
$$f^{(4)}(0) = 4!, \quad \ldots, \quad f^{(n)}(0) = (-1)^n n!$$

Therefore, the required nth Taylor polynomial is

$$P_n(x) = 1 + \frac{(-1)}{1!}x + \frac{2}{2!}x^2 + \frac{-3!}{3!}x^3$$
$$+ \frac{4!}{4!}x^4 + \cdots + \frac{(-1)^n n!}{n!}x^n$$
$$= 1 - x + x^2 - x^3 + x^4 - \cdots + (-1)^n x^n$$

b. The third Taylor polynomial of f at $x = 0$ is

$$P_3(x) = 1 - x + x^2 - x^3$$

So

$$P_3(0.1) = 1 - 0.1 + (0.1)^2 - (0.1)^3$$
$$= 0.909$$

The actual value of $f(x)$ at $x = 0.1$ is

$$f(0.1) = \frac{1}{1 + 0.1} = \frac{1}{1.1} = 0.909090\ldots$$

Thus, the error incurred in approximating $f(0.1)$ by $P_3(0.1)$ is $0.00009090\ldots$.

2. First we need to find a number M such that

$$|f^{(4)}(t)| \le M$$

for all t lying in the interval $[0, 0.1]$. From the solution of Exercise 1, we see that

$$f^{(5)}(x) = (-1)^5 5! (1 + x)^{-6} = -5!(1 + x)^{-6}$$

Since $f^{(5)}(x)$ is negative for x in $[0, 0.1]$, $f^{(4)}(x)$ is decreasing on $[0, 0.1]$. Therefore, the maximum of the function

$$f^{(4)}(x) = 4!(1 + x)^{-5}$$

is attained at $x = 0$, and its value is

$$f^{(4)}(0) = 4! = 24$$

So we can take $M = 24$. Next, with $a = 0$, $x = 0.1$, and $M = 24$, formula (3) gives

$$|R_3(0.1)| \leq \frac{24}{(3+1)!} |0.1 - 0|^{3+1}$$

$$= \frac{24}{24}(0.1)^4 = 0.0001$$

the required error bound.

11.2 INFINITE SEQUENCES

Infinite Sequences of Real Numbers

In Section 11.1 we saw that the approximation of a function f near a point $x = a$ by the nth Taylor polynomial of f at $x = a$ generally improves as n increases. This leads us to wonder what happens when the degree of a Taylor polynomial of a function at a point $x = a$ is allowed to increase without bound. Letting n go to infinity, however, leads to an expression involving the sum of "infinitely" many terms:

$$f(a) + f'(a)(x - a) + \frac{f''(a)}{2!}(x - a)^2 + \cdots$$

Clearly, this "sum" cannot be found by simply adding all the terms. Before we can define such a sum, we need to discuss sequences. (We will return to the problem posed here in Section 11.4.)

An **infinite sequence** of real numbers is a function whose domain is the set of positive integers. For example, the infinite sequence defined by

$$f(n) = \frac{1}{n}$$

is the ordered list of numbers

$$1, \frac{1}{2}, \frac{1}{3}, \ldots, \frac{1}{n}, \ldots$$

It is customary to represent the functional values called the **terms** of the sequence

$$f(1), f(2), f(3), \ldots, f(n), \ldots$$

by the notation

$$a_1, a_2, a_3, \ldots, a_n, \ldots$$

The term $a_n = f(n)$ is called the **nth term,** or **general term,** of the sequence. The sequence itself is abbreviated $\{a_n\}$. Observe that we can describe a sequence by specifying its nth term. For the example under consideration, the sequence is denoted by $\{1/n\}$, and its terms are

$$a_1 = 1, \quad a_2 = \frac{1}{2}, \quad a_3 = \frac{1}{3}, \quad \ldots, \quad a_n = \frac{1}{n}, \quad \ldots$$

INFINITE SEQUENCE

An infinite sequence is a function whose domain is the set of positive integers. If f is a function defining the sequence $\{a_n\}$, then the functional values $a_1, a_2, a_3, \ldots, a_n, \ldots$, where $a_n = f(n)$ $(n = 1, 2, 3, \ldots)$, are called the **terms** of the sequence, and a_n is called the **nth term** of the sequence.

EXAMPLE 1   Write down the terms of each of the following infinite sequences.

a. $\left\{ \dfrac{1}{2^{n-1}} \right\}$ **b.** $\left\{ \dfrac{1+n}{1+n^2} \right\}$ **c.** $\left\{ \dfrac{\pi^n}{n!} \right\}$

Solution

a. Taking $n = 1, 2, 3, 4, \ldots$ in succession, we see that

$$a_1 = \frac{1}{2^0} = 1, \quad a_2 = \frac{1}{2^1} = \frac{1}{2}, \quad a_3 = \frac{1}{2^2} = \frac{1}{4},$$

$$a_4 = \frac{1}{2^3} = \frac{1}{8}, \quad \ldots,$$

so the required sequence is

$$1, \frac{1}{2}, \frac{1}{4}, \frac{1}{8}, \ldots, \frac{1}{2^{n-1}}, \ldots$$

b. The required sequence is

$$a_1 = \frac{1+1}{1+1^2} = 1, \quad a_2 = \frac{1+2}{1+2^2} = \frac{3}{5}, \quad a_3 = \frac{1+3}{1+3^2} = \frac{4}{10},$$

$$a_4 = \frac{1+4}{1+4^2} = \frac{5}{17}, \quad \ldots, \quad a_n = \frac{1+n}{1+n^2}, \quad \ldots$$

or

$$1, \frac{3}{5}, \frac{4}{10}, \frac{5}{17}, \ldots, \frac{1+n}{1+n^2}, \ldots$$

c. Here

$$a_1 = \frac{\pi}{1!} = \pi, \quad a_2 = \frac{\pi^2}{2!} = \frac{\pi^2}{2}, \quad a_3 = \frac{\pi^3}{3!} = \frac{\pi^3}{6}, \quad a_4 = \frac{\pi^4}{4!} = \frac{\pi^4}{24}, \quad \ldots$$

so the required sequence is

$$\pi, \frac{\pi^2}{2!}, \frac{\pi^3}{3!}, \frac{\pi^4}{4!}, \ldots, \frac{\pi^n}{n!}, \ldots$$

EXAMPLE 2 Find the general term of each sequence:

a. $1, \dfrac{1}{4}, \dfrac{1}{9}, \dfrac{1}{16}, \dfrac{1}{25}, \ldots$ **b.** $-1, 1, -1, 1, -1, 1, \ldots$

Solution

a. Observe that the terms may be written as

$$a_1 = \frac{1}{1^2}, \quad a_2 = \frac{1}{2^2}, \quad a_3 = \frac{1}{3^2}, \quad a_4 = \frac{1}{4^2}, \quad a_5 = \frac{1}{5^2}, \quad \cdots$$

and we conclude that the nth term is $a_n = \dfrac{1}{n^2}$.

b. Here

$$a_1 = -1, \quad a_2 = 1, \quad a_3 = -1, \quad a_4 = 1, \quad a_5 = -1, \quad \cdots$$

which may also be written as

$$a_1 = (-1)^1, \quad a_2 = (-1)^2, \quad a_3 = (-1)^3,$$
$$a_4 = (-1)^4, \quad a_5 = (-1)^5, \quad \cdots$$

and so $a_n = (-1)^n$.

Graphs of Infinite Sequences

Since an infinite sequence is a function, we can sketch its graph. Because the domain of the function is the set of positive integers, the graph of a sequence consists of an infinite collection of points in the xy-plane.

EXAMPLE 3 Sketch the graph of each sequence:

a. $\{n + 1\}$ **b.** $\left\{\dfrac{n+1}{n}\right\}$ **c.** $\{(-1)^n\}$

Solution

a. The terms of the sequence $\{n + 1\}$ are $2, 3, 4, 5, \ldots$. Recalling that these numbers are precisely the functional values of $f(n) = n + 1$ for $n = 1, 2, 3, \ldots$, we obtain the following table of values of f:

n	1	2	3	4	$\cdots$
$f(n)$	2	3	4	5	$\cdots$

from which we construct the graph of $\{n + 1\}$ shown in Figure 11.3a.

Figure 11.3

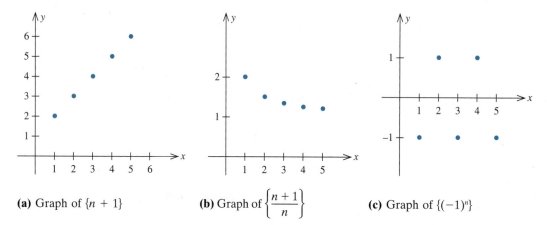

(a) Graph of $\{n + 1\}$ **(b)** Graph of $\left\{\dfrac{n+1}{n}\right\}$ **(c)** Graph of $\{(-1)^n\}$

b. From the following table of values of f:

n	1	2	3	4	$\cdots$
$f(n)$	2	$\frac{3}{2}$	$\frac{4}{3}$	$\frac{5}{4}$	$\cdots$

we sketch the graph shown in Figure 11.3b.

c. We use the following table of values for the sequence

n	1	2	3	4	$\cdots$
$f(n)$	-1	1	-1	1	$\cdots$

to sketch the graph of $\{(-1)^n\}$ shown in Figure 11.3c.   o o o

REMARK   Notice that the function $f(n) = n + 1$ defining the sequence $\{n + 1\}$ may be viewed as the function $f(x) = x + 1$ $(-\infty < x < \infty)$ with x *restricted* to the set of positive integers. Similarly, the function defining the sequence $\{(n + 1)/n\}$ is just the function $f(x) = (x + 1)/x$ with a similar restriction on its domain. Finally, the function defining the sequence $\{(-1)^n\}$ is the function $f(x) = \cos \pi x$ with x a positive integer.* These observations suggest that it is possible to study the properties of a sequence by analyzing the properties of a corresponding function defined for *all* values of x in some suitable interval.   o o o

The Limit of a Sequence

Given a sequence $\{a_n\}$, we may ask whether the terms a_n of the sequence approach some specific number L as n gets larger and larger. If they do, we say

o o o

* The trigonometric function $f(x) = \cos \pi x$ will be defined in Chapter 12.

that the sequence a_n *converges* to L. More specifically, we have the following informal definition.

LIMIT OF A SEQUENCE	Let $\{a_n\}$ be a given sequence. We say that the sequence $\{a_n\}$ **converges** and has the limit L, written $$\lim_{n \to \infty} a_n = L$$ if the terms of the sequence, a_n, can be made as close to L as we please by taking n sufficiently large. If a sequence is not convergent, it is said to be **divergent.**

EXAMPLE 4 Determine whether each of the following infinite sequences converges or diverges.

a. $\left\{\dfrac{1}{n}\right\}$ **b.** $\{\sqrt{n}\}$ **c.** $\{(-1)^n\}$

Solution

a. The terms of the sequence,

$$1, \frac{1}{2}, \frac{1}{3}, \frac{1}{4}, \frac{1}{5}, \ldots, \frac{1}{n}, \ldots$$

approach the number $L = 0$ as n gets larger and larger. We conclude that

$$\lim_{n \to \infty} a_n = \lim_{n \to \infty} \frac{1}{n} = 0$$

Compare this with

$$\lim_{x \to \infty} \frac{1}{x} = 0$$

b. The terms of the sequence,

$$1, \sqrt{2}, \sqrt{3}, \sqrt{4}, \sqrt{5}, \ldots, \sqrt{n}, \ldots$$

get larger and larger as n gets larger and larger. Consequently, they do not approach any finite number L as n tends to infinity. Therefore, the sequence is divergent. Compare this with

$$\lim_{x \to \infty} \sqrt{x}$$

c. The terms of the sequence are

$$a_1 = -1, \quad a_2 = 1, \quad a_3 = -1, \quad a_4 = 1,$$
$$a_5 = -1, \quad \ldots, \quad a_n = (-1)^n, \quad \ldots$$

Thus, no matter how large n is, there are terms that are equal to -1 (those with odd-numbered subscripts) and also terms that are equal to 1 (those with even-numbered subscripts). This implies that there cannot be a *unique* real number L such that a_n is arbitrarily close to L no matter how large n is. Therefore, the sequence is divergent.

○ ○ ○

EXPLORING WITH TECHNOLOGY

Refer to Example 4 and the remark on page 770.

a. Plot the graph of $f(x) = 1/x$ using the viewing rectangle $[0, 10] \times [0, 3]$ and thus verify graphically that $\lim_{n \to \infty} (1/n) = 0$.

b. Plot the graph of $f(x) = \sqrt{x}$ using an appropriate viewing rectangle and thus verify graphically that $\lim_{n \to \infty} \sqrt{n}$ does not exist.

c. Plot the graph of $f(x) = \cos \pi x$ using the viewing rectangle $[0, 10] \times [-2, 2]$ and thus verify graphically that $\lim_{n \to \infty} (-1)^n$ does not exist. [*Note:* Trigonometric functions will be studied in Chapter 12.]

○   ○   ○

The following properties of sequences, which parallel those of the limit of $f(x)$ at infinity, are helpful in computing limits of sequences.

LIMIT PROPERTIES OF SEQUENCES

Suppose that

$$\lim_{n \to \infty} a_n = A \quad \text{and} \quad \lim_{n \to \infty} b_n = B$$

Then

1. $\lim_{n \to \infty} ca_n = c \lim_{n \to \infty} a_n = cA \qquad (c, \text{a constant})$

2. $\lim_{n \to \infty} (a_n \pm b_n) = \lim_{n \to \infty} a_n \pm \lim_{n \to \infty} b_n = A \pm B \qquad$ (Sum Rule)

3. $\lim_{n \to \infty} a_n b_n = \left(\lim_{n \to \infty} a_n \right)\left(\lim_{n \to \infty} b_n \right) = AB \qquad$ (Product Rule)

4. $\lim_{n \to \infty} \dfrac{a_n}{b_n} = \dfrac{\lim_{n \to \infty} a_n}{\lim_{n \to \infty} b_n} = \dfrac{A}{B} \quad$ provided $B \neq 0 \qquad$ (Quotient Rule)

EXAMPLE 5   Evaluate:

a. $\lim\limits_{n\to\infty} \left(\dfrac{1}{2}\right)^n$ **b.** $\lim\limits_{n\to\infty} \dfrac{2n^2+1}{3n^2+n+2}$

Solution

a.
$$\lim\limits_{n\to\infty} \left(\frac{1}{2}\right)^n = \lim\limits_{n\to\infty} \frac{1}{2^n}$$
$$= \frac{1}{\lim\limits_{n\to\infty} 2^n} = 0$$

b. Dividing the numerator and denominator by n^2, we find

$$\lim\limits_{n\to\infty} \frac{2n^2+1}{3n^2+n+2} = \lim\limits_{n\to\infty} \frac{2+\dfrac{1}{n^2}}{3+\dfrac{1}{n}+\dfrac{2}{n^2}}$$

$$= \frac{\lim\limits_{n\to\infty}\left(2+\dfrac{1}{n^2}\right)}{\lim\limits_{n\to\infty}\left(3+\dfrac{1}{n}+\dfrac{2}{n^2}\right)}$$

$$= \frac{2}{3}$$

● ● ●

Consider the sequences $\{a_n\}$ and $\{b_n\}$ defined by $a_n = (-1)^{n+1}$ and $b_n = (-1)^n$.

a. Show that $\lim\limits_{n\to\infty} a_n$ and $\lim\limits_{n\to\infty} b_n$ do not exist.

b. Show that $\lim\limits_{n\to\infty} (a_n + b_n) = 0$.

c. Do the results of parts (a) and (b) contradict the limit properties of sequences listed on page 772? Explain your answer.

EXAMPLE 6   Two percent of the spark plugs manufactured by the Parts Division of United Motors Corporation are defective. It can be shown that the probability of getting at least one defective plug in a random sample of n spark plugs is $f(n) = 1 - (0.98)^n$. Consider the sequence $\{a_n\}$ defined by $a_n = f(n)$.

a. Write down the terms a_5, a_{10}, a_{25}, a_{100}, and a_{200} of the sequence $\{a_n\}$.

b. Evaluate

$$\lim\limits_{n\to\infty} a_n$$

and interpret your results.

Solution

a. The required terms of the sequence are

$$0.10, 0.18, 0.40, 0.87, 0.98$$

For example, the probability of getting at least one defective plug in a random sample of 25 is 0.4—that is, a 40% chance.

b.

$$\lim_{n\to\infty} a_n = \lim_{n\to\infty} [1 - (0.98)^n]$$

$$= \lim_{n\to\infty} 1 - \lim_{n\to\infty} (0.98)^n$$

$$= 1 - 0$$

$$= 1$$

The result tells us that if the sample is large enough, we will almost certainly pick at least one defective plug!   ○ ○ ○

SELF-CHECK EXERCISES 11.2

1. Determine whether the sequence $\left\{\dfrac{3n^2 + 2n + 1}{n^2 + 4}\right\}$ converges or diverges. If it converges, find its limit.

2. Consider the sequence $\left\{\dfrac{n}{n^2 + 1}\right\}$.

    a. Sketch the graph of the sequence.
    b. Show that the sequence is decreasing—that is, $a_1 > a_2 > a_3 > \cdots$. [*Hint:* Consider $f(x) = x/(x^2 + 1)$ and show that f is decreasing by computing f'.]

Solutions to Self-Check Exercises 11.2 can be found on page 775.

11.2 EXERCISES

In exercises 1–9, write down the first five terms of each of the given sequences.

1. $\{a_n\} = \{2^{n-1}\}$

2. $\{a_n\} = \left\{\dfrac{2n}{1 + n^2}\right\}$

3. $\{a_n\} = \left\{\dfrac{n-1}{n+1}\right\}$

4. $\{a_n\} = \left\{\left(-\dfrac{1}{3}\right)^n\right\}$

5. $\{a_n\} = \left\{\dfrac{2^{n-1}}{n!}\right\}$

6. $\{a_n\} = \left\{\dfrac{(-1)^n}{(2n)!}\right\}$

7. $\{a_n\} = \left\{\dfrac{e^n}{n^3}\right\}$

8. $\{a_n\} = \left\{\dfrac{\sqrt{n}}{\sqrt{n}+1}\right\}$

9. $\{a_n\} = \left\{\dfrac{3n^2 - n + 1}{2n^2 + 1}\right\}$

In exercises 10–21, find the general terms of each of the given sequences.

10. $\dfrac{1}{2}, \dfrac{1}{4}, \dfrac{1}{6}, \dfrac{1}{8}, \ldots$

11. $1, 4, 7, 10, \ldots$

12. $\dfrac{1}{3}, \dfrac{1}{7}, \dfrac{1}{11}, \dfrac{1}{15}, \ldots$

13. $1, \dfrac{1}{8}, \dfrac{1}{27}, \dfrac{1}{64}, \ldots$

14. $1, \dfrac{2}{3}, \dfrac{4}{9}, \dfrac{8}{27}, \ldots$

15. $2, \dfrac{8}{5}, \dfrac{32}{25}, \dfrac{128}{125}, \ldots$

16. $1, \dfrac{5}{4}, \dfrac{7}{5}, \dfrac{9}{6}, \ldots$

17. $1, -\dfrac{1}{2}, \dfrac{1}{4}, -\dfrac{1}{8}, \ldots$

18. $1 + \dfrac{1}{2}, 1 + \dfrac{1}{3}, 1 + \dfrac{1}{4}, 1 + \dfrac{1}{5}, \ldots$

19. $\dfrac{1}{2 \cdot 3}, \dfrac{2}{3 \cdot 4}, \dfrac{3}{4 \cdot 5}, \dfrac{4}{5 \cdot 6}, \ldots$

20. $1, \dfrac{2}{1 \cdot 3}, \dfrac{4}{1 \cdot 3 \cdot 5}, \dfrac{8}{1 \cdot 3 \cdot 5 \cdot 7}, \ldots$

21. $1, e, \dfrac{e^2}{2}, \dfrac{e^3}{6}, \dfrac{e^4}{24}, \dfrac{e^5}{120}, \ldots$

In exercises 22–29, sketch the graph of each of the given sequences.

22. $\{n^2\}$

23. $\left\{\dfrac{2n}{n+1}\right\}$

24. $\left\{\dfrac{(-1)^n}{n}\right\}$

25. $\{\sqrt{n}\}$

26. $\{\ln n\}$

27. $\{e^n\}$

28. $\{ne^{-n}\}$

29. $\{n - \sqrt{n}\}$

In exercises 30–44, determine the convergence or divergence of each given sequence $\{a_n\}$. If the sequence converges, find its limit.

30. $a_n = \dfrac{n}{n^2 + 1}$

31. $a_n = \dfrac{n+1}{2n}$

32. $a_n = \sqrt[3]{n}$

33. $a_n = \dfrac{(-1)^n}{\sqrt{n}}$

34. $a_n = \dfrac{1}{n+1} - \dfrac{1}{n+2}$

35. $a_n = \dfrac{\sqrt{n}-1}{\sqrt{n}+1}$

36. $a_n = \dfrac{3n^2 + n - 1}{6n^2 + n + 1}$

37. $a_n = \dfrac{2n^3 - 1}{n^3 + 2n + 1}$

38. $a_n = \dfrac{1 + (-1)^n}{3^n}$

39. $a_n = 2 - \dfrac{1}{2^n}$

40. $a_n = \dfrac{2n}{n!}$

41. $a_n = \dfrac{2^n}{3^n}$

42. $a_n = \dfrac{2^n - 1}{2^n}$

43. $a_n = \dfrac{n}{\sqrt{2n^2 + 3}}$

44. $a_n = \dfrac{2^n}{n!}$

45. Automobile Microprocessors   One and a half percent of the microprocessors manufactured by a microelectronics firm for use in regulating fuel consumption in automobiles are defective. It can be shown that the probability of getting at least one defective microprocessor in a random sample of n microprocessors is $f(n) = 1 - (0.985)^n$. Consider the sequence $\{a_n\}$ defined by $a_n = f(n)$.
  a. Write down the terms a_1, a_{10}, a_{100}, and a_{1000} of the sequence $\{a_n\}$.
  b. Evaluate $\lim\limits_{n \to \infty} a_n$ and interpret your results.

46. Savings Accounts   An amount of \$1000 is deposited in a bank that pays 8% interest per year compounded daily (take the number of days in a year to be 365). Let a_n denote the total amount on deposit after n days, assuming no deposits or withdrawals are made during the period in question.
  a. Find the formula for a_n. [*Hint:* See Section 5.3.]
  b. Compute a_1, a_{10}, a_{50}, and a_{100}.
  c. What amount is on deposit after one year?

47. Accumulated Amount   Suppose \$100 is deposited at the beginning of each month into an account earning interest at the rate of 12% per year compounded monthly. Let a_n denote the amount on deposit at the end of the nth month. This is called the accumulated amount.
  a. Show that $a_1 = 100(1.01)$, $a_2 = 100(1.01)^2$, and $a_3 = 100(1.01)^3$.
  b. Find the accumulated amount.
  c. Find the 24th term of the sequence $\{a_n\}$ and interpret your result.

SOLUTIONS TO SELF-CHECK EXERCISES 11.2

1. We compute

$$\lim_{n \to \infty} \frac{3n^2 + 2n + 1}{n^2 + 4} = \lim_{n \to \infty} \frac{3 + \dfrac{2}{n} + \dfrac{1}{n^2}}{1 + \dfrac{4}{n^2}} \qquad \text{(Divide numerator and denominator by } n^2.)$$

$$= \frac{3}{1} = 3$$

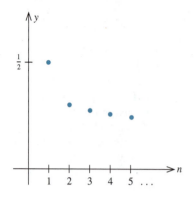

2. a. From the table

n	1	2	3	4	5	$\cdots$
$f(n)$	$\frac{1}{2}$	$\frac{2}{5}$	$\frac{3}{10}$	$\frac{4}{17}$	$\frac{5}{26}$	$\cdots$

we obtain the graph shown in the figure.

b. We compute

$$f'(x) = \frac{(x^2 + 1)(1) - x(2x)}{(x^2 + 1)^2} = \frac{1 - x^2}{(x^2 + 1)^2} < 0$$

for $x > 1$, and so f is decreasing on $(1, \infty)$. We conclude that $\{a_n\}$ is decreasing.

11.3 INFINITE SERIES

The Sum of an Infinite Series

In Section 11.2, we raised a question concerning the "sum" of a series involving infinitely many terms. In this section, we will show how we define such a sum. Consider the **infinite series**

$$a_1 + a_2 + \cdots + a_n + \cdots$$

which may be abbreviated through the use of sigma notation as

$$\sum_{n=1}^{\infty} a_n = a_1 + a_2 + a_3 + \cdots + a_n + \cdots \tag{4}$$

and is read "the sum of the numbers a_n for n running from 1 to infinity."

Now, given an infinite series (4) whose terms are drawn from the sequence $\{a_n\}$, let us define the sums

$$S_1 = \sum_{n=1}^{1} a_n = a_1$$

$$S_2 = \sum_{n=1}^{2} a_n = a_1 + a_2$$

$$S_3 = \sum_{n=1}^{3} a_n = a_1 + a_2 + a_3$$

$$S_N = \sum_{n=1}^{N} a_n = a_1 + a_2 + a_3 + \cdots + a_N$$

Observe that each of these sums exists since each is obtained by adding together finitely many numbers. For each N, S_N is called the **Nth partial sum** of the series (4), and the sequence $\{S_n\}$ is called the **sequence of partial sums** of the series (4). We are now in a position to define the sum of an infinite series.

SUM OF AN INFINITE SERIES

Let

$$\sum_{n=1}^{\infty} a_n = a_1 + a_2 + a_3 + \cdots$$

be an infinite series, and let $\{S_n\}$ be the sequence of partial sums of the infinite series. If

$$\lim_{n \to \infty} S_n = S$$

we say that the infinite series $\sum_{n=1}^{\infty} a_n$ **converges** to S and write

$$\sum_{n=1}^{\infty} a_n = \lim_{n \to \infty} S_n = S$$

In this case, S is called the **sum of the series.** If $\{S_n\}$ does not converge, we say that the infinite series **diverges** and has no sum.

REMARK   Simply stated, this definition tells us that the partial sums of a convergent series ultimately form a sequence of increasingly accurate approximations to the sum of the series.*   ◦ ◦ ◦

EXAMPLE 1   **a.** Show that the following infinite series diverges:

$$\sum_{n=1}^{\infty} (-1)^{n+1} = 1 - 1 + 1 - 1 + 1 - 1 + \cdots$$

b. Show that the following infinite series converges:

$$\sum_{n=1}^{\infty} \frac{1}{n(n+1)} = \frac{1}{1 \cdot 2} + \frac{1}{2 \cdot 3} + \frac{1}{3 \cdot 4} + \cdots$$

Solution

a. The partial sums of the given infinite series are

$$S_1 = 1, \quad S_2 = 1 - 1 = 0, \quad S_3 = 1 - 1 + 1 = 1,$$
$$S_4 = 1 - 1 + 1 - 1 = 0, \quad \ldots$$

The sequence of partial sums $\{S_n\}$ evidently diverges, and so the given infinite series diverges.

b. Let us write

$$\frac{1}{n(n+1)} = \frac{1}{n} - \frac{1}{n+1}$$

◦ ◦ ◦

* This is true for most of the convergent series that arise in applications.

an equality that is easily verified. The Nth partial sum of the given series is

$$S_N = \sum_{n=1}^{N} \frac{1}{n(n+1)} = \sum_{n=1}^{N} \left(\frac{1}{n} - \frac{1}{n+1} \right)$$

$$= \left(1 - \frac{1}{2} \right) + \left(\frac{1}{2} - \frac{1}{3} \right) + \left(\frac{1}{3} - \frac{1}{4} \right) + \cdots + \left(\frac{1}{N} - \frac{1}{N+1} \right)$$

$$= 1 + \left(-\frac{1}{2} + \frac{1}{2} \right) + \left(-\frac{1}{3} + \frac{1}{3} \right) + \cdots + \left(-\frac{1}{N} + \frac{1}{N} \right) - \frac{1}{N+1}$$

$$= 1 - \frac{1}{N+1}$$

Since

$$\lim_{N \to \infty} S_N = \lim_{N \to \infty} \left(1 - \frac{1}{N+1} \right) = 1$$

we conclude that the given series converges and has a sum equal to 1; that is,

$$\sum_{n=1}^{\infty} \frac{1}{n(n+1)} = 1$$

○ ○ ○

EXPLORING WITH TECHNOLOGY

Refer to Example 1b.

a. Verify graphically the results of Example 1b by plotting the graphs of the partial sums

$$S_1 = \frac{1}{1 \cdot 2}, \quad S_2 = \frac{1}{1 \cdot 2} + \frac{1}{2 \cdot 3}, \quad \ldots, \quad S_6 = \frac{1}{1 \cdot 2} + \frac{1}{2 \cdot 3} + \cdots + \frac{1}{6 \cdot 7}$$

and $S = 1$ in the viewing rectangle $[0, 3] \times [0, 1]$.

b. Refer to the abbreviated form for S_N—namely, $S_N = \left(1 - \frac{1}{N+1} \right)$. By plotting the graphs of $y_1 = \left(1 - \frac{1}{x+1} \right)$ and $y_2 = 1$ in the viewing rectangle $[0, 50] \times [0, 1.1]$, verify graphically that $\lim_{N \to \infty} S_N = 1$

and thus the result $\sum_{n=1}^{\infty} \frac{1}{n(n+1)} = 1$, as obtained in Example 1b.

○ ○ ○

a. An example of a divergent infinite series is $\sum_{n=0}^{\infty} 1 = 1 + 1 + 1 + \cdots$. Show that the se-

ries is divergent by establishing the following:

(i) The nth partial sum of the infinite series is $S_n = n$.

(ii) $\lim_{n \to \infty} S_n = \infty$ so that $\{S_n\}$ is divergent and the desired result follows.

b. Since the *terms* of the infinite series in part (a) do not decrease, it is evident that the partial sums of the series must grow without bound. Therefore, it is not difficult to see that the infinite series cannot converge. Now consider the infinite series

$$\sum_{n=1}^{\infty} \frac{1}{n} = 1 + \frac{1}{2} + \frac{1}{3} + \cdots$$

Even though the terms of this series (called the *harmonic series*) approach zero as n goes to infinity, it can be shown that the harmonic series is divergent. Show that this result is intuitively true by establishing the following:

(i) Observe that

$$S_2 = 1 + \frac{1}{2} > \frac{1}{2} + \frac{1}{2} = 1$$

and

$$S_4 = 1 + \frac{1}{2} + \frac{1}{3} + \frac{1}{4} = S_2 + \frac{1}{3} + \frac{1}{4} > 1 + \left(\frac{1}{4} + \frac{1}{4}\right) = \frac{3}{2}$$

Using a similar argument, show that $S_8 > 4/2$ and $S_{16} > 5/2$. Conclude that in general

$$S_{2^n} > \frac{n+1}{2}$$

(ii) Explain why $S_1 < S_2 < S_3 < \cdots < S_n < \cdots$.

(iii) Use the results of (i) and (ii) to explain why the harmonic series is divergent.

Geometric Series

In general, it is no easy task to determine whether a given infinite series is convergent or divergent. It is an even more difficult problem to determine the sum of an infinite series that is known to be convergent. But there is an important and useful series whose sum, when it exists, is easy to find.

GEOMETRIC SERIES   A **geometric series** with ratio r is a series of the form

$$\sum_{n=0}^{\infty} ar^n = a + ar + ar^2 + ar^3 + \cdots + ar^n + \cdots \tag{5}$$

The ratio here refers to the ratio of two consecutive terms.

Note that, for this series, we begin the summation with $n = 0$ instead of $n = 1$. To determine the conditions under which the geometric series (5) converges and find its sum, let us consider the nth partial sum of the infinite series

$$S_n = a + ar + ar^2 + \cdots + ar^n$$

Multiplying both sides of the equation by r gives

$$rS_n = ar + ar^2 + ar^3 + \cdots + ar^{n+1}$$

Subtracting the second equation from the first yields

$$S_n - rS_n = a - ar^{n+1}$$
$$(1 - r)S_n = a(1 - r^{n+1})$$

or

$$S_n = \frac{a(1 - r^{n+1})}{1 - r}$$

provided $r \neq 1$. You are asked to show that if $|r| \geq 1$, then the series (5) diverges (exercise 34). On the other hand, observe that if $|r| < 1$ (that is, if $-1 < r < 1$), then

$$\lim_{n \to \infty} r^{n+1} = 0$$

For example, if $r = \frac{1}{2}$, then

$$\lim_{n \to \infty} r^{n+1} = \lim_{n \to \infty} \left(\frac{1}{2}\right)^{n+1} = \lim_{n \to \infty} \frac{1}{2^{n+1}} = 0$$

Using this fact, together with the properties of limits stated earlier, we see that

$$\lim_{n \to \infty} S_n = \lim_{n \to \infty} \frac{a(1 - r^{n+1})}{1 - r}$$

$$= \frac{a}{1 - r} \lim_{n \to \infty} (1 - r^{n+1})$$

$$= \frac{a}{1 - r}$$

These results are summarized in Theorem 1.

THEOREM 1

The geometric series

$$\sum_{n=0}^{\infty} ar^n = a + ar + ar^2 + \cdots$$

converges and its sum is $a/(1 - r)$; that is,

$$\sum_{n=0}^{\infty} ar^n = a + ar + ar^2 + \cdots = \frac{a}{1 - r} \qquad (6)$$

if $|r| < 1$. The series diverges if $|r| \geq 1$.

EXAMPLE 2 Show that each of the following infinite series is a geometric series, and find its sum if it is convergent.

a. $\displaystyle\sum_{n=0}^{\infty} \frac{1}{2^n}$ **b.** $\displaystyle\sum_{n=0}^{\infty} 3\left(\frac{5}{2}\right)^n$ **c.** $\displaystyle\sum_{n=1}^{\infty} 5\left(-\frac{3}{4}\right)^n$

Solution

a. Observe that

$$\frac{1}{2^n} = \left(\frac{1}{2}\right)^n$$

so that

$$\sum_{n=0}^{\infty} \frac{1}{2^n} = \sum_{n=0}^{\infty} \left(\frac{1}{2}\right)^n$$

is a geometric series with $a = 1$ and $r = \frac{1}{2}$. Since $|r| = \frac{1}{2} < 1$, we conclude that the series is convergent. Finally, using (6), we have

$$\sum_{n=0}^{\infty} \left(\frac{1}{2}\right)^n = 1 + \frac{1}{2} + \frac{1}{4} + \cdots = \frac{1}{1 - \dfrac{1}{2}} = 2$$

b. This is a geometric series with $a = 3$ and $r = \frac{5}{2}$. Since

$$|r| = \left|\frac{5}{2}\right| = \frac{5}{2} > 1$$

we deduce that the series is divergent.

c. The summation here begins with $n = 1$. However, we may rewrite the series as

$$\sum_{n=1}^{\infty} 5\left(-\frac{3}{4}\right)^n = 5\left(-\frac{3}{4}\right) + 5\left(-\frac{3}{4}\right)^2 + 5\left(-\frac{3}{4}\right)^3 + \cdots$$

$$= 5\left(-\frac{3}{4}\right)\left[1 + \left(-\frac{3}{4}\right) + \left(-\frac{3}{4}\right)^2 + \cdots\right]$$

$$= \sum_{n=0}^{\infty} \left(\frac{-15}{4}\right)\left(-\frac{3}{4}\right)^n$$

which is just a geometric series with $a = -\frac{15}{4}$ and $r = -\frac{3}{4}$. Since $|r| = |-\frac{3}{4}| = \frac{3}{4} < 1$, the series is convergent. Using (6), we find

$$\sum_{n=1}^{\infty} 5\left(-\frac{3}{4}\right)^n = \sum_{n=0}^{\infty} \left(-\frac{15}{4}\right)\left(-\frac{3}{4}\right)^n = \frac{-\dfrac{15}{4}}{1 - \left(-\dfrac{3}{4}\right)} = -\frac{\dfrac{15}{4}}{\dfrac{7}{4}} = -\frac{15}{7}$$

Properties of Infinite Series

The following properties of infinite series enable us to perform algebraic operations on convergent series.

PROPERTIES OF INFINITE SERIES	If $\sum\limits_{n=1}^{\infty} a_n$ and $\sum\limits_{n=1}^{\infty} b_n$ are convergent infinite series and c is a constant, then
	1. $\sum\limits_{n=1}^{\infty} ca_n = c \sum\limits_{n=1}^{\infty} a_n$
	2. $\sum\limits_{n=1}^{\infty} (a_n \pm b_n) = \sum\limits_{n=1}^{\infty} a_n \pm \sum\limits_{n=1}^{\infty} b_n$

Thus, we may multiply each term of a convergent series by a constant c, which results in a convergent series whose sum is c times the sum of the original series. We may also add (subtract) the corresponding terms of two convergent series, which gives a convergent series whose sum is the sum (difference) of the sums of the original series.

EXAMPLE 3   Find the sum of the following series if it exists:

$$\sum_{n=0}^{\infty} \frac{2 \cdot 3^n - 2^n}{5^n} = 1 + \frac{4}{5} + \frac{14}{25} + \cdots$$

Solution   We can write

$$\sum_{n=0}^{\infty} \frac{2 \cdot 3^n - 2^n}{5^n} = \sum_{n=0}^{\infty} \left(\frac{2 \cdot 3^n}{5^n} - \frac{2^n}{5^n} \right)$$

Now observe that

$$\sum_{n=0}^{\infty} \frac{2 \cdot 3^n}{5^n} \quad \text{and} \quad \sum_{n=0}^{\infty} \frac{2^n}{5^n}$$

are convergent geometric series with ratios $r = \frac{3}{5} < 1$ and $r = \frac{2}{5} < 1$, respectively. Therefore, using property 2 of infinite series, we have

$$\sum_{n=0}^{\infty} \frac{2 \cdot 3^n - 2^n}{5^n} = \sum_{n=0}^{\infty} \frac{2 \cdot 3^n}{5^n} - \sum_{n=0}^{\infty} \frac{2^n}{5^n}$$

$$= 2 \sum_{n=0}^{\infty} \frac{3^n}{5^n} - \sum_{n=0}^{\infty} \frac{2^n}{5^n} \qquad \text{(Using property 1 on the first sum)}$$

$$= 2 \sum_{n=0}^{\infty} \left(\frac{3}{5} \right)^n - \sum_{n=0}^{\infty} \left(\frac{2}{5} \right)^n$$

$$= 2 \left(\frac{1}{1 - \dfrac{3}{5}} \right) - \left(\frac{1}{1 - \dfrac{2}{5}} \right)$$

$$= 2 \left(\frac{5}{2} \right) - \frac{5}{3}$$

$$= \frac{10}{3}$$

Applications of Geometric Series

We now consider some applications of geometric series.

EXAMPLE 4 Find the rational number that has the repeated decimal representation 0.222. . . .

Solution By definition, the decimal representation

$$0.222\ldots = \frac{2}{10} + \frac{2}{100} + \frac{2}{1000} + \cdots$$

$$= \frac{2}{10}\left(1 + \frac{1}{10} + \frac{1}{100} + \cdots\right)$$

$$= \sum_{n=0}^{\infty}\left(\frac{2}{10}\right)\left(\frac{1}{10}\right)^n$$

which is a geometric series with $a = \frac{2}{10}$ and $r = \frac{1}{10}$. Since $|r| = r = \frac{1}{10} < 1$, the series converges. In fact, using formula (6), we have

$$0.222\ldots = \frac{\left(\dfrac{2}{10}\right)}{1 - \dfrac{1}{10}} = \frac{\dfrac{2}{10}}{\dfrac{9}{10}} = \frac{2}{9}$$

○ ○ ○

The following example illustrates a phenomenon in economics known as the **multiplier effect**.

EXAMPLE 5 Suppose that the average wage earner saves 10% of his or her take-home pay and spends the other 90%. Estimate the impact that a proposed $20 billion tax cut will have on the economy over the long run in terms of the additional spending generated.

Solution Of the $20 billion received by the original beneficiaries of the proposed tax cut, $(0.9)(20)$ billion dollars will be spent. Of the $(0.9)(20)$ billion dollars reinjected into the economy, 90% of it, or $(0.9)(0.9)(20)$ billion dollars, will find its way into the economy again. This process will go on ad infinitum, so this one-time proposed tax cut will result in additional spending over the years in the amount of

$$(0.9)(20) + (0.9)^2(20) + (0.9)^3(20) + \cdots = (0.9)(20)[1 + 0.9 + 0.9^2 + 0.9^3 + \cdots]$$

$$= 18\left[\frac{1}{1 - 0.9}\right]$$

$$= 180$$

or $180 billion.

○ ○ ○

A **perpetuity** is a sequence of payments made at regular time intervals and continuing on forever. The **capital value** of a perpetuity is the sum of the present values of all future payments. The following example illustrates these concepts.

EXAMPLE 6   The Robinson family wishes to create a scholarship fund at a college. If a scholarship in the amount of $5000 is to be awarded on an annual basis beginning next year, find the amount of the endowment they are required to make now. Assume that this fund will earn interest at a rate of 10% per year compounded continuously.

Solution   The amount of the endowment, A, is given by the sum of the present values of the amounts awarded annually in perpetuity. Now, the present value of the amount of the first award is equal to

$$5000e^{-0.1(1)}$$

(see Section 6.7). The present value of the amount of the second award is

$$5000e^{-0.1(2)}$$

and so on. Continuing, we see that the present value of the amount of the nth award is

$$5000e^{-0.1(n)}$$

Therefore, the amount of the endowment is

$$A = 5000e^{-0.1(1)} + 5000e^{-0.1(2)} + \cdots + 5000e^{-0.1(n)} + \cdots$$

To find the sum of the infinite series on the right-hand side, let

$$r = e^{-0.1}$$

Then

$$A = 5000\, r^1 + 5000\, r^2 + \cdots + 5000\, r^n + \cdots$$
$$= 5000\, r(1 + r + r^2 + \cdots + r^n + \cdots)$$

The series inside the brackets is a geometric series with $r = e^{-0.1} \approx 0.905 < 1$, so that, using (6), we find

$$A = 5000\, r\left(\frac{1}{1-r}\right) = \frac{5000e^{-0.1}}{1 - e^{-0.1}} = 47{,}541.66$$

Thus, the amount of the endowment is $47,541.66.   ◦ ◦ ◦

Our final example is an application of a geometric series in the field of medicine.

EXAMPLE 7   A patient is to be given 5 units of a certain drug daily for an indefinite period of time. For this particular drug, it is known that the fraction of a dose that remains in the patient's body after t days is given by $e^{-0.3t}$. Determine the residual amount of the drug that may be expected to be in the patient's body after an extended treatment.

Solution The amount of the drug in the patient's body one day after the first dose is administered, and prior to administration of the second dose, is $5e^{-0.3}$ units. The amount of the drug in the patient's body two days later, and prior to administration of the third dose, consists of the residuals from the first two doses. Of the first dose, $5e^{-(0.3)2}$ units of the drug are left in the patient's body, and of the second dose, $5e^{-0.3}$ units of the drug are left. Thus, the amount of the drug two days later is given by

$$5e^{-0.3} + 5e^{-0.3(2)}$$

units. Continuing, we see that the amount of the drug left in the patient's body in the long run, and prior to administration of a fresh dose, is given by

$$R = 5e^{-0.3} + 5e^{-0.3(2)} + 5e^{-0.3(3)} + \cdots$$

To find the sum of the infinite series, we let

$$r = e^{-0.3}$$

Then

$$A = 5r + 5r^2 + 5r^3 + \cdots = 5r(1 + r + r^2 + \cdots)$$
$$= \frac{5r}{1 - r} = \frac{5e^{-0.3}}{1 - e^{-0.3}} \approx 14.29$$

Therefore, after an extended treatment, the residual amount of drug in the patient's body is approximately 14.29 units. ○ ○ ○

SELF-CHECK EXERCISES 11.3

1. Determine whether the geometric series

$$\sum_{n=0}^{\infty} 5\left(-\frac{1}{3}\right)^n = 5 - 5\left(\frac{1}{3}\right) + 5\left(\frac{1}{9}\right) - \cdots$$

is convergent or divergent. If it is convergent, find its sum.

2. Suppose that the average wage earner in a certain country saves 12% of his or her take-home pay and spends the other 88%. Estimate the impact that a proposed $10 billion tax cut will have on the economy over the long run due to the additional spending generated.

Solutions to Self-Check Exercises 11.3 can be found on page 787.

11.3 EXERCISES

In exercises 1–4, find the Nth partial sum of the infinite series and evaluate its limit to determine whether the series converges or diverges. If the series is convergent, find its sum.

1. $\displaystyle\sum_{n=1}^{\infty} (-2)^n$

2. $\displaystyle\sum_{n=1}^{\infty} \left(\frac{1}{n+1} - \frac{1}{n+2}\right)$

3. $\displaystyle\sum_{n=1}^{\infty} \frac{1}{n^2 + 3n + 2}$

$$\left[\textit{Hint:}\; \frac{1}{n^2 + 3n + 2} = \frac{1}{n+1} - \frac{1}{n+2}.\right]$$

4. $\displaystyle\sum_{n=2}^{\infty} \left(\frac{1}{\ln n} - \frac{1}{\ln(n+1)}\right)$

In exercises 5–14, determine whether the geometric series converges or diverges. If it converges, find its sum.

5. $\displaystyle\sum_{n=0}^{\infty} \left(\frac{1}{3}\right)^n$

6. $\displaystyle\sum_{n=0}^{\infty} 4\left(-\frac{2}{3}\right)^n$

7. $\displaystyle\sum_{n=0}^{\infty} 2(1.01)^n$

8. $\displaystyle\sum_{n=0}^{\infty} 3(0.9)^n$

9. $\displaystyle\sum_{n=0}^{\infty} \frac{(-2)^n}{3^n}$

10. $\displaystyle\sum_{n=0}^{\infty} \frac{3}{2^n}$

11. $\displaystyle\sum_{n=0}^{\infty} \frac{2^n}{3^{n+2}}$

12. $\displaystyle\sum_{n=0}^{\infty} \frac{3^{n+1}}{4^{n-1}}$

13. $\displaystyle\sum_{n=0}^{\infty} e^{-0.2n}$

14. $\displaystyle\sum_{n=0}^{\infty} 2e^{-0.1n}$

In exercises 15–22, determine whether the series converges or diverges. If it converges, find its sum.

15. $8 + 4 + \dfrac{1}{2} + \dfrac{1}{4} + \dfrac{1}{8} + \cdots$

16. $1 + 0.2 + 0.04 + 0.0016 + \cdots$

17. $3 - \dfrac{1}{3} + \dfrac{1}{9} - \dfrac{1}{27} + \cdots$

18. $5 - 1.01 + (1.01)^2 - (1.01)^3 + \cdots$

19. $\displaystyle\sum_{n=0}^{\infty} \frac{3 + 2^n}{3^n}$

20. $\displaystyle\sum_{n=0}^{\infty} \frac{2^n - 3^n}{4^n}$

21. $\displaystyle\sum_{n=0}^{\infty} \frac{3 \cdot 2^n + 4^n}{3^n}$

22. $\displaystyle\sum_{n=0}^{\infty} \frac{2 \cdot 3^n - 3 \cdot 5^n}{7^n}$

In exercises 23–26, express each of the given decimals as a rational number.

23. $0.3333\ldots$

24. $0.121212\ldots$

25. $1.213213213\ldots$

26. $6.2314314314\ldots$

27. Effect of a Tax Cut on Spending Suppose that the average wage earner saves 9% of his or her take-home pay and spends the other 91%. Estimate the impact that a proposed $30 billion tax cut will have on the economy over the long run due to the additional spending generated.

28. A Bouncing Ball A ball is dropped from a height of 10 meters. After hitting the ground, it rebounds to a height of 5 meters and then continues to rebound at one-half of its former height thereafter. Find the total distance traveled by the ball before it comes to rest.

29. Endowments Hal Corporation wants to establish a fund to provide the art center of a large metropolitan area with an annual grant of $250,000 beginning next year. If the fund will earn interest at a rate of 10% per year compounded continuously, find the amount of endowment the corporation must make at this time.

30. Capital Value of a Perpetuity Find a formula for the capital value of a perpetuity involving payments of P dollars each, paid at the end of each of m periods per year into a fund that earns interest at the nominal rate of r% per year compounded m times a year by verifying the following:

a. The present value of the nth payment is

$$P\left(1 + \frac{r}{m}\right)^{-n}$$

b. The capital value is

$$A = P\left(1 + \frac{r}{m}\right)^{-1} + P\left(1 + \frac{r}{m}\right)^{-2}$$
$$+ P\left(1 + \frac{r}{m}\right)^{-3} + \cdots$$

c. Using the fact that the series in (b) is a geometric series, its sum is

$$A = \frac{mP}{r}$$

31. Capital Value of a Perpetuity Find a formula for the capital value of a perpetuity involving payments of P dollars paid at the end of each investment period into a fund that earns interest at the rate of $r\%$ per year compounded continuously.

[*Hint:* Study exercise 30.]

32. Residual Drug in the Bloodstream Ten units of a certain drug are administered to a patient on a daily basis. The fraction of this drug that remains in the patient's bloodstream after t days is given by

$$f(t) = e^{(-1/4)t}$$

Determine the residual amount of the drug in the patient's bloodstream after a period of extended treatment with the drug.

33. Residual Drug in the Bloodstream Suppose that a dose of C units of a certain drug is administered to a patient and that the fraction of the dose remaining in the patient's bloodstream t hours after the dose is administered is given by Ce^{-kt}, where k is a positive constant.

a. Show that the residual concentration of the drug in the bloodstream after extended treatment when a dose of C units is administered at intervals of t hours is given by

$$R = \frac{Ce^{-kt}}{1 - e^{-kt}}$$

b. If the highest concentration of this particular drug that is considered safe is S units, find the minimal time that must exist between doses.

[*Hint:* $C + R \leq S$.]

34. Let

$$\sum_{n=0}^{\infty} ar^n = a + ar + ar^2 + ar^3 + \cdots + ar^n + \cdots$$

be a geometric series with common ratio r. Show that if $|r| \geq 1$, then the series diverges.

SOLUTIONS TO SELF-CHECK EXERCISES 11.3

1. This is a geometric series with $r = -\frac{1}{3}$ and $a = 5$. Since $0 < |-\frac{1}{3}| < 1$, we see that the series is convergent. Its sum is

$$\frac{5}{1 - \left(-\frac{1}{3}\right)} = \frac{5}{\frac{4}{3}} = \frac{15}{4} = 3\frac{3}{4}$$

2. Of the $10 billion received by the original beneficiaries of the proposed cut, $(0.88)(10)$ billion dollars will be spent. Of the $(0.88)(10)$ billion dollars reinjected into the economy, 88% of it, or $(0.88)(0.88)(10)$ billion dollars, will find its way into the economy again. This process will go on ad infinitum, so this one-time proposed tax cut will result in additional spending over the years in the amount of

$$(0.88)(10) + (0.88)^2(10) + (0.88)^3(10) + \cdots$$
$$= (0.88)(10)[1 + 0.88 + (0.88)^2 + \cdots]$$
$$= 8.8\left[\frac{1}{1 - 0.88}\right]$$
$$= 73.3$$

or $73.3 billion.

11.4 POWER SERIES AND TAYLOR SERIES

Power Series and Intervals of Convergence

Recall that one of our goals in this chapter is to see what happens when the number of terms of a Taylor polynomial is allowed to increase without bound. In other words, we wish to study the expression

$$f(a) + f'(a)(x - a) + \frac{f''(a)}{2!}(x - a)^2 + \cdots + \frac{f^{(n)}(a)}{n!}(x - a)^n + \cdots \quad \text{(7)}$$

called the **Taylor series of $f(x)$ at $x = a$.** If the series (7) is truncated after $(n + 1)$ terms, the result is a Taylor polynomial of degree n of $f(x)$ at $x = a$ (see Section 11.1). Note that when $a = 0$, (7) reduces to

$$f(0) + f'(0)x + \frac{f''(0)}{2!}x^2 + \cdots + \frac{f^{(n)}(0)}{n!}x^n + \cdots \quad \text{(8)}$$

which is referred to as the **Maclaurin series of $f(x)$.** Thus, the Maclaurin series is the special case of the Taylor series when $a = 0$.

Observe that the Taylor series (7) has the form

$$\sum_{n=0}^{\infty} a_n(x - x_0)^n = a_0 + a_1(x - x_0) + a_2(x - x_0)^2 + \cdots$$
$$+ a_n(x - x_0)^n + \cdots \quad \text{(9)}$$

In fact, comparing the Taylor series (7) with the infinite series (9), we see that

$$a_0 = f(a), \quad a_1 = f'(a), \quad a_2 = \frac{f''(a)}{2!}, \quad \ldots, \quad a_n = \frac{f^{(n)}(a)}{n!}, \quad \ldots$$

The infinite series (9) is called a **power series centered at $x = a$.** Thus, the Taylor series is just a power series with coefficients that involve the values of some function f and its derivatives at the point $x = a$.

Let us examine some important properties of the power series given in (9). Observe that when x is assigned a value, then this power series becomes an infinite series with constant terms. Accordingly, we can determine, at least theoretically, whether this infinite series converges or diverges. Now the totality of all values of x for which this power series *converges* comprises a set of points in the real line called the **interval of convergence** of the power series. This observation suggests that we may view the power series (9) as a function f whose domain coincides with the interval of convergence of the series and whose functional values are the sums of the infinite series obtained by allowing x to take on all values in the interval of convergence. In this case, we also say the function f is **represented by the power series** (9).

Given such a power series, how does one determine its interval of convergence? The following theorem, whose proof we will omit, provides the answer to this question.

THEOREM 2

Suppose that we are given the power series

$$\sum_{n=0}^{\infty} a_n(x - a)^n$$

Let

$$R = \lim_{n \to \infty} \left| \frac{a_n}{a_{n+1}} \right|$$

1. If $R = 0$, the series converges only for $x = a$.
2. If $0 < R < \infty$, the series converges for x in the interval $(a - R, a + R)$ and diverges for x outside this interval (Figure 11.4).

Figure 11.4

3. If $R = \infty$, the series converges for all x.

REMARKS

1. The domain of convergence of a power series is an interval called the *interval of convergence.* This interval of convergence is determined by R, the **radius of convergence.** Depending on whether $R = 0, 0 < R < \infty$, or $R = \infty$, the "interval" of convergence may just be the degenerate interval consisting of the point a, a bona fide interval $(a - R, a + R)$, or the entire real line.

2. As mentioned earlier, a power series, inside its interval of convergence, represents a function f whose domain of definition coincides with the interval of convergence; that is,

$$f(x) = \sum_{n=0}^{\infty} a_n(x - a)^n, \qquad x \in (a - R, a + R)$$

3. The power series may or may not converge at an endpoint. In general, it is difficult to determine whether a power series is convergent at an endpoint. For this reason, we will restrict our attention to points inside the interval of convergence of a power series.   ◦ ◦ ◦

EXAMPLE I Find the radius of convergence and the interval of convergence of the power series

$$\sum_{n=0}^{\infty} \frac{(x - 1)^n}{2^n}$$

Show that $f(2)$ exists and find its value, where

$$f(x) = \sum_{n=0}^{\infty} \frac{(x - 1)^n}{2^n}$$

Solution Since $a_n = 1/2^n$, we have

$$R = \lim_{n \to \infty} \left| \frac{a_n}{a_{n+1}} \right| = \lim_{n \to \infty} \left| \frac{\dfrac{1}{2^n}}{\dfrac{1}{2^{n+1}}} \right|$$

$$= \lim_{n \to \infty} \frac{2^{n+1}}{2^n} = \lim_{n \to \infty} 2 = 2$$

Therefore, bearing in mind that $a = 1$, we see that the interval of convergence of the given series is $(-1, 3)$. Thus, in the interval $(-1, 3)$, the given power series defines a function

$$f(x) = \sum_{n=0}^{\infty} \frac{(x-1)^n}{2^n}, \qquad x \in (-1, 3)$$

Since $2 \in (-1, 3)$, $f(2)$ exists. In fact,

$$f(2) = \sum_{n=0}^{\infty} \frac{(2-1)^n}{2^n} = \sum_{n=0}^{\infty} \frac{1}{2^n}$$

which is a geometric series with $a = 1$ and $r = \frac{1}{2}$, so that

$$f(2) = \frac{1}{1 - \dfrac{1}{2}} = 2$$

○ ○ ○

EXAMPLE 2 Find the interval of convergence of each of the following power series:

a. $\displaystyle\sum_{n=0}^{\infty} n^3(x+2)^n$ **b.** $\displaystyle\sum_{n=0}^{\infty} n!(x-1)^n$ **c.** $\displaystyle\sum_{n=0}^{\infty} \frac{x^n}{n!}$

Solution
a. Here $a_n = n^3$, so

$$R = \lim_{n \to \infty} \left| \frac{a_n}{a_{n+1}} \right| = \lim_{n \to \infty} \frac{n^3}{(n+1)^3}$$

$$= \lim_{n \to \infty} \frac{1}{\left(1 + \dfrac{1}{n}\right)^3} = 1 \qquad \text{(Dividing numerator and denominator by } n^3\text{)}$$

Since $a = -2$, we find that the interval of convergence of the series is $(-3, -1)$.

b. Here $a_n = n!$, so

$$R = \lim_{n \to \infty} \left| \frac{a_n}{a_{n+1}} \right| = \lim_{n \to \infty} \left| \frac{n!}{(n+1)!} \right|$$

$$= \lim_{n \to \infty} \frac{1}{n+1} = 0$$

Therefore, the series converges only at the point $a = 1$.

c. Here $a_n = 1/n!$ and

$$R = \lim_{n\to\infty}\left|\frac{a_n}{a_{n+1}}\right| = \lim_{n\to\infty}\left|\frac{\dfrac{1}{n!}}{\dfrac{1}{(n+1)!}}\right| = \lim_{n\to\infty}(n+1) = \infty$$

Therefore, the interval of convergence of the series is $(-\infty, \infty)$; that is, it converges for any value of x.   ◑ ◑ ◑

Suppose the power series $\displaystyle\sum_{n=0}^{\infty} a_n(x-a)^n$ has a radius of convergence R. What can you deduce about the radius of convergence of the series $\displaystyle\sum_{n=0}^{\infty} a_n(x-a)^{2n}$?

Finding a Taylor Series

Theorem 2 guarantees that a power series represents a function whose domain is precisely the interval of convergence of the series. Let us examine the converse problem. More precisely, suppose we are given a function f that has derivatives of *all* orders at a point $x = a$. Can we find a power series centered at $x = a$ that represents the function f; that is, can we find a power series that *coincides* with f at each point x in some suitable subset of the real line? The answer to this question is supplied by Theorem 3.

THEOREM 3

Suppose that the function f has derivatives of all orders in an interval I containing the point $x = a$. Then

$$f(x) = \sum_{n=0}^{\infty} \frac{f^{(n)}(a)}{n!}(x-a)^n \qquad\qquad \textbf{(10)}$$

if and only if $\lim\limits_{n\to\infty} R_n(x) = 0$ for every x in I, where $R_n(x) = f(x) - P_n(x)$, as defined on page 761.

Theorem 3 states that under the stated hypothesis a function f has a power series representation that is just the Taylor series of f at $x = a$. Observe that this representation is restricted to the interval of convergence of the series. To derive this result, we need the following theorem, which we state without proof.

THEOREM 4

Suppose that

$$f(x) = \sum_{n=0}^{\infty} a_n(x - a)^n, \qquad x \in (a - R, a + R)$$

Then f may be differentiated term by term and

$$f'(x) = \sum_{n=1}^{\infty} na_n(x - a)^{n-1}, \qquad x \in (a - R, a + R)$$

$$= a_1 + 2a_2(x - a) + 3a_3(x - a)^2 + \cdots$$
$$+ na_n(x - a)^{n-1} + \cdots$$

Thus, the derivative of a function represented by a power series may be found by differentiating the power series term by term. The power series representing the derivative f' of f has the same interval of convergence as f.

Now suppose that f is represented by a power series centered about $x = a$; that is,

$$f(x) = a_0 + a_1(x - a) + a_2(x - a)^2 + a_3(x - a)^3 + \cdots$$
$$+ a_n(x - a)^n + \cdots$$

Then, applying Theorem 4 repeatedly, we find

$$f'(x) = a_1 + 2a_2(x - a) + 3a_3(x - a)^2 + \cdots + na_n(x - a)^{n-1} + \cdots$$
$$f''(x) = 2a_2 + 3 \cdot 2a_3(x - a) + 4 \cdot 3a_4(x - a)^2 + \cdots$$
$$+ n(n - 1)a_n(x - a)^{n-2} + \cdots$$
$$f'''(x) = 3 \cdot 2a_3 + 4 \cdot 3 \cdot 2a_4(x - a) + \cdots$$
$$+ n(n - 1)(n - 2)a_n(x - a)^{n-3} + \cdots$$
$$= 3!a_3 + 4!a_4(x - a) + \cdots + n(n - 1)(n - 2)a_n(x - a)^{n-3} + \cdots$$
$$\vdots$$
$$f^{(n)}(x) = n!a_n + (n + 1)!a_{n+1}(x - a) + \cdots$$

Evaluating $f(x)$ and each of these derivatives at $x = a$ yields

$$f(a) = a_0$$
$$f'(a) = a_1$$
$$f''(a) = 2a_2 = 2!a_2$$
$$f'''(a) = 3 \cdot 2a_3 = 3!a_3$$
$$\vdots$$
$$f^{(n)}(a) = n!a_n$$

Thus,

$$a_0 = f(a), \quad a_1 = f'(a), \quad a_2 = \frac{f''(a)}{2!},$$

$$a_3 = \frac{f'''(a)}{3!}, \quad \ldots, \quad a_n = \frac{f^{(n)}(a)}{n!}, \quad \ldots$$

as we set out to show.

EXAMPLE 3   Find the Taylor series of the function $f(x) = \dfrac{1}{x-1}$ at $x = 2$.

Solution   Here $a = 2$, and

$$f(x) = \frac{1}{x-1}$$

$$f'(x) = -\frac{1}{(x-1)^2}$$

$$f''(x) = \frac{2}{(x-1)^3}$$

$$f'''(x) = -\frac{3 \cdot 2}{(x-1)^4}$$

$$\vdots$$

$$f^{(n)}(x) = (-1)^n \frac{n!}{(x-1)^{n+1}}$$

$$\vdots$$

so that

$$f(2) = 1, \quad f'(2) = -1, \quad f''(2) = 2,$$
$$f'''(2) = -6 = -3!, \quad \ldots, \quad f^{(n)}(2) = (-1)^n n!, \quad \ldots$$

Therefore, the required Taylor series is

$$\frac{1}{x-1} = f(2) + f'(2)(x-2) + \frac{f''(2)}{2!}(x-2)^2 + \frac{f'''(2)}{3!}(x-2)^3 + \cdots$$

$$+ \frac{f^{(n)}(2)}{n!}(x-2)^n + \cdots$$

$$= 1 - (x-2) + \frac{2!}{2!}(x-2)^2 - \frac{3!}{3!}(x-2)^3 + \cdots$$

$$+ (-1)^n \frac{n!}{n!}(x-2)^n + \cdots$$

$$= 1 - (x-2) + (x-2)^2 - (x-2)^3 + \cdots + (-1)^n(x-2)^n + \cdots$$

$$= \sum_{n=0}^{\infty} (-1)^n (x-2)^n$$

To find the radius of convergence of the series, we compute

$$R = \lim_{n \to \infty} \left| \frac{a_n}{a_{n+1}} \right| = \lim_{n \to \infty} \left| \frac{(-1)^n}{(-1)^{n+1}} \right|$$

$$= \lim_{n \to \infty} 1 = 1$$

Thus, the series converges in the interval $(1, 3)$. By Theorem 3, we see that the function $f(x) = 1/(x - 1)$ is represented by the Taylor series in the interval $(1, 3)$.   ○ ○ ○

 The representation of the function $f(x) = 1/(x - 1)$ by its Taylor series at $x = 2$ is valid only in the interval $(1, 3)$, in spite of the fact that f itself has a domain that is the set of all real numbers except $x = 1$. This serves to remind us of the local nature of this representation.

EXPLORING WITH TECHNOLOGY

Refer to Example 3, where it was shown that the Taylor series at $x = 2$ representing $f(x) = 1/(x - 1)$ is

$$P(x) = \sum_{n=0}^{\infty} (-1)^n (x - 2)^n = 1 - (x - 2) + (x - 2)^2 - (x - 2)^3 + \cdots$$

for $1 < x < 3$. This means that if c is any number satisfying $1 < c < 3$, then $f(c) = 1/(c - 1) = P(c)$. In particular, this means that $\lim_{n \to \infty} |P_n(c) - f(c)| = 0$, where $\{P_n(x)\}$ is the sequence of partial sums of $P(x)$.

a. Plot the graphs of f, P_0, P_1, P_2, P_3, P_4, P_5, and P_6 on the same set of axes using the viewing rectangle $[2.5, 3.1] \times [-0.1, 1.1]$.

b. Do the results of part (a) give a visual confirmation of the statement $\lim_{n \to \infty} P_n(c) = f(c)$ for the special case where $c = 2.8$? Explain what happens when $c = 3$.

○ ○ ○

EXAMPLE 4   Find the Taylor series of the function $f(x) = \ln x$ at $x = 1$.

Solution   Here $a = 1$, and

$$f(x) = \ln x$$

$$f'(x) = \frac{1}{x}$$

$$f''(x) = -\frac{1}{x^2}$$

$$f'''(x) = \frac{2}{x^3} = \frac{2!}{x^3}$$

$$f^{(4)}(x) = -\frac{3 \cdot 2}{x^4} = -\frac{3!}{x^4}$$

$$\vdots$$

$$f^{(n)}(x) = (-1)^{n+1}\frac{(n-1)!}{x^n}$$

so that

$$f(1) = 0, \quad f'(1) = 1, \quad f''(1) = -1, \quad f'''(1) = 2!,$$
$$f^{(4)}(1) = -3!, \quad \ldots, \quad f^{(n)}(1) = (-1)^{n+1}(n-1)!, \quad \ldots$$

Therefore, the required Taylor series is

$$f(x) = \ln x = f(1) + f'(1)(x-1) + \frac{f''(1)}{2!}(x-1)^2 + \frac{f'''(1)}{3!}(x-1)^3$$

$$+ \cdots + \frac{f^{(n)}(1)}{n!}(x-1)^n + \cdots$$

$$= (x-1) - \frac{1}{2!}(x-1)^2 + \frac{2!}{3!}(x-1)^3 - \frac{3!}{4!}(x-1)^4 + \cdots$$

$$+ (-1)^{n+1}\frac{(n-1)!}{n!}(x-1)^n + \cdots$$

$$= (x-1) - \frac{1}{2}(x-1)^2 + \frac{1}{3}(x-1)^3 - \frac{1}{4}(x-1)^4 + \cdots$$

$$+ \frac{(-1)^{n+1}}{n}(x-1)^n + \cdots$$

$$= \sum_{n=1}^{\infty} \frac{(-1)^{n+1}}{n}(x-1)^n$$

Since

$$R = \lim_{n \to \infty}\left|\frac{a_n}{a_{n+1}}\right| = \lim_{n \to \infty}\left|\frac{\frac{(-1)^{n+1}}{n}}{\frac{(-1)^{n+2}}{n+1}}\right|$$

$$= \lim_{n \to \infty}\frac{n+1}{n} = \lim_{n \to \infty}\left(1 + \frac{1}{n}\right) = 1$$

we see that the power series representation of $f(x) = \ln x$ is valid in the interval $(0, 2)$. It can be shown that the representation is valid at $x = 2$ as well.

○ ○ ○

EXAMPLE 5 Find the power series representation about $x = 0$ for the function $f(x) = e^x$.

Solution Here $a = 0$, and since

$$f(x) = f'(x) = f''(x) = \cdots = e^x$$

we see that

$$f(0) = f'(0) = f''(0) = \cdots = e^0 = 1$$

Therefore, the required power series (Taylor series) is

$$f(x) = e^x = f(0) + f'(0)(x - 0) + \frac{f''(0)}{2!}(x - 0)^2 + \cdots + \frac{f^{(n)}}{n!}(x - 0)^n + \cdots$$

$$= 1 + x + \frac{x^2}{2!} + \frac{x^3}{3!} + \cdots + \frac{x^n}{n!} + \cdots$$

$$= \sum_{n=0}^{\infty} \frac{x^n}{n!}$$

Since

$$R = \lim_{n \to \infty} \left| \frac{a_n}{a_{n+1}} \right| = \lim_{n \to \infty} \left| \frac{\dfrac{1}{n!}}{\dfrac{1}{(n+1)!}} \right|$$

$$= \lim_{n \to \infty} (n + 1) = \infty$$

we see that the series representation is valid for all x.   ○ ○ ○

SELF-CHECK EXERCISES 11.4

1. Find the interval of convergence of the power series $\displaystyle\sum_{n=0}^{\infty} \frac{x^n}{n^2 + 1}$ (disregard the endpoints).

2. Find the Taylor series of the function $f(x) = e^{-x}$ at $x = 1$ and determine its interval of convergence.

Solutions to Self-Check Exercises 11.4 can be found on page 797.

11.4 EXERCISES

In exercises 1–20, find the radius of convergence and the interval of convergence of each of the given power series.

1. $\displaystyle\sum_{n=0}^{\infty} (x - 1)^n$

2. $\displaystyle\sum_{n=0}^{\infty} \left(\frac{x}{2} \right)^n$

3. $\displaystyle\sum_{n=1}^{\infty} n^2 x^n$

4. $\displaystyle\sum_{n=0}^{\infty} \frac{(n + 1)(x + 2)^n}{2^n}$

5. $\displaystyle\sum_{n=0}^{\infty} \frac{(-1)^n x^n}{4^n}$

6. $\displaystyle\sum_{n=0}^{\infty} \frac{(2x)^n}{3^n}$

7. $\displaystyle\sum_{n=0}^{\infty} \frac{(x - 1)^n}{n! 2^n}$

8. $\displaystyle\sum_{n=0}^{\infty} (2n)! x^n$

9. $\displaystyle\sum_{n=0}^{\infty} \frac{(-1)^n n! (x + 2)^n}{2^n}$

10. $\displaystyle\sum_{n=2}^{\infty} \frac{x^n}{n(n + 1)}$

11. $\displaystyle\sum_{n=2}^{\infty} \frac{(x + 3)^n}{(n + 1)^2}$

12. $\displaystyle\sum_{n=0}^{\infty} \frac{n!(x + 1)^n}{(3n)!}$

13. $\displaystyle\sum_{n=1}^{\infty} \frac{2n(x-3)^n}{(n+1)!}$

14. $\displaystyle\sum_{n=0}^{\infty} \frac{(-2x)^{2n}}{4^n(n+1)}$

15. $\displaystyle\sum_{n=1}^{\infty} \frac{n(-2x)^n}{n+1}$

16. $\displaystyle\sum_{n=0}^{\infty} \frac{x^{2n+1}}{(2n+1)!}$

17. $\displaystyle\sum_{n=0}^{\infty} \frac{n!(x+1)^n}{3^n}$

18. $\displaystyle\sum_{n=0}^{\infty} (n+1)(x+2)^n$

19. $\displaystyle\sum_{n=0}^{\infty} \frac{n^3(x-3)^n}{3^n}$

20. $\displaystyle\sum_{n=0}^{\infty} \frac{(-1)^{n+1}(x-2)^n}{n2^n}$

23. $f(x) = \dfrac{1}{x+1}; x = 2$

24. $f(x) = \dfrac{1}{1-x}; x = 0$

25. $f(x) = \dfrac{1}{1-x}; x = 2$

26. $f(x) = \ln(x+1); x = 0$

27. $f(x) = \sqrt{x}; x = 1$

28. $f(x) = \sqrt{1-x}; x = 0$

29. $f(x) = e^{2x}; x = 0$

30. $f(x) = e^{2x}; x = 1$

31. $f(x) = \dfrac{1}{\sqrt{x+1}}; x = 0$

32. $f(x) = \sqrt{x+1}; x = 1$

In exercises 21–32, find the Taylor series of each of the given functions at the indicated point and give its radius and interval of convergence (disregard the endpoints).

21. $f(x) = \dfrac{1}{x}; x = 1$

22. $f(x) = \dfrac{1}{x+1}; x = 0$

SOLUTIONS TO SELF-CHECK EXERCISES 11.4

1. We first find the radius of convergence of the power series. Since $a_n = \dfrac{1}{n^2+1}$, we have

$$R = \lim_{n\to\infty} \left| \frac{a_n}{a_{n+1}} \right| = \lim_{n\to\infty} \frac{\dfrac{1}{n^2+1}}{\dfrac{1}{(n+1)^2+1}}$$

$$= \lim_{n\to\infty} \frac{(n+1)^2+1}{n^2+1}$$

$$= \lim_{n\to\infty} \frac{\left(1+\dfrac{1}{n}\right)^2 + \dfrac{1}{n^2}}{1+\dfrac{1}{n^2}} \quad \text{(Dividing numerator and denominator by } n^2\text{)}$$

$$= 1$$

Therefore, the interval of convergence of the series is $(-1, 1)$,

2. Here $a = 1$ and

$$f(x) = e^{-x}$$
$$f'(x) = -e^{-x}$$
$$f''(x) = e^{-x}$$
$$f'''(x) = -e^{-x}$$
$$\vdots$$
$$f^{(n)}(x) = (-1)^n e^{-x}$$

so that

$$f(1) = e^{-1}, \quad f'(1) = -e^{-1}, \quad f''(1) = e^{-1}$$
$$f'''(1) = -e^{-1}, \quad \ldots, \quad f^{(n)}(1) = (-1)^n e^{-1}$$

Therefore, the required Taylor series is

$$e^{-x} = f(1) + f'(1)(x - 1) + \frac{f''(1)}{2!}(x - 1)^2$$

$$+ \frac{f'''(1)}{3!}(x - 1)^3 + \cdots + \frac{f^{(n)}(1)}{n!}(x - 1)^n + \cdots$$

$$= e^{-1} - \frac{e^{-1}}{1!}(x - 1) + \frac{e^{-1}}{2!}(x - 1)^2$$

$$- \frac{e^{-1}}{3!}(x - 1)^3 + \cdots + \frac{(-1)^n e^{-1}}{n!}(x - 1)^n + \cdots$$

$$= \frac{1}{e} - \frac{1}{e}(x - 1) + \frac{1}{2!e}(x - 1)^2 - \frac{1}{3!e}(x - 1)^3 + \cdots$$

$$+ \frac{(-1)^n}{n!e}(x - 1)^n + \cdots$$

$$= \sum_{n=0}^{\infty} \frac{(-1)^n}{n!e}(x - 1)^n$$

11.5 MORE ON TAYLOR SERIES

A Useful Technique for Finding Taylor Series

In the last section, we showed how to find the power series representation of certain functions. This representation turned out to be the Taylor series of f at $x = a$. The method we used to compute the series relies solely on our ability to find the higher-order derivatives of the function f. This method, however, is rather tedious.

In this section, we will show how the Taylor series of a function can often be found by manipulating some well-known power series. For this purpose, we first catalog the power series of some of the most commonly used functions (see Table 11.1). Recall that each of these representations was derived in Sections 11.1 and 11.4.

Table 11.1

Power series representations for some common functions.

1. $\dfrac{1}{1 - x} = 1 + x + x^2 + \cdots + x^n + \cdots$ $(-1 < x < 1)$

2. $e^x = 1 + x + \dfrac{1}{2!}x^2 + \dfrac{1}{3!}x^3 + \cdots + \dfrac{1}{n!}x^n + \cdots$ $(-\infty < x < \infty)$

3. $\ln x = (x - 1) - \dfrac{1}{2}(x - 1)^2 + \dfrac{1}{3}(x - 1)^3 - \cdots$

   $\quad + \dfrac{(-1)^{n+1}}{n}(x - 1)^n + \cdots$ $(0 < x \le 2)$

EXAMPLE 1   Find the Taylor series of each function at the indicated point.

a. $f(x) = \dfrac{1}{1+x}; x = 0$ **b.** $f(x) = \dfrac{1}{1+x}; x = 2$

c. $f(x) = \dfrac{1}{1-3x}; x = 0$ **d.** $f(x) = \dfrac{x}{1+x^2}; x = 0$

Solution

a. We write

$$f(x) = \frac{1}{1-(-x)}$$

and use the power series representation of $1/(1-x)$ with x replaced by $-x$ to obtain

$$f(x) = \frac{1}{1+x}$$
$$= 1 + (-x) + (-x)^2 + \cdots + (-x)^n + \cdots$$
$$= 1 - x + x^2 - x^3 + \cdots + (-1)^n x^n + \cdots \qquad (-1 < x < 1)$$

b. We write

$$f(x) = \frac{1}{1+x} = \frac{1}{3+(x-2)} = \frac{1}{3\left[1 + \left(\dfrac{x-2}{3}\right)\right]} = \frac{1}{3}\left[\frac{1}{1 + \left(\dfrac{x-2}{3}\right)}\right]$$

From the result of (a), we have

$$\frac{1}{1+u} = 1 - u + u^2 - u^3 + \cdots + (-1)^n u^n + \cdots \qquad (-1 < u < 1)$$

Thus, with $u = \dfrac{x-2}{3}$, we find

$$f(x) = \frac{1}{3}\left[\frac{1}{1 + \left(\dfrac{x-2}{3}\right)}\right]$$

$$= \frac{1}{3}\left[1 - \left(\frac{x-2}{3}\right) + \left(\frac{x-2}{3}\right)^2 - \left(\frac{x-2}{3}\right)^3 + \cdots \right.$$
$$\left. + (-1)^n \left(\frac{x-2}{3}\right)^n + \cdots \right]$$

$$= \frac{1}{3} - \frac{1}{3^2}(x-2) + \frac{1}{3^3}(x-2)^2 - \frac{1}{3^4}(x-2)^3 + \cdots$$

$$+ \frac{(-1)^n}{3^{n+1}}(x-2)^n + \cdots$$

which converges when

$$-1 < \frac{x-2}{3} < 1 \quad \text{or} \quad -1 < x < 5$$

c. Here we simply replace x in the power series representation of $1/(1 - x)$ by $3x$ to get

$$f(x) = \frac{1}{1 - 3x} = 1 + 3x + (3x)^2 + (3x)^3 + \cdots + (3x)^n + \cdots$$

$$= 1 + 3x + 3^2x^2 + 3^3x^3 + \cdots + 3^nx^n + \cdots$$

The series converges for $-1 < 3x < 1$, or $-\frac{1}{3} < x < \frac{1}{3}$.

d. We write

$$f(x) = \frac{x}{1 + x^2} = x\left[\frac{1}{1 - (-x^2)}\right]$$

$$= x[1 + (-x^2) + (-x^2)^2 + (-x^2)^3 + \cdots + (-x^2)^n + \cdots]$$

$$= x - x^3 + x^5 - x^7 + \cdots + (-1)^nx^{2n+1} + \cdots$$

The series converges for $-1 < x < 1$.   ○ ○ ○

EXAMPLE 2   Find the Taylor series of each of the following functions at the indicated point.

a. $f(x) = xe^{-x}; x = 0$ **b.** $f(x) = \ln(1 + x); x = 0$

Solution

a. First, we replace x with $-x$ in the expression

$$e^x = 1 + x + \frac{x^2}{2!} + \frac{x^3}{3!} + \cdots + \frac{x^n}{n!} + \cdots \qquad \text{(see Table 11.1)}$$

to obtain

$$e^{-x} = 1 + (-x) + \frac{(-x)^2}{2!} + \frac{(-x)^3}{3!} + \cdots + \frac{(-x)^n}{n!} + \cdots$$

$$= 1 - x + \frac{x^2}{2!} - \frac{x^3}{3!} + \cdots + \frac{(-1)^nx^n}{n!} + \cdots$$

Then, multiplying both sides of this expression by x gives the required expression

$$f(x) = xe^{-x} = x - x^2 + \frac{x^3}{2!} - \frac{x^4}{3!} + \cdots + \frac{(-1)^nx^{n+1}}{n!} + \cdots$$

b. From Table 11.1 we have

$$\ln x = (x - 1) - \frac{1}{2}(x - 1)^2 + \frac{1}{3}(x - 1)^3 - \cdots$$

$$+ \frac{(-1)^{n+1}}{n}(x - 1)^n \cdots \qquad (0 < x \le 2)$$

Replacing x with $1 + x$ in this expression, we obtain the required expression

$$f(x) = \ln(1 + x) = x - \frac{1}{2}x^2 + \frac{1}{3}x^3 - \cdots + \frac{(-1)^{n+1}}{n}x^n + \cdots$$

valid for $-1 < x \le 1$. (Why?)   ○ ○ ○

The formula

$$\ln x = (x - 1) - \frac{1}{2}(x - 1)^2 + \frac{1}{3}(x - 1)^3 - \cdots \qquad (0 < x \le 2) \qquad \textbf{(A)}$$

from Table 11.1 can be used to compute the value of ln x for $0 < x \le 2$. However, the restriction on x and the slow convergence of the series limit its effectiveness from the computational point of view. A more effective formula, first obtained by the Scottish mathematician James Gregory (1638–1675), follows:

a. Using formula (A), derive the formulas

$$\ln(1 + x) = x - \frac{x^2}{2} + \frac{x^3}{3} - \frac{x^4}{4} - \cdots \qquad (-1 < x \le 1)$$

and

$$\ln(1 - x) = -x - \frac{x^2}{2} - \frac{x^3}{3} - \frac{x^4}{4} - \cdots \qquad (-1 \le x < 1)$$

b. Use part (a) to show that

$$\ln\left(\frac{1 + x}{1 - x}\right) = 2\left(x + \frac{x^3}{3} + \frac{x^5}{5} + \frac{x^7}{7} + \cdots\right) \qquad (-1 < x < 1)$$

c. To compute the natural logarithm of a positive number p, let

$$p = \frac{1 + x}{1 - x}$$

and show that

$$x = \frac{p - 1}{p + 1} \qquad (-1 < x < 1)$$

d. Use parts (b) and (c) to show that

$$\ln 2 = 2\left[\left(\frac{1}{3}\right) + \frac{(\frac{1}{3})^3}{3} + \frac{(\frac{1}{3})^5}{5} + \frac{(\frac{1}{3})^7}{7} + \cdots\right]$$

and this yields ln 2 ≈ 0.6931 when we add the first four terms of the series. This approximation of ln 2 is accurate to four decimal places.

e. Compare this method of computing ln 2 with that of using formula (A) directly.

In Theorem 4 in the last section we saw that a power series may be differentiated term by term to yield another power series whose interval of convergence coincides with that of the original series. The latter is, of course, the power series representation of the derivative f' of the function f represented by the first series. The following theorem tells us that we may integrate a power series term by term.

THEOREM 5

Suppose that

$$f(x) = \sum_{n=0}^{\infty} a_n(x - a)^n$$

$$= a_0 + a_1(x - a) + a_2(x - a)^2 + a_3(x - a)^3 + \cdots$$
$$+ a_n(x - a)^n + \cdots, \qquad x \in (a - R, a + R)$$

Then

$$\int f(x)\, dx = \sum_{n=0}^{\infty} \frac{a_n}{n + 1}(x - a)^{n+1}$$

$$= a_0(x - a) + \frac{a_1}{2}(x - a)^2 + \frac{a_2}{3}(x - a)^3 + \cdots$$

$$+ \frac{a_n}{n + 1}(x - a)^{n+1} + \cdots, \qquad x \in (a - R, a + R)$$

Theorems 4 and 5 can also be used to help us find the power series representation of a function starting from the series representation of some appropriate function, as the next two examples show.

EXAMPLE 3 Differentiate the power series for the function $1/(1 - x)$ at $x = 0$ (see Table 11.1) to obtain a Taylor series representation of the function $f(x) = 1/(1 - x)^2$ at $x = 0$.

Solution From Table 11.1, we have

$$\frac{1}{1 - x} = 1 + x + x^2 + x^3 + \cdots + x^n + \cdots \qquad (-1 < x < 1)$$

Differentiating both sides of the equation with respect to x, and using Theorem 4, we obtain

$$\frac{d}{dx}\left(\frac{1}{1 - x}\right) = \frac{d}{dx}(1 + x + x^2 + x^3 + \cdots + x^n + \cdots)$$

or

$$f(x) = \frac{1}{(1 - x)^2}$$

$$= 1 + 2x + 3x^2 + \cdots + nx^{n-1} + \cdots \qquad (-1 < x < 1) \qquad \circ \; \circ \; \circ$$

EXAMPLE 4 Integrate the Taylor series for the function $1/(1 + x)$ at $x = 0$ (see Table 11.1) to obtain a power series representation for the function $f(x) = \ln(1 + x)$ centered at $x = 0$. Compare your result with that of Example 2b.

Solution   From Table 11.1 we see that

$$\frac{1}{1-x} = 1 + x + x^2 + x^3 + \cdots + x^n + \cdots \qquad (-1 < x < 1)$$

Replacing x by $-x$ gives

$$\frac{1}{1-(-x)} = 1 + (-x) + (-x)^2 + (-x)^3 + \cdots + (-x)^n + \cdots$$

or

$$\frac{1}{1+x} = 1 - x + x^2 - x^3 + \cdots + (-1)^n x^n + \cdots \qquad (-1 < x < 1)$$

Finally, integrating both sides of this equation with respect to x and using Theorem 5, we obtain

$$\int \frac{1}{1+x}\, dx = \int [1 - x + x^2 - x^3 + \cdots + (-1)^n x^n + \cdots]\, dx$$

or

$$f(x) = \ln(1 + x) = x - \frac{1}{2}x^2 + \frac{1}{3}x^3 - \cdots + \frac{(-1)^n}{n+1}x^{n+1} + \cdots$$

$$= x - \frac{1}{2}x^2 + \frac{1}{3}x^3 - \cdots + \frac{(-1)^{n+1}}{n}x^n + \cdots \qquad (-1 < x < 1)$$

This result is the same as that of Example 2b, as expected.   ◦ ◦ ◦

Application

In many practical applications, all that is required is an approximation of the actual solution to a problem. Thus, rather than working with the Taylor series of a function at a point that is *equal* to $f(x)$ inside its interval of convergence, one often works with the truncated Taylor series. The truncated Taylor series is a Taylor polynomial that gives an acceptable approximation to the values of $f(x)$ in the neighborhood of the point about which the function is expanded, provided the degree of the polynomial or, equivalently, the number of terms of the Taylor series retained is large enough.

Before looking at an example, we want to point out the comparative ease with which one can obtain the Taylor polynomial approximation of a function using the method of this section, rather than obtaining it directly as was done in Section 11.1.

EXAMPLE 5   The serum cholesterol levels (in mg/dl) in a current Mediterranean population are found to be normally distributed with a probability density function given by

$$f(x) = \frac{1}{50\sqrt{2\pi}} e^{-1/2[(x-160)/50]^2}$$

Scientists at the National Heart, Lung, and Blood Institute consider this pattern ideal for a minimal risk of heart attacks. Find the percentage of the population who have blood cholesterol levels between 160 and 180 mg/dl.

Solution The required probability is given by

$$P(160 \le x \le 180) = \frac{1}{50\sqrt{2\pi}} \int_{160}^{180} e^{-1/2[(x-160)/50]^2} \, dx$$

(see Section 10.3). Let us approximate the integrand by a sixth-degree Taylor polynomial about $x = 160$. From Table 11.1, we have

$$e^x = 1 + x + \frac{1}{2!}x^2 + \frac{1}{3!}x^3 + \cdots$$

Replacing x with $-\frac{1}{2}[(x - 160)/50]^2$, we obtain

$$e^{-1/2[(x-160)/50]^2} \approx 1 - \frac{1}{2}[(x-160)/50]^2 + \frac{1}{2!}\left\{-\frac{1}{2}[(x-160)/50]^2\right\}^2$$

$$+ \frac{1}{3!}\left\{-\frac{1}{2}[(x-160)/50]^2\right\}^3$$

$$= 1 - \frac{(x-160)^2}{5000} + \frac{(x-160)^4}{5 \cdot 10^7} - \frac{(x-160)^6}{7.5 \cdot 10^{11}}$$

Therefore,

$$P(160 \le x \le 180)$$

$$\approx \frac{1}{50\sqrt{2\pi}} \int_{160}^{180} \left[1 - \frac{(x-160)^2}{5000} + \frac{(x-160)^4}{5 \cdot 10^7} - \frac{(x-160)^6}{7.5 \cdot 10^{11}}\right] dx$$

$$\approx \frac{1}{50\sqrt{2\pi}} \left[x - \frac{(x-160)^3}{15,000} + \frac{(x-160)^5}{2.5 \cdot 10^8} - \frac{(x-160)^7}{5.25 \cdot 10^{12}}\right]\Bigg|_{160}^{180}$$

$$\approx \frac{1}{50\sqrt{2\pi}} [(180 - 0.53333 + 0.0128 - 0.00024) - 160]$$

$$\approx 0.1554$$

and so approximately 15.5% of the population has blood cholesterol levels between 160 and 180 mg/dl.   ◖ ◖ ◖

SELF-CHECK EXERCISES 11.5

1. Find the Taylor series of the function $f(x) = xe^{-x^2}$ at $x = 0$.

2. Use the result from Exercise 1 to write the seventh Taylor polynomial of f about $x = 0$, and use this polynomial to approximate

$$\int_0^{0.5} xe^{-x^2} \, dx$$

Compare this result with the exact value of the integral.

Solutions to Self-Check Exercises 11.5 can be found on page 806.

11.5 EXERCISES

In exercises 1–20, find the Taylor series of each function at the indicated point. Give the interval of convergence for each series.

1. $f(x) = \dfrac{1}{1 - x}; x = 2$

2. $f(x) = \dfrac{1}{1 + x}; x = 1$

3. $f(x) = \dfrac{1}{1 + 3x}; x = 0$

4. $f(x) = \dfrac{x}{1 - 2x}; x = 0$

5. $f(x) = \dfrac{1}{4 - 3x}; x = 0$

6. $f(x) = \dfrac{1}{4 - 3x}; x = 1$

7. $f(x) = \dfrac{1}{1 - x^2}; x = 0$

8. $f(x) = \dfrac{x^2}{1 + x^3}; x = 0$

9. $f(x) = e^{-x}; x = 0$

10. $f(x) = e^x; x = 1$

11. $f(x) = xe^{-x^2}; x = 0$

12. $f(x) = xe^{x/2}; x = 0$

13. $f(x) = \dfrac{1}{2}(e^x + e^{-x}); x = 0$

14. $f(x) = \dfrac{1}{2}(e^x - e^{-x}); x = 0$

15. $f(x) = \ln(1 + 2x); x = 0$

16. $f(x) = \ln\left(1 + \dfrac{x}{2}\right); x = 0$

17. $f(x) = \ln(1 + x^2); x = 0$

18. $f(x) = \ln(1 + 2x); x = 2$

19. $f(x) = (x - 2)\ln x; x = 2$

20. $f(x) = x^2 \ln\left(1 + \dfrac{x}{2}\right); x = 0$

21. Differentiate the power series for $\ln(1 + x)$ at $x = 0$ to obtain a series representation for the function $f(x) = 1/(1 + x)$.

22. Differentiate the power series for $1/(1 + x)$ at $x = 0$ to obtain a series representation for the function $f(x) = 1/(1 + x)^2$.

23. Integrate the power series for $1/(1 + x)$ to obtain a power series representation for the function $f(x) = \ln(1 + x)$.

24. Integrate the power series for $2x/(1 + x^2)$ to obtain a power series representation for the function $f(x) = \ln(1 + x^2)$.

25. Use the sixth-degree Taylor polynomial to approximate

$$\int_0^{0.5} \frac{1}{\sqrt{1 + x^2}}\, dx$$

[Hint: $P_6(x) = 1 - \frac{1}{2}x^2 + \frac{3}{8}x^4 - \frac{5}{16}x^6$]

26. Use the sixth-degree Taylor polynomial to approximate

$$\int_0^{0.4} \ln(1 + x^2)\, dx$$

27. Use the eighth-degree Taylor polynomial to approximate

$$\int_0^1 e^{-x^2}\, dx$$

28. Use the sixth-degree Taylor polynomial to approximate

$$\int_0^{0.5} \frac{\ln(1 + x)}{x}\, dx$$

29. Use the eighth-degree Taylor polynomial of $f(x) = 1/(1 + x^2)$ at $x = 0$ and the relationship

$$\pi = 4 \int_0^1 \frac{dx}{1 + x^2}$$

to obtain an approximation of π.

30. Factory Worker Wages According to data released by a city's Chamber of Commerce, the weekly wages of factory workers are normally distributed according to the probability density function

$$f(x) = \frac{1}{50\sqrt{2\pi}} e^{-1/2[(x - 400)/50]^2}$$

Find the probability that a worker selected at random from the city has a weekly wage of $350–$450.

31. TV Set Reliability The General Manager of the Service Department of MCA Television Company has estimated that the time that elapses between the dates of purchase and the dates on which the 19-inch sets manufactured by the company first require service is normally distributed according to the probability density function

$$f(x) = \frac{1}{10\sqrt{2\pi}} e^{-1/2[(x - 30)/10]^2}$$

where x is measured in months. Determine the percentage of sets manufactured and sold by MCA that may require service 28–32 months after purchase.

32. Demand for Wristwatches   The demand equation for the "Tempus" quartz wristwatch is given by

$$p = 50e^{-0.1(x+1)^2}$$

where x (measured in units of a thousand) is the quantity demanded per week and p is the unit wholesale price in dollars. National Importers, the supplier of the watches, will make x units (in thousands) available in the market if the unit wholesale price is

$$p = 10 + 5x^2$$

dollars. Use an appropriate Taylor polynomial approximation of the function

$$p = f(x) = 50e^{-0.1(x+1)^2}$$

to find the consumers' surplus.

[*Hint:* The equilibrium quantity is 1689 units.]

SOLUTIONS TO SELF-CHECK EXERCISES 11.5

1. First, we replace x in the expression

$$e^x = 1 + x + \frac{x^2}{2!} + \frac{x^3}{3!} + \cdots + \frac{x^n}{n!} + \cdots$$

(see Table 11.1) with $-x^2$ to obtain

$$e^{-x^2} = 1 + (-x^2) + \frac{(-x^2)^2}{2!}$$

$$+ \frac{(-x^2)^3}{3!} + \cdots + \frac{(-x^2)^n}{n!} + \cdots$$

$$= 1 - x^2 + \frac{x^4}{2!} - \frac{x^6}{3!} + \cdots + \frac{(-1)^n x^{2n}}{n!} + \cdots$$

Then multiplying both sides of this expression by x gives the required expression

$$f(x) = xe^{-x^2} = x - x^3 + \frac{x^5}{2!}$$

$$- \frac{x^7}{3!} + \cdots + \frac{(-1)^n x^{2n+1}}{n!} + \cdots$$

$$= \sum_{n=0}^{\infty} \frac{(-1)^n x^{2n+1}}{n!}$$

2. Using the result from Exercise 1, we see that the seventh Taylor polynomial of f about $x = 0$ is

$$P_7(x) = x - x^3 + \frac{x^5}{2!} - \frac{x^7}{3!}$$

$$= x - x^3 + \frac{1}{2}x^5 - \frac{1}{6}x^7$$

Therefore,

$$\int_0^{0.5} xe^{-x^2}\, dx \approx \int_0^{0.5} \left(x - x^3 + \frac{1}{2}x^5 - \frac{1}{6}x^7 \right) dx$$

$$= \frac{1}{2}x^2 - \frac{1}{4}x^4 + \frac{1}{12}x^6 - \frac{1}{48}x^8 \Big|_0^{0.5}$$

$$\approx 0.1105957$$

The exact value of the integral is

$$\int_0^{0.5} xe^{-x^2}\, dx = -\frac{1}{2}e^{-x^2}\Big|_0^{0.5} \qquad \text{(Use the substitution } u = x^2.\text{)}$$

$$= -\frac{1}{2}(e^{-0.25} - 1)$$

which is approximately 0.1105996. Thus, the error is 0.0000039.

11.6 THE NEWTON-RAPHSON METHOD

The Newton-Raphson Method

Recall that in much of our previous work we had occasion to find the zeros of a function f or, equivalently, the **roots** of the equation $f(x) = 0$. For example, the x-intercepts of a function f are precisely the values of x when $f(x) = 0$; the critical points of f include the roots of the equation $f'(x) = 0$; and the candidates for the inflection points of f include the roots of the equation $f''(x) = 0$.

If $f(x)$ is a linear or quadratic function or $f(x)$ is a polynomial that is easily factored, the roots of f are readily found. In practice, however, we often encounter functions with zeros that cannot be found as readily. For example, the function

$$f(t) = -t^3 + 96t^2 + 195t + 5$$

of Example 8, page 179, which gives the altitude (in feet) of a rocket t seconds into flight, is not easily factored, so its zeros cannot be found by elementary algebraic methods. (We will find the zeros of this function in Example 2.) Another example of a function with roots that are not easily found is

$$g(x) = e^{2x} - 3x - 2$$

In this section we will develop an algorithm, based on the approximation of a function $f(x)$ by a first-degree Taylor polynomial, to approximate the value of a zero of $f(x)$ to any desired degree of accuracy.

Suppose that the function f has a zero at $x = c$ (Figure 11.5). Let x_0 be an initial estimate of the actual zero c of $f(x)$. Now, the first-degree Taylor polynomial of $f(x)$ at $x = x_0$ is

$$P_1(x) = f(x_0) + f'(x_0)(x - x_0)$$

and, as observed earlier, is just the equation of the tangent line to the graph of $y = f(x)$ at the point $(x_0, f(x_0))$. Since the tangent line T is an approximation of the graph of $y = f(x)$ near $x = x_0$ (and x_0 is assumed to be close to c!), it may be expected that the zero of $P_1(x)$ is close to the zero $x = c$ of $f(x)$. But

Figure 11.5
x_0 is our first guess in estimating the value of c, the zero of $f(x)$.

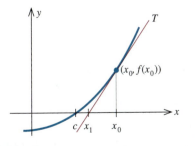

the zero of $P_1(x)$, a linear function, is found by setting $P_1(x) = 0$. Thus,

$$f(x_0) + f'(x_0)(x - x_0) = 0$$

$$f'(x_0)(x - x_0) = -f(x_0)$$

$$x - x_0 = -\frac{f(x_0)}{f'(x_0)}$$

and

$$x = x_0 - \frac{f(x_0)}{f'(x_0)}$$

This number provides us with another estimate

$$x_1 = x_0 - \frac{f(x_0)}{f'(x_0)}$$

of the zero of $f(x)$. In general, the estimate x_1 is better than the initial estimate x_0.

This process may be repeated with the initial estimate x_0 replaced by the recent estimate x_1. This leads to yet another estimate

$$x_2 = x_1 - \frac{f(x_1)}{f'(x_1)}$$

which is usually better than x_1. In this manner, we generate a sequence of approximations $x_0, x_1, x_2, \ldots, x_n, x_{n+1}, \ldots$, with

$$x_{n+1} = x_n - \frac{f(x_n)}{f'(x_n)}$$

which, in most instances, approaches the zero $x = c$ of $f(x)$.

A summary of this algorithm follows.

THE NEWTON-RAPHSON ALGORITHM

1. Pick an initial estimate x_0 of the root c.
2. Find a new estimate using the iterative formula

$$x_{n+1} = x_n - \frac{f(x_n)}{f'(x_n)} \qquad (n = 0, 1, 2, \ldots) \tag{11}$$

3. Compute $|x_n - x_{n+1}|$. If this number is less than a prescribed positive number, stop. The required approximation to the root $x = c$ is $x = x_{n+1}$.

Solving Equations Using the Newton-Raphson Method

EXAMPLE 1   Use the Newton-Raphson algorithm to approximate the zero of $f(x) = x^2 - 2$. Start the iteration with initial guess $x_0 = 1$ and terminate the process when two successive approximations differ by less than 0.00001.

Solution   We have

$$f(x) = x^2 - 2$$

and

$$f'(x) = 2x$$

so that, by (11), the required iterative formula is

$$x_{n+1} = x_n - \frac{x_n^2 - 2}{2x_n} = \frac{x_n^2 + 2}{2x_n}$$

With $x_0 = 1$, we find

$$x_1 = \frac{1^2 + 2}{2(1)} = 1.5$$

$$x_2 = \frac{(1.5)^2 + 2}{2(1.5)} \approx 1.416667$$

$$x_3 = \frac{(1.416667)^2 + 2}{2(1.416667)} \approx 1.414216$$

$$x_4 = \frac{(1.414216)^2 + 2}{2(1.414216)} \approx 1.414214$$

Since $x_3 - x_4 = 0.000002 < 0.00001$, we terminate the process. The sequence generated converges to $\sqrt{2}$, which is one of the two roots of the equation $x^2 - 2 = 0$. Note that, to six places, $\sqrt{2} = 1.414214$!   ○ ○ ○

EXAMPLE 2   Refer to Example 8, page 179. The altitude in feet of a rocket t seconds into flight is given by

$$s = f(t) = -t^3 + 96t^2 + 195t + 5 \qquad (t \geq 0)$$

Find the time T when the rocket hits Earth.

Solution   The rocket hits Earth when the altitude is equal to zero. So we are required to solve the equation

$$s = f(t) = -t^3 + 96t^2 + 195t + 5 = 0$$

Let us use the Newton-Raphson algorithm with initial guess $t_0 = 100$ (Figure 11.6). Here

$$f'(t) = -3t^2 + 192t + 195$$

so that the required iterative formula takes the form

$$t_{n+1} = t_n - \frac{-t_n^3 + 96t_n^2 + 195t_n + 5}{-3t_n^2 + 192t_n + 195}$$

$$= t_n - \frac{t_n^3 - 96t_n^2 - 195t_n - 5}{3t_n^2 - 192t_n - 195}$$

$$= \frac{3t_n^3 - 192t_n^2 - 195t_n - t_n^3 + 96t_n^2 + 195t_n + 5}{3t_n^2 - 192t_n - 195}$$

$$= \frac{2t_n^3 - 96t_n^2 + 5}{3t_n^2 - 192t_n - 195}$$

Figure 11.6

The rocket's altitude t seconds into flight is given by f(t).

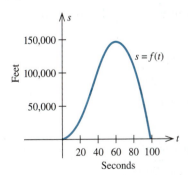

We have

$$t_1 = \frac{2(100)^3 - 96(100)^2 + 5}{3(100)^2 - 192(100) - 195} \approx 98.0674$$

$$t_2 = \frac{2(98.0674)^3 - 96(98.0674)^2 + 5}{3(98.0674)^2 - 192(98.0674) - 195} \approx 97.9906$$

$$t_3 = \frac{2(97.9906)^3 - 96(97.9906)^2 + 5}{3(97.9906)^2 - 192(97.9906) - 195} \approx 97.9905$$

Thus, we see that $T \approx 97.99$, so the rocket hits Earth approximately 98 seconds after lift-off.

The next example shows how the Newton-Raphson Method may be used to help determine a certain commodity's equilibrium quantity and price.

EXAMPLE 3 The demand equation for the "Tempus" quartz wristwatch is given by

$$p = 50e^{-0.1(x+1)^2}$$

where x (measured in units of a thousand) is the quantity demanded per week and p is the unit wholesale price in dollars. National Importers, the supplier of the watches, will make x units (in thousands) available in the market if the unit wholesale price is

$$p = 10 + 5x^2$$

dollars. Find the equilibrium quantity and price.

Solution We determine the equilibrium point by finding the point of intersection of the demand curve and the supply curve (Figure 11.7). To solve the system of equations comprising

$$p = 50e^{-0.1(x+1)^2}$$

and

$$p = 10 + 5x^2$$

we substitute the second equation into the first, obtaining

$$10 + 5x^2 = 50e^{-0.1(x+1)^2}$$

or

$$10 + 5x^2 - 50e^{-0.1(x+1)^2} = 0$$

To solve the last equation, we use the Newton-Raphson Method with

$$f(x) = 10 + 5x^2 - 50e^{-0.1(x+1)^2}$$
$$= 5[2 + x^2 - 10e^{-0.1(x+1)^2}]$$

and

$$f'(x) = 10x - 50e^{-0.1(x+1)^2} \cdot [-0.2(x+1)]$$
$$= 10[x + (x+1)e^{-0.1(x+1)^2}]$$

leading to the iterative formula

$$x_{n+1} = x_n - \frac{2 + x_n^2 - 10e^{-0.1(x_n+1)^2}}{2[x_n + (x_n+1)e^{-0.1(x_n+1)^2}]}$$

Figure 11.7
The equilibrium point is the point of intersection of the demand curve and the supply curve.

Referring to Figure 11.7, we see that a reasonable initial estimate is $x_0 = 2$. We have

$$x_1 = 2 - \frac{2 + 2^2 - 10e^{-0.1(3)^2}}{2[2 + 3e^{-0.1(3)^2}]} \approx 1.69962$$

$$x_2 = 1.69962 - \frac{2 + (1.69962)^2 - 10e^{-0.1(2.69962)^2}}{2[1.69962 + 2.69962e^{-0.1(2.69962)^2}]} \approx 1.68899$$

$$x_3 = 1.68899 - \frac{2 + (1.68899)^2 - 10e^{-0.1(2.68899)^2}}{2[1.68899 + 2.68899e^{-0.1(2.68899)^2}]} \approx 1.68898$$

Thus, the equilibrium quantity is approximately 1.689 units, and the equilibrium wholesale price is given by

$$p = 10 + 5(1.68898)^2 \approx 24.263$$

or approximately \$24.26 per watch.

○ ○ ○

The Internal Rate of Return on an Investment

Yet another use of the Newton-Raphson algorithm is in finding a quantity called the *internal rate of return* on an investment. Suppose a company has an initial outlay of C dollars in an investment that yields returns of R_1, $R_2, \ldots, R_n$ dollars at the end of the first, second, $\ldots$, nth periods, respectively. Then the *net present value* of the investment is

$$\frac{R_1}{1 + r} + \frac{R_2}{(1 + r)^2} + \frac{R_3}{(1 + r)^3} + \cdots + \frac{R_n}{(1 + r)^n} - C$$

where r denotes the interest rate per period earned on the investment. Now the **internal rate of return** on the investment is defined as the rate of return for which the net present value of the investment is equal to zero; that is, it is the value of r that satisfies the equation

$$\frac{R_1}{1 + r} + \frac{R_2}{(1 + r)^2} + \frac{R_3}{(1 + r)^3} + \cdots + \frac{R_n}{(1 + r)^n} - C = 0$$

or, equivalently, the equation

$$C(1 + r)^n - R_1(1 + r)^{n-1} - R_2(1 + r)^{n-2} - R_3(1 + r)^{n-3} - \cdots - R_n = 0$$

obtained by multiplying both sides of the former by $(1 + r)^n$. The internal rate of return is used by management to decide on the worthiness or profitability of an investment.

EXAMPLE 4   The management of A-1 Rental—a tool and equipment rental service for industry, contractors, and homeowners—is contemplating purchasing new equipment. The initial outlay for the equipment, which has a useful life of 4 years, is \$45,000, and it is expected that the

investment will yield returns of $15,000 at the end of the first year, $18,000 at the end of the second year, $14,000 at the end of the third year, and $10,000 at the end of the fourth year. Find the internal rate of return on this investment.

Solution   Here $n = 4$, $C = 45{,}000$, $R_1 = 15{,}000$, $R_2 = 18{,}000$, $R_3 = 14{,}000$, and $R_4 = 10{,}000$. So we are required to solve the equation

$$45{,}000(1 + r)^4 - 15{,}000(1 + r)^3 - 18{,}000(1 + r)^2 - 14{,}000(1 + r) - 10{,}000 = 0$$

for r. The equation may be written more simply by letting $x = 1 + r$. Thus,

$$f(x) = 45{,}000x^4 - 15{,}000x^3 - 18{,}000x^2 - 14{,}000x - 10{,}000 = 0$$

To solve the equation $f(x) = 0$ using the Newton-Raphson algorithm, we first compute

$$f'(x) = 180{,}000x^3 - 45{,}000x^2 - 36{,}000x - 14{,}000$$

The required iterative formula is

$$x_{n+1} = x_n - \frac{45{,}000x_n^4 - 15{,}000x_n^3 - 18{,}000x_n^2 - 14{,}000x_n - 10{,}000}{180{,}000x_n^3 - 45{,}000x_n^2 - 36{,}000x_n - 14{,}000}$$

Starting with the initial estimate $x_0 = 1.1$, we find

$$x_1 = 1.1 - \frac{45{,}000(1.1)^4 - 15{,}000(1.1)^3 - 18{,}000(1.1)^2 - 14{,}000(1.1) - 10{,}000}{180{,}000(1.1)^3 - 45{,}000(1.1)^2 - 36{,}000(1.1) - 14{,}000}$$

$$\approx 1.10958$$

$$x_2 \approx 1.10941$$

$$x_3 \approx 1.10941$$

Therefore, we may take $x \approx 1.1094$, in which case we see that $r \approx 0.1094$. Thus, the rate of return on the investment is approximately 10.94% per year.

◐ ◐ ◐

Having seen how effective the Newton-Raphson Method can be in finding the zeros of a function, we want to close this section by reminding you that there are situations in which the method fails and that care must be exercised in applying it. Figure 11.8a illustrates a situation where $f'(x_n) = 0$ for some n (in this case $n = 2$). Since the iterative formula (11) involves division by $f'(x_n)$, it should be clear why the method fails to work in this case. However, if you choose a different initial estimate x_0, the situation may yet be salvaged (Figure 11.8b).

Figure 11.8
In (a), the Newton-Raphson Method fails to work because $f'(x_2) = 0$, but this situation is remedied in (b) by selecting a different initial estimate x_0.

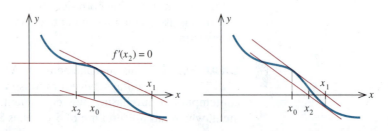

For a concrete example of a situation similar to that depicted in Figure 11.8, consider the function

$$f(x) = x^3 - 1.5x^2 - 6x + 2$$

a. Show that the Newton-Raphson Method fails to work if we choose $x_0 = -1$ or $x_0 = 2$ for an initial estimate.
b. Using the initial estimates $x_0 = -2.5$, $x_0 = 1$, and $x_0 = 2.5$, show that the three roots of $f(x) = 0$ are -2, 0.313859, and 3.186141, respectively.
c. Using a graphing utility, plot the graph of f in the viewing rectangle $[-3, 4] \times [-10, 7]$. Verify the results of part (b) using TRACE and ZOOM or the root-finding function of your calculator.

The next situation, shown in Figure 11.9, is more serious, and the method will not work for any choice of the initial estimate x_0 other than the actual zero of the function $f(x)$ because the sequence $x_1, x_2, \ldots x_n$ diverges.

Figure 11.9
The Newton-Raphson Method fails here because the sequence of estimates diverges.

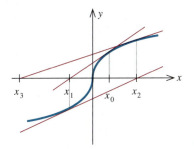

For a concrete example of a situation similar to that depicted in Figure 11.9, consider the function $f(x) = x^{1/3}$.

a. Show that the Newton-Raphson iteration for solving the equation $f(x) = 0$ is $x_{n+1} = -2x_n$, with x_0 being an initial guess.
b. Show that the sequence $x_0, x_1, x_2, \ldots$ diverges for any choice of x_0 other than zero and thus the Newton-Raphson Method does not lead to the unique solution $x = 0$ for the problem under consideration.
c. Illustrate this situation geometrically.

SELF-CHECK EXERCISES 11.6

1. Use three iterations of the Newton-Raphson algorithm on an appropriate function $f(x)$ and initial guess x_0 to obtain an estimate of $\sqrt[3]{10}$.

2. A study prepared for a certain Sunbelt town's Chamber of Commerce projected that the town's population in the next 3 years will grow according to the rule

$$P(t) = 50{,}000 + 30t^{3/2} + 20t$$

where $P(t)$ denotes the population t months from now. Using this model, estimate the time when the population will reach 55,000.

Solutions to Self-Check Exercises 11.6 can be found on page 816.

Solutions to Self-Check Exercises 11.6 can be found on page 816.

11.6 EXERCISES

A calculator is recommended for this exercise set. In exercises 1–6, estimate the value of each radical by using three iterations of the Newton-Raphson Method with the indicated initial guess on each function.

1. $\sqrt{3}$; $f(x) = x^2 - 3$; $x_0 = 1.5$

2. $\sqrt{5}$; $f(x) = x^2 - 5$; $x_0 = 2$

3. $\sqrt{7}$; $f(x) = x^2 - 7$; $x_0 = 2.5$

4. $\sqrt[3]{6}$; $f(x) = x^3 - 6$; $x_0 = 2$

5. $\sqrt[3]{14}$; $f(x) = x^3 - 14$; $x_0 = 2.5$

6. $\sqrt[4]{50}$; $f(x) = x^4 - 50$; $x_0 = 2.5$

In exercises 7–14, use the Newton-Raphson Method to approximate the indicated zero of each function. Continue with the iteration until two successive approximations differ by less than 0.0001.

7. The zero of $f(x) = x^2 - x - 3$ between $x = 2$ and $x = 3$

8. The zero of $f(x) = x^3 + x - 3$ between $x = 1$ and $x = 2$

9. The zero of $f(x) = x^3 + 2x^2 + x - 5$ between $x = 1$ and $x = 2$

10. The zero of $f(x) = x^5 + x - 1$ between $x = 0$ and $x = 1$

11. The zero of $f(x) = \sqrt{x + 1} - x$ between $x = 0$ and $x = 2$

12. The zero of $f(x) = e^{-x} - x$ between $x = 0$ and $x = 1$

13. The zero of $f(x) = e^x - (1/x)$ between $x = 0$ and $x = 1$

14. The zero of $f(x) = \ln x^2 - 0.7x + 1$ between $x = 6$ and $x = 7$

15. Let $f(x) = 2x^3 - 9x^2 + 12x - 2$.
   a. Show that $f(x) = 0$ has a root between $x = 0$ and $x = 1$.
   [*Hint:* Compute $f(0)$ and $f(1)$ and use the fact that f is continuous.]

   b. Use the Newton-Raphson Method to find the zero of f in the interval $(0, 1)$.

16. Let $f(x) = x^3 - x - 1$.
   a. Show that $f(x) = 0$ has a root between $x = 1$ and $x = 2$.
   [*Hint:* See exercise 15.]
   b. Use the Newton-Raphson Method to find the zero of f in the interval $(1, 2)$.

17. Let $f(x) = x^3 - 3x - 1$.
   a. Show that f has a zero between $x = 1$ and $x = 2$.
   [*Hint:* See exercise 15.]
   b. Use the Newton-Raphson Method to find the zero of f.

18. Let $f(x) = x^4 - 4x^3 + 10$.
   a. Show that $f(x) = 0$ has a root between $x = 1$ and $x = 2$.
   b. Use the Newton-Raphson Method to find the zero of f.

In exercises 19–24, make a rough sketch of the graphs of each of the given pairs of functions. Use your sketch to approximate the point(s) of intersection of the two graphs, and then apply the Newton-Raphson Method to refine the approximation of the x-coordinate of the point of intersection.

19. $f(x) = 2\sqrt{x + 3}$; $g(x) = 2x - 1$

20. $f(x) = e^{-x^2}$; $g(x) = x^2$

21. $f(x) = e^{-x}$; $g(x) = x - 1$

22. $f(x) = \ln x$; $g(x) = 2 - x$

23. $f(x) = \sqrt{x}$; $g(x) = e^{-x}$

24. $f(x) = 2 - x^2$; $g(x) = \ln x$

25. Minimizing Average Cost   A division of Ditton Industries manufactures the "Futura" model microwave oven.

Given that the daily cost of producing these microwave ovens (in dollars) obeys the rule

$$C(x) = 0.0002x^3 - 0.06x^2 + 120x + 5000$$

where x stands for the number of units produced, find the level of production that minimizes the daily average cost per unit.

26. Altitude of a Rocket The altitude (in feet) of a rocket t seconds into flight is given by

$$s = f(t) = -2t^3 + 114t^2 + 480t + 1 \qquad (t \geq 0)$$

Find the time T when the rocket hits Earth.

27. Temperature The temperature at 6 A.M. on a certain December day was measured at 15.6°F. In the next t hours the temperature was given by the function

$$T = -0.05t^3 + 0.4t^2 + 3.8t + 15.6 \qquad (0 \leq t \leq 15)$$

where T is measured in degrees Fahrenheit. At what time was the temperature 0°F?

28. Internal Rate of Return on an Investment The proprietor of Qwik Film Lab, Inc., recently purchased $12,000 of new film-processing equipment. She expects that this investment, which has a useful life of 4 years, will yield returns of $4000 at the end of the first year, $5000 at the end of the second year, $4000 at the end of the third year, and $3000 at the end of the fourth year. Find the internal rate of return on the investment.

29. Internal Rate of Return on an Investment Executive Limousine Service recently acquired limousines worth $120,000. The projected returns over the next 3 years, the time period the limousines will be in service, are $80,000 at the end of the first year, $60,000 at the end of the second year, and $40,000 at the end of the third year. Find the internal rate of return on the investment.

30. Internal Rate of Return on an Investment Suppose that an initial outlay of $\$C$ in an investment yields returns of $\$R$ at the end of each period over N periods.
  a. Show that the internal rate of return on the investment, r, may be obtained by solving the equation

$$Cr + R[(1 + r)^{-N} - 1] = 0$$

$$\left[Hint: 1 + x + x^2 + \cdots + x^{N-1} = \frac{1 - x^{N+1}}{1 - x} \right]$$

  b. Show that r can be found by performing the iteration

$$r_{n+1} = r_n - \frac{Cr_n + R[(1 + r_n)^{-N} - 1]}{C - NR(1 + r_n)^{-N-1}}$$

  [r_0 (positive), an initial guess]

[*Hint:* Apply the Newton-Raphson Method to the function $f(r) = Cr + R[(1 + r)^{-N} - 1]$.]

31. Home Mortgages Refer to exercise 30. The Flemings secured a loan of $100,000 from a bank to finance the purchase of a house. They have agreed to repay the loan in equal monthly installments of $1053 over 25 years. The bank charges interest at the rate of $12r$ per year on the unpaid balance, and interest computations are made at the end of each month. Find r.

32. Home Mortgages The Blakelys borrowed a sum of $80,000 from a bank to help finance the purchase of a house. The bank charges interest at the rate of $12r$ per year on the unpaid balance, with interest being computed at the end of each month. The Blakelys have agreed to repay the loan in equal monthly installments of $643.70 over 30 years. What is the true rate of interest charged by the bank?

[*Hint:* See exercise 30.]

33. Car Loans The price of a certain new car is $8000. Suppose that an individual makes a down payment of 25% toward the purchase of the car and secures financing for the balance over 4 years. If the monthly payment is $152.18, what is the true rate of interest charged by the finance company?

[*Hint:* See exercise 30.]

34. Real Estate Investment Groups Refer to exercise 30. A group of private investors purchased a condominium complex for $2 million. They made an initial down payment of 10% and have obtained financing for the balance. If the loan is amortized over 15 years with quarterly repayments of $65,039, determine the interest rate charged by the bank. Assume that interest is calculated at the end of each quarter and is based on the unpaid balance.

35. Demand for Wristwatches The quantity of "Sicard" wristwatches demanded per month is related to the unit price by the equation

$$p = d(x) = \frac{50}{0.01x^2 + 1} \qquad (1 \leq x \leq 20)$$

where p is measured in dollars and x is measured in units of a thousand. The supplier is willing to make x thousand wristwatches available per month when the price per watch is given by $p = s(x) = 0.1x + 20$ dollars. Find the equilibrium quantity and price.

36. Demand for TV Sets The weekly demand for the Pulsar 25-inch color console television is given by the demand equation

$$p = -0.05x + 600 \qquad (0 \leq x \leq 12,000)$$

where p denotes the wholesale unit price in dollars and x denotes the quantity demanded. The weekly total cost function associated with the manufacture of these sets is given by

$$C = 0.000002x^3 - 0.03x^2 + 400x + 80,000$$

where $C(x)$ denotes the total cost incurred in the production of x sets. Find the break-even level(s) of operation for the company.

[*Hint:* Solve the equation $P(x) = 0$, where P is the total profit function.]

SOLUTIONS TO SELF-CHECK EXERCISES 11.6

1. Since $\sqrt[3]{10}$ is one of the cube roots of the equation $x^3 - 10 = 0$, let us take $f(x) = x^3 - 10$. We have

$$f'(x) = 3x^2$$

and so the required iteration formula is

$$x_{n+1} = x_n - \frac{f(x_n)}{f'(x_n)}$$

$$= x_n - \frac{x_n^3 - 10}{3x_n^2}$$

$$= \frac{2x_n^3 + 10}{3x_n^2}$$

Taking $x_0 = 2$, we find

$$x_1 = \frac{2(2^3) + 10}{3(2^2)} \approx 2.166667$$

$$x_2 = \frac{2(2.166667)^3 + 10}{3(2.166667)^2} \approx 2.154504$$

$$x_3 = \frac{2(2.154504)^3 + 10}{3(2.154504)^2} \approx 2.154435$$

Therefore, $\sqrt[3]{10} \approx 2.1544$.

2. We need to solve the equation $P(t) = 55,000$ or

$$50,000 + 30t^{3/2} + 20t = 55,000$$

Rewriting, we have

$$30t^{3/2} + 20t - 5000 = 0$$

or $$3t^{3/2} + 2t - 500 = 0$$

Thus, the problem reduces to that of finding the zero of the function

$$f(t) = 3t^{3/2} + 2t - 500$$

Using the Newton-Raphson Method, we have

$$f'(t) = \frac{9}{2}t^{1/2} + 2 = \frac{9t^{1/2} + 4}{2}$$

So the required iteration formula is

$$t_{n+1} = t_n - \frac{\dfrac{3t_n^{3/2} + 2t_n - 500}{9t_n^{1/2} + 4}}{2}$$

$$= t_n - \frac{6t_n^{3/2} + 4t_n - 1000}{9t_n^{1/2} + 4}$$

$$= \frac{9t_n^{3/2} + 4t_n - (6t_n^{3/2} + 4t_n - 1000)}{9t_n^{1/2} + 4}$$

$$= \frac{3t_n^{3/2} + 1000}{9t_n^{1/2} + 4}$$

Using the initial guess $t_0 = 24$, we find

$$t_1 = \frac{3(24)^{3/2} + 1000}{9(24)^{1/2} + 4} \approx 28.128584$$

$$t_2 = \frac{3(28.128584)^{3/2} + 1000}{9(28.128584)^{1/2} + 4} \approx 27.981338$$

and

$$t_3 = \frac{3(27.981338)^{3/2} + 1000}{9(27.981338)^{1/2} + 4} \approx 27.981160$$

So we may take $t \approx 27.98$. Thus, the population will reach 55,000 approximately 28 months from now.

CHAPTER 11 SUMMARY OF PRINCIPAL FORMULAS AND TERMS

Formulas

1. nth Taylor polynomial of f at $x = a$

$$P_n(x) = f(a) + f'(a)(x-a) + \frac{f''(a)}{2!}(x-a)^2 + \cdots + \frac{f^{(n)}(a)}{n!}(x-a)^n$$

2. Infinite series

$$\sum_{n=1}^{\infty} a_n = a_1 + a_2 + a_3 + \cdots$$

3. Geometric series

$$\sum_{n=0}^{\infty} ar^n = a + ar + \cdots$$

$$= \frac{a}{1-r}, \quad |r| < 1$$

4. Taylor series expansion of $f(x)$ at $x = a$

$$f(x) = f(a) + f'(a)(x-a) + \frac{f''(a)}{2!}(x-a)^2 + \frac{f'''(a)}{3!}(x-a)^3 + \cdots$$

5. Iterative formula for the
Newton-Raphson algorithm

$$x_{n+1} = x_n - \frac{f(x_n)}{f'(x_n)}$$

$$(n = 0, 1, 2, \ldots)$$

Terms

Error bound for Taylor
  polynomials
Infinite sequence
Perpetuity
Capital value of a perpetuity
Limit of a sequence

Convergent series
Divergent series
Partial sum of an infinite series
Maclaurin series of $f(x)$
Interval of convergence
Internal rate of return

CHAPTER 11 REVIEW EXERCISES

In exercises 1–4, find the fourth Taylor polynomial of each of the given functions f at the indicated point.

1. $f(x) = \dfrac{1}{x+2}$ at $x = -1$

2. $f(x) = e^{-x}$ at $x = 1$

3. $f(x) = \ln(1 + x^2)$ at $x = 0$

4. $f(x) = \dfrac{1}{(1+x)^2}$ at $x = 0$

5. Find the second Taylor polynomial of $f(x) = \sqrt[3]{x}$ at $x = 8$ and use it to estimate the value of $\sqrt[3]{7.8}$.

6. Find the sixth Taylor polynomial of $f(x) = e^{-x^2}$ at $x = 0$ and use it to estimate the value of

$$\int_0^1 e^{-x^2}\, dx$$

7. Use the second Taylor polynomial of $f(x) = \sqrt[3]{x}$ at $x = 27$ to approximate $\sqrt[3]{26.98}$ and find a bound for the error in the approximation.

8. Use the third Taylor polynomial of $f(x) = 1/(1 + x)$ to approximate $f(0.1)$. Find a bound for the error in the approximation and compare your results to the exact value of $f(0.1)$.

9. Find an estimate of e^{-1} with an error less than 10^{-2}.

10. Estimate $\int_0^{0.2} e^{-(1/2)x^2}\, dx$ with an error less than 10^{-2}.

In exercises 11–14, the nth term of a sequence is given. Determine whether the sequence converges or diverges. If the sequence converges, find its limit.

11. $a_n = \dfrac{2n^2 + 1}{3n^2 - 1}$

12. $a_n = \dfrac{(-1)^{n-1}n}{n+1}$

13. $a_n = 1 - \dfrac{1}{2^n}$

14. $a_n = \dfrac{1 + \sqrt{n}}{1 - \sqrt{n}}$

In exercises 15–18, find the sum of the geometric series if it converges.

15. $\displaystyle\sum_{n=1}^{\infty} \frac{2^n}{3^n}$

16. $\displaystyle\sum_{n=1}^{\infty} 2^{-n}3^{-n+1}$

17. $\displaystyle\sum_{n=1}^{\infty} (-1)^{n-1}\left(\frac{1}{\sqrt{2}}\right)^n$

18. $\displaystyle\sum_{n=1}^{\infty} \left(\frac{1}{e}\right)^n$

19. Express the repeating decimal $1.424242\ldots$ as a rational number.

20. Express the repeating decimal $3.142142142\ldots$ as a rational number.

In exercises 21–24, find the radius of convergence and the interval of convergence of each power series.

21. $\displaystyle\sum_{n=0}^{\infty} \frac{x^n}{n^2 + 2}$

22. $\displaystyle\sum_{n=1}^{\infty} \frac{(-1)^{n-1}}{\sqrt{n}}x^n$

23. $\displaystyle\sum_{n=1}^{\infty} \frac{(x-1)^n}{n(n+1)}$

24. $\displaystyle\sum_{n=2}^{\infty} \frac{e^n}{n^2}(x-2)^n$

In exercises 25–28, find the Taylor series of each function at the indicated point. Give the interval of convergence for each series.

25. $f(x) = \dfrac{1}{2x - 1}; x = 0$

26. $f(x) = e^{-x}; x = 1$

27. $f(x) = \ln(1 + 2x); x = 0$

28. $f(x) = x^2 e^{-2x}; x = 0$

29. Apply the Newton-Raphson Method to an appropriate function $f(x)$ to obtain an estimate of $\sqrt[3]{12}$.

30. Use the Newton-Raphson Method to find the root of the equation $x^3 + x^2 - 1 = 0$ between $x = 0$ and $x = 1$.

31. Use the Newton-Raphson Method to find the point of intersection of the graphs of the functions $y = f(x) = 2x$ and $y = g(x) = e^{-x}$.

32. A suitcase released from rest at the top of a plane metal slide moves a distance of

$$x = f(t) = 27(t + 3e^{-t/3} - 3)$$

feet in t seconds. If the metal slide is 24 feet long, how long does it take the suitcase to reach the bottom?

[*Hint:* Use the Newton-Raphson Method.]

33. Mr. Simpson wishes to establish a trust fund that will provide each of his two children with an annual income of $10,000 beginning next year and continuing throughout their lifetimes. Find the amount of money he is required to place in trust now if the fund will earn interest at the rate of 9% per year compounded continuously.

34. Suppose that the average wage earner in a certain country saves 8% of his or her take-home pay and spends the other 92%.
   a. Estimate the impact that a proposed $10 billion tax cut will have on the economy over the long run in terms of the additional spending generated.
   b. Estimate the impact that a $10 billion tax cut will have on the economy over the long run if, at the same time, legislation is enacted to boost the rate of savings of the average taxpayer from 8% to 10%.

35. The heights of 4000 women who participated in a recent survey were found to be normally distributed according to the probability density function

$$f(x) = \dfrac{1}{2.5\sqrt{2\pi}} e^{-1/2[(x-64.5)/2.5]^2}$$

What percentage of these women have heights between 63.5 and 65.5 inches?

During the golden period of Greek civilization, Apollonius (262–200 B.C.) developed the trigonometric techniques necessary for calculating the radii of various circles. This was done in an effort to describe planetary motion, a physical process that exhibits a cyclical, or periodic, mode of behavior. Many other real-life phenomena, such as business cycles, earthquake vibrations, respiratory cycles, sales trends, and sound waves, exhibit cyclical behavior patterns.

In this chapter, we extend our study of calculus to an important class of functions called the trigonometric functions. These functions are periodic and hence lend themselves readily to describing the natural phenomena previously mentioned.

The revenue of McMenamy's Fish Shanty follows a cyclical pattern. When is the revenue of the restaurant increasing most rapidly? In Example 6, page 844, we attempt to answer this question.

TRIGONOMETRIC FUNCTIONS

12.1 MEASUREMENT OF ANGLES

Angles

Recall that an angle consists of two rays that intersect at a common endpoint. If we rotate the ray l_1 in a *counterclockwise* direction about the point O, the angle generated is *positive* (Figure 12.1a). On the other hand, if we rotate the ray l_1 in a *clockwise* direction about the point O, the angle generated is *negative* (Figure 12.1b). We refer to the ray l_1 as the **initial ray,** the ray l_2 as the **terminal ray,** and the endpoint O as the **vertex** of the angle. If A and B are points on l_2 and l_1, respectively, then we refer to the angle as angle AOB (Figure 12.1c).

Figure 12.1

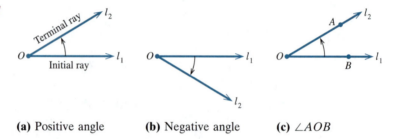

(a) Positive angle   **(b)** Negative angle   **(c)** $\angle AOB$

We can represent an angle in the rectangular coordinate system. An angle is in **standard position** if the vertex of the angle is centered at the origin and its initial side coincides with the positive x-axis (Figures 12.2a and b).

Figure 12.2

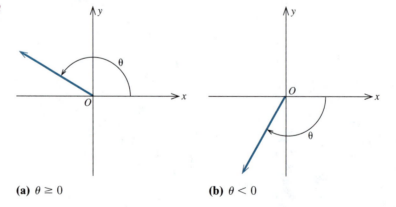

(a) $\theta \geq 0$ **(b)** $\theta < 0$

When we say that an angle lies in a certain quadrant, we are referring to the quadrant in which the terminal ray lies. The angle shown in Figure 12.3a lies in Quadrant II, and the angles shown in Figures 12.3b and c lie in Quadrant IV.

Figure 12.3

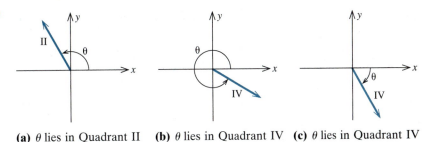

(a) θ lies in Quadrant II   **(b)** θ lies in Quadrant IV   **(c)** θ lies in Quadrant IV

Figure 12.4

$\theta = \dfrac{s}{r} \ radians$

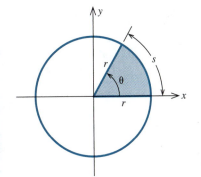

Degree and Radian Measure

An angle may be measured in either degrees or radians. A **degree** is the measure of the angle formed by $\frac{1}{360}$ of one complete revolution. If an initial ray in standard position is rotated through one complete revolution, an angle of 360° is the result.

In our study of the trigonometric functions, we will most often be using radians. A **radian** is the measure of the central angle subtended by an arc equal in length to the radius of the circle. In Figure 12.4, if s is the length of the arc subtended by a central angle θ in a circle of radius r, then

$$\theta = \frac{s}{r} \text{ radians} \qquad (1)$$

Figure 12.5

$\theta = \dfrac{s}{r} = 1 \ radian$

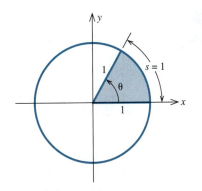

For convenience, we will consider a circle of radius 1 centered at the origin. We refer to this circle as the **unit circle.** Then an angle of 1 radian is subtended by an arc of length 1 (Figure 12.5). The circumference of the unit circle is 2π and, consequently, the angle subtended by one complete revolution is 2π radians. In terms of degrees, the angle subtended by one complete revolution is 360°. Therefore,

$$360 \text{ degrees} = 2\pi \text{ radians}$$

so that

$$1 \text{ degree} = \frac{\pi}{180} \text{ radians}$$

and

$$x \text{ degrees} = \frac{\pi}{180} x \text{ radians} \qquad (2)$$

Since this relationship is linear, we may specify the rule for converting degree measure to radian measure in terms of the following linear function:

CONVERTING DEGREES TO RADIANS

$$f(x) = \frac{\pi}{180} x \text{ radians} \qquad (3)$$

where x is the number of degrees and $f(x)$ is the number of radians.

EXAMPLE 1   Convert each angle to radian measure:

a. 30°   **b.** 45°   **c.** 300°   **d.** 450°   **e.** −240°

Solution   Using formula (3), we have

a. $f(30) = \dfrac{\pi}{180}(30)$ radians, or $\dfrac{\pi}{6}$ radians

b. $f(45) = \dfrac{\pi}{180}(45)$ radians, or $\dfrac{\pi}{4}$ radians

c. $f(300) = \dfrac{\pi}{180}(300)$ radians, or $\dfrac{5\pi}{3}$ radians

d. $f(450) = \dfrac{\pi}{180}(450)$ radians, or $\dfrac{5\pi}{2}$ radians

e. $f(-240) = \dfrac{\pi}{180}(-240)$ radians, or $-\dfrac{4\pi}{3}$ radians

By multiplying both sides of equation (2) by $180/\pi$, we have

$$x \text{ radians} = \frac{180}{\pi} x \text{ degrees}$$

Thus, we have the following rule for converting radian measure to degree measure:

CONVERTING RADIANS TO DEGREES

$$g(x) = \frac{180}{\pi} x \text{ degrees} \tag{4}$$

where x is the number of radians and $g(x)$ is the number of degrees.

EXAMPLE 2   Convert each angle to degree measure:

a. $\dfrac{\pi}{2}$ radians   **b.** $\dfrac{5\pi}{4}$ radians   **c.** $-\dfrac{3\pi}{4}$ radians   **d.** $\dfrac{7\pi}{2}$ radians

Solution   Using formula (4), we have

a. $g\left(\dfrac{\pi}{2}\right) = \dfrac{180}{\pi}\left(\dfrac{\pi}{2}\right)$ degrees, or 90 degrees

b. $g\left(\dfrac{5\pi}{4}\right) = \dfrac{180}{\pi}\left(\dfrac{5\pi}{4}\right)$ degrees, or 225 degrees

c. $g\left(-\dfrac{3\pi}{4}\right) = \dfrac{180}{\pi}\left(-\dfrac{3\pi}{4}\right)$ degrees, or −135 degrees

d. $g\left(\dfrac{7\pi}{2}\right) = \dfrac{180}{\pi}\left(\dfrac{7\pi}{2}\right)$ degrees, or 630 degrees

Take time to familiarize yourself with the radian and degree measures of the common angles given in Table 12.1.

Table 12.1

Degrees	0°	30°	45°	60°	90°	120°	135°	150°	180°	270°	360°
Radians	0	$\dfrac{\pi}{6}$	$\dfrac{\pi}{4}$	$\dfrac{\pi}{3}$	$\dfrac{\pi}{2}$	$\dfrac{2\pi}{3}$	$\dfrac{3\pi}{4}$	$\dfrac{5\pi}{6}$	π	$\dfrac{3\pi}{2}$	2π

You may have observed by now that more than one angle may be described by the same initial and terminal rays. The two angles in Figures 12.6a and b illustrate this case. The angle $\theta = 5\pi/4$ radians is generated by rotating the initial ray in a counterclockwise direction, and the angle $\theta = -3\pi/4$ radians is generated by rotating the initial ray in a clockwise direction. We refer to such angles as **coterminal** angles.

Figure 12.6
Coterminal angles.

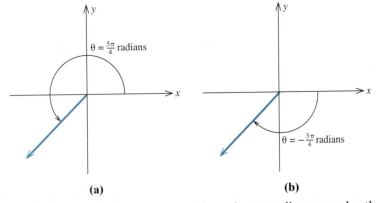

(a)   (b)

You may also have observed from the preceding examples that an angle may be greater than 2π radians. For example, an angle of 3π radians is generated by rotating a ray through one and a half revolutions. As illustrated in Figure 12.7, an angle of 3π radians has its initial and terminal rays in the same position as an angle of π radians. A similar statement is true for the negative angles $-\pi$ and -3π radians. Thus, in identifying an angle, we must be careful to specify the *direction* of the rotation and the *number of revolutions* through which the angle has gone.

Figure 12.7
$\theta = 3\pi$ radians has its initial and ter-minal rays in the same position as $\theta = \pi$ radians.

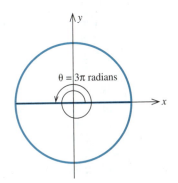

SELF-CHECK EXERCISES 12.1

1. a. Convert 315° to radian measure.

   b. Convert $-\dfrac{5\pi}{4}$ radians to degree measure.

2. Make a sketch of the angle $-\dfrac{2\pi}{3}$ radians.

Solutions to Self-Check Exercises 12.1 can be found on page 827.

12.1 EXERCISES

In exercises 1–4, express each of the angles shown in the diagrams in radian measure.

1.

2.

3.

4.

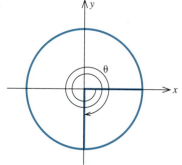

5. Identify the quadrant in which each of the angles lies.
   a. 220°   **b.** −110°
   c. 460°   **d.** −310°

6. Identify the quadrant in which each of the angles lies.
   a. $\dfrac{13}{6}\pi$ radians   **b.** $-\dfrac{11}{4}\pi$ radians

   c. $\dfrac{17}{3}\pi$ radians   **d.** $-\dfrac{25}{12}\pi$ radians

In exercises 7–12, convert each of the given angles to radian measure.

7. 75°   **8.** 330°   **9.** 160°
10. −210°   **11.** 630°   **12.** −420°

In exercises 13–18, convert each of the given angles to degree measure.

13. $\dfrac{2}{3}\pi$ radians   **14.** $\dfrac{7}{6}\pi$ radians   **15.** $-\dfrac{3}{2}\pi$ radians

16. $-\dfrac{13}{12}\pi$ radians   **17.** $\dfrac{22}{18}\pi$ radians   **18.** $-\dfrac{21}{6}\pi$ radians

In exercises 19–22, make a sketch of each of the given angles on a unit circle centered at the origin.

19. 225°   **20.** −120°

21. $\dfrac{7}{3}\pi$ radians   **22.** $-\dfrac{13}{6}\pi$ radians

In exercises 23–26, determine a positive angle and a negative angle that are coterminal angles of the angle θ. Use degree measure.

23.

$\theta = \frac{5\pi}{6}$

24.

$\theta = \frac{5\pi}{3}$

25.

$\theta = -\frac{\pi}{4}$

26.

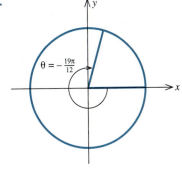

$\theta = -\frac{19\pi}{12}$

SOLUTIONS TO SELF-CHECK EXERCISES 12.1

1. a. Using formula (3), we find that the required radian measure is

$$f(315) = \frac{\pi}{180}(315) = \frac{7\pi}{4}, \quad \text{or} \quad \frac{7\pi}{4}\text{ radians}$$

b. Using formula (4), we find that the required degree measure is

$$g\left(-\frac{5\pi}{4}\right) = \frac{180}{\pi}\left(-\frac{5\pi}{4}\right) = -225, \quad \text{or} \quad -225°$$

2.

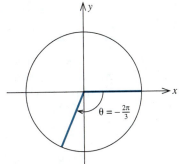

$\theta = -\frac{2\pi}{3}$

THE TRIGONOMETRIC FUNCTIONS

The Trigonometric Functions and Their Graphs

Let $P(x, y)$ be a point on the unit circle so that the radius OP forms an angle of θ radians ($0 \le \theta < 2\pi$) with respect to the positive x-axis (see Figure 12.8). We define the **sine** of the angle θ, written $\sin \theta$, to be the y-coordinate of P. Similarly, the **cosine** of the angle θ, written $\cos \theta$, is defined to be the x-coordinate of P. The other trigonometric functions, tangent, cosecant, secant, and cotangent of θ—written $\tan \theta$, $\csc \theta$, $\sec \theta$, and $\cot \theta$, respectively—are defined in terms of the sine and cosine functions.

Figure 12.8
P is a point on the unit circle with coordinates $x = \cos \theta$ and $y = \sin \theta$.

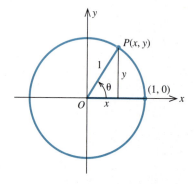

THE TRIGONOMETRIC FUNCTIONS	If P is a point on the unit circle and the coordinates of P are (x, y), then

$$\cos \theta = x \quad \text{and} \quad \sin \theta = y$$

$$\tan \theta = \frac{y}{x} = \frac{\sin \theta}{\cos \theta} \quad (x \ne 0)$$

$$\csc \theta = \frac{1}{y} = \frac{1}{\sin \theta} \quad (y \ne 0)$$

$$\sec \theta = \frac{1}{x} = \frac{1}{\cos \theta} \quad (x \ne 0)$$

$$\cot \theta = \frac{x}{y} = \frac{\cos \theta}{\sin \theta} \quad (y \ne 0)$$

As you work with trigonometric functions, it is useful to remember the values of the sine, cosine, and tangent of some important angles, such as $\theta = 0$, $\pi/6$, $\pi/4$, $\pi/3$, $\pi/2$, and so on. These values may be found using elementary trigonometry. For example, if $\theta = 0$, then the point P has coordinates $(1, 0)$ [see Figure 12.9a], and we see that

$$\sin 0 = y = 0, \quad \cos 0 = x = 1, \quad \tan 0 = \frac{y}{x} = 0$$

As another example, the diagram for $\theta = \pi/4$ (Figure 12.9b) suggests that $x = y$, so that, by the Pythagorean Theorem, we have

$$x^2 + y^2 = 2x^2 = 1$$

and $x = y = \sqrt{2}/2$. Therefore,

$$\sin\frac{\pi}{4} = y = \frac{\sqrt{2}}{2}, \quad \cos\frac{\pi}{4} = x = \frac{\sqrt{2}}{2}, \quad \tan\frac{\pi}{4} = \frac{y}{x} = 1$$

Figure 12.9

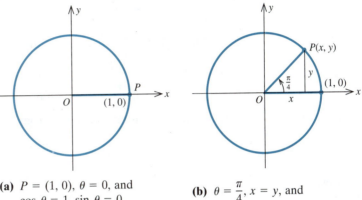

(a) $P = (1, 0)$, $\theta = 0$, and $\cos\theta = 1$, $\sin\theta = 0$

(b) $\theta = \dfrac{\pi}{4}$, $x = y$, and

$$\cos\theta = \sin\theta = \frac{\sqrt{2}}{2}$$

The values of the sine, cosine, and tangent of some common angles are given in Table 12.2.

Table 12.2

θ in radians	0	$\dfrac{\pi}{6}$	$\dfrac{\pi}{4}$	$\dfrac{\pi}{3}$	$\dfrac{\pi}{2}$	$\dfrac{2}{3}\pi$	$\dfrac{3}{4}\pi$	$\dfrac{5}{6}\pi$	π	$\dfrac{7}{6}\pi$	$\dfrac{5}{4}\pi$	$\dfrac{4}{3}\pi$	$\dfrac{3}{2}\pi$	$\dfrac{5}{3}\pi$	$\dfrac{7}{4}\pi$	$\dfrac{11}{6}\pi$	2π
$\sin\theta$	0	$\dfrac{1}{2}$	$\dfrac{\sqrt{2}}{2}$	$\dfrac{\sqrt{3}}{2}$	1	$\dfrac{\sqrt{3}}{2}$	$\dfrac{\sqrt{2}}{2}$	$\dfrac{1}{2}$	0	$-\dfrac{1}{2}$	$-\dfrac{\sqrt{2}}{2}$	$-\dfrac{\sqrt{3}}{2}$	-1	$-\dfrac{\sqrt{3}}{2}$	$-\dfrac{\sqrt{2}}{2}$	$-\dfrac{1}{2}$	0
$\cos\theta$	1	$\dfrac{\sqrt{3}}{2}$	$\dfrac{\sqrt{2}}{2}$	$\dfrac{1}{2}$	0	$-\dfrac{1}{2}$	$-\dfrac{\sqrt{2}}{2}$	$-\dfrac{\sqrt{3}}{2}$	-1	$-\dfrac{\sqrt{3}}{2}$	$-\dfrac{\sqrt{2}}{2}$	$-\dfrac{1}{2}$	0	$\dfrac{1}{2}$	$\dfrac{\sqrt{2}}{2}$	$\dfrac{\sqrt{3}}{2}$	1
$\tan\theta$	0	$\dfrac{1}{\sqrt{3}}$	1	$\sqrt{3}$	∞	$-\sqrt{3}$	-1	$-\dfrac{1}{\sqrt{3}}$	0	$\dfrac{1}{\sqrt{3}}$	1	$\sqrt{3}$	$-\infty$	$-\sqrt{3}$	-1	$-\dfrac{1}{\sqrt{3}}$	0

Since the rotation of the radius OP by 2π radians leaves it in its original configuration (see Figure 12.10), we see that

$$\sin(\theta + 2\pi) = \sin\theta \quad \text{and} \quad \cos(\theta + 2\pi) = \cos\theta$$

It can be shown that the number 2π is the smallest positive number such that the preceding equation holds true. That is, the sine and cosine functions are

Figure 12.10

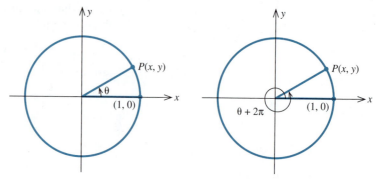

(a) The angle θ **(b)** The angle $(\theta + 2\pi)$

periodic with **period 2π.** It follows that

$$\sin(\theta + 2n\pi) = \sin\theta \quad \text{and} \quad \cos(\theta + 2n\pi) = \cos\theta \tag{5}$$

whenever n is an integer. Also, from Figure 12.11, we see that

$$\sin(-\theta) = -\sin\theta \quad \text{and} \quad \cos(-\theta) = \cos\theta \tag{6}$$

Figure 12.11
$\mathrm{Sin}(-\theta) = -\sin\theta$ *and*
$\cos(-\theta) = \cos\theta.$

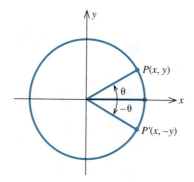

EXAMPLE 1   Evaluate

a. $\sin\dfrac{7\pi}{2}$ **b.** $\cos 5\pi$ **c.** $\sin\left(-\dfrac{5\pi}{2}\right)$ **d.** $\cos\left(-\dfrac{11\pi}{4}\right)$

Solution

a. Using (5) and Table 12.2, we find

$$\sin\left(\frac{7\pi}{2}\right) = \sin\left(2\pi + \frac{3\pi}{2}\right) = \sin\frac{3\pi}{2} = -1 \cdot$$

b. Using (5) and Table 12.2, we have

$$\cos 5\pi = \cos(4\pi + \pi) = \cos\pi = -1$$

c. Using (5), (6), and Table 12.2, we have

$$\sin\left(-\frac{5\pi}{2}\right) = -\sin\left(\frac{5\pi}{2}\right) = -\sin\left(2\pi + \frac{\pi}{2}\right) = -\sin\frac{\pi}{2} = -1$$

d. Using (5), (6), and Table 12.2, we have

$$\cos\left(-\frac{11\pi}{4}\right) = \cos\frac{11\pi}{4} = \cos\left(2\pi + \frac{3\pi}{4}\right) = \cos\frac{3\pi}{4} = -\frac{\sqrt{2}}{2} \qquad \text{o o o}$$

To draw the graph of the function $y = f(x) = \sin x$, we first note that $\sin x$ is defined for every real number x so that the domain of the sine function is $(-\infty, \infty)$. Next, since the sine function is periodic with period 2π, it suffices to concentrate on sketching that part of the graph of $y = \sin x$ on the interval $[0, 2\pi]$ and repeating it as necessary. With the help of Table 12.2, we sketch Figure 12.12, the graph of $y = \sin x$.

Figure 12.12
The graph of $y = \sin x$.

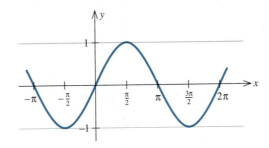

In a similar manner, we can sketch the graph of $y = \cos x$ (Figure 12.13).

Figure 12.13
The graph of $y = \cos x$.

EXPLORING WITH TECHNOLOGY

Plot the graphs of $f(x) = \sin x$ and $g(x) = \cos x$ in the viewing rectangle $[-10, 10] \times [-1.1, 1.1]$.

a. What do the graphs suggest about the relationship between the two trigonometric functions?

b. Confirm your observation by plotting the graphs of $h(x) = \sin\left(x + \frac{\pi}{2}\right)$ and $g(x) = \cos x$ in the viewing rectangle $[-10, 10] \times [-1.1, 1.1]$.

o o o

To sketch the graph of $y = \tan x$, note that $\tan x = \sin x/\cos x$, and so $\tan x$ is not defined when $\cos x = 0$—that is, when $x = (\pi/2) \pm n\pi$ $(n = 0, 1, 2, 3, \ldots)$. The function is defined at all other points so that the domain of the tangent function is the set of all real numbers with the exception of the points just noted. Next, we can show that the vertical lines with equation $x = (\pi/2) \pm n\pi(n = 0, 1, 2, 3, \ldots)$ are vertical asymptotes of the function $f(x) = \tan x$. For example, since

$$\lim_{x \to \frac{\pi}{2}^-} \tan x = \infty \quad \text{and} \quad \lim_{x \to \frac{\pi}{2}^+} \tan x = -\infty$$

which we readily verify with the help of a calculator, we conclude that $x = \pi/2$ is a vertical asymptote of $y = \tan x$. Finally, using Table 12.2, we can sketch the graph of $y = \tan x$ (Figure 12.14).

Figure 12.14
The graph of $y = \tan x$.

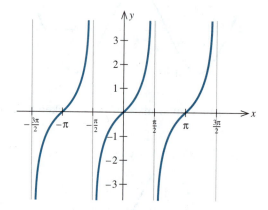

Observe that the tangent function is periodic with period π. It follows that

$$\tan(x + n\pi) = \tan x \tag{7}$$

whenever n is an integer.

The graphs of $y = \sec x$, $y = \csc x$, and $y = \cot x$ may be sketched in a similar manner (see exercises 23–25).

The Predator-Prey Population Model

We will now look at a specific mathematical model of a phenomenon exhibiting cyclical behavior—the so-called **predator-prey population model.**

EXAMPLE 2   The population of owls (predators) in a certain region over a two-year period is estimated to be

$$P(t) = 1000 + 100 \sin\left(\frac{\pi t}{12}\right)$$

in month t, and the population of mice (prey) in the same area at time t is given by

$$p(t) = 20{,}000 + 4000 \cos\left(\frac{\pi t}{12}\right)$$

Sketch the graphs of these two functions and explain the relationship between the sizes of the two populations.

Solution   We first observe that both of the given functions are periodic with period 24 (months). To see this, recall that both the sine and cosine functions are periodic with period 2π. Now the smallest value of $t > 0$ such that $\sin(\pi t/12) = 0$ is obtained by solving the equation

$$\frac{\pi t}{12} = 2\pi$$

giving $t = 24$ as the period of $\sin(\pi t/12)$. Since $P(t + 24) = P(t)$, we see that the function P is periodic with period 24. Similarly, one verifies that the function p is also periodic with period 24, as asserted. Next, we recall that both the sine and cosine functions oscillate between -1 and $+1$ so that $P(t)$ is seen to oscillate between $[1000 + 100(-1)]$, or 900, and $[1000 + 100(1)]$, or 1100, while $p(t)$ oscillates between $[20{,}000 + 4000(-1)]$, or 16,000, and $[20{,}000 + 4000(1)]$, or 24,000. Finally, plotting a few points on each graph for, say, $t = 0, 2, 3$, and so on, we obtain the graphs of the functions P and p as shown in Figure 12.15.

Figure 12.15

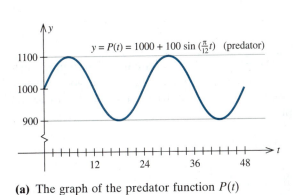

(a) The graph of the predator function $P(t)$

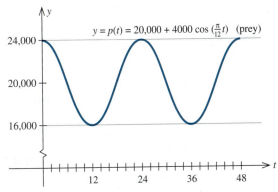

(b) The graph of the prey function $p(t)$

From the graphs, we see that at time $t = 0$ the predator population stands at 1000 owls. As it increases, the prey population decreases from 24,000 mice at that instant of time. Eventually, this decrease in the food supply causes the predator population to decrease, which in turn causes an increase in the prey population. But as the prey population increases, resulting in an increase in food supply, the predator population once again increases. The cycle is complete and starts all over again.

○ ○ ○

Trigonometric Identities

Equations expressing the relationships between trigonometric functions, such as

$$\sin(-\theta) = -\sin \theta \quad \text{and} \quad \cos(-\theta) = \cos \theta$$

are called **trigonometric identities.** Some other important trigonometric identities are listed in Table 12.3. Each identity holds true for every value of θ in the domain of the specified function. The proofs of these identities may be found in any elementary trigonometry book.

Table 12.3
Trigonometric identities.

Pythagorean identities	Half-angle formulas	Sum and difference formulas
$\sin^2 \theta + \cos^2 \theta = 1$	$\cos^2 \theta = \dfrac{1}{2}(1 + \cos 2\theta)$	$\sin(A \pm B) = \sin A \cos B \pm \cos A \sin B$
$\tan^2 \theta + 1 = \sec^2 \theta$	$\sin^2 \theta = \dfrac{1}{2}(1 - \cos 2\theta)$	$\cos(A \pm B) = \cos A \cos B \mp \sin A \sin B$
$\cot^2 \theta + 1 = \csc^2 \theta$		

Double-angle formulas	Cofunctions of complementary angles
$\sin 2A = 2 \sin A \cos A$	$\sin \theta = \cos\left(\dfrac{\pi}{2} - \theta\right)$
$\cos 2A = \cos^2 A - \sin^2 A$	$\cos \theta = \sin\left(\dfrac{\pi}{2} - \theta\right)$

As we will see later, these identities are useful in simplifying trigonometric expressions and equations and in deriving other trigonometric relationships. They are also used to verify other trigonometric identities, as illustrated in the next example.

EXAMPLE 3   Verify the identity

$$\sin \theta(\csc \theta - \sin \theta) = \cos^2 \theta$$

Solution   We verify this identity by showing that the expression on the left side of the equation can be transformed into the expression on the right side. Thus,

$$\sin \theta(\csc \theta - \sin \theta) = \sin \theta \csc \theta - \sin^2 \theta$$
$$= \sin \theta \frac{1}{\sin \theta} - \sin^2 \theta$$
$$= 1 - \sin^2 \theta$$
$$= \cos^2 \theta$$

○ ○ ○

EXPLORING WITH TECHNOLOGY

a. You can confirm some of the trigonometric identities graphically in many ways. For example, explain why an identity given in the form $f(\theta) = g(\theta)$ over some domain D is equivalent to the following statements:

  (i) The graphs of f and g coincide when viewed in any appropriate viewing rectangle.
  (ii) The graphs of f and $-g$ are reflections of each other with respect to the θ-axis when viewed in any appropriate viewing rectangle.
  (iii) The graph of $h = f - g$ is the graph of the zero function $h(\theta) = 0$ for all θ in the domain of f (and g).

b. Use the observations made in part (a) to verify graphically the identity $\sin^2 \theta + \cos^2 \theta = 1$ by taking $f(x) = \sin^2 x + \cos^2 x$ and $g(x) = 1$. [*Hint:* Enter $f(x)$ as $y_1 = (\sin x)^{\wedge}2 + (\cos x)^{\wedge}2$ and use the viewing rectangle $[-10, 10] \times [-1.1, 1.1]$.]

c. Verify graphically the half-angle formula $\cos^2 \theta = \frac{1}{2}(1 + \cos 2\theta)$.

d. Verify graphically the formula $\sin \theta = \cos\left(\dfrac{\pi}{2} - \theta\right)$.

○   ○   ○

SELF-CHECK EXERCISES 12.2

  1. Evaluate $\cos\left(-\dfrac{13\pi}{3}\right)$.

  2. Solve the equation $\cos \theta = -\dfrac{\sqrt{2}}{2}$ for $0 \le \theta \le 2\pi$.

  3. Sketch the graph of $y = 2 \cos x$.

  Solutions to Self-Check Exercises 12.2 can be found on page 837.

12.2 EXERCISES

In exercises 1–10, evaluate the given trigonometric functions.

1. $\sin 3\pi$

2. $\cos\left(-\dfrac{3}{2}\pi\right)$

3. $\sin \dfrac{9}{2}\pi$

4. $\cos \dfrac{13}{6}\pi$

5. $\sin\left(-\dfrac{4}{3}\pi\right)$

6. $\cos\left(-\dfrac{5}{4}\pi\right)$

7. $\tan \dfrac{\pi}{6}$

8. $\cot\left(-\dfrac{\pi}{3}\right)$

9. $\sec\left(-\dfrac{5}{8}\pi\right)$

10. $\csc \dfrac{9}{4}\pi$

In exercises 11–14, find the six trigonometric functions of each of the given angles.

11. $\dfrac{\pi}{2}$

12. $-\dfrac{\pi}{6}$

13. $\dfrac{5}{3}\pi$

14. $-\dfrac{3}{4}\pi$

In exercises 15–22, find all values of θ that satisfy the given equation over the interval [0, 2π].

15. $\sin \theta = -\dfrac{1}{2}$

16. $\tan \theta = 1$

17. $\cot \theta = -\sqrt{3}$

18. $\csc \theta = \sqrt{2}$

19. $\sec \theta = -1$

20. $\cos \theta = \sin \theta$

21. $\sin \theta = \sin\left(-\dfrac{4}{3}\pi\right)$

22. $\cos \theta = \cos\left(-\dfrac{\pi}{6}\right)$

In exercises 23–26, sketch the graph of the given function over the interval [0, 2π].

23. $y = \csc x$

24. $y = \sec x$

25. $y = \cot x$

26. $y = \tan 2x$

In exercises 27–30, sketch the graph of each of the given functions over the interval [0, 2π] by comparing it to the graph of y = sin x.

27. $y = \sin 2x$

28. $y = 2 \sin x$

29. $y = -\sin x$

30. $y = -2 \sin 2x$

In exercises 31–38, verify each of the given identities.

31. $\cos^2 \theta - \sin^2 \theta = 2 \cos^2 \theta - 1$

32. $1 - 2 \sin^2 \theta = 2 \cos^2 \theta - 1$

33. $(\sec \theta + \tan \theta)(1 - \sin \theta) = \cos \theta$

34. $\dfrac{\sin^2 \theta}{1 + \cos^2 \theta} = \dfrac{1 - \cos^2 \theta}{2 - \sin^2 \theta}$

35. $(1 + \cot^2 \theta)\tan^2 \theta = \sec^2 \theta$

36. $\dfrac{\sec \theta - \cos \theta}{\tan \theta} = \sin \theta$

37. $\dfrac{\csc \theta}{\tan \theta + \cot \theta} = \cos \theta$

38. $\tan(A + B) = \dfrac{\tan A + \tan B}{1 - \tan A \tan B}$

39. The accompanying figure shows a right triangle ABC superimposed over a unit circle in the xy-coordinate

system. By considering similar triangles, show that

$$\sin \theta = \frac{BC}{AC} = \frac{\text{Opposite side}}{\text{Hypotenuse}}$$

$$\cos \theta = \frac{AB}{AC} = \frac{\text{Adjacent side}}{\text{Hypotenuse}}$$

$$\tan \theta = \frac{BC}{AB} = \frac{\text{Opposite side}}{\text{Adjacent side}}$$

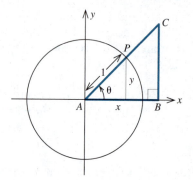

40. Refer to exercise 39. Find the values of the six trigonometric functions of θ, where θ is the angle shown in the right triangle in the accompanying figure.

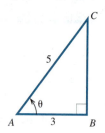

41. Refer to exercise 39. Find the values of the six trigonometric functions of θ, where θ is the angle shown in the right triangle in the accompanying figure.

42. **Predator-Prey Population** The population of foxes (predators) in a certain region over a two-year period is estimated to be

$$P(t) = 400 + 50 \sin\left(\frac{\pi t}{12}\right)$$

in month t, and the population of rabbits (prey) in the same region at time t is given by

$$p(t) = 3000 + 500 \cos\left(\frac{\pi t}{12}\right)$$

Sketch the graphs of each of these two functions and explain the relationship between the sizes of the two populations.

43. Blood Pressure   The arterial blood pressure of an individual in a state of relaxation is given by

$$P(t) = 100 + 20 \sin 6t$$

where $P(t)$ is measured in mm of mercury (Hg) and t is the time in seconds.

a. Show that the individual's systolic pressure (maximum blood pressure) is 120 and his diastolic pressure (minimum blood pressure) is 80.
b. Find the values of t when the individual's blood pressure is highest and lowest.

SOLUTIONS TO SELF-CHECK EXERCISES 12.2

1. $\cos\left(-\dfrac{13\pi}{3}\right) = \cos\dfrac{13\pi}{3}$

$$= \cos\left(4\pi + \frac{\pi}{3}\right)$$

$$= \cos\frac{\pi}{3} = \frac{1}{2}$$

2. Cos θ is negative for θ in Quadrants II and III. From Table 12.2, we see that the required values of θ are $3\pi/4$ and $5\pi/4$.

3. By comparison with the graph of $y = \cos x$, we obtain the accompanying graph of $y = 2 \cos x$.

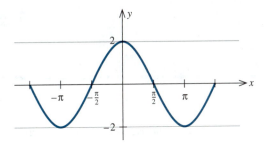

12.3 DIFFERENTIATION OF TRIGONOMETRIC FUNCTIONS

In this section we will develop rules for differentiating trigonometric functions. Knowledge of the derivatives of the trigonometric functions will enable us to study the properties of functions involving the trigonometric functions in much the same way we analyzed the algebraic, exponential, and logarithmic functions in our earlier work.

The Derivatives of the Sine and Cosine Functions

We begin by deriving the rule for differentiating the sine function. Then, using the usual rules of differentiation, we will be able to derive rules for differentiating the other trigonometric functions. For a start, we need the following two limits, whose proofs we postpone until the end of the section:

$$\lim_{h \to 0} \frac{\sin h}{h} = 1 \quad \text{and} \quad \lim_{h \to 0} \frac{\cos h - 1}{h} = 0 \tag{8}$$

EXPLORING WITH TECHNOLOGY

You can obtain a visual confirmation of the limits $\lim\limits_{h \to 0} \dfrac{\sin h}{h} = 1$ and $\lim\limits_{h \to 0} \dfrac{\cos h - 1}{h} = 0$ given in equation (8) as follows:

a. Plot the graph of $f(x) = (\sin x)/x$ using the viewing rectangle $[-1, 1] \times [0, 2]$. Then use ZOOM and TRACE to find the values of $f(x)$ for values of x close to $x = 0$. What happens when you try to evaluate $f(0)$? Explain your answer.

b. Plot the graph of $g(x) = (\cos x - 1)/x$ using the viewing rectangle $[-1, 1] \times [-0.5, 0.5]$. Then use ZOOM and TRACE to show that $g(x)$ is close to zero if x is close to zero.

Caution: Convincing as these arguments are, they do not establish the stated results. As noted earlier, the proofs are given at the end of the section.

Now suppose $f(x) = \sin x$. Then

$$f'(x) = \lim_{h \to 0} \frac{f(x + h) - f(x)}{h}$$

$$= \lim_{h \to 0} \frac{\sin(x + h) - \sin x}{h}$$

$$= \lim_{h \to 0} \frac{\sin x \cos h + \cos x \sin h - \sin x}{h} \qquad \text{(Using the Sine Rule for sums)}$$

$$= \lim_{h \to 0} \frac{\sin x (\cos h - 1) + \cos x \sin h}{h}$$

$$= \lim_{h \to 0} \sin x \left[\frac{\cos h - 1}{h} \right] + \lim_{h \to 0} \cos x \left[\frac{\sin h}{h} \right]$$

$$= \sin x \lim_{h \to 0} \left[\frac{\cos h - 1}{h} \right] + \cos x \lim_{h \to 0} \left[\frac{\sin h}{h} \right]$$

$$= (\sin x)(0) + (\cos x)(1) \qquad \text{[Using equation (8)]}$$

$$= \cos x$$

With the help of the General Power Rule, this result can be generalized as follows: Suppose

$$y = h(x) = \sin f(x)$$

where $f(x)$ is a differentiable function of x. Then putting

$$u = f(x)$$

so that $y = \sin u$, we find, upon applying the Chain Rule,

$$h'(x) = \frac{dy}{du} \cdot \frac{du}{dx} = (\cos u) \frac{du}{dx} = [\cos f(x)] f'(x)$$

THE DERIVATIVES OF THE SINE AND GENERALIZED SINE FUNCTIONS

Rule 1a: $\dfrac{d}{dx} (\sin x) = \cos x$

Rule 1b: $\dfrac{d}{dx} (\sin f(x)) = [\cos f(x)] f'(x)$

Observe that Rule 1b reduces to Rule 1a when $f(x) = x$, as expected.

EXAMPLE 1   Differentiate each of the following functions:

a. $f(x) = x^2 \sin x$ **b.** $g(x) = \sin(2x + 1)$ **c.** $h(x) = (x + \sin x^2)^{10}$

Solution

a. Using the Product Rule followed by Rule 1a, we find

$$f'(x) = 2x \sin x + x^2 \frac{d}{dx} (\sin x)$$

$$= 2x \sin x + x^2 \cos x = x(2 \sin x + x \cos x)$$

b. Using Rule 1b, we find

$$g'(x) = [\cos(2x + 1)] \cdot \frac{d}{dx} (2x + 1) = 2 \cos(2x + 1)$$

c. We first use the General Power Rule followed by Rule 1b. We obtain

$$h'(x) = 10(x + \sin x^2)^9 \cdot \frac{d}{dx} (x + \sin x^2)$$

$$= 10(x + \sin x^2)^9 \left[1 + \cos x^2 \cdot \frac{d}{dx} (x^2) \right]$$

$$= 10(x + \sin x^2)^9 (1 + 2x \cos x^2) \qquad \circ\ \circ\ \circ$$

To derive the rule for differentiating the cosine function, we make use of the following relationships between the cosine function and the sine function:

$$\cos x = \sin\left(\frac{\pi}{2} - x\right) \quad \text{and} \quad \sin x = \cos\left(\frac{\pi}{2} - x\right)$$

(see Table 12.3). Now, with the help of Rule 1b, we see that if $f(x) = \cos x$, then

$$f'(x) = \frac{d}{dx}\cos x$$

$$= \frac{d}{dx}\sin\left(\frac{\pi}{2} - x\right)$$

$$= \cos\left(\frac{\pi}{2} - x\right) \cdot \frac{d}{dx}\left(\frac{\pi}{2} - x\right)$$

$$= (\sin x)(-1)$$

$$= -\sin x$$

This result may be generalized immediately using the General Power Rule. In fact, if

$$y = h(x) = \cos f(x)$$

where $f(x)$ is a differentiable function of x, then with $u = f(x)$, we have $y = \cos u$ so that

$$\frac{dy}{dx} = \frac{dy}{du} \cdot \frac{du}{dx} = (-\sin u)\frac{du}{dx} = -[\sin f(x)]f'(x)$$

THE DERIVATIVES OF THE COSINE AND GENERALIZED COSINE FUNCTIONS

Rule 2a: $\dfrac{d}{dx}(\cos x) = -\sin x$

Rule 2b: $\dfrac{d}{dx}(\cos f(x)) = -[\sin f(x)]f'(x)$

Observe that Rule 2b reduces to Rule 2a when $f(x) = x$, as expected.

EXPLORING WITH TECHNOLOGY

Derive Rule 2a: $\dfrac{d}{dx}(\cos x) = -\sin x$ directly from the definition of the derivative and the relationships given in equation (8).

● ● ●

EXAMPLE 2 Find the derivative of each function:

a. $f(x) = \cos(2x^2 - 1)$ **b.** $g(x) = \sqrt{\cos 2x}$

c. $h(x) = e^{\sin 2x + \cos 3x}$

Solution

a. Using Rule 2b, we find

$$f'(x) = -\sin(2x^2 - 1) \cdot \frac{d}{dx}(2x^2 - 1)$$

$$= -[\sin(2x^2 - 1)]4x$$

$$= -4x \sin(2x^2 - 1)$$

b. We first rewrite $g(x)$ as $g(x) = (\cos 2x)^{1/2}$. Using the General Power Rule followed by Rule 2b, we find

$$g'(x) = \frac{1}{2}(\cos 2x)^{-1/2} \cdot \frac{d}{dx}(\cos 2x)$$

$$= \frac{1}{2}(\cos 2x)^{-1/2}(-\sin 2x) \cdot \frac{d}{dx}(2x)$$

$$= \frac{1}{2}(\cos 2x)^{-1/2}(-\sin 2x)(2)$$

$$= -\frac{\sin 2x}{\sqrt{\cos 2x}}$$

c. Using the corollary to the Chain Rule for exponential functions, we find

$$h'(x) = e^{\sin 2x + \cos 3x} \cdot \frac{d}{dx}(\sin 2x + \cos 3x)$$

$$= e^{\sin 2x + \cos 3x}\left[(\cos 2x)\frac{d}{dx}(2x) - (\sin 3x)\frac{d}{dx}(3x)\right]$$

$$= (2 \cos 2x - 3 \sin 3x)e^{\sin 2x + \cos 3x} \qquad \text{◦ ◦ ◦}$$

The Derivatives of the Other Trigonometric Functions

We are now in a position to find the rule for differentiating the tangent function. In fact, using the Quotient Rule, we see that if $f(x) = \tan x$, then

$$f'(x) = \frac{d}{dx}\tan x = \frac{d}{dx}\left(\frac{\sin x}{\cos x}\right)$$

$$= \frac{(\cos x)\dfrac{d}{dx}\sin x - (\sin x)\dfrac{d}{dx}\cos x}{\cos^2 x}$$

$$= \frac{(\cos x)(\cos x) - (\sin x)(-\sin x)}{\cos^2 x}$$

$$= \frac{\cos^2 x + \sin^2 x}{\cos^2 x} = \frac{1}{\cos^2 x} = \sec^2 x$$

Once again, this result may be generalized with the help of the General Power Rule, and we are led to the next two rules.

THE DERIVATIVES OF THE TANGENT AND THE GENERALIZED TANGENT FUNCTIONS

Rule 3a: $\dfrac{d}{dx}(\tan x) = \sec^2 x$

Rule 3b: $\dfrac{d}{dx}(\tan f(x)) = [\sec^2 f(x)]f'(x)$

The rules for differentiating the remaining three trigonometric functions are derived in a similar manner. We state the generalized versions of these rules. Here $f(x)$ is a differentiable function of x.

DERIVATIVES OF THE COSECANT, SECANT, AND COTANGENT FUNCTIONS

Rule 4: $\dfrac{d}{dx}(\csc f(x)) = -[\csc f(x)][\cot f(x)]f'(x)$

Rule 5: $\dfrac{d}{dx}(\sec f(x)) = [\sec f(x)][\tan f(x)]f'(x)$

Rule 5: $\dfrac{d}{dx}(\cot f(x)) = -[\csc^2 f(x)]f'(x)$

EXAMPLE 3 Find an equation of the tangent line to the graph of the function $f(x) = \tan 2x$ at the point $(\pi/8, 1)$.

Solution The slope of the tangent line at any point on the graph of f is given by

$$f'(x) = 2\sec^2 2x$$

In particular, the slope of the tangent line at the point $(\pi/8, 1)$ is given by

$$f'\left(\frac{\pi}{8}\right) = 2\sec^2\frac{\pi}{4}$$

$$= 2\left(\frac{2}{\sqrt{2}}\right)^2 = 4$$

Therefore, a required equation is given by

$$y - 1 = 4\left(x - \frac{\pi}{8}\right)$$

or

$$y = 4x + \left(1 - \frac{\pi}{2}\right)$$

Applications

The techniques developed in Chapter 4 may be used to study the properties of functions involving trigonometric functions, as the next examples show.

EXAMPLE 4   The owl population in a certain area is estimated to be

$$P(t) = 1000 + 100 \sin\left(\frac{\pi t}{12}\right)$$

in month t, and the mouse population in the same area at time t is given by

$$p(t) = 20{,}000 + 4000 \cos\left(\frac{\pi t}{12}\right)$$

Find the rate of change of the population when $t = 2$.

Solution   The rate of change of the owl population at any time t is given by

$$P'(t) = 100\left[\cos\left(\frac{\pi t}{12}\right)\right]\left(\frac{\pi}{12}\right)$$

$$= \frac{25\pi}{3} \cos\left(\frac{\pi t}{12}\right)$$

and the rate of change of the mouse population at any time t is given by

$$p'(t) = 4000\left[-\sin\left(\frac{\pi t}{12}\right)\right]\left(\frac{\pi}{12}\right)$$

$$= -\frac{1000\pi}{3} \sin\left(\frac{\pi t}{12}\right)$$

In particular, when $t = 2$ the rate of change of the owl population is

$$P'(2) = \frac{25\pi}{3} \cos\frac{\pi}{6}$$

$$= \frac{25\pi}{3}\left(\frac{\sqrt{3}}{2}\right) \approx 22.7$$

That is, the predator population is increasing at the rate of approximately 22.7 owls per month, and the rate of change of the mouse population is

$$p'(2) = -\frac{1000\pi}{3} \sin\frac{\pi}{6}$$

$$= -\frac{1000\pi}{3}\left(\frac{1}{2}\right) \approx -523.6$$

That is, the prey population is decreasing at the rate of approximately 523.6 mice per month.   ০ ০ ০

EXAMPLE 5 Sketch the graph of the function $y = f(x) = \sin^2 x$ on the interval $[0, \pi]$ by first obtaining the following information:

a. the intervals where f is increasing and where it is decreasing
b. the relative extrema of f
c. the concavity of f
d. the inflection points of f

Figure 12.16
Sign diagram for f'.

Solution

a. $f'(x) = 2 \sin x \cos x = \sin 2x$. Setting $f'(x) = 0$ gives $x = 0$, $\pi/2$, or π. The sign diagram for f' (Figure 12.16) shows that f is increasing on $(0, \pi/2)$ and decreasing on $(\pi/2, \pi)$.

b. From the result of (a) and the following table

x	0	$\dfrac{\pi}{2}$	π
$f(x)$	0	1	0

we see that the endpoints $x = 0$ and $x = \pi$ yield the absolute minimum of f, while the absolute maximum of f is 1.

c. We compute $f''(x) = 2 \cos 2x$. Setting $f''(x) = 0$ gives $2x = \pi/2$ or $3\pi/2$; that is, $x = \pi/4$ or $3\pi/4$. From the sign diagram for f'', shown in Figure 12.17, we conclude that f is concave upward on the intervals $(0, \pi/4)$ and $(3\pi/4, \pi)$ and concave downward on the interval $(\pi/4, 3\pi/4)$.

d. From the results of (c), we see that $(\pi/4, 1/2)$ and $(3\pi/4, 1/2)$ are inflection points of f.

Finally, the graph of f is sketched in Figure 12.18.

Figure 12.17
Sign diagram for f''.

Figure 12.18
The graph of $y = \sin^2 x$.

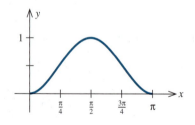

EXAMPLE 6 The revenue of McMenamy's Fish Shanty, located at a popular summer resort, is approximately

$$R(t) = 2\left(5 - 4\cos\frac{\pi}{6}t\right) \qquad (0 \le t \le 12)$$

during the tth week ($t = 1$ corresponds to the first week of June), where R is measured in thousands of dollars. When is the weekly revenue increasing most rapidly?

Solution The revenue function R is increasing at the rate of

$$R'(t) = -8\left[-\sin\frac{\pi t}{6}\right]\left(\frac{\pi}{6}\right)$$

$$= \frac{4\pi}{3}\sin\frac{\pi t}{6}$$

Figure 12.19
The graph of

$$R(t) = 2\left(5 - 4\cos\left(\frac{\pi}{6}t\right)\right).$$

thousand dollars per week. We want to maximize R'. Thus, we compute

$$R''(t) = \frac{4\pi}{3}\left[\cos\frac{\pi t}{6}\right]\left(\frac{\pi}{6}\right)$$

$$= \frac{2\pi^2}{9}\cos\frac{\pi t}{6}$$

Setting $R''(t) = 0$ gives

$$\cos\left(\frac{\pi t}{6}\right) = 0$$

from which we see that $t = 3$ and $t = 9$ are critical points of $R'(t)$. Since $R'(t)$ is a continuous function on the closed interval $[0, 12]$, the absolute maximum must occur at a critical point or at an endpoint of the interval. From the following table we see that the weekly revenue is increasing most rapidly when $t = 3$—that is, in the third week of June (Figure 12.19).

t	0	3	9	12
$R'(t)$	0	$\dfrac{4\pi}{3}$	$-\dfrac{4\pi}{3}$	0

EXAMPLE 7 A trough with a trapezoidal cross section is to be constructed with a 1-foot base and sides that are 6 feet long (Figure 12.20). Find the angle of inclination θ that maximizes the capacity of the trough.

Figure 12.20

(a) We want to maximize the capacity of the trough with the given dimensions.

(b) A cross section of the trough has the shape of a trapezoid.

Solution   The volume of the trough is the product of the area of its cross section and its length. Since the length is constant, it suffices to maximize the area of its cross section. But the cross section is a trapezoid with area given by one-half the sum of the parallel sides times its height, or

$$A = \frac{1}{2}[1 + (1 + 2\cos\theta)]\sin\theta$$

$$= (1 + \cos\theta)\sin\theta$$

To find the absolute maximum of the continuous function A over the closed interval $[0, \pi/2]$, we first compute the derivative of A, obtaining

$$A' = -\sin^2\theta + (1 + \cos\theta)\cos\theta$$

$$= -\sin^2\theta + \cos\theta + \cos^2\theta$$

$$= (\cos^2\theta - 1) + \cos\theta + \cos^2\theta \quad \text{(Using the identity}$$
$$\quad \sin^2\theta + \cos^2\theta = 1)$$

$$= 2\cos^2\theta + \cos\theta - 1$$

$$= (2\cos\theta - 1)(\cos\theta + 1)$$

Setting $A' = 0$ gives $\cos\theta = 1/2$ or $\cos\theta = -1$; that is, $\theta = \pi/3$ [the other values of θ lie outside the interval $(0, \pi/2)$]. Evaluating the function A at the critical point $\theta = \pi/3$ of A and at the endpoints of the interval, we have

θ	0	$\dfrac{\pi}{3}$	$\dfrac{\pi}{2}$
A	0	$\dfrac{3\sqrt{3}}{4}$	1

from which we see that the volume of the trough is maximized when $\theta = \pi/3$, or $60°$.   ◦ ◦ ◦

Verification of Limits in Equation (8)

We close this section by deriving the limits in equation (8). We begin by showing that

$$\lim_{h \to 0} \frac{\sin h}{h} = 1$$

or, equivalently,

$$\lim_{\theta \to 0} \frac{\sin\theta}{\theta} = 1$$

To see this, consider the areas of $\triangle OPQ$, the sector OPR, and $\triangle ORS$ in Figure 12.21. Now the area of $\triangle OPQ$ is equal to

$$\frac{1}{2}|OQ||PQ| = \frac{1}{2}\cos\theta\sin\theta$$

Figure 12.21
Area of $\triangle OPQ = \frac{1}{2}\cos\theta\sin\theta$, area of sector $OPR = \frac{1}{2}\theta$, and area of $\triangle ORS = \dfrac{1}{2}\dfrac{\sin\theta}{\cos\theta}$.

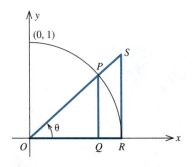

since $|OQ| = \cos\theta$ and $|PQ| = \sin\theta$. Next, using the fact that the area of a circular sector in a circle of radius r subtended by an angle of θ radians is $\frac{1}{2}r^2\theta$, we find that the area of the sector OPR is equal to

$$\frac{1}{2}\theta$$

since $r = 1$. Also, the area of $\triangle ORS$ is equal to

$$\frac{1}{2}|OR||RS| = \frac{1}{2}|RS| \qquad \text{(Since } |OR| = 1\text{)}$$

But, by similar triangles,

$$\frac{|RS|}{|PQ|} = \frac{|OR|}{|OQ|} = \frac{1}{|OQ|} \qquad \text{(Since } |OR| = 1\text{)}$$

and

$$|RS| = \frac{|PQ|}{|OQ|}$$

so that the area of $\triangle ORS$ is, in fact, equal to

$$\frac{1}{2}|RS| = \frac{1}{2}\frac{\sin\theta}{\cos\theta}$$

In view of these computations and the relationship

$$\text{Area of } \triangle OPQ < \text{Area of sector } OPR < \text{Area of } \triangle ORS$$

we see that

$$\frac{1}{2}\cos\theta\sin\theta < \frac{1}{2}\theta < \frac{1}{2}\frac{\sin\theta}{\cos\theta}$$

Dividing each member of the inequality by $\frac{1}{2}\sin\theta$, we have

$$\cos\theta < \frac{\theta}{\sin\theta} < \frac{1}{\cos\theta}$$

or, upon taking reciprocals,

$$\frac{1}{\cos\theta} > \frac{\sin\theta}{\theta} > \cos\theta$$

Finally, observing that $(\sin\theta)/\theta$ is sandwiched between two terms that both tend to 1 as θ tends to 0, we conclude that

$$\lim_{\theta\to 0}\frac{\sin\theta}{\theta} = 1$$

Although the result was derived under the assumption that $\theta > 0$, the identity $\sin(-\theta) = -\sin\theta$ may be used to show that the same result holds when θ approaches 0 through negative values of θ.

To prove

$$\lim_{h\to 0}\frac{\cos h - 1}{h} = 0$$

we compute

$$\lim_{h \to 0} \frac{\cos h - 1}{h} = \lim_{h \to 0} \frac{\cos h - 1}{h} \cdot \frac{\cos h + 1}{\cos h + 1}$$

$$= \lim_{h \to 0} \frac{\cos^2 h - 1}{h(\cos h + 1)}$$

$$= \lim_{h \to 0} \frac{-\sin^2 h}{h(\cos h + 1)}$$

$$= -\left[\lim_{h \to 0} \frac{\sin h}{h}\right]\left[\lim_{h \to 0} \frac{\sin h}{\cos h + 1}\right]$$

$$= -(1)(0) = 0$$

Self-Check Exercises 12.3

1. Find the derivative of $f(x) = x \cos 2x$.

2. If $f(x) = x + \tan x$, find $f'(\pi/4)$.

3. Suppose that the volume of air inhaled by a person during respiration is given by

$$V(t) = \frac{6}{5\pi}\left(1 - \sin\frac{\pi t}{2}\right) \qquad (t \geq 0)$$

liters at time t (in seconds). When is the volume of inhaled air at a maximum? What is the maximum volume?

Solutions to Self-Check Exercises 12.3 can be found on page 851.

12.3 EXERCISES

In exercises 1–30, find the derivative of the given function.

1. $f(x) = \cos 3x$

2. $f(x) = \sin 5x$

3. $f(x) = 2 \cos \pi x$

4. $f(x) = \pi \sin 2x$

5. $f(x) = \sin(x^2 + 1)$

6. $f(x) = \cos \pi x^2$

7. $f(x) = \tan 2x^2$

8. $f(x) = \cot \sqrt{x}$

9. $f(x) = x \sin x$

10. $f(x) = x^2 \cos x$

11. $f(x) = 2 \sin 3x + 3 \cos 2x$

12. $f(x) = 2 \cot 2x + \sec 3x$

13. $f(x) = x^2 \cos 2x$

14. $f(x) = \sqrt{x} \sin \pi x$

15. $f(x) = \sin\sqrt{x^2 - 1}$

16. $f(x) = \csc(x^2 + 1)$

17. $f(x) = e^x \sec x$

18. $f(x) = e^{-x} \csc x$

19. $f(x) = x \cos \dfrac{1}{x}$

20. $f(x) = x^2 \sin \dfrac{1}{x}$

21. $f(x) = \dfrac{x - \sin x}{1 + \cos x}$

22. $f(x) = \dfrac{\sin 2x}{1 + \cos 3x}$

23. $f(x) = \sqrt{\tan x}$

24. $f(x) = \sqrt{\cos x + \sin x}$

25. $f(x) = \dfrac{\sin x}{x}$

26. $f(x) = \dfrac{\cos x}{x^2 + 1}$

27. $f(x) = \tan^2 x$

28. $f(x) = \cot \sqrt{x}$

29. $f(x) = e^{\cot x}$

30. $f(x) = e^{\tan x + \sec x}$

31. Find the equation of the tangent line to the graph of the function $f(x) = \cot 2x$ at the point $(\pi/4, 0)$.

32. Find an equation of the tangent line to the graph of the function $f(x) = e^{\sec x}$ at the point $(\pi/4, e^{\sqrt{2}})$.

33. Determine the intervals where the function

$$f(x) = e^x \cos x \qquad (0 \le x \le 2\pi)$$

is increasing and where it is decreasing.

34. Determine the point(s) of inflection of the function

$$f(x) = x + \sin x \qquad (-2\pi \le x \le 2\pi)$$

In exercises 35–38, sketch the graph of the given function over the specified interval by obtaining the following information:

  a. the intervals where f is increasing and where it is decreasing

  b. the relative extrema of f

  c. the concavity of f

  d. the inflection points of f

35. $f(x) = \sin x + \cos x \quad (0 \le x \le 2\pi)$

36. $f(x) = x - \sin x \quad (0 \le x \le 2\pi)$

37. $f(x) = 2 \sin x + \sin 2x \quad (0 \le x \le 2\pi)$

38. $f(x) = e^x \cos x \quad (0 \le x \le 2\pi)$

39. a. Find the Taylor series about $x = 0$ for the function $f(x) = \sin x$.

  b. Use the result of (a) to show that

$$\lim_{x \to 0} \frac{\sin x}{x} = 1$$

40. a. Find the Taylor series about $x = 0$ for the function $f(x) = \cos x$.

  b. Use the result of (a) to show that

$$\lim_{x \to 0} \frac{\cos x - 1}{x} = 0$$

41. Predator-Prey Population   The wolf population in a certain northern region is estimated to be

$$P(t) = 8000 + 1000 \sin\left(\frac{\pi t}{24}\right)$$

in month t, and the caribou population in the same region is given by

$$p(t) = 40{,}000 + 12{,}000 \cos\left(\frac{\pi t}{24}\right)$$

Find the rate of change of each population when $t = 12$.

42. Stock Prices   The closing price (in dollars) per share of stock of Tempco Electronics on the tth day it was traded is approximated by

$$P(t) = 20 + 12 \sin \frac{\pi t}{30} - 6 \sin \frac{\pi t}{15} + 4 \sin \frac{\pi t}{10}$$

$$- 3 \sin \frac{2\pi t}{15} \qquad (0 \le t \le 20)$$

where $t = 0$ corresponds to the time the stock was first listed in a major stock exchange. What was the rate of change of the stock's price at the close of the fifteenth day of trading? What was the closing price on that day?

43. Average Daily Temperature   The average daily temperature (in degrees Fahrenheit) at a tourist resort in Cameron Highlands is approximated by

$$T = 62 - 18 \cos \frac{2\pi(t - 23)}{365}$$

on the tth day ($t = 1$ corresponds to January 1). Which day was the warmest day of the year? The coldest day?

44. Water Level in a Harbor   The water level (in feet) in Boston Harbor during a certain 24-hour period is approximated by the formula

$$H = 4.8 \sin \frac{\pi}{6}(t - 10) + 7.6 \qquad (0 \le t \le 24)$$

where $t = 0$ corresponds to 12 A.M. When is the water level rising and when is it falling? Find the relative extrema of H and interpret your results.

45. Restaurant Revenue   The revenue of McMenamy's Fish Shanty located at a popular summer resort is approximately

$$R(t) = 2\left(5 - 4\cos\frac{\pi}{6}t\right) \qquad (0 \le t \le 12)$$

during the tth week ($t = 1$ corresponds to the first week of June), where R is measured in thousands of dollars. On what week does the restaurant realize the greatest revenue?

46. Volume of Air Inhaled During Respiration   Suppose the volume of air inhaled by a person during respiration is given by

$$V(t) = \frac{6}{5\pi}\left(1 - \cos\frac{\pi t}{2}\right)$$

liters at time t (in seconds). When is the volume of inhaled air at a maximum? What is the maximum volume?

47. Maximum Air Flow During Respiration   Refer to exercise 46. Find the rate of flow of air in and out of the lungs. At what time is the rate of flow of air at a maximum? A minimum?

48. Televising a Rocket Launch A major network is televising the launching of a rocket. A camera tracking the lift-off of the rocket is located at point A, as shown in accompanying figure, where ϕ denotes the angle of elevation of the camera at A. How fast is ϕ changing at the instant when the rocket is at a distance of 13,000 ft from the camera and this distance is increasing at the rate of 480 ft/sec?

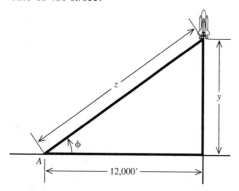

49. Tracking a Suspect A police cruiser hunting for a suspect pulls over and stops at a point 20 ft from a straight wall. The flasher on top of the cruiser revolves at a constant rate of 90° per second, and the light beam casts a spot of light as it strikes the wall. How fast is the spot of light moving along the wall at a point 30 ft from the point on the wall closest to the cruiser?

50. Percentage Error in Measuring Height From a point on level ground 150 ft from the base of a derrick, Jose measures the angle of elevation to the top of the derrick as 60°. If Jose's measurement is subject to an error of ±1%, find the percentage error in the measured height of the derrick. (See the accompanying figure.)

51. Maximizing Drainage Capacity The cross section of a drain is a trapezoid, as shown in the figure. The sides and the bottom of the trapezoid are each 5 ft long. Determine the angle θ so that the drain will have a maximal cross-sectional area.

52. Surface Area of a Honeycomb The accompanying figure depicts a prism-shaped single cell in a honeycomb. The front end of the prism is a regular hexagon, and the back is formed by the sides of the cell coming together at a point. It can be shown that the surface area of a cell is given by

$$S(\theta) = 6ab + \frac{3}{2}b^2 \left(\frac{\sqrt{3} - \cos \theta}{\sin \theta} \right) \qquad \left(0 < \theta < \frac{\pi}{2} \right)$$

where θ is the angle between one of the (three) upper surfaces and the altitude. Show that the surface area is minimized if $\cos \theta = 1/\sqrt{3}$, or $\theta \approx 54.7°$. Measurements of actual honeycombs have confirmed that this is, in fact, the angle found in beehives.

53. Finding the Position of a Planet As shown in the figure on page 851, the position of a planet that revolves about the Sun with an elliptical orbit can be located by calculating the central angle θ. Suppose the central angle sustained by a planet on a certain day satisfies the equation

$$\theta - 0.5 \sin \theta = 1$$

Use the Newton-Raphson method to find θ.

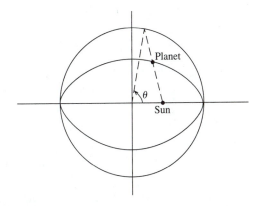

54. Flow of Blood Suppose that some of the fluid flowing along a pipe of radius R is diverted to a pipe of smaller radius r attached to the former at an angle θ (see the figure). Such is the case when blood flowing along an artery is pumped into an arteriole. What should the angle θ be so that the energy loss due to friction in moving the fluid is minimal? Assume that the minimum exists and solve the problem using the following steps:

a. Use Poiseuille's Law, which states that the loss of

energy due to friction in nonturbulent flow is proportional to the length of the path and inversely proportional to the fourth power of the radius, to show that the energy loss in moving the fluid from P to S via Q is

$$L = \frac{kd_1}{R^4} + \frac{kd_2}{r^4}$$

where k is a constant.

b. Suppose a and b are fixed. Find d_1 and d_2 in terms of a and b. Then use this result together with the result from (a) to show that

$$E = k\left[\frac{a - b\cot\theta}{R^4} + \frac{b\csc\theta}{r^4}\right]$$

c. Using the technique of this section, show that E is minimized when

$$\theta = \cos^{-1}\frac{r^4}{R^4}$$

55. Prove Rule 4:

If $h(x) = \csc f(x)$,

then $h'(x) = -[\csc f(x)][\cot f(x)]f'(x)$

[*Hint:* $\csc x = 1/\sin x$]

56. Prove Rule 5:

If $h(x) = \sec f(x)$

then $h'(x) = [\sec f(x)][\tan f(x)]f'(x)$

[*Hint:* $\sec x = 1/\cos x$]

57. Prove Rule 6:

If $h(x) = \cot f(x)$

then $h'(x) = -[\csc^2 f(x)]f'(x)$

[*Hint:* $\cot x = 1/\tan x = (\cos x)/(\sin x)$]

SOLUTIONS TO SELF-CHECK EXERCISES 12.3

1. $f'(x) = x\dfrac{d}{dx}\cos 2x + (\cos 2x)\dfrac{d}{dx}(x)$

$= x(-\sin 2x)(2) + (\cos 2x)(1)$

$= \cos 2x - 2x\sin 2x$

2. $f'(x) = 1 + \sec^2 x$. Therefore,

$$f'\left(\frac{\pi}{4}\right) = 1 + \sec^2\left(\frac{\pi}{4}\right) = 1 + (\sqrt{2})^2 = 3$$

USING TECHNOLOGY

ANALYZING TRIGONOMETRIC FUNCTIONS

Graphing utilities can be used to analyze complicated trigonometric functions, as this example shows.

EXAMPLE 1 Let $f(x) = \dfrac{\sqrt{x} \cos 2x}{\sqrt{x^2 + \sin 3x}}$, where $0 < x \leq 2$.

a. Use a graphing utility to plot the graph of f in the viewing rectangle $[0, 2] \times [-2, 1]$.
b. Use the numerical derivative operation of a graphing utility to find the rate of change of $f(x)$ at $x = 1$.
c. Find the inflection points of f.
d. Evaluate $\lim\limits_{x \to 0^+} f(x)$. Explain why $f(0)$ is not defined. What happens if you use the evaluation function of the graphing utility to find $f(0)$?
e. Find the absolute maximum and absolute minimum values of f on the interval $[0.5, 2]$.

Solution
a. The graph of f in the viewing rectangle $[0, 2] \times [-2, 1]$ is shown in Figure T1.

Figure T1

b. We find $f'(1) \approx -2.0628$.
c. The inflection points are $(1.0868, -0.5734)$ and $(1.9244, -0.5881)$. [*Note:* Move the cursor near $x = 1$ to locate the first inflection point and near $x = 1.5$ to locate the other.]
d. Using ZOOM-IN, we find $\lim\limits_{x \to 0^+} f(x) \approx 0.5773$.

e. Using the operation to find the absolute minimum and the absolute maximum in the viewing window $[0.5, 2] \times [-2, 1]$, we find the absolute minimum value to be -1.0823 and the absolute maximum value to be 0.3421.

○ ○ ○

EXERCISES

In exercises 1–6, use the numerical derivative operation of a graphing utility to find the rate of change of $f(x)$ at the given value of x. Give your answer accurate to four decimal places.

1. $f(x) = x^2 \sqrt{1 + \sin^2 x}; x = 0.5$

2. $f(x) = \dfrac{1 + \cos x}{\sqrt{1 + \sin 3x}}; x = \dfrac{\pi}{4}$

3. $f(x) = \dfrac{x \tan x}{\sqrt{1 + x^2}}; x = 0.4$

4. $f(x) = e^{\cot x}(1 + x^2)^{3/2}; x = 1$

5. $f(x) = \cos[\csc(\sqrt{x} + 1)]; x = \dfrac{\pi}{3}$

6. $f(x) = \dfrac{\ln(\cos x)}{(1 + e^{-x})^{3/2}}; x = 0.8$

In exercises 7–10, use a graphing utility to find the absolute maximum and the absolute minimum values of f in the given interval. Express your answers accurate to four decimal places.

7. $f(x) = \dfrac{\sin x}{x}; [1, 6]$

8. $f(x) = \dfrac{\cos x^2}{\sqrt{x}}; [1, 3]$

9. $f(x) = \dfrac{x \sec x}{1 + \cot x}; [0.5, 1]$

10. $f(x) = \dfrac{x + \cos x}{1 + 0.5 \sin x}; [0, 2]$

11. Refer to exercise 42, page 849.
    a. Plot the graph of P using the viewing rectangle $[0, 20] \times [15, 35]$.
    b. What was the rate of change of the stock's price at the close of the 23rd day of trading?
    c. What was the closing price on that day?

12. Refer to exercise 43, page 849.
    a. Plot the graph of T using the viewing rectangle $[0, 365] \times [20, 80]$.
    b. Using TRACE and ZOOM or using the graphing utility's operation for finding the absolute extrema of a function, verify the solution to the problem.

13. Refer to exercise 46, page 849.
    a. Plot the graph of V using the viewing rectangle $[0, 12] \times [0, 2]$.
    b. Verify the solution to the problem using a graphing utility.

14. **Effect of an Earthquake on a Structure** In order to study the effect an earthquake has on a structure, engineers look at the way a beam bends when subjected to an earth tremor. The equation

$$D = a - a \cos\left(\frac{\pi h}{2L}\right) \qquad (0 \le h \le L)$$

where L is the length of a beam and a is the maximum deflection from the vertical, has been used by engineers to calculate the deflection D at a point on the beam h feet from the ground (see the figure). Suppose a 10-ft vertical beam has a maximum deflection of $\frac{1}{2}$ foot when subjected to an external force. Using differentials, estimate the difference in the deflection between the point midway on the beam and the point $\frac{1}{10}$ foot above it.

3. We compute

$$V'(t) = \frac{6}{5\pi}\left(-\cos\frac{\pi t}{2}\right)\left(\frac{\pi}{2}\right) = -\frac{3}{5}\cos\frac{\pi t}{2}$$

Setting $V'(t) = 0$ and solving the resulting equation give $t = 1 + 2n$ ($n = 1, 2, \ldots$) as critical points of V. The points $t = 3, 5, 7, \ldots$ give rise to the relative maxima of V, which coincide with the absolute maximum of V with value $12/(5\pi)$. Thus, the volume of inhaled air is at a maximum when $t = 3, 5, 7, \ldots$, and its value is $12/(5\pi)$ liters.

12.4 INTEGRATION OF TRIGONOMETRIC FUNCTIONS

Integrating Trigonometric Functions

Each of the six equations pertaining to a rule of differentiation given in the last section may be integrated to yield a corresponding integration formula. For example, integrating the equation $f'(x) = \cos x$ of Rule 1 with respect to x yields

$$\int \cos x \, dx = \int f'(x) \, dx = f(x) + C = \sin x + C$$

where C is a constant of integration. The six rules of integration obtainable in this manner follow.

TRIGONOMETRIC INTEGRATION FORMULAS

1. $\displaystyle\int \sin x \, dx = -\cos x + C$

2. $\displaystyle\int \cos x \, dx = \sin x + C$

3. $\displaystyle\int \sec^2 x \, dx = \tan x + C$

4. $\displaystyle\int \csc^2 x \, dx = -\cot x + C$

5. $\displaystyle\int \sec x \tan x \, dx = \sec x + C$

6. $\displaystyle\int \csc x \cot x \, dx = -\csc x + C$

EXAMPLE 1   Evaluate $\int \cos 3x \, dx$.

Solution   Let us put $u = 3x$, so that $du = 3 \, dx$ and $dx = \frac{1}{3} \, du$. Then

$$\int \cos 3x \, dx = \frac{1}{3}\int \cos u \, du$$

$$= \frac{1}{3}\sin u + C \qquad \text{(Using Rule 2)}$$

$$= \frac{1}{3}\sin 3x + C$$

EXAMPLE 2 Evaluate $\int \sec(2x + 1)\tan(2x + 1)\, dx$.

Solution Put $u = 2x + 1$, so that $du = 2\, dx$, or $dx = \frac{1}{2}\, du$. Then

$$\int \sec(2x + 1)\tan(2x + 1)\, dx = \frac{1}{2} \int \sec u \tan u \, du$$

$$= \frac{1}{2} \sec u + C$$

$$= \frac{1}{2} \sec(2x + 1) + C \qquad \text{\small◦ ◦ ◦}$$

EXAMPLE 3 Evaluate $\int \dfrac{\sin x \, dx}{1 + \cos x}$.

Solution Put $u = 1 + \cos x$, so that $du = -\sin x \, dx$, or $\sin x \, dx = -du$. Then

$$\int \frac{\sin x \, dx}{1 + \cos x} = -\int \frac{du}{u}$$

$$= -\ln |u| + C$$

$$= -\ln |1 + \cos x| + C \qquad \text{\small◦ ◦ ◦}$$

EXAMPLE 4 Evaluate $\int_0^{\pi/2} x \cos 2x \, dx$.

Solution We first integrate the indefinite integral

$$\int x \cos 2x \, dx$$

by parts with

$$u = x \quad \text{and} \quad dv = \cos 2x \, dx$$

so that

$$du = dx \quad \text{and} \quad v = \frac{1}{2} \sin 2x$$

Therefore,

$$\int x \cos 2x \, dx = uv - \int v \, du$$

$$= \frac{1}{2} x \sin 2x - \frac{1}{2} \int \sin 2x \, dx$$

$$= \frac{1}{2} x \sin 2x + \frac{1}{4} \cos 2x + C$$

We have used the method of substitution to evaluate the integral on the right. Therefore,

$$\int_0^{\pi/2} x \cos 2x \, dx = \frac{1}{2} x \sin 2x + \frac{1}{4} \cos 2x \, \Big|_0^{\pi/2}$$

$$= \left[\frac{1}{2} \left(\frac{\pi}{2} \right) \sin \pi + \frac{1}{4} \cos \pi \right] - \left[0 + \frac{1}{4} \cos 0 \right]$$

$$= -\frac{1}{4} - \frac{1}{4} = -\frac{1}{2} \qquad \text{\small◦ ◦ ◦}$$

Refer to the Group Discussion question on page 493.

a. Show that if f is even and g is odd, then the function fg is odd.
b. Refer to Example 4, page 855, and use the result of part (a) to show that

$$\int_{-\pi/2}^{\pi/2} x \cos 2x \, dx = 0$$

Figure 12.22
The area under the curve $y = \sin 2t$
from $t = 0$ to $t = \dfrac{\pi}{4}$.

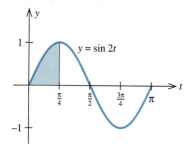

EXAMPLE 5 Find the area under the curve of $y = \sin 2t$ from $t = 0$ to $t = \pi/4$.

Solution The required area, shown in Figure 12.22, is given by

$$\int_0^{\pi/4} \sin 2t \, dt = -\frac{1}{2} \cos 2t \, \Big|_0^{\pi/4} = -\frac{1}{2} \cos \frac{\pi}{2} + \frac{1}{2} \cos 0 = \frac{1}{2}$$

or $\frac{1}{2}$ square unit.

● ● ●

Application

EXAMPLE 6 The weekly closing price of HAL Corporation stock in week t is approximated by the rule

$$f(t) = 30 + t \sin \frac{\pi}{6} t \qquad (0 \le t \le 15)$$

where $f(t)$ is the price (in dollars) per share. Find the average weekly closing price of the stock over the 15-week period.

Solution The average weekly closing price of the stock over the 15-week period in question is given by

$$A = \frac{1}{15 - 0} \int_0^{15} \left[30 + t \sin \frac{\pi}{6} t \right] dt$$

$$= \frac{1}{15} \int_0^{15} 30 \, dt + \frac{1}{15} \int_0^{15} t \sin \frac{\pi}{6} t \, dt$$

$$= 30 + \frac{1}{15} \int_0^{15} t \sin \frac{\pi}{6} t \, dt$$

Integrating by parts with

$$u = t \quad \text{and} \quad dv = \sin \frac{\pi}{6} t \, dt$$

so that

$$du = dt \quad \text{and} \quad v = -\frac{6}{\pi} \cos \frac{\pi}{6} t$$

we have

$$A = 30 + \frac{1}{15} \left[-\frac{6}{\pi} t \cos \frac{\pi}{6} t \Big|_0^{15} + \frac{6}{\pi} \int_0^{15} \cos \frac{\pi}{6} t \, dt \right]$$

$$= 30 + \frac{1}{15} \left[-\left(\frac{6}{\pi}\right)(15) \cos \frac{15\pi}{6} + \frac{6}{\pi}(0) \cos 0 + \left(\frac{6}{\pi}\right)^2 \sin \frac{\pi}{6} t \Big|_0^{15} \right]$$

$$= 30 + \frac{1}{15} \left[-\left(\frac{6}{\pi}\right)(15)(0) + \left(\frac{6}{\pi}\right)^2 \sin \frac{15\pi}{6} - \left(\frac{6}{\pi}\right)^2 \sin 0 \right]$$

$$= 30 + \frac{1}{15} \left(\frac{6}{\pi}\right)^2 (1) \approx 30.24$$

or approximately \$30.24 per share.   ◑ ◑ ◑

Without actually evaluating the integral, explain the following using the properties of the trigonometric functions:

a. $\displaystyle \int_0^{\pi/2} \sin x \, dx = \int_{10\pi}^{21\pi/2} \sin x \, dx$

b. $\displaystyle \int_{-3}^{3} (2 + \cos 3x - \sin 5x) \, dx \geq 0$

Additional Trigonometric Integration Formulas

The six integration formulas listed at the beginning of this section were immediate consequences of the six corresponding rules of differentiation of the previous section. We will now complete the list of integration formulas for the basic trigonometric functions by giving the integration formulas for the tangent, secant, cosecant, and cotangent functions.

ADDITIONAL TRIGONOMETRIC INTEGRATION FORMULAS	
7.	$\displaystyle \int \tan x \, dx = -\ln |\cos x| + C$
8.	$\displaystyle \int \sec x \, dx = \ln |\sec x + \tan x| + C$
9.	$\displaystyle \int \csc x \, dx = \ln |\csc x - \cot x| + C$
10.	$\displaystyle \int \cot x \, dx = \ln |\sin x| + C$

To prove Formula 7, we write

$$\int \tan x \, dx = \int \frac{\sin x}{\cos x} \, dx$$

Putting $u = \cos x$, so that $du = -\sin x \, dx$, we find

$$\int \tan x \, dx = -\int \frac{du}{u} = -\ln |u| + C$$
$$= -\ln |\cos x| + C$$

as we set out to show.

SELF-CHECK EXERCISES 12.4

1. Evaluate $\int_0^{\pi/2} \frac{\cos x \, dx}{1 + \sin x}$.

2. The rate of change of the population of foxes in a certain region in month t is

$$\frac{25\pi}{6} \cos\left(\frac{\pi t}{12}\right)$$

foxes per month. If the population at $t = 0$ is 400, find an expression giving the number of foxes in month t.

Solutions to Self-Check Exercises 12.4 can be found on page 862.

12.4 EXERCISES

In exercises 1–32, evaluate each integral.

1. $\int \sin 3x \, dx$

2. $\int \cos(x + \pi) \, dx$

3. $\int (3 \sin x + 4 \cos x) \, dx$

4. $\int (x^2 - \cos 2x) \, dx$

5. $\int \sec^2 2x \, dx$

6. $\int \csc^2 3x \, dx$

7. $\int x \cos x^2 \, dx$

8. $\int x \sec x^2 \tan x^2 \, dx$

9. $\int \csc \pi x \cot \pi x \, dx$

10. $\int \sec 2x \tan 2x \, dx$

11. $\int_{-\pi/2}^{\pi/2} (\sin x + \cos x) \, dx$

12. $\int_0^{\pi} (2 \cos x + 3 \sin x) \, dx$

13. $\int_0^{\pi/6} \tan 2x \, dx$

14. $\int_{\pi/8}^{\pi/4} \cot 2x \, dx$

15. $\int \sin^3 x \cos x \, dx$

16. $\int \cos^2 x \sin x \, dx$

17. $\int \sec \pi x \, dx$

18. $\int \csc(1 - x) \, dx$

19. $\int_0^{\pi/12} \sec 3x \, dx$

20. $\int_0^{\pi/8} \tan 2x \, dx$

21. $\int \sqrt{\cos x} \sin x \, dx$

22. $\int (\sin x)^{1/3} \cos x \, dx$

23. $\int \cos 3x \sqrt{1 - 2 \sin 3x} \, dx$

24. $\displaystyle\int \frac{\cos x\, dx}{\sqrt{1 + \sin x}}$

25. $\displaystyle\int \tan^3 x \sec^2 x\, dx$

26. $\displaystyle\int \sec^5 x \tan x\, dx$

27. $\displaystyle\int \csc^2 x (\cot x - 1)^3\, dx$

28. $\displaystyle\int \sec^2 x \sqrt{1 + \tan x}\, dx$

29. $\displaystyle\int_1^{e^\pi} \frac{\sin(\ln x)}{x}\, dx$

30. $\displaystyle\int_0^{\pi/2} \frac{\cos x}{1 + \sin x}\, dx$

31. $\displaystyle\int \sin(\ln x)\, dx$

[*Hint:* Integrate by parts.]

32. $\displaystyle\int_1^{e^{\pi/2}} \cos(\ln x)\, dx$

[*Hint:* Integrate by parts.]

In exercises 33–36, find the area of the region under the graph of the function f from x = a to x = b.

33. $f(x) = \cos\dfrac{x}{4}$; $a = 0$, $b = \pi$

34. $f(x) = x + \sin x$; $a = 0$, $b = \dfrac{\pi}{4}$

35. $f(x) = \tan x$; $a = 0$, $b = \dfrac{\pi}{4}$

36. $f(x) = \cot x + \csc x$; $a = \dfrac{\pi}{4}$, $b = \dfrac{\pi}{2}$

37. Find the area of the region bounded above by the graph of $y = x$ and below by the graph of $y = \sin x$ from $x = 0$ to $x = \pi$.

38. Find the area of the region bounded above by the graph of $y = \cos x$ and below by the graph of $y = x$ from $x = 0$ to $x = \pi/4$. Make a rough sketch of the region.

39. Stock Prices The weekly closing price of TMA Corporation stock in week t is approximated by the rule

$$f(t) = 80 + 3t \cos\frac{\pi t}{6} \qquad (0 \le t \le 15)$$

where $f(t)$ is the price (in dollars) per share. Find the average weekly closing price of the stock over the 15-week period.

40. Average Daily Temperature The average daily temperature in degrees Fahrenheit at a tourist resort in the

Cameron Highlands is approximately

$$T = 62 - 18 \cos\frac{2\pi(t - 23)}{365}$$

on the tth day ($t = 1$ corresponds to January 1). What is the average temperature at the resort in the month of January?

[*Hint:* Find the average value of T over the interval $[0, 31]$.]

41. Restaurant Revenue The revenue of McMenamy's Fish Shanty, located at a popular summer resort, is approximately

$$R(t) = 2\left(5 - 4\cos\frac{\pi}{6}t\right) \qquad (0 \le t \le 12)$$

during the tth week ($t = 1$ corresponds to the first week of June), where R is measured in thousands of dollars. What is the total revenue realized by the restaurant over the 12-week period starting June 1?

42. Stock Prices Refer to exercise 42, Section 12.3. What was the average price per share of Tempco Electronics stock over the first 20 days that it was traded on the stock exchange?

43. Volume of Air Inhaled During Respiration Suppose that the rate of air flow in and out of a person's lungs during respiration is

$$R(t) = 0.6 \sin\frac{\pi t}{2}$$

liters per second, where t is the time in seconds. Find an expression for the volume of air in the person's lungs at any time t.

44. Average Volume of Inhaled Air During Respiration Refer to exercise 43. What is the average volume of inspired air in the person's lungs during one respiratory cycle?

[*Hint:* Find the average value of the function giving the volume of inhaled air (obtained in the solution to exercise 43) over the interval $[0, 4]$.]

45. Growth of a Fruit Fly Colony A biologist has determined that by varying the food supply available to a population of *Drosophila* (fruit flies) in a controlled experiment, the rate at which the population grows obeys the rule

$$\frac{dQ}{dt} = 0.0001(4 + 5\cos 2t)Q(400 - Q)$$

USING TECHNOLOGY

EVALUATING INTEGRALS OF TRIGONOMETRIC FUNCTIONS

As we saw in Chapter 6, the numerical integration operation of a graphing utility can be used to evaluate definite integrals involving trigonometric functions.

EXAMPLE I   Use a graphing utility to find the area of the region R that is completely enclosed by the graphs of the functions

$$f(x) = \cos x + \sin 2x \quad \text{and} \quad g(x) = 1 - x^2$$

Solution   The graphs of f and g in the viewing rectangle $[-2, 2] \times [-2, 2]$ are shown in Figure T1. Using the intersection operation of a graphing utility,

Figure T1

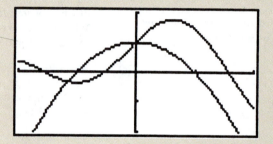

we find that the x-coordinates of the points of intersection of the two graphs are ≈ -1.1554 and 0. Since the graph of g lies above that of f on the interval $[-1.1554, 0]$, we see that the area of R is given by

$$A = \int_{-1.1554}^{0} [(1 - x^2) - (\cos x + \sin 2x)] \, dx$$

$$= \int_{-1.1554}^{0} (1 - x^2 - \cos x - \sin 2x) \, dx$$

$$\approx 0.5635$$

or approximately 0.5635 square unit.   ○ ○ ○

EXERCISES

In exercises 1–4, use a graphing utility to find the area of the region under the graph of f on the interval $[a, b]$. Express your answer accurate to four decimal places.

1. $f(x) = \sqrt{x} \sin 2x$; $[0, 1]$

2. $f(x) = \dfrac{x}{\sin x}$; $[1, 2]$

3. $f(x) = \dfrac{\cos x}{\sqrt{1 + x^2}}$; $[0, 1]$

4. $f(x) = (\cos x) \ln\left(\dfrac{1 + x}{1 - x}\right)$; $[0, \frac{1}{2}]$

In exercises 5–12, use a graphing calculator to evaluate the definite integral. Express your answers accurate to four decimal places.

5. $\displaystyle\int_0^{0.3} (0.2 \cos x - 0.3 \sin 2x + 1.2 \tan 3x)\, dx$

6. $\displaystyle\int_0^{1.1} \sqrt{1 + x^2} \sec x\, dx$

7. $\displaystyle\int_0^{2\pi} \dfrac{dx}{10 + 3 \cos x}$

8. $\displaystyle\int_0^{\pi/2} \sqrt{1 + \frac{1}{2} \sin^2 x}\, dx$

9. $\displaystyle\int_{0.5}^1 \ln \sin x\, dx$

10. $\displaystyle\int_1^2 e^{\cot x}\, dx$

11. $\displaystyle\int_0^1 \cos x^2\, dx$

12. $\displaystyle\int_0^3 \dfrac{\cos x}{(1 + \sin^2 x)^{3/2}}\, dx$

In exercises 13–16, use a graphing utility (a) to plot the graphs of the functions f and g and (b) to find the area of the region enclosed by these graphs and the vertical lines $x = a$ and $x = b$. Express your answers accurate to four decimal places.

13. $f(x) = x \sin 2x$; $g(x) = \frac{1}{2}x^2 - 2$; $a = 0$, $b = 1$

14. $f(x) = \tan^2 x$; $g(x) = 5 - \sqrt{1 + \sqrt{x}}$; $a = 0, b = 1$

15. $f(x) = e^{\cos x}$; $g(x) = \ln \tan x$; $a = 0.2$, $b = 0.8$

16. $f(x) = \cot x$; $g(x) = \sqrt{1 + 2 \sin x}$; $a = 1$, $b = 2$

In exercises 17–20, use a graphing utility (a) to plot the graphs of the functions f and g and (b) to find the area of the region completely enclosed by the graphs of these functions.

17. $f(x) = 2 \cos 3x$ and $g(x) = x^2$

18. $f(x) = \cos^2 x$ and $g(x) = e^{x^2} - 1$

19. $f(x) = \sin 2x$ and $g(x) = x - \sqrt{x}$

20. $f(x) = e^{\sin x}$ and $g(x) = 0.5x^2$

21. Boston Harbor Water Level The water level (in feet) in Boston Harbor during a certain 24-hour period is approximated by the formula

$$H = 4.8 \sin \frac{\pi}{6}(t - 10) + 7.6 \qquad (0 \le t \le 24)$$

where $t = 0$ corresponds to 12 A.M. What is the average water level in Boston Harbor over the 24-hour period on that day?

22. Number of Daylight Hours in Chicago The number of hours of daylight at any time t in Chicago is approximated by

$$L(t) = 2.8 \sin \frac{2\pi}{365}(t - 79) + 12$$

where t is measured in days and $t = 0$ corresponds to January 1. What is the daily average number of hours of daylight in Chicago over the year? Over the summer months from June 21 ($t = 171$) through September 20 ($t = 262$)?

where Q denotes the population of *Drosophila* at any time t. If there were 10 fruit flies initially, determine the number of fruit flies in the colony after 20 days.

Hint: Solve the separable differential equation using the following identity:

$$\frac{1}{Q(C - Q)} = \frac{1}{C}\left[\frac{1}{Q} + \frac{1}{C - Q}\right]$$

46. Use the sixth Taylor polynomial approximation of $(\sin x)/x$ (see exercise 39, Section 12.3)

$$\frac{\sin x}{x} \approx 1 - \frac{x^2}{6} + \frac{x^4}{120} - \frac{x^6}{5040}$$

to obtain an approximation of the integral

$$\int_0^{\pi/2} \frac{\sin x}{x}\, dx$$

47. Use the twelfth Taylor polynomial approximation of $\cos x^2$ (see exercise 40, Section 12.3)

$$\cos x^2 \approx 1 - \frac{x^4}{2} + \frac{x^8}{24} - \frac{x^{12}}{720}$$

to obtain an approximation of the Fresnel cosine integral

$$\int_0^x \cos t^2\, dt$$

(It is important in the study of the theory of diffraction.) In particular, evaluate

$$\int_0^1 \cos t^2\, dt$$

SOLUTIONS TO SELF-CHECK EXERCISES 12.4

1. Let $u = 1 + \sin x$, so that $du = \cos x\, dx$. Then

$$\int \frac{\cos x\, dx}{1 + \sin x} = \int \frac{du}{u} = \ln |u| + C$$

$$= \ln |1 + \sin x| + C$$

Therefore,

$$\int_0^{\pi/2} \frac{\cos x\, dx}{1 + \sin x} = \ln |1 + \sin x|\, \bigg|_0^{\pi/2} = \ln 2$$

2. The population of foxes in month t is

$$P = \int \frac{25\pi}{6} \cos\left(\frac{\pi t}{12}\right) dt$$

$$= \frac{25\pi}{6} \int \cos\left(\frac{\pi t}{12}\right) dt$$

Let $u = \dfrac{\pi t}{12}$, so that $du = \dfrac{\pi}{12}\, dt$. Therefore,

$$P = \left(\frac{25\pi}{6}\right)\left(\frac{12}{\pi}\right) \int \cos u\, du = 50 \sin u + C$$

$$= 50 \sin\left(\frac{\pi t}{12}\right) + C$$

To evaluate C, note that $P = 400$ when $t = 0$. That is,

$$400 = 50(0) + C \quad \text{or} \quad C = 400$$

Therefore,

$$P(t) = 400 + 50 \sin\left(\frac{\pi t}{12}\right)$$

CHAPTER 12 SUMMARY OF PRINCIPAL FORMULAS AND TERMS

Formulas

1. Degree–radian conversion   360 degrees $= 2\pi$ radians

2. Derivatives of the trigonometric functions:

   Sine $\dfrac{d}{dx}(\sin u) = \cos u \dfrac{du}{dx}$

   Cosine $\dfrac{d}{dx}(\cos u) = -\sin u \dfrac{du}{dx}$

   Tangent $\dfrac{d}{dx}(\tan u) = \sec^2 u \dfrac{du}{dx}$

   Cosecant $\dfrac{d}{dx}(\csc u) = -\csc u \cot u \dfrac{du}{dx}$

   Secant $\dfrac{d}{dx}(\sec u) = \sec u \tan u \dfrac{du}{dx}$

   Cotangent $\dfrac{d}{dx}(\cot u) = -\csc^2 u \dfrac{du}{dx}$

3. Integrals of the trigonometric functions:

$$\int \sin u \, du = -\cos u + C \qquad \int \csc u \cot u \, du = -\csc u + C$$

$$\int \cos u \, du = \sin u + C \qquad \int \tan u \, du = -\ln|\cos u| + C$$

$$\int \sec^2 u \, du = \tan u + C \qquad \int \sec u \, du = \ln|\sec u + \tan u| + C$$

$$\int \csc^2 u \, du = -\cot u + C \qquad \int \csc u \, du = \ln|\csc u - \cot u| + C$$

$$\int \sec u \tan u \, du = \sec u + C \qquad \int \cot u \, du = \ln|\sin u| + C$$

Terms

Angle   Coterminal angles
Initial ray   Period of a trigonometric function
Terminal ray   Predator-prey population model
Degree   Trigonometric identity
Radian

CHAPTER 12 REVIEW EXERCISES

In exercises 1–3, convert each angle to radian measure.

1. $120°$ **2.** $450°$ **3.** $-225°$

In exercises 4–6, convert each angle to degree measure.

4. $\dfrac{11}{6}\pi$ radians   **5.** $-\dfrac{5}{2}\pi$ radians   **6.** $-\dfrac{7}{4}\pi$ radians

In exercises 7 and 8, find all values of θ that satisfy the given equation over the interval $[0, 2\pi]$.

7. $\cos\theta = \dfrac{1}{2}$ **8.** $\cot\theta = -\sqrt{3}$

In exercises 9–20, find the derivative of the function f.

9. $f(x) = \sin 3x$ **10.** $f(x) = 2\cos\dfrac{x}{2}$

11. $f(x) = 2\sin x - 3\cos 2x$

12. $f(x) = \sec^2\sqrt{x}$ **13.** $f(x) = e^{-x}\tan 3x$

14. $f(x) = (1 - \csc 2x)^2$ **15.** $f(x) = 4\sin x\cos x$

16. $f(x) = \dfrac{\cos x}{1 - \cos x}$ **17.** $f(x) = \dfrac{1 - \tan x}{1 - \cot x}$

18. $f(x) = \ln(\cos^2 x)$ **19.** $f(x) = \sin(\sin x)$

20. $f(x) = e^{\sin x}\cos x$

21. Find the equation of the tangent line to the graph of the function $f(x) = \tan^2 x$ at the point $(\pi/4, 1)$.

22. Sketch the graph of the function $y = f(x) = \cos^2 x$ on the interval $[0, \pi]$ by obtaining the following information:
  a. the intervals where f is increasing and where it is decreasing
  b. the relative extrema of f
  c. the concavity of f
  d. the inflection points of f

In exercises 23–32, evaluate each of the given integrals.

23. $\displaystyle\int \cos\dfrac{2}{3}x\,dx$ **24.** $\displaystyle\int \sin^2 2x\,dx$

25. $\displaystyle\int x\csc x^2\cot x^2\,dx$ **26.** $\displaystyle\int x\cos x\,dx$

27. $\displaystyle\int \sin^2 x\cos x\,dx$ **28.** $\displaystyle\int e^x\sin e^x\,dx$

29. $\displaystyle\int \dfrac{\cos x}{\sin^2 x}\,dx$ **30.** $\displaystyle\int_0^{\pi/4} \tan x\,dx$

31. $\displaystyle\int_{\pi/6}^{\pi/2} \dfrac{\cos x\,dx}{1 - \cos^2 x}$

32. $\displaystyle\int_0^{\pi} \sin x(1 + \cos^3 x)\,dx$

33. Find the area of the region bounded by the curves $f(x) = \sin x$ and $f(x) = \cos x$ from $x = \pi/4$ to $x = (5/4)\pi$.

34. The population of foxes in a certain region over a two-year period is estimated to be

$$P(t) = 400 + 50\sin\left(\dfrac{\pi t}{12}\right)$$

in month t, and the population of rabbits in the same region at time t is given by

$$p(t) = 3000 + 500\cos\left(\dfrac{\pi t}{12}\right)$$

Find the rate of change of the populations when $t = 4$.

35. The occupancy rate of a large hotel in Maui in month t is described by the function

$$R(t) = 60 + 37\sin^2\left(\dfrac{\pi t}{12}\right) \qquad (0 \le t \le 12)$$

where $t = 0$ corresponds to the beginning of June. When is the occupancy rate highest? When is the occupancy rate increasing most rapidly?

36. Refer to exercise 35. Determine the average occupancy rate of the hotel over a one-year period from June 1 through May 31 of the following year.

[*Hint:* $\sin^2 x = (1 - \cos 2x)/2$.]

TABLES

Table 1 *Exponential functions.*

x	e^x	e^{-x}	x	e^x	e^{-x}
0.00	1.0000	1.0000	1.5	4.4817	0.2231
0.01	1.0101	0.9901	1.6	4.9530	0.2019
0.02	1.0202	0.9802	1.7	5.4739	0.1827
0.03	1.0305	0.9705	1.8	6.0496	0.1653
0.04	1.0408	0.9608	1.9	6.6859	0.1496
0.05	1.0513	0.9512	2.0	7.3891	0.1353
0.06	1.0618	0.9418	2.1	8.1662	0.1225
0.07	1.0725	0.9324	2.2	9.0250	0.1108
0.08	1.0833	0.9331	2.3	9.9742	0.1003
0.09	1.0942	0.9139	2.4	11.023	0.0907
0.10	1.1052	0.9048	2.5	12.182	0.0821
0.11	1.1163	0.8958	2.6	13.464	0.0743
0.12	1.1275	0.8869	2.7	14.880	0.0672
0.13	1.1388	0.8781	2.8	16.445	0.0608
0.14	1.1503	0.8694	2.9	18.174	0.0550
0.15	1.1618	0.8607	3.0	20.086	0.0498
0.16	1.1735	0.8521	3.1	22.198	0.0450
0.17	1.1853	0.8437	3.2	24.533	0.0408
0.18	1.1972	0.8353	3.3	27.113	0.0369
0.19	1.2092	0.8270	3.4	29.964	0.0334
0.20	1.2214	0.8187	3.5	33.115	0.0302
0.21	1.2337	0.8106	3.6	36.598	0.0273
0.22	1.2461	0.8025	3.7	40.447	0.0247
0.23	1.2586	0.7945	3.8	44.701	0.0224
0.24	1.2712	0.7866	3.9	49.402	0.0202
0.25	1.2840	0.7788	4.0	54.598	0.0183
0.30	1.3499	0.7408	4.1	60.340	0.0166
0.35	1.4191	0.7047	4.2	66.686	0.0150
0.40	1.4918	0.6703	4.3	73.700	0.0136
0.45	1.5683	0.6376	4.4	81.451	0.0123
0.50	1.6487	0.6065	4.5	90.017	0.0111
0.55	1.7333	0.5769	4.6	99.484	0.0101
0.60	1.8221	0.5488	4.7	109.95	0.0091
0.65	1.9155	0.5220	4.8	121.51	0.0082
0.70	2.0138	0.4966	4.9	134.29	0.0074
0.75	2.1170	0.4724	5.0	148.41	0.0067
0.80	2.2255	0.4493	5.5	244.69	0.0041
0.85	2.3396	0.4274	6.0	403.43	0.0025
0.90	2.4596	0.4066	6.5	665.14	0.0015
0.95	2.5857	0.3867	7.0	1096.6	0.0009
1.0	2.7183	0.3679	7.5	1808.0	0.0006
1.1	3.0042	0.3329	8.0	2981.0	0.0003
1.2	3.3201	0.3012	8.5	4914.8	0.0002
1.3	3.6693	0.2725	9.0	8103.1	0.0001
1.4	4.0552	0.2466	10.0	22026	0.00005

From MODERN COLLEGE ALGEBRA AND TRIGONOMETRY, Fourth Edition, by Edwin F. Beckenbach, Irving Drooyan, and Michael D. Grady. © 1981 by Wadsworth, Inc. Reprinted by permission.

Table 2 Common logarithms.

x	0	1	2	3	4	5	6	7	8	9
1.0	.0000	.0043	.0086	.0128	.0170	.0212	.0253	.0294	.0334	.0374
1.1	.0414	.0453	.0492	.0531	.0569	.0607	.0645	.0682	.0719	.0755
1.2	.0792	.0828	.0864	.0899	.0934	.0969	.1004	.1038	.1072	.1106
1.3	.1139	.1173	.1206	.1239	.1271	.1303	.1335	.1367	.1399	.1430
1.4	.1461	.1492	.1523	.1553	.1584	.1614	.1644	.1673	.1703	.1732
1.5	.1761	.1790	.1818	.1847	.1875	.1903	.1931	.1959	.1987	.2014
1.6	.2041	.2068	.2095	.2122	.2148	.2175	.2201	.2227	.2253	.2279
1.7	.2304	.2330	.2355	.2380	.2405	.2430	.2455	.2480	.2504	.2529
1.8	.2553	.2577	.2601	.2625	.2648	.2672	.2695	.2718	.2742	.2765
1.9	.2788	.2810	.2833	.2856	.2878	.2900	.2923	.2945	.2967	.2989
2.0	.3010	.3032	.3054	.3075	.3096	.3118	.3139	.3160	.3181	.3201
2.1	.3222	.3243	.3263	.3284	.3304	.3324	.3345	.3365	.3385	.3404
2.2	.3424	.3444	.3464	.3483	.3502	.3522	.3541	.3560	.3579	.3598
2.3	.3617	.3636	.3655	.3674	.3692	.3711	.3729	.3747	.3766	.3784
2.4	.3802	.3820	.3838	.3856	.3874	.3892	.3909	.3927	.3945	.3962
2.5	.3979	.3997	.4014	.4031	.4048	.4065	.4082	.4099	.4116	.4133
2.6	.4150	.4166	.4183	.4200	.4216	.4232	.4249	.4265	.4281	.4298
2.7	.4314	.4330	.4346	.4362	.4378	.4393	.4409	.4425	.4440	.4456
2.8	.4472	.4487	.4502	.4518	.4533	.4548	.4564	.4579	.4594	.4609
2.9	.4624	.4639	.4654	.4669	.4683	.4698	.4713	.4728	.4742	.4757
3.0	.4771	.4786	.4800	.4814	.4829	.4843	.4857	.4871	.4886	.4900
3.1	.4914	.4928	.4942	.4955	.4969	.4983	.4997	.5011	.5024	.5038
3.2	.5051	.5065	.5079	.5092	.5105	.5119	.5132	.5145	.5159	.5172
3.3	.5185	.5198	.5211	.5224	.5237	.5250	.5263	.5276	.5289	.5302
3.4	.5315	.5328	.5340	.5353	.5366	.5378	.5391	.5403	.5416	.5428
3.5	.5441	.5453	.5465	.5478	.5490	.5502	.5514	.5527	.5539	.5551
3.6	.5563	.5575	.5587	.5599	.5611	.5623	.5635	.5647	.5658	.5670
3.7	.5682	.5694	.5705	.5717	.5729	.5740	.5752	.5763	.5775	.5786
3.8	.5798	.5809	.5821	.5832	.5843	.5855	.5866	.5877	.5888	.5899
3.9	.5911	.5922	.5933	.5944	.5955	.5966	.5977	.5988	.5999	.6010
4.0	.6021	.6031	.6042	.6053	.6064	.6075	.6085	.6096	.6107	.6117
4.1	.6128	.6138	.6149	.6160	.6170	.6180	.6191	.6201	.6212	.6222
4.2	.6232	.6243	.6253	.6263	.6274	.6284	.6294	.6304	.6314	.6325
4.3	.6335	.6345	.6355	.6365	.6375	.6385	.6395	.6405	.6415	.6425
4.4	.6435	.6444	.6454	.6464	.6474	.6484	.6493	.6503	.6513	.6522
4.5	.6532	.6542	.6551	.6561	.6571	.6580	.6590	.6599	.6609	.6618
4.6	.6628	.6637	.6646	.6656	.6665	.6675	.6684	.6693	.6702	.6712
4.7	.6721	.6730	.6739	.6749	.6758	.6767	.6776	.6785	.6794	.6803
4.8	.6812	.6821	.6830	.6839	.6848	.6857	.6866	.6875	.6884	.6893
4.9	.6902	.6911	.6920	.6928	.6937	.6946	.6955	.6964	.6972	.6981
5.0	.6990	.6998	.7007	.7016	.7024	.7033	.7042	.7050	.7059	.7067
5.1	.7076	.7084	.7093	.7101	.7110	.7118	.7126	.7135	.7143	.7152
5.2	.7160	.7168	.7177	.7185	.7193	.7202	.7210	.7218	.7226	.7235
5.3	.7243	.7251	.7259	.7267	.7275	.7284	.7292	.7300	.7308	.7316
5.4	.7324	.7332	.7340	.7348	.7356	.7364	.7372	.7380	.7388	.7396
x	0	1	2	3	4	5	6	7	8	9

From *MODERN COLLEGE ALGEBRA AND TRIGONOMETRY, Fourth Edition,* by Edwin F. Beckenbach, Irving Drooyan, and Michael D. Grady. © 1981 by Wadsworth, Inc. Reprinted by permission.

Table 2 (*continued*)

x	0	1	2	3	4	5	6	7	8	9
5.5	.7404	.7412	.7419	.7427	.7435	.7443	.7451	.7459	.7466	.7474
5.6	.7482	.7490	.7497	.7505	.7513	.7520	.7528	.7536	.7543	.7551
5.7	.7559	.7566	.7574	.7582	.7589	.7597	.7604	.7612	.7619	.7627
5.8	.7634	.7642	.7649	.7657	.7664	.7672	.7679	.7686	.7694	.7701
5.9	.7709	.7716	.7723	.7731	.7738	.7745	.7752	.7760	.7767	.7774
6.0	.7782	.7789	.7796	.7803	.7810	.7818	.7825	.7832	.7839	.7846
6.1	.7853	.7860	.7868	.7875	.7882	.7889	.7896	.7903	.7910	.7917
6.2	.7924	.7931	.7938	.7945	.7952	.7959	.7966	.7973	.7980	.7987
6.3	.7993	.8000	.8007	.8014	.8021	.8028	.8035	.8041	.8048	.8055
6.4	.8062	.8069	.8075	.8082	.8089	.8096	.8102	.8109	.8116	.8122
6.5	.8129	.8136	.8142	.8149	.8156	.8162	.8169	.8176	.8182	.8189
6.6	.8195	.8202	.8209	.8215	.8222	.8228	.8235	.8241	.8248	.8254
6.7	.8261	.8267	.8274	.8280	.8287	.8293	.8299	.8306	.8312	.8319
6.8	.8325	.8331	.8338	.8344	.8351	.8357	.8363	.8370	.8376	.8382
6.9	.8388	.8395	.8401	.8407	.8414	.8420	.8426	.8432	.8439	.8445
7.0	.8451	.8457	.8463	.8470	.8476	.8482	.8488	.8494	.8500	.8506
7.1	.8513	.8519	.8525	.8531	.8537	.8543	.8549	.8555	.8561	.8567
7.2	.8573	.8579	.8585	.8591	.8597	.8603	.8609	.8615	.8621	.8627
7.3	.8633	.8639	.8645	.8651	.8657	.8663	.8669	.8675	.8681	.8686
7.4	.8692	.8698	.8704	.8710	.8716	.8722	.8727	.8733	.8739	.8745
7.5	.8751	.8756	.8762	.8768	.8774	.8779	.8785	.8791	.8797	.8802
7.6	.8808	.8814	.8820	.8825	.8831	.8837	.8842	.8848	.8854	.8859
7.7	.8865	.8871	.8876	.8882	.8887	.8893	.8899	.8904	.8910	.8915
7.8	.8921	.8927	.8932	.8938	.8943	.8949	.8954	.8960	.8965	.8971
7.9	.8976	.8982	.8987	.8993	.8998	.9004	.9009	.9015	.9020	.9025
8.0	.9031	.9036	.9042	.9047	.9053	.9058	.9063	.9069	.9074	.9079
8.1	.9085	.9090	.9096	.9101	.9106	.9112	.9117	.9122	.9128	.9133
8.2	.9138	.9143	.9149	.9154	.9159	.9165	.9170	.9175	.9180	.9186
8.3	.9191	.9196	.9201	.9206	.9212	.9217	.9222	.9227	.9232	.9238
8.4	.9243	.9248	.9253	.9258	.9263	.9269	.9274	.9279	.9284	.9289
8.5	.9294	.9299	.9304	.9309	.9315	.9320	.9325	.9330	.9335	.9340
8.6	.9345	.9350	.9355	.9360	.9365	.9370	.9375	.9380	.9385	.9390
8.7	.9395	.9400	.9405	.9410	.9415	.9420	.9425	.9430	.9435	.9440
8.8	.9445	.9450	.9455	.9460	.9465	.9469	.9474	.9479	.9484	.9489
8.9	.9494	.9499	.9504	.9509	.9513	.9518	.9523	.9528	.9533	.9538
9.0	.9542	.9547	.9552	.9557	.9562	.9566	.9571	.9576	.9581	.9586
9.1	.9590	.9595	.9600	.9605	.9609	.9614	.9619	.9624	.9628	.9633
9.2	.9638	.9643	.9647	.9652	.9657	.9661	.9666	.9671	.9675	.9680
9.3	.9685	.9689	.9694	.9699	.9703	.9708	.9713	.9717	.9722	.9727
9.4	.9731	.9736	.9741	.9745	.9750	.9754	.9759	.9763	.9768	.9773
9.5	.9777	.9782	.9786	.9791	.9795	.9800	.9805	.9809	.9814	.9818
9.6	.9823	.9827	.9832	.9836	.9841	.9845	.9850	.9854	.9859	.9863
9.7	.9868	.9872	.9877	.9881	.9886	.9890	.9894	.9899	.9903	.9908
9.8	.9912	.9917	.9921	.9926	.9930	.9934	.9939	.9943	.9948	.9952
9.9	.9956	.9961	.9965	.9969	.9974	.9978	.9983	.9987	.9991	.9996
x	0	1	2	3	4	5	6	7	8	9

Table 3 *Natural logarithms.*

n	$\log_e n$	n	$\log_e n$	n	$\log_e n$
0.1	7.6974 *	4.5	1.5041	9.0	2.1972
0.2	8.3906	4.6	1.5261	9.1	2.2083
0.3	8.7960	4.7	1.5476	9.2	2.2192
0.4	9.0837	4.8	1.5686	9.3	2.2300
		4.9	1.5892	9.4	2.2407
0.5	9.3069	5.0	1.6094	9.5	2.2513
0.6	9.4892	5.1	1.6292	9.6	2.2618
0.7	9.6433	5.2	1.6487	9.7	2.2721
0.8	9.7769	5.3	1.6677	9.8	2.2824
0.9	9.8946	5.4	1.6864	9.9	2.2925
1.0	0.0000	5.5	1.7047	10	2.3026
1.1	0.0953	5.6	1.7228	11	2.3979
1.2	0.1823	5.7	1.7405	12	2.4849
1.3	0.2624	5.8	1.7579	13	2.5649
1.4	0.3365	5.9	1.7750	14	2.6391
1.5	0.4055	6.0	1.7918	15	2.7081
1.6	0.4700	6.1	1.8083	16	2.7726
1.7	0.5306	6.2	1.8245	17	2.8332
1.8	0.5878	6.3	1.8405	18	2.8904
1.9	0.6419	6.4	1.8563	19	2.9444
2.0	0.6931	6.5	1.8718	20	2.9957
2.1	0.7419	6.6	1.8871	25	3.2189
2.2	0.7885	6.7	1.9021	30	3.4012
2.3	0.8329	6.8	1.9169	35	3.5553
2.4	0.8755	6.9	1.9315	40	3.6889
2.5	0.9163	7.0	1.9459	45	3.8067
2.6	0.9555	7.1	1.9601	50	3.9120
2.7	0.9933	7.2	1.9741	55	4.0073
2.8	1.0296	7.3	1.9879	60	4.0943
2.9	1.0647	7.4	2.0015	65	4.1744
3.0	1.0986	7.5	2.0149	70	4.2485
3.1	1.1314	7.6	2.0281	75	4.3175
3.2	1.1632	7.7	2.0412	80	4.3820
3.3	1.1939	7.8	2.0541	85	4.4427
3.4	1.2238	7.9	2.0669	90	4.4998
3.5	1.2528	8.0	2.0794	100	4.6052
3.6	1.2809	8.1	2.0919	110	4.7005
3.7	1.3083	8.2	2.1041	120	4.7875
3.8	1.3350	8.3	2.1163	130	4.8676
3.9	1.3610	8.4	2.1282	140	4.9416
4.0	1.3863	8.5	2.1401	150	5.0106
4.1	1.4110	8.6	2.1518	160	5.0752
4.2	1.4351	8.7	2.1633	170	5.1358
4.3	1.4586	8.8	2.1748	180	5.1930
4.4	1.4816	8.9	2.1861	190	5.2470

* Subtract 10 for $n < 1$. Thus, $\log_e 0.1 = 7.6974 - 10 = -2.3026$.

From *MODERN COLLEGE ALGEBRA AND TRIGONOMETRY*, Fourth Edition, by Edwin F. Beckenbach, Irving Drooyan, and Michael D. Grady. © 1981 by Wadsworth, Inc. Reprinted by permission.

Table 4 *The standard normal distribution.*

$$F_z(z) = P[Z \le z]$$

z	0.00	0.01	0.02	0.03	0.04	0.05	0.06	0.07	0.08	0.09
−3.4	0.0003	0.0003	0.0003	0.0003	0.0003	0.0003	0.0003	0.0003	0.0003	0.0002
−3.3	0.0005	0.0005	0.0005	0.0004	0.0004	0.0004	0.0004	0.0004	0.0004	0.0003
−3.2	0.0007	0.0007	0.0006	0.0006	0.0006	0.0006	0.0006	0.0005	0.0005	0.0005
−3.1	0.0010	0.0009	0.0009	0.0009	0.0008	0.0008	0.0008	0.0008	0.0007	0.0007
−3.0	0.0013	0.0013	0.0013	0.0012	0.0012	0.0011	0.0011	0.0011	0.0010	0.0010
−2.9	0.0019	0.0018	0.0017	0.0017	0.0016	0.0016	0.0015	0.0015	0.0014	0.0014
−2.8	0.0026	0.0025	0.0024	0.0023	0.0023	0.0022	0.0021	0.0021	0.0020	0.0019
−2.7	0.0035	0.0034	0.0033	0.0032	0.0031	0.0030	0.0029	0.0028	0.0027	0.0026
−2.6	0.0047	0.0045	0.0044	0.0043	0.0041	0.0040	0.0039	0.0038	0.0037	0.0036
−2.5	0.0062	0.0060	0.0059	0.0057	0.0055	0.0054	0.0052	0.0051	0.0049	0.0048
−2.4	0.0082	0.0080	0.0078	0.0075	0.0073	0.0071	0.0069	0.0068	0.0066	0.0064
−2.3	0.0107	0.0104	0.0102	0.0099	0.0096	0.0094	0.0091	0.0089	0.0087	0.0084
−2.2	0.0139	0.0136	0.0132	0.0129	0.0125	0.0122	0.0119	0.0116	0.0113	0.0110
−2.1	0.0179	0.0174	0.0170	0.0166	0.0162	0.0158	0.0154	0.0150	0.0146	0.0143
−2.0	0.0228	0.0222	0.0217	0.0212	0.0207	0.0202	0.0197	0.0192	0.0188	0.0183
−1.9	0.0287	0.0281	0.0274	0.0268	0.0262	0.0256	0.0250	0.0244	0.0239	0.0233
−1.8	0.0359	0.0352	0.0344	0.0336	0.0329	0.0322	0.0314	0.0307	0.0301	0.0294
−1.7	0.0446	0.0436	0.0427	0.0418	0.0409	0.0401	0.0392	0.0384	0.0375	0.0367
−1.6	0.0548	0.0537	0.0526	0.0516	0.0505	0.0495	0.0485	0.0475	0.0465	0.0455
−1.5	0.0668	0.0655	0.0643	0.0630	0.0618	0.0606	0.0594	0.0582	0.0571	0.0559
−1.4	0.0808	0.0793	0.0778	0.0764	0.0749	0.0735	0.0722	0.0708	0.0694	0.0681
−1.3	0.0968	0.0951	0.0934	0.0918	0.0901	0.0885	0.0869	0.0853	0.0838	0.0823
−1.2	0.1151	0.1131	0.1112	0.1093	0.1075	0.1056	0.1038	0.1020	0.1003	0.0985
−1.1	0.1357	0.1335	0.1314	0.1292	0.1271	0.1251	0.1230	0.1210	0.1190	0.1170
−1.0	0.1587	0.1562	0.1539	0.1515	0.1492	0.1469	0.1446	0.1423	0.1401	0.1379
−0.9	0.1841	0.1814	0.1788	0.1762	0.1736	0.1711	0.1685	0.1660	0.1635	0.1611
−0.8	0.2119	0.2090	0.2061	0.2033	0.2005	0.1977	0.1949	0.1922	0.1894	0.1867
−0.7	0.2420	0.2389	0.2358	0.2327	0.2296	0.2266	0.2236	0.2206	0.2177	0.2148
−0.6	0.2743	0.2709	0.2676	0.2643	0.2611	0.2578	0.2546	0.2514	0.2483	0.2451
−0.5	0.3085	0.3050	0.3015	0.2981	0.2946	0.2912	0.2877	0.2843	0.2810	0.2776

Table 4 (*continued*)

$$F_z(z) = P[Z \leq z]$$

z	0.00	0.01	0.02	0.03	0.04	0.05	0.06	0.07	0.08	0.09
−0.4	0.3446	0.3409	0.3372	0.3336	0.3300	0.3264	0.3228	0.3192	0.3156	0.3121
−0.3	0.3821	0.3783	0.3745	0.3707	0.3669	0.3632	0.3594	0.3557	0.3520	0.3483
−0.2	0.4207	0.4168	0.4129	0.4090	0.4052	0.4013	0.3974	0.3936	0.3897	0.3859
−0.1	0.4602	0.4562	0.4522	0.4483	0.4443	0.4404	0.4364	0.4325	0.4286	0.4247
−0.0	0.5000	0.4960	0.4920	0.4880	0.4840	0.4801	0.4761	0.4721	0.4681	0.4641
0.0	0.5000	0.5040	0.5080	0.5120	0.5160	0.5199	0.5239	0.5279	0.5319	0.5359
0.1	0.5398	0.5438	0.5478	0.5517	0.5557	0.5596	0.5636	0.5675	0.5714	0.5753
0.2	0.5793	0.5832	0.5871	0.5910	0.5948	0.5987	0.6026	0.6064	0.6103	0.6141
0.3	0.6179	0.6217	0.6255	0.6293	0.6331	0.6368	0.6406	0.6443	0.6480	0.6517
0.4	0.6554	0.6591	0.6628	0.6664	0.6700	0.6736	0.6772	0.6808	0.6844	0.6879
0.5	0.6915	0.6950	0.6985	0.7019	0.7054	0.7088	0.7123	0.7157	0.7190	0.7224
0.6	0.7257	0.7291	0.7324	0.7357	0.7389	0.7422	0.7454	0.7486	0.7517	0.7549
0.7	0.7580	0.7611	0.7642	0.7673	0.7704	0.7734	0.7764	0.7794	0.7823	0.7852
0.8	0.7881	0.7910	0.7939	0.7967	0.7995	0.8023	0.8051	0.8078	0.8106	0.8133
0.9	0.8159	0.8186	0.8212	0.8238	0.8264	0.8289	0.8315	0.8340	0.8365	0.8389
1.0	0.8413	0.8438	0.8461	0.8485	0.8508	0.8531	0.8554	0.8577	0.8599	0.8621
1.1	0.8643	0.8665	0.8686	0.8708	0.8729	0.8749	0.8770	0.8790	0.8810	0.8830
1.2	0.8849	0.8869	0.8888	0.8907	0.8925	0.8944	0.8962	0.8980	0.8997	0.9015
1.3	0.9032	0.9049	0.9066	0.9082	0.9099	0.9115	0.9131	0.9147	0.9162	0.9177
1.4	0.9192	0.9207	0.9222	0.9236	0.9251	0.9265	0.9278	0.9292	0.9306	0.9319
1.5	0.9332	0.9345	0.9357	0.9370	0.9382	0.9394	0.9406	0.9418	0.9429	0.9441
1.6	0.9452	0.9463	0.9474	0.9484	0.9495	0.9505	0.9515	0.9525	0.9535	0.9545
1.7	0.9554	0.9564	0.9573	0.9582	0.9591	0.9599	0.9608	0.9616	0.9625	0.9633
1.8	0.9641	0.9649	0.9656	0.9664	0.9671	0.9678	0.9686	0.9693	0.9699	0.9706
1.9	0.9713	0.9719	0.9726	0.9732	0.9738	0.9744	0.9750	0.9756	0.9761	0.9767
2.0	0.9772	0.9778	0.9783	0.9788	0.9793	0.9798	0.9803	0.9808	0.9812	0.9817
2.1	0.9821	0.9826	0.9830	0.9834	0.9838	0.9842	0.9846	0.9850	0.9854	0.9857
2.2	0.9861	0.9864	0.9868	0.9871	0.9875	0.9878	0.9881	0.9884	0.9887	0.9890
2.3	0.9893	0.9896	0.9898	0.9901	0.9904	0.9906	0.9909	0.9911	0.9913	0.9916
2.4	0.9918	0.9920	0.9922	0.9925	0.9927	0.9929	0.9931	0.9932	0.9934	0.9936
2.5	0.9938	0.9940	0.9951	0.9943	0.9945	0.9946	0.9948	0.9949	0.9951	0.9952
2.6	0.9953	0.9955	0.9956	0.9957	0.9959	0.9960	0.9961	0.9962	0.9963	0.9964
2.7	0.9965	0.9966	0.9967	0.9968	0.9969	0.9970	0.9971	0.9972	0.9973	0.9974
2.8	0.9974	0.9975	0.9976	0.9977	0.9977	0.9978	0.9979	0.9979	0.9980	0.9981
2.9	0.9981	0.9982	0.9982	0.9983	0.9984	0.9984	0.9985	0.9985	0.9986	0.9986
3.0	0.9987	0.9987	0.9987	0.9988	0.9988	0.9989	0.9989	0.9989	0.9990	0.9990
3.1	0.9990	0.9991	0.9991	0.9991	0.9992	0.9992	0.9992	0.9992	0.9993	0.9993
3.2	0.9993	0.9993	0.9994	0.9994	0.9994	0.9994	0.9994	0.9995	0.9995	0.9995
3.3	0.9995	0.9995	0.9995	0.9996	0.9996	0.9996	0.9996	0.9996	0.9996	0.9997
3.4	0.9997	0.9997	0.9997	0.9997	0.9997	0.9997	0.9997	0.9997	0.9997	0.9998

Table 5 *Trigonometric functions.*

Degrees	Radians	sin	tan	cot	cos	Radians	Degrees
0	0.000	0.000	0.000		1.000	1.571	90
1	0.017	0.017	0.017	57.29	1.000	1.553	89
2	0.035	0.035	0.035	28.64	0.999	1.536	88
3	0.052	0.052	0.052	19.081	0.999	1.518	87
4	0.070	0.070	0.070	14.301	0.998	1.501	86
5	0.087	0.087	0.087	11.430	0.996	1.484	85
6	0.105	0.105	0.105	9.514	0.995	1.460	84
7	0.122	0.122	0.123	8.144	0.993	1.449	83
8	0.140	0.139	0.141	7.115	0.990	1.431	82
9	0.157	0.156	0.158	6.314	0.988	1.414	81
10	0.175	0.174	0.176	5.671	0.985	1.396	80
11	0.192	0.191	0.194	5.145	0.982	1.379	79
12	0.209	0.208	0.213	4.705	0.978	1.361	78
13	0.227	0.225	0.231	4.331	0.974	1.344	77
14	0.244	0.242	0.249	4.011	0.970	1.326	76
15	0.262	0.259	0.268	3.732	0.966	1.309	75
16	0.279	0.276	0.287	3.487	0.961	1.292	74
17	0.297	0.292	0.306	3.271	0.956	1.274	73
18	0.314	0.309	0.325	3.078	0.951	1.257	72
19	0.332	0.326	0.344	2.904	0.946	1.239	71
20	0.349	0.342	0.364	2.747	0.940	1.222	70
21	0.367	0.358	0.384	2.605	0.934	1.204	69
22	0.384	0.375	0.404	2.475	0.927	1.187	68
23	0.401	0.391	0.424	2.356	0.921	1.169	67
24	0.419	0.407	0.445	2.246	0.914	1.152	66
25	0.436	0.423	0.466	2.144	0.906	1.134	65
26	0.454	0.438	0.488	2.050	0.899	1.117	64
27	0.471	0.454	0.510	1.963	0.891	1.100	63
28	0.489	0.469	0.532	1.881	0.883	1.082	62
29	0.506	0.485	0.554	1.804	0.875	1.065	61
30	0.524	0.500	0.577	1.732	0.866	1.047	60
31	0.541	0.515	0.601	1.664	0.857	1.030	59
32	0.559	0.530	0.625	1.600	0.848	1.012	58
33	0.576	0.545	0.649	1.540	0.839	0.995	57
34	0.593	0.559	0.675	1.483	0.829	0.977	56
35	0.611	0.574	0.700	1.428	0.819	0.960	55
36	0.628	0.588	0.727	1.376	0.809	0.942	54
37	0.646	0.602	0.754	1.327	0.799	0.925	53
38	0.663	0.616	0.781	1.280	0.788	0.908	52
39	0.681	0.629	0.810	1.235	0.777	0.890	51
40	0.698	0.643	0.839	1.192	0.766	0.873	50
41	0.716	0.656	0.869	1.150	0.755	0.855	49
42	0.733	0.669	0.900	1.111	0.743	0.838	48
43	0.750	0.682	0.933	1.072	0.731	0.820	47
44	0.768	0.695	0.966	1.036	0.719	0.803	46
45	0.785	0.707	1.000	1.000	0.707	0.785	45
Degrees	**Radians**	cos	cot	tan	sin	**Radians**	**Degrees**

From CALCULUS WITH ANALYTIC GEOMETRY, Second Alternate Edition, by Earl W. Swokowski. © 1988 by PWS-KENT Publishing. Reprinted by permission.

ANSWERS TO ODD-NUMBERED EXERCISES

CHAPTER 1

Exercises 1.1, page 12

1. False **3.** False

5.
(number line with open interval between 3 and 6, marked at 0, 3, 6)

7.
(number line with interval from −1 to 4, marked at −1, 0, 4)

9.
(number line with ray from 0, marked at 0)

11. $(-\infty, 2)$ **13.** $(-\infty, -5]$ **15.** $(-4, 6)$

17. $(-\infty, -3) \cup (3, \infty)$ **19.** $(-2, 3)$ **21.** 4

23. 2 **25.** $5\sqrt{3}$ **27.** $\pi + 1$ **29.** 2

31. False **33.** False **35.** True **37.** False

39. True **41.** False **43.** 9 **45.** 1 **47.** 4

49. 7 **51.** $\frac{1}{5}$ **53.** 2 **55.** 2 **57.** 1

59. True **61.** False **63.** False **65.** False

67. False **69.** $\dfrac{1}{(xy)^2}$ **71.** $\dfrac{1}{x^{5/6}}$ **73.** $\dfrac{1}{(s+t)^3}$

75. $x^{13/3}$ **77.** $\dfrac{1}{x^3}$ **79.** x **81.** $\dfrac{9}{x^2 y^4}$ **83.** $\dfrac{y^8}{x^{10}}$

85. $2x^{11/6}$ **87.** $-2xy^2$ **89.** $2x^{4/3} y^{1/2}$ **91.** 2.828

93. 5.196 **95.** 31.62 **97.** 316.2 **99.** $\dfrac{3\sqrt{x}}{2x}$

101. $\dfrac{2\sqrt{3y}}{3}$ **103.** $\dfrac{\sqrt[3]{x^2}}{x}$ **105.** $\dfrac{2x}{3\sqrt{x}}$ **107.** $\dfrac{2y}{\sqrt{2xy}}$

109. $\dfrac{xz}{y\sqrt[3]{xz^2}}$ **111.** [362, 488.7] **113.** \$12,300

115. \$18,666.67 **117.** $|x - 0.5| < 0.01$

119. Between 1000 and 4000 units

Exercises 1.2, page 26

1. $9x^2 + 3x + 1$ **3.** $4y^2 + y + 8$ **5.** $-x - 1$

7. $\frac{2}{3} + e - e^{-1}$ **9.** $6\sqrt{2} + 8 + \frac{1}{2}\sqrt{x} - \frac{11}{4}\sqrt{y}$

11. $x^2 + 6x - 16$ **13.** $a^2 + 10a + 25$

15. $x^2 + 4xy + 4y^2$ **17.** $4x^2 - y^2$ **19.** $-2x$

21. $2t(2\sqrt{t} + 1)$ **23.** $2x^3(2x^2 - 6x - 3)$

25. $7a^2(a^2 + 7ab - 6b^2)$ **27.** $e^{-x}(1 - x)$

29. $\frac{1}{2}x^{-5/2}(4 - 3x)$ **31.** $(2a + b)(3c - 2d)$

33. $(2a + b)(2a - b)$ **35.** $-2(3x + 5)(2x - 1)$

37. $3(x - 4)(x + 2)$ **39.** $2(3x - 5)(2x + 3)$

41. $(3x - 4y)(3x + 4y)$ **43.** $(x^2 + 5)(x^4 - 5x^2 + 25)$

45. $x^3 - xy^2$ **47.** $4(x - 1)(3x - 1)(2x + 2)^3$

49. $4(x - 1)(3x - 1)(2x + 2)^3$

51. $2x(x^2 + 2)^2(5x^4 + 20x^2 + 17)$

53. -4 and 3 **55.** -1 and $\frac{1}{2}$ **57.** 2 and 2

59. -2 and $\frac{3}{4}$ **61.** $\frac{1}{2} + \frac{1}{4}\sqrt{10}$ and $\frac{1}{2} - \frac{1}{4}\sqrt{10}$

63. $-1 + \frac{1}{2}\sqrt{10}$ and $-1 - \frac{1}{2}\sqrt{10}$ **65.** $\dfrac{x - 1}{x - 2}$

67. $\dfrac{3(2t + 1)}{2t - 1}$ **69.** $-\dfrac{7}{(4x - 1)^2}$ **71.** -8

73. $\dfrac{3x - 1}{2}$ **75.** $\dfrac{t + 20}{3t + 2}$ **77.** $-\dfrac{x(2x - 13)}{(2x - 1)(2x + 5)}$

79. $-\dfrac{x + 27}{(x - 3)^2(x + 3)}$ **81.** $\dfrac{x + 1}{x - 1}$ **83.** $\dfrac{4x^2 + 7}{\sqrt{2x^2 + 7}}$

85. $\dfrac{x - 1}{x^2\sqrt{x + 1}}$ **87.** $\dfrac{x - 1}{(2x + 1)^{3/2}}$ **89.** $\dfrac{\sqrt{3} + 1}{2}$

91. $\dfrac{\sqrt{x} + \sqrt{y}}{x - y}$ **93.** $\dfrac{(\sqrt{a} + \sqrt{b})^2}{a - b}$ **95.** $\dfrac{x}{3\sqrt{x}}$

97. $-\dfrac{2}{3(1 + \sqrt{3})}$ **99.** $-\dfrac{x + 1}{\sqrt{x + 2}(1 - \sqrt{x + 2})}$

Exercises 1.3, page 32

1. (3, 3); Quadrant I   **3.** (2, −2); Quadrant IV

5. (−4, −6); Quadrant III   **7.** *A*   **9.** *E*, *F*, and *G*

11. *F*   **13.–19.** See accompanying figure.

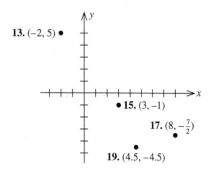

13. (−2, 5) •

• **15.** (3, −1)

17. $(8, -\frac{7}{2})$

19. (4.5, −4.5)

21. 5   **23.** $\sqrt{61}$ **25.** (−8, −6) and (8, −6)

29. 3400 miles   **31.** Route 1   **33.** Model C

Exercises 1.4, page 46

1. e   **3.** a   **5.** f   **7.** 1/2   **9.** Not defined

11. 5   **13.** 5/6   **15.** $\dfrac{d - b}{c - a}$ $(a \neq c)$

17. a. 4   **b.** −8   **19.** Parallel   **21.** Perpendicular

23. −5   **25.** $y = -3$ **27.** $y = 2x - 10$

29. $y = 2$ **31.** $y = 3x - 2$ **33.** $y = x + 1$

35. $y = 3x + 4$ **37.** $y = 5$

39. $y = \frac{1}{2}x$; $m = \frac{1}{2}$; $b = 0$

41. $y = \frac{2}{3}x - 3$; $m = \frac{2}{3}$; $b = -3$

43. $y = -\frac{1}{2}x + \frac{7}{2}$; $m = -\frac{1}{2}$; $b = \frac{7}{2}$

45. $y = \frac{1}{2}x + 3$ **47.** $y = -6$ **49.** $y = b$

51. $y = \frac{2}{3}x - \frac{2}{3}$ **53.** $k = 8$

55.

57.

59.

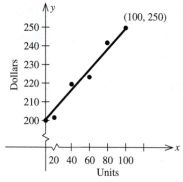

63. $y = -2x - 4$ **65.** $y = \frac{1}{8}x - \frac{1}{2}$ **67.** Yes

69. a. $y = 0.55x$ **b.** 2000

71. a. and **b.**

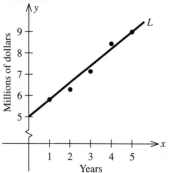

(100, 250)

Dollars

250
240
230
220
210
200

20 40 60 80 100
Units

c. $y = \frac{1}{2}x + 200$ **d.** $227

73. a. and **b.**

L

Millions of dollars

9
8
7
6
5

1 2 3 4 5
Years

c. $y = 0.8x + 5$ **d.** $12.2 million

Chapter I Review Exercises, page 52

1. $[-2, \infty)$ **3.** $(-\infty, -4) \cup (5, \infty)$ **5.** 4

7. $\pi - 6$ **9.** $\frac{27}{8}$ **11.** $\frac{1}{144}$ **13.** $\frac{1}{4}$

15. $4(x^2 + y)^2$ **17.** $\dfrac{2x}{3z}$ **19.** $6xy^7$

21. $2vw(v^2 + w^2 + u^2)$ **23.** $6t(t + 1)(2t - 3)$

25. -2 and $\frac{1}{3}$ **27.** $\dfrac{\sqrt{2}}{2}$ and $-\dfrac{\sqrt{2}}{2}$

29. $-2 + \dfrac{\sqrt{2}}{2}$ and $-2 - \dfrac{\sqrt{2}}{2}$ **31.** $\dfrac{15x^2 + 24x + 2}{4(x + 2)(3x^2 + 2)}$

33. $\dfrac{2(x + 2)}{\sqrt{x + 1}}$ **35.** $\dfrac{x - \sqrt{x}}{2x}$ **37.** 2   **39.** $y = 4$

41. $y = -\frac{4}{5}x + \frac{12}{5}$ **43.** $y = \frac{3}{4}x + \frac{11}{2}$

45. $y = -\frac{3}{5}x + \frac{12}{5}$

47.

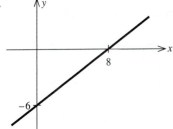

49. $100

CHAPTER 2

Exercises 2.1, page 64

1. $21, -9, 5a + 6, -5a + 6, 5a + 21$

3. $-3, 6, 3a^2 - 6a - 3, 3a^2 + 6a - 3, 3x^2 - 6$

5. $\dfrac{8}{15}, 0, \dfrac{2a}{a^2 - 1}, \dfrac{2(2 + a)}{a^2 + 4a + 3}, \dfrac{2(t + 1)}{t(t + 2)}$

7. $8, \dfrac{2a^2}{\sqrt{a - 1}}, \dfrac{2(x + 1)^2}{\sqrt{x}}, \dfrac{2(x - 1)^2}{\sqrt{x - 2}}$

9. $5, 1, 1$ **11.** $\frac{5}{2}, 3, 3, 9$ **13.** Yes   **15.** Yes

17. $(-\infty, \infty)$ **19.** $(-\infty, 0) \cup (0, \infty)$ **21.** $(-\infty, \infty)$

23. $(-\infty, 5]$ **25.** $(-\infty, -1) \cup (-1, 1) \cup (1, \infty)$

27. $[-3, \infty)$ **29.** $(-\infty, -2) \cup (-2, 1]$

31. a. $(-\infty, \infty)$
   b. $6, 0, -4, -6, -\frac{25}{4}, -6, -4, 0$

c.

33.

35.

37.

39.

41.

43.

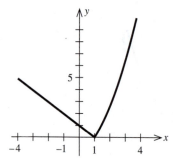

45. Yes   **47.** No   **49.** Yes   **51.** Yes

53. 10π in.   **55. a.** From 1985 to 1990
                               b. From 1990 on
                               c. 1990; \$3.5 billion

57. a.
$$f(t) = \begin{cases} 0.0185t + 0.58 & \text{if } 0 \le t \le 20 \\ 0.015t + 0.65 & \text{if } 20 < t \le 30 \end{cases}$$
  b. 0.0185/yr from 1960 through 1980; 0.015/yr from 1980 through 1990
  c. 1980

59. a. $0.06x$ **b.** \$12.00; \$0.34   **61.** 160 mg

63. a. $30 + 0.15x$; $25 + 0.20x$
  b.

  c. Acme

65. \$620,000; \$540,000

67. a. $0 \le r \le 0.2$
  b. 40; 30; 0; as the distance r increases, the velocity of the blood decreases.

69. 20; 26

71. 0.77. When the proportion of popular votes won by the Democratic presidential candidate is 0.60, the proportion of seats in the House of Representatives won by Democratic candidates is 0.77.

73. a. $(0, 12]$
  b.

Using Technology Exercises 2.1, page 73

1.

3.

5.

7.

9. a.

b.

11. a.

b.

13. a.

b.

15. a.

b.

17. a.

b.

19. a.

b.

21.

23.

25.

27.

29.

31. 18   **33.** 2   **35.** 18.5505   **37.** 17.3850

39. 4.1616   **41.** 1.7214

43. a.

b. 2.1762%; 1.9095%

45. a.

b. 44.7; 52.7; 129.2

Exercises 2.2, page 81

1. $f(x) + g(x) = x^3 + x^2 + 3$

3. $f(x)g(x) = x^5 - 2x^3 + 5x^2 - 10$

5. $\dfrac{f(x)}{g(x)} = \dfrac{x^3 + 5}{x^2 - 2}$

7. $\dfrac{f(x)g(x)}{h(x)} = \dfrac{x^5 - 2x^3 + 5x^2 - 10}{2x + 4}$

9. $f(x) + g(x) = x - 1 + \sqrt{x + 1}$

11. $f(x)g(x) = (x - 1)\sqrt{x + 1}$

13. $\dfrac{g(x)}{h(x)} = \dfrac{\sqrt{x + 1}}{2x^3 - 1}$

15. $\dfrac{f(x)g(x)}{h(x)} = \dfrac{(x - 1)\sqrt{x + 1}}{2x^3 - 1}$

17. $\dfrac{f(x) - h(x)}{g(x)} = \dfrac{x - 2x^3}{\sqrt{x + 1}}$

19. $f(x) + g(x) = x^2 + \sqrt{x} + 3$;
$f(x) - g(x) = x^2 - \sqrt{x} + 7$;
$f(x)g(x) = (x^2 + 5)(\sqrt{x} - 2)$; $\dfrac{f(x)}{g(x)} = \dfrac{x^2 + 5}{\sqrt{x} - 2}$

21. $f(x) + g(x) = \dfrac{(x - 1)\sqrt{x + 3} + 1}{x - 1}$;
$f(x) - g(x) = \dfrac{(x - 1)\sqrt{x + 3} - 1}{x - 1}$;
$f(x)g(x) = \dfrac{\sqrt{x + 3}}{x - 1}$; $\dfrac{f(x)}{g(x)} = (x - 1)\sqrt{x + 3}$

23. $f(x) + g(x) = \dfrac{2(x^2 - 2)}{(x - 1)(x - 2)}$;

$f(x) - g(x) = \dfrac{-2x}{(x - 1)(x - 2)}$;

$f(x)g(x) = \dfrac{(x + 1)(x + 2)}{(x - 1)(x - 2)}$; $\dfrac{f(x)}{g(x)} = \dfrac{(x + 1)(x - 2)}{(x - 1)(x + 2)}$

25. $f(g(x)) = x^4 + x^2 + 1$; $g(f(x)) = (x^2 + x + 1)^2$

27. $f(g(x)) = \sqrt{x^2 - 1} + 1$; $g(f(x)) = x + 2\sqrt{x}$

29. $f(g(x)) = \dfrac{x}{x^2 + 1}$; $g(f(x)) = \dfrac{x^2 + 1}{x}$

31. 49   **33.** $\dfrac{\sqrt{5}}{5}$

35. $f(x) = 2x^3 + x^2 + 1$ and $g(x) = x^5$

37. $f(x) = x^2 - 1$ and $g(x) = \sqrt{x}$

39. $f(x) = x^2 - 1$ and $g(x) = \dfrac{1}{x}$

41. $f(x) = 3x^2 + 2$ and $g(x) = \dfrac{1}{x^{3/2}}$

43. $3h$ **45.** $-h(2a + h)$ **47.** $2a + h$

49. $C(x) = 0.6x + 12,100$

51. a. $P(x) = -0.000003x^3 - 0.07x^2 + 300x - 100,000$
b. $182,375

53. a. $N(t) = \dfrac{7}{1 + 0.02\left[\dfrac{10t + 150}{t + 10}\right]^2}$
b. 1.27 million units; 1.74 million units; 1.85 million units

55. $N(x(t)) = 9.94\left[\dfrac{(t + 10)^2}{(t + 10)^2 + 2(t + 15)^2}\right]$; 2.24 million jobs; 2.48 million jobs

Exercises 2.3, page 92

1. Yes; $y = -\tfrac{2}{3}x + 2$ **3.** Yes, $y = \tfrac{1}{2}x + 2$

5. Yes; $y = \tfrac{1}{2}x + \tfrac{9}{4}$ **7.** No

9. Polynomial function; degree 6

11. Polynomial function; degree 6

13. Some other function   **15.** $m = -1$, $b = 2$

17. a. $C(x) = 8x + 40,000$
b. $R(x) = 12x$

c. $P(x) = 4x - 40,000$
d. A loss of $8000; a profit of $8000

19. $28,800   **21.** 104 mg   **23.** $400,000

25. a. $R(x) = \dfrac{100x}{40 + x}$ **b.** 60%

27. a. $T = \tfrac{1}{4}N + 40$ **b.** $N = 4T - 160$; 248 times/min

29. $72,000

31. $\dfrac{110}{\frac{1}{2}t + 1} - 26(\tfrac{1}{4}t^2 - 1)^2 - 52$; $32, $6.71, $3; the gap was closing.

33. a.

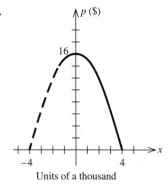

Units of a thousand

b. 3000 units

35. a.

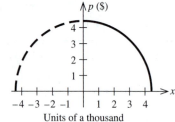

Units of a thousand

b. 3000

37. a.

Units of a thousand

b. $76

39. a.

Units of a thousand

b. $15

41. L_2; for each dollar increase in the price of a clock radio, more model A clock radios than model B clock radios will be made available in the marketplace.

43. $p = \sqrt{-x^2 + 100}$; 6614 units

45. $p = \frac{1}{10}\sqrt{x} + 10$; $30   **47.** 2500; $67.50

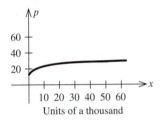

Units of a thousand

49. 11,000; $3   **51.** 5000 tents; $32.50

Using Technology Exercises 2.3, page 97

1. $(-3.0414, 0.1503)$; $(3.0414, 7.4497)$

3. $(-2.3371, 2.4117)$; $(6.0514, -2.5015)$

5. $(-1.0219, -6.3461)$; $(1.2414, -1.5931)$; $(5.7805, 7.9391)$

7. a.

b. 438 wall clocks; $40.92

Exercises 2.4, page 116

1. $\lim_{x \to -2} f(x) = 3$ **3.** $\lim_{x \to 3} f(x) = 3$ **5.** $\lim_{x \to -2} f(x) = 3$

7. The limit does not exist.

9.

x	1.9	1.99	1.999
$f(x)$	4.61	4.9601	4.9960

x	2.001	2.01	2.1
$f(x)$	5.004	5.0401	5.41

$\lim_{x \to 2} (x^2 + 1) = 5$

11.

x	-0.1	-0.01	-0.001
$f(x)$	-1	-1	-1

x	0.001	0.01	0.1
$f(x)$	1	1	1

The limit does not exist.

13.

x	0.9	0.99	0.999
$f(x)$	100	10,000	1,000,000

x	1.001	1.01	1.1
$f(x)$	1,000,000	10,000	100

The limit does not exist.

15.

x	0.9	0.99	0.999	1.001	1.01	1.1
$f(x)$	2.9	2.99	2.999	3.001	3.01	3.1

$\lim_{x \to 1} \dfrac{x^2 + x - 2}{x - 1} = 3$

17.

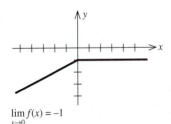

$\lim_{x \to 0} f(x) = -1$

19.

$\lim_{x \to 1} f(x) = 1$

21.

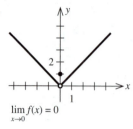

$\lim\limits_{x\to 0} f(x) = 0$

23. 3

25. 3   **27.** −1   **29.** 2   **31.** −4   **33.** $\frac{5}{4}$

35. 2   **37.** $\sqrt{171} = 3\sqrt{19}$ **39.** $\frac{3}{2}$ **41.** −1

43. −6   **45.** 2   **47.** $\frac{1}{6}$ **49.** 2   **51.** −1

53. −10   **55.** The limit does not exist.   **57.** $\frac{5}{3}$

59. $\frac{1}{2}$ **61.** $\frac{1}{3}$ **63.** $\lim\limits_{x\to\infty} f(x) = \infty$; $\lim\limits_{x\to-\infty} f(x) = \infty$

65. 0; 0   **67.** $\lim\limits_{x\to\infty} f(x) = -\infty$; $\lim\limits_{x\to-\infty} f(x) = -\infty$

69.

x	1	10	100	1000
$f(x)$	0.5	0.009901	0.0001	0.000001

x	−1	−10	−100	−1000
$f(x)$	0.5	0.009901	0.0001	0.000001

$\lim\limits_{x\to\infty} f(x) = 0$ and $\lim\limits_{x\to-\infty} f(x) = 0$

71.

x	1	5	10	100
$f(x)$	12	360	2910	2.99×10^6

x	1000	−1	−5
$f(x)$	2.999×10^9	6	−390

x	−10	−100	−1000
$f(x)$	−3090	-3.01×10^6	-3.0×10^9

$\lim\limits_{x\to\infty} f(x) = \infty$ and $\lim\limits_{x\to-\infty} f(x) = -\infty$

73. 3   **75.** 3   **77.** $\lim\limits_{x\to-\infty} f(x) = -\infty$ **79.** 0

81. a. $0.5 million; $0.75 million; $1,166,667; $2 million; $4.5 million; $9.5 million
   b. The limit does not exist; as the percentage of pollutant to be removed approaches 100, the cost becomes astronomical.

83. $2.20; the average cost of producing x video discs will approach $2.20/disc in the long run.

85. a. $24 million; $60 million; $83.1 million
   b. $120 million

87. a. 76.1 cents/mile; 30.5 cents/mile; 23 cents/mile; 20.6 cents/mile; 19.5 cents/mile
   b.

   c. 17.8 cents/mile

89. No

Using Technology Exercises 2.4, page 122

1. 5   **3.** 3   **5.** $\frac{2}{3}$ **7.** $\frac{1}{2}$ **9.** e^2

13. a.

   b. 25,000

Exercises 2.5, page 134

1. 3; 2; the limit does not exist.

3. The limit does not exist; 2; the limit does not exist.

5. 0; 2; the limit does not exist.

7. −2; 2; the limit does not exist.

9. True   **11.** True   **13.** False   **15.** True

17. False   **19.** True   **21.** 6   **23.** $-\frac{1}{4}$

25. The limit does not exist.   **27.** −1   **29.** 0

31. −4   **33.** The limit does not exist.   **35.** 4

37. 0   **39.** 0; 0   **41.** 2; 3

43. $x = 0$; conditions 2 and 3

45. Continuous everywhere   **47.** $x = 0$; condition 3

49. $x = 0$; condition 3   **51.** $(-\infty, \infty)$ **53.** $(-\infty, \infty)$

55. $(-\infty, \frac{1}{2}) \cup (\frac{1}{2}, \infty)$ **57.** $(-\infty, -2) \cup (-2, 1) \cup (1, \infty)$

59. $(-\infty, \infty)$ **61.** $(-\infty, \infty)$ **63.** $(-\infty, \infty)$

65. $(-\infty, \infty)$ **67.** -1 and 1 **69.** 1 and 2

71. f is discontinuous at $x = 1, 2, \ldots, 11$

73. Michael makes progress toward solving the problem until $x = x_1$. Between $x = x_1$ and $x = x_2$, he makes no further progress. But at $x = x_2$ he suddenly achieves a breakthrough, and at $x = x_3$ he proceeds to complete the problem.

75. Conditions 2 and 3 are not satisfied at each of these points.

77.

79.

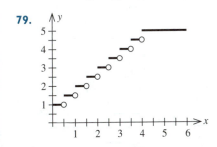

f is discontinuous at $x = \frac{1}{2}, 1, 1\frac{1}{2}, \ldots, 4$.

81. a. ∞; if the speed of the fish is very close to that of the current, the energy expended by the fish is enormous.
b. ∞; if the speed of the fish increases greatly, so does the amount of energy required to swim L feet.

83. $k = -4$ **85. a.** No   **b.** No

87. a. f is a polynomial of degree 3.
b. $f(-1) = -4$ and $f(1) = 4$

89. f is continuous on $[14, 16]$ and $f(14) \approx -6.06$ and $f(16) \approx 1.60$.

91. $x = 2$ **93.** ≈ 1.34 **95.** False   **97. c.** No

Using Technology Exercises 2.5, page 142

1. $x = 0, 1$ **3.** $x = 2$ **5.** $x = 0, \frac{1}{2}$

7. $x = -\frac{1}{2}, 2$ **9.** $x = -2, 1$

11.

13.

15.

Exercises 2.6, page 159

1. 1.5; 0.5833; 1.25   **3.** 3.075; -21.15

5. a. Car A
b. They are traveling at the same speed.
c. Car B
d. Both cars covered the same distance.

7. a. P_2 **b.** P_1 **c.** Bactericide B; bactericide A

9. 0   **11.** 2   **13.** $6x$ **15.** $-2x + 3$

17. 2; $y = 2x + 7$ **19.** 6; $y = 6x - 3$

21. 1/9; $y = \frac{1}{9}x - \frac{2}{3}$

23. a. $4x$
b. $y = 4x - 1$
c.

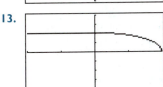

25. a. $2x - 2$

b. $(1, 0)$

c.

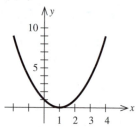

d. 0

27. a. 6; 5.5; 5.1

b. 5

c. The computations in (a) show that as h approaches zero, the average velocity approaches the instantaneous velocity.

29. a. 130 ft/sec; 128.2 ft/sec; 128.02 ft/sec

b. 128 ft/sec

c. The computations in (a) show that as the time intervals over which the average velocity are computed become smaller and smaller, the average velocity approaches the instantaneous velocity of the car at $t = 20$.

31. a. 5 sec   **b.** 80 ft/sec   **c.** 160 ft/sec

33. a. $-\frac{1}{6}$ liter/atmosphere   **b.** $-\frac{1}{4}$ liter/atmosphere

35. a. $-\frac{2}{3}x + 7$ **b.** $333/quarter; $-$13,000/quarter

37. $6 billion/yr; $10 billion/yr

39. Average rate of change of the seal population over $[a, a + h]$; instantaneous rate of change of the seal population at $x = a$

41. Average rate of change of the country's industrial production over $[a, a + h]$; instantaneous rate of change of the country's industrial production at $x = a$

43. Average rate of change of atmospheric pressure over $[a, a + h]$; instantaneous rate of change of atmospheric pressure at $x = a$

45. a. Yes   **b.** No   **c.** No

47. a. Yes   **b.** Yes   **c.** No

49. a. No   **b.** No   **c.** No

51. 5.06060; 5.06006; 5.060006; 5.0600006; 5.06000006; $5.06/case

53.

55.

Yes; no; the graph of f has a kink at $x = 0$.

Using Technology Exercises 2.6, page 167

1. a. $y = 4x - 3$

b.

3. a. $y = -7x - 8$

b.

5. a. $y = 9x - 11$

b.

7. a. $y = 2$

b.

9. a. $y = \frac{1}{4}x + 1$

b.

11. a. 4

b. $y = 4x - 1$

c.

13. a. 20
b. $y = 20x - 35$
c.

15. a. 0.75
b. $y = 0.75x - 1$
c.

17. a. -0.25
b. $y = -0.25x + 0.75$
c.

19. a. 4.02
b. $y = 4.02x - 3.57$
c.

Chapter 2 Review Exercises, page 169

1. a. $(-\infty, 9]$
b. $(-\infty, -1) \cup (-1, \frac{3}{2}) \cup (\frac{3}{2}, \infty)$

3. a.

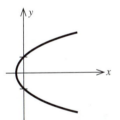

b. No
c. Yes

5. a. $\dfrac{2x + 3}{x}$ **b.** $\dfrac{1}{x(2x + 3)}$ **c.** $\dfrac{1}{2x + 3}$ **d.** $\dfrac{2}{x} + 3$

7. 2 **9.** 0 **11.** The limit does not exist.

13. $\frac{9}{2}$ **15.** $\frac{1}{2}$ **17.** 1 **19.** The limit does not exist.

21.

4; 2; the limit does not exist.

23. $x = -\frac{1}{2}, 1$ **25.** $x = 0$ **27.** 3

29. $\frac{3}{2}$; $y = \frac{3}{2}x + 5$ **31. a.** Yes **b.** No

33. a. $S(t) = t + 2.4$ **b.** \$5.4 million

35. $(6, \frac{21}{2})$ **37.** 6000; \$22 **39.** \$45,000

41.

As the length of the list increases, the time taken to learn the list increases by a very large amount.

43. $C(x) = \begin{cases} 5 & \text{if} \quad 1 \le x \le 100 \\ 9 & \text{if} \quad 100 < x \le 200 \\ 12.50 & \text{if} \quad 200 < x \le 300 \\ 15.00 & \text{if} \quad 300 < x \le 400 \\ 7 + 0.02x & \text{if} \quad x > 400 \end{cases}$

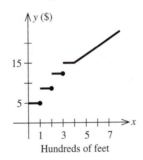

The function is discontinuous at $x = 100, 200,$ and 300.

CHAPTER 3
Exercises 3.1, page 181

1. 0 **3.** $5x^4$ **5.** $2.1x^{1.1}$ **7.** $6x$ **9.** $2\pi r$

11. $\dfrac{3}{x^{2/3}}$ **13.** $\dfrac{3}{2\sqrt{x}}$ **15.** $-84x^{-13}$ **17.** $10x - 3$

19. $-3x^2 + 4x$ **21.** $0.06x - 0.4$ **23.** $2x - 4 - \dfrac{3}{x^2}$

25. $16x^3 - 7.5x^{3/2}$ **27.** $-\dfrac{3}{x^2} - \dfrac{8}{x^3}$ **29.** $-\dfrac{16}{t^5} + \dfrac{9}{t^4} - \dfrac{2}{t^2}$

31. $2 - \dfrac{5}{2\sqrt{x}}$ **33.** $-\dfrac{4}{x^3} + \dfrac{1}{x^{4/3}}$

35. a. 20   **b.** -4 **c.** 20   **37.** 3   **39.** 11

41. $m = 5$; $y = 5x - 4$ **43.** $m = -2$; $y = -2x + 2$

45. a. $(0, 0)$ **b.**

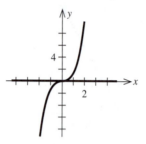

47. a. $(-2, -7), (2, 9)$
b. $y = 12x + 17$ and $y = 12x - 15$
c.

49. a. $(0, 0)$; $(1, -\frac{13}{12})$
b. $(0, 0)$; $(2, -\frac{8}{3})$; $(-1, -\frac{5}{12})$
c. $(0, 0)$; $(4, \frac{80}{3})$; $(-3, \frac{81}{4})$

51. a. $\dfrac{16\pi}{9}$ cm³/cm   **b.** $\dfrac{25\pi}{4}$ cm³/cm

53. -89.01; -4.36; if you make 0.25 stop per mile, your average speed will decrease at the rate of approximately 89.01 mph/stop/mile. If you make 2 stops per mile, your average speed will decrease at the rate of approximately 4.36 mph/stop/mile.

55. a. 15 pts/yr; 12.6 pts/yr; 0 pts/yr   **b.** 10 pts/yr

57. 155/month; 200/month

59. 32 turtles/yr; 428 turtles/yr; 3260 turtles

61. a. $120 - 30t$ **b.** 120 ft/sec   **c.** 240 ft

63. a. 2.495 million   **b.** 405,000/year

65. a. $20\left(1 - \dfrac{1}{\sqrt{t}}\right)$

b. 50 mph; 30 mph; 33.43 mph
c. -8.28; 0; 5.86; at 6:30 A.M., the average velocity is decreasing at the rate of 8.28 mph/hr; at 7 A.M., it is unchanged; and at 8 A.M., it is increasing at the rate of 5.86 mph.

Using Technology Exercises 3.1, page 183

1. 1   **3.** 0.4226   **5.** 0.1613

7. a.

b. 3.4295 parts/million; 105.4332 parts/million

9. a.

b. Decreasing at the rate of 9 days/yr; increasing at the rate of 13 days/yr

Exercises 3.2, page 195

1. $6x^2 + 2$ **3.** $4t - 1$ **5.** $9x^2 + 2x - 6$

7. $4x^3 + 3x^2 - 1$ **9.** $5w^4 - 4w^3 + 9w^2 - 6w + 2$

11. $\dfrac{25x^2 - 10x^{3/2} + 1}{x^{1/2}}$ **13.** $\dfrac{3x^4 - 10x^3 + 4}{x^2}$

15. $\dfrac{-1}{(x - 2)^2}$ **17.** $\dfrac{3}{(2x + 1)^2}$ **19.** $-\dfrac{2x}{(x^2 + 1)^2}$

21. $\dfrac{s^2 + 2s + 4}{(s + 1)^2}$ **23.** $\dfrac{1 - 3x^2}{2\sqrt{x}(x^2 + 1)^2}$ **25.** $\dfrac{x^2 - 2x - 2}{(x^2 + x + 1)^2}$

27. $\dfrac{2x^3 - 5x^2 - 4x - 3}{(x - 2)^2}$

29. $\dfrac{-2(x^5 + 12x^4 - 8x^3 + 16x + 64)}{(x^4 - 16)^2}$ **31.** 8   **33.** -9

35. $2(3x^2 - x + 3)$; 10   **37.** $\dfrac{-3x^4 + 2x^2 - 1}{(x^4 - 2x^2 - 1)^2}$; $-\dfrac{1}{2}$

39. 60; $y = 60x - 102$ **41.** $-\frac{1}{2}$; $y = -\frac{1}{2}x + \frac{3}{2}$

43. $y = 7x - 5$ **45.** $(\frac{1}{3}, \frac{50}{27})$; $(1, 2)$

47. $(\frac{4}{3}, -\frac{770}{27}); (2, -30)$ **49.** $y = -\frac{1}{2}x + 1; y = 2x - \frac{3}{2}$

51. 0.125; 0.5; 2; 50

53. $-5000/\text{min}; -1600/\text{min}; 7000; 4000$

55. a. $\dfrac{180}{(t+6)^2}$ **b.** 3.7; 2.2; 1.8; 1.1

c.

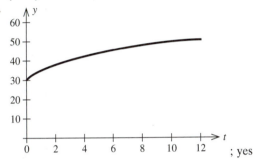

; yes

d. 50 words/min

57. Dropping at the rate of 0.0375 part/million/year; dropping at the rate of 0.006 part/million/year

Using Technology Exercises 3.2, page 199

1. 0.8750   **3.** 0.0774   **5.** -0.5000

7. 87,322 per year

Exercises 3.3, page 210

1. $8(2x - 1)^3$ **3.** $10x(x^2 + 2)^4$

5. $6x^2(1 - x)(2 - x)^2$ **7.** $\dfrac{-4}{(2x + 1)^3}$ **9.** $3x\sqrt{x^2 - 4}$

11. $\dfrac{3}{2\sqrt{3x - 2}}$ **13.** $\dfrac{-2x}{3(1 - x^2)^{2/3}}$ **15.** $-\dfrac{6}{(2x + 3)^4}$

17. $\dfrac{-1}{(2t - 3)^{3/2}}$ **19.** $-\dfrac{3(16x^3 + 1)}{2(4x^4 + x)^{5/2}}$

21. $\dfrac{-4(3x + 1)}{(3x^2 + 2x + 1)^3}$ **23.** $6x(2x^2 - x + 1)$

25. $3(t^{-1} - t^{-2})^2(-t^{-2} + 2t^{-3})$ **27.** $\dfrac{1}{2\sqrt{x - 1}} + \dfrac{1}{2\sqrt{x + 1}}$

29. $-12x(4x - 1)(3 - 4x)^3$

31. $6(x - 1)(2x - 1)(2x + 1)^3$ **33.** $-\dfrac{15(x + 3)^2}{(x - 2)^4}$

35. $\dfrac{3\sqrt{t}}{2(2t + 1)^{5/2}}$ **37.** $\dfrac{-1}{2\sqrt{u + 1}\,(3u + 2)^{3/2}}$

39. $-\dfrac{2x(3x^2 + 1)}{(x^2 - 1)^5}$ **41.** $-\dfrac{2x(3x^2 + 1)^2(3x^2 + 13)}{(x^2 - 1)^5}$

43. $-\dfrac{3x^2 + 2x + 1}{\sqrt{2x + 1}\,(x^2 - 1)^2}$ **45.** $-\dfrac{t^2 + 2t - 1}{2\sqrt{t + 1}\,(t^2 + 1)^{3/2}}$

47. $\frac{4}{3}u^{1/3}; 6x; 8x(3x^2 - 1)^{1/3}$

49. $-\dfrac{2}{3u^{5/3}}; 6x^2 - 1; -\dfrac{2(6x^2 - 1)}{3(2x^3 - x + 1)^{5/3}}$

51. $\frac{1}{2}u^{-1/2} - \frac{1}{2}u^{-3/2}; 3x^2 - 1; \dfrac{(3x^2 - 1)(x^3 - x - 1)}{2(x^3 - x)^{3/2}}$

53. $y = -33x + 57$ **55.** $y = \frac{43}{5}x - \frac{54}{5}$

57. 0.333 million/wk; 0.304 million/wk; 16 million; 22.7 million

59. a. $0.027(0.2t^2 + 4t + 64)^{-1/3}(0.1t + 1)$
b. 0.009 part per million

61. a. $0.21t^2(t - 3)(t - 7)^3$
b. 90.72; 0; -90.72; at 8 A.M., the level of nitrogen dioxide is increasing; at 10 A.M., the level stops increasing; and at 11 A.M., the level is decreasing.

63. $\dfrac{3450t}{(t + 25)^2\sqrt{\frac{1}{2}t^2 + 2t + 25}}$; 2.9 beats/min²; 0.7 beat/min²; 0.2 beat/min²; 179 beats/min

65. 160π ft²/sec   **67.** -27 mph/decade; 19 mph

69. $\dfrac{1.42(140t^2 + 3500t + 21,000)}{(3t^2 + 80t + 550)^2}$; 87,322 jobs/year

71. Decreasing at the rate of $5.86/passenger/year

Using Technology Exercises 3.3, page 209

1. 0.5774   **3.** 0.9390   **5.** -4.9498

7. 10,146,200/decade; 7,810,520/decade

Exercises 3.4, page 226

1. a. $C(x)$ is always increasing because as the number of units x produced increases, the greater the amount of money that must be spent on production.
b. 4000

3. a. $114; $120.16; $138.12   **b.** $114; $120; $138

5. a. $\dfrac{5000}{x} + 2$ **b.** $-\dfrac{5000}{x^2}$

7. $0.0002x^2 - 0.06x + 120 + \dfrac{5000}{x}$; $0.0004x - 0.06 - \dfrac{5000}{x^2}$

9. a. $-0.04x^2 + 800x$ **b.** $-0.08x + 800$
c. 400; when the level of production is 5000 units, the production of the next speaker system will bring in additional revenue of $400.

11. a. $-0.04x^2 + 600x - 300,000$ **b.** $-0.08x + 600$ **c.** 200; -40
d. The profit increases as production increases, peaking at 7500 units; beyond this level, profit falls.

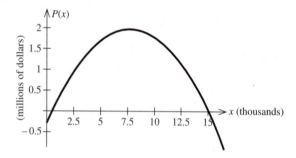

13. a. $-0.006x^2 + 180x$; $-0.000002x^3 + 0.014x^2 + 60x - 60,000$
b. $0.000006x^2 - 0.04x + 120$; $-0.012x + 180$; $-0.000006x^2 + 0.028x + 60$
c. 64; 156; 92   **d.**

 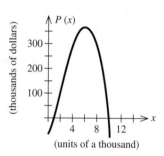

15. a. $0.000004x - 0.02 - \dfrac{60,000}{x^2}$

b. -0.0024; 0.0194; the average cost is decreasing when 5000 TV sets are produced and increasing when 10,000 units are produced.

17. 0.712 billion/billion dollars

21. $-\$0.209$ billion/billion dollars

23. $\frac{5}{3}$; elastic   **25.** 1; unitary   **27.** 0.104; inelastic

29. a. Inelastic; elastic   **b.** When $p = 8.66$
c. Increase   **d.** Increase

31. a. Inelastic   **b.** Increase

33. $\dfrac{25}{50 - p}$; for $p > 25$, demand is elastic; for $p = 25$, demand is unitary; and for $p < 25$, demand is inelastic.

Exercises 3.5, page 235

1. $8x - 2$; 8   **3.** $6x^2 - 6x$; $6(2x - 1)$

5. $4t^3 - 6t^2 + 12t - 3$; $12(t^2 - t + 1)$

7. $10x(x^2 + 2)^4$; $10(x^2 + 2)^3(9x^2 + 2)$

9. $6t(2t^2 - 1)(6t^2 - 1)$; $6(60t^4 - 24t^2 + 1)$

11. $14x(2x^2 + 2)^{5/2}$; $28(2x^2 + 2)^{3/2}(6x^2 + 1)$

13. $(x^2 + 1)(5x^2 + 1)$; $4x(5x^2 + 3)$

15. $\dfrac{1}{(2x + 1)^2}$; $-\dfrac{4}{(2x + 1)^3}$

17. $\dfrac{2}{(s + 1)^2}$; $-\dfrac{4}{(s + 1)^3}$

19. $-\dfrac{3}{2(4-3u)^{1/2}}; -\dfrac{9}{4(4-3u)^{3/2}}$

21. $72x - 24$

23. $-\dfrac{6}{x^4}$

25. $\frac{81}{8}(3s - 2)^{-5/2}$

27. $192(2x - 3)$

29. 128 ft/sec; 32 ft/sec^2

31. a. and **b.**

t	0	1	2	3	4	5	6	7
$N'(t)$	0	2.7	4.8	6.3	7.2	7.5	7.2	6.3
$N''(t)$				0.6	0	−0.6	−1.2	

33. a. $\frac{1}{4}t(t^2 - 12t + 32)$ **b.** 0 ft/sec; 0 ft/sec; 0 ft/sec
c. $\frac{3}{4}t^2 - 6t + 8$ **d.** 8 ft/sec^2; −4 ft/sec^2; 8 ft/sec^2
e. 0 ft; 16 ft; 0 ft

35. −3.09; 0.35; ten minutes after the test began, the smoke was being removed at the rate of 3%/min and the rate of the rate of change of the amount of smoke remaining was .35%/min^2.

37. $f(x) = x^{n+1/2}$

Using Technology Exercises 3.5, page 237

1. −18 **3.** 15.2762 **5.** −0.6255 **7.** 0.1973

9. −68.46214; at the beginning of 1988, the rate of the rate of the rate at which banks were failing was 68 banks/yr/yr/yr.

Exercises 3.6, page 247

1. a. $-\frac{1}{2}$ **b.** $-\frac{1}{2}$ **3. a.** $-\dfrac{1}{x^2}$ **b.** $-\dfrac{y}{x}$

5. a. $2x - 1 + \dfrac{4}{x^2}$ **b.** $3x - 2 - \dfrac{y}{x}$

7. a. $\dfrac{1 - x^2}{(1 + x^2)^2}$ **b.** $-2y^2 + \dfrac{y}{x}$

9. $-\dfrac{x}{y}$ **11.** $\dfrac{x}{2y}$ **13.** $1 - \dfrac{y}{x}$ **15.** $-\dfrac{y}{x}$

17. $-\dfrac{\sqrt{y}}{\sqrt{x}}$ **19.** $2\sqrt{x + y} - 1$ **21.** $-\dfrac{y^3}{x^3}$

23. $\dfrac{2\sqrt{xy} - y}{x - 2\sqrt{xy}}$ **25.** $\dfrac{6x - 3y - 1}{3x + 1}$ **27.** $\dfrac{2(2x - y^{3/2})}{3x\sqrt{y} - 4y}$

29. $-\dfrac{2x^2 + 2xy + y^2}{x^2 + 2xy + 2y^2}$ **31.** $y = 2$ **33.** $y = -\frac{3}{2}x + \frac{5}{2}$

35. $\dfrac{2y}{x^2}$ **37.** $\dfrac{2y(y - x)}{(2y - x)^3}$

39. a. $\dfrac{dV}{dt} = \pi r\left(r\dfrac{dh}{dt} + 2h\dfrac{dr}{dt}\right)$ **b.** 3.6π cu in./sec

41. Dropping at the rate of 111 tires/wk

43. Increasing at the rate of 44 10-packs/week

45. Dropping at the rate of 3.7 cents/carton/week

47. 0.37; inelastic **49.** 160π ft^2/sec **51.** 17 ft/sec

53. 7.69 ft/sec

55. Sliding down at the rate of 1.5 ft/sec

Exercises 3.7, page 258

1. $4x\,dx$ **3.** $(3x^2 - 1)\,dx$ **5.** $\dfrac{dx}{2\sqrt{x + 1}}$

7. $\dfrac{6x + 1}{2\sqrt{x}}dx$ **9.** $\dfrac{x^2 - 2}{x^2}dx$

11. $\dfrac{-x^2 + 2x + 1}{(x^2 + 1)^2}dx$ **13.** $\dfrac{6x - 1}{2\sqrt{3x^2 - x}}dx$

15. a. $2x\,dx$ **b.** 0.04 **c.** 0.0404

17. a. $-\dfrac{dx}{x^2}$ **b.** −0.05 **c.** −0.05263

19. 3.167 **21.** 7.0358 **23.** 1.983 **25.** 0.298

27. 2.50146 **29.** ±8.64 cm^3 **31.** ±0.076 cm^3

33. $7.9 billion **35.** $2000 **37.** 75 cents

39. 0.133% **41.** ±1.91%

43. a. $\dfrac{10{,}000\left\{\left[1 - \left(1 + \dfrac{r}{12}\right)^{-360}\right] - 30r\left(1 + \dfrac{r}{12}\right)^{-361}\right\}}{\left[1 - \left(1 + \dfrac{r}{12}\right)^{-360}\right]^2}dr$

b. $17.27; $25.91; $34.54; $43.17

Using Technology Exercises 3.7, page 261

1. 7.5787 **3.** 0.031220185778 **5.** −0.0198761598

7. $26.60/month; $35.47/month; $44.34/month

9. 625

Chapter 3 Review Exercises, page 264

1. $15x^4 - 8x^3 + 6x - 2$ **3.** $\dfrac{6}{x^4} - \dfrac{3}{x^2}$

5. $-\dfrac{1}{t^{3/2}} - \dfrac{6}{t^{5/2}}$ **7.** $1 - \dfrac{2}{t^2} - \dfrac{6}{t^3}$

9. $2x + \dfrac{3}{x^{5/2}}$ **11.** $\dfrac{2t}{(2t^2 + 1)^2}$

13. $\dfrac{1}{\sqrt{x}(\sqrt{x} + 1)^2}$ **15.** $\dfrac{2x(x^4 - 2x^2 - 1)}{(x^2 - 1)^2}$

17. $72x^2(3x^3 - 2)^7$ **19.** $\dfrac{2t}{\sqrt{2t^2 + 1}}$

21. $\dfrac{-4(3t - 1)}{(3t^2 - 2t + 5)^3}$ **23.** $\dfrac{2(x^2 + 1)(x^2 - 1)}{x^3}$

25. $4t^2(5t + 3)(t^2 + t)^3$, or $4t^5(5t + 3)(t + 1)^3$

27. $\dfrac{1}{2\sqrt{x}}(x^2 - 1)^2(13x^2 - 1)$

29. $-\dfrac{12x + 25}{2\sqrt{3x + 2}\,(4x - 3)^2}$

31. $2(12x^2 - 9x + 2)$ **33.** $\dfrac{2t(t^2 - 12)}{(t^2 + 4)^3}$

35. $\dfrac{2}{(2x^2 + 1)^{3/2}}$ **37.** $\dfrac{2x}{y}$

39. $-\dfrac{2x}{y^2 - 1}$ **41.** $\dfrac{x - 2y}{2x + y}$

43. a. $(2, -25)$ and $(-1, 14)$
    b. $y = -4x - 17;\ y = -4x + 10$

45. $y = -\dfrac{\sqrt{3}}{3}x + \dfrac{4}{3}\sqrt{3}$

47. $-\dfrac{48}{(2x - 1)^4};\ (-\infty; \tfrac{1}{2}) \cup (\tfrac{1}{2}, \infty)$

49. a. \$2.20; \$2.20   **b.** $\dfrac{2500}{x} + 2.2;\ -\dfrac{2500}{x^2}$

    c. $\displaystyle\lim_{x \to \infty}\left(\dfrac{2500}{x} + 2.2\right) = 2.2$

51. a. $2000x - 0.04x^2;$
$-0.000002x^3 - 0.02x^2 + 1000x - 120{,}000;$
$0.000002x^2 - 0.02x + 1000 + \dfrac{120{,}000}{x}$
    b. $0.000006x^2 - 0.04x + 1000;\ 2000 - 0.08x;$
$-0.000006x^2 - 0.04x + 1000;$
$0.000004x - 0.02 - \dfrac{120{,}000}{x^2}$
    c. 934; 1760; 826

d. -0.0048; 0.010125; at a production level of 5000, the average cost is decreasing by 0.48 cent/unit. At a production level of 8000, the average cost is increasing by 1.0125 cents/unit.

CHAPTER 4

Exercises 4.1, page 281

1. Decreasing on $(-\infty, 0)$ and increasing on $(0, \infty)$

3. Increasing on $(-\infty, -1) \cup (1, \infty)$ and decreasing on $(-1, 1)$

5. Decreasing on $(-\infty, 0) \cup (2, \infty)$ and increasing on $(0, 2)$

7. Decreasing on $(-\infty, -1) \cup (1, \infty)$ and increasing on $(-1, 1)$

9. Increasing on $(20.2, 20.6) \cup (21.7, 21.8)$, constant on $(19.6, 20.2) \cup (20.6, 21.1)$, and decreasing on $(21.1, 21.7) \cup (21.8, 22.7)$

11. Increasing on $(-\infty, \infty)$

13. Decreasing on $(-\infty, \tfrac{3}{2})$ and increasing on $(\tfrac{3}{2}, \infty)$

15. Decreasing on $\left(-\infty, -\dfrac{\sqrt{3}}{3}\right) \cup \left(\dfrac{\sqrt{3}}{3}, \infty\right)$ and increasing on $\left(-\dfrac{\sqrt{3}}{3}, \dfrac{\sqrt{3}}{3}\right)$

17. Increasing on $(-\infty, -2) \cup (0, \infty)$ and decreasing on $(-2, 0)$

19. Increasing on $(-\infty, 3) \cup (3, \infty)$

21. Decreasing on $(-\infty, 0) \cup (0, 3)$ and increasing on $(3, \infty)$

23. Decreasing on $(-\infty, 2) \cup (2, \infty)$

25. Decreasing on $(-\infty, 1) \cup (1, \infty)$

27. Increasing on $(-\infty, 0) \cup (0, \infty)$

29. Increasing on $(-1, \infty)$

31. Increasing on $(-4, 0)$; decreasing on $(0, 4)$

33. Increasing on $(-\infty, 0) \cup (0, \infty)$

35. Increasing on $(-\infty, 1)$; decreasing on $(1, \infty)$

37. Relative maximum: $f(0) = 1$; relative minima: $f(-1) = 0$ and $f(1) = 0$

39. Relative maximum: $f(-1) = 2$; relative minimum: $f(1) = -2$

41. Relative maximum: $f(1) = 3$; relative minimum: $f(2) = 2$

43. Relative minimum: $f(0) = 2$

45. a   **47.** d

49. Relative minimum: $f(2) = -4$

51. Relative minimum: $f(2) = 2$

53. Relative minimum: $f(0) = 2$

55. Relative maximum: $g(0) = 4$; relative minimum: $g(2) = 0$

57. Relative minimum: $F(3) = -5$; relative maximum: $F(-1) = \frac{17}{3}$

59. Relative minimum: $g(3) = -19$

61. Relative minimum: $f(\frac{1}{2}) = \frac{63}{16}$

63. None

65. Relative maximum: $f(-3) = -4$; relative minimum: $f(3) = 8$

67. Relative maximum: $f(1) = \frac{1}{2}$; relative minimum: $f(-1) = -\frac{1}{2}$

69. Relative maximum: $f(0) = 0$

71. Relative minimum: $f(1) = 0$

73. Increasing on $(0, 4000)$ and decreasing on $(4000, \infty)$

75. Decreasing on $(0, 5)$ and increasing on $(5, 10)$; after declining from 1984 through 1989, the index began to increase after 1989.

79. Increasing on $(0, 4)$ and decreasing on $(4, 5)$; the cash in the Central Provident Fund will be increasing from 1990 to 2010 and decreasing from 2010 to 2020.

81. Increasing on $(0, 1)$ and decreasing on $(1, 4)$

85. If $a > 0$, f is decreasing on $\left(-\infty, -\frac{b}{2a}\right)$ and increasing on $\left(-\frac{b}{2a}, \infty\right)$. If $a < 0$, f is increasing on $\left(-\infty, -\frac{b}{2a}\right)$ and decreasing on $\left(-\frac{b}{2a}, \infty\right)$.

87. a. $f'(x) = -2x$ if $x \neq 0$ and is not defined if $x = 0$.
b. No

91. $f'(x) = 0$ for all x, so f is a constant function.

Using Technology Exercises 4.1, page 288

1. a. f is decreasing on $(-\infty, -0.2934)$ and increasing on $(-0.2934, \infty)$
b. Relative minimum: $f(-0.2934) = -2.5435$

3. a. f is increasing on $(-\infty, -1.6144) \cup (0.2390, \infty)$ and decreasing on $(-1.6144, 0.2390)$
b. Relative maximum: $f(-1.6144) = 26.7991$; relative minimum: $f(0.2390) = 1.6733$

5. a. f is decreasing on $(-\infty, -1) \cup (0.33, \infty)$ and increasing on $(-1, 0.33)$
b. Relative maximum: $f(0.33) = 1.11$; relative minimum: $f(-1) = -0.63$

7. a. f is decreasing on $(-1, -0.71)$ and increasing on $(-0.71, 1)$
b. Relative minimum: $f(-0.71) = -1.41$

9. a.

b. f is decreasing on $(0, 0.2398) \cup (6.8758, 12)$ and increasing on $(0.2398, 6.8758)$
c. $(6.8758, 200.14)$

11. a.

b. f is decreasing on $(0, 0.8343) \cup (7.6726, 12)$ and increasing on $(0.8343, 7.6726)$

13. Increasing from 7 A.M. to 11:30 A.M. and decreasing from 11:30 A.M. to 6 P.M.; highest at 11:30 A.M. when it reaches 164 PSI.

Exercises 4.2, page 301

1. Concave downward on $(-\infty, 0)$ and concave upward on $(0, \infty)$; inflection point: $(0, 0)$

3. Concave downward on $(-\infty, 0) \cup (0, \infty)$

5. Concave upward on $(-\infty, 0) \cup (1, \infty)$ and concave downward on $(0, 1)$; inflection points: $(0, 0)$ and $(1, -1)$

7. Concave downward on $(-\infty, -2) \cup (-2, 2) \cup (2, \infty)$

9. a   **11.** b

13. a. $D_1'(t) > 0$, $D_2'(t) > 0$, $D_1''(t) > 0$, and $D_2''(t) < 0$ on $(0, 12)$

b. With or without the proposed promotional campaign, the deposits will increase, but with the promotion, the deposits will increase at an increasing rate whereas without the promotion, the deposits will increase at a decreasing rate.

15. At the time t_0, corresponding to its t-coordinate, the restoration process is working at its peak.

23. Concave upward on $(-\infty, \infty)$

25. Concave downward on $(-\infty, 0)$; concave upward on $(0, \infty)$

27. Concave upward on $(-\infty, 0) \cup (3, \infty)$; concave downward on $(0, 3)$

29. Concave downward on $(-\infty, 0) \cup (0, \infty)$

31. Concave downward on $(-\infty, 4)$

33. Concave downward on $(-\infty, 2)$; concave upward on $(2, \infty)$

35. Concave upward on $\left(-\infty, -\dfrac{\sqrt{6}}{3}\right) \cup \left(\dfrac{\sqrt{6}}{3}, \infty\right)$; concave downward on $\left(-\dfrac{\sqrt{6}}{3}, \dfrac{\sqrt{6}}{3}\right)$

37. Concave downward on $(-\infty, 1)$; concave upward on $(1, \infty)$

39. Concave upward on $(-\infty, 0) \cup (0, \infty)$

41. Concave upward on $(-\infty, 2)$; concave downward on $(2, \infty)$

43. Concave upward on $(-\infty, -1) \cup (1, \infty)$; concave downward on $(-1, 1)$

45. $(0, -2)$ **47.** $(1, -15)$ **49.** $(0, 1)$ and $\left(\tfrac{2}{3}, \tfrac{11}{27}\right)$

51. $(0, 0)$ **53.** $(1, 2)$ **55.** $\left(-\dfrac{\sqrt{3}}{3}, \dfrac{3}{2}\right)$ and $\left(\dfrac{\sqrt{3}}{3}, \dfrac{3}{2}\right)$

57. Relative maximum: $f(1) = 5$

59. None

61. Relative maximum: $f(-1) = -\tfrac{22}{3}$; relative minimum: $f(5) = -\tfrac{130}{3}$

63. Relative maximum: $f(-3) = -6$; relative minimum: $f(3) = 6$

65. None

67. Relative minimum: $f(-2) = 12$

69. Relative maximum: $g(1) = \tfrac{1}{2}$; relative minimum: $g(-1) = -\tfrac{1}{2}$

71. Relative maximum: $f(0) = 0$; relative minimum: $f\left(\tfrac{4}{3}\right) = \tfrac{256}{27}$

73. Relative minimum: $g(4) = -\tfrac{1}{108}$

75. a. N is increasing on $(0, 12)$.
b. $N''(t) < 0$ on $(0, 6)$ and $N''(t) > 0$ on $(6, 12)$
c. The rate of growth of the number of help-wanted advertisements was decreasing over the first 6 months of the year and increasing over the last 6 months.

77. $(100, 4600)$; the sales increase rapidly until $\$100,000$ is spent on advertising; after that, any additional expenditure results in increased sales but at a slower rate of increase.

79. 10 A.M.   **81.** $(18, 20{,}310)$; 1452 ft/sec

85. $\left(-\dfrac{b}{3a}, f\left(-\dfrac{b}{3a}\right)\right)$

Using Technology Exercises 4.2, page 307

1. a. f is concave upward on $(-\infty, 0) \cup (1.1667, \infty)$ and concave downward on $(0, 1.1667)$
b. $(1.1667, 1.1153)$; $(0, 2)$

3. a. f is concave downward on $(-\infty, 0)$ and concave upward on $(0, \infty)$
b. $(0, 2)$

5. a. f is concave downward on $(-\infty, 0)$ and concave upward on $(0, \infty)$
b. $(0, 0)$

7. a. f is concave downward on $(-\infty, -2.4495) \cup (0, 2.4495)$ and concave upward on $(-2.4495, 0) \cup (2.4495, \infty)$
b. $(2.4495, 0.3402)$; $(-2.4495, -0.3402)$

9. a.

11. a.

b. $(5.5318, 35.9483)$
c. $t = 5.5318$

b. $(3.9024, 77.0919)$

Exercises 4.3, page 319

1. Horizontal asymptote: $y = 0$

3. Horizontal asymptote: $y = 0$; vertical asymptote: $x = 0$

5. Horizontal asymptote: $y = 0$; vertical asymptotes: $x = -1$ and $x = 1$

7. Horizontal asymptote: $y = 3$; vertical asymptote: $x = 0$

9. Horizontal asymptotes: $y = 1$ and $y = -1$

11. Horizontal asymptote: $y = 0$; vertical asymptote: $x = 0$

13. Horizontal asymptote: $y = 0$; vertical asymptote: $x = 0$

15. Horizontal asymptote: $y = 1$; vertical asymptote: $x = -1$

17. None

19. Horizontal asymptote: $y = 1$: vertical asymptotes: $t = -3$ and $t = 3$

21. Horizontal asymptote: $y = 0$; vertical asymptotes: $x = -2$ and $x = 3$

23. Horizontal asymptote: $y = 2$; vertical asymptote: $t = 2$

25. Horizontal asymptote: $y = 1$; vertical asymptotes: $x = -2$ and $x = 2$

27. None

29. f is the derivative function of the function g

31.

33.

35.

37.

39.

41.

43.

45.

47.

49.

51.

53.

55.

57.

59.

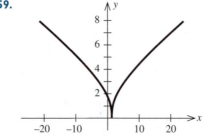

61. a. $y = 2.2$ **b.** \$2.20/disc

63. a. a

  b. The initial speed of the reaction approaches a moles/liter/sec as the amount of substrate becomes arbitrarily large.

65.

67.

69.

Using Technology Exercises 4.3, page 325

1.

3.

5. −0.9733; 2.3165, 4.6569   **7.** −1.1301; 2.9267

9. 1.5142

Exercises 4.4, page 338

1. None

3. Absolute minimum value: 0

5. Absolute maximum value: 3; absolute minimum value: −2

7. Absolute maximum value: 3; absolute minimum value: $-\frac{27}{16}$

9. Absolute minimum value: $-\frac{41}{8}$

11. No absolute extrema

13. Absolute maximum value: 1

15. Absolute maximum value: 5; absolute minimum value: −4

17. Absolute maximum value: 10; absolute minimum value: 1

19. Absolute maximum value: 19; absolute minimum value: −1

21. Absolute maximum value: 16; absolute minimum value: −1

23. Absolute maximum value: 3; absolute minimum value: $\frac{5}{3}$

25. Absolute maximum value: $\frac{37}{3}$; absolute minimum value: 5

27. Absolute maximum value ≈ 1.04; absolute minimum value: −1.5

29. No absolute extrema

31. Absolute maximum value: 1; absolute minimum value: 0

33. Absolute maximum value: 0; absolute minimum value: −3

35. Absolute maximum value: $\dfrac{\sqrt{2}}{4} \approx 0.35$; absolute minimum value: $-\frac{1}{3}$

37. Absolute maximum value: $\dfrac{\sqrt{2}}{2}$; absolute minimum value: $-\dfrac{\sqrt{2}}{2}$

39. 144 ft   **41.** 3000

43. $f(6) = 3.60$, $f(0.5) = 1.13$; the number of nonfarm, full-time, self-employed women over the time interval from 1963 to 1993 reached its highest level, 3.6 million, in 1993.

45. 5000   **47.** 3333

49. a. $0.0025x + 80 + \dfrac{10,000}{x}$ **b.** 2000

c. 2000   **d.** Same

51. 533

53. a. 2 days after the organic waste was dumped into the pond
b. 3.5 days after the organic waste was dumped into the pond

63. c.

Using Technology Exercises 4.4, page 343

1. Absolute maximum value: 145.8985; absolute minimum value: −4.3834

3. Absolute maximum value: 16; absolute minimum value: −0.1257

5. Absolute maximum value: 2.8889; absolute minimum value: 0

7. a.

b. 200.1410 banks per year

9. a.

b. Absolute maximum value: 108.8756; absolute minimum value: 49.7773

Exercises 4.5, page 353

1. 750 yd × 1500 yd; 1,125,000 sq yd

3. $10\sqrt{2}$ ft × $40\sqrt{2}$ ft

5. $\frac{16}{3}$ inches × $\frac{16}{3}$ inches × $\frac{4}{3}$ inches

7. 5.04 inches × 5.04 inches × 5.04 inches

9. 18 inches × 18 inches × 36 inches; 11,664 cu in.

11. $r = \frac{36}{\pi}$ inches; $l = 36$ inches; $\frac{46{,}656}{\pi}$ cu in.

13. $\frac{2}{3}\sqrt[3]{9}$ ft × $\sqrt[3]{9}$ ft × $\frac{2}{5}\sqrt[3]{9}$ ft　　**15.** 250; $62,500; $250

17. $w \approx 13.86$ inches; $h \approx 19.60$ inches

19. $x = 750\sqrt{2}$ ft　　**21.** 56.57 mph　　**23.** 45; 44,445

Chapter 4 Review Exercises, page 358

1. a. f is increasing on $(-\infty, 1) \cup (1, \infty)$
b. No relative extrema
c. Concave down on $(-\infty, 1)$; concave up on $(1, \infty)$
d. $\left(1, -\frac{17}{3}\right)$

3. a. f is increasing on $(-1, 0) \cup (1, \infty)$; decreasing on $(-\infty, -1) \cup (0, 1)$
b. Relative maximum: 0; relative minimum: −1
c. Concave up on $\left(-\infty, -\frac{\sqrt{3}}{3}\right) \cup \left(\frac{\sqrt{3}}{3}, \infty\right)$; concave down on $\left(-\frac{\sqrt{3}}{3}, \frac{\sqrt{3}}{3}\right)$
d. $\left(-\frac{\sqrt{3}}{3}, -\frac{5}{9}\right); \left(\frac{\sqrt{3}}{3}, -\frac{5}{9}\right)$

5. a. f is increasing on $(-\infty, 0) \cup (2, \infty)$; decreasing on $(0, 1) \cup (1, 2)$
b. Relative maximum: 0; relative minimum: 4
c. Concave up on $(1, \infty)$; concave down on $(-\infty, 1)$
d. None

7. a. f is decreasing on $(-\infty, 1) \cup (1, \infty)$
b. No relative extrema
c. Concave down on $(-\infty, 1)$; concave up on $(1, \infty)$
d. $(1, 0)$

9. a. f is increasing on $(-\infty, -1) \cup (-1, \infty)$
b. No relative extrema
c. Concave down on $(-1, \infty)$; concave up on $(-\infty, -1)$
d. No inflection point

11.

13.

15.

17.

27.

29.

19. Vertical asymptote: $x = -\frac{3}{2}$; horizontal asymptote: $y = 0$

21. Vertical asymptotes: $x = -2$, $x = 4$; horizontal asymptote: $y = 0$

23. Absolute minimum value: $-\frac{25}{8}$

25. Absolute maximum value: 5; absolute minimum value: 0

27. Absolute maximum value: -16; absolute minimum value: -32

29. Absolute maximum value: $\frac{8}{3}$; absolute minimum value: 0

31. Absolute maximum value: $\frac{1}{2}$; absolute minimum value: $-\frac{1}{2}$

33. $4000   **35.** 168   **37.** 10 A.M.

39. 20,000 cases

31.

33.

35.

Using Technology Exercises 5.1, page 369

1.

3.

5.

7.

9.

CHAPTER 5

Exercises 5.1, page 367

1. a. 16   **b.** 27   **3. a.** 3   **b.** $\sqrt{5}$

5. a. -3 **b.** 8   **7. a.** 25   **b.** $4^{1.8}$

9. a. $4x^3$ **b.** $5xy^2\sqrt{x}$ **11. a.** $\dfrac{2}{a^2}$ **b.** $\frac{1}{3}b^2$

13. a. $8x^9y^6$ **b.** $16x^4y^4z^6$

15. a. $\dfrac{64x^6}{y^4}$ **b.** $(x - y)(x + y)$

17. 2   **19.** 3   **21.** 3   **23.** $\frac{5}{4}$ **25.** 1 or 2

Exercises 5.2, page 377

1. $\log_2 64 = 6$ **3.** $\log_3 \frac{1}{9} = -2$ **5.** $\log_{1/3} \frac{1}{3} = 1$

7. $\log_{32} 8 = \frac{3}{5}$ **9.** $\log_{10} 0.001 = -3$ **11.** 1.0792

13. 1.2042 **15.** 1.6813 **17.** $\log x + 4 \log (x + 1)$

19. $\frac{1}{2} \log (x + 1) - \log (x^2 + 1)$ **21.** $\ln x - x^2$

23. $-\frac{3}{2} \ln x - \frac{1}{2} \ln (1 + x^2)$ **25.** $x \ln x$

27.

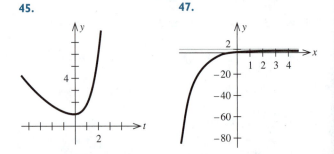

29.

31.

33. 5.1986 **35.** -0.0912 **37.** -8.0472

39. -4.9041 **41.** $-2 \ln \left(\dfrac{A}{B}\right)$ **43.** 105.7 mm

45. a. $10^3 I_0$
   b. 100,000 times greater
   c. 10,000,000 times greater

47. $34\frac{1}{2}$ hr earlier, at 1:30 P.M.

Exercises 5.3, page 389

1. $4974.47   **3.** $223,403.11

5. a. 10.25% per year   **b.** 9.31% per year

7. a. $29,227.61   **b.** $29,137.83   **9.** $6885.64

11. $112,926.52   **13.** $3.795 million   **15.** $23,329.48

17. 9.58 years   **19.** 12.75%   **21.** $40,000

23. a. $33,885.14   **b.** $33,565.38

25. a. $16,262.79   **b.** $12,047.77   **c.** $6611.96

29. Bank B   **31.** 9.531%

Exercises 5.4, page 400

1. $3e^{3x}$ **3.** $-e^{-t}$ **5.** $e^x + 1$ **7.** $x^2 e^x(x + 3)$

9. $\dfrac{2e^x(x - 1)}{x^2}$ **11.** $3(e^x - e^{-x})$ **13.** $-\dfrac{1}{e^w}$

15. $6e^{3x-1}$ **17.** $-2xe^{-x^2}$ **19.** $\dfrac{3e^{-1/x}}{x^2}$

21. $25e^x(e^x + 1)^{24}$ **23.** $\dfrac{e^{\sqrt{x}}}{2\sqrt{x}}$ **25.** $e^{3x+2}(3x - 2)$

27. $\dfrac{2e^x}{(e^x + 1)^2}$ **29.** $2(8e^{-4x} + 9e^{3x})$ **31.** $6e^{3x}(3x + 2)$

33. $y = 2x - 2$

35. f is increasing on $(-\infty, 0)$ and decreasing on $(0, \infty)$

37. Concave downward on $(-\infty, 0)$; concave upward on $(0, \infty)$

39. $(1, e^{-2})$

41. Absolute maximum value: 1; absolute minimum value: e^{-1}

43. Absolute minimum value: -1; absolute maximum value: $2e^{-3/2}$

45.

47.

49. −$6065/day; −$3679/day; −$2231/day; −$1353/day

51. **b.** 4505 per year; 273 cases per year

53. **a.** −1.68 cents/case   **b.** $40.36 per case

55. Tenth year

57. 1.8; −0.1084; −0.2286; −0.1279; initially, the number of barrels of oil needed to fuel productivity ($1000 worth of economic output) was increasing, but after one decade the number of barrels required was decreasing.

Using Technology Exercises 5.4, page 402

1. 5.4366   3. 12.3929   5. 0.1861

7. **a.** 50   **c.**

9. **a.**

b. 4.2720 billion/half century

11. **a.** 196 aphids/stem; 180 aphids/stem
b. Increasing at the rate of 226 aphids/month; decreasing at the rate of 238 aphids/month

Exercises 5.5, page 411

1. $\dfrac{5}{x}$ 3. $\dfrac{1}{x+1}$ 5. $\dfrac{8}{x}$ 7. $\dfrac{1}{2x}$ 9. $\dfrac{-2}{x}$

11. $\dfrac{2(4x-3)}{4x^2-6x+3}$ 13. $\dfrac{1}{x(x+1)}$ 15. $x(1+2\ln x)$

17. $\dfrac{2(1-\ln x)}{x^2}$ 19. $\dfrac{3}{u-2}$ 21. $\dfrac{1}{2x\sqrt{\ln x}}$

23. $\dfrac{3(\ln x)^2}{x}$ 25. $\dfrac{3x^2}{x^3+1}$ 27. $\dfrac{(x\ln x+1)e^x}{x}$

29. $\dfrac{e^{2t}[2(t+1)\ln(t+1)+1]}{t+1}$ 31. $\dfrac{1-\ln x}{x^2}$ 33. $-\dfrac{1}{x^2}$

35. $\dfrac{2(2-x^2)}{(x^2+2)^2}$ 37. $(x+1)(5x+7)(x+2)^2$

39. $(x-1)(x+1)^2(x+3)^3(9x^2+14x-7)$

41. $\dfrac{(2x^2-1)^4(38x^2+40x+1)}{2(x+1)^{3/2}}$ 43. $3^x\ln 3$

45. $(x^2+1)^{x-1}[2x^2+(x^2+1)\ln(x^2+1)]$

47. $y=x-1$

49. f is decreasing on $(-\infty, 0)$ and increasing on $(0, \infty)$

51. Concave up: $(-\infty, -1)\cup(1, \infty)$; concave down: $(-1, 0)\cup(0, 1)$

53. $(-1, \ln 2)$ and $(1, \ln 2)$

55. Absolute minimum value: 1; absolute maximum value: $3-\ln 3$

57.

Exercises 5.6, page 421

1. **a.** 0.05
b. 400
c.

t	0	10	20	100	1000
Q	400	660	1087	59,365	2.07×10^{24}

3. **a.** $Q(t)=100e^{0.035t}$
b. 266 minutes
c. $Q(t)=1000e^{0.035t}$

5. **a.** 55.5 years   **b.** 14.25 billion   7. 8.7 lb/sq in.

9. $Q(t)=100e^{-0.049t}$; 70.6 grams   11. 13,412 years ago

13.

a. 60 words per minute
b. 107 words per minute
c. 136 words per minute

15.

a. 573 computers; 1177 computers; 1548 computers; 1925 computers
b. 2000 computers
c. 46 computers/month

17. a. 11   **b.** 937   **c.** 1000   **19.** 3%; 65%

21. 1080   **23. b.** $\dfrac{1}{k}\ln A$

Chapter 5 Review Exercises, page 425

1. a.–b.

3. $\log_{16} 0.125 = -\frac{3}{4}$ **5.** 2   **7.** $x + 2y - z$

9.

11. $(2x + 1)e^{2x}$ **13.** $\dfrac{(1 - 4t)}{2\sqrt{te^{2t}}}$ **15.** $\dfrac{2(e^{2x} + 2)}{(1 + e^{-2x})^2}$

17. $(1 - 2x^2)e^{-x^2}$ **19.** $(x + 1)^2 e^x$ **21.** $\dfrac{2xe^{x^2}}{e^{x^2} + 1}$

23. $\dfrac{x + 1 - x\ln x}{x(x + 1)^2}$ **25.** $\dfrac{4e^{4x}}{e^{4x} + 3}$ **27.** $\dfrac{1 + e^x(1 - x\ln x)}{x(1 + e^x)^2}$

29. $-\dfrac{9}{(3x + 1)^2}$ **31.** 0

33. $6x(x^2 + 2)^2(3x^3 + 2x + 1)$ **35.** $y = -(2x - 3)e^{-2}$

37.

39. Absolute maximum value: $\dfrac{1}{e}$ **41.** 12% per year

43. a. $Q(t) = 2000e^{0.01831t}$ **b.** 161,992

45.

a. 1175; 2540; 3289   **b.** 4000

CHAPTER 6
Exercises 6.1, page 438

5. b. $y = 2x + C$

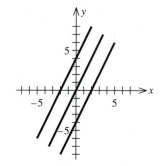

7. b. $y = \frac{1}{3}x^3 + C$

9. $6x + C$ **11.** $\frac{1}{4}x^4 + C$ **13.** $-\dfrac{1}{3x^3} + C$

15. $\frac{3}{5}x^{5/3} + C$ **17.** $-\dfrac{4}{x^{1/4}} + C$ **19.** $-\dfrac{2}{x} + C$

21. $\frac{2}{3}\pi t^{3/2} + C$ **23.** $3x - x^2 + C$

25. $\dfrac{1}{3}x^3 + \dfrac{1}{2}x^2 - \dfrac{1}{2x^2} + C$ **27.** $4e^x + C$

29. $x + \frac{1}{2}x^2 + e^x + C$ **31.** $x^4 + \dfrac{2}{x} - x + C$

33. $\frac{2}{7}x^{7/2} + \frac{4}{5}x^{5/2} - \frac{1}{2}x^2 + C$ **35.** $\frac{2}{3}x^{3/2} + 6\sqrt{x} + C$

37. $\frac{1}{6}u^3 + \frac{1}{3}u^2 - \frac{1}{3}u + C$ **39.** $\frac{2}{3}t^3 - \frac{3}{2}t^2 - 2t + C$

41. $\dfrac{1}{3}x^3 - 2x - \dfrac{1}{x} + C$ **43.** $\frac{1}{3}s^3 + s^2 + s + C$

45. $e^t + \dfrac{t^{e+1}}{e+1} + C$ **47.** $\dfrac{1}{2}x^2 + x - \ln|x| - \dfrac{1}{x} + C$

49. $\ln|x| + \dfrac{4}{\sqrt{x}} - \dfrac{1}{x} + C$ **51.** $x^2 + x + 1$

53. $x^3 + 2x^2 - x - 5$ **55.** $x - \dfrac{1}{x} + 2$ **57.** $x + \ln|x|$

59. $\sqrt{x}$ **61.** $e^x + \frac{1}{2}x^2 + 2$ **63.** $f(t) = \frac{4}{3}t^{3/2}$

65. \$3370   **67.** 5000 units; \$34,000

69. a. $-t^3 + 6t^2 + 45t$ **b.** 212

71. 21,960   **73.** 120,000

75. 1.9424 sq m   **77.** $\frac{1}{2}k(R^2 - r^2)$

79. $9\frac{7}{9}$ ft/sec²; 396 ft   **81.** 36 ft/sec²

Exercises 6.2, page 451

1. $\frac{1}{5}(4x + 3)^5 + C$ **3.** $\frac{1}{3}(x^3 - 2x)^3 + C$

5. $-\dfrac{1}{2(2x^2 + 3)^2} + C$ **7.** $\frac{2}{3}(t^3 + 2)^{3/2} + C$

9. $\frac{1}{20}(x^2 - 1)^{10} + C$ **11.** $-\frac{1}{5}\ln|1 - x^5| + C$

13. $\ln(x - 2)^2 + C$ **15.** $\frac{1}{2}\ln(0.3x^2 - 0.4x + 2) + C$

17. $\frac{1}{6}\ln|3x^2 - 1| + C$ **19.** $-\frac{1}{2}e^{-2x} + C$

21. $-e^{2-x} + C$ **23.** $-\frac{1}{2}e^{-x^2} + C$ **25.** $e^x + e^{-x} + C$

27. $\ln(1 + e^x) + C$ **29.** $2e^{\sqrt{x}} + C$

31. $-\dfrac{1}{6(e^{3x} + x^3)^2} + C$ **33.** $\frac{1}{8}(e^{2x} + 1)^4 + C$

35. $\frac{1}{2}(\ln 5x)^2 + C$ **37.** $\ln|\ln x| + C$

39. $\frac{2}{3}(\ln x)^{3/2} + C$ **41.** $\frac{1}{2}e^{x^2} - \frac{1}{2}\ln(x^2 + 2) + C$

43. $\frac{2}{3}(\sqrt{x} - 1)^3 + 3(\sqrt{x} - 1)^2 + 8(\sqrt{x} - 1)$
    $+ 4\ln|\sqrt{x} - 1| + C$

45. $\dfrac{(6x + 1)(x - 1)^6}{42} + C$

47. $5 + 4\sqrt{x} - x - 4\ln(1 + \sqrt{x}) + C$

49. $-\frac{1}{252}(1 - v)^7(28v^2 + 7v + 1) + C$

51. $\frac{1}{2}[(2x - 1)^5 + 5]$ **53.** $e^{-x^2+1} - 1$

55. $21,000 - \dfrac{20,000}{\sqrt{1 + 0.2t}}$; 6858   **57.** $p(x) = \dfrac{250}{\sqrt{16 + x^2}}$

59. $30(\sqrt{2t + 4} - 2)$; $14,400\pi$ sq ft

61. $\dfrac{71.86887}{1 + 2.449e^{-0.3277t}} - 1.43760$; 59.6 inches

63. 24,555 pairs

Exercises 6.3, page 463

1. 4.27 sq units

3. a. 6 sq units

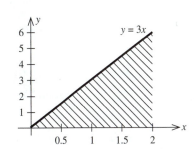

b. 4.5 sq units   **c.** 5.25 sq units   **d.** Yes

5. a. 4 sq units

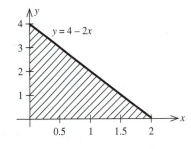

b. 4.8 sq units   **c.** 4.4 sq units   **d.** Yes

7. a. 18.5 sq units   **b.** 18.64 sq units
    c. 18.66 sq units   **d.** $\approx$18.7 sq units

9. a. 25 sq units   **b.** 21.12 sq units
c. 19.88 sq units   **d.** $\approx$19.9 sq units

11. a. 0.0625 sq unit   **b.** 0.16 sq unit
c. 0.2025 sq unit   **d.** $\approx$0.2 sq unit

13. 4.64 sq units   **15.** 0.95 sq unit   **17.** 9400 sq ft

Exercises 6.4, page 474

1. 6 sq units   **3.** 8 sq units   **5.** 12 sq units

7. 9 sq units   **9.** ln 2 sq units   **11.** $17\frac{1}{3}$ sq units

13. $18\frac{1}{4}$ sq units   **15.** $(e^2 - 1)$ sq units   **17.** 6

19. 14   **21.** $18\frac{2}{3}$ **23.** $\frac{4}{3}$ **25.** 45   **27.** $\frac{7}{12}$

29. ln 2   **31.** 56   **33.** $\frac{256}{15}$ **35.** $\frac{2}{3}$ **37.** $2\frac{2}{3}$

39. $19\frac{1}{2}$ **41. a.** \$4100   **b.** \$900

43. a. \$2800   **b.** \$219.20   **45.** $10{,}133\frac{1}{3}$ ft

47. 46% of the smoke; 24% of the smoke

Using Technology Exercises 6.4, page 477

1. 6.1787   **3.** 0.7873   **5.** −0.5888   **7.** 2.7044

9. 3.9973   **11.** 37.7 million

Exercises 6.5, page 486

1. 10   **3.** $\frac{19}{15}$ **5.** $32\frac{4}{15}$ **7.** $\sqrt{3} - 1$ **9.** $24\frac{1}{5}$

11. $\frac{32}{15}$ **13.** $18\frac{2}{15}$ **15.** $\frac{1}{2}(e^4 - 1)$ **17.** $\frac{1}{2}e^2 + \frac{5}{6}$

19. 0   **21.** 2 ln 4   **23.** $\frac{1}{3}(\ln 19 - \ln 3)$

25. $2e^4 - 2e^2 - \ln 2$ **27.** $\frac{1}{2}(e^{-4} - e^{-8} - 1)$ **29.** 5

31. $\frac{17}{3}$ **33.** −1   **35.** $\frac{13}{6}$ **37.** $\frac{1}{4}(e^4 - 1)$

39. 120.3 billion metric tons   **41.** $\approx$\$2.24 million

43. \$40,339   **45.** 26°F   **47.** 16,863

49. 0.071 mg/cm³   **51.** $\dfrac{2k}{3} R^2$ cm/sec

Exercises 6.6, page 497

1. 108 sq units   **3.** $\frac{2}{3}$ sq unit   **5.** $2\frac{2}{3}$ sq units

7. $1\frac{1}{2}$ sq units   **9.** 3   **11.** $3\frac{1}{3}$ **13.** 27

15. $2(e^2 - e^{-1})$ **17.** $12\frac{2}{3}$ **19.** $3\frac{1}{3}$ **21.** $4\frac{3}{4}$

23. 12 − ln 4   **25.** $e^2 - e - \ln 2$ **27.** $2\frac{1}{2}$ **29.** $7\frac{1}{3}$

31. $\frac{3}{2}$ **33.** $e^3 - 4 + \dfrac{1}{e}$ **35.** $20\frac{5}{6}$ **37.** $\frac{1}{12}$ **39.** $\frac{1}{3}$

41. S is the additional revenue that Odyssey Travel could realize by switching to the new agency;

$$S = \int_0^b [g(x) - f(x)]\, dx$$

43. a. S is the additional amount of smoke that brand B will remove over brand A in the time interval $[a, b]$.

b. $S = \int_a^b [f(t) - g(t)]\, dt$

45. 42.79 million metric tons   **47.** 57,179 people

Using Technology Exercises 6.6, page 503

1. a.

3. a.

b. 1074.2857   **b.** 0.9961

5. a.   **7. a.**

b. 5.4603   **b.** 25.8549

9. a.   **11. a.**

b. 10.5144   **b.** 3.5799

Exercises 6.7, page 515

1. \$11,667   **3.** \$6667   **5.** \$11,667

7. Consumers' surplus: \$13,333;
producers' surplus: \$11,667

9. \$824,200   **11.** \$148,239   **13.** \$52,203

15. \$76,615   **17.** \$111,869   **19.** \$20,964

21. a.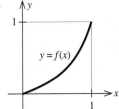
$y = f(x)$

b. 0.175; 0.816

23. a.
$y = f(x)$

b. 0.104; 0.504

Using Technology Exercises 6.7, page 518

1. Consumers' surplus: \$18,000,000;
producers' surplus: \$11,700,000

3. Consumers' surplus: \$33,120;
producers' surplus: \$2,880

Exercises 6.8, page 526

1. 3π cu units **3.** $\dfrac{15\pi}{2}$ cu units

5. $\dfrac{4\pi}{3}$ cu units **7.** $\dfrac{16\pi}{15}$ cu units

9. $\frac{1}{2}(e^2 - 1)$ cu units **11.** $\dfrac{2\pi}{15}$ cu units

13. $\dfrac{136\pi}{15}$ cu units **15.** $\dfrac{64\sqrt{2}\pi}{3}$ cu units

17. $\dfrac{\pi}{2}(e^2 - 2 + e^{-2})$ cu units

19. $\dfrac{\pi}{6}$ cu units **21.** $\dfrac{6517\pi}{240}$ cu units

23. $\dfrac{64\sqrt{2}\pi}{3}$ cu units **25.** $\dfrac{8(\sqrt{2}-1)\pi}{3}$ cu units

27. $\dfrac{4\pi}{3}r^3$ cu units **29.** $50,000\pi$ cu ft

Chapter 6 Review Exercises, page 529

1. $\frac{1}{4}x^4 + \frac{2}{3}x^3 - \frac{1}{2}x^2 + C$ **3.** $\dfrac{1}{5}x^5 - \dfrac{1}{2}x^4 - \dfrac{1}{x} + C$

5. $\frac{1}{2}x^4 + \frac{2}{5}x^{5/2} + C$ **7.** $\frac{1}{3}x^3 - \frac{1}{2}x^2 + 2\ln|x| + 5x + C$

9. $\frac{3}{8}(3x^2 - 2x + 1)^{4/3} + C$

11. $\frac{1}{2}\ln(x^2 - 2x + 5) + C$ **13.** $\frac{1}{2}e^{x^2+x+1} + C$

15. $\frac{1}{6}(\ln x)^6 + C$ **17.** $\dfrac{(11x^2 - 1)(x^2 + 1)^{11}}{264} + C$

19. $\frac{2}{3}(x + 4)\sqrt{x - 2} + C$ **21.** $\frac{1}{2}$ **23.** $5\frac{2}{3}$

25. -80 **27.** $\frac{1}{2}\ln 5$ **29.** 4 **31.** $\dfrac{e - 1}{2(1 + e)}$

33. $f(x) = x^3 - 2x^2 + x + 1$ **35.** $f(x) = x + e^{-x} + 1$

37. -4.28

39. a. $-0.015x^2 + 60x$ **b.** $p = -0.015x + 60$

41. $3000t - 50,000(1 - e^{-0.04t})$; 16,939

43. \$3100 **45.** 15 sq units **47.** $\frac{2}{3}$ sq unit

49. $e^2 - 3$ sq units **51.** $\frac{1}{2}$ sq unit **53.** $\frac{1}{3}$

55. \$2083; \$3333 **57.** \$98,973

59. a.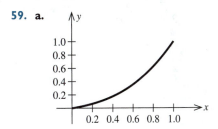

b. 0.1017; 0.3733 **c.** 0.315

61. $\dfrac{2\pi}{3}$ cu units

CHAPTER 7

Exercises 7.1, page 539

1. $\frac{1}{4}e^{2x}(2x - 1) + C$ **3.** $4(x - 4)e^{x/4} + C$

5. $\frac{1}{2}e^{2x} - 2(x - 1)e^x + \frac{1}{3}x^3 + C$ **7.** $xe^x + C$

9. $\dfrac{2(x + 2)}{\sqrt{x + 1}} + C$ **11.** $\frac{2}{3}x(x - 5)^{3/2} - \frac{4}{15}(x - 5)^{5/2} + C$

13. $\dfrac{x^2}{4}(2\ln 2x - 1) + C$ **15.** $\dfrac{x^4}{16}(4\ln x - 1) + C$

17. $\frac{2}{9}x^{3/2}(3\ln\sqrt{x} - 1) + C$ **19.** $-\dfrac{1}{x}(\ln x + 1) + C$

21. $x(\ln x - 1) + C$ **23.** $-(x^2 + 2x + 2)e^{-x} + C$

25. $\frac{1}{4}x^2[2(\ln x)^2 - 2 \ln x + 1] + C$ 27. $2 \ln 2 - 1$

29. $4 \ln 4 - 3$ 31. $\frac{1}{4}(3e^4 + 1)$

33. $-\frac{1}{2}xe^{-2x} - \frac{1}{4}e^{-2x} + \frac{13}{4}$ 35. $5 \ln 5 - 4$

37. 1485 ft   39. 2.04 mg/ml

41. $-20e^{-0.1t}(t + 10) + 200$ 43. $521,087

Exercises 7.2, page 547

1. $\frac{2}{9}[2 + 3x - 2 \ln|2 + 3x|] + C$

3. $\frac{3}{32}[(1 + 2x)^2 - 4(1 + 2x) + 2 \ln|1 + 2x|] + C$

5. $2\left[\frac{x}{8}\left(\frac{9}{4} + 2x^2\right)\sqrt{\frac{9}{4} + x^2} - \frac{81}{128}\ln\left(x + \sqrt{\frac{9}{4} + x^2}\right)\right] + C$

7. $\ln\left(\frac{\sqrt{1 + 4x} - 1}{\sqrt{1 + 4x} + 1}\right) + C$ 9. $\frac{1}{2}\ln 3$

11. $\dfrac{x}{9\sqrt{9 - x^2}} + C$

13. $\frac{x}{8}(2x^2 - 4)\sqrt{x^2 - 4} - 2 \ln|x + \sqrt{x^2 - 4}| + C$

15. $\sqrt{4 - x^2} - 2 \ln\left|\dfrac{2 + \sqrt{4 - x^2}}{x}\right| + C$

17. $\frac{1}{4}(2x - 1)e^{2x} + C$ 19. $\ln|\ln(1 + x)| + C$

21. $\frac{1}{9}\left[\dfrac{1}{1 + 3e^x} + \ln(1 + 3e^x)\right] + C$

23. $6[e^{(1/2)x} - \ln(1 + e^{(1/2)x})] + C$

25. $\frac{1}{9}(2 + 3 \ln x - 2 \ln|2 + 3 \ln x|) + C$

27. $e - 2$ 29. $\dfrac{x^3}{9}(3 \ln x - 1) + C$

31. $x[(\ln x)^3 - 3(\ln x)^2 + 6 \ln x - 6] + C$

33. $\approx$$2329   35. 27,136   37. 44; 49   39. 26,157

41. $418,444

Exercises 7.3, page 562

1. 2.7037; 2.6667; $2\frac{2}{3}$ 3. 0.2656; 0.2500; $\frac{1}{4}$

5. 0.6970; 0.6933; $\approx$0.6931   7. 0.5090; 0.5004; $\frac{1}{2}$

9. 5.2650; 5.3046; $\frac{16}{3}$ 11. 0.6336; 0.6321; $\approx$0.6321

13. 0.3837; 0.3863; $\approx$0.3863   15. 1.1170; 1.1114

17. 1.3973; 1.4052   19. 0.8806; 0.8818

21. 3.7757; 3.7625   23. a. 3.6   b. 0.0324

25. a. 0.013   b. 0.00043

27. a. 0.0078125   b. 0.0002848

29. 52.84 miles   31. 21.65 mpg

33. a. $142,374   b. $142,698

35. 103.9 PSI   37. $\approx$30%   39. $\approx$6.42 1/min

Exercises 7.4, page 574

1. $\frac{2}{3}$ sq unit   3. 1 sq unit   5. 2 sq units

7. $\frac{2}{3}$ sq unit   9. $\frac{1}{2}e^4$ sq units   11. 1 sq unit

13. a. $\frac{2}{3}b^{3/2}$ 15. 1   17. 2   19. Divergent

21. $-\frac{1}{8}$ 23. 1   25. 1   27. $\frac{1}{2}$ 29. Divergent

31. -1 33. Divergent   35. 0   37. 0

39. Divergent   41. $18,750

43. $\dfrac{10,000r + 4000}{r^2}$ dollars

45. b. $83,333

Chapter 7 Review Exercises, page 578

1. $-2(1 + x)e^{-x} + C$ 3. $x(\ln 5x - 1) + C$

5. $\frac{1}{4}(1 - 3e^{-2})$ 7. $2\sqrt{x}(\ln x - 2) + 2$

9. $\dfrac{1}{8}\left[3 + 2x - \dfrac{9}{3 + 2x} - 6 \ln|3 + 2x|\right] + C$

11. $\frac{1}{32}e^{4x}(8x^2 - 4x + 1) + C$

13. $\dfrac{1}{4}\dfrac{\sqrt{x^2 - 4}}{x} + C$ 15. $\frac{1}{2}$

17. Divergent   19. $\frac{1}{10}$

21. 0.8421; 0.8404   23. 2.2379; 2.1791

25. $1,157,641   27. $41,100

29. $111,111

CHAPTER 8

Exercises 8.1, page 589

1. $f(0, 0) = -4$; $f(1, 0) = -2$; $f(0, 1) = -1$; $f(1, 2) = 4$; $f(2, -1) = -3$

3. $f(1, 2) = 7$; $f(2, 1) = 9$; $f(-1, 2) = 1$; $f(2, -1) = 1$

5. $g(1, 2) = 4 + 3\sqrt{2}$; $g(2, 1) = 8 + \sqrt{2}$; $g(0, 4) = 2$; $g(4, 9) = 56$

7. $h(1, e) = 1$; $h(e, 1) = -1$; $h(e, e) = 0$

9. $g(1, 1, 1) = e$; $g(1, 0, 1) = 1$; $g(-1, -1, -1) = -e$

11. All real values of x and y

13. All real values of u and v except those satisfying the equation $u = v$

15. All real values of r and s satisfying $rs \geq 0$

17. All real values of x and y satisfying $x + y > 5$

19.

21.

23.

25. 9π cu ft   **27.** $40{,}000k$ dynes

29. $25{,}500$; $23{,}580$ **31.** 8310; 7910

33. a. The set of all nonnegative values of W and H
   b. 1.871 sq meters

35. $13{,}498.59$ **37.** $76{,}704.11$; $50{,}814.62$ **39.** 18

Exercises 8.2, page 604

1. 2; 3 **3.** $4x$; 4 **5.** $-\dfrac{4y}{x^3}$; $\dfrac{2}{x^2}$

7. $\dfrac{2v}{(u+v)^2}$; $-\dfrac{2u}{(u+v)^2}$

9. $3(2s-t)(s^2-st+t^2)^2$; $3(2t-s)(s^2-st+t^2)^2$

11. $\dfrac{4x}{3(x^2+y^2)^{1/3}}$; $\dfrac{4y}{3(x^2+y^2)^{1/3}}$ **13.** ye^{xy+1}; xe^{xy+1}

15. $\ln y + \dfrac{y}{x}$; $\dfrac{x}{y} + \ln x$ **17.** $e^u \ln v$; $\dfrac{e^u}{v}$

19. $yz + y^2 + 2xz$, $xz + 2xy + z^2$; $xy + 2yz + x^2$

21. ste^{rst}; rte^{rst}; rse^{rst} **23.** $f_x(1, 2) = 8$; $f_y(1, 2) = 5$

25. $f_x(2, 1) = 1$; $f_y(2, 1) = 3$

27. $f_x(1, 2) = \frac{1}{2}$; $f_y(1, 2) = -\frac{1}{4}$

29. $f_x(1, 1) = e$; $f_y(1, 1) = e$

31. $f_x(1, 0, 2) = 0$; $f_y(1, 0, 2) = 8$; $f_z(1, 0, 2) = 0$

33. $f_{xx} = 2y$; $f_{xy} = 2x + 3y^2 = f_{yx}$; $f_{yy} = 6xy$

35. $f_{xx} = 2$; $f_{xy} = f_{yx} = -2$; $f_{yy} = 4$

37. $f_{xx} = \dfrac{y^2}{(x^2+y^2)^{3/2}}$; $f_{xy} = f_{yx} = -\dfrac{xy}{(x^2+y^2)^{3/2}}$;

   $f_{yy} = \dfrac{x^2}{(x^2+y^2)^{3/2}}$

39. $f_{xx} = \dfrac{1}{y^2}e^{-x/y}$; $f_{xy} = \dfrac{y-x}{y^3}e^{-x/y} = f_{yx}$;

   $f_{yy} = \dfrac{x}{y^3}\left(\dfrac{x}{y} - 2\right)e^{-x/y}$

41. a. $f_x = 7.5$; $f_y = 40$ **b.** Yes

43. $P_x = 10$—at $(0, 1)$, the price of land is changing at the rate of \$10 per sq ft per mile change to the right; $P_y = 0$—at $(0, 1)$, the price of land is constant per mile change upward.

45. Complementary commodities

47. \$30/unit change in finished desks; $-$\$25/unit change in unfinished desks. The weekly revenue increases by \$30/unit for each additional finished desk produced (beyond 300) when the level of production of unfinished desks remains fixed at 250; the revenue decreases by \$25/unit when each additional unfinished desk (beyond 250) is produced and the level of production of finished desks remains fixed at 300.

49. 0.039 l/degree; $-$0.015 l/mm of mercury. The volume increases by 0.039 l when the temperature increases by 1 degree (beyond 300 K) and the pressure is fixed at 800 mm of mercury. The volume decreases by 0.015 l when the pressure increases by 1 mm of mercury (beyond 800 mm) and the temperature is fixed at 300 K.

Using Technology Exercises 8.2, page 607

1. 1.3124; 0.4038   **3.** $-$1.8889; 0.7778

5. $-$0.3863; $-$0.8497

Exercises 8.3, page 618

1. $(0, 0)$; relative maximum value: $f(0, 0) = 1$

3. $(1, 2)$; saddle point: $f(1, 2) = 4$

5. $(8, -6)$; relative minimum value: $f(8, -6) = -41$

7. $(1, 2)$ and $(2, 2)$; saddle point: $f(1, 2) = -1$; relative minimum value: $f(2, 2), = -2$

9. $(-\frac{1}{3}, \frac{11}{3})$ and $(1, 5)$; saddle point: $f(-\frac{1}{3}, \frac{11}{3}) = -\frac{319}{27}$; relative minimum value: $f(1, 5) = -13$

11. $(0, 0)$ and $(1, 1)$; saddle point: $f(0, 0) = -2$; relative minimum value: $f(1, 1) = -3$

13. $(2, 1)$; relative minimum value: $f(2, 1) = 6$

15. $(0, 0)$; saddle point: $f(0, 0) = -1$

17. $(0, 0)$; relative minimum value: $f(0, 0) = 1$

19. $(0, 0)$; relative minimum value: $f(0, 0) = 0$

21. 200 finished units and 100 unfinished units; $10,500

23. Price of land ($200/sq ft) is highest at $(\frac{1}{2}, 1)$

25. $(0, 1)$ gives desired location

27. 18 inches $\times$ 36 inches $\times$ 18 inches

29. 6 inches $\times$ 4 inches $\times$ 2 inches

Exercises 8.4, page 628

1. a. $y = 2.3x + 1.5$

b.

3. a. $y = -0.77x + 5.74$

b.

5. a. $y = 1.2x + 2$

b.
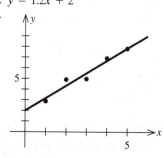

7. a. $y = 0.34x - 0.9$

b.

c. 1276 applications

9. a. $y = -2.8x + 440$

b.

c. 420

11. a. $y = 6.23x + 167.6$ **b.** 541 acres

13. a. $y = 2.8x + 17.6$ **b.** $40,000,000

15. a. $y = 4.842x + 11.842$ **b.** 103.8 billion cans

Using Technology Exercises 8.4, page 631

1. $y = 2.3596x + 3.8639$ **3.** $y = -1.1948x + 3.5525$

5. a. $y = 13.321x + 72.571$ **b.** 192 million tons

Exercises 8.5, page 643

1. Min of $\frac{3}{4}$ at $(\frac{3}{4}, \frac{1}{4})$ **3.** Max of $-\frac{7}{4}$ at $(2, \frac{7}{2})$

5. Min of 4 at $\left(\sqrt{2}, \frac{\sqrt{2}}{2}\right)$ and $\left(-\sqrt{2}, -\frac{\sqrt{2}}{2}\right)$

7. Max of $-\frac{3}{4}$ at $(\frac{3}{2}, 1)$

9. Max of $2\sqrt{3}$ at $\left(\frac{\sqrt{3}}{3}, -\sqrt{6}\right)$ and $\left(\frac{\sqrt{3}}{3}, \sqrt{6}\right)$

11. Max of 8 at $(2\sqrt{2}, 2\sqrt{2})$ and $(-2\sqrt{2}, -2\sqrt{2})$; min of -8 at $(2\sqrt{2}, -2\sqrt{2})$ and $(-2\sqrt{2}, 2\sqrt{2})$

13. Max: $\frac{2\sqrt{3}}{9}$; min: $-\frac{2\sqrt{3}}{9}$

15. Min of $\frac{18}{7}$ at $(\frac{9}{7}, \frac{6}{7}, \frac{3}{7})$

17. 140 finished and 60 unfinished units

19. $10\sqrt{2}$ ft $\times 40\sqrt{2}$ ft   **21.** 2 ft $\times$ 2 ft $\times$ 3 ft

23. $15,000 on newspaper and $45,000 on TV

25. 30 ft $\times$ 40 ft $\times$ 10 ft; $7200

Exercises 8.6, page 650

1. $2x\,dx + 2\,dy$

3. $(4x - 3y + 4)\,dx - 3x\,dy$

5. $\dfrac{x}{\sqrt{x^2 + y^2}}\,dx + \dfrac{y}{\sqrt{x^2 + y^2}}\,dy$

7. $-\dfrac{5y}{(x - y)^2}\,dx + \dfrac{5x}{(x - y)^2}\,dy$

9. $(10x^4 + 3ye^{-3x})\,dx - e^{-3x}\,dy$

11. $\left(2xe^y + \dfrac{y}{x}\right) dx + (x^2 e^y + \ln x)\,dy$

13. $y^2 z^3\,dx + 2xyz^3\,dy + 3xy^2 z^2\,dz$

15. $\dfrac{1}{y + z}\,dx - \dfrac{x}{(y + z)^2}\,dy - \dfrac{x}{(y + z)^2}\,dz$

17. $(yz + e^{yz})\,dx + xz(1 + e^{yz})\,dy + xy(1 + e^{yz})\,dz$

19. 0.04   **21.** −0.01   **23.** −0.10

25. −0.18   **27.** 0.06   **29.** −0.1401

31. An increase of $19,250/month

33. An increase of $5000/month

35. 38.4π cu cm   **37.** 0.55 ohm

Exercises 8.7, page 658

1. $\frac{7}{2}$ **3.** 0   **5.** $4\frac{1}{2}$

7. $(e^2 - 1)(1 - e^{-2})$

9. 1   **11.** $\frac{2}{3}$ **13.** $\frac{188}{3}$

15. $\frac{84}{5}$ **17.** $2\frac{2}{3}$ **19.** 1

21. $\frac{1}{2}(3 - e)$ **23.** $\frac{1}{4}(e^4 - 1)$ **25.** $\frac{2}{3}(e - 1)$

Exercises 8.8, page 665

1. 42 cu units   **3.** 20 cu units   **5.** $\frac{64}{3}$ cu units

7. $\frac{1}{3}$ cu unit   **9.** 4 cu units   **11.** $3\frac{1}{3}$ cu units

13. $2(e^2 - 1)$ cu units

15. $\frac{2}{35}$ cu unit   **17.** 54   **19.** $\frac{1}{3}$

21. 1   **23.** 43,329   **25.** $10,460

Chapter 8 Review Exercises, page 669

1. $f(0, 1) = 0, f(1, 0) = 0, f(1, 1) = \frac{1}{2}$; no

3. The set of all ordered pairs of real numbers (u, v) such that $u \geq 0$ and $u \neq v$.

5.

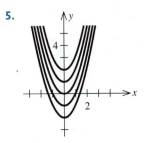

7. $2xy^3 + 3y^2 + \dfrac{1}{y}; 3x^2 y^2 + 6xy - \dfrac{x}{y^2}$

9. $\dfrac{y(y^2 - x^2)}{(x^2 + y^2)^2}, \dfrac{x(x^2 - y^2)}{(x^2 + y^2)^2}$

11. $2x(1 + x^2 + y^2)e^{x^2+y^2}, 2y(1 + x^2 + y^2)e^{x^2+y^2}$

13. $f_{xx} = 6x - 4y, f_{xy} = f_{yx} = -4x, f_{yy} = 2$

15. $g_{xx} = 2(1 + 2x^2)e^{x^2+y^2}, g_{xy} = g_{yx} = 4xye^{x^2+y^2},$
$g_{yy} = 2(1 + 2y^2)e^{x^2+y^2}$

17. $f_x(1, 1, 0) = 3, f_y(1, 1, 0) = 3, f_z(1, 1, 0) = -2$

19. $2\,dx$ **21.** $\frac{227}{54}$

23. $(8, -2)$; saddle point at $(8, -2)$

25. $(0, 0)$; rel min of 1 at $(0, 0)$

27. $y = 0.5x + 1.5$ **29.** $\frac{175}{44}$ at $\left(-\frac{1}{22}, \frac{23}{22}\right)$

31. Max of $e^{\sqrt{2}}$ at $\left(\dfrac{\sqrt{2}}{2}, -\dfrac{\sqrt{2}}{2}\right)$; min of $e^{-\sqrt{2}}$

at $\left(-\dfrac{\sqrt{2}}{2}, \dfrac{\sqrt{2}}{2}\right)$

33. $\frac{1}{2}(e^{-2} - 1)^2$ **35.** $\frac{1}{4}(3 - 2\ln 2)$

37. $10\frac{4}{5}$ cu units

39. a. $R(x, y) = -0.02x^2 - 0.05y^2 - 0.2xy + 80x + 60y$
    b. The set of all points (x, y) satisfying

$$0.02x + 0.1y \leq 80$$
$$0.1x + 0.05y \leq 60$$
$$x \geq 0, y \geq 0$$

    c. 15,300

41. $490

43. a. $y = 8.2x + 361.2$; ≈ 7 hr 48 min

45. 75 units on labor and 25 units on capital

CHAPTER 9

Exercises 9.1, page 679

13. $y = 12x^2 - 2x$

15. $y = \dfrac{1}{x}$

17. $y = -\dfrac{e^x}{x} + \dfrac{1}{2} xe^x$

19. $\dfrac{dQ}{dt} = -kQ$; $Q(0) = Q_0$

21. $\dfrac{dA}{dt} = k(C - A)$

23. $\dfrac{dC}{dt} = -kC$; $C(0) = C_0$

Exercises 9.2, page 686

1. $\frac{1}{3}y^3 = \frac{1}{2}x^2 + x + C$

3. $\frac{1}{3}y^3 = e^x + C$

5. $y = ce^{2x}$

7. $-\dfrac{1}{y} = \dfrac{1}{2}x^2 + C$

9. $y = -\frac{4}{3} + ce^{-6x}$

11. $y^3 = \frac{1}{3}x^3 + x + C$

13. $y^{1/2} - x^{1/2} = C$

15. $\ln |y| = \frac{1}{2}(\ln x)^2 + C$

17. $y^2 = 2x^2 + 2$

19. $y = 2 - e^{-x}$

21. $y = \frac{2}{3} + \frac{1}{3}e^{(3/2)x^2}$

23. $y = \sqrt{x^2 + 1}$

25. $\ln |y| = (x - 1)e^x$

27. $y = \ln (x^3 + e)$

29. $y^2 = x^3 + 8$

31. $Q(t) = Q_0 e^{-kt}$

33. $A = \dfrac{C}{k} - d_2 e^{-kt}$

35. $S(t) = D - (D - S_0)e^{-kt}$

Exercises 9.3, page 695

1. $y = y_0 e^{-kt}$

3. $Q(t) = 4.5e^{0.02t}$; 7.4 billion

5. $\frac{1}{2}$ in.

7. $Q(t) = \dfrac{50}{4t + 1}$; 5.56 gm

9. 3.6 min

11. 23

13. 290.5 million

15. 395

17. $Q(t) = e^{C - (C - \ln Q_0)e^{-kt}}$

19. $x(t) = 40(1 - e^{-(3/20)t})$; 38 lb; 40 lb

Exercises 9.4, page 705

1. a. $y(1) = \frac{369}{128} \approx 2.8828$ **b.** $y(1) = \frac{70{,}993}{23{,}328} \approx 3.043$

3. a. $y(2) = \frac{51}{16} \approx 3.1875$ **b.** $y(2) = \frac{793}{243} \approx 3.2634$

5. a. $y(0.5) \approx 0.8324$ **b.** $y(0.5) \approx 0.8207$

7. a. $y(1.5) \approx 1.7831$ **b.** $y(1.5) \approx 1.7920$

9. a. $y(1) \approx 1.3390$ **b.** $y(1) \approx 1.3654$

11.

x	0.0	0.2	0.4	0.6	0.8	1
$\tilde{y}_n$	1	1	1.02	1.0608	1.1245	1.2144

13.

x	0.0	0.2	0.4	0.6	0.8	1
$\tilde{y}_n$	2	1.8	1.72	1.736	1.8288	1.9830

15.

x	0.0	0.1	0.2	0.3	0.4	0.5
$\tilde{y}_n$	1	1.1	1.211	1.3361	1.4787	1.6426

Chapter 9 Review Exercises, page 706

5. $y = (9x + 8)^{-1/3}$

7. $y = 4 - Ce^{-2t}$

9. $y = -\dfrac{2}{2x^3 + 2x + 1}$

11. $y = 3e^{-x^3/2}$

13. a. $y(1) \approx 0.3849$ **b.** $y(1) \approx 0.4361$

15. a. $y(1) \approx 1.3258$ **b.** $y(1) \approx 1.4570$

17.

x	0.0	0.2	0.4	0.6	0.8	1
$\tilde{y}_n$	1	1	1.08	1.2528	1.5535	2.0506

19. a. $S = 50{,}000(0.8)^t$ **b.** \$16,384

21. $A = \dfrac{P}{r}(e^{rt} - 1)$; \$342,549.50

23. a. $\approx$7:30 P.M.   **25. a.** 183

CHAPTER 10

Exercises 10.1, page 717

11. $k = \frac{1}{3}$ **13.** $k = \frac{1}{8}$ **15.** $k = \frac{3}{16}$ **17.** $k = 2$

19. a. $\frac{1}{2}$ **b.** $\frac{5}{8}$ **c.** $\frac{7}{8}$ **d.** 0

21. a. $\frac{11}{16}$ **b.** $\frac{1}{2}$ **c.** $\frac{27}{32}$ **d.** 0

23. a. $\frac{1}{2}$ **b.** $\frac{1}{2}(2\sqrt{2} - 1)$ **c.** 0   **d.** $\frac{1}{2}$

25. a. 0.63   **b.** 0.30   **c.** 0.30

27. a. 0.10   **b.** 0.30   **29.** 0.9355

31. 0.4815; 0.7407   **33.** $\frac{1}{6}$

Using Technology Exercises 10.1, page 720

1.

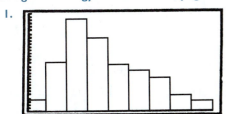

3. a.

x	0	1	2	3	4
$P(X = x)$	.017	.067	.033	.117	.233

x	5	6	7	8	9	10
$P(X = x)$	.133	.167	.1	.05	.067	.017

b.

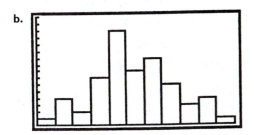

Exercises 10.2, page 733

1. $\mu = \frac{9}{2}$; Var $(x) = \frac{3}{4}$; $\sigma \approx 0.866$

3. $\mu = \frac{15}{4}$; Var $(x) = \frac{15}{16}$; $\sigma \approx 0.9682$

5. $\mu = 3$; Var $(x) = 0.8$; $\sigma = 0.8944$

7. $\mu \approx 2.3765$; Var $(x) \approx 2.3522$; $\sigma = 1.534$

9. $\mu = \frac{93}{35}$; Var $(x) \approx 0.7151$; $\sigma \approx 0.846$

11. $\mu = \frac{3}{2}$; Var $(x) = \frac{3}{4}$; $\sigma \approx \frac{1}{2}\sqrt{3}$

13. $\mu = 4$; Var $(x) = 16$; $\sigma = 4$

15. 100 days   **17.** $3\frac{1}{3}$ minutes   **19.** 1.5 ft

21. 2500 lb/wk   **23.** $m = 5$

25. $m \approx 2.52$ **27.** $m = 2$

Using Technology Exercises 10.2, page 736

1. a.

b. $\mu = 4$ and $\sigma = 1.40$

3. a. X gives the minimum age requirement for a regular driver's license.

b.

x	15	16	17	18	19	21
$P(X = x)$	.02	.30	.08	.56	.02	.02

c.

d. $\mu = 17.34$ and $\sigma = 1.11$

5. a. Let X denote the random variable that gives the weight of a carton of sugar.

b.

x	4.96	4.97	4.98	4.99	5.00	5.01
$P(X = x)$	$\frac{3}{30}$	$\frac{4}{30}$	$\frac{4}{30}$	$\frac{1}{30}$	$\frac{1}{30}$	$\frac{5}{30}$

x	5.02	5.03	5.04	5.05	5.06
$P(X = x)$	$\frac{3}{30}$	$\frac{3}{30}$	$\frac{4}{30}$	$\frac{1}{30}$	$\frac{1}{30}$

c. $\mu \approx 5.00$; Var $(X) \approx 0.0009$; $\sigma \approx 0.03$

Exercises 10.3, page 746

1. 0.9265   **3.** 0.0401   **5.** 0.8657

7. a.   **b.** 0.9147

1.37

9. a.

−0.65

b. 0.2578

11. a.

−1.25

b. 0.8944

13. a.

0.68   2.02

b. 0.2266

15. a. 1.23   **b.** −0.81   **17. a.** 1.9   **b.** −1.9

19. a. 0.9772   **b.** 0.9192   **c.** 0.7333

21. a. 0.2206   **b.** 0.2206   **c.** 0.3034

23. a. 0.0228   **b.** 0.0228   **c.** 0.4772   **d.** 0.7258

25. a. 0.0038   **b.** 0.0918   **c.** 0.4082   **d.** 0.2514

27. 0.6247   **29.** 0.62%

31. A:80; B:73; C:62; D:54

Chapter 10 Review Exercises, page 750

5. $\frac{1}{243}$ **7.** $\frac{3}{2}$ **9. a.** $\frac{4}{7}$ **b.** 0   **c.** $\frac{1}{3}$

11. a. ≈ 0.52 **b.** ≈ 0.65 **c.** 0

13. $\mu = \frac{9}{2}$; Var $(x) \approx 2.083$; $\sigma \approx 1.44$

15. $\mu = 0$; Var $(x) \approx \frac{7}{15}$; $\sigma \approx 0.6831$

17. 0.9875   **19.** 0.3049

21. a. 0.6915   **b.** 0.8944   **c.** 0.4681

23. a. 0.22   **b.** 0.39   **c.** 4 days

CHAPTER 11

Exercises 11.1, page 764

1. $P_1(x) = 1 - x$; $P_2(x) = 1 - x + \frac{1}{2}x^2$;
$P_3(x) = 1 - x + \frac{1}{2}x^2 - \frac{1}{6}x^3$

3. $P_1(x) = 1 - x$; $P_2(x) = 1 - x + x^2$;
$P_3(x) = 1 - x + x^2 - x^3$

5. $P_1(x) = 1 - (x - 1)$; $P_2(x) = 1 - (x - 1) + (x - 1)^2$;
$P_3(x) = 1 - (x - 1) + (x - 1)^2 - (x - 1)^3$

7. $P_1(x) = 1 - \frac{1}{2}x$; $P_2(x) = 1 - \frac{1}{2}x - \frac{1}{8}x^2$;
$P_3(x) = 1 - \frac{1}{2}x - \frac{1}{8}x^2 - \frac{1}{16}x^3$

9. $P_1(x) = -x$; $P_2(x) = -x - \frac{1}{2}x^2$;
$P_3(x) = -x - \frac{1}{2}x^2 - \frac{1}{3}x^3$

11. $P_2(x) = 16 + 32(x - 2) + 24(x - 2)^2$

13. $P_4(x) = (x - 1) - \frac{1}{2}(x - 1)^2 + \frac{1}{3}(x - 1)^3 - \frac{1}{4}(x - 1)^4$

15. $P_4(x) = e + e(x - 1) + \frac{1}{2}e(x - 1)^2 + \frac{1}{6}e(x - 1)^3 + \frac{1}{24}e(x - 1)^4$

17. $P_3(x) = 1 - \frac{1}{2}x - \frac{1}{8}x^2 - \frac{1}{16}x^3$

19. $P_3(x) = \frac{1}{3} - \frac{2}{9}x + \frac{4}{27}x^2 - \frac{8}{81}x^3$

21. $P_n(x) = 1 - x + x^2 - x^3 + \cdots + (-1)^n x^n$; 0.9091; 0.909090 ...

23. $P_4(x) = 1 - \frac{1}{2}x + \frac{1}{8}x^2 - \frac{1}{48}x^3 + \frac{1}{384}x^4$; 0.90484

25. $P_2(x) = 4 + \frac{1}{8}(x - 16) - \frac{1}{512}(x - 16)^2$; ≈ 3.94969

27. $P_3(x) = x - \frac{1}{2}x^2 + \frac{1}{3}x^3$; 0.109; 0.108

29. 2.01494375; 0.00000042

31. 1.248; 0.00493; 1.25   **33.** 0.095; 0.00033

35. a. 0.04167   **b.** 0.625   **c.** 0.04167
d. 0.007121

37. 0.47995   **39.** 48%   **41.** 2600

Exercises 11.2, page 774

1. 1, 2, 4, 8, 16   **3.** 0, $\frac{1}{3}$, $\frac{2}{4}$, $\frac{3}{5}$, $\frac{4}{6}$

5. 1, 1, $\frac{4}{6}$, $\frac{8}{24}$, $\frac{16}{120}$ **7.** $e, \frac{e^2}{8}, \frac{e^2}{27}, \frac{e^4}{64}, \frac{e^5}{125}$

9. 1, $\frac{11}{9}$, $\frac{25}{19}$, $\frac{45}{33}$, $\frac{71}{51}$ **11.** $a_n = 3n - 2$

13. $a_n = \dfrac{1}{n^3}$ **15.** $a_n = \dfrac{2^{2n-1}}{5^{n-1}}$

17. $a_n = \dfrac{(-1)^{n+1}}{2^{n-1}}$ **19.** $a_n = \dfrac{n}{(n+1)(n+2)}$

21. $a_n = \dfrac{e^{n-1}}{(n-1)!}$

23.

25.

27.

29.

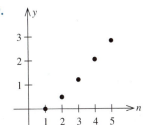

31. Converges; $\frac{1}{2}$

33. Converges; 0

35. Converges; 1

37. Converges; 2

39. Converges; 2

41. Converges; 0

43. Converges; $\dfrac{\sqrt{2}}{2}$

45. a. $a_1 = 0.015$, $a_{10} = 0.140$, $a_{100} = 0.77939$,
$a_{1000} = 0.999999727$
b. 1

47. b. $a_n = 100(1.01)^n$
c. $a_{24} = 126.97$. The accumulated amount at the end of 2 years is \$126.97.

Exercises 11.3, page 786

1. $\lim\limits_{N\to\infty} S_N$ does not exist; divergent

3. $S_N = \dfrac{1}{2} - \dfrac{1}{N+2}; \dfrac{1}{2}$

5. Converges; $\frac{3}{2}$

7. Diverges

9. Converges; $\frac{3}{5}$

11. Converges; $\frac{1}{3}$

13. Converges; 5.52

15. Converges; 13

17. Converges; $2\frac{3}{4}$

19. Converges; $7\frac{1}{2}$

21. Diverges

23. $\frac{1}{3}$

25. $\frac{404}{333}$

27. \$303 billion

29. \$2,377,083

31. $\dfrac{P}{e^r - 1}$

33. b. $\dfrac{1}{k} \ln \dfrac{S}{S - C} \, \text{hr}$

Exercises 11.4, page 796

1. $R = 1$; $(0, 2)$

3. $R = 1$; $(-1, 1)$

5. $R = 4$; $(-4, 4)$

7. $R = \infty$; $(-\infty, \infty)$

9. $R = 0$; $x = -2$

11. $R = 1$; $(-4, -2)$

13. $R = \infty$; $(-\infty, \infty)$

15. $R = 1$; $(-\frac{1}{2}, \frac{1}{2})$

17. $R = 0$; $x = -1$

19. $R = 3$; $(0, 6)$

21. $\displaystyle\sum_{n=0}^{\infty} (-1)^n (x - 1)^n$; $R = 1$; $(0, 2)$

23. $\displaystyle\sum_{n=0}^{\infty} (-1)^n \dfrac{(x - 2)^n}{3^{n+1}}$; $R = 3$; $(-1, 5)$

25. $\displaystyle\sum_{n=0}^{\infty} (-1)^{n+1}(x - 2)^n$; $R = 1$; $(1, 3)$

27. $1 + \dfrac{1}{2}(x - 1) + \displaystyle\sum_{n=2}^{\infty} (-1)^{n+1} \dfrac{1 \cdot 3 \cdot 5 \cdots (2n - 3)}{n! 2^n}(x - 1)^n$;
$R = 1$; $(0, 2)$

29. $\displaystyle\sum_{n=0}^{\infty} \dfrac{2^n}{n!} x^n$; $R = \infty$; $(-\infty, \infty)$

31. $\displaystyle\sum_{n=0}^{\infty} (-1)^n \dfrac{1 \cdot 3 \cdot 5 \cdots (2n - 1)}{n! 2^n} x^n$; $R = 1$; $(-1, 1)$

Exercises 11.5, page 805

1. $\displaystyle\sum_{n=0}^{\infty} (-1)^{n+1}(x - 2)^n$; $(1, 3)$

3. $\displaystyle\sum_{n=0}^{\infty} (-1)^n 3^n x^n$; $(-\frac{1}{3}, \frac{1}{3})$

5. $\displaystyle\sum_{n=0}^{\infty} \dfrac{3^n}{4^{n+1}} x^n$; $(-\frac{4}{3}, \frac{4}{3})$

7. $\displaystyle\sum_{n=0}^{\infty} x^{2n}$; $(-1, 1)$

9. $\displaystyle\sum_{n=0}^{\infty} (-1)^n \dfrac{x^n}{n!}$; $(-\infty, \infty)$

11. $\displaystyle\sum_{n=0}^{\infty} (-1)^n \dfrac{x^{2n+1}}{n}$; $(-\infty, \infty)$

13. $f(x) = 1 + \dfrac{x^2}{2!} + \dfrac{x^4}{4!} + \dfrac{x^6}{6!} + \cdots + \dfrac{x^{2n}}{2n!} + \cdots$; $(-\infty, \infty)$

15. $\sum_{n=1}^{\infty} (-1)^{n-1} \dfrac{2^n x^n}{n}$; $(-\frac{1}{2}, \frac{1}{2}]$

17. $\sum_{n=1}^{\infty} (-1)^{n+1} \dfrac{x^{2n}}{n}$; $(-1, 1)$

19. $(\ln 2)(x - 2) + \sum_{n=1}^{\infty} (-1)^{n-1}\left(\dfrac{1}{n2^n}\right)(x - 2)^n$; $(0, 4]$

21. $f'(x) = 1 - x + x^2 + \cdots + (-1)^n x^n + \cdots$

23. $f(x) = x - \dfrac{1}{2}x^2 + \dfrac{1}{3}x^3 + \cdots + \dfrac{(-1)^{n+1}}{n}x^n + \cdots$

25. 0.4812   **27.** 0.7475

29. 3.34   **31.** 15.85%

Exercises 11.6, page 814

1. 1.732051   **3.** 2.645751   **5.** 2.410142

7. 2.30278   **9.** 1.11634   **11.** 1.61803

13. 0.5671   **15. b.** 0.19356   **17. b.** 1.87939

19. 2.9365   **21.** 1.2785   **23.** 0.4263

25. 294 units/day   **27.** 8:39 P.M.

29. 26.82%/yr   **31.** 12%/yr   **33.** 10%/yr

35. 11,671 units; $21.17 per unit

Chapter 11 Review Exercises, page 818

1. $f(x) = 1 - (x + 1) + (x + 1)^2 - (x + 1)^3 + (x + 1)^4$

3. $f(x) = x^2 - \frac{1}{2}x^4$

5. $f(x) = 2 + \frac{1}{12}(x - 8) - \frac{1}{288}(x - 8)^2$; 1.983

7. 2.9992591; 8×10^{-11} **9.** 0.37

11. Converges; $\frac{2}{3}$ **13.** Converges; 1

15. 2   **17.** $\dfrac{1}{\sqrt{2} + 1}$ **19.** $\frac{141}{99}$

21. $R = 1$; $(-1, 1)$ **23.** $R = 1$; $(0, 2)$

25. $f(x) = -1 - 2x - 4x^2 - 8x^3 - \cdots - 2^n x^n - \cdots$; $(-\frac{1}{2}, \frac{1}{2})$

27. $f(x) = 2x - 2x^2 + \frac{8}{3}x^3 - \cdots + \dfrac{(-1)^{n+1}2^n}{n}x^n + \cdots$; $(-\frac{1}{2}, \frac{1}{2}]$

29. 2.28943   **31.** (0.35173, 0.70346)

33. $106,186.10   **35.** 31.08%

CHAPTER 12

Exercises 12.1, page 826

1. $\dfrac{5\pi}{2}$ radians   **3.** $-\dfrac{3\pi}{2}$ radians

5. a. III   **b.** III   **c.** II   **d.** I

7. $\dfrac{5\pi}{12}$ radians   **9.** $\dfrac{8\pi}{9}$ radians   **11.** $\dfrac{7\pi}{2}$ radians

13. 120°   **15.** −270°   **17.** 220°

19.   **21.**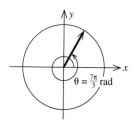

23. 150°, −210°   **25.** 315°, −45°

Exercises 12.2, page 835

1. 0   **3.** 1   **5.** $\dfrac{\sqrt{3}}{2}$

7. $\dfrac{\sqrt{3}}{3}$ **9.** −2.6131

11. $\operatorname{Sin} \dfrac{\pi}{2} = 1$, $\cos \dfrac{\pi}{2} = 0$, $\tan \dfrac{\pi}{2}$ is undefined, $\csc \dfrac{\pi}{2} = 1$, $\sec \dfrac{\pi}{2}$ is undefined, $\cot \dfrac{\pi}{2} = 0$

13. $\operatorname{Sin} \dfrac{5\pi}{3} = -\dfrac{\sqrt{3}}{2}$, $\cos \dfrac{5\pi}{3} = \dfrac{1}{2}$, $\tan \dfrac{5\pi}{3} = -\sqrt{3}$, $\csc \dfrac{5\pi}{3} = -\dfrac{2\sqrt{3}}{3}$, $\sec \dfrac{5\pi}{3} = 2$, $\cot \dfrac{5\pi}{3} = -\dfrac{\sqrt{3}}{3}$

15. $\dfrac{7\pi}{6}, \dfrac{11\pi}{6}$ **17.** $\dfrac{5\pi}{6}, \dfrac{11\pi}{6}$

19. π **21.** $\dfrac{\pi}{3}, \dfrac{2\pi}{3}$

23.

25.

27.

29.

41. Sin $\theta = \frac{5}{13}$, cos $\theta = \frac{12}{13}$, tan $\theta = \frac{5}{12}$, csc $\theta = \frac{13}{5}$ sec $\theta =$ $\frac{13}{12}$, cot $\theta = \frac{12}{5}$

43. b. $\dfrac{\pi(4n + 1)}{12}$ $(n = 0, 1, 2, \ldots)$;

$\dfrac{\pi(4n + 3)}{12}$ $(n = 0, 1, 2, \ldots)$

Exercises 12.3, page 848

1. $-3 \sin 3x$

3. $-2\pi \sin \pi x$

5. $2x \cos(x^2 + 1)$

7. $4x \sec^2 2x^2$

9. $x \cos x + \sin x$

11. $6(\cos 3x - \sin 2x)$

13. $2x(\cos 2x - x \sin 2x)$

15. $\dfrac{x \cos \sqrt{x^2 - 1}}{\sqrt{x^2 - 1}}$

17. $e^x \sec x(1 + \tan x)$

19. $\cos \dfrac{1}{x} + \dfrac{1}{x} \sin \dfrac{1}{x}$

21. $\dfrac{x \sin x}{(1 + \cos x)^2}$

23. $\dfrac{\sec^2 x}{2\sqrt{\tan x}}$

25. $\dfrac{x \cos x - \sin x}{x^2}$

27. $2 \tan x \sec^2 x$

29. $-\csc^2 x \cdot e^{\cot x}$

31. $y = -2x + \dfrac{\pi}{2}$

33. Increasing on $\left(0, \dfrac{\pi}{4}\right) \cup \left(\dfrac{5\pi}{4}, 2\pi\right)$; decreasing on $\left(\dfrac{\pi}{4}, \dfrac{5\pi}{4}\right)$

35. $f(x) = \sin x + \cos x$

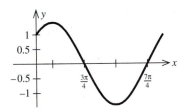

37. $f(x) = 2 \sin x + \sin 2x$

39. a. $f(x) = \sin x = x - \dfrac{x^3}{3!} + \dfrac{x^5}{5!} - \dfrac{x^7}{7!} - \cdots$

$+ (-1)^{n+1} \dfrac{x^{2n+1}}{(2n + 1)!} + \cdots$

41. Zero wolves/month; -1571 caribou/month

43. Warmest day is July 25; coldest day is January 23

45. Sixth week

47. Maximum when $t = 1, 5, 9, 13, \ldots$; minimum when $t = 3, 7, 11, 15, \ldots$

49. 70.7 ft/sec

51. 60°

53. 1.4987 radians

Using Technology Exercises 12.3, page 853

1. 1.2038

3. 0.7762

5. -0.2368

7. 0.8415; -0.2172

9. 1.1271; 0.2013

11. a.

b. ≈ $0.63   **c.** ≈ $27.79

13. a.

b. 2.2687 sq units

13. a.

15. a.

b. 1.8239 sq units

Exercises 12.4, page 858

1. $-\frac{1}{3}\cos 3x + C$ **3.** $-3\cos x + 4\sin x + C$

5. $\frac{1}{2}\tan 2x + C$ **7.** $\frac{1}{2}\sin x^2 + C$

9. $-\dfrac{1}{\pi}\csc \pi x + C$ **11.** 2

13. $-\frac{1}{2}\ln\frac{1}{2}$ **15.** $\frac{1}{4}\sin^4 x + C$

17. $\dfrac{1}{\pi}\ln|\sec \pi x + \tan \pi x| + C$

19. $\frac{1}{3}\ln(1 + \sqrt{2})$ **21.** $-\frac{2}{3}(\cos x)^{3/2} + C$

23. $-\frac{1}{9}(1 - 2\sin 3x)^{3/2} + C$ **25.** $\frac{1}{4}\tan^4 x + C$

27. $-\frac{1}{4}(\cot x - 1)^4 + C$ **29.** 2

31. $\frac{1}{2}x[\sin(\ln x) - \cos(\ln x)] + C$

33. $2\sqrt{2}$ sq units   **35.** $\frac{1}{2}\ln 2$ sq units

37. $\frac{1}{2}(\pi^2 - 4)$ sq units   **39.** $85

41. $120,000   **43.** $\dfrac{1.2}{\pi}\left(1 - \cos\dfrac{\pi t}{2}\right)$

45. 162 fruit flies   **47.** ≈0.9

Using Technology Exercises 12.4, page 861

1. 0.5419   **3.** 0.7544   **5.** 0.2231

7. 0.6587   **9.** −0.2032   **11.** 0.9045

17. a.

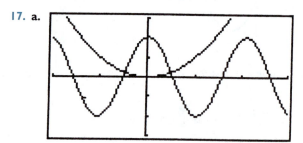

b. 1.2484 sq units

19. a.

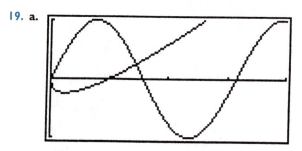

b. 1.0983 sq units

21. 7.6 ft

Chapter 12 Review Exercises, page 864

1. $\frac{2\pi}{3}$ radians

3. $-\frac{5\pi}{4}$ radians

5. $-450°$

7. $\frac{\pi}{3}$ or $\frac{5\pi}{3}$

9. $3 \cos 3x$

11. $2 \cos x + 6 \sin 2x$

13. $e^{-x}(3 \sec^2 3x - \tan 3x)$

15. $4 \cos 2x$ or $4(\cos^2 x - \sin^2 x)$

17. $\dfrac{(\cot x - 1)\sec^2 x - (1 - \tan x)\csc^2 x}{(1 - \cot x)^2}$

19. $\cos(\sin x)\cos x$

21. $y = 4x + 1 - \pi$

23. $\frac{3}{2} \sin \frac{2}{3}x + C$

25. $-\frac{1}{2} \csc x^2 + C$

27. $\frac{1}{3} \sin^3 x + C$

29. $-\csc x + C$

31. 1

33. $2\sqrt{2}$ sq units

35. December 1; September 1

INDEX

Take the Journey

Journey *Through*™ Calculus

A CD-ROM unlike any other, *Journey Through Calculus* is a program so exciting, so visual, so affordable, it is destined to change the way you learn calculus.

Seven modules provide a learning expedition like no other

Journey is the only calculus CD-ROM that brings together activities, tutorials, testing, a computer algebra system, and calculus content into one, unified environment.

Journey's seven modules:

▲ Teach concepts through interactive explorations, animations, and applications.

▲ Give you unlimited practice with automatic grading and feedback via algorithmically generated tests and quizzes.

▲ Include interactive, real-world applications that bring relevance to abstract and often difficult concepts.

▲ Help you visualize the concepts through vivid animations.

▲ Introduce concepts, often in gamelike environments, with interactive activities that call upon student intuition and interest to develop a concrete conceptual understanding.

▲ Utilize the power of the Maple® kernel throughout, in any computation—both symbolic and numeric—or in graphing.

Brooks/Cole Publishing Company
511 Forest Lodge Road
Pacific Grove, CA 93950-5098

To order, call toll-free (800) 423-0563
Fax: (606) 647-5020
email: info@brookscole.com
http://www.brookscole.com/math

Source Code 9BCMABK1
© 1999 Brooks/Cole Publishing Company

The Seven Modules

1. A Survey of Limits
2. Limits of Functions
3. Derivatives
4. Derivative Calculations
5. Derivative Applications
6. Integration
7. Integration Applications